CW00704342

MODERN MAGNETIC MATERIALS

MODERN MAGNETIC MATERIALS
Principles and Applications

ROBERT C. O'HANDLEY
Massachusetts Institute of Technology

A Wiley-Interscience Publication
JOHN WILEY & SONS, INC.

This book is printed on acid-free paper. ♾

Published simultaneously in Canada.

For ordering and customer service, call 1-800-CALL WILEY.

Library of Congress Cataloging-in-Publication Data:

O'Handley, Robert C., 1942–
Modern magnetic materials : principles and applications / Robert C. O'Handley.
p. cm.
Includes bibliographical references and index.
ISBN 978-0-471-15566-9
1. Magnetic materials. 2. Magnetism. I. Title.
TK454.4.M30353 1999
621.34--dc21 99-21372

To Carol,
who knows much more than magnetism, l'aimant.

CONTENTS

PREFACE

Magnetism is an open field in which physicists, electrical engineers, materials scientists, chemists, metallurgists, and others practice together. This book is intended as a modern text for an advanced undergraduate or graduate level course in magnetic materials. It should appeal to readers with a broad range of backgrounds. It begins by laying a foundation of the most widely used concepts of magnetic materials and proceeds to describe their application in a number of areas of current importance. The text leads, hopefully, to a point from which the current literature in much of the field can be read with a critical appreciation.

As the title indicates, the text makes an attempt to be current. This is evident in the chapters on amorphous magnetism (Chapter 11) and nonocrystalline materials (Chapter 12), and perhaps more so in the last three chapters: charge and spin transport (Chapter 15), surface and thin-film magnetism (Chapter 16), and magnetic recording (Chapter 17). Part of the price paid for these chapters of current interest is the volatility of their content. An attempt has been made at distillation in order to cover those aspects that are more enduring. The rest will soon be of historical value only. Some of the examples and problems in the text are presented with computer solutions that allow the student to explore and learn more actively and hence, more thoroughly. There has been a deliberate attempt to rely heavily on figures and to include extensive data from the current literature.

The prominent place of the word "materials" in the title indicates that this is not a book about the physics of magnetism, although there is some of that. It is about understanding magnetic materials, particularly those that are of technical importance. Thus, there is little coverage of many interesting materials and phenomena that are limited to extremely high fields or low temperatures. The emphasis is on the science and engineering of materials that are subjectively regarded as useful.

An understanding of modern magnetic materials cannot be achieved without elements of metallurgy, the physics and chemistry of solids, as well as concepts from electrical engineering. For example, the most advanced developments in magnetic recording, or in permanent magnets, are based squarely on careful control of microstructure. Recent electronic structure calculations shed new light on the interplay of magnetic moment formation and chemical binding that is changing the ways magnetic materials are designed and processed. The design of high-performance magnetic devices is not possible without micromagnetic computer calculations of magnetization and field distributions. The modern magnetician must be facile in appreciating and integrating concepts and data that cut across tranditional disciplinary boundaries.

The treatment of magnetism and magnetic materials presented here reflects the importance of *process* in addition to *product* or *result*. In general, a result often depends on the route by which it was reached. This is as true for the derivation of a formula as it is for the properties of a fabricated material or device. Most calculated results are *relative* in the sense that they depend strongly on the assumptions and methods used to derive them. The more fundamental the result, the more paths lead there. But with approximations, different assumptions and different methods lead to different results. Experimentalists work with real materials that are approximations to the ideal, pure crystals treated in many texts. Hence, the world of the materials scientist—the theories, models or experimental measurements of real material behavior—is rarely a "conservative" one; the endpoint or conclusion is generally a function of the path taken. The challenge is to map out the shortest paths to the most important places.

The selection of topics in a text such as this is a humbling exercise. Authors struggle under competing tensions: seeking broad coverage on one hand and acknowledging personal limitations on the other. Even after selection of the material to be covered, it is impossible in any one text to shape the material into a sequence suited to a variety of learning styles. Most people learn from the particular to the general, from examples to principles. An attempt has been made, therefore, to begin each chapter with a relevant experimental observation to motivate the treatment. However, the book as a whole cannot follow this paradigm; one cannot thoroughly treat the results of magnetism, such as soft or hard magnetic materials or magnetic recording, without first covering the fundamentals of magnetostatics, anisotropy, domain walls, and so on. The first chapter is a modest effort to motivate the reader with a foretaste of some of the exciting topics to come while at the same time establishing a common platform of preparation. One of the major compromises in subject organization is the interruption of the largely phenomenological approach with three chapters (3–5) intended to answer the questions "Where do magnetic moments come from and how do they interact?" as well as "How does magnetism reflect the different types of bonding in different materials?" These chapters may be passed over by those already well prepared in the physics of magnetism or by those interested only in the phenomenology of magnetism. Otherwise, this is an appropriate place to cover such fundamental material.

The sequence of chapters is chosen to give the student an appreciation of the energies appropriate to understanding the behavior of magnetic materials. After a coverage of magnetostatics (Chapter 2), exchange interactions (Chapters 3–5), magnetic anisotropy (Chapter 6), and magnetostriction (Chapter 7), the reader is in a position to understand quantitatively the concepts of magnetic domains and domain walls (Chapter 8). This provides adequate preparation to grasp the details of the magnetization process (Chapter 9). The spectrum of magnetic behavior from soft to hard magnetic materials is treated in the sequence of chapters from 10 to 13. After treating the technical properties of soft magnetic materials, both crystalline and amorphous, in Chapter 10, more aspects of the physics of amorphous magnetism are covered in Chapter 11. Nanocrystalline materials (commonly derived from amorphous precursors) and issues related to the effects of magnetic exchange interactions in small structures are treated in Chapter 12. Hard magnetic materials, often based on isolated nanocrystalline magnetic particles, are then covered in Chapter 13. Magnetic ordering and field annealing, important in many soft and hard materials, are covered in Chapter 14. Electronic and spin transport in magnetic materials, a field presently experiencing explosive growth in fundamental understanding and applications, is treated in Chapter 15. Some of the aspects of magnetism unique to surfaces and thin films are described in Chapter 16. Perhaps the single most important application driving modern magnetic materials development—namely, magnetic information storage, both media and transducers—provides an important opportunity to apply much of the material in earlier chapters. Magnetic recording physics and materials are covered in the final chapter, 17.

The text uses mks/SI (meter–kilogram–second/Système Internationale) units primarily with many important equations and quantities given also in cgs (centimeter–gram–second) units. It was not possible to convert all of the data borrowed from the literature to mks units. This may be a reminder to the reader that in this field, two (or more) languages are still spoken, often in the same sentence.

It is hoped that this text will contribute in some measure to the growing appreciation of the field of magnetic materials as a paradigm of the new scientific order in which the stubborn disciplinary barriers of physics, chemistry, mathematics, and metallurgy are not only breached but also reformed as bridges for scientific understanding and technological development.

ROBERT C. O'HANDLEY

Massachusetts Institute of Technology
June 1999

ACKNOWLEDGMENT

This text would not have been realized without input from numerous friends and colleagues. I would like to begin by thanking my thesis advisor, Hellmut Juretschke, who introduced me to the exciting field of magnetism and magnetic materials. I am most grateful to my colleagues at Allied Corporation, especially Souk Hasegawa, and IBM, most notably Dick Gambino and Tom McGuire, who shared with me the riches of their magnetic experiences. Particular gratitude goes to my students, postdocs and other colleagues over the years at MIT who have helped me learn various aspects of this fascinating field. I would especially like to thank Nick Grant, who saw a need for returning magnetism to the Department of Materials Science and Engineering, and mentored me in how to make it work here. I have been privileged to teach much of the material in this book to graduate students at MIT since 1989. Also, with generous support from the National Magnet Lab through Jack Crow, Reza Abbascian hosted my sabbatical at the University of Florida in 1994, where much of this material was given as a graduate course. Hans-Joachim Güntherodt graciously hosted my 1997 visit to the University of Basel, where, with the support of the Swiss National Science Foundation, much of the material was presented to the diploma and graduate students. I am truly grateful for these valuable experiences.

Thanks go sincerely to my editors at Wiley, Greg Franklin, John Falcone, and Rosalyn Farkas who had faith in the project as well as patient hope for its delivery before the new millennium. The anonymous, Wiley-selected reviewers of the early drafts were extremely helpful in their specific comments and broader recommendations.

Particular and heartfelt thanks go to those friends and colleagues who selflessly read, corrected, and commented on drafts of various chapters. These include Ami Berkowitz, Neal Bertram, Chia-Ling Chien, Miguel Ciria, Arthur Clark, Jeff Childress, Dennis Clougherty, Jim Cullen, Dick Gambino, C. D. Graham, Kin Ha, Hans Hug, Dave Lambeth, Laura Henderson Lewis, Mike McHenry, Tadashi Mizoguchi, Jagadeesh Moodera, Agustin del Moral, David Paul, Dan Pierce, Fred Pinkerton, Caroline Ross, John Unguris, and Hongru Zhai. I deeply appreciate the time they took to improve the treatment of various topics and search out errors. While much of the material presented here has appeared elsewhere, the selection, arrangement, and presentation of the material is, for better or worse, my own. I apologize in advance to those whose important contributions have been overlooked here because of my own unfamiliarity with them or my inability fully to appreciate their significance.

I am extremely grateful to Mrs. Lee Ward, who worked tirelessly on the drawings and patiently accepted my suggestions and reversals of opinion on many of the figures.

Special thanks go to Mr. Robin Lippincott, who worked with the manuscript as it evolved, through several generations of hardware and software, from class notes to its present form. The formidable task of keeping track of the references, figures, tables, and permissions as material was rearranged, was accomplished only by his tireless work. Robin dealt calmly with my unending revisions and changes of notation. He also applied his own professional writing skills to editing my drafts. Although his literary style soars above the dry technical genre, he was still able to detect and expunge many of my awkward and incorrect constructions. Thank you so much, Robin.

Finally, I would like to thank my wife, Carol, and children, Kevin, Meghan, and Kara, who shared my time with this book for too many nights, weekends, and years. Their love and encouragement made it easier to persevere when the task grew bigger than I ever would have imagined.

ROBERT C. O'HANDLEY

Massachusetts Institute of Technology
June 1999

SYMBOLS

A	Exchange stiffness constant, area
a, a_0	Lattice constant, first Bohr radius
B, $\mathbf{B}$	Flux density or magnetic induction, $B = \phi/\text{area}$
B_i	Magnetoelastic coupling coefficient
C	Curie constant
c	Speed of light
c_{ij}	Elastic stiffness constant
D	Grain size
d	Sample size, film thickness
E, E_F	Energy, Fermi energy
E_K	Kinetic energy
e	Electronic charge, strain
F	Helmholtz free energy, force
f	Helmholtz free energy per unit volume, frequency
G	Gibbs free energy
g	Gibbs free energy per unit volume, Landé g factor
H, $\mathbf{H}$	Magnetic field intensity due to macroscopic currents
$\mathscr{H}$	Hamiltonian
H_c	Coercive field, coercivity (used when ${}_BH_c = {}_iH_c$)
${}_BH_c$	Flux density coercivity, field at which $B = 0$
${}_iH_c$	Intrinsic coercivity, field at which magnetization vanishes
h, $\hbar$	Planck's constant, Planck's constant/2π
I	Current, moment of inertia
i	The imaginary number $(-1)^{1/2}$
J, $\mathbf{J}$	Current density or total angular momentum; scalar and vector
$\mathscr{J}$	Magnetic exchange integral

j, $\boldsymbol{J}$ Total angular momentum quantum number of electron, atom
K_u, K Uniaxial magnetic anisotropy energy coefficient, surface current density (current per unit length)
k, $\boldsymbol{k}$ Wavenumber, wavevector, of magnitude $2\pi/\lambda$, magnetomechanical coupling coefficient
k_B Boltzmann constant
$\boldsymbol{L}$, L Orbital angular momentum operator, quantum number
l Orbit angular momentum quantum number, correlation length
l_{ex} Exchange length
$l(r)$ Dipole coefficient
M, $\boldsymbol{M}$ Magnetization density, total magnetic moment per unit volume
m Mass
m_l, m_s Quantum number for the z component of orbital or spin angular momentum
m_j Quantum number for the z component of total angular momentum
N Number of atoms, domain walls, particles, and other quantities
N_v, n Number per unit volume
n_B Magneton number = number of Bohr magnetons per atom or molecule
p Momentum
Q Activation energy
q Charge
$q(r)$ Quadrupole coefficient
R Radius
r Radius
$\boldsymbol{S}$, S Spin operator, spin quantum number of atom
s Spin quantum number of electron
T Temperature, torque, kinetic energy
T_C Curie temperature
T_N Néel temperature
t Time, thickness
U Internal energy
u Internal energy per unit volume
V Volume
v Velocity
W Work done in a given process
w Width of sample or region
x, y, z Cartesian coordinates
α Direction cosine of magnetization vector
β Direction cosine of strain direction
γ Gyromagnetic ratio
δ_{dw} Domain wall thickness
δ_{cl} Classical skin depth
ε_0 Dielectric constant of free space
ν Frequency

π	Circular constant, 3.14159
π, π^*	Twofold rotationally symmetric bonding, antibonding molecular orbitals
κ	Symmetry invariant magnetic anisotropy constant
λ	Wavelength, molecular field coefficient
λ, λ_s	Magnetostriction, saturation magnetostriction
μ, μ_i	Permeability, initial permeability
μ_B	Bohr magneton, magnetic moment of the electron
μ_m	Local magnetic moment
χ	Paramagnetic susceptibility, spin wavefunction
ρ, ρ_m	Electrical resistivity, mass density
σ	Electrical conductivity, surface energy density, specific magnetization
σ, σ^*	Axially symmetric bonding, antibonding molecular orbitals
τ	Relaxation time
υ	Poisson's ratio
φ	Azimuthal polar angle
ϕ	Orbital wave function
ψ	Spatial wavefunction
Ψ	Spatial and spin wavefunction
ω, $\boldsymbol{\omega}$	Angular frequency $2\pi\nu$, angular velocity vector

ABOUT THE AUTHOR

Robert C. O'Handley is a Senior Research Scientist in the Department of Materials Science and Engineering at the Massachusetts Institute of Technology, where he has taught several undergraduate and graduate courses and conducted research on magnetic materials, superconductors, and a variety of rapidly solidified materials. Prior to M.I.T., he was a Research Staff Member at the IBM Thomas J. Watson Research Center in Yorktown Heights, New York and a staff physicist at Allied Corporation's Materials Research Center Morristown, New Jersey. He held a National Research Council Postdoctoral Research Associate position at Michelson Laboratory, China Lake, California. He earned his Ph.D. and M.S. in physics at the Polytechnic Institute of Brooklyn and taught for two years at a secondary school in New York City. He received his B.A. in physics from Marist College, Poughkeepsie, New York.

Dr. O'Handley has authored more than 200 technical and scientific publications, including several book chapters and review articles. He holds more than ten U.S. and international patents. He has been active in the annual Conference on Magnetism and Magnetic Materials since serving as Publications Co-chairman from 1979 to 1981. Web site: http://web.mit.edu/bobohand/www/

CHAPTER 1

INTRODUCTION AND OVERVIEW

This chapter serves as an introduction to magnetism and an overview of the aspects of magnetic materials to be covered in this text. It reviews many elementary concepts (Maxwell's equations in differential and integral form, units, concepts of magnetic fields and magnetic moments, types of magnetism, and generic applications) that may be familiar to some readers.

1.1 INTRODUCTION

The Chinese are believed to have first used a lodestone compass, what they call *shao shih* or *tzhu shih*, meaning loving stone (hence the French *l'aimant*, attraction or friendship, for magnet), more than four millennia ago. The shao shih is a ladle-shaped magnet that balances and pivots on a brass plate. The handle of the ladle is the north-seeking pole of the compass. The English word *magnet* came from Magnesia, the name of a region of the ancient Middle East, in what is now Turkey, where magnetic ores were found.

The Chinese knew abstract binary concepts such as yin/yang and male/female, as well as understanding the concrete binary process of counting by the presence or absence of a bead in an abacus. But the development of a magnetic abacus, that is, a computer with binary magnetic information storage, took thousands more years to achieve. The late Dr. An Wang pioneered the use of magnetic core memories in his early Wang computers. These magnetic memories (see Section 17.6.1) were tiny toroids of ferrite that could be individually magnetized clockwise or counterclockwise by simultaneous current pulses passing through two orthogonal wires defining a core address on a grid. After a current pulse of critical magnitude (related to the coercive field of the core) at a given address, the core remained magnetized (a remanent magnetization in a given direction persists in zero field). The core could be read or overwritten by later pulses.

Today, information technologies ranging from personal computers to mainframes use magnetic materials to store information on tapes, floppy diskettes, and hard disks. The dollar value of magnetic components coming out of Silicon Valley is greater than that of the semiconductor components made there. Our seemingly insatiable appetite for more computer memory will probably be met by a variety of magnetic recording technologies based on nanocrystalline thin-film media and magnetooptic materials. Personal computers and many of our consumer and industrial electronics components are now powered largely by lightweight switch-mode power supplies using new magnetic materials technology that was unavailable 20 years ago. Magnetic materials touch many other aspects of our lives. Each automobile contains dozens of motors, actuators, sensors, inductors, and other electromagnetic and magnetomechanical components using hard (permanent) as well as soft magnetic materials. Electric power generation, transformation, and distribution systems rely on hundreds of millions of transformers and generators that use various magnetic materials ranging from the standard 3% SiFe alloys to new amorphous magnetic alloys. Finally, magnetic materials are the backbone of expanding businesses: electronic article surveillance, asset protection, and access control. Tiny strips or films of specially processed magnetic materials store one or more bits of information about an item or about the owner of an identification badge. Access to secure areas or inappropriate removal of merchandise or property can be monitored and controlled.

The purpose of this text is to introduce readers to the basic concepts needed to understand magnetism, magnetic materials, and their applications and bring readers to a point from which they can appreciate the technical literature. The text moves from an exposition of the principles of magnetism to a consideration of some of the major classes and applications of magnetic materials. The introductory chapters consider magnetism from three perspectives:

1. What requirements do *Maxwell's equations* place on magnetism?
2. And does *classical electron theory* indicate?
3. Why is *quantum mechanics* needed to understand magnetism?

The major energies controlling magnetic processes, domain wall formation, and technical magnetism are considered next. These principles are then applied to specific magnetic materials used in various devices. Several chapters are dedicated to the properties of specific magnetic materials in three classes: soft, nanocrystalline, and hard magnetic materials. Processing and annealing are covered. A number of more advanced topics are included to meet the needs of specific audiences. Throughout the text the treatment is intended to be empirical, moving from observation to understanding. An effort is made to build new concepts on familiar ones.

It is worth reviewing what may be three fairly familiar observations related to magnetism. Their quantitative understanding will lead to a discussion of Maxwell's equations.

1.2 OBSERVATIONS RELATED TO MAGNETIC FIELDS

1.2.1 Field of a Current-Carrying Wire

When a current passes through a length of wire, a magnetic field having a direction indicated by the right-hand rule is generated (Fig. 1.1). Here the thumb indicates the direction of the positive current I and the fingers indicate the direction of the magnetic field lines B. It is important to know the magnitude and direction of this field as well as its dependence on current and on distance from the wire.

1.2.2 Field of a Solenoid

If a length of current-carrying wire is formed into a solenoid, a field can be observed about the solenoid (Fig. 1.2). The topology of the field has the cylindrical symmetry of a torus, and its sense is again given by a right-hand rule. With the fingers this time indicating the direction of the current, the thumb gives the direction of the magnetic field *inside* the solenoid. The field outside follows from the symmetry of a torus.

Because solenoids are often used to provide fields for testing magnetic materials, it is important to be able to calculate the strength of the field along the axis of the solenoid.

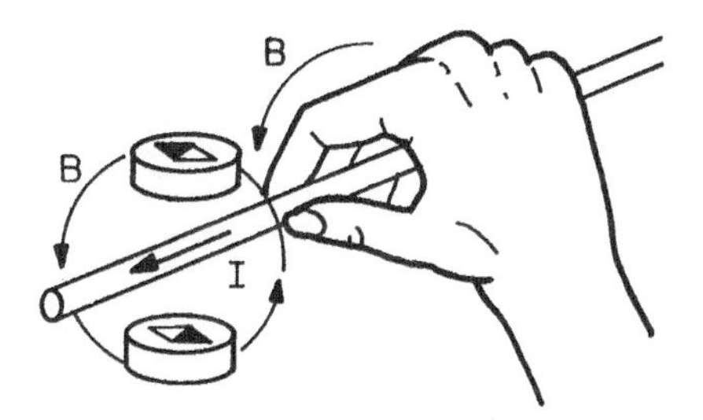

Figure 1.1 With thumb in direction of current, direction of magnetic B field about a current-carrying wire is given by the direction of the fingers according to the right-hand rule.

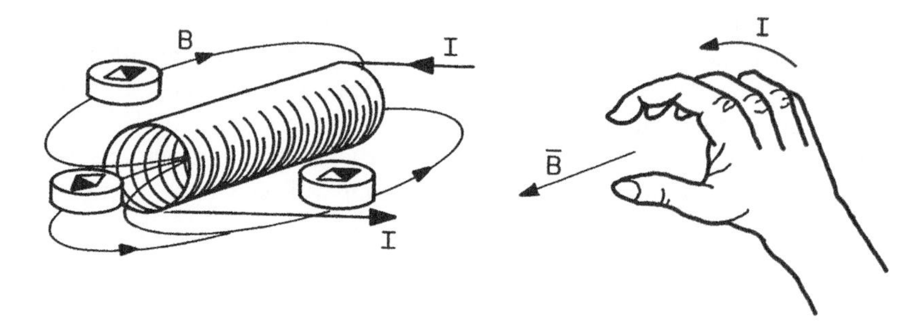

Figure 1.2 Direction of magnetic B field about a current-carrying solenoid is given by the right-hand rule.

It is believed that the magnetic field of the earth results from a current in its molten iron core. What would the direction of such a current have to be for the black end of a compass needle like that in Figure 1.2 to point to the geographic north pole?

1.2.3 Voltage Induced in a Coil

If a bar magnet is removed from inside a coil as shown in Figure 1.3, a voltage pulse will be detected across the windings of the coil. The voltage results from a change in magnetic flux inside the coil. This induced voltage is a result of Lenz' law—if there is a change in flux in a coil, a voltage is induced in the coil with a sense that would produce a current whose magnetic field opposes the initial change. Note that the voltage in Figure 1.3 is such that its current would create a field coming out of the coil (in the direction opposite the change caused by the motion of the bar magnet). The sense of the voltage is also given by a right-hand rule with assignments similar to those in Figure 1.2, except now B is replaced by $-dB/dt$.

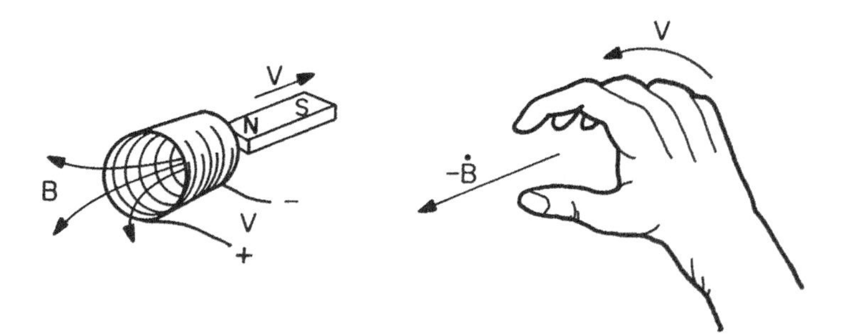

Figure 1.3 A change in flux through a coil results in a voltage in that coil whose sense is such that its current would create a field opposing the initial change.

The voltage induced by a change in flux is responsible for the operation of generators and transformers as well as for the material behavior known as *diamagnetism.*

Observations similar to 1 and 2 above were first reported by Hans Christian Oersted, and in the early nineteenth century André Ampère was able to describe them mathematically. Observation 3 was first recorded by Michael Faraday, who used it to write the mathematical form of the law of magnetic induction.

In order to be able to calculate the magnitude and direction of the fields described here qualitatively, the set of magnetic and electric fields must be defined and the equations that relate them to each other and to charge and current distributions, must bę understood.

1.3 QUANTITATIVE DESCRIPTION OF MAGNETIC FIELDS

Primarily mks (or Systéme International, SI) units will be used in the text. However, because the literature abounds with data in cgs and other units, cgs equations and quantities will often be given along with mks. An Appendix to this chapter summarizes the equations and conversion factors for the important magnetic quantities in mks and cgs units.

1.3.1 Constitutive Relations

The simplest and most intuitive equations needed are the constitutive relations that describe how a material responds to electric or magnetic fields. Ohm's law, $I = V/R$, relates current I and voltage V by the resistance R. Converting to intensive variables gives $\boldsymbol{J} = \sigma \boldsymbol{E}$, where $\boldsymbol{J}$ is the current density (amperes per unit area), $\boldsymbol{E}$ the electric field (volts per unit length), and σ is the conductivity of the material (the conductivity is the inverse of the electrical resistivity, $\sigma = \rho^{-1} = l/RA$, where l is the conductor length and A is its cross-sectional area).

It is known that an electric field can cause positive and negative charges in a material to be displaced relative to each other creating an *electric* dipole moment $\boldsymbol{p}$ (coulomb-meter). The definition of $\boldsymbol{p}_e$ is charge ($+$ and $-q$) times separation distance, $\boldsymbol{p}_e = qd$. The macroscopic dipole moment density is given by $P = np_e$ C/m^2 (coulombs per square meter), where $n = N/V$ = number/volume. $\boldsymbol{P}$ is related to $\boldsymbol{E}$ by an electric susceptibility χ_e:

$$\boldsymbol{P} = \chi_e \boldsymbol{E} \tag{1.1}$$

and the electric displacement field $\boldsymbol{D}$ is related to $\boldsymbol{E}$ and $\boldsymbol{P}$ through the permittivity tensor $\varepsilon = \varepsilon_r \varepsilon_0$, $\varepsilon_0 = 8.85 \times 10^{-12}$F/m (Farads per meter):

$$\boldsymbol{D} = \varepsilon_0 \boldsymbol{E} + \boldsymbol{P} = \varepsilon_0 \boldsymbol{E} + \chi_e \boldsymbol{E} = (\varepsilon_0 + \chi_e)\boldsymbol{E} = \varepsilon \boldsymbol{E} \tag{1.2}$$

Similarly, materials generally respond to an applied magnetic field H with a change in their *magnetic* dipole moment p_m. The macroscopic magnetic dipole density or magnetization, $M = np_m$, is given by

$$M = \chi_m H \tag{1.3}$$

where χ_m is the magnetic susceptibility.

The magnetic flux density B is related to M and H by the permeability $\mu = \mu_r \mu_0$ with $\mu_0 = 4\pi \times 10^{-7}$ henry/m

$$\begin{aligned} B &= \mu_0(H + M) = \mu_0(H + \chi_m H) = \mu_0(1 + \chi_m)H = \mu H \qquad \text{(mks)} \\ B &= H + 4\pi M = H + 4\pi\chi_m H = (1 + 4\pi\chi_m)H \qquad \text{(cgs)} \end{aligned} \tag{1.4}$$

Therefore $\mu_r = 1 + \chi_m$. The parameters μ_r and χ_m are different ways of describing the response of a material to magnetic fields. In general, the permeability and susceptibility are tensors because they relate two vector quantities that need not be parallel. In SI units, $\mu_0 M$ has the same units as B (tesla) and in cgs units, $4\pi M$ has the same units as B (gauss).

The magnetic response M of the sample to H causes B/μ_0 to differ from H inside the material. Thus H is the cause and M as the material effect. B is a field that includes both the external field $\mu_0 H$, due to macroscopic currents, and the material response, $\mu_0 M$, due to microscopic currents. B is important technically because it is the magnetic flux density ϕ/A Wb/m^2 (webers per square meter) and the change with time flux density gives rise to an electric field or to a voltage (Faraday's law, explained below). Ferromagnets represent a low reluctance path for magnetic field lines; hence they draw in the flux of a nearby field and add to it by their magnetization (Fig. 1.4).

A better understanding is needed of the atomic magnetic moment, $p_m = M/n$ analogous to the electric dipole moment, $p_e = P_e/n$. An explanation will be sought for the strength of the atomic magnetic dipole moment strength after a review of Maxwell's equations and their consequences.

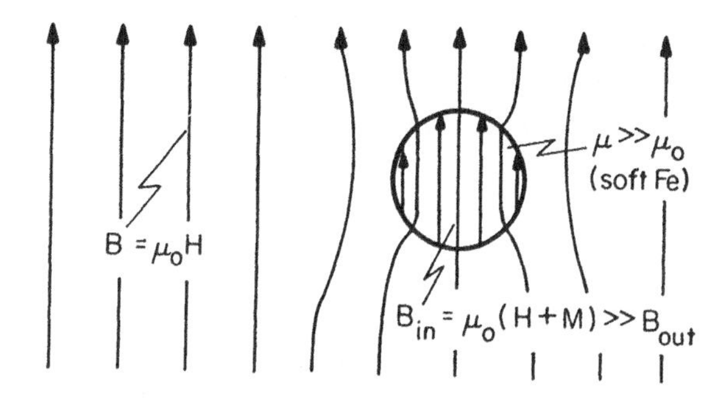

Figure 1.4 The B field in a vacuum is proportional to H. However, inside a material there is an additional contribution to B from the sample magnetization M.

1.3.2 Maxwell's Equations

The ***B***, ***H***, and ***E*** fields are related to each other and to charge and current densities ρ and J by a fundamental set of differential equations described by James Clerk Maxwell in 1865:

SI:

$$\nabla \cdot \boldsymbol{E} = \frac{\rho}{\varepsilon} \qquad \nabla \cdot \boldsymbol{B} = 0$$

$$\nabla \times \boldsymbol{E} = -\frac{\partial \boldsymbol{B}}{\partial t} \qquad \nabla \times \boldsymbol{B} = \mu_0 \boldsymbol{J} + \frac{\mu_0 \varepsilon \partial \boldsymbol{E}}{\partial t}$$

cgs:

$$\nabla \cdot \boldsymbol{E} = \frac{4\pi\rho}{\varepsilon} \qquad \nabla \cdot \boldsymbol{B} = 0 \tag{1.5}$$

$$\nabla \times \boldsymbol{E} = -\frac{c^{-1}\partial \boldsymbol{B}}{\partial t} \qquad \nabla \times \boldsymbol{H} = \left(\frac{4\pi}{c}\right)\boldsymbol{J} + \left(\frac{\varepsilon}{c}\right)\frac{\partial \boldsymbol{E}}{\partial t}$$

SI units are defined such that ***B*** is the more important field [as will be seen later, it is the ***B*** field that determines the energy of a magnetic moment, $U = -p_m \cdot \boldsymbol{B}$ (joule)] whereas cgs units are defined with ***H*** playing the major role [$U = -p_m \cdot \boldsymbol{H}$ (erg)].

The general form of these equations is such that a characteristic of a given field, such as its divergence or its rotational quality, is equated to a source term, namely, charge or current density or to a time change in a complementary field. Figure 1.5 illustrates diverging and curling vector fields as well as fields that have neither a divergence nor a curl. Comparing Eqs. (1.5) and Figure 1.5, it is clear that ***B*** fields can never terminate at a source, $\nabla \cdot \boldsymbol{B} = 0$ (unless a magnetic monopole were discovered) but they may show a curling character in the presence of a current density, ***J***.

The Maxwell–Ampère equation, $\nabla \times \boldsymbol{B} = \mu_0 \boldsymbol{J} + \mu_0 \varepsilon_0 \partial \boldsymbol{E}/\partial t$ indicates that a circulating ***B*** field results from a free current density or from an electric

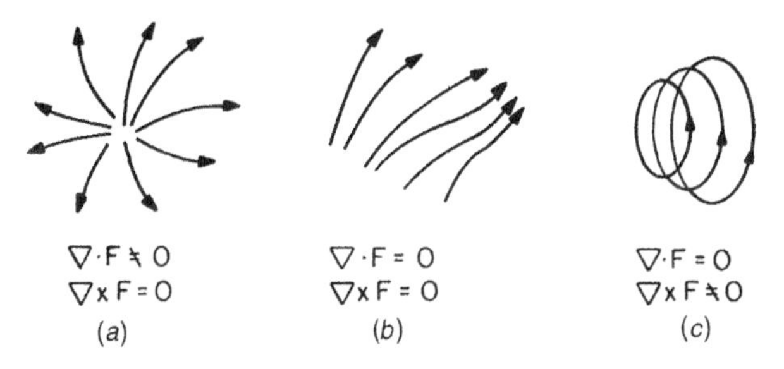

Figure 1.5 Topology of three different classes of vector field, **F**: (*a*) diverging (having a source) but not curling; (*b*) neither diverging nor curling; (*c*) divergenceless but curling.

polarization current. The $\boldsymbol{B}$ field curls around $\boldsymbol{J}$ in a right-hand sense (Figs. 1.1 and 1.2). The Maxwell–Faraday equation, curl $\boldsymbol{E} = \nabla \times \boldsymbol{E} = -\partial \boldsymbol{B}/\partial t$ requires that a time dependent $\boldsymbol{B}$ field gives rise to a spatially rotating $\boldsymbol{E}$ field normal to the direction of the change in $\boldsymbol{B}$, for instance, $-(\partial B/\partial t)\boldsymbol{e}_x = (\partial E_y/\partial z - \partial E_z/\partial y)\boldsymbol{e}_x$, where $\boldsymbol{e}_x$ is a unit vector in the x direction. The electric field curls around the direction of $-dB/dt$ in a right-hand sense (Fig. 1.3). The negative sign in the Maxwell–Faraday equation is a manifestation of Lenz' law; a changing $\boldsymbol{B}$ field induces a back electromotive force (EMF) opposing the current change that gave rise to the $\boldsymbol{B}$ field change. Alternatively, a changing $\boldsymbol{B}$ field induces an electric field whose current generates a magnetic field that opposes the change in the first $\boldsymbol{B}$ field.

There will often be interest in the use of Eqs. (1.5) to describe fields due to a given charge or current distribution. In these cases, the integral forms of Maxwell's equations are more useful.

Div $\boldsymbol{E} = \nabla \cdot \boldsymbol{E} = \rho/\varepsilon$ implies that charge density is the source of a diverging $\boldsymbol{E}$ field (Fig. 1.5). This can be seen more clearly on integrating this differential equation over a volume containing the charge density. Integrating both divergence Eqs. (1.5) over a volume enclosing the source charge distribution gives

$$\int \nabla \cdot E(x - x')d^3x' = \frac{1}{\varepsilon_0} \int \rho(x - x')d^3x'$$

$$\int \nabla \cdot B(x - x')d^3x' = 0 \tag{1.6}$$

A theorem due to Gauss, $\int (\nabla \cdot \boldsymbol{F})d^3x = \int \boldsymbol{F} \cdot d\boldsymbol{A}$, where $\boldsymbol{F}$ is a vector field, converts the volume integrals on the left-hand side (LHS) of Eqs. (1.6) to the integral of the normal component of the vector field over the surface enclosing that volume:

$$\oint E \cdot dA = \frac{1}{\varepsilon_0} \int \rho(x - x')d^3x'$$

$$\oint B \cdot dA = 0 \tag{1.7}$$

These equations say that the normal component of a field leaving a closed surface integrated over that surface (i.e., the field emerging from a volume) is equal to the total amount of source charge inside that volume. $\nabla \cdot \boldsymbol{B} = 0$ implies there can be no net outflow of $\boldsymbol{B}$ over any closed surface, therefore no sources of $\boldsymbol{B}$, no magnetic monopoles. Magnetic poles always come in pairs, usually designated north and south, called *dipoles*.

To get the integral forms of the curl equations (1.5), integrate them over a surface chosen to change the magnetic and current flux densities $\boldsymbol{B}$ and $\boldsymbol{J}$ to flux, $\phi = B \cdot A$, and current, $I = J \cdot A$ (neglect the electric displacement term):

$$\begin{aligned} \int (\nabla \times E) \cdot dA &= -\int \frac{\partial B}{\partial t} \cdot dA \\ \int (\nabla \times B) \cdot dA &= \mu_0 \int J \cdot dA \end{aligned} \tag{1.8}$$

A theorem due to Stokes, $\int (\nabla \times \boldsymbol{F}) \cdot \boldsymbol{dA} = \int \boldsymbol{F} \cdot \boldsymbol{dl}$, converts the area integrals of the curls to line integrals of the vector field, $\boldsymbol{F}$ along the path enclosing the original area:

$$\begin{aligned} \oint E \cdot dl &= -\frac{\partial}{\partial t} \int B \cdot dA = -\frac{\partial \phi}{\partial t} \\ \oint B \cdot dl &= \mu_0 \int J \cdot dA = \mu_0 I \end{aligned} \tag{1.9}$$

Equations (1.7) and (1.9) are integral forms of Maxwell's equations. The latter two, originating with Faraday and Ampère, respectively, are at the foundation of our traditional understanding and application of magnetism. Equations (1.9) say that the line integral of an $\boldsymbol{E}$ or $\boldsymbol{B}$ field about a closed path depends on the normal component of the complementary field passing through the surface area defined by that path:

Faraday—the normal component of a time-dependent $\boldsymbol{B}$ field passing through an area A induces a voltage along the closed path about that area.

Ampère—the normal component of a current density $\boldsymbol{J}$ passing through an area A gives rise to a $\boldsymbol{B}$ field circulating around that area.

It is often of interest to consider magnetostatic situations where $\partial \boldsymbol{E}/\partial t = 0$ and $\partial \boldsymbol{B}/\partial t = 0$ (i.e., $\nabla \times \boldsymbol{E} = 0$). But allowance should be made for the existence of currents, $\boldsymbol{J} \neq 0$ (that is why this case is sometimes called *magnetoquasistatics*; charges can still move, but the fields are independent of time). In this case, Maxwell's differential equations become

$$\begin{aligned} &\text{SI:} \quad &\nabla \cdot \boldsymbol{E} &= \frac{\rho}{\varepsilon} \quad &\nabla \cdot \boldsymbol{B} &= 0 \\ & &\nabla \times \boldsymbol{E} &= 0 \quad &\nabla \times \boldsymbol{B} &= \mu_0 \boldsymbol{J} \\ &\text{cgs:} \quad &\nabla \cdot \boldsymbol{E} &= \frac{4\pi\rho}{\varepsilon} \quad &\nabla \cdot \boldsymbol{B} &= 0 \\ & &\nabla \times \boldsymbol{E} &= 0 \quad &\nabla \times \boldsymbol{H} &= \left(\frac{4\pi}{c}\right) \boldsymbol{J} \end{aligned} \tag{1.10}$$

The integral forms of Maxwell's *magnetoquasistatic* equations are given by Eqs. (1.7 and 1.9) with $\partial f/\partial t = 0$:

$$\text{SI:} \quad \oint E \cdot dA = \frac{1}{\varepsilon_0} \int \rho(x - x')d^3x' \qquad \oint B \cdot dA = 0$$
$$\oint E \cdot dl = 0 \qquad \oint B \cdot dl = \mu_0 \int J \cdot dA = \mu_0 I \tag{1.11}$$

$$\text{cgs:} \quad \oint E \cdot dA = \frac{4\pi}{\varepsilon_0} \int \rho(x - x')d^3x' \qquad \oint B \cdot dA = 0$$
$$\oint E \cdot dl = 0 \qquad \oint H \cdot dl = \frac{4\pi}{c} \int J \cdot dA = \frac{4\pi}{c} I$$

This last set of equations (1.11) indicate that there is no net voltage along a path enclosing a *static* magnetic flux. Complete magnetostatics apply in the limit that there is no current flowing. Equations (1.10) apply with $J = 0$. The vanishing of $\nabla \times \boldsymbol{B}$ suggests that $\boldsymbol{B}$ can be derived from a magnetostatic scalar potential, ϕ_m: $\boldsymbol{B} = -\nabla\phi_m$, because the curl of a gradient is always zero. This case is treated in Chapter 2. These forms of Maxwell's equations form the basis of electrostatics and magnetostatics.

The most fundamental applications of these equations to magnetism are now reviewed.

1.4 MAGNETISM AND CURRENTS

1.4.1 Magnetic Field about a Straight Current-Carrying Wire

Consider the situation described in Figure 1.1. To get a quantitative measure of the $\boldsymbol{B}$ field due to the current in the wire, go to the last of Maxwell's equations $\nabla \times \boldsymbol{B} = \mu_0 \boldsymbol{J}$. Symmetry suggests that $\boldsymbol{B}$ circulates around the wire, so it can be assumed that $\boldsymbol{B}$ is circular and has a constant value at a distance R from the wire (Fig. 1.6).

Therefore, construct a circular surface of radius $R > r$ normal to the direction of current I and use the integral form of the Maxwell–Ampère equation, (1.9). From Figure 1.6, the integral of $\boldsymbol{B}$ around the circle of radius R is $2\pi RB$. The area integral of the RHS gives $\mu_0 I$. Hence

$$B = \frac{\mu_0 I}{2\pi R} \tag{1.12}$$

where B has units henry-amperes per square meter (HA/m^2, which is defined as Wb/m^2 or tesla). With units Wb/m^2, it is natural to define B as ϕ/A where

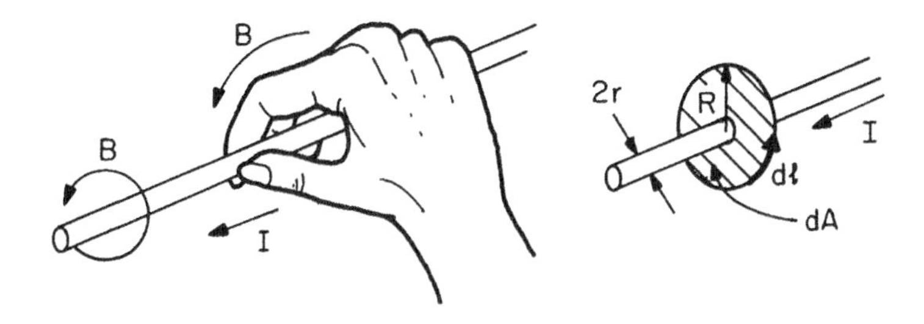

Figure 1.6 Right-hand rule for direction of magnetic field circulating about a current-carrying wire. Construction for calculation of strength of ***B*** field is shown on right.

ϕ is the number of magnetic flux lines and A is the perpendicular area through which they pass. Alternatively, it is possible to speak in terms of the magnetic field $H = B/\mu_0$ (A/m) generated by the current

$$H = \frac{I}{2\pi R} \tag{1.13}$$

In this text, the symbol H will be used for the magnetic field created exclusively by *macroscopic* currents or, as will be seen in the next chapter, by magnetic charges. The B field will be used for the magnetic field when *microscopic* currents contribute to the flux density, specfically, situations for which the contribution of the magnetization is important: $B = \mu_0(H + M)$.

The same result as Eq. (1.12) follows by integration of Ampère's differential equation, $\nabla \times B = \mu_0 J$. It is done most conveniently in cylindrical coordinates.

$$\nabla \times B = \frac{1}{r}\frac{\partial}{\partial r}[rB_\theta(r)] = \mu_0 J$$

Integrating from 0 to R

$$\int_0^R \left\{\frac{\partial}{\partial r}[rB_\theta(r)] = \mu_0 Jr\right\} dr$$

gives

$$B_\theta(R) = \mu_0 \frac{J_z R}{2} + C$$

Noting that J_z is the current *per unit area* πR^2 gives Eq. (1.12) ($C = 0$ because B vanishes at $R = \infty$.)

A more useful form of Eq. (1.12) is the differential vector solution to Maxwell's Ampère equation, the Biot–Savart law:

$$d\boldsymbol{B} = \frac{\mu_0 I}{4\pi} \frac{\boldsymbol{dl} \times \boldsymbol{r}}{r^3} \tag{1.14}$$

Note that Eq. (1.14) gives the correct direction for the $\boldsymbol{B}$ field relative to the current density segment and the distance from that segment to the observation point.

1.4.2 Moment of a Current Loop

The straight wire of the previous example (Fig. 1.6) is now formed in a loop. The magnetic field shape about the current loop is toroidal as can be deduced from the right-hand rule for a current-carrying wire (Fig. 1.6). A closed current loop has a *far field* given by

$$B = \frac{\mu_0}{4\pi}\left(\frac{2IA\cos\theta}{r^3}\mathbf{e}_r + \frac{IA\sin\theta}{r^3}\mathbf{e}_\theta\right) \tag{1.15}$$

where $\boldsymbol{e}_r$ and $\boldsymbol{e}_\theta$ are unit vectors in the $\boldsymbol{r}$ and θ directions, respectively, and A is the area of the loop. (This equation will be derived later.) At large distances, the field of a current loop is the same as that of a small bar magnet, a magnetic dipole (Fig. 1.7).

A solenoid is now formed consisting of N of these loops, each carrying a current I. The field inside the solenoid can be calculated simply by integrating the Maxwell–Ampère equation over an area normal to several adjacent turns (Fig. 1.8):

$$\oint B \cdot dl = \mu_0 \int J \cdot dA = \mu_0 NI \tag{1.16}$$

The sense of the line integral along the boundary of the area element is determined to be clockwise by the right-hand rule with respect to enclosed current.

Inside the solenoid the B field is strong because the field lines are compressed; outside the solenoid the B field is weak because the field lines are spread out. The return field just outside the middle of an infinite solenoid is zero; along the outside branch of the rectangular path in Fig. 1.8, $\boldsymbol{B}\cdot\boldsymbol{dl}$ is zero. Therefore, for the closed line integral of $\boldsymbol{B}$, it follows that

$$\int \boldsymbol{B}\cdot\boldsymbol{dl} = B_{\text{in}} l$$

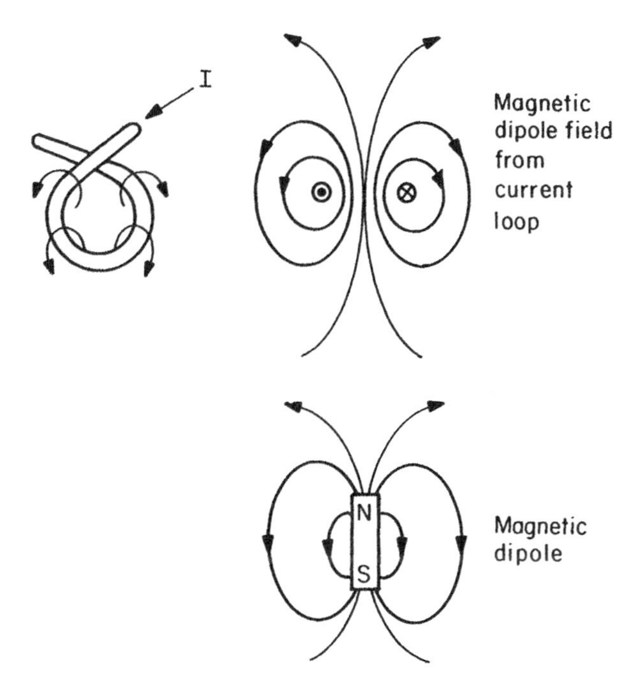

Figure 1.7 Upper left, magnetic field about a loop of current-carrying wire. Right, shape of dipole field about a current loop (above) and about a permanent magnet (below).

Finally, this suggests $Bl = \mu_0 NI$ or

$$B = \frac{\mu_0 NI}{l} \tag{1.17}$$

$$H = \frac{NI}{l}$$

($B = H = 0.4\pi NI/l$, practical units: H given in oersteds for I in ampère, l in centimeters).

Solenoids are often used as sources of magnetic fields in the laboratory and are sometimes filled with a soft iron core ($\mu \gg \mu_0$) to increase the magnetic field they produce.

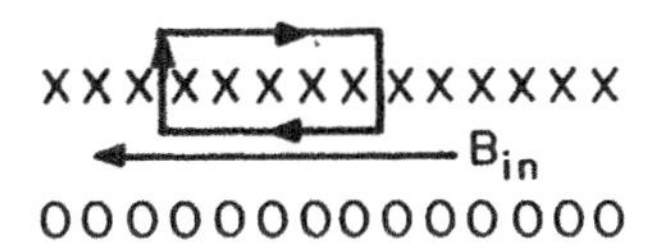

Figure 1.8 Cross section through a long solenoid (as in Fig. 1.2) showing path and sense of integration for calculating field inside solenoid using Eq. (1.16).

Equations (1.11)–(1.17) give the field, $H = B/\mu_0$, generated by *macroscopic* currents in different configurations. They may also apply to the fields generated by microscopic currents to give $B = \mu_0 H$. The fact that the magnetic moments of materials arise from *microscopic*, atomic-scale, currents is now justified.

1.4.3 Origin of Atomic Moments

What is the microscopic origin of magnetism in materials? What is the atomic magnetic moment p_m? It will be shown that atomic magnetic moments come from microscopic current loops.

Consider a line of N circular atomic current loops with a common axis (Fig. 1.9). This represents a number of atoms with their atomic orbitals aligned along a given direction in a material. The solenoid equation, Eq. (1.17), can be applied to this model where N/l is the number of atomic current loops per unit length. With no field applied by any external currents $H = 0$, so the B field along the axis of this atomic-scale solenoid is $B = \mu_0(H + M) = \mu_0 M$. Thus, Eq. (1.17) gives the field inside this "material":

$$B = \mu_0 M = \mu_0 \left(\frac{N}{l}\right) I \tag{1.18}$$

But $M = np_m$ and $n = N/Al$, where A is the area of the atomic current loop. Thus $M = NI/l = (N/Al)p_m$ implies

$$p_m = IA$$
$$\left(p_m = \frac{4\pi IA}{c} \text{ cgs}, \quad p_m = \frac{IA}{10} \text{ practical}\right) \tag{1.19}$$

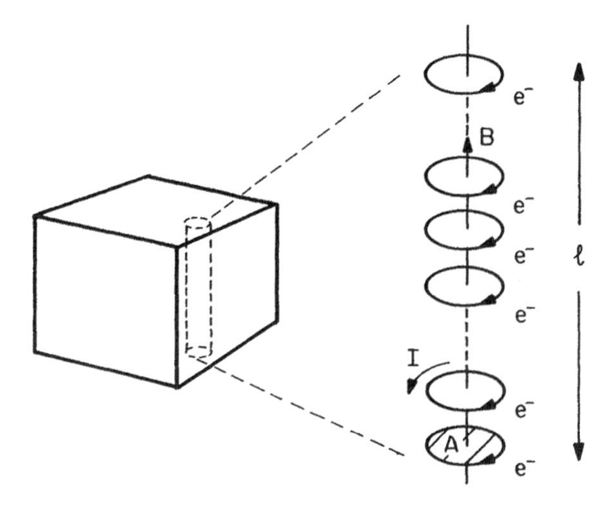

Figure 1.9 Schematic representation of a number of atomic current loops, each of area A.

with units Am^2. This is a crude plausibility argument for a very important relation. Thus, atomic magnetism has its origin in microscopic currents and the atomic magnetic moments can be calculated if the currents and the areas they enclose are known. Another microscopic current, in addition to electron orbital motion, also important to magnetism, is one due to the intrinsic angular momentum or spin of electrons. Spin will be discussed in Chapter 3. For now, note that IA can be replaced by p_m in Eq. (1.15) for the field of a current dipole:

$$\boldsymbol{B} = \frac{\mu_0}{4\pi}\left(\frac{2p_m \cos\theta}{r^3}\boldsymbol{e}_r + \frac{p_m \sin\theta}{r^3}\boldsymbol{e}_\theta\right) \tag{1.20}$$

From here on, the symbol μ_m will be used instead of p_m for the atomic magnetic moment. The p_m notation was used till now in order to avoid confusion with the permeability and also to draw attention to the analogy between electric, p_e, and magnetic, p_m, dipole moments.

The magnitude of $\mu_m = p_m$ can be estimated for a hydrogen atom in the Bohr model. $IA \approx e(\omega/2\pi)\pi r_0^2$ and take $\omega = v/r_0$ with the electronic velocity given by $v = (2E/m)^{1/2}$ (where E is the electronic energy of the $1s$ electron in hydrogen). This gives for the orbital magnetic moment, $\mu_m = IA \approx 9.27 \times 10^{-24} (\mathrm{Am}^2)$.

Now consider a magnetic material to be made up of about $n \approx 10^{29}/\mathrm{m}^3$ atoms per unit volume, each with a circulating current I which makes the atom behave like a magnetic dipole. The magnetization $M = n\mu_m$ of this assembly is thus of order 10^6 A/m or $B = \mu_0 M \approx 1$ T [in cgs, $B_s \approx 4\pi M_s = 10$ kilogauss (kG)]. For comparison, the saturation magnetizations, $B_s = \mu_0 M_s$ (all atomic moments aligned) of metallic Fe, Co, and Ni are about 2.2, 1.7, and 0.6 T, respectively.

1.5 TYPES OF MAGNETISM

This value $B \approx 1$ T represents an upper limit to the magnetization density for our hypothetical hydrogenic material. $B = 1$ T is typical for a magnetic material in which all atomic moments are aligned. But atomic magnetic moments are not necessarily aligned in all materials.

The way the local atomic moments couple to each other, parallel, antiparallel, or not at all, provides the first way of classifying magnetic materials. The individual atomic moments μ_m may be randomly oriented if they do not interact with each other. In this case $\Sigma\,\mu_m = 0$ in zero field. Such uncoupled magnetic moments may be aligned partially (depending on thermally induced agitation) in an applied magnetic field H. This weak field-induced magnetization behavior defines a paramagnet. Alternatively, the atomic dipoles may couple somehow to each other and cooperatively align so $\Sigma\,\mu_m \neq 0$ even in the absence of an applied field. This defines an ordered magnetic material, examples of which are ferromagnets, antiferromagnets, and ferrimagnets.

1.5.1 Weak Magnetism

The magnetic susceptibility χ_m is usually used to describe weak magnetic responses to H (Fig. 1.10) as in *paramagnetic* and diamagnetic materials. The magnitude of χ_m is usually $\pm 10^{-4}$ to 10^{-6} (dimensionless in SI). (It will be seen later that *diamagnetism* is not a matter of *aligning* preexisting atomic magnetic moments but rather an electronic response to B that creates a new atomic or molecular magnetic moment.) A material with a paramagnetic susceptibility of 10^{-5} would show a magnetization $M = 1$ A/m in a field of 10^5 A/m (applied field $B_0 \approx 0.1$ T). This value of magnetization corresponds to a flux density $\mu_0 M$ of order 10^{-6} T, which is much less than the 1 T that our hydrogenic model suggests for fully aligned moments. What is keeping the moments in a paramagnet from aligning in an external field? Perhaps it is thermal energy, $k_B T$.

The degree to which a paramagnetic moment will respond to a field can be appreciated by considering the potential energy U of the moment μ_m in an applied field B:

$$U = -\mu_m \cdot \mathbf{B}$$
$$(U = -\mu_m \cdot H \text{ cgs}) \tag{1.21}$$

[In mks units the energy is written as $-\mu_m \cdot \mathbf{B}$ because the factor μ_0 needed to

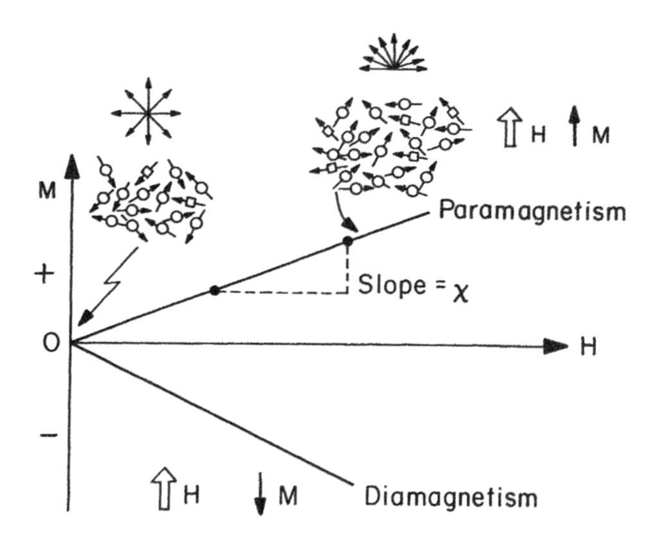

Figure 1.10 Field dependence of magnetization response in paramagnets and diamagnets. Inset shows a schematic of the distribution of local moments in the paramagnetic case.

relate μ_m to a flux density has been put with the field H. If μ_m were defined as $\mu_0 IA$ (Wb/m^2), then the energy would be $-\mu_m \cdot H$.]

Using the value $\mu_m \approx 10^{-23}$ A m^2 calculated above for a hydrogenic orbit gives $U \approx 9.3 \times 10^{-24}$ J $= 6 \times 10^{-5}$ eV (for $B = 1$ T, i.e., $H = 8 \times 10^5$ A/m). This energy is very small compared to thermal energy at room temperature, $k_B T \approx 4 \times 10^{-21}$ J $= \frac{1}{40}$ eV, so it is not expected that a 1-tesla field would produce much magnetic response in a paramagnet. One can even crudely estimate the fraction of saturation that would be measured in a thermally disordered paramagnet by multiplying $n\mu_m$ by the scale factor $(U/k_B T) = \mu_m B/k_B T$: $\chi_m = M/H \approx n\mu_m(U/k_B T)/(B/\mu_0) = \mu_0 n \mu_m^2/k_B T \approx 9.3 \times 10^{-4}$. So it can be seen why paramagnetic susceptibilities are so small. The field has only a weak linear effect in aligning the moments because thermal energy is large relative to the magnetic energy. (Proper derivations of susceptibilities will be given in Chapters 3 and 4.)

1.5.2 Ferromagnetism

Ferromagnetic materials are characterized by a long-range ordering of their atomic moments, even in the absence of an external field. The spontaneous, long-range magnetization of a ferromagnet is observed to vanish above an ordering temperature called the Curie temperature T_C (Fig. 1.11*b*).

The relative magnetic permeability $\mu_r = \mu/\mu_0 (= 1 + \chi_m)$ is used more often than susceptibility to describe the magnetic response of ferromagnetic materials to H. This is because ferromagnets are useful in electromagnetic devices where it is B, or more specifically $\partial B/\partial t$, that is important in producing a voltage (see Faraday's law, explained below).

Ferromagnets are useful because a large $B \approx 1$–2 T is produced by a fairly small field, $H \approx 100$ A/m ($B \approx 10^{-4}$ T $= 1$ gauss, e.g., 10 turns/cm carrying only 0.1 A!) (Fig. 1.11). The full magnetization of ferromagnets $M \approx B/\mu_0$ of

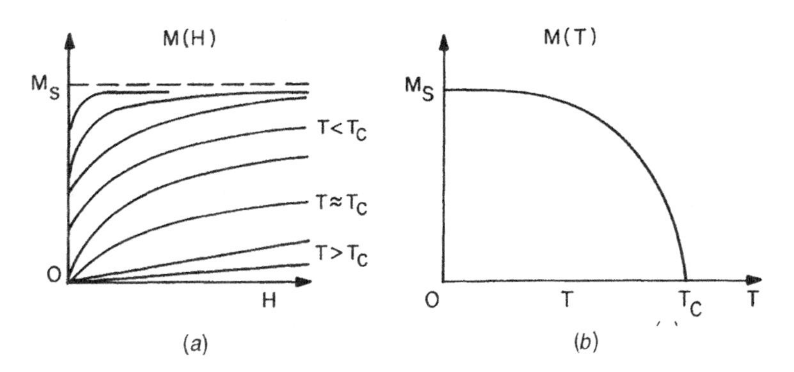

Figure 1.11 (*a*) Magnetization of a strongly magnetic material (e.g., a ferromagnet) versus field; (*b*) temperature dependence of the saturation magnetization. T_C is the Curie temperature.

order 10^6 A/m is just what our simple atomic model estimated for fully aligned moments. Where does the energy for magnetic alignment in a ferromagnet come from? For a paramagnet, M was found to be of order $n\mu_m(U/k_BT)$. Is not the moment-disordering effect of thermal energy just as strong in a ferromagnetic solid; could there be something other than an external field contributing to the tendency to align the local moments?

Pierre Weiss thought long and hard about this problem and eventually concluded that in ferromagnetic materials there must exist a giant "molecular" field, H_{molec} that is present even when no external field H_{ext} is applied: $H = H_{\text{ext}} + H_{\text{molec}}$. H_{molec} overcomes the thermal agitation and essentially aligns all the atomic moments in a ferromagnet so that $M \approx n\mu_m = M_s$ instead of $M \approx n\mu_m(U/k_BT)$ as in a paramagnet. The molecular field will be examined in Chapters 4 and 5.

But if ferromagnets have such strong magnetizations, why do two pieces of iron not attract each other the way they are attracted to a permanent magnet? How is iron magnetized or demagnetized? Weiss came up with an hypothesis for that, too. He postulated the existence of magnetic domains, regions (ranging in size upwards from approximately 0.1 μm) in a ferromagnetic material over which all moments are essentially parallel (Fig. 1.12). Domains are separated from each other by domain walls, surfaces over which the orientation of μ_m changes relatively abruptly (within about 10–100 nm). The domain walls will be the subject of Chapter 8 and domains will be examined more closely in Chapter 9.

The magnetizations in different domains have different directions so that over the whole sample their vector sum may vanish. Figure 1.13 is an image of the magnetic domains at the surface of a body-centered cubic (BCC) 3% SiFe crystal. (The image was taken with a scanning electron microscope fitted with a special detector to reveal magnetization direction; see Chapter 16.) In this figure the $\langle 100 \rangle$ directions of the crystal are parallel to the figure edges. It is not a coincidence that M inside the domains is parallel to these crystallographic directions. The coupling of the direction of M to the crystal axes is

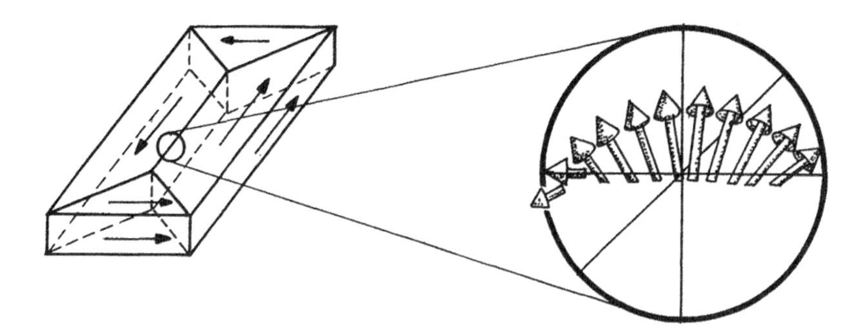

Figure 1.12 Simplified picture of magnetic domains in an iron crystal and the spatial variation of the atomic moments within the wall shown in an expanded view.

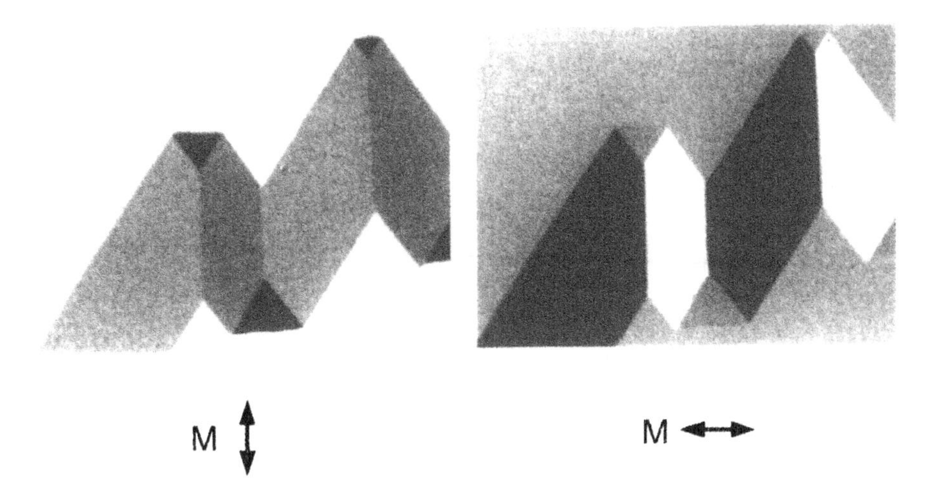

Figure 1.13 Magnetic domains at the surface of a 3% Si-Fe crystal taken by scanning electron microscopy with spin polarization analysis (SEMPA; see Chapter 16). Crystallographic ⟨100⟩ axes lie in the image plane along the horizontal and vertical directions. Left panel shows magnetic contrast when the instrument is sensitive to the horizontal component of magnetization; dark is magnetized to the left, light to the right. In the right panel, the contrast is sensitive to vertical component of magnetization; dark is magnetized down, light is magnetized up. (Courtesy of R. S. Celotta et al.)

called magnetocrystalline anisotropy (Chapter 6). Note also that domains often form so as to create closed flux loops. This minimizes magnetostatic energy (Chapter 2).

In an ideal magnet, does it matter much *where* the domain wall lies? If the potential energy of a domain wall were independent of its position, it would only take a relatively weak field to move the wall, much as it is relatively easy to move a ripple in a carpet. In some soft magnetic materials, domain walls can be moved with fields of order 0.1 A/m. However, defects such as grain boundaries and precipitates cause the wall energy to depend on position, so in most materials, higher fields are required to move domain walls (Chapter 9). It is now possible to understand, at least qualitatively, the B–H loops of ferromagnets and connect them with domain wall motion and other magnetization processes. But before describing the B–H loop of a ferromagnet, the classification of strongly magnetic materials should be completed.

1.5.3 Antiferromagnetism and Ferrimagnetism

In some magnetically ordered materials (i.e., materials with atomic magnetic moments having long-range correlation in their orientations) the atomic moments couple in antiparallel arrangements with zero net moment, rather than parallel as in a ferromagnet. Such materials are called *antiferromagnets.*

In a crystalline material, the structure dictates how the antiparallel spins are arranged. Many transition metal monoxides assume the rocksalt structure and are antiferromagnetic with alternate $\{111\}$ planes oppositely magnetized. Chromium has the BCC structure with the body-center atoms having one direction of spin (parallel to $\langle 100 \rangle$) and the corner atoms having the opposite direction. (Actually, the period of the magnetic fluctuation is incommensurate with the crystal lattice by about 5%.) Antiferromagnets have limited technical application because their net magnetization is zero; they produce no external field and the direction of the atomic moments is not easily affected by an external field.

When two antiferromagnetically coupled sublattices in a material have unequal moments—usually because different species are found on the different sites—the net moment is not zero. Such materials are called *ferrimagnets.* An example is magnetite, Fe_3O_4, for which iron ions of different valence are found at sites of different coordination. Ferrimagnets are very important technically because of their good high-frequency properties. They will be covered in Chapters 4 and 11.

1.6 TECHNICAL MAGNETIC MATERIALS

1.6.1 *B-H* Loops and Magnetic Domains

Consider a magnetic material in the demagnetized state ($B = 0$, $H = 0$ in Fig. 1.14).

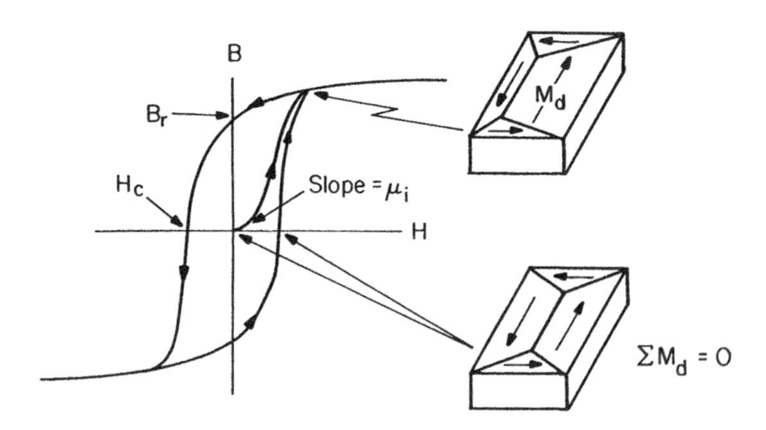

Figure 1.14 Hysteresis loop of a magnetic material showing the variation of B with changing H. Initial magnetization curve from the demagnetized state is shown with the initial permeability μ_i indicated. The remanence B_r and coercive field H_c are indicated. The approximate domain structures are indicated at right for demagnetized state and for approach to saturation.

Application of a weak field produces motion of domain walls such as to expand the volume of those domains having the largest component of M along H. (The macroscopic form of the energy of an atomic moment μ_m in a $\boldsymbol{B}$ field, $U = -\mu_m \cdot \boldsymbol{B}$, is $NU = -\boldsymbol{M} \cdot \boldsymbol{B}V$, where V is the volume of the domain of magnetization density M.) The initial induction B produced in response to a small field H defines the initial permeability $\mu_i = (B/H)_{H \approx 0}$. (The initial permeability can be as great as $10^5 \mu_0 \approx 10^{-1}$ in some materials. This is quite a contrast to a paramagnet where $1 + \chi_m > 1$ and typically of order 10^{-6} to 10^{-4}.) At higher fields B increases sharply and the permeability increases to its maximum value $\mu_{\max}$. When most domain wall motion has been completed there often remains domains with nonzero components of magnetization at right angles to the applied field direction. The magnetization in these domains must be rotated into the field direction to minimize the potential energy $-\boldsymbol{M} \cdot \boldsymbol{B}$. This process generally costs more energy than wall motion because it involves rotating the magnetization away from an "easy" direction [which may be fixed in the sample by sample shape (Chapter 2), crystallography (Chapter 6), stress (Chapter 7), or atomic pair ordering (Chapter 14)]. When the applied field is of sufficient magnitude that these two processes, wall motion and magnetization rotation, are complete, the sample is in a state of magnetic saturation $B_s = \mu_0(H + M_s)$ (in cgs units $B_s = H + 4\pi M_s$).

On decreasing the magnitude of the applied field, the magnetization rotates back toward its "easy" directions, generally without hysteresis (i.e., rotation is a largely reversible, lossless process). As the applied field decreases further, domain walls begin moving back across the sample. Because energy is lost when a domain wall jumps abruptly from one local energy minimum to the next (Barkhausen jumps), wall motion is an irreversible, lossy process. The B–H loop opens up, that is, it shows hysteresis, when lossy magnetization processes are involved. The induction and magnetization remaining in the sample when the applied field is zero are called the residual induction B_r and remanence M_r, respectively. The reverse field needed to restore B to zero is called the *coercivity* or *coercive field* H_c. It is a good measure of the ease or difficulty of magnetizing a material. (The field needed to restore M to zero is called the *intrinsic coercivity*, ${}_iH_c$. The distinction between H_c and ${}_iH_c$ is of importance only in permanent magnets because in a soft magnetic material $H_c \ll M$ so that $M = 0$ for essentially the same field that gives $B = 0$.) Cyclic application of an applied field causes the material to respond in the way described by its B–H loop. The magnetization process, both static and dynamic, is described in Chapter 10, and applications to soft and hard magnetic materials are presented in Chapters 11 and 12, respectively.

1.6.2 Soft Magnetic Materials

When the magnetization processes (domain wall motion and domain magnetization rotation) occur in weak fields, $H_c \leqslant 10^3$ A/m (readily generated by a modest current though a few turns of wire), the material is called a *soft magnet.* In some very soft magnetic materials such as certain crystalline NiFe alloys (permalloys) or amorphous metallic alloys, H_c can be as low as 1.0 A/m (12 mOe). The earth's magnetic field is about 0.4 Oe or 30 A/m. Other soft magnetic materials include pure Fe, Fe with up to 3 or even 6% Si, Ni, many FeNiCo alloys, and ferrites such as $(MnZnO)Fe_2O_3$ or $(NiZnO)Fe_2O_3$. Applications of soft magnetic materials exploit the large flux changes (proportional to $\partial B/\partial t \approx \omega B_s$) that occur in these materials with relatively weak changes in applied field. Soft magnetic materials are used in transformers both large (tens of tons) and small (ounces), in inductors, motors, and generators and as field sensors in magnetic recording or as stress/strain gauges.

The area inside the B–H loop, of order $4B_rH_c$, is the energy per unit volume lost per cycle in magnetizing the material. It is called the *hysteresis loss.* (The product BH has units $\text{Wb} \cdot \text{A/m}^3 = \text{J/m}^3$.) In AC (alternating-current) applications the power loss is the frequency times the AC loop area ($\text{J/m}^3 \cdot \text{s}$). Soft materials are covered in Chapter 10.

1.6.3 Hard Magnetic Materials

In some magnetic materials the coercivity can be as high as 2×10^6 A/m (25,000 G). This is the case if the material has defects that strongly impede wall motion or if it consists of single domain particles with such high magnetic anisotropy that the direction of magnetization is changed only by very large fields. Such materials resist demagnetization (once magnetized) and therefore are called *hard or permanent magnets.* Hard magnets are generally used in applications where they resist a negative applied field that pushes them back along the demagnetization part (second quadrant) of their M–H loop. Because they resist demagnetization in the presence of a negative field, hard magnets exert a repelling force against the source of a negative imposed field. Thus, permanent magnets store energy like a spring; they exert a restoring force *without contact.* The energy stored in a permanent magnet is related to the area inside the second quadrant of the B–H loop (Fig. 1.14). The $(BH)_{max}$ quality factor commonly used to rate permanent magnets is the maximum value of the BH product along the B–H curve in second quadrant.

Hard magnets are used in many motors and actuators; they also find applications in frictionless bearings, microwave generators, and lenses for charged particle machines. Common permanent magnets are Alnico, hexagonal ferrites such as $BaO \cdot 6Fe_2O_3$ and the newer rare earth magnets based on Co_5 Sm or $Nd_2Fe_{14}B$. Permanent magnetic materials are covered in Chapter 13.

1.6.4 Recording Media

In between these two extremes of soft and hard magnets lie magnetic recording media. They will remain magnetized in the face of ambient fields from nearby components or electrical currents but can be reversed by application of a suitable field $H > H_c$ with $10^4 < H_c \leqslant 10^5$ A/m (which is of order several hundred to several thousands of oersteds). In analog recording the degree of magnetization changes with position in proportion to the amplitude of the signal to be stored. In digital recording, use is made of the binary quality of the magnetization when there is a preferred axis of magnetization. A magnetic recording medium is usually made of an array of microscopic, independently magnetizable elements. The γ phase of Fe_2O_3, often coated or impregnated with cobalt, has been one of the most widely used materials for magnetic recording. This material can be processed to give rod-shaped particles about 10 nm in diameter and 100 nm in length. These particles are coated onto Mylar tape or onto aluminum disks (Winchester® technology) to make magnetic recording tapes and hard disks. More recently, high-density magnetic recording materials have been made of thin films, including CoP alloys, doped CoCr, metallic CrO_2, and $(FeCo)_{80}Tb_{20}$. Digital or analog information can be written on the surface of a magnetic medium using a small electromagnet capable of magnetizing the recording medium on a scale measured in micrometers. Reading the information can also be done with the same write head or with a special read head using the magnetoresistance effect (Chapter 14). Keep in mind that these heads write and read magnetic fields on a scale of micrometers! This microscopic control of magnetization is amazing, especially when you compare it with electrical power transformers weighing tons and handling megawatts of power. Both of these extremes of magnetic material application operate on the principles contained in Maxwell's Ampère and Faraday equations. Magnetic recording is treated in Chapter 17.

1.7 GENERIC APPLICATIONS

This introductory chapter began with a description of three simple and fundamental magnetic observations: the magnetic field about a wire, that inside a solenoid and the voltage induced in a coil by a flux change. So far discussed are mostly the first two effects, related to the magnetic fields produced by currents (Ampère's law). Attention is now given to the third observation, the appearance of a voltage in a conducting loop through which the magnetic flux changes. This is Faraday's law of induction.

1.7.1 Inductors and Transformers

Inductors are circuit elements that resist a change in current ($L dI/dt = L d^2q/dt^2$) just as inertial mass resists an acceleration $m\, d^2x/dt^2$. Inductors are most commonly found as multiple windings of a current-carrying wire (Figure 1.15).

For a sinusoidal current $I = I_0 \sin \omega t$, the electrical equation of motion $L\partial I/\partial t + RI = V(t)$ gives for the voltage across an inductor $V_L = LI_0\omega \cos \omega t$. Thus I and V_L are 90° out of phase. The current in the windings can also give the flux density in the core of the inductor [Eq. (1.6)] $B = \mu NI/l$, where both

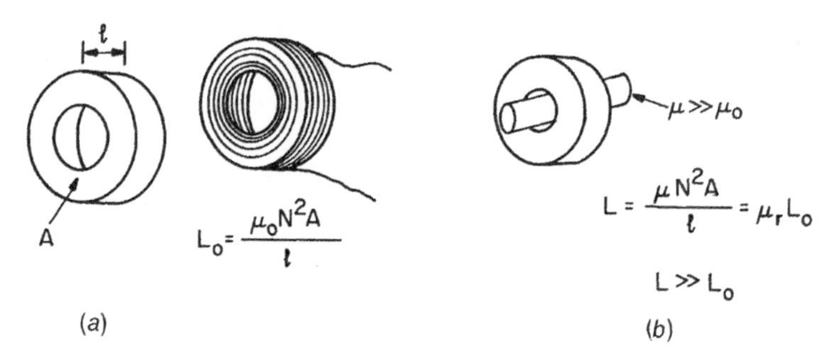

Figure 1.15 (*a*) Inductance of a hollow inductor; (*b*) inductance of a coil filled with a magnetic core.

I and B are time-dependent. From this expression for induction, an alternate expression can be derived for the voltage in the circuit using Faraday's law: $V = -NA\, dB/dt = -N^2 A\omega\mu I_0 \cos(\omega t)/l$. Comparing the two expressions for the voltage, $LI_0\omega \cos \omega t$ and $\mu N^2 A\omega I_0 \cos(\omega t)/l$, it can be seen that the value of the inductance for an air-filled core is given by

$$L_0 = \mu_0 N^2 \frac{A}{l} \tag{1.22}$$

where N is the number of turns, A the winding area, and l the length as illustrated in Figure 1.15. Comparison with Figure 1.8 shows that as current goes from zero to I, the number of magnetic flux lines inside the coil increases. By the Maxwell–Faraday equation, $\partial\phi/\partial t$ induces a voltage in each of the N windings that opposes (by Lenz' law) the current change producing the flux change. Thus inductors oppose changes in current; they provide electrical inertia.

Clearly, if the inductor is filled with a high-permeability magnetic core, then μ_0 in Eqs. (1.21) becomes $\mu = \mu_0 \mu_r$ with $\mu_r \gg 1$. In this way the inertial effect can be enhanced by orders of magnitude or the same inductance can be maintained with fewer turns and/or smaller inductor area.

A transformer is a pair of inductors coupled only by each other's flux changes (Fig. 1.16); a change in current in one set of turns induces a voltage in the other. When the inductors are linked by a soft, high-permeability core (Fig. 1.16*b*), the linkage or mutual inductance is strongly enhanced. Electrical signals can be transferred through the transformer with their DC (direct-current) components blocked. Also, large differences in AC voltage or current can be generated across a transformer depending on the ratio of turns N_1/N_2. As was the case for the inductor, the primary current can be related by Eq. (1.16) to the flux density linking the two sets of turns $B = \mu N_1 I/l$. This time dependent flux density induces an EMF in each set of turns $V_i = -N_i A\, dB/dt$, where $i = 1$ or 2 for primary or secondary windings. From these equations it is clear that

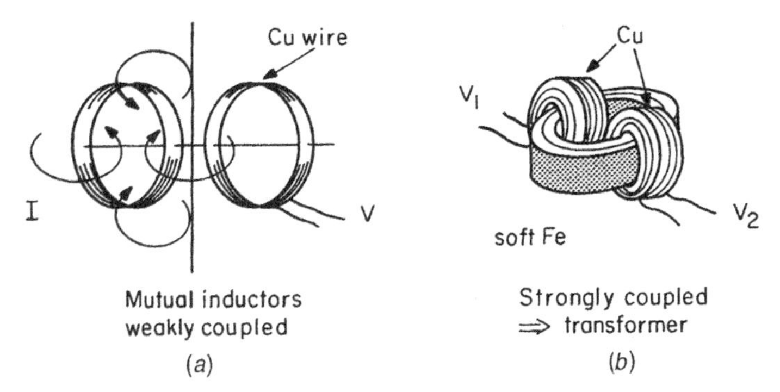

Figure 1.16 (*a*) Two coils near each other are coupled by the flux associated with a current through one of the coils—changes in the flux induce a voltage in the second coil; (*b*) the coupling between two coils is strongly enhanced if they are filled with a magnetic material that concentrates the flux linking the two coils—this is a transformer.

$V_2/V_1 = N_2/N_1$. If the transformer is purely inductive [i.e., $Ld^2q/dt^2 + Rdq/dt + Cq = V(t)$, with $R = C = 0$], there is a 90° phase shift between the current and voltage. In this case the energy lost in the transfer $\int IVdt$ over a complete cycle, is zero. In reality not only do the electrical windings have resistance but the magnetic core material itself does not perfectly link N_1 and N_2. The core also contributes to the loss of the equivalent circuit of the transformer. The phase lag then differs from 90°, and $\int IVdt$ gives the loss per cycle in the transformer. The loss is made up of core loss and coil loss. The core loss is related to the B–H loop area at the frequency of operation of the transformer. Transformer core loss represents a significant fraction of the national electricity bill. It has been estimated at tens of billions of dollars per year for distribution transformers alone (and they are among the most efficient).

1.7.2 Shielding

Finally, a word about shielding. It is known that good metals, $\sigma \approx 10^7(\Omega \cdot \text{m})$ ($\rho = 10\mu\Omega \cdot \text{cm}$), can shield high-frequency electromagnetic fields by the skin effect $\delta = [2/(\mu_0 \sigma \omega)]^{1/2}$. If a magnetic material is used as a shield, μ_0 becomes $\mu_0 \mu_r$, $\mu_r \gg 1$ and the shielding can be much more effective. In addition, magnetic materials can block DC or low-frequency magnetic fields, which nonmagnetic metals cannot do. They do this by channeling flux lines through the highly permeable material and reducing the flux density in the adjacent space. Soft magnetic materials (often a Ni-rich FeNi alloy called *mu-metal*, or amorphous magnetic alloys) are used to shield the cathode ray tubes in televisions, computer monitors and oscilloscopes from the earth's magnetic field $H \approx 30$ A/m (≈ 0.4 Oe) and from fields created by other current-carrying elements.

TABLE 1.1 MKS/SI and CGS Units and Conversion Factors for Common Magnetic Quantities

MKS, SI	CGS	Conversion
$H = NI/l$ (A/m)	$0.4\pi NI/l$ (Oe)	79.6 A/m = 1 Oe
$B = B = \mu_0(H + M)$ (T)	$B = H + 4\pi M$ (G)	$1\,T = 10^4 G$
$M = M = \chi_m H$ (A/m)	$M = \chi_m H$ (A/m)	79.6 A/m = 1 Oe
$\mu = \mu_0(1 + \chi_m)$ (H/m)	$1 + 4\pi\chi_m$ (none)	
$p_m = \mu_m = iA$ (Am^2)	iA (emu, $G \cdot cm^3$)	
$U = -\mu_m \cdot \boldsymbol{B}$ (J)	$-\mu_m \cdot \boldsymbol{H}$ (erg)	10^7 erg = 1 J
$u = -\boldsymbol{M} \cdot \boldsymbol{B}$ (J/m^3)	$-\boldsymbol{M} \cdot \boldsymbol{H}$ (erg/cm^3)	$10\, erg/cm^3 = 1\, J/m^3$

It is hoped that this chapter has given the reader a survey of the broad range of issues involved in the study of magnetism and magnetic materials. The next chapter takes a more careful look into the issues and concepts touched on here. Chapter 2 begins by examining the macroscopic fields produced by magnetized materials. This is the field of magnetostatics.

1.8 SUMMARY

Table 1.1 gives the equations for some of the important magnetic quantities in mks and cgs units, as well as the conversion factors between them. Note that units derived from proper names (e.g., gauss or tesla from Karl Friedrich Gauss and Nicola Tesla) are abbreviated with capital letters (G or T, respectively) while simple quantities such as meters and seconds are not. The value of the permeability of free space is $\mu_0 = 4\pi \times 10^{-7}$ (H/m).

Some other important relations are given in Table 1.2.

Maxwell's equations in differential and integral form are given in the two systems of units given in Table 1.3.

TABLE 1.2 Equations for Field About a Wire and a Solenoid in MKS and CGS Units

	MKS, SI	CGS
Field about wire	$B = \mu_0 i/2\pi R$ (T)	$H = 2i/cR$ (Oe)
Solenoid	$B = \mu_0 Ni/l$ (T)	$H = 0.4\pi Ni/l$ (Oe)

TABLE 1.3 Maxwell's Equations in Differential and Integral Form in MKS and CGS Units

MKS, SI	CGS
$\nabla \cdot \boldsymbol{E} = \rho/\varepsilon$	$\nabla \cdot \boldsymbol{E} = 4\pi\rho/\varepsilon$
$\nabla \cdot \boldsymbol{B} = 0$	$\nabla \cdot \boldsymbol{B} = 0$
$\nabla \times \boldsymbol{E} = -\partial \boldsymbol{B}/\partial t$	$\nabla \times \boldsymbol{E} = -c^{-1}\partial \boldsymbol{B}/\partial t$
$\nabla \times \boldsymbol{B} = \mu_0 \boldsymbol{J} + \mu_0 \varepsilon \partial \boldsymbol{E}/\partial t$	$\nabla \times \boldsymbol{H} = (4\pi/c)\boldsymbol{J} + (\varepsilon/c)\partial \boldsymbol{E}/\partial t$
$\int \boldsymbol{E} \cdot \boldsymbol{dA} = (1/\varepsilon) \int \rho(x - x']d^3x' = q/\varepsilon$	$\int \boldsymbol{E} \cdot \boldsymbol{dA} = (4\pi/\varepsilon) \int \rho(x - x']d^3x' = 4\pi q/\varepsilon$
$\int \boldsymbol{B} \cdot \boldsymbol{dA} = 0$	$\int \boldsymbol{B} \cdot \boldsymbol{dA} = 0$
$\int \boldsymbol{E} \cdot \boldsymbol{dl} = -\partial/\partial t \int \boldsymbol{B} \cdot \boldsymbol{dA}' = -\partial\phi/\partial t$	$\int \boldsymbol{E} \cdot \boldsymbol{dl} = -(1/c)\partial/\partial t \int \boldsymbol{B} \cdot \boldsymbol{dA}' = -(1/c)\partial\phi/\partial t$
$\int \boldsymbol{B} \cdot \boldsymbol{dl} = \mu_0 \int \boldsymbol{J} \cdot \boldsymbol{dA}' = \mu_0 I$	$\int \boldsymbol{B} \cdot \boldsymbol{dl} = (4\pi\mu_0/c) \int \boldsymbol{J} \cdot \boldsymbol{dA}' = (4\pi\mu_0/c)I$

APPENDIX: GAUSS AND STOKES THEOREMS

A theorem due to Gauss, $\int (\nabla \cdot \boldsymbol{F}) d^3 x = \int \boldsymbol{F} \cdot dA$, provides a useful way to convert the volume integral of the divergence of a vector field $\boldsymbol{F}$ to the integral of the normal component of the vector field over the surface enclosing that volume. Figure 1A.1 illustrates this in Cartesian coordinates.

A theorem due to Stokes, $\int (\nabla \times \boldsymbol{F}) \cdot \boldsymbol{dA} = \int \boldsymbol{F} \cdot \boldsymbol{dl}$, converts the area integral of the curl of $\boldsymbol{F}$ to a line integral of $\boldsymbol{F}$ along the path enclosing the original bounded area (Fig. 1A.2).

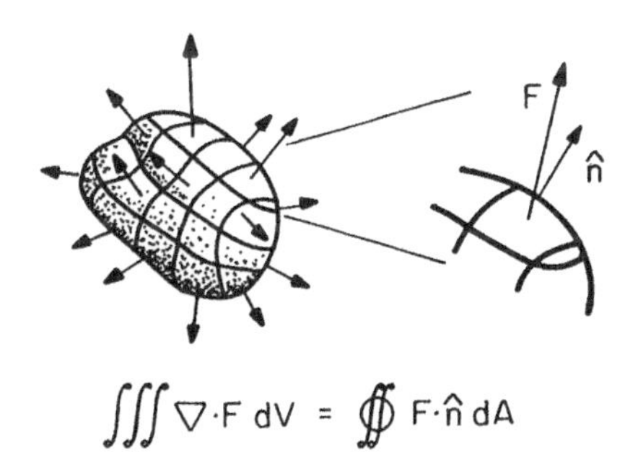

Figure A1.1 Graphical illustration of the meaning of Gauss' theorem.

PROBLEMS

1.1 What is the field in the middle and at the end of a 20-cm-long solenoid (2 cm in diameter) uniformly wound with 200 turns of wire carrying 0.5 A. Give fields B and H in SI and cgs/emu units.

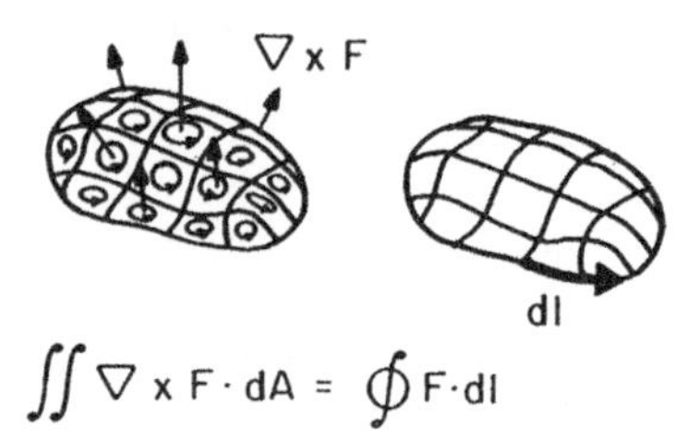

Figure A1.2 Graphical illustration of the meaning of Stokes' theorem.

1.2 Consider the magnetic field inside a toroid of circular cross section (see diagram) with inner radius $r_1 = 3$ cm and outer radius $r_2 = 4$ cm wound uniformly with 100 turns of wire carrying a current of 1.0 A. Calculate the field at $r = 3.5$ cm. Plot the field as a function of r for $3.1 \leq r \leq 3.9$ cm.

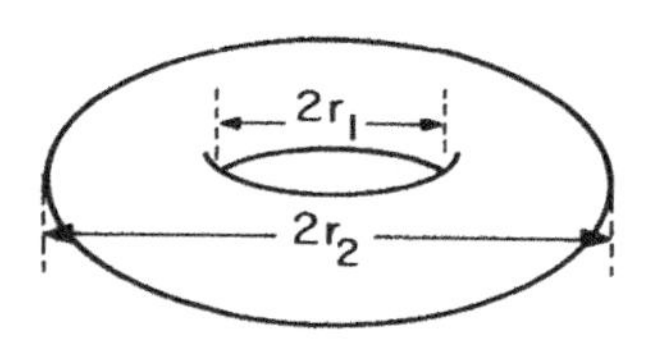

Figure P1.2

1.3 For what values of $\Delta r = r_2 - r_1$ and $r_{av} = (r_1 + r_2)/2$ is the variation of the field less than 10% across the inside of a toroid?

1.4 Carry out the steps described after Figure 1.9 to estimate the magnetic moment of a hydrogenic orbit in the Bohr model.

1.5 Use Eq. (1.14) to sketch the vector components of the magnetic field along the axis, 45° off the axis and on the plane of a current loop at a distance of twice the radius of the loop.

BIBLIOGRAPHY

Bozorth, R. M., *Ferromagnetic Materials*, Van Nostrand, New York, 1955; IEEE Press, New York, 1993.

Chikazumi, S., *Physics of Magnetism*, Wiley, New York, 1965.

Chikazumi, S., *Physics of Ferromagnetism*, Oxford Univ. Press, Oxford, 1997.

Craik, D., *Magnetism: Principles and Applications*, Wiley, New York, 1996.

Cullity, B. D., *Introduction to Magnetic Materials*, Addison Wesley, New York, 1973.

Jiles, D. C., *Introduction to Magnetism and Magnetic Materials*, Chapman & Hall, London, 1991.

Livingston, J. D., *Driving Force: The Natural Magic of Magnets*, Harvard Univ. Press, Cambridge, MA, 1996.

Mattis, D. C., *Theory of Magnetism*, Harper & Row, New York, 1965.

White, R. M., *Quantum Theory of Magnetism*, McGraw Hill, New York, 1970.

CHAPTER 2

MAGNETOSTATICS

This chapter treats the magnetostatic fields about magnetized samples, the energy associated with those fields, and how they combine with externally applied fields to influence the magnetization process. An appendix is included to help the student in designing simple magnets and estimating the fields they produce.

2.1 OBSERVATIONS OF MAGNETOSTATIC EFFECTS

If you measure the magnetization of a polycrystalline ferromagnetic sample measuring $1 \times 1 \times 5$ mm, you notice very different behavior for application of the field along the long axis as opposed to the short direction (Fig. 2.1). A greater external field is needed to achieve the same degree of magnetization for fields applied in the short direction compared to the long direction. What is it about the shape or aspect ratio of a sample that makes the magnetization process easier or harder? How does one determine a true material parameter from such different experimental results?

Before answering these questions, let us look at another related observation.

Two cylindrical samples of equal volume and comprised of the same permanent magnet composition are tested in an application. One is shorter and wider; the other, longer and narrower (Fig. 2.2). One end of the permanent magnet is to be placed at a fixed distance from an object that is to be magnetized. Space is a limitation. It is found that the long, thin magnet is more

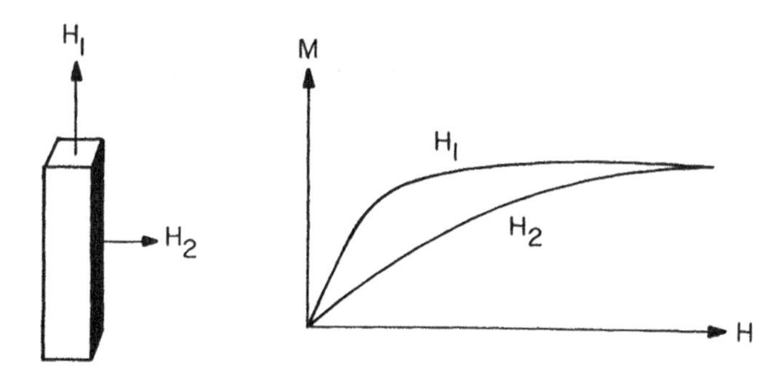

Figure 2.1 Magnetization curves for a polycrystalline ferromagnetic sample with field applied in different directions.

effective in magnetizing the object to the required level; the shorter one fails to provide a sufficient field. How can you calculate the appropriate shape that will provide the required field?

One more example of magnetostatics is given. When the B–H response for a ring or toroid of a polycrystalline magnetic material is measured as in Figure 2.3*a*, the hysteresis loop may resemble that shown in Figure 2.3*b*. The sides of the loop are steep where they cross $B = 0$ and then may round gradually toward saturation. Because this sample configuration is a closed path, the flux is contained almost entirely within the sample, there are no stray fields outside the sample. However, if a gap is cut in the toroid (Fig. 2.3*c*), the B–H loop

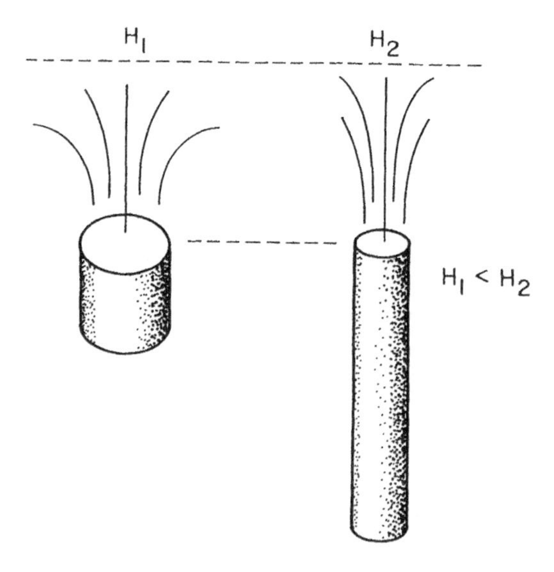

Figure 2.2 Dipole fields at a fixed distance from the ends of two permanent magnets of the same volume but different shapes.

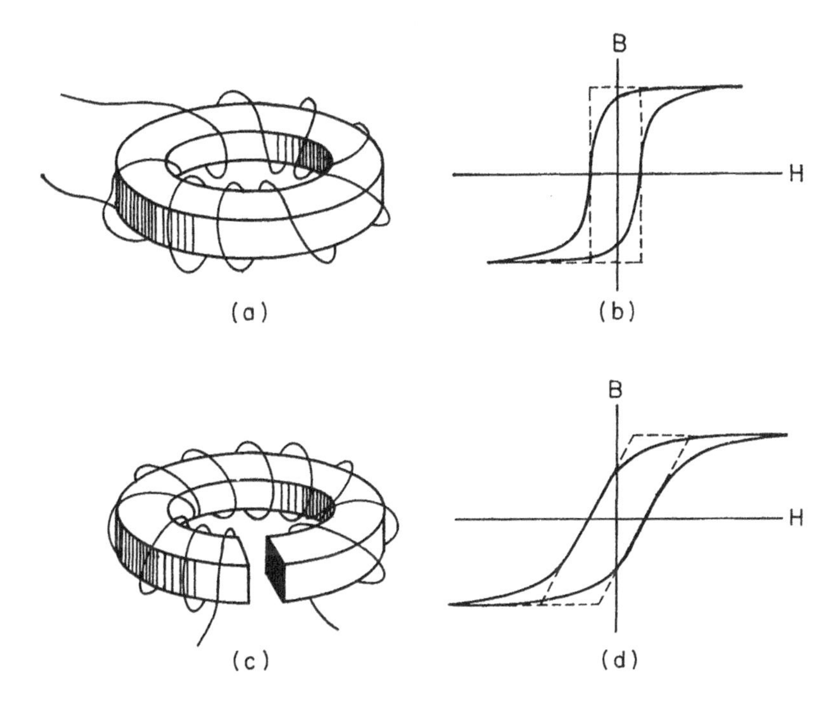

Figure 2.3 Demagnetization observed: (*a*) a closed toroidal magnetic sample and (*b*) its B–H loop; (*c*) a toroidal magnetic sample with a gap; (*d*) its B–H loop.

changes dramatically. It appears to be sheared over as shown by the curve in Figure 2.3*d*. The slopes of the branches near $B = 0$ are decreased and a higher field is required to achieve a given degree of magnetization. The "shearing" of the B–H loop is more dramatic for a wider gap in a toroid sample or for a shorter sample aspect ratio in a straight length of material.

This shearing effect on the B–H loop is related to the surfaces created on introduction of the gap in the toroid. When a magnetized sample has surfaces through which flux lines emerge with a normal component, "free poles" exist at the end surfaces (see Fig. 2.4). A magnetic field emanates from the north poles and terminates at the south poles. Depending on the shape and aspect ratio of the sample, the closing path of least energy for part of this dipole field *is through the sample.* To the extent that this field passes through the sample, it opposes the magnetization that set up the surface poles in the first place. The magnetic pole strength per unit surface area σ is given by the component of the magnetization at the surface that is normal to the surface:

$$\sigma = \boldsymbol{M} \cdot \boldsymbol{n} \tag{2.1}$$

where $\boldsymbol{n}$ is a unit vector normal to the surface. The surface pole density σ has

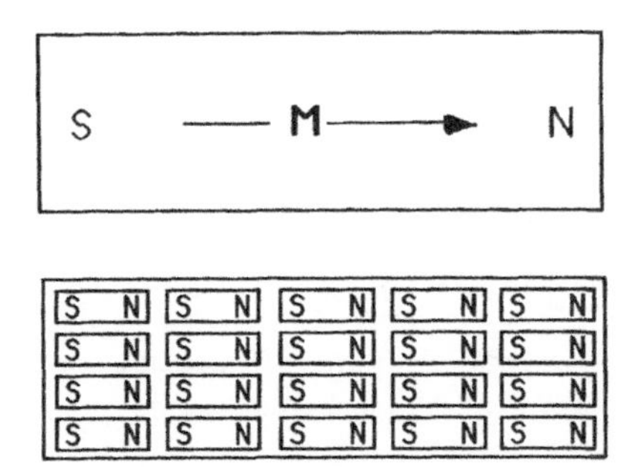

Figure 2.4 Schematic of magnetized sample as a composite of microscopic dipoles.

units of monopoles q_m per unit surface area or, equivalently, dipoles $q_m \cdot d$ per unit volume. σ gives rise to a vector magnetic field on *both* sides of a surface. The H field lines emanate from north (+) poles and terminate on south (−) poles (Fig. 2.5).

The field from the surface poles that passes through the interior of the sample is called the *demagnetizing field*. Its strength and direction generally vary with position inside the sample but are often assumed to be constant (they are constant only inside an ellipsoid).

For a flat, charged surface of infinite extent in one direction (perpendicular to the paper in Fig. 2.5) and finite in its other direction (bounded by r_1 and

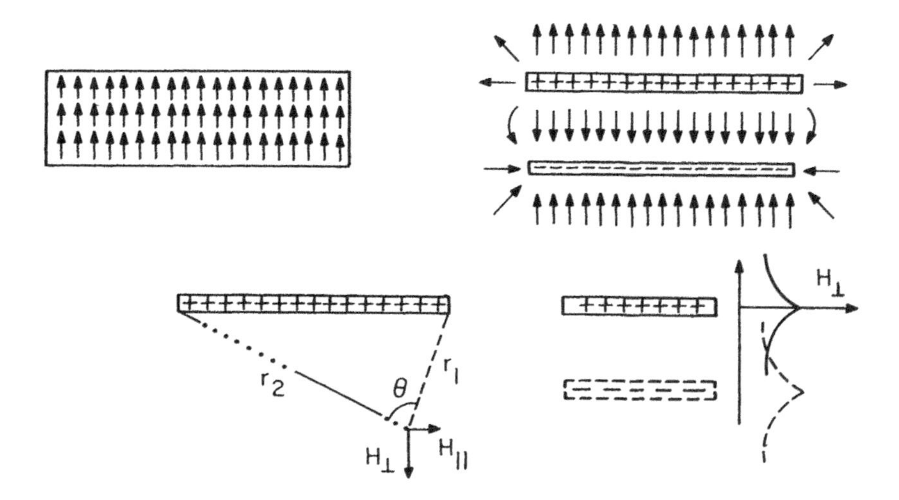

Figure 2.5 Magnetic dipoles inside a magnetized sample, upper left. Upper right, free poles at surfaces and field lines that they give rise to. Lower left, geometry for calculation of field components due to surface charge. Lower right, variation of the magnitude of the internal field with position inside a thicker sample.

r_2), the field components parallel and perpendicular to the charged surface are given in SI units by

$$H_{\|} = \frac{\sigma}{2\pi} \ln \frac{r_2}{r_1} \tag{2.2}$$

$$H_{\perp} = \frac{\sigma}{2\pi} \theta \tag{2.3}$$

where r_1, r_2, and θ are as defined in Figure 2.5. In cgs units the right-hand side expressions in Eqs. (2.2) and (2.3) must be multiplied by 4π, and M is in electromagnetic units per unit volume (emu/cm^3). [Equations (2.2) and (2.3) are exact two-dimensional forms of the 3D field due to a surface charge which will be given later.] Note that the parallel field component vanishes along a line perpendicular to the surface through its middle ($r_2 = r_1$). For a very thin dipole sheet, the field $H_{\perp}$ from *each* surface right at the surface ($\theta = \pi$) is given by $-\boldsymbol{M} \cdot \boldsymbol{n}/2$, so the internal or demagnetizing field from *both* surfaces is $-M(-4\pi M$ in cgs). For a thicker sample, the field from each surface drops off with distance from the surface, so the internal field varies approximately as sketched in the lower right of Figure 2.5; the demagnetization field is strongest near the charged surfaces of the sample and weaker in the middle.

2.2 BOUNDARY CONDITIONS ON *B* AND *H*

It is very useful to be able to relate the strength and orientation of the $\boldsymbol{B}$ and $\boldsymbol{H}$ fields across an interface. The boundary conditions on $\boldsymbol{B}$ and $\boldsymbol{H}$ are derived from Maxwell's equations.

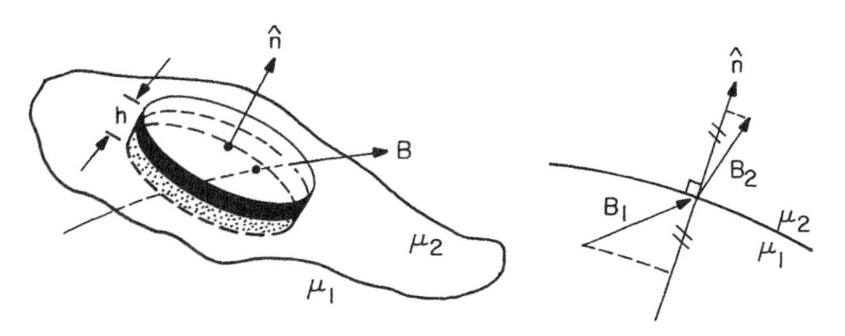

Figure 2.6 Construction for determining boundary conditions on the flux density $\boldsymbol{B}$ from Maxwell's equations: (left) perspective view of a pillbox with top and bottom surfaces on opposite sides of an interface between two magnetic media characterized by μ_1 and μ_2; (right) boundary conditions on B demand that normal component of $\boldsymbol{B}$ be continuous across an interface.

2.2.1 Flux Density

For the flux density $\boldsymbol{B}$, the integral form of $\nabla \cdot \boldsymbol{B} = 0$ and Gauss' theorem can be used to write the following, as in Eq. (1.7):

$$\int \boldsymbol{B} \cdot d\boldsymbol{A} = 0 \tag{2.4}$$

We construct a pillbox that intersects the interface between media 1 and 2 (Fig. 2.6*a*) and whose area elements dA have, by convention, normals directed outward from the pillbox. Equation (2.4) says that the net flux emerging from the pillbox is zero. In the limit that the height h of the pillbox shrinks to zero, the only contributions to the surface integral come from the normal component of $\boldsymbol{B}$ passing through the top and bottom surfaces:

$$-B_1 \cos \theta_1 + B_2 \cos \theta_2 = 0 \tag{2.5}$$

Here θ_i is the angle between the $\boldsymbol{B}$ field in each medium and the surface normal $\boldsymbol{n}$, not the pillbox normal. Equation (2.5) can be written

$$(\boldsymbol{B}_2 - \boldsymbol{B}_1) \cdot \boldsymbol{n} = 0 \tag{2.6}$$

In other words, the normal component of the $\boldsymbol{B}$ field is continuous from one medium to an adjacent one, regardless of their permeabilities (Fig. 2.6*b*). Also from Eq. (2.6), the normal components of $\boldsymbol{H}_1$ and $\boldsymbol{H}_2$ are related by

$$(\boldsymbol{H}_2 - \boldsymbol{H}_1) \cdot \boldsymbol{n} = (\boldsymbol{M}_1 - \boldsymbol{M}_2) \cdot \boldsymbol{n}$$

That is, the normal component of the $\boldsymbol{H}$ field is discontinuous across a surface at which a normal component of $\boldsymbol{M}$ has a discontinuity. Near the center of a magnetically charged surface, the perpendicular components of the field due to the surface charge are equal and opposite (Eq. 2.3), so the normal component of the H field outside a sample is equal to half the normal component of magnetization inside:

$$H_2^z = \frac{M_1^z}{2}$$

This is consistent with Eq. (2.3).

2.2.2 Field Intensity

For another boundary condition on the $\boldsymbol{H}$ field, the integral form of $\nabla \times \boldsymbol{H} = \boldsymbol{J}$ and Stokes theorem are used to give [cf. Eq. (1.10)]:

$$\int \boldsymbol{H} \cdot d\boldsymbol{l} = \iint \boldsymbol{J} \cdot d\boldsymbol{A} \tag{2.7}$$

with the integration path crossing the material interface (Fig. 2.6*a*).

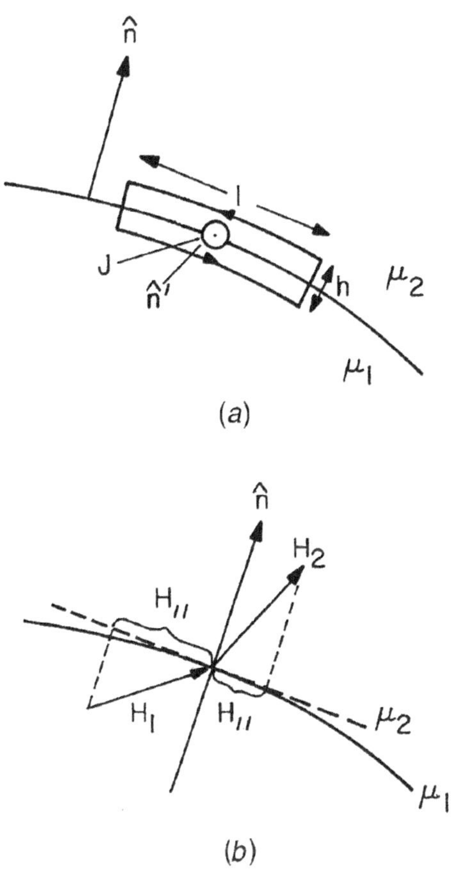

Figure 2.7 (*a*) Construction for determining boundary conditions on $\boldsymbol{H}$ from Maxwell's equations. An edge view of a right-handed, closed line element with long segments on opposite sides of the interface. Note the difference between $\boldsymbol{n}$, the interface normal, and $\boldsymbol{n}'$, the normal to the area enclosed by the path of integration. (*b*) Boundary conditions on $\boldsymbol{H}$ deduced from Maxwell's equations demand that the tangential component of $\boldsymbol{H}$ be continuous across an interface unless there are surface currents present.

The normal $\boldsymbol{n}'$ to the area enclosed by this path is determined by a right-hand rule; $\boldsymbol{n}'$ is tangent to the interface. Again, as the dimension of the rectangular path normal to the material interface shrinks to zero, Eq. (2.7) becomes

$$l(H_1 \sin\theta_1 - H_2 \sin\theta_2) = \iint \boldsymbol{J} \cdot \boldsymbol{dA} = I \tag{2.8}$$

where θ_i is the angle between the $\boldsymbol{H}_i$ field in medium i and the interface normal

$\boldsymbol{n}$ and l is the length of the rectangle along the surface. We can write Eq. (2.8) as

$$\boldsymbol{n} \times (\boldsymbol{H}_1 - \boldsymbol{H}_2) = \boldsymbol{K} \tag{2.9}$$

where $\boldsymbol{K}$ is a surface current density, i/l (A/m), in the interface plane parallel to $\boldsymbol{n}'$ and $\boldsymbol{K}$ has the magnitude

$$K = l^{-1} \int \boldsymbol{J} \cdot dA \tag{2.10}$$

Equation (2.9) states that the tangential components of $\boldsymbol{H}$ are continuous across the interface in the absence of an interfacial current density (Fig. 2.7*b*); otherwise they differ by the amount of that current density.

We summarize these boundary conditions on $\boldsymbol{B}$ and $\boldsymbol{H}$. Across an interface between two media with different magnetic properties the normal component of $\boldsymbol{B} = \mu_0(\boldsymbol{H} + \boldsymbol{M})$ is always continuous and the tangential component of $\boldsymbol{H}$ is discontinuous by the magnitude of the transverse surface currents. These boundary conditions will help us to understand the phenomena in Figures 2.1–2.3.

Let us be clear about the nature of the surface current K defined in Eq. (2.10). The constructions in Figure 2.7 are drawn for convenience to express the mathematical process of limitation of the integration path in the direction normal to the interface. Expressing the RHS of Eq. (2.8) in Cartesian coordinates gives

$$\iint J(x, y)\boldsymbol{e}_z dxdy \tag{2.11}$$

where $\boldsymbol{e}_z$ is a unit vector in the z direction. Choose the y direction normal to the interface so that the limits of y integration go to zero. If $\boldsymbol{J}$ were concentrated like a Dirac delta function, $\delta(y)$ at $y = 0$, then $\boldsymbol{J}$ could be expressed as

$$J(x, y) = \kappa_z(x)\delta(y) \tag{2.12}$$

where $\kappa_z(x)$ has dimensions of current per unit length in the x direction and the Dirac delta function has dimensions of inverse length. In this case the integral in Eq. (2.11) reduces to

$$\iint \kappa_z \delta(z) dxdy = lK$$

and Eq. (2.10) follows. Thus, K is the average current along the surface per unit length transverse to the current path. It is concentrated right at the surface. If the

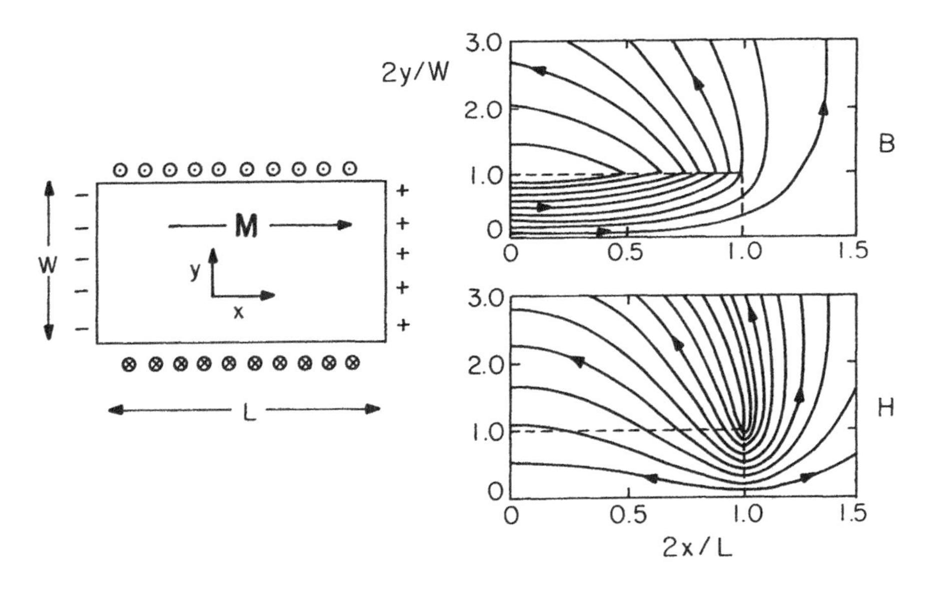

Figure 2.8 ***B*** and ***H*** fields in and around a uniformly magnetized bar of length L and width W (infinite extent out of paper). At the left is sketched the magnetized bar and its surface poles (or the equivalent surface current around the bar), which are the sources for the ***H*** and ***B*** fields, respectively. Only one quarter of the bar is sketched at right because of the symmetry of the situation. [After Bertram (1994).]

surface current were distributed over a depth dy, then the tangential field H_x would change gradually over that depth (see Problem 2.1).

Figure 2.8 (Bertram 1994) provides an excellent pedagogical summary of the important issues in magnetostatics that have been presented. This figure is the output of micromagnetic calculations of the field distribution inside and around a uniformly magnetized bar. Only the upper right quadrant of the bar, of finite extent out of the paper, is shown. Note that the surface charges are sources for the ***H*** field inside and outside the sample. An equivalent current through the surface windings is the source of the ***B*** field inside and outside the sample. Further, outside the sample the B and H fields are equivalent to each other: $B = \mu_0 H$ (SI) and $B = H$ (cgs). Note also that the boundary conditions on ***B*** (normal component continuous across surface) and ***H*** (tangential component continuous across surface) are properly satisfied in these calculated fields. ***M*** is proportional to $\boldsymbol{B} - \boldsymbol{H}$, which in this case is uniform inside the bar. The boundary conditions on ***B*** and ***H*** are properly satisfied.

The H field inside the magnetized sample in Figure 2.8 opposes the state of magnetization in which the sample is held. In a sample that is not so constrained, this internal field would tend to demagnetize the sample. We now consider such demagnetizing fields separately.

2.3 DEMAGNETIZING FIELDS AND FACTORS

Demagnetizing fields hold the key to an understanding of the observations described in Figures 2.1–2.3.

At the end of a sample that is assumed to be magnetized perpendicular to the ends without application of an external field, Eq. (2.7) indicates that fields exist inside $H_i = -M_i/2$, and outside, $H_0 = M_i/2$, the sample. Just inside the surface, $B_i = \mu_0(H_i + M_i) = \mu_0 M_i/2$. Just outside the surface $B_0 = \mu_0(H_0 + M_0) = \mu_0 M_i/2$ with $M_0 = 0$. [Note that $B_i = \mu_0(H_i + M_i) = \mu_r\mu_0 H_i$, i.e., either $\mu_0 M_i$ is added to $\mu_0 H_i$ to give B_i or the permeability of the medium, $\mu_r\mu_0$ times H_i gives B_i.] If an external field is now applied to change M_i, its magnitude adds to H_i and H_0:

$$H_0 = H_{\text{appl}} + \frac{M_i}{2} \quad \text{and} \quad H_i = H_{\text{appl}} - \frac{M_i}{2} \qquad \text{(SI)} \qquad (2.13a)$$

$$H_0 = H_{\text{appl}} + 2\pi M_i \quad \text{and} \quad H_i = H_{\text{appl}} + 2\pi M_i \qquad \text{(cgs)} \qquad (2.13b)$$

In other words, the H field inside the material is equal to the applied field reduced by half the amount of magnetization normal to the surface. (If the sample is thin in the direction of M_i, then there are contributions to the internal field of $-M_i/2$ from *each* surface.) It is now shown how this explains the introductory examples.

What happens in Figures 2.1 and 2.3 is that the field from the poles at the surfaces (Fig. 2.4) opposes the magnetizing field so the internal field that M responds to is less than the applied field. This surface pole field or dipole field is the demagnetizing field H_d mentioned earlier. The internal field is sometimes written

$$H_i = H_{\text{appl}} + H_d \qquad (2.14)$$

Comparison with Eq. (2.13) shows that $H_d = -M_i/2$ for each surface normal to M_i. If the magnetization is zero $H_d = 0$, but as M increases H_d becomes more negative.

We have so far assumed $\boldsymbol{M}$ is normal to the surface. More generally, $H_d = -\boldsymbol{M}\cdot\boldsymbol{n}$, where $\boldsymbol{n}$ is the surface normal. For an arbitrarily shaped sample, the demagnetizing field for a given direction of $\boldsymbol{M}$ relative to the sample axes may be approximated as

$$H_d = -NM \qquad (2.15)$$

The constant of proportionality N is called the demagnetization factor and, in general, it is a tensor function of sample shape. If the ends are flat and M_i is normal to the surface, $N = 1$. For ellipsoids N is a diagonal tensor and can be calculated because for those shapes, the internal field turns out to be uniform.

N can be measured fairly well, or at least approximated, for most other shapes. It tells us the component of flux density normal to the surface for a given shape and direction of magnetization. It expresses the effect of the unit vector $\boldsymbol{n}$ in Eq. (2.1).

Equation (2.15) says that the greater the magnetization of the sample, the more the field from the surface poles opposes the external field. Thus, for soft magnetic materials, where a large magnetization results from a relatively weak external field, the internal field can be much less than the applied field even if the shape factor N is very much less than unity. However, for permanent-magnet materials, where very large external fields are required to achieve appreciable magnetization, shape effects become important only for much smaller-aspect-ratio, larger N, samples (Fig. 2.2).

We can *quantitatively* interpret Figure 2.3 using Eq. (2.14) and the constitutive relation $B_i = \mu_0(H_i + M)$, where B_i is the induction inside the material and H_i is the internal field, $H_i = H_{\text{appl}} - NM$. Eliminating M between these equations gives

$$\frac{B_i}{\mu_0} = \frac{1}{N} H_{\text{appl}} - \frac{1-N}{N} H_i \tag{2.16}$$

On a familiar B_i/H_{appl} plot (e.g., Fig. 2.3*d* or 2.9*a*), Eq. (2.16) is a straight line with slope μ_0/N. Along the straight line of slope μ_0/N that passes through the origin, $H_i = 0$. The coercivity is defined as the external field needed to reduce B_i to zero. From Eq. (2.16), this gives $H_{\text{appl}} = H_c = (1 - N)H_i$. If $H_c \neq 0$, the remanent flux density is given by Eq. (2.16) with $H_{\text{appl}} = 0$, namely, $B_r = -\mu_0(1 - N)H_i/N = -\mu_0 H_c/N$. A plot of B versus H_i can be generated from the B versus H_{appl} curve by shearing the latter so the line of slope μ_0/N in Figure 2.9*a* becomes vertical and the vertical in Figure 2.9*a* takes on a slope of $\mu_0(1 - N)/N$ (Fig. 2.9*b*). [This slope is given by Eq. (2.16) with $H_{\text{appl}} = 0$.] The intersection of this "load line" with the B_i–H_i loop indicates the remanence that actually would be measured in the B_i–H_{appl} loop for a sample of a given geometry.

It should be clear from Figure 2.8 and from Eqs. (2.2) and (2.3) that the demagnetization field is generally a function of position and magnetization orientation inside a sample. Hence, as was stated above, N should be a tensor function of position and orientation in a given sample. However, N is a double-valued or triple-valued diagonal tensor for magnetization along the principal axes of an ellipsoid. For a truly ellipsoidal specimen in a uniform field oriented along one of the major axes, the magnetization is uniform throughout the sample. This is not true for samples of other shapes where the magnetic response is greatly reduced by dipole fields near corners of rectangular specimens.

For an infinitely long cylinder of arbitrary cross-sectional shape, magnetized along its length, $N = 0$. If it is magnetized perpendicular to its length, $N = \frac{1}{2}$.

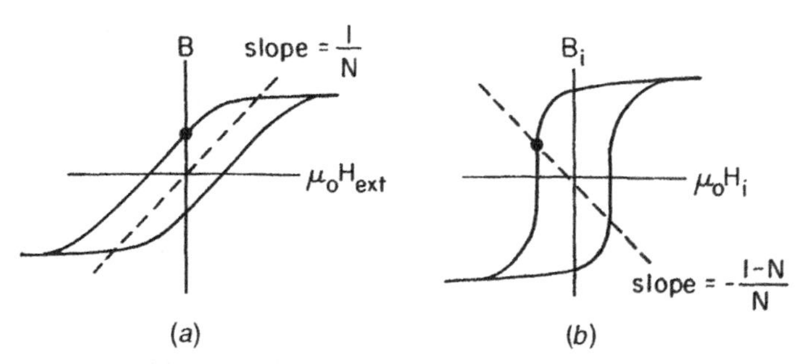

Figure 2.9 Schematic representation of demagnetization effect on B-H_{appl} loops (a) and on B–H_i loops (b). The dashed lines rotated into the vertical axis in each case relate one loop to the other.

A sphere has $N = \frac{1}{3}$ along each of its three orthogonal axes, as will be shown later. An infinite sheet has $N = 1$ if it is magnetized normal to its surface and $N = 0$ if it is magnetized in plane. Note that the demagnetization factors in three principal orthogonal directions add up to unity: $N_x + N_y + N_z = 1$ ($\Sigma N_i = 4\pi$ in cgs units); the trace of the demagnetization tensor is 1 (or 4π in cgs units). Clearly, it costs less field energy to magnetize ferromagnetic materials along their long directions, preferably in closed circuits, because then there are no surface poles to cause opposing fields.

The exact demagnetization factors for various ellipsoids have been calculated for magnetization along the three axes, for arbitrary aspect ratios (Osborn 1945). Here, as for the simple cases cited above, $N_x + N_y + N_z = 1$. These equations may be used as approximations for samples of similar geometries.

For a *prolate ellipsoid* with major axis m times as long as either of its two minor axes and magnetized along the major axis

$$N_z = \frac{1}{m^2 - 1}\left\{\frac{m}{(m^2-1)^{1/2}} \ln\left[\frac{m + (m^2-1)^{1/2}}{1}\right] - 1\right\} \tag{2.17}$$

For a *prolate ellipsoid* magnetized along a minor axis

$$N_x = N_y = \frac{m}{2(m^2-1)}\left\{m - \frac{1}{m(m^2-1)^{1/2}} \ln\left[\frac{m + (m^2-1)^{1/2}}{m - (m^2-1)^{1/2}}\right]\right\} \tag{2.18}$$

For $m \gg 1$, this becomes $N_x = N_y = \frac{1}{2}\{1 - (\ln[2m - 1])/m^3\}$.

For *oblate ellipsoids* having two long axes m times the length of the axis of symmetry and magnetized parallel to a long axis

$$N_x = N_y = \frac{1}{2}\left\{\frac{m^2}{(m^2-1)^{3/2}} \arcsin\left[\frac{(m^2-1)^{1/2}}{m}\right] - \frac{1}{m^2-1}\right\} \tag{2.19}$$

and if $m \gg 1$, this becomes $N_x = N_y = (\pi/4m)(1 - 4/\pi m)$.

For an *oblate ellipsoid* magnetized along its short axis

$$N_z = \frac{m^2}{m^2 - 1}\left\{1 - \frac{1}{(m^2 - 1)^{1/2}} \arcsin\left[\frac{(m^2 - 1)^{1/2}}{m}\right]\right\} \tag{2.20}$$

and if $m \gg 1$, this becomes $N_z = 1 - \pi/2m + 2/m^2$.

For a more *general ellipsoid* of principal axis lengths a, b, and c with $a \gg b, c$, and for magnetization in the a direction, N is given by

$$N_a \approx \left(\frac{bc}{a^2}\right)\left\{\ln\left(\frac{4a}{b + c}\right) - 1\right\} \tag{2.21}$$

Table 2.1 summarizes results of these equations for various aspect ratios. Figure 2.10 shows the demagnetization factors calculated for ellipsoids and cylinders magnetized along their long axes for various dimensional ratios. The curves for nonellipsoidal bodies depend on the permeability of the material.

There is an approximation for a *two-dimensional* problem where the field is essentially uniform in the third direction. The demagnetization factors are taken from the demagnetization fields calculated from Eqs. (2.2) and (2.3). For example, in Figure 2.11, for in-plane magnetization, it may be assumed that the field due to the end poles is independent of the coordinate parallel to the length of the fact on which the poles lie. For magnetization along the z direction, the

TABLE 2.1 Demagnetizing Factors N for Rods and Ellipsoids Magnetized Parallel to Long Axis

Dimensional Ratio (Length/Diameter)	Rod	Prolate Ellipsoid	Oblate Ellipsoid
0	1.0	1.0	1.0
1	0.27	0.3333	0.3333
1.5	—	0.233	0.329
2	0.14	0.1735	0.2364
5	0.040	0.558	0.1248
10	0.0172	0.0203	0.0696
20	0.00617	0.00675	0.0369
50	0.00129	0.00144	0.01472
100	0.00036	0.000430	0.00776
200	0.00090	0.000125	0.00390
500	0.000014	0.0000236	0.001567
1000	0.0000036	0.0000066	0.000784
2000	0.0000009	0.0000019	0.000392

Source: Bozorth, IEEE Press, 1993, p. 849.

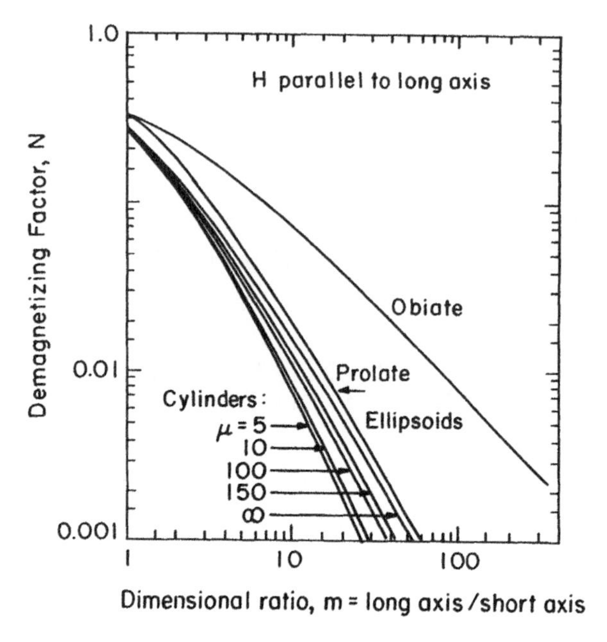

Figure 2.10 Demagnetization factors for ellipsoids and cylinders with field applied parallel to long axis, with aspect ratios closer to unity. [Bozorth, © IEEE Press (1993)].

y-independent demagnetization factor derived from Eq. (2.3) is

$$N_w \approx \frac{2t}{\pi w} \tag{2.22}$$

and for magnetization along the y direction, the z-independent factor is

$$N_h \approx \frac{2t}{\pi h} \tag{2.23}$$

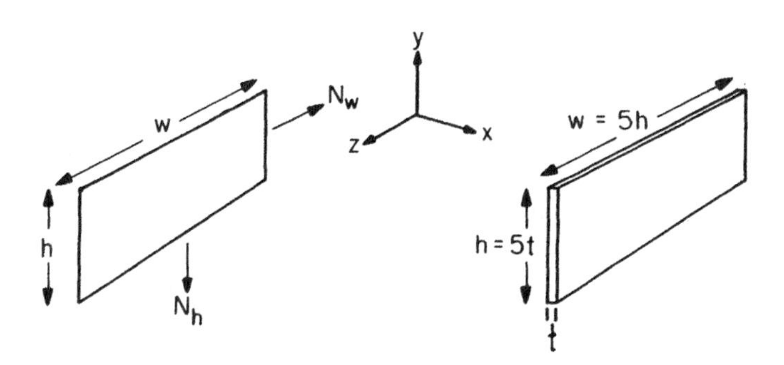

Figure 2.11 Two-dimensional approximation for demagnetization factors, left, and schematic of thin film, right.

If the dependence on the third dimension becomes significant, additional factors of $(2h/w)^{1/2}$ and $(2w/h)^{1/2}$ should appear in Eqs. (2.22) and (2.23), respectively.

As a cautionary note, it is emphasized that N is not a constant inside any magnetized sample that is not an ellipsoid; N is merely approximated. Consider a thin film of dimensions $w:h:t$ $25:5:1$ (Fig. 2.11, right). Equation (2.21), for magnetization along the z direction of a triaxial ellipsoid indicates a demagnetization factor of $N_w \approx 0.0145$. The two-dimensional approximation, Eq. (2.22), gives $N_w \approx 0.025$. Use of Eq. (2.22) with the factor $(2h/w)^{1/2}$ gives $N_w \approx 0.0155$. If the film is assumed to be a prolate ellipsoid of aspect ratio $m = w/(th)^{1/2} \approx 11.2$, Table 2.1 indicates $N_w \approx 0.02$. The large disparity in these various approximations is due to the square shape and small aspect ratio of this thin film.

The preference for the magnetization to lie in a particular direction in a polycrystalline sample is given by the shape anisotropy. The shape anisotropy field is proportional to the difference in the demagnetization fields in two orthogonal directions:

$$H_{\text{shape}} = M(N_w - N_h) = M\,\Delta N \tag{2.24}$$

In the two-dimensional approximation for the dimensions given above, if $M = 64\,\text{kA/m} = (800\,\text{G})$, then $H_{\text{shape}} \approx 10.2\,\text{kA/m}$ (128 Oe).

2.4 MAGNETIZATION CURVES

It is possible to calculate the shape of the M–H curve for magnetizing a *single-domain* sample against its demagnetizing field (i.e., magnetizing in a hard direction) if a value is known for N. The magnetostatic energy density is

$$\frac{U_{ms}}{V} = u_{ms} = -\mu_0 \boldsymbol{M} \cdot \boldsymbol{H}_d = -\mu_0 M_s H_d \cos\theta = +\left(\frac{\Delta N}{2}\right)\mu_0 M_s^2 \cos^2\theta \tag{2.25}$$

because H_d is parallel to the short sample dimension and is given by $H_d = -\Delta N M_s \cos\theta$; θ is the angle between $\boldsymbol{M}$ and the direction of the applied field. The factor of $\frac{1}{2}$ reflects the fact that the energy is actually the result of a moment-by-moment assembly of the sample in the sense of $\int M dM = \frac{1}{2}M_s^2$. This energy is minimum for $\boldsymbol{M}$ in the easy (long) direction ($\theta = 90°$) and maximum for $\boldsymbol{M}$ in the hard direction, $\theta = 0°$. See Figure 2.12, $H = 0$ curve. The Zeeman energy density due to the magnetization orientation in the applied field is

$$u_H = -\mu_0 M_s \cdot H_0 = -\mu_0 M_s H_0 \cos\theta \tag{2.26}$$

Figure 2.12 plots the sum of the two energies, $u = u_{ms} + u_H$ versus θ for various values of applied field.

The magnetization takes on an orientation that minimizes the total energy density u. The minimum energy is found at a value of θ for which $du/d\theta$, the negative of the torque on M, is zero and u has positive curvature. The torque is zero when the energy is an extremum:

$$\frac{du}{d\theta} = 0 = -\Delta N M_s^2 \sin\theta\cos\theta + M_s H \sin\theta \tag{2.27}$$

Divide both sides of Eq. (2.27) by $\sin\theta$ except when $\sin\theta = 0$. But $\sin\theta = 0$ for M parallel to H, that is, at and above the saturation field H_s, where it is known that the shape of M–H is flat. Below saturation Eq. (2.27) gives $\Delta N M_s \cos\theta = H$. The component of M_s of interest is that parallel to H, namely, $M(H) = M_s \cos\theta$; hence

$$M(H) = M_s \cos\theta = \frac{H}{\Delta N} \tag{2.28}$$

The stability condition, $u'' > 0$, is $\Delta N M_s^2 \cos^2\theta + M_s H \cos\theta > 0$. It cannot be met for $135° < \theta \leqslant 180°$ and can always be met for $45° < \theta \leqslant 90°$. The stable range extends down to $\theta = 0$ for $H > \Delta N M_s$. This is clear by considering Figure 2.12.

Equation (2.28) is the equation for the M–H loop in the hard shape direction. For $H < H_{\text{saturation}}$, the magnetization increases linearly with H. M

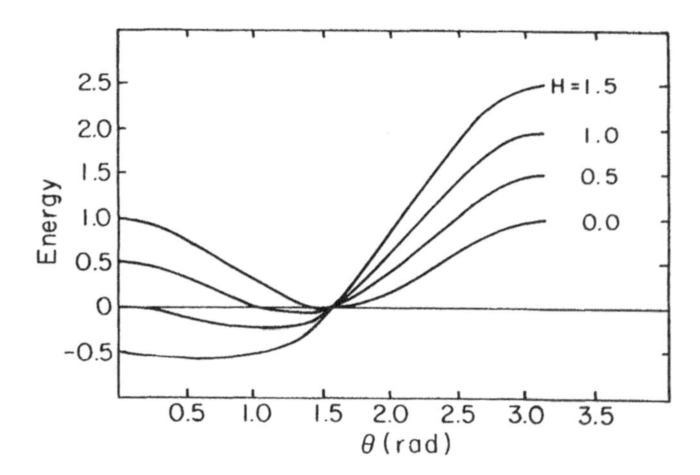

Figure 2.12 Variation of magnetostatic plus Zeeman energy density with θ for increasing values of applied field (arbitrary units). Note how the stable energy minimum moves from $\pi/2$ toward zero as applied field increases.

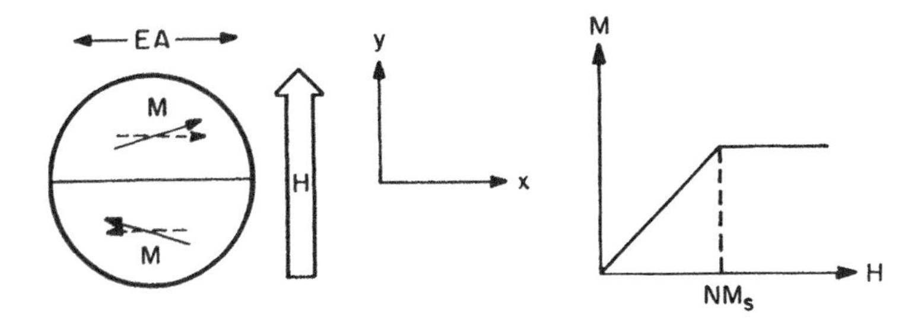

Figure 2.13 Schematic representation of a magnetic material having purely uniaxial anisotropy in the direction of the easy axis (EA). Dashed lines indicate magnetization configurations for $H = 0$. Application of a field H transverse to the EA results in rotation of the domain magnetizations but no wall motion.

reaches the value M_s when $H = \Delta N M_s = H_s$ (Fig. 2.13). Note that the demagnetizing factor can be determined from the field at the knee of the M–H curve if M_s is known and if there is no anisotropy other than shape. Alternatively, M_s can be determined from the saturation field if ΔN is known. The simplest case is magnetization normal to a thin film, $N = 1$ ($N = 4\pi$ in cgs). In that case the material does not saturate until $H = M_s$ (or $4\pi M_s$, cgs). This field is about 0.6 T for pure Ni, 1.0 T for $Ni_{81}Fe_{19}$, and 2.2 T for Fe.

If domain walls parallel to the easy direction are present, this simple result is unchanged because the hard-axis field causes no wall motion and rotates the domain magnetization by equal amounts (Fig. 2.13).

Even for application of a field along an axis of relatively easy magnetization (Fig. 2.14), in a *multidomain* sample, the demagnetizing factor can shear over

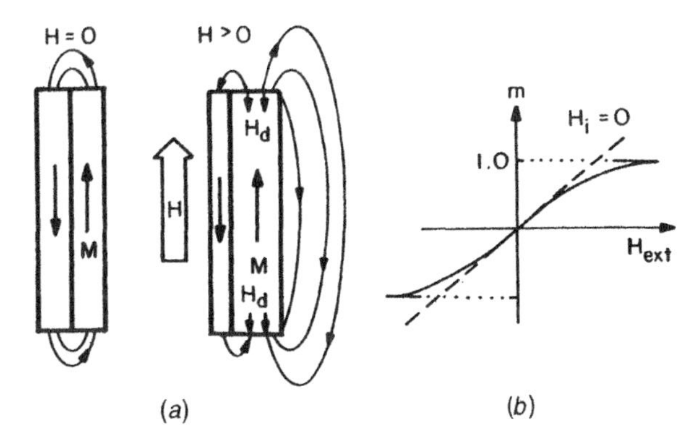

Figure 2.14 (*a*) A demagnetized sample for which shape is a factor responds to an applied field at the cost of increased demagnetization factor and increased magnetostatic energy; (*b*), the changing demagnetization factor causes the M–H loop to be less than linear in the external field.

an $M-H$ loop. The wall motion that results from application of the field will be justified in Chapter 10. The internal field seen by the material is the applied field plus the demagnetizing field [Eq. (2.14)]. Note that as the shape of each domain changes, its demagnetizing factor changes (Fig. 2.14*a*). The $M-H$ curve is no longer linear in H but is sublinear (Fig. 2.14*b*) due to the increasing demagnetizing factor of the larger domain, which represents most of the sample.

Equation (2.15) may be written as $M(H) = -H_d/N = -(H_i - H_{\mathrm{appl}})/N$, so the slope of the magnetization line at zero internal field is $1/N$ (which is vertical if there are no shape effects). [A similar result applies for the $B-H$ curve, Eq. (2.16).] $M(H)$ increases with increasing field to the extent that $H_i = 0$ is maintained during the magnetization process, that is, the demagnetization field exactly cancels out the applied field until saturation is reached. Thus the loop is sheared over by the sample's shape as indicated in Figure 2.14*b*. Note that it is not possible from this measurement alone to distinguish shape anisotropy from some other source of anisotropy.

Example 2.1 In soft magnetic materials where the anisotropy is weak, even small demagnetizing effects, $N \approx 10^{-4}$ to 10^{-5}, can have a strong effect on measured $M-H$ response or susceptibility. In the case of thin ribbons of amorphous magnetic material (25 μm thick and 3–6 mm wide), changes in length from 5–10 cm dramatically alter the $M-H$ curves (Fig. 2.15) and can be readily accounted for.

We can calculate the demagnetization effects on the $M-H_{\mathrm{ext}}$ loop by expressing the measured magnetization in terms of the external and demagnetizing fields

$$M = \chi H_i = \chi(H_{\mathrm{ext}} - NM) \tag{2.29}$$

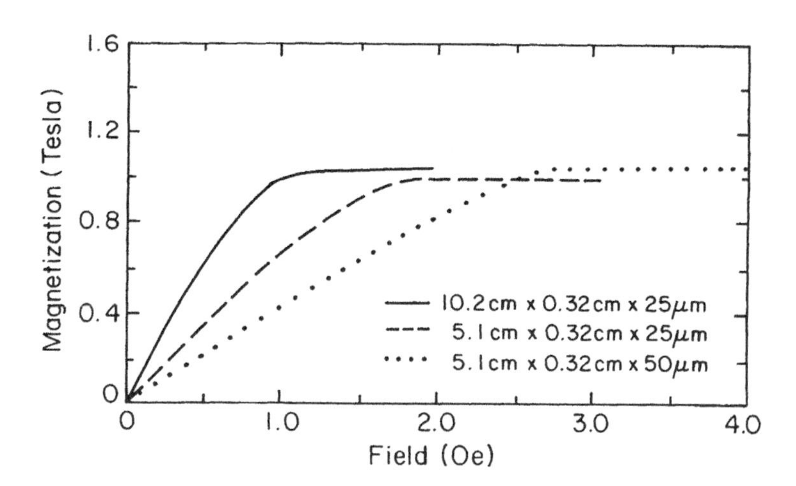

Figure 2.15 Magnetization for transversely annealed amorphous alloy ribbons of various aspect ratios. [After Clark and Wun-Fogel, 1989.]

Note that the magnetization results from the internal field, not the applied field. Equation (2.29) leads to the definition of an effective or apparent susceptibility χ_e for a sample of a given shape:

$$\frac{M}{H_{\text{ext}}} = \frac{\chi}{1 + N\chi} = \chi_e \tag{2.30}$$

This equation expresses the same effect that is described by Eq. (2.16), but in different parameters. See Problem 2.5.

Figure 2.15 shows how sensitive the observed M–H characteristics of high-permeability or high-susceptibility materials are to very small changes in N. The effective susceptibility, which measures the initial slope of M–H_{ext}, is sharply reduced in shorter samples even though N is of order 10^{-4}!

The data for the longest sample in Figure 2.15 give a slope of order $1.0\,\text{T}/\mu_0\,80\,\text{A/m} = 10^4$ or, from Eq. (2.16), $N_a = 6.3 \times 10^{-5}$. Approximating the ribbon as an extremely prolate ellipsoid, Eq. (2.21) gives $N_a = 3.6 \times 10^{-5}$. An assumption of prolate ellipsoidal shape with $m = 100/(4 \cdot 0.025)^{1/2} = 316$ gives $N_a = 5 \times 10^{-5}$ from Table 2.1. Cutting the ribbon length from 10 to 5 cm increases N derived from the data by a factor of 3.5.

2.5 MAGNETOSTATIC ENERGY AND THERMODYNAMICS

The shape effects or demagnetizing effects described in this chapter can be thought of as resulting from the dipole energy of the magnetized sample, that is, the energy needed to assemble the atomic dipoles that constitute the macroscopic magnetic dipole that is the sample. It costs energy to place dipoles *adjacent* to each other if they have the same orientation as in Figure 2.16*a*. This energy is stored in the fields about the dipole configuration. The energy cost is less when you assemble dipoles head to tail as in Figure 2.16*b*. Two end-to-end dipoles have a negative energy that gets more negative the closer they are. Let us make these concepts quantitative.

It is a familiar result that the potential energy of interaction of a magnetic dipole μ_m with an external field $\mathbf{B}_0 = \mu_0 H$ is given by the quantity $U = -\mu_m \cdot \mathbf{B}_0$. The same result can be extended to a rigid assembly of dipoles. The potential energy per unit volume for a macroscopic sample of magnetization $\mathbf{M} = N\mu_m/V$ in an external field may be written as

$$\left(\frac{N}{V}\right) U = u = -\mathbf{M} \cdot \mathbf{B}_0 \tag{2.31}$$

When the magnetic field of interest is not external, but is due to the magnetization itself, H_d, the factor of one-half (which occurs in all self-energy problems) must be included in Eq. (2.31). The factor of $\frac{1}{2}$ enters because dipole

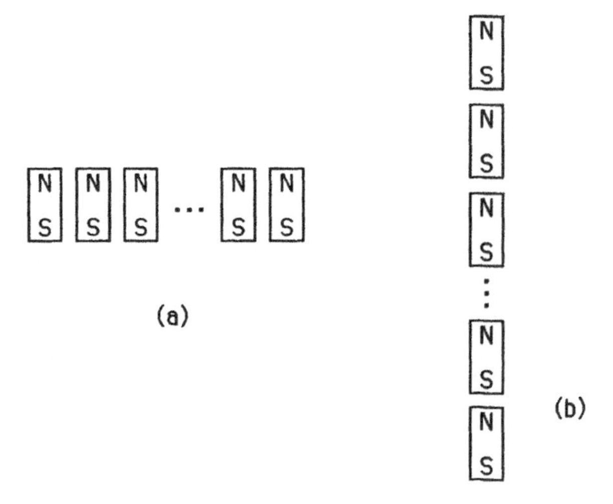

Figure 2.16 Assembly of dipoles in high-energy configuration (*a*) and low-energy configuration (*b*).

pair interactions must not be counted twice; application of Eq. (2.31) to self-energy effectively counts each dipole once as a source of field and once as a magnet in the field. The correct magnetostatic energy density is

$$u = -\left(\frac{\mu_0}{2}\right)\boldsymbol{M}\cdot\boldsymbol{H}_d$$

or

$$u = \left(\frac{\mu_0}{2}\right)NM^2 \tag{2.32}$$

$$u = 2\pi NM^2 \text{ (cgs)}$$

Equation (2.32) represents the work done in assembling a given state of magnetization in a sample. For example, an infinite sheet magnetized perpendicular to its plane has energy density $u = \mu_0 M^2/2$ ($u = 2\pi M^2$ in cgs). These relations are very useful.

It is appropriate to examine more carefully the process of magnetizing a sample. Consider the magnetization curve in Figure 2.17. For the process of magnetizing the sample from the demagnetized state to any point (M_1, B_1), three energy densities can be defined: the potential energy of the magnetized sample in B_1, $-M_1B_1$, as well as the energies A_1 and A_2.

It should be clear from Figure 2.17 that integral expressions for A_1 and A_2 can be written as

$$A_1 = \int_0^{M_1} B(M)dM \tag{2.33}$$

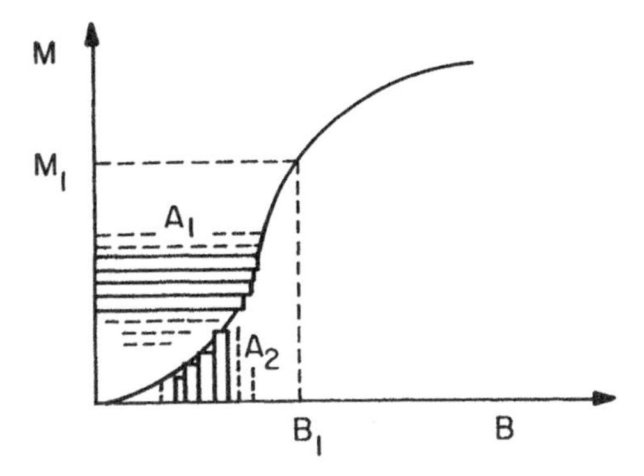

Figure 2.17 Magnetization curve for sample brought from demagnetized state to M_1 in field increasing from $B = 0$ to $B = B_1$.

and

$$A_2 = \int_0^{B_1} M(B)dB \tag{2.34}$$

where $M(B)$ and $B(M)$ are functions defining the magnetization curve in terms of the independent variables B and M, respectively. What is the physical significance of these two contributions to the energy density? If A_1/A_2 is small, the material is considered to be easily magnetized, thus not much work needs to be done to magnetize the material. A_1 is the work done *by the field* to bring the sample to the state of magnetization M_1.

Consider a magnetic sample of cross section A and volume Al inside a solenoid of N turns over the length l of the sample. The electrical work dw per unit volume expended to magnetize the sample quasistatically to a given level of magnetization is given by

$$\frac{dW}{Al} = dw = i\,\frac{V\,dt}{Al} \quad \text{(energy/vol)}$$

The current i produces an H field in the solenoid [Eq. (1.8)] of $H = Ni/l$ (A/m) [$H = 0.4\pi Ni/l$ (Oe)]. The back EMF induced in the solenoid by the changing flux density in the sample is given by $V = -NA\,dB/dt$ ($V = -10^{-8}NA\,dB/dt$ in cgs units). Hence, the incremental work per unit volume done in magnetizing the sample becomes

$$dw = H\,dB = \mu_0 H\,dM \tag{2.35}$$

where attention has been focused on only the material effect of the applied field H, namely, the change, $dB = \mu_0\,dM$. This *differential* energy is the area inside of the rectangular elements between the $M-B$ curve and the M axis in Figure 2.17. Its integral is A_1 as defined in Eq. (2.33).

If there is no sample present, or if only the field energy aside from the sample energy is of interest, then

$$w = \mu_0 \int \boldsymbol{H} \cdot d\boldsymbol{M} = \frac{B^2}{2\mu_0} \quad (\mathrm{J/m^3}) \qquad \left[w = \frac{BH}{8\pi} \quad (\mathrm{erg/cm^3}) \right]$$

This is just the energy per unit volume stored in the field, $B = \mu_0 H$.

The term A_2 is proportional to the energy given up *by the magnetized material* as it is drawn into a field. Thus, A_1 represents work done on the sample and A_2 describes work done by the sample. Note that A_1 and A_2 are functions of the path by which the sample is magnetized.

The internal energy of the sample is its potential energy, $-\mu_0 M_1 H_1$, plus the work done to magnetize it:

$$u = -M_1 B_1 + A_1 = -A_2 = -\mu_0 \int_0^{H_1} M(B) dH \tag{2.36}$$

If a sample is already magnetized in the absence of a field (i.e., $A_1 = 0$) and properly aligned, when the field is turned on, the sample is drawn into the field, lowering its energy by $A_2 = - = M_1 B_1$. The internal energy of a magnetic sample is decreased in the presence of a field; a magnetic sample can do work when exposed to a magnetic field.

In Chapter 6 the values of these integrals A_1 and A_2 will be shown to depend on the direction in which a crystal is magnetized. Their variation with direction defines the anisotropy energies.

The second law of thermodynamics says that if an amount of heat dQ is added to a system, the result is an increase in its internal energy dU and/or the system can expand, $dV > 0$, doing work on its environment:

$$dQ = T\,dS = dU + p\,dV \tag{2.37}$$

where dS is the change in entropy and T is the temperature. For a magnetic material, the added heat may show up as a decrease in magnetization, giving

$$T\,dS = dU + p\,dV - \mu_0 HV dM \tag{2.38}$$

Note that a decrease in magnetization, $dM < 0$, is equivalent to an increase in entropy in terms of providing a channel for heat input, TdS. Hence the change in internal energy for a magnetic system should be written

$$dU = T\,dS - p\,dV + \mu_0 HV\,dM \tag{2.39}$$

The *internal* energy of a material *increases* as it is magnetized by a frield (A_1). In situations for which T, V, and M are the independent variables, it is the Helmholtz free energy, $F = U - TS$ that is minimized

$$dF = -S\,dT - p\,dV + \mu_0 HV\,dM$$

In the chapters ahead, expressions for the Helmholtz free energy will be minimized with respect to the direction of magnetization to find the equilibrium magnetization direction. When T, p, and H are the independent variables, it is the Gibbs free energy, $G = F + pV - MHV$, that is minimized:

$$dG = -S\,dT + V\,dp - \mu_0 MV\,dH$$

This free energy decreases as a sample of magnetization M is placed in a field (A_2).

It is important in many cases to consider also the *work done by the material* as it strains in response to being magnetized (see Chapter 7). This strain, driven by a magnetic stress σ_M will be seen to add a term to dU analogous to $-p\,dV$: $-\sigma_M(dl/l)V = -\sigma_M eV$. The energy of the material decreases if the strain is of the same sign as the magnetoelastic stress.

2.6 ANALYTIC MAGNETOSTATICS

2.6.1 The Magnetostatic Potential

The existence of a magnetostatic potential is justified in the same way the electrostatic potential is. Thus, one way to satisfy $\nabla \times \boldsymbol{E} = 0$ is to allow the electric field to be derived from a potential gradient, $\boldsymbol{E} = -\nabla\phi$ [because $\nabla \times (\nabla \text{ scalar}) = 0$, always]. If macroscopic current densities $\boldsymbol{J}$ vanish, then $\nabla \times \boldsymbol{B} = 0$. The vanishing curl implies that $\boldsymbol{B}$ also can be derived from the gradient, of a scalar, $\boldsymbol{B} = -\nabla\phi'_m$. This definition of the scalar magnetic potential is used by Jackson (1965). In this text the practice more common in the engineering literature is followed. The magnetic potential may be defined from $\nabla \times H = 0$ as

$$\boldsymbol{H} = -\nabla\phi_m \tag{2.40}$$

Because $\nabla \cdot \boldsymbol{B} = \mu_0 \nabla \cdot (H + M) = 0$, ϕ_m satisfies Poisson's equation in the absence of current densities:

$$\nabla^2\phi_m = \nabla \cdot M \tag{2.41}$$

where $\nabla \cdot \boldsymbol{M}$ defines a volume magnetic charge density ρ_m, again in analogy with electrostatics. Solutions for ϕ_m may be obtained from the differential Eq.

(2.43) or, knowing the magnetic charge distribution, from the integral

$$\phi_m(r) = -\frac{1}{4\pi} \iiint \frac{\rho_m}{r - r'} \, dV' \tag{2.42}$$

In regions where there are no magnetic charges present, ϕ_m satisfies Laplace's equation

$$\nabla^2 \phi_m = 0 \tag{2.43}$$

It has been shown that in addition to volume "charges," $\nabla \cdot M$, surface charges, $\boldsymbol{M} \cdot \boldsymbol{n}$ may also contribute to the potential. The most general and useful form, then, of the magnetic potential is the following:

$$\phi_m = -\frac{1}{4\pi} \int \frac{\nabla' \cdot \boldsymbol{M}(\boldsymbol{x}')}{|\boldsymbol{x} - \boldsymbol{x}'|} d^3\boldsymbol{x}' + \frac{1}{4\pi} \int \frac{\boldsymbol{M}(\boldsymbol{x}') \cdot \boldsymbol{n}}{|\boldsymbol{x} - \boldsymbol{x}'|} dA' \tag{2.44}$$

Because $H(r) = -\nabla_r \phi_m(r, r')$, it follows that

$$\begin{aligned} H(r) = &-\frac{1}{4\pi} \iiint d^3r' \nabla \cdot \boldsymbol{M}(r') \frac{(r - r')}{|r - r'|^3} \\ &+ \frac{1}{4\pi} \iint d^2r' \boldsymbol{n}' \cdot \boldsymbol{M}(r') \frac{(r - r')}{|r - r'|^3} \end{aligned} \tag{2.45}$$

The sources of demagnetization fields are surface poles, $\boldsymbol{M} \cdot \boldsymbol{n}$, and volume magnetic charges associated with a divergence of magnetization. Brown (1963) has shown that Eq. (2.45) is the solution to Maxwell's equations in the magnetostatic limit. The 2D equations used for surface poles [Eqs. (2.2) and (2.3)] are special cases of the surface integral in Eq. (2.45).

It was shown in Section 2.2 that surface charge, $\boldsymbol{M} \cdot \boldsymbol{n}$, is a source of $\boldsymbol{H}$ fields. The volume term in Eq. (2.45) also satisfies the magnetostatic equation, Eq. (2.43). To show this, apply the Laplacian operator to the $\boldsymbol{H}$ field potential ϕ_m:

$$\nabla^2 \phi_m = \frac{1}{4\pi} \nabla^2 \int \frac{\nabla' \cdot \boldsymbol{M}(x')}{|\boldsymbol{x} - \boldsymbol{x}'|} d^3\boldsymbol{x}'$$

Because the Laplacian here operates only on the observation coordinates x and not the source coordinates x', it may be written

$$\nabla^2 \phi_m = \frac{1}{4\pi} \int \nabla' \cdot \boldsymbol{M}(\boldsymbol{x}') \nabla^2 \frac{1}{|\boldsymbol{x} - \boldsymbol{x}'|} d^3\boldsymbol{x}'.$$

It is a consequence of Gauss' theorem [see Jackson (1965), p.13] that

$$\nabla^2 \frac{1}{|\boldsymbol{x} - \boldsymbol{x}'|} = -4\pi\delta(\boldsymbol{x} - \boldsymbol{x}')$$

Equations (2.41) and (2.43) follow from this.

The process of creating a magnetic charge distribution costs energy; the work done per unit of magnetic charge is the magnetostatic potential ϕ_m, and hence the energy of a given pole distribution can be expressed as follows:

$$U_m = \tfrac{1}{2} \iiint \rho_m \phi_m dV \tag{2.46}$$

The electrostatic solutions to Laplace's equation for various geometries can be carried over to magnetostatics and those of Poisson's equation apply when magnetic charge is present. Equation (2.43) is now used to derive the form of the potential and hence the $\boldsymbol{H}$ field due to two classical charge distributions: (1) that on a uniformly magnetized sphere and (2) that of a periodic, alternating magnetic pole density.

2.6.2 Uniformly Magnetized Sphere

Consider a sphere of radius a that is uniformly magnetized with no external field applied (Fig. 2.18). Inside the sphere, assume the magnetization to be parallel to the z axis, $\boldsymbol{M} = M_0 \boldsymbol{e}_z$. Because the sphere is uniformly magnetized, take $\boldsymbol{B}^{\text{in}} = B_0^{\text{in}} \boldsymbol{e}_z$. Using $\boldsymbol{B}^{\text{in}} = \mu_0(\boldsymbol{H}^{\text{in}} + \boldsymbol{M})$, then $\boldsymbol{H}^{\text{in}} = (B_0^{\text{in}}/\mu_0 - M_0)\boldsymbol{e}_z$, which is also a constant.

Outside the sphere there is no magnetization so Laplace's equation, Eq. (2.43) applies. Assuming $B = 0$ at $r = \infty$, the solution to Eq. (2.43) has the form of a sum of spherical harmonics (familiar from electrostatics):

$$\phi_m^{\text{out}} = \sum_{l=0}^{\infty} C_l \frac{P_l(\cos\theta)}{r^{l+1}}$$

$$\phi_m^{\text{out}} = C_0 \frac{1}{r} + C_1 \frac{\cos\theta}{r^2} + C_2 \frac{3\cos^2\theta - 1}{2r^3} + \cdots \tag{2.47}$$

The C values are determined by matching the normal ($\boldsymbol{e}_r$) and tangential ($\boldsymbol{e}_\theta$) components of the $\boldsymbol{B}$ and $\boldsymbol{H}$ fields inside and out (derived from ϕ^{in} and ϕ^{out}) at $r = a$: $B_r^{\text{in}} = B_r^{\text{out}}$ and $H_\theta^{\text{in}} = H_\theta^{\text{out}}$.

Using the cylindrically symmetric form of the Del operator, we obtain

$$\nabla = \frac{\partial}{\partial r}\boldsymbol{e}_r + \frac{1}{r}\frac{\partial}{\partial \theta}\boldsymbol{e}_\theta$$

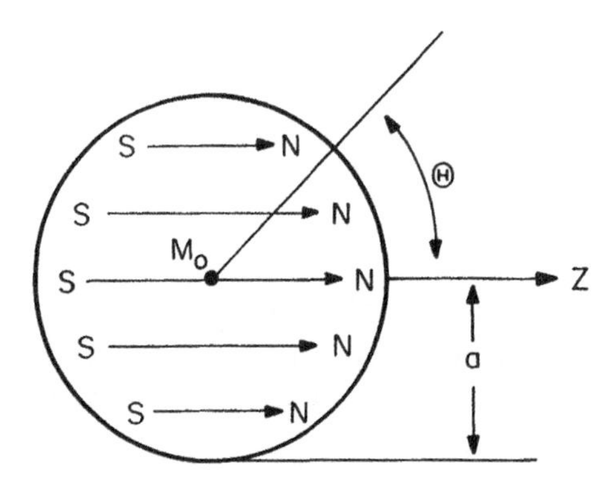

Figure 2.18 Uniformly magnetized sphere of radius a showing coordinate θ.

Equation (2.47) gives the following for the boundary conditions:

$$B_r^{\text{out}} = \left[\frac{C_0}{r^2} + 2\frac{C_1 \cos\theta}{r^3} + \cdots\right]_a = B_r^{\text{in}} = B_0^{\text{in}} \cos\theta \tag{2.48}$$

and

$$H_\theta^{\text{out}} = \left[\frac{C_1 \sin\theta}{\mu_0 r^3} + \cdots\right]_a = H_\theta^{\text{in}} = \left(\frac{B_0^{\text{in}}}{\mu_0} - M_0\right)(-\sin\theta) \tag{2.49}$$

Clearly, from Eq. (2.48), $C_0 = 0$, $C_1 = a^3 B_0^{\text{in}}/2$, and all higher-order coefficients vanish. From Eq. (2.49) and the value determined for C_1, it follows that $B_0^{\text{in}} = (\frac{2}{3})\mu_0 M_0$ (in cgs units $B_0^{\text{in}} = 8\pi M_0/3$) and $C_1 = \mu_0 a^3 M_0/3$ ($4\pi a^3 M_0/3$ in cgs). Thus from Eq. (2.47), the fields inside the sphere are

$$\boldsymbol{B}^{\text{in}} = \left(\frac{2\mu_0}{3}\right)\boldsymbol{M}$$

$$\boldsymbol{H}^{\text{in}} = -\frac{\boldsymbol{M}}{3} \tag{2.50}$$

The latter is the demagnetizing field and it is due to the free poles at the surface. Thus, what was merely stated before, namely that $N = 1/3$ for a sphere, has now been calculated. The internal B^{in} field is reduced from $\mu_0 M$ by the demagnetizing field H_d.

Outside the sphere the result is

$$\boldsymbol{H}^{\text{out}} = \frac{\boldsymbol{B}^{\text{out}}}{\mu_0} = 2\frac{a^3 M_0}{3r^3}\cos\theta \boldsymbol{e}_r + \frac{a^3 M_0}{3r^3}\sin\theta \boldsymbol{e}_\theta \tag{2.51}$$

which is a dipole field for a magnetic moment $\mu_m = a^3 M_0/3$. Figure 2.19 shows

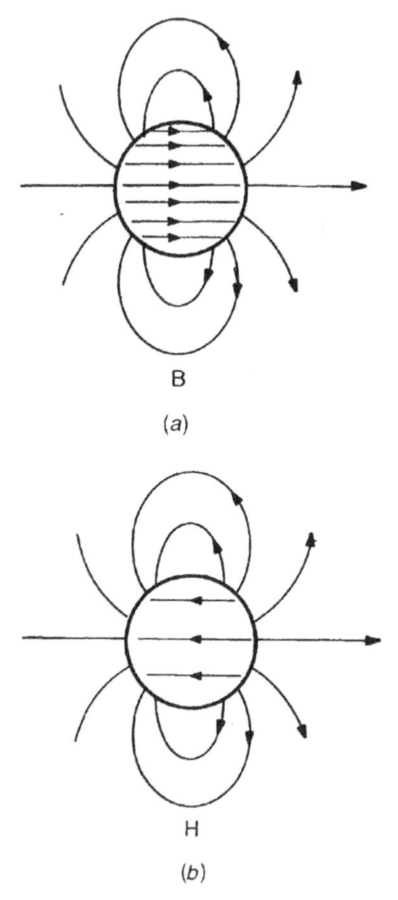

Figure 2.19 Fields inside and outside a uniformly magnetized sphere: (*a*) ***B*** field whose lines form continuous loops inside and outside the material; (*b*) ***H*** field, whose lines are not continuous; some may terminate at the surface poles.

these fields. The ***B*** field lines are continuous; those of ***H*** originate and terminate on the "poles" at the surface.

The results in Figure 2.19 should be compared with those for the 2D results for the rectangular cross section shown in Figure 2.8.

2.6.3 Field Due to Periodic Surface Poles

A useful application of the magnetic potential is to calculate the ***H*** field due to a periodic array of alternating surface poles such as exist at the surface of a multidomain sample (Fig. 2.20). The existence of such domain structures will be justified in Chapter 9 after treating the other energies involved in determining domain structure.

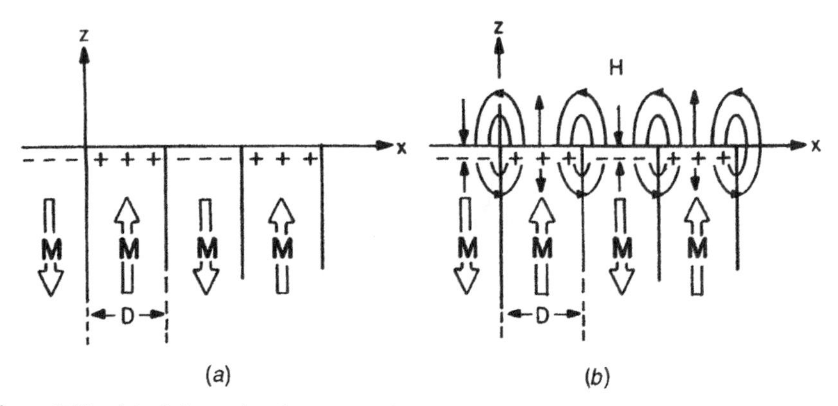

Figure 2.20 (*a*), Schematic of cross section of a semi-infinite sample with a periodic domain structure; (*b*), field distribution due to magnetic surface charges.

The periodic charge density can be expressed as a Fourier sine or cosine series. If the surface charge density is sinusoidal, only the term $n = 1$ need be kept. For a square surface pole density, more terms in the expansion are required.

It is possible to solve for ϕ_m by integrating over the charge density

$$\rho_m = +M_s \quad \text{for} \quad 2nD < x < (2n+1)D \tag{2.52a}$$

$$\rho_m = -M_s \quad \text{for} \quad 2(n+1) \quad D < x < (2n+1)D \tag{2.52b}$$

Alternatively, general expressions for the magnetic potential can be written by inspection:

$$\phi_m(x, z) = X(x)Z(z)$$

$$\nabla^2\phi = 0 \rightarrow \frac{X''}{X} + \frac{Z''}{Z} = 0 \tag{2.53}$$

thus

$$\phi^{\text{in}} = \sum_n A_n^{\text{in}} e^{nkz} \sin(nkx) \tag{2.54a}$$

$$\phi^{\text{out}} = \sum_n A_n^{\text{out}} e^{-nkz} \sin(nkx) \tag{2.54b}$$

The values of A_n^{in} and A_n^{out} can be determined by applying the boundary conditions for continuity of $H_x = (\partial\phi/\partial x)_0$ and $B_z = \mu_0(-\partial\phi/\partial x \pm M)_0$ across the interface [Eqs. (2.9) and (2.6), respectively]

$$-\sum nkA_n^{\text{in}} \cos(nkx) = -\sum nkA_n^{\text{out}} \cos(nkx)$$

thus $A_n^{\text{in}} = A_n^{\text{out}}$, and

$$-\sum nkA_n \sin(nkx) \pm M_s = \sum nkA_n \sin(nkx)$$

thus

$$\sum nkA_n \sin(nkx) = \pm \frac{M_s}{2} \tag{2.55}$$

with the plus or minus applying depending on the value of x as in Figure 2.20.

The value of A_n can be determined from the second boundary condition, (Eq. 2.55), by multiplying both sides by $\sin(mkx)$ and integrating over one period in x:

$$\sum_n A_n nk \int \sin(mkx) \sin(nkx) dx = \frac{M_s}{2} \left\{ \int_0^D \sin(nkx) dx - \int_D^{2D} \sin(nhx) dx \right\}$$

which gives

$$A_n = \begin{cases} \dfrac{2M}{n^2 k\pi} & \text{for } n \text{ odd} \\ 0 & \text{for } n \text{ even} \end{cases} \tag{2.56}$$

Because of the n^2 in the denominator of A_n, it is usually sufficient to keep only one or two terms in the expansion of the field. The reader should verify that the H fields derived from ϕ_m inside and outside the material have the form shown in Figure 2.20*b*. Note that the exponential decay length scales with the wavelength of the periodic pole distribution.

The magnetostatic energy per unit surface area of the periodic distribution can be calculated from Eq. (2.54), evaluated at $z = 0$, and from Eqs. (2.46), (2.52), and (2.51) to give

$$U_{\text{area}} = \frac{1}{2} \int_0^D \sum_{n\,\text{odd}} \frac{2M_s}{n^2 k} \sin(nkx) M_s dx = 4 \frac{2M_s^2 D}{\pi^2} \sum \frac{1}{n^3} = 2.13 \frac{M_s^2 D}{\pi^2}$$

For a two-dimensional square pole pattern the magnetostatic energy is reduced to 62% of the value for one dimension and for a circular domain pattern the energy is reduced by another factor of 72% [see Kittel (1949)]. We will have more to say about the magnetostatic fields of various domain configurations in Chapter 9.

2.7 SUMMARY AND EXAMPLES

This chapter contains descriptions of some common ways in which magnetostatic effects are observed. Samples are harder to magnetize along their shorter directions. These effects were considered first in terms of a single demagnetizing factor N related to sample shape, then in terms of the three diagonal components of the demagnetization tensor. A phenomenological approach was taken to understanding the effects of sample shape on magnetization curves, indicating that the field at which hard-axis saturation occurs gives the demagnetization factor: $H_a \approx NM_s$.

The boundary conditions on $\boldsymbol{B}$ and $\boldsymbol{H}$ fields were helpful in understanding magnetostatic effects; the normal component of $\boldsymbol{B}$ must be continuous across an interface and the tangential component of $\boldsymbol{H}$ must be discontinuous by the amount of surface current.

After reviewing the phenomenology of magnetostatic energy, attention was given to a more analytic approach solving Laplace's equation for a uniformly charged sphere and for a periodic charged surface. The sphere has a demagnetizing field equal to $-\frac{1}{3}M_s$, and the periodic charged surface produces a field that drops off exponentially with distance from the surface.

APPENDIX: MAGNETIC CIRCUITS

Many flow or flux phenomena obey a diffusion equation: $-D\, dc/dx = \boldsymbol{J}$, where, in general, $\boldsymbol{J}$ is the flux, that is, quantity of heat, mass or charge flowing per unit time through a unit area; dc/dx is the concentration gradient that drives the flux; and D is the constant that describes the ease with which the medium permits the flux.

The following table compares some values of the constant D for electrical and magnetic fluxes in air and in representative media:

$\boldsymbol{J} = \sigma \boldsymbol{E}$	$\sigma_0(\mu_0/\varepsilon_0 = 377\,\Omega)$	$\sigma_{\text{Fe}} \approx 10 \times 10^6 (\Omega \cdot \text{m})^{-1}$
$\boldsymbol{D} = \varepsilon \boldsymbol{E}$	$\varepsilon_0 = 1/\mu_0 c^2 = 8.85 \times 10^{-12}$ F/m	$\varepsilon_r = \varepsilon/\varepsilon_0 \approx 1$–$10$ for dielectric
$\boldsymbol{B} = \mu \boldsymbol{H}$	$\mu_0 = 4\pi \times 10^{-7}$ H/m	$\mu_r \approx 10$–10^6 for Fe

In the relation $\boldsymbol{J} = \sigma \boldsymbol{E}$, $\boldsymbol{J}$ is the electrical flux or current per unit area, and $\boldsymbol{E}$ is the potential gradient (analogous to $\partial c/\partial x$) having units of volts per unit length, so σ represents for electrical conductivity what D describes for diffusion. Thus, $J = I/A$ is equivalent to $V = IR$, which is Ohm's law.

Figure 2A.1 illustrates a flux between two bodies at different concentrations, along a path of low impedance or high conductivity.

In electrostatics $\int \boldsymbol{E} \cdot d\boldsymbol{l} = V$ defines the electromotive force (EMF) which satisfies Ohm's law. Similarly, in magnetostatics, the integral of the $\boldsymbol{H}$ field around a closed path, $\int \boldsymbol{H} \cdot d\boldsymbol{l} = NI$ defines the magnetomotive force (MMF) or "pressure" that creates a magnetic response. This MMF also satisfies a magnetic Ohm's law:

$$V_m = NI = \phi R_m \tag{2A.1}$$

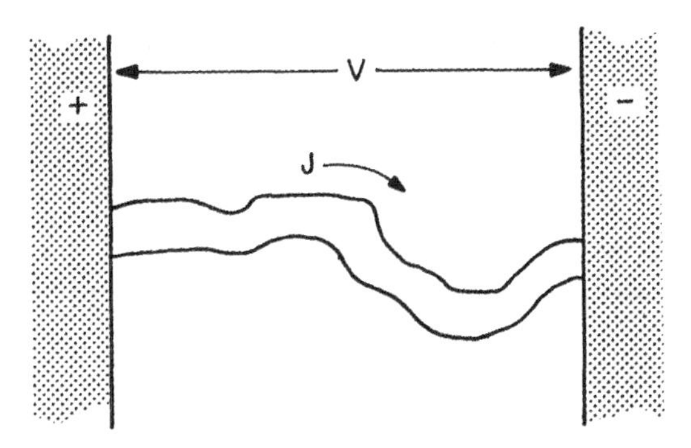

Figure 2A.1 Representation of flux between two bodies at different concentrations or potentials. The flow is shown along a path of low impedance.

where ϕ is the magnetic flux ($\phi = \int \boldsymbol{B} \cdot \boldsymbol{dA}$) that results from a magnetomotive force NI just as current is the electrical flux or flow that results from an electromotive force. Equation (2A.1) defines the reluctance, which is the magnetic analog of resistance.

What is the reluctance R_m? It should be related inversely to the permeability of a material, its ability to draw in magnetic flux. It can be defined from the integral form of NI:

$$NI = \oint H \cdot dl = \oint \frac{\phi}{\mu A} dl \tag{2A.2}$$

A single component magnetic circuit gives

$$NI = \frac{\phi l}{\mu A} = \phi R_m$$

Thus

$$R_m = \frac{l}{(\mu A)} \tag{2A.3}$$

If the medium through which the flux flows is inhomogeneous then the reluctance is

$$R_m = \oint \frac{dl}{\mu A}$$

just as

$$R_{\text{elec}} = \oint \frac{dl}{\sigma A} \tag{2A.4}$$

Thus, from Eq. (2A.1), and writing R_m as a discrete sum over various circuit elements

$$NI = \phi \sum_i \frac{dl_i}{\mu_i A_i} \tag{2A.5}$$

This last form is most useful for magnetic circuit analysis. It is important to note that in a magnetic circuit with the N turns confined to a portion of the circuit, the right-hand side (RHS) of (2A.5) or (2A.2) sums over all elements of the circuit, *including* the length containing the turns. This part of the circuit provides a back MMF on the magnetic power source, NI.

The concept of magnetic circuits also applies if the MMF, NI, is replaced by a permanent magnet of magnetization per unit area. $M_r l$ (SI units, A amperes). This will be covered in Chapter 14 on permanent magnets.

Examples are as follows:

1. What field is produced by one ampere of current passing through 400 turns on a hollow toroid ($r_1 = 2$ cm $r_2 = 3$ cm) of circular cross section (Fig. A2.2)? The magnetic path length is $2\pi(r_1 + r_2)/2 = 0.157$ m. *Answer:* $H = NI/l = 400 \times 1/0.157 = 2548$ A/m (or $H = 32$ Oe). This result serves as a reference for the second exercise.
2. A similar toroid with 400 turns is now prepared with a soft iron core ($\mu_r = \mu/\mu_0 \approx 10^2$) and a 1-cm gap (Fig. 2A.3). How much current is needed to produce a field of 0.1 T ($H = 1000$ Oe or 79,618 A/m) in the gap? *Answer:* We first determine the path length in the magnet. From Example 1 (above): $l = 0.157 - 0.01 = 0.147$ m. For the core, the cross-sectional area is given by $A = \pi[(r_2 - r_1)/2]^2 = 7.85 \times 10^{-5}\,\text{m}^2$. Use $NI = 400I = \Sigma(\phi/A)(l/\mu)$ and keep in mind the B field desired, $0.1\,\text{T} = \phi/A$:

$$400I = \frac{(\phi/A)}{\mu_0}\sum \frac{l_i}{\mu_r} = \frac{0.1}{4\pi \times 10^{-7}}\left(\frac{0.01}{1} + \frac{0.147}{100}\right)^2$$

$$= 913\ \text{At (ampere-turns)}\ I = 2.29\ \text{A}$$

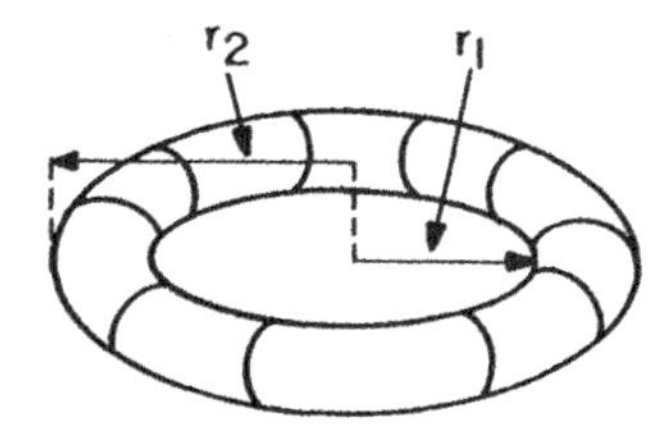

Figure 2A.2 A hollow toroid of inner and outer radius r_1 and r_2 has 400 turns of wire carrying a current of 1 A.

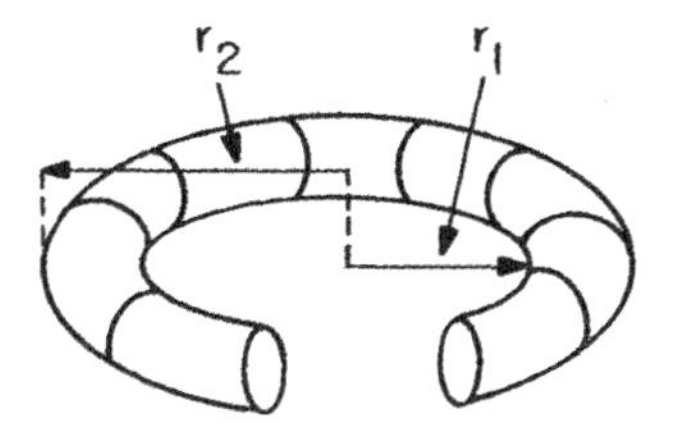

Figure 2A.3 The toroid of Figure 2A.2 has a 1 cm gap and is otherwise filled with an iron core having $\mu = 100\,\mu_0$.

Here, the flux in the gap is assumed to fill an area equal to that filled by the flux in the core. Actually, the flux in the gap flares out to a larger area. Note that for the hollow toroid (example 1) the same current of 2.3 A would give $H = 5830$ A/m (73 Oe) and the addition of a magnetic core with a gap (example 2) gives a field of nearly 80,000 A/m (1000 Oe) for the same current.

It is important to note that if the N turns around the core are bunched up over a length shorter than the length of the core ($l_0 = 0.147$ m in this case), the resulting gap field is generally the same. This can be seen by breaking l_0 into two parts l_1 and $l_2 = l_0 - l_1$, with the N turns confined to l_1. The circuit equation term l_0/μ_{Fe} is replaced by $l_1/\mu_{Fe} + l_2/\mu_{Fe}$, which is equal to l_0/μ_{Fe}. The result would be different if the permeability or cross-sectional area of the unwrapped part of the core were different from the wrapped portion.

PROBLEMS

2.1 Assume that an interface, similar to that sketched in Figure 2.7*b* but with $\mu_1/10 = \mu_2 = \mu_0$, is carrying a current $[0, 0, J(y)] = 2J_0(y/h + \frac{1}{2})$ and h is not necessarily small:

(a) Sketch the dependence of the current density on y.

(b) What are the boundary conditions on an $\boldsymbol{H}$ field $[H_x(y), H_y(y), 0]$ across the interface.

(c) Sketch your result and justify it in terms of the field generated by the current distribution. Make sure that your result behaves properly as h approaches zero.

2.2 You need to make an electromagnet that produces a field of 0.6 T in a 1-cm gap. You use a rectangular core design (20 × 10 cm with the gap on one of the 20-cm legs) and make the core out of soft iron with $\mu_r \approx 100$. The yoke has cross-sectional area = 10 cm^2.

(a) If you are able to conveniently wind 600 turns of copper wire on the 20-cm leg, how much current will you need to achieve $H = 0.6\,\mathrm{T}$ in the gap? (Neglect edge effects.)

(b) What is the flux density in the iron core at this current?

2.3 Consider a permanent dipole magnet with its north pole a distance h away from a sheet of copper. Calculate the magnetic field in the copper, and sketch the field in and near the copper sheet when it carries no current and when it carries a uniform current I_0 out of the plane of the figure as sketched below.

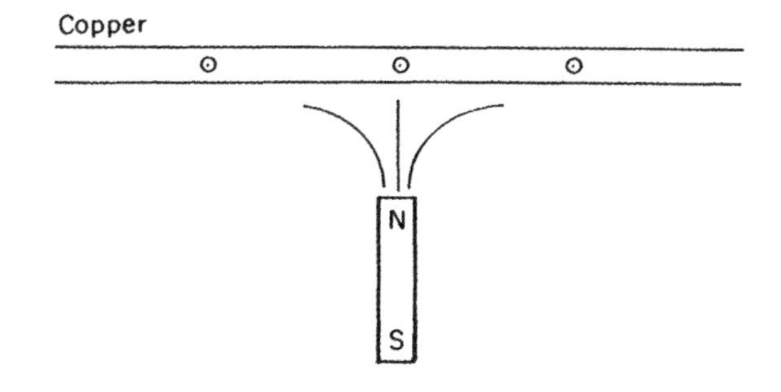

Figure P2.3

2.4 Describe the energy required to move a *demagnetized* ferromagnetic sample from negative infinity toward a permanent magnetic dipole at the origin having its north pole pointed toward the ferromagnet. Include the effects of magnetization of the sample and its energy in the dipole field. Then, assuming you can hold the magnetization fixed, remove the magnetized sample back to negative infinity considering only its energy in the dipole field is changing. What is the energy change in this second process? How do these energies relate to A_1 and A_2?

2.5 Derive Eq. (2.30) by eliminating H_i from the two expressions that led to Eq. (2.16).

2.6 Verify that Eq. (2.45) satisfies Laplace equation, Eq. (2.48).

2.7 You formulate a new soft ferrite composition and want to test its low-field magnetic properties. You sinter a toroid and a rod from your new proprietary composition. The figures below depict the B–H loops you observe on each.

(a) What is the remanent induction in each case?

(b) What is the coercivity in each case?

(c) Approximate the effective permeability in each case for a drive field of 2 Oe.

(d) Assume $B = H + 4\pi M \approx 4\pi M$ and use $M = \chi H$ with Eq. (2.29) to estimate the demagnetization factor, N, for the rod.

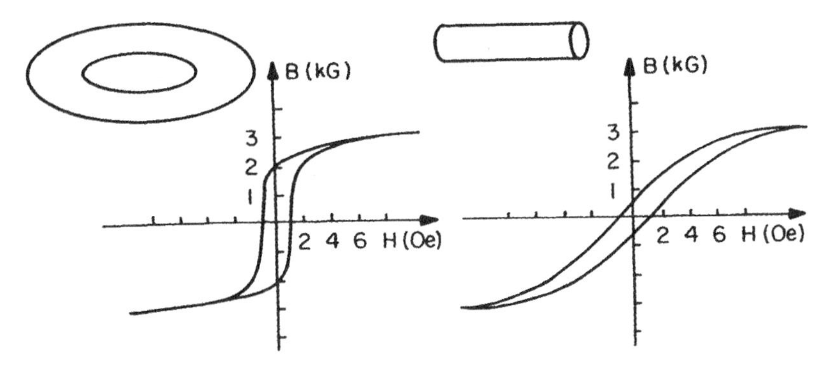

Figure P2.7

(e) Use Table 2.1 to estimate the demagnetization factor of the rod-shaped sample which you measure to have an aspect ratio of 10:1.

2.8 You need to calculate the magnetic field parallel to and outside a strip of a permanent magnet of length l. The strip is $l/5$ wide (perpendicular to the plane of the figure below) and $l/200$ thick (in the y-direction, below) and you need to know the field at heights of $0.05l$ and $0.3l$ for $0 < x < l$. Use a three-dimensional formula [Eq. (2.45)] for the field assuming that the width is small relative to the length (strip acts like a rod) and also use Eq. (2.3) for a two-dimensional approximation (ignoring field spread perpendicular to the plane of the paper). The form of the x dependence of $H(x)$ is more important to find than the magnitude of the field.

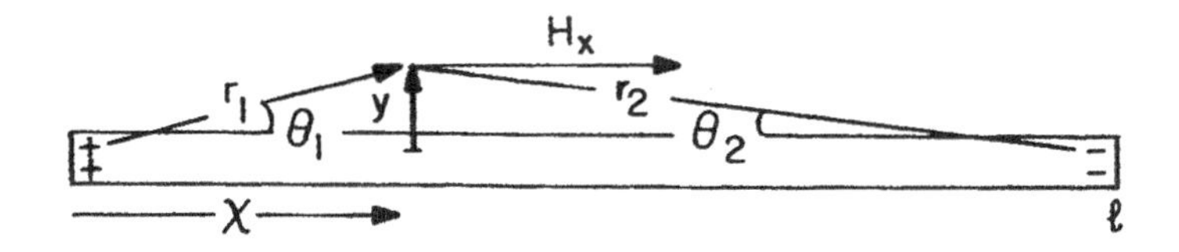

Figure P2.8

BIBLIOGRAPHY

Aharoni, A., "Magnetostatic energy calculations," *IEEE Trans.* **MAG-27**, 3539 (1991).

Aharoni, A., *Introduction to the Theory of Ferromagnetism* Clarendon Press, Oxford, England, 1996.

Bertram, H. N., *Theory of the Magnetic Recording*, Cambridge Univ. Press, 1994.

Bozorth, R. M., *Ferromagnetism*, van Nostrand, 1954; IEEE Press, 1993.

Brown, W. F., *Magnetostatic Principles in Ferromagnetism*, Interscience, New York, 1962; *Micromagnetics*, Interscience, New York, 1963.

Clarke, A. E., and Wun-Fogel, M., *IEEE Trans.* **MAG-25**, 3611 (1989).

Jackson, J. D., *Electricity and Magnetism,* Wiley, New York, 1965.

Kittel, C., "Physical theory of ferromagnetic domains," *Rev. Mod. Phys.* **21**, 541 (1949).

Osborn, J. A., "Demagnetizing factors of the general ellipsoid," *Phys. Rev.* **67**, 351 (1945).

CHAPTER 3

CLASSICAL AND QUANTUM PHENOMENOLOGY OF MAGNETISM

This chapter begins with a classical treatment of the origin and behavior of magnetic moments. Then paramagnetism, diamagnetism, and atomic spectra are described, as best they can be, in this simple picture. The quantization of electron orbital motion and spin are then introduced, as well as the interaction between these two angular momenta. These concepts lead to a more accurate description of the phenomena already treated classically and provide a solid basis for understanding ferromagnetism in simple systems. The consequences of bonding in oxides and metals is covered in Chapters 4 and 5.

3.1 ORBITAL ANGULAR MOMENTUM AND ORBITAL MAGNETIC MOMENTS

It was pointed out that an atomic magnetic dipole moment is given by $\mu_m = IA$. Once it was realized that magnetic moments were associated with circulating charges, several people, including Einstein, became curious about the connection between magnetic moments and angular momentum. (It turns out that if a rod of copper is rotated about its cylindrical axis, it is more strongly affected by a magnetic field than if it is not spinning. An angular momentum is imparted to the electrons that otherwise would have none. This angular momentum produces magnetism where there was none before.) It was of interest to determine whether the ratio of the magnetic moment μ_m to the mechanical angular momentum L of a system, $\gamma = \mu_m/L$, is a constant; γ is called the *gyromagnetic ratio.*

Let us make an estimate of γ for the $n = 1$ electron in hydrogen using the Bohr model (Fig. 3.1). The angular momentum of the electron is $\boldsymbol{L} = \boldsymbol{r} \times \boldsymbol{p} = mvr_0$. The magnetic moment is given by $\mu_m = IA = er_0v/2$, where r_0 is the Bohr radius and the relation $\omega = v/r_0$ is used. Thus the gyromagnetic ratio for orbital motion is given by

$$\gamma = \frac{\mu_m}{L} = \frac{e}{2m} \tag{3.1}$$

$$= -8.78 \times 10^{10}\,(\text{C/kg})$$

Note that because e is a negative number, $\boldsymbol{L}$ and μ_m are in opposite directions for electron orbits. This value for γ turns out to be exact for materials where the magnetism comes from the orbital motion of electrons. It will be seen that this γ is not valid for many other materials and quantum mechanics is needed to fix it.

The x component of force on an object is given by $F_x = -dU/dx$, where U is the potential energy. Similarly, the torque on an axial vector such as the magnetic moment μ_m is given by $T = -dU/d\theta$, where U is the potential energy of the moment in a B field [Eq. (1.21)]. Thus

$$T = -\mu_m B \sin\theta \tag{3.2}$$

This has the magnitude of $\mu_m \times B$ or $B \times \mu_m$, but it may not be immediately clear what direction to assign to the torque. Note that the torque causes the magnetic moment to precess around B in the same way the angular momentum axis of a top spinning on a smooth surface precesses around the gravitational force.

A classical picture of this precession can be developed by considering the Lorentz force

$$\boldsymbol{F} = q(\boldsymbol{v} \times \boldsymbol{B}) \tag{3.3}$$

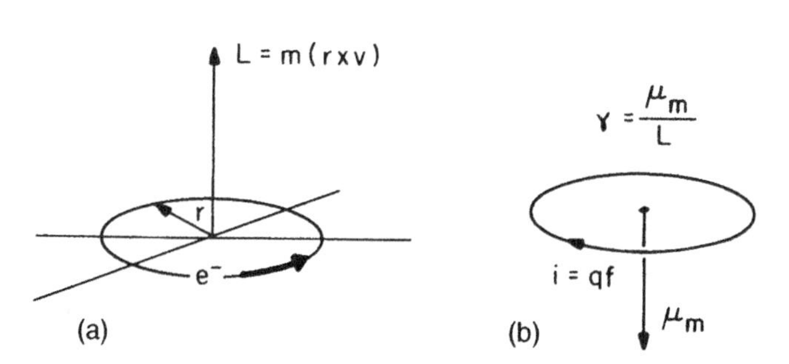

Figure 3.1 (*a*), Angular momentum of a circulating electron; (*b*), magnetic moment of a circulating electron.

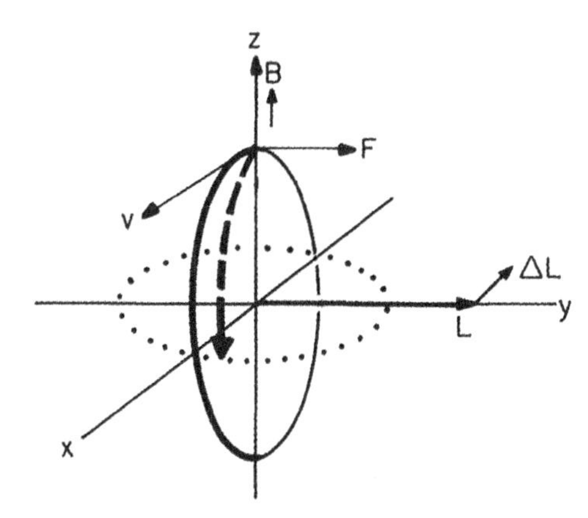

Figure 3.2 Classical picture of the Lorentz force effect of the B field on the trajectory of a circulating electron. The effect is to rotate the angular momentum vector of the initial orbit about $\boldsymbol{B}$.

on an electron in circular orbit of radius r in the x–z plane (Fig. 3.2). This orbit has angular momentum $\boldsymbol{L}$ in the y direction. Consider the motion at the apex of the orbit, $v = v_x$, where $F = qv_xB_z$ is in the positive y direction (because $q < 0$ and $v_x > 0$). Following this Lorentz force around the orbit shows that $\boldsymbol{L}$ changes by $\Delta\boldsymbol{L}$ in the negative x direction; that is, $\boldsymbol{L}$ precesses with angular frequency ω in the field direction. The magnetic moment vector is oriented opposite to $\boldsymbol{L}$; $\mu_m = \gamma\boldsymbol{L} = -|e/2m|\boldsymbol{L}$, and it also precesses with an angular frequency ω in the field direction. Similar arguments can be made for orbits in the other two planes; y–z orbits precess as x–z orbits do, but x–y orbits do not.

Let us calculate the frequency of this precession, called the *Larmor frequency* ω_L. The torque causes the orientation of the angular momentum vector $\boldsymbol{L}$ to change by $d\boldsymbol{L}$ perpendicular to both $\boldsymbol{B}$ and μ_m. From Fig. 3.3, $dL = L\sin\theta d\phi$, where $d\phi = \omega_L dt$, thus $dL/dt = \omega_L L\sin\theta$. But the definition of torque is $\boldsymbol{T} = d\boldsymbol{L}/dt$ (analogous to $\boldsymbol{F} = d\boldsymbol{p}/dt$), so it is possible to write $\omega_L L\sin\theta = -\mu_m B\sin\theta$ [Eq. (3.2)]. Therefore

$$\omega_L = \frac{-\mu_m \boldsymbol{B}}{L} = -\gamma\boldsymbol{B} \tag{3.4}$$

$$= -\frac{e}{2m}\boldsymbol{B}$$

This is called the *Larmor frequency*.

So the sign of the torque is now known; it is such as to cause an angular rotation of the magnetic moment vector with ω_L parallel to $\boldsymbol{B}$; ω_L would be antiparallel to $\boldsymbol{B}$ for a positively charged particle. The expression $\omega_L = -\gamma B$ is

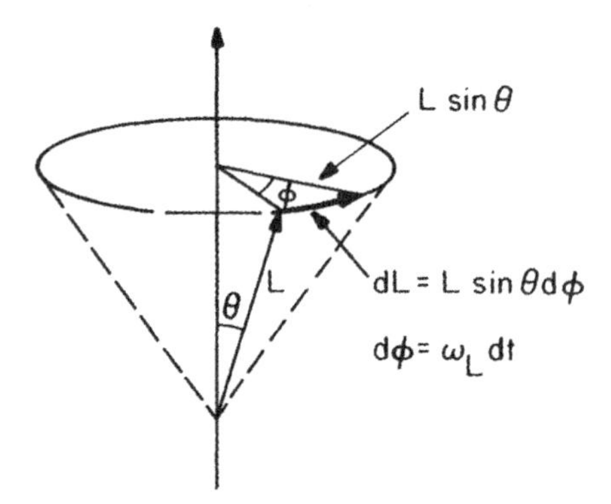

Figure 3.3 Construction for relating the change in angular momentum to the precession frequency.

the basis for magnetic resonance. A radiofrequency (RF) field incident on a precessing moment in a quasistatic magnetic field B can be absorbed if the RF frequency satisfies $\omega = \gamma B$. The value of γ [Eq. (3.1)] corresponds to a frequency $f = 14\,\text{GHz/T}$. For an applied field $B = 1.0\,\text{T}$ (10 kOe), this implies that magnetic resonance occurs for microwaves of a few centimers in length. As an exercise, compare the physics of magnetic spin resonance with that of cyclotron resonance $\omega = (e/m)B$.

3.2 CLASSICAL DIAMAGNETISM

A classical explanation of the origin of diamagnetism will now be given (see also Fig. 1.5). Diamagnetism is present in all matter but is often obscured by paramagnetism or ferromagnetism. Diamagnetism does not require that the atoms have orbital moments or unpaired spins; it occurs for filled or partially filled orbitals.

It was shown above that the Larmor precession is the additional angular frequency acquired by an orbiting electron when a magnetic field is turned on. It was possible to explain it by resorting to the classical Lorentz force. The Larmor frequency ω_L for an electron orbit is always in the direction of $\boldsymbol{B}$. Thus, ω_L corresponds to a new, field-induced angular momentum, $d\boldsymbol{L} = \boldsymbol{I} \cdot \omega_L$ (here $\boldsymbol{I}$ is the moment of inertia tensor, which, in simplest terms, has magnitude $r^2 m$). (Here is where this classical explanation of diamagnetism is clearly wrong. Where does the extra angular motion come from in diamagnetism? If angular momentum is conserved, the net must be the same before and after $\boldsymbol{B}$ is turned on. Classical arguments cannot answer this question; quantum mechanics is needed.) Because the magnetic moment is related to the angular momentum of an orbiting electron, $d\boldsymbol{L}$ gives rise to a field-induced change in the moment $d\mu_m = \gamma d\boldsymbol{L}$ (opposite to $d\boldsymbol{L}$ because $\gamma < 0$). Thus, $d\mu_m = (e/2m)(r^2 m)|e/$

$2m|B = -(e^2r^2/4m)\boldsymbol{B}$. But not all components of the moment of inertia $\boldsymbol{I}$ respond to the field; only those corresponding to motions in the plane perpendicular to $\boldsymbol{B}$ do. Hence, only two components of $r^2 = x^2 + y^2 + z^2 = 3x^2$ are of importance, namely, those that contribute to the moment of inertia about the field axis $2r^2/3$. Therefore, $d\mu_m = -e^2\boldsymbol{B}\langle r^2\rangle/(6m)$ and for a number n_v of these atomic oscillators per unit volume

$$\chi_d = \frac{n_v\mu_m}{H} = \frac{-n_v e^2\mu_0\Sigma\langle r^2\rangle}{6m} \tag{3.5}$$

A summation over the different orbitals in each atom has been included in Eq. (3.5). Note that $\langle r^2\rangle$ is the average value of the orbital area divided by π.

Figure 3.4 shows the atomic magnetic susceptibilities of most elements; the negative values are the susceptibilities of materials for which diamagnetism

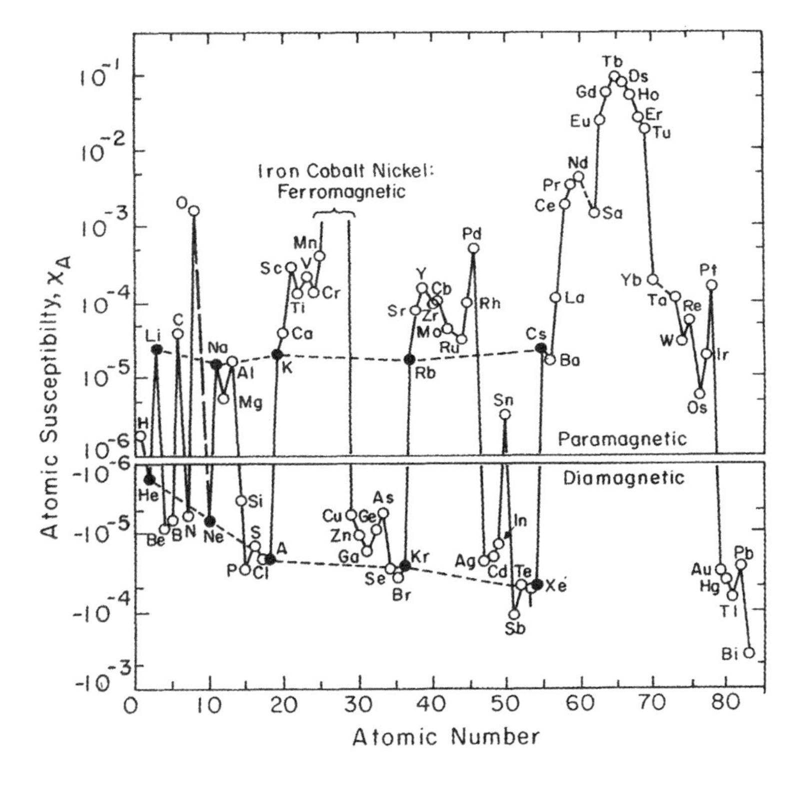

Figure 3.4 Magnetic susceptibilities of the elements in atomic units. Negative values of χ indicate that the diamagnetic part of the susceptibility is greater than the paramagnetic part [After Bozorth, copyright IEEE Press (1993)].

dominates paramagnetism. (The atomic susceptibility χ is related to the experimental volume susceptibility, $\kappa = M/H$ (moment per volume per field) by $\chi = \kappa A/\rho_m$, where A is the atomic weight and ρ_m is the mass density; χH is the magnetic moment per gram atomic weight, which, when divided by Avogadro's number, gives the component along the field direction of the magnetic moment per atom.) Table 3.1 lists the diamagnetic susceptibilities of rare gases and alkali metal ions. [The units in this table are 10^{-6} cm^3/mol, so you must convert to units of m^3/mol *and* remember that in cgs units it is $4\pi\chi$ that corresponds to the SI χ. So, for helium, the table indicates $\chi_d = -2.4 \times 10^{-11}$ m^3/mol). If $\langle r^2 \rangle \approx a_0^2$ is assumed for the helium atom, then Eq. (3.5) gives $\chi_d = -2 \times 10^{-11}$ (m^3/mole). The accuracy of this result is impressive for a simple model. Clearly, for atoms of larger radius, χ_d increases in magnitude.

An alternate classical derivation of diamagnetic susceptibility is given by Feynman et al. (1964). He considers a quasistatic switching on of the B field and calculates the voltage $\int \boldsymbol{E} \cdot \boldsymbol{dl}$ applied to an electron orbit in a plane normal to $\boldsymbol{B}$. Integrate the Maxwell–Faraday equation over the area of the orbit:

$$-\int \frac{\partial B}{\partial t} \cdot dA = \int (\nabla \times E) \cdot dA = \oint E \cdot dl$$

giving:

$$-\left(\frac{\partial B}{\partial t}\right) \pi r^2 = 2\pi r E \tag{3.6}$$

The tangential electric field $E = -(\partial B/\partial t)r/2$ gives a torque $\boldsymbol{r} \times \boldsymbol{F} = -|e|Er$ parallel to $\boldsymbol{B}$ which changes the angular momentum $\boldsymbol{L}$ (later it will be shown that

TABLE 3.1 Diamagnetic Susceptibilities ξ_m, of Various Atoms and Ions ^{a} in CGS Units

Element	$-\chi_m$ (10^{-6} cm^3/mol)	Element	$-\chi_m$ (10^{-6} cm^3/mol)	Element	$-\chi_m$ (10^{-6} cm^3/mol)
F^-	9.4	Ne	7.2 (5.6)	Na^+	6.1 (5.5)
Cl^-	24.2	A	19.4 (19.5)	K^+	14.6 (17.5)
Br^-	34.5	Kr	28	Rb^+	22 (30.1)
I^-	50.6	Xe	43	Cs^+	35.1
He	1.9 (1.7)	Li^+	0.7		

^{a}Ions in each row have the same electronic configuration. Values from Kubo and Nagamiya, McGraw-Hill, New York, 1969, p. 439, except where indicated by asterisk, which are from Condon and Odishaw, *Handbook of Physics*, McGraw-Hill, New York, 1958, pp. 4–130. Calculated values are given in parentheses.

angular momentum can arise from sources other than orbital motion):

$$\boldsymbol{r} \times \boldsymbol{F} = \frac{d\boldsymbol{L}}{dt} = |e| \frac{(\partial B/\partial t) r^2}{2} \tag{3.7}$$

Integrating over the time to turn on the B field gives the additional angular momentum $\Delta L = |e| r^2 B/2$ which describes a field-induced orbital magnetic moment $\Delta\mu_m = \gamma \Delta L = -e^2 r^2 B/4m$. From which, as above, $\chi_d = -\mu_0 N e^2 \langle r \rangle^2 / 4m = -\mu_0 N e^2 (x^2 + y^2)/6m$. The minus sign is important.

This treatment makes it clear that if the B field is *decreased* from a finite value to 0, the diamagnetic response is to create a *positive* moment that opposes the field change. Thus diamagnetism is seen simply as a manifestation of Lenz law: moving charges respond to a change in field by changing their motion so as to set up a field (response) that opposes the initial change.

The lower panel of Figure 3.5 shows the temperature dependence of the atomic or molecular magnetic susceptibility in a number of important diamagnetic materials. It should be noted that the diamagnetic susceptibility is largely independent of temperature; paramagnetic susceptibility, covered in the next section (and shown in the top panel of Fig. 3.5), generally varies inversely with temperature.

To try to understand the lack of temperature dependence exhibited by the diamagnetic susceptibility, let us first compare the potential energy of the diamagnetic moment with the thermal energy. Using for the diamagnetic moment $\mu_m = \chi B/(\mu_0 N_v)$, leads to the energy of a diamagnetic moment in a 1-T field

$$|U| \approx \mu_m B = \frac{e^2 \langle r^2 \rangle B^2}{6m} \approx 10^{-28}\ \text{J}$$

But $k_B T$ is of order 4×10^{-21} J at room temperature (RT) (6×10^{-23} J at 4.2 K). (In cgs units $\mu_m H = \chi H^2/N_v \approx 1.7 \times 10^{-21}$ erg and $k_B T$, of course, is $0.025\ \text{eV} = 4 \times 10^{-14}$ ergs at RT). Clearly, $k_B T \gg \mu_m H$ at most temperatures. But this has nothing to do with why diamagnetism is independent of temperature. The diamagnetic susceptibility does not come from alignment of a preexisting moment that can be thermally disordered. Diamagnetism is temperature independent because it results from an interaction between a magnetic field and the velocity of electronic charge. The electron velocity is a function of the energy of the electronic states and hence is essentially independent of temperature. Nevertheless, it is interesting to note that diamagnetism persists even though it is much weaker than $k_B T$. Most solids are not made of independent, diamagnetic atoms.

Simply put, diamagnetism is a manifestation of Lenz law, namely, that if you apply a field to a system of moving charges, their motions change in such a

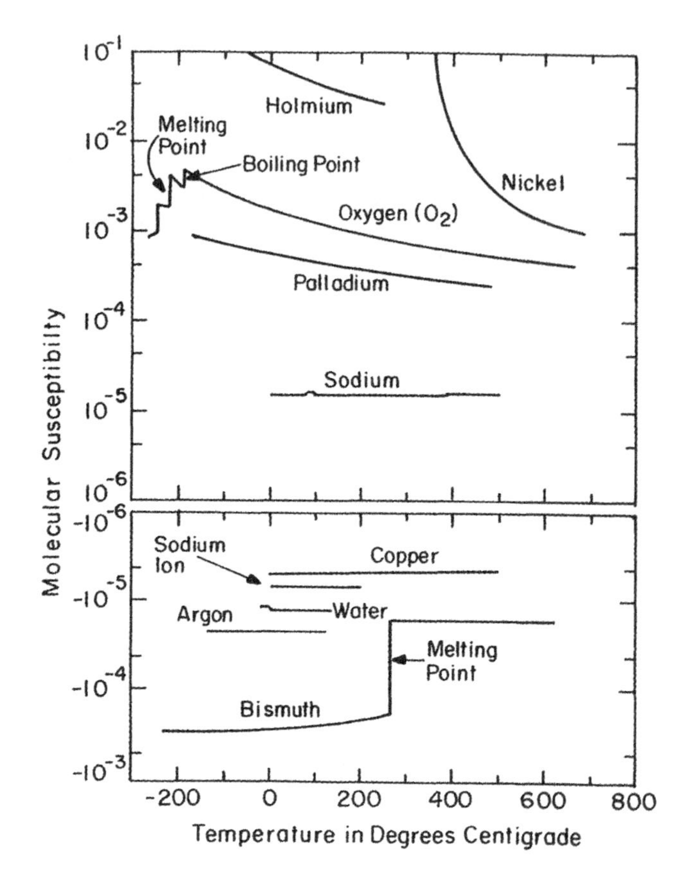

Figure 3.5 Temperature dependence of paramagnetic and diamagnetic susceptibility in some materials. [After Bozorth, copyright IEEE Press (1993)].

way that they create a magnetization that opposes the original field. Diamagnetism is stronger for large electron orbits uninterrupted by scattering. A superconductor is characterized by macroscopic orbits and is the strongest possible diamagnet. For a superconductor below its lower critical field, $\chi_d = -1/4\pi$ (cgs). The new magnetization is the diamagnetic response of the system $M = \chi_d H$, with $\chi_d < 0$. (An electron does not have to have spin or angular momentum to show diamagnetism; only charge and a component of velocity perpendicular to B are necessary. Even a linear electron trajectory will acquire angular momentum in a B field according to the classical argument.) Are you curious where the added angular momentum comes from? It does not come from the field because B is *not* weakened, only opposed, and angular momentum is conserved. If this appears unusual, good! Unfortunately, the classical model so far described cannot explain it; the treatment is naive even though it gives the correct numerical answer.

Nonzero susceptibilities cannot be proved classically for an ensemble of atoms in thermal equilibrium with their surroundings. The reason for this is that the magnetic field does not change the motion of a particle in a way that affects the energy. This is because energy is changed by work $\boldsymbol{F} \cdot d\boldsymbol{x}$, and a magnetic field classically affects electron orbits by the Lorentz force, $\boldsymbol{F} = q[\boldsymbol{E} + (\boldsymbol{v} \times \boldsymbol{B})]$; the magnetic force on a classical particle is always perpendicular to its velocity. The classical statistical average of a particle's variables of motion is given by

$$\langle v \rangle = \frac{\int v \cdot e^{-(E/kT)} d\Omega}{\int e^{-(E/kT)} d\Omega} \tag{3.8}$$

Since the energy E is independent of B (other than effects of a preexisting magnetic moment), $\langle v \rangle$ also must be independent of B. What this means is that the angular momentum imparted to a particle by a magnetic field does not come from work done by that field. Classically, the angular momentum comes from the surroundings of the particle so the surroundings are left with an equal and opposite angular momentum. The net effect is zero. This problem was identified by Niels Bohr in his PhD thesis and independently a few years later by a Miss Van Leeuwen in Leiden. It is only resolved by a quantum mechanical treatment of magnetism. A quantum mechanical derivation of diamagnetism is outlined in an appendix to this chapter.

3.3 CLASSICAL PARAMAGNETISM

It has been shown that an applied B field perturbs the orbit of a moving electric charge in such a way as to create a diamagnetic response. Attention is now given to the effect of a magnetic field on a *preexisting* orbital or spin moment. (Even though spin has not yet been discussed as a source of magnetism, it should be appreciated that magnetic dipole moments can arise from orbital *or* spin angular momentum.) Experience indicates that magnets align with fields. Yet the torque $\boldsymbol{T} = \mu_m \times \boldsymbol{B}$ is orthogonal to $\boldsymbol{B}$ and μ_m. How, then, can the field align a magnetic moment that is already present in order to reduce its potential energy $U = -\mu_m \cdot \boldsymbol{B}$? As it turns out, the *field alone* is not enough. Scattering processes interrupt the precession of the magnetic moment and allow it to relax to a lower energy orientation relative to $\boldsymbol{B}$.

The angular momentum $\boldsymbol{L}$ of a top precesses about the gravitational field and becomes parallel to it *only if there is a frictional force* (Fig. 3.6); otherwise the torque never lowers the energy of the top, it continues to precess. In magnetic metals, angular momentum scattering takes place every few nanoseconds. This is the frictional effect that allows μ_m to align with $\boldsymbol{B}$. (Electrons get scattered, i.e.,

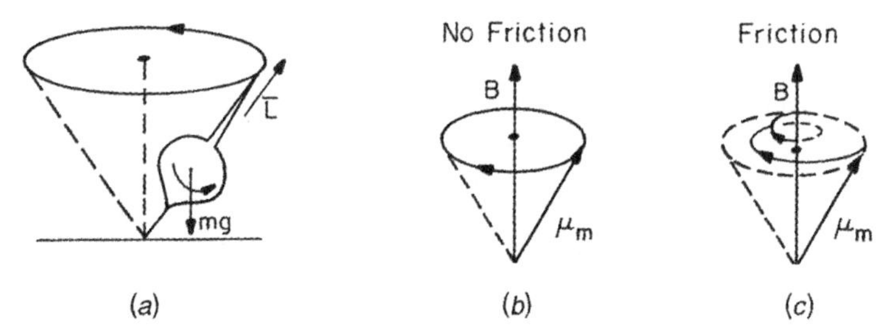

Figure 3.6 (*a*) A top precesses under the torque produced by the action of a gravitational force; (*b*) precession of magnetic moment under the action of a magnetic field; (*c*), same as center but with scattering present.

change their linear momentum, every 10^{-14} s in Cu at RT, so from these numbers you can conclude that only about one in 10^5 scattering events alters the spin.) On the timescale of unaided human observations, magnetic moments quickly align with applied fields rather than precess about them indefinitely. If you do an experiment fast enough, $f \geqslant (1 \text{ nanosecond})^{-1} \approx \text{GHz}$, you can detect the precession; this is done in magnetic resonance. For experiments at longer time scales, μ_m or $\boldsymbol{M}$ appears to align with the applied field.

Paramagnetism describes the behavior of materials that have local magnetic moments but no strong magnetic interaction between those moments (energy of interaction $\ll k_B T$). Examples of materials exhibiting classical paramagnetism for noninteracting moments include the diatomic molecules B_2 (the highest occupied molecular orbital is π^2), $O_2(\pi^{*2})$, some salts such as iron ammonium alum [Fe^{3+} in $Al_2O_3(NH_2)$], and ferromagnetic materials well above their Curie temperatures. All of these materials have unpaired electrons (a nonzero spin magnetic moment) and/or a net orbital angular momentum. The important question is this: "How do noninteracting magnetic moments respond to the field B at low frequencies when thermal effects are considered?"

The treatment for magnetic response to a B field is exactly parallel to dielectric response to an E field and involves thermodynamics and statistical averages. It is worth repeating even though it is a classical result. First, the potential energy U of a magnetic dipole μ_m in a $\boldsymbol{B}$ field is given by Eq. (1.6).

In a classical solid, the atomic moments can take on any possible orientation in space relative to $\boldsymbol{B}$ (Fig. 3.7), that is, θ is a continuous variable. The probability of occupying one of these energy states $E_i = -\mu_m B \cos\theta_i$ at temperature T is given by the Boltzmann factor:

$$P = C \exp\left[\frac{-E_i}{k_B T}\right] = C \exp\left(\frac{\mu_m B \cos\theta_i}{k_B T}\right) \tag{3.9}$$

(Even though the electrons *within* each atom or molecule obey Fermi-Dirac

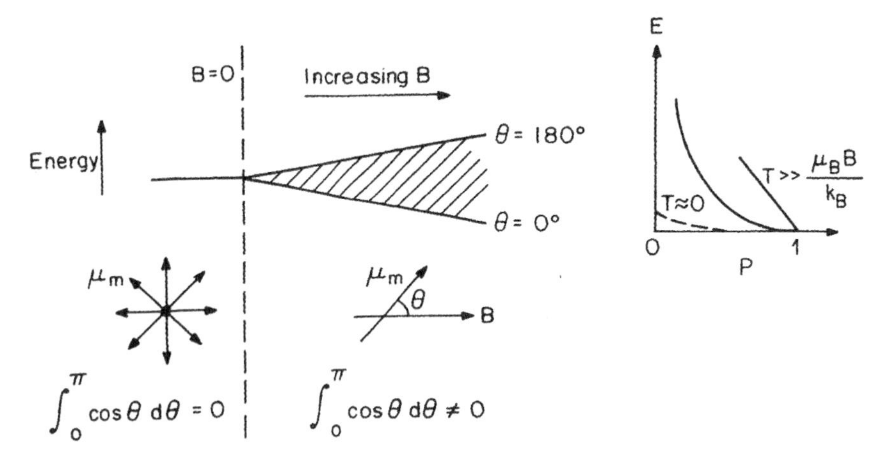

Figure 3.7 Left, a degenerate energy level ($B = 0$) for a random distribution of spins broadens in energy as a magnetic field is applied. In the presence of a field the moment orientation relative to $\boldsymbol{B}$ defines the energy. Far right, the probability of occupation of the distribution of energy states is a function of energy and temperature; probability should be normalized to unity at $E = 0$.

statistics in filling the states that give the magnetic moment, the moments of the different molecules are separate and distinguishable, so Maxwell-Boltzmann statistics apply).

The component of μ_m in the field direction, $\mu_m \cdot \boldsymbol{e}_z$, is given by $\langle \mu_m \rangle = \mu_m \langle \cos\theta \rangle$, hence the probability of observing a particular orientation of μ_m relative to $\boldsymbol{B}$ is given by

$$\langle \mu_m \rangle = \frac{\displaystyle\int \mu_m \cos\theta \left[\exp \frac{\mu_m B \cos\theta}{k_B T} \right] d\Omega}{\displaystyle\int \left[\exp \frac{\mu_m B \cos\theta}{k_B T} \right] d\Omega} \tag{3.10}$$

where $d\Omega = \sin\theta\, d\theta\, d\phi$. Carrying out the $d\phi$ integral and making the substitutions $s = \mu_m B / k_B T$ and $x = \cos\theta$, Eq. (3.10) gives

$$\langle \mu_m \rangle = \mu_m \frac{\displaystyle\int_{-1}^{1} e^{sx} x\, dx}{\displaystyle\int_{-1}^{1} e^{sx} dx} = \mu_m \frac{s(e^s + e^{-s}) - (e^s - e^{-s})}{s(e^s - e^{-s})}$$

This can be expressed as

$$\langle \mu_m \rangle = \mu_m \left(\coth s - \frac{1}{s} \right) = \mu_m L(s) \tag{3.11}$$

where $L(s)$ is the Langevin function for $s = \mu_m B/k_B T$ and is shown in Figure 3.8. For a number of moments per unit volume given by N_v, the magnetization density is

$$M = N_v \langle \mu_m \rangle = N_v \mu_m L(s) \tag{3.12}$$

When $s \approx 0$, by L'Hôpital's rule $L(s) \approx s/3 = \mu_m B/3k_B T$:

$$\chi_m = \frac{\mu_m^2 N_v \mu_0}{3k_B T} = \frac{C}{T} \tag{3.13}$$

where $C = \mu_0 \mu_m^2 N_v / 3k_B$ is called the *Curie constant* because it was Pierre Curie who first observed that the susceptibility χ_m goes as $1/T$ at sufficiently low fields ($s \approx 0$) for many paramagnets. Also note that in this regime M is linear in H (cf. Figs. 3.8 and 1.6). The top panel of Figure 3.5 plots the paramagnetic susceptibilities of several elements.

Example 3.1 It is possible to calculate χ_m at room temperature for diatomic oxygen for which the magnetic moment is $2\mu_B$ (due to spin) from the unpaired electrons in the doubly degenerate π^* highest occupied molecular orbital. The result is $\chi = 3.5 \times 10^{-3}$ (dimensionless) or 2.1×10^{-8} mol^{-1} 5.4×10^{-9} m^3/mol. Figure 3.4 gives $\mu_m = 2 \times 10^{-3}$ corresponding to $\chi = 1.2 \times 10^{-8} mol^{-1}$. The result of the calculation is too large by a factor of nearly 2. The diamagnetism that is included in the experimental data of Figure 3.4 have not been subtracted from the calculated result.

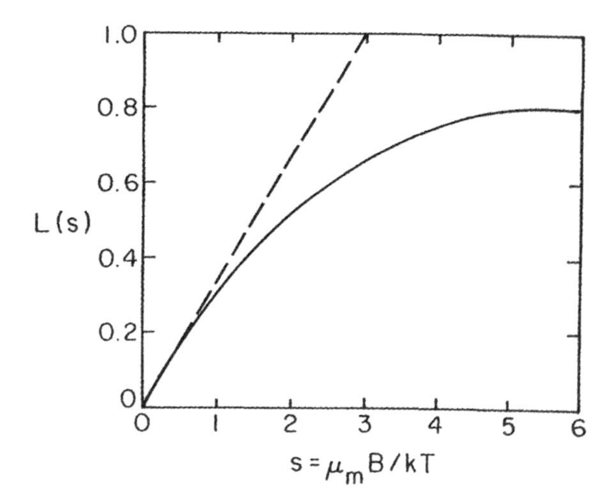

Figure 3.8 Langevin function versus s describes the universal behavior of classical paramagnets.

In very strong fields or at low temperatures such that $s \approx \infty$, the Langevin function becomes $L(s) \approx 1$, so $M \approx N_v \mu_m$, which defines the saturation magnetization.

Our simple classical model accounts reasonably well for some of the most important observations (T and H dependence) of paramagnetism. Further, classical paramagnetism $M = \chi_m H$ is similar to classical dielectric polarization $P = \chi_e E$ in the form of its temperature and field dependence. More accurate calculations, which can be done in many cases, involve corrections beyond the scope of a treatment at this level.

3.4 MAGNETISM AND SPECTROSCOPY

Much of our understanding of atoms and their quantized energy levels was known in the nineteenth century from analysis of atomic spectra, well before the birth of quantum mechanics. It was appreciated that the spacings of the major groups of lines in atomic spectra were fairly well described by two sets of integers n and k related to 2 degrees of freedom of the "atomic oscillators," as they were called.

Balmer observed that the visible spectrum of hydrogen followed a frequency rule $v = A(1 - 4/n^2)$. Later it was realized that this was the rule $v \propto (1/n_f^2 - 1/n_i^2)$, where n_f and n_i are the integer numbers for the final and initial states, respectively. For the Balmer series, $n_f^2 = 4$ (for the Lyman series, $n_f = 1$ which is in the ultraviolet; for Paschen, $n_f = 3$; Blacket, $n_f = 4$; and Pfund, $n_f = 5$, in the infrared).

In 1905, Henri Poincare observed:

Our first glance at the distribution of the [spectral] lines makes us think of the harmonics that are met with in acoustics, but the difference is great; not only are the [spectral] wave numbers *not* successive multiples of the same number, but we do not find anything analogous to the roots of those transcendental equations to which we are often led in physical mathematics.... The laws [of spectra] are simpler, but they are of an entirely different nature.... Of that, we have not taken account, and I believe that therein lies one of the most important secrets of nature.

A planetary model was unable to explain fixed spectral energies, fixed orbital energies.

Niels Bohr borrowed Planck's quantization of action $\int \mathbf{p} \cdot d\mathbf{l} = nh$ giving $mvr = nh/2\pi$, which, when combined with the classical mechanics relations, $mv^2/r = ke^2/r^2$ and $E = T + V$, allows elimination of v and r to give $E = -K/n^2$ and $r = n^2 r_0$, with $K = 2.18 \times 10^{-18}$ J and $r_0 = 0.53 \times 10^{-10}$ m. Thus there are only certain energy levels that can exist, governed by the integers n. This energy quantization will be pursued and its implications for magnetism will be described in the next two chapters.

A quarter of a century before quantum mechanics revealed the significance of these integers (principal quantum number n and orbital angular momentum quantum number $l = k - 1$), their significance was being studied by Michael Faraday in England and a group in Leiden by applying static magnetic fields to the gas discharge tubes from which the spectra were excited. The race was won by the same laboratory that was to discover superconductivity a little more than a decade later, the Leiden group.

Peter Zeeman, a student of Hendrik Lorentz in Leiden, observed (in 1896) that the lines of the optical spectra emitted from a gas in a strong magnetic field showed field-dependent splitting that differed when viewed along the field axis or normal to it (see Fig. 3.9, where the axes are rotated for convenience from the orientation shown in the related Fig. 3.2). The B field is applied along the z axis. The Zeeman split lines had frequencies $\omega = \omega_0 \pm \gamma B$. The characteristic directions of polarization observed for the Zeeman lines are also shown in Figure 3.9.

Lorentz was able to give a simple explanation of the splittings in terms of the classical theory of oscillating electrons. [Lorentz and Zeeman shared the Nobel Physics prize in 1902 for this work. Bohr's theory of the atom did not come until 1913, but Lorentz knew about J. J. Thompson's experiments with electrons (1897)]. Lorentz' classical explanation followed that given above (Section 3.3) for precession. Consider an orbit in the x–y plane. Whether the angular frequency ω of such an electron orbit has a positive or negative projection along the z axis, $\Delta\omega$ is in the direction of the field. Hence, for $\omega < 0$, $|\omega|$ decreases and for $\omega > 0$, $|\omega|$ increases. When viewed along $\boldsymbol{B}$, the line upshifted in frequency is circularly polarized counterclockwise about $\boldsymbol{B}$ and vice versa for the downshifted line. So both the line shift and polarization viewed along $\boldsymbol{B}$ were accounted for by classical free electron theory using the Lorentz force.

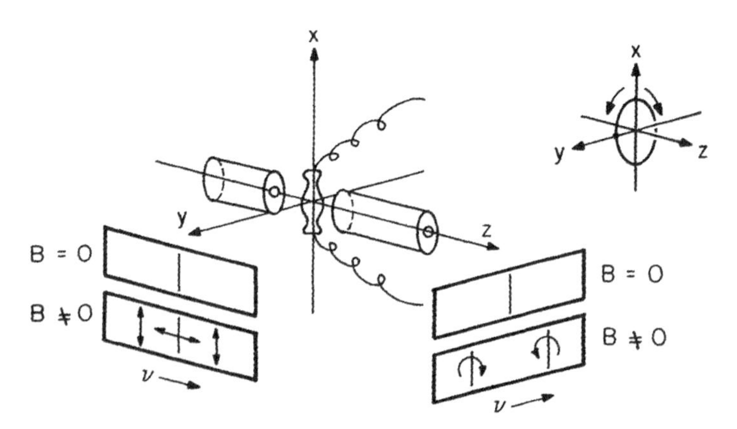

Figure 3.9 Simplified sketch of Zeeman's experimental observation of the effects of a magnetic field B_z on optical spectra. Only the effects on an x–z orbit are shown.

When this same orbit is viewed perpendicular to $\boldsymbol{B}$, the same shifts result, but from the sense of oscillation, only vertical polarization is observed. Oscillations in the z direction have not been considered because they are unaffected by $\boldsymbol{B}$; this explains the unshifted line viewed perpendicular to $\boldsymbol{B}$. The frequency shift is the Larmor frequency $\omega_L = (e/2m)\boldsymbol{B}$:

$$\omega = \omega_0 \pm \omega_L = \omega_0 \pm \frac{eB}{2m} \tag{3.14}$$

Here ω_L is the *additional* frequency acquired by an orbit having angular momentum $\boldsymbol{L}$ in a magnetic field and it shows up as a precession of the orbital angular momentum about $\boldsymbol{B}$ if $\boldsymbol{L}$ is not parallel to $\boldsymbol{B}$. Note that the added frequency, $\pm 14\,\text{GHz/T} = 1.4 \times 10^{10}$ s/T, is a small shift (for laboratory values of B) on a visible optical spectral line for which $\nu \approx 6 \times 10^{14}\,\text{s}^{-1}$. (It was recognized by Sommerfeld that this orbital angular momentum of the electrons was the additional degree of freedom that necessitated a new quantum number k in addition to the principal quantum number n of the Bohr model. Sommerfeld's definition of k is no longer used; instead, the letter l, which has the value $k - 1$, is commonly used to describe the orbital angular momentum quantum number.)

Figure 3.10 shows the level splitting scheme for a $d \rightarrow p$ transition in the vector model of the atom. Each level splits in a B field into $2l + 1$ components depending on its magnetic quantum number m_l. The usual dipole selection rule $\Delta m_l = 0, \pm 1$ indicates three allowed lines at $\nu - \Delta\nu$, ν, and $\nu + \Delta\nu$. Note that the sense of polarization, which is readily understood from the classical picture above, now comes from the conservation of angular momentum during the

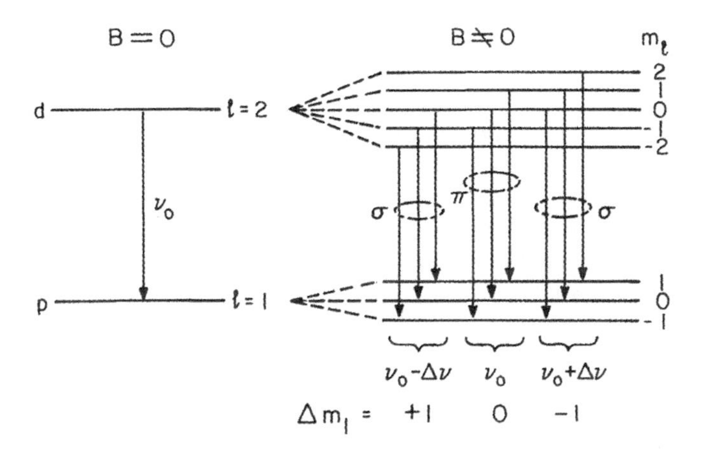

Figure 3.10 Electronic level structure for $d \rightarrow p$ transitions in zero field and in the presence of a magnetic field. Allowed transitions and their polarizations are shown.

transition. When $\Delta m_l = 0$, lines result that are linearly polarized parallel to the field (π). When $\Delta m_l = \pm 1$, observation reveals right or left circularly polarized light when viewed along the field axis. The change in angular momentum from initial to final electron states is balanced by the angular momentum of the emitted photons. Of course the circular polarization seen when viewed along $\boldsymbol{B}$ appears as linear polarization perpendicular to $\boldsymbol{B}$ (σ) when viewing perpendicular to $\boldsymbol{B}$ (Figs. 3.9 and 3.10). (These small changes in the state of polarization of light emitted in a magnetic field give us a preview of the microscopic mechanisms, selection rules, and transitions, involved in the Faraday and Kerr effects that are of technical importance.)

But there were many spectral lines whose field-induced splittings were more complex than the simple three-line patterns Lorentz was able to explain. These more complicated magnetic-field-induced splittings became known as the anomalous Zeeman effect, and their explanation lies in quantum mechanics, which is treated next.

3.5 QUANTUM MECHANICS AND MAGNETISM

Earlier in this chapter the magnetic moment of hydrogen ($1s^1$) was calculated in the classical, Bohr model to be $\mu_m = IA = 9.27 \times 10^{-24}\ \mathrm{A \cdot m^2}$. It was assumed that the position, $r = a_0 = 0.52$ Å, of the point electronic charge, $q = -e = -1.6 \times 10^{-19}$ C, was known as it traveled, presumably in a circular orbit, with a velocity given exactly by $v = (2E/m)^{1/2}$. At the atomic scale, the two preceding sentences cannot be accurate: position and momentum cannt be known simultaneously with arbitrary accuracy (Heisenberg's uncertainty principle; see Appendix, Chapter 4), the electronic charge is distributed in space, not localized at a point, and to the extent that the $1s^1$ electron can be described classically, it travels in a straight line through the nucleus (an s electron has zero angular momentum), not in a circle. Further, at the atomic or quantum level, variables such as position, momentum, and energy cannot take on a continuum of values; rather a set of discrete values is found to describe what is observed.

Replacing the classical angular momentum $\boldsymbol{L} = \boldsymbol{r} \times \boldsymbol{p}$ for an electron (which is needed to calculate the orbital magnetic moment, $\mu_m = \gamma \boldsymbol{L}$) with the appropriate quantum mechanical form must be justified. The way this is done is given by Schrödinger's equation. The part of the Schrödinger equation that describes the angular variables of a quantum mechanical wave/particle in a central potential is given by

$$\left[\frac{1}{\sin\theta}\frac{\partial}{\partial\theta}\left(\sin\theta\frac{\partial}{\partial\theta}\right) + \frac{1}{\sin^2\theta}\frac{\partial^2}{\partial\varphi^2}\right] Y_l^{m_l} = -l(l+1)Y_l^{m_l} \tag{3.15}$$

This equation separates into one equation for the magnitude of the angular momentum, $|\boldsymbol{L}|$ and another for its z component L_z:

$$\boldsymbol{L}^2 Y_l^m = \hbar^2 l(l+1) Y_l^m$$

and

$$L_z\Phi = \hbar m_l \Phi \tag{3.16}$$

Here, $\hbar = 1.05 \times 10^{-34}$ J/s is Planck's constant divided by 2π. The $Y_l^{m_l} = A_l^{m_l}\Phi_{m_l}$ are spherical harmonics, $A_l^{m_l}$ are related to the associated Legendre polynomials, and $\Phi_{m_l} = e^{im_l\phi}$. The convention that capital (uppercase) letters, L, designate operators and small (lowercase) letters, l, designate eigenvalues will be followed.

The first important point in Eq. (3.16) is that the classical expression for the value of angular momentum, $\boldsymbol{L} = \boldsymbol{r} \times \boldsymbol{p}$, is now replaced by a discrete set of allowed values, $\hbar[l(l+1)]^{1/2}$, where l is the orbital angular momentum quantum number. Further, the component of $\boldsymbol{L}$ in some direction (usually set by a field) is also quantized and has the values $\hbar m_l$ where m_l takes on the integer values in the range $-l \leqslant m_l \leqslant l$. The appearance of Planck's constant sets the lower limit to the magnitude of angular momentum (of order 10^{-34} J/s). In a central force potential, the quantum mechanical angular momentum is a constant of the motion, a good quantum number; it commutes with the Hamiltonian, and it is conserved.

The magnetic moment due to orbital motion is no longer the classical value, $\mu_m = \gamma(\boldsymbol{r} \times \boldsymbol{p})$, but instead

$$|\mu_m| = \gamma\hbar[l(l+1)]^{1/2} \tag{3.17a}$$

and

$$\mu_m|_z = \gamma\hbar m_l \tag{3.17b}$$

These formulas indicate that the magnitude of the magnetic moment is of order $\gamma\hbar$, which is about 10^{-23} A · m^2 (essentially what was calculated classically).

One of the key factors that necessitated the development of quantum mechanics was the accumulation of spectral data such as the Zeeman splittings described earlier in this chapter, and the more complicated aspects of those spectra that became apparent as higher-resolution spectrometers were available. This leads now to a different kind of angular momentum that is known to play a major role in many magnetic materials: spin.

3.5.1 Spin

Earlier in this chapter, it was shown that certain lines in the optical spectra of many materials split in a magnetic field into three lines characterized by frequencies $\omega = \omega_0$ and $\omega_0 \pm \gamma B$ with $\gamma = e/2m$. This could be explained by assuming that the three-fold degeneracy of a p state ($l = 1$ and degeneracy of $2l + 1$) was lifted by the magnetic field.

The Zeeman contributions to the spectral line energies E_0, were $\Delta E = \hbar\Delta\omega = (eB/2m)\hbar m_l$, where $m_l = 0 \pm 1$. But this picture of the Zeeman effect was not complete. It turned out that many lines showed more complex splittings. These cases were known collectively as the "anomalous" Zeeman effect because they could not be explained by the orbital angular momentum of the classical electron.

The anomalous Zeeman effect demanded a new quantum number, but its meaning was a matter of considerable controversy. Uhlenbeck (a graduate student of Ehrenfest's in Leiden) recognized that the new quantum number had to describe an extra electronic degree of freedom; specifically, the electron must have some dynamics in addition to translational motion (r) with angular momentum (θ, ϕ). In 1924, Uhlenbeck and his friend Sam Goudsmit (an expert in atomic spectra) proposed that the electron has an intrinsic angular momentum called *spin*. It turned out that the new angular momentum of the electron has a gyromagnetic ratio that is twice that determined above for orbital motion $\mu_L/L = \gamma_L = e/2m$:

$$\begin{aligned} \frac{\mu_s}{S} = \gamma_S = \frac{e}{m} &= 1.76 \times 10^{11} \qquad [\text{C/kg} = (\text{s} \cdot \text{T})^{-1}] \\ &= 1.76 \times 10^{7} \qquad (\text{s} \cdot \text{Oe})^{-1} \end{aligned} \tag{3.18}$$

Thus, by analogy with Eqs. (3.17), the result for an atom of spin S is

$$|\mu_m| = \gamma_S \hbar [s(s+1)]^{1/2} \tag{3.19a}$$

and

$$\mu_m|_z = \gamma_S \hbar m_s \tag{3.19b}$$

The concept of spin has been introduced. But Schrödinger's equation says nothing about spin. It took Dirac to incorporate Lorentz covariance into a wave equation (now bearing Dirac's name) that predicts electron spin. The Dirac equation will not be considered here. It is enough to know that our previous wavefunctions $\psi(r)$ can have a new multiplicative factor $\chi(s)$ to describe the probability amplitude for occupation of different spin states: $\Psi(r, s) = \psi(r)\chi(s)$. The spin wavefunction $\chi(s)$, in complete analogy with Eq. (3.16), satisfies

$$S^2\chi = \hbar^2 s(s+1)\chi \quad \text{and} \quad S_z\chi = \hbar m_s \chi \tag{3.20}$$

where m_s can take on discrete values from $+s$ to $-s$, e.g., for $s = \frac{1}{2}$, $m_s = \pm\frac{1}{2}$ (unlike m_l, which can take on only the integer values in the range $-1 \leqslant m_l \leqslant 1$).

According to Eq. (3.20), when the component of intrinsic (spin) angular momentum of an electron is measured along a field direction, the result is $\hbar m_s = \pm\hbar/2$. The magnetic moment associated with this spin is $\mu_m = \pm\gamma_s\hbar/2$, where $\gamma_s = 2\gamma = e/m$. Thus the spin magnetic moment of a single electron is

$\mu_s = \pm e\hbar/2m$, which shows up so often it is given the name Bohr magneton:

$$\mu_B = \frac{e\hbar}{2m}$$
$$= 0.927 \times 10^{-23}\,(\text{J/T or A}\cdot\text{m}^2) \qquad (3.21)$$
$$= 0.927 \times 10^{-20}\,(\text{erg/Oe})$$

The number that was estimated earlier for the *orbital* moment of the 1*s* electron in hydrogen is identical to this value because the quantum of orbital angular momentum is twice that for spin angular momentum and the orbital gyromagnetic ratio is half that for spin, $\gamma_s = 2\gamma_L$.

A spin magnetic moment precesses in a B field just as does an orbital moment except again the gyromagnetic ratio must be that for the spin: $\omega_p = \gamma_S B = eB/m$. For an electron in a 0.1 T (1 kOe) field $\omega_p/2\pi = 2.8 \times 10^9$ Hz (3 GHz) which is a microwave of about 10 cm wavelength. Nuclear resonances are nearly 2000 times slower (MHz, radio waves) than electron resonances because the mass of a proton is 1836 times that of an electron.

Because both kinds of moments, orbital and spin, can contribute to the total magnetism of an atom, a weighting factor is needed that gives their relative contributions. It is called the g factor: $g = 1$ for purely orbital magnetism ($s = 0$) and $g = 2$ for purely spin magnetism ($l = 0$).

If the spin–orbit interaction (a very important interaction that will be described later) were neglected, the spin and orbital angular momenta of the various electrons would be independent of each other. L and S have fixed magnitudes, and their projections on a specified axis are constant. Under this assumption, the orbital and spin angular momenta can be measured independently; they obey Eqs. (3.16) and (3.20), respectively. The appropriate quantum numbers would be n, l, m_l, s, and m_s.

However, when considering the effect of spin–orbit coupling, the spin and orbital angular momenta of the individual electrons are no longer independent, but instead their vector sum, $\boldsymbol{J} = \boldsymbol{L} + \boldsymbol{S}$, becomes a constant of the motion; $\boldsymbol{J}$ precesses about the field direction. $\boldsymbol{J}$ is quantized and has the following expectation (measured) values:

$$\langle J \rangle = \hbar[j(j+1)]^{1/2} \qquad (3.22)$$

$$\langle J_z \rangle = \hbar m_J \qquad (3.23)$$

Thus four quantum numbers are now needed to fully describe the state of a quantum system. In many cases the appropriate quantum numbers are n, j, m_j, and m_s.

The coupling of the motion of $\boldsymbol{L}$ and $\boldsymbol{S}$ by the spin–orbit interaction makes it more difficult to determine the relative spin and orbital contributions to the total magnetic moment μ_J. The difficulty can be appreciated by considering Figure 3.11. Clearly, the total angular momentum, $\boldsymbol{J} = \boldsymbol{L} + \boldsymbol{S}$, is conserved. It is now $\boldsymbol{J}$, not $\boldsymbol{L}$, that is acted on by a torque: $\boldsymbol{T} = d\boldsymbol{J}/dt$. It has been shown that the magnetic moments for L and S have different scale factors $e/2m$ and e/m, respectively; thus the total magnetic moment is given by $\mu_m = [e\hbar/2m][l_z + 2s_z]$. In general $\mu_J = \mu_L + \mu_s$ is *not* collinear with $\boldsymbol{J}$. Because $\boldsymbol{L}$ and $\boldsymbol{S}$ are coupled, they have fixed projections on $\boldsymbol{J}$. Therefore, μ_s and μ_L, also must have constant projections on $\boldsymbol{J}$ and so must their vector sum μ_J. All that can be measured is the component of μ_J on $\boldsymbol{J}$, designated in Figure 3.11 as μ'_J. In general, $\mu'_J = g\mu_B m_j$ with (as you may confirm in Problem 3.4) the Landé g factor given by

$$g = 1 + \frac{j(j+1) + s(s+1) - l(l+1)}{2j(j+1)} \tag{3.24}$$

This quantum mechanical treatment of electronic states is now applied to the problem of the "anomalous" Zeeman spectra that was introduced in Chapter 3. The Zeeman splitting of different states proceeds as in Figure 3.12 with field-induced splittings given by $\Delta E = g\gamma B\langle J_z\rangle = g\mu_B m_j B$.

The level scheme for sodium ($3s^1$ ground state: $^2S_{1/2}$) is shown in Figure 3.13. Note the differences here from the normal Zeeman effect (Fig. 3.10). The simplest expression for the energy of a moment (orbiting electron) in a field is $U = -\mu_m \cdot \boldsymbol{B}$. But now $\mu_m = (ge/2m)\boldsymbol{J} = (ge\hbar/2m)m_j$, where $\hbar m_j$ is the expectation value of $\boldsymbol{J}$ in the field direction. Values for gm_j are shown for the various Zeeman-split lines. At the left of the figure is shown the spectroscopic notation for each state; the definition is in the box at the lower left.

The longer wavelength D_1 line splits into four components while D_2 splits into six. Because the magnitude of the energy splittings ($\Delta E \sim gm_j$) of $^2S_{1/2}$

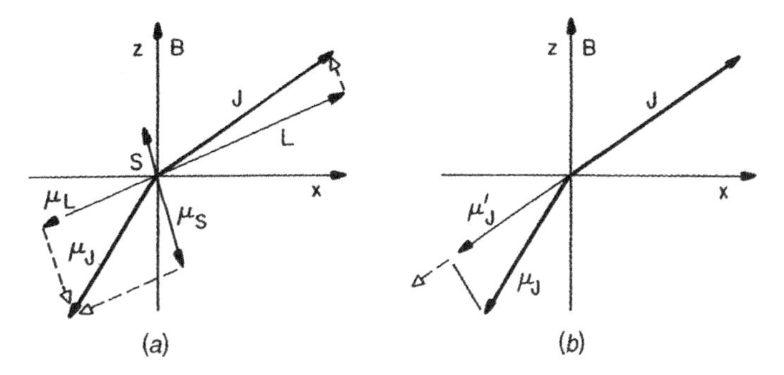

Figure 3.11 (*a*) Vector diagram of total angular momentum $\boldsymbol{J}$ and its component spin and orbital momenta. These mechanical moments scaled with their respective g factors give rise to the magnetic moments, μ_J, μ_L, μ_S. (*b*) because μ_L, and μ_S precess about $\boldsymbol{J}$, μ_J precesses about $\boldsymbol{J}$, and what is measured is μ'_j.

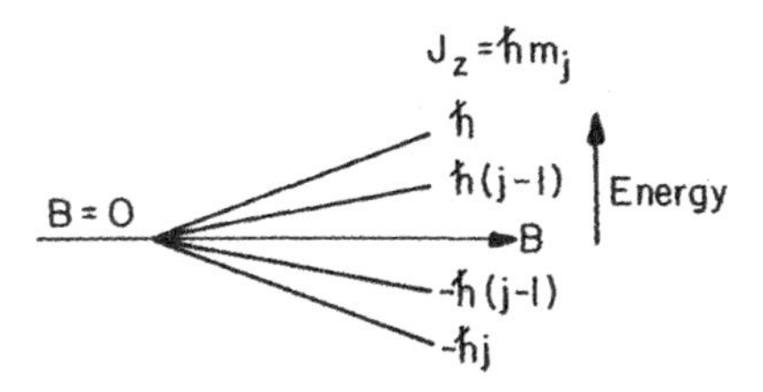

Figure 3.12 Zeeman splitting of a total angular momentum state in a B field.

are different from those of the $^2P_{1/2}$ and $^2P_{3/2}$ states, these lines all have different energies. The polarizations shown in Figure 3.13 are for viewing perpendicular to B. When viewed parallel to B the σ components are circularly polarized and the π components are not observed. The polarization is readily understood from the classical description of the normal Zeeman effect earlier in this chapter. The Larmor precession frequency measures the energy difference between any two adjacent split states for which $m_j = 1$: $\Delta E = \hbar\omega_p = \gamma\hbar B = g\mu_B B = ge\hbar B/2m$.

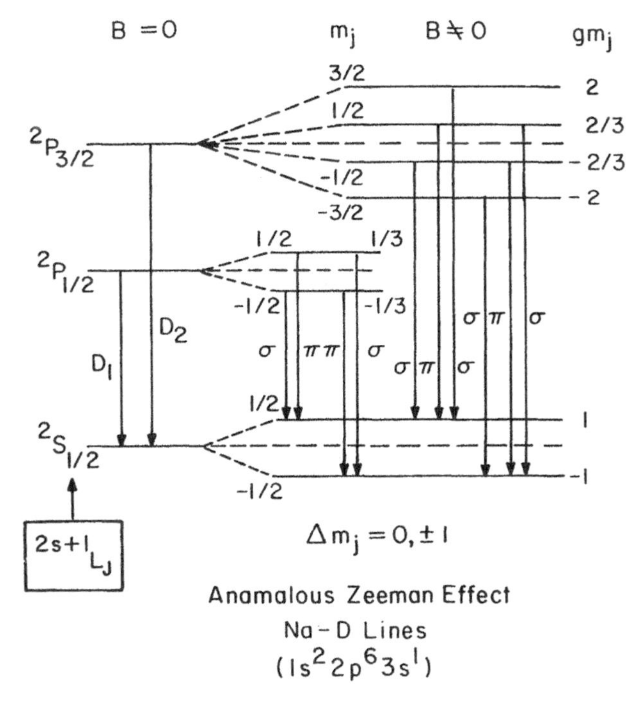

Figure 3.13 Fine structure for sodium in $B = 0$, (left), and $B \neq 0$, (right). Allowed transitions and their polarizations are also shown.

3.5.2 Spin–Orbit Interaction

Attention was drawn to the role of the spin–orbit interaction in coupling $\boldsymbol{L}$ and $\boldsymbol{S}$ so that the total angular momentum $\boldsymbol{J}$ becomes the conserved quantity. Spin–orbit coupling is of temendous importance not just for the interpretation of atomic spectra but also for a host of magnetic phenomena that are key to most applications of magnetic materials. Spin–orbit coupling is at the root of magnetocrystalline anisotropy, magnetostriction, magnetooptic effects (Kerr and Faraday), anisotropic magnetoresistance, ferromagnetic Hall effect and magnetic resonance damping. For these reasons, the origin of this ubiquitous interaction will be reviewed.

In simplest terms, the spin–orbit interaction describes the effects of an electron's orbital motion on the orientation of its spin. From the electron rest frame (moving about the charged nucleus (Fig. 3.14) it "sees" a positive charge in motion just as the sun appears to be in motion about a casual observer on the earth. Hence the electron is situated near the center of a current loop, which generates a magnetic field that causes a preferred direction of orientation for the spin magnetic moment of the electron.

The B field due to the apparent orbital motion of the nuclear charge as seen by the electron is first calculated. According to the Biot–Savart form of Ampère's law [Eq. (1.14)], the magnetic field due to the relative nuclear motion is given by

$$\boldsymbol{B} = \left(\frac{\mu_0}{4\pi}\right)\frac{\boldsymbol{j} \times \boldsymbol{r}}{r^3} = -\frac{Ze\mu_0}{4\pi}\left(\frac{\boldsymbol{v} \times \boldsymbol{r}}{r^3}\right) \tag{3.25}$$

This B field can be expressed in terms of the Coulomb field of the nucleus, in which the electron moves:

$$\boldsymbol{E} = \frac{Ze}{4\pi\varepsilon_0}\frac{\boldsymbol{r}}{r^3} = \frac{1}{e}\frac{\partial V}{\partial r}\frac{\boldsymbol{r}}{r} \tag{3.26}$$

Hence

$$\boldsymbol{B} = -\mu_0\varepsilon_0(\boldsymbol{v} \times \boldsymbol{E}) = -\frac{1}{c^2}(\boldsymbol{v} \times \boldsymbol{E}) \tag{3.27}$$

$$= -\frac{1}{ec^2 r}(\boldsymbol{v} \times \boldsymbol{r})\left(\frac{\partial V}{\partial r}\right) \tag{3.28}$$

Making use of the definition $\boldsymbol{L} = \boldsymbol{r} \times \boldsymbol{p}$ leads to

$$\boldsymbol{B} = \frac{1}{emc^2}\frac{1}{r}\frac{\partial V}{\partial r}\boldsymbol{L} \tag{3.29}$$

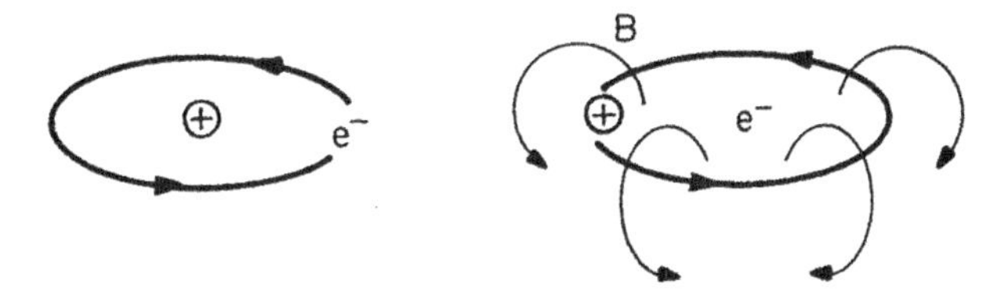

Figure 3.14 Diagram on left, depicts an electron orbiting about the nucleus. In the diagram on the right, from the electron's rest frame, the nucleus appears to be orbiting around the electron and hence producing a magnetic field in the sense indicated. The interaction of the electron spin with its orbitally induced magnetic field is the spin–orbit interaction.

The potential energy of the spin moment $\mu_s = -g\mu_B m_s = -g\mu_B S/\hbar$ in this field is

$$U = -\mu_s \cdot B = \frac{g\mu_B}{emc^2\hbar}\frac{1}{r}\frac{\partial V}{\partial r} L \cdot S \tag{3.30}$$

Using $g\mu_B = e\hbar/m$ for spin, and returning to the frame of reference in which the nucleus is at rest, finally gives

$$\Delta E_{SO} = \frac{1}{2m^2c^2}\frac{1}{r}\frac{\partial V}{\partial r}\boldsymbol{L} \cdot \boldsymbol{S} \tag{3.31}$$

[The relativistic correction factor of $\frac{1}{2}$, the Thomas precession, has been included here; see, e.g., Leighton (1959).]

By expressing the relative electron–nuclear motion in terms of $\boldsymbol{L}$, it is clear that for an s electron that is classically traveling on a linear trajectory through the nucleus, no B field is experienced. However, when $L \neq 0$, the B field created by this relative motion is stronger the more open is the electronic orbit, $l \gg 1$. The coefficient that describes the strength of the spin–orbit interaction makes this effect stronger in heavier atoms [implicit atomic number Z in $V(r)$].

Exercise The magnitude of ΔE_{SO} is estimated for a p electron in a hydrogenic potential, $V = -(1/4\pi\varepsilon_0)(e^2/r)$:

$$\Delta E_{SO} = \frac{e^2}{8\pi\varepsilon_0 m^2c^2r^3}\boldsymbol{L} \cdot \boldsymbol{S}$$

Using $\langle r^{-3}\rangle \approx 1/(3a_0)^3$, and remembering that $\boldsymbol{L}$ and $\boldsymbol{S}$ are operators so that their product is of order $\hbar^2$, leads to

$$\Delta E_{SO} \approx 10^{-23}\,\text{J} = 10^{-4}\,\text{eV/atom}$$

The B field seen by the $2p$ electron in this case is

$$B \approx \frac{\Delta E_{sO}}{\mu_B} \approx 1\,\mathrm{T}$$

Thus, spin–orbit effects are of a strength that can be sensed by typical applied fields.

3.5.3 Hund's Rules

Having added a fourth quantum number to the previous three, n, l, and m_l, it is now necessary to expand the *aufbau* principle accordingly. In what sequence are states filled? No two electrons in an atom can have the same set of quantum numbers (Pauli's exclusion principle). Clearly the states of lowest energy are occupied first. In a purely central potential the energy depends only on n, so electrons first fill states of lowest n. But in many-electron atoms, the electrons see more than just the nuclear charge; they also see the charge of the other electrons. This makes orbitals of different angular momentum take on different energies; the orbital degeneracy is partially lifted by electron–electron interactions. Because low angular momentum electrons spend more time near the nucleus than high-angular momentum electrons do, the l levels [$l = 0(s)$, $1(p)$, $2(d), \ldots, n-1$] are filled lowest l first for each n. But each l orbital has $2(2l+1)$ different states corresponding to $2l+1$ different values of m_l; each one has two possible orientations of its spin value, $m_s = \pm\frac{1}{2}$. What subtle interactions can lift these degeneracies, namely, cause one m_l state to be favored over another? The answer will become clear below. Hund noticed a pattern in atomic spectra indicating that there is a preferred sequence of filling the $2(2l+1)$ orbital states. He did not know why because he did not know quantum mechanics, but what he observed led to these rules that bear his name:

1. Quantum states are filled so as to maximize $s = \Sigma\, m_s$. This rule forces electrons into different orbital states thereby tending to minimize their Coulomb repulsion. The energies here can be up to a few electronvolts (eV).
2. If rule 1 does not determine the occupancy, the state filling is such that $l = \Sigma\, m_l$ is maximized. This also keeps electrons in orbits which circulate in the same sense and have lower probability near the nucleus, again tending to minimize their Coulomb repulsion. This correlation among the motions of different electrons is particularly strong for atomic (as opposed to molecular orbital or free electron) wavefunctions. The energies involved can be up to 0.5 eV.
3. Finally, l and s combine by subtraction when a spin sub-band is less than half filled, $j = |l - s|$, and by addition if more than half filled. This condition tends to minimize the spin–orbit energy, $\xi l_i \cdot s_i$, specifically,

$\xi > 0$ for first-half shell and $\xi < 0$ for second-half shell. The spin–orbit interaction energy can be of order of 10^{-4} eV. The understanding of the spin–orbit interactions that partially lift the $2(2l + 1)$-fold degeneracy of the lth orbital came long after Hund's rules empirically outlined their implications.

Example 3.2 As examples of Hund's rules consider an Fe atom (or Ni^{2+} ion) that has 26 electrons, 8 outside a filled Ar core. These valence electrons fill first the two 4s states (lower in energy than 3d because their zero angular momentum gives them a strong interaction with the nuclear potential). Five electrons fill the spin up states $m_l = 2, 1, 0, -1, -2$ (this gives $l \approx 0$, $s = \frac{5}{2}$ so far), and the remaining electron occupies $s = -\frac{1}{2}$, $l = -2$, giving for the atom $l = 2$, $s = 2$. The spectroscopic notation (see Fig. 3.13) is $^5D_4$ for multiplicity $2s + 1 = 5$, $\Sigma\, l_i = 2(D)$ and $j = 4$. If the iron atom were doubly ionized, it would lose its $4s^2$ electrons and remain $^5D_4$; if triply ionized, it would be $^6S_{5/2}$. A Cr atom has two less electrons than Fe, so $s = +2$, giving for the atom $l = +2$, but $j = 0$ by Hund's third rule. The spectroscopic notation is $^5D_0$.

In the case of Sm^{3+} there are six electrons outside a Xe core, to distribute among 14 different 4f states. Clearly the states with $s = +\frac{1}{2}$ are filled in sequence $-3, -2, -1, 0, 1, 2$ giving $s = 3$, $l = 3$, and $j = 0$ or $^7F_0$.

It should be clear that Hund's rules (1) are based primarily on Coulomb repulsion (of order 1 eV per atom) and secondarily on spin–orbit interactions ($\approx 10^{-4}$ eV/atom) and (2) account for the existence of atomic magnetic moments, even in some atoms with an *even* number of valence electrons. Hence Hund's rules make up what is sometimes called *intraatomic exchange*, the effect responsible for atomic moment formation.

The next chapter will consider what happens to Hund's rules in a solid. For now it is instructive to consider diatomic oxygen, O_2 (see Fig. 5.3 for its molecular orbital structure). Remember that its eight 2p electrons filled the doubly degenerate π^4 molecular orbital (MO), the single $3\sigma^2$ MO, and half of the doubly degenerate π^{*2} MO. The partially filled π^* orbital allows Hund's rule to come into play so the two electrons have the *same* spin and O_2 has a net magnetic moment equivalent to two unpaired spins.

3.6 QUANTUM PARAMAGNETISM AND DIAMAGNETISM

The basic question here concerns the response to an applied field at various temperatures for a large number N_v per unit volume of local magnetic moments μ_m, which can now include both orbital and spin components. This sort of problem always involves writing the Boltzmann factor for the fractional occupation of the various states of energy E_i as a function of temperature. Before going on you might do well to review the approach and method used in Section 3.3 for the classical paramagnet.

First, it is important to know the energy of the magnetic moment μ_m in the $\boldsymbol{B}$ field: $E = -\mu_m \cdot \boldsymbol{B} = -\mu_m B \cos\theta$. The net magnetization of the assembly will be $N_v \mu_m \langle\cos\theta\rangle$. The average of all possible orientations of the μ_m in this field must be determined. Here is where the classical and quantum treatments part ways.

Classical moments can take any orientation relative to $\boldsymbol{B}$, so (Fig. 3.15) θ is a continuous variable. Quantum mechanics limits the orientations of a spin $s = j = \frac{1}{2}$ ($\mu_m = g\mu_B m_j$) to one of two orientations up or down, specifically, $\cos\theta = \pm 1$ [Eq. (3.20)]. The magnetic moment must be averaged over the various states, weighted according to the probability of their occupancy.

If there are a total of N particles, each with spin $\frac{1}{2}$, the fractional populations of the upper and lower energy states N_2/N and N_1/N, are written as

$$n_2 = \frac{N_2}{N} = \frac{e^{-E_2/kT}}{e^{-E_1/kT} + e^{-E_2/kT}} \tag{3.32a}$$

$$n_1 = \frac{N_1}{N} = \frac{e^{-E_1/kT}}{e^{-E_1/kT} + e^{-E_2/kT}} \tag{3.32b}$$

with $E_2 = +\mu_m B$, $E_1 = -\mu_m B$, $\mu_m = g\mu_B m_s$, and $N_1 + N_2 = N$. The denominators in Eqs. (3.32) are the partition function or sum of states (Zustandsumme) that normalizes the Boltzmann factors to net unit probability.

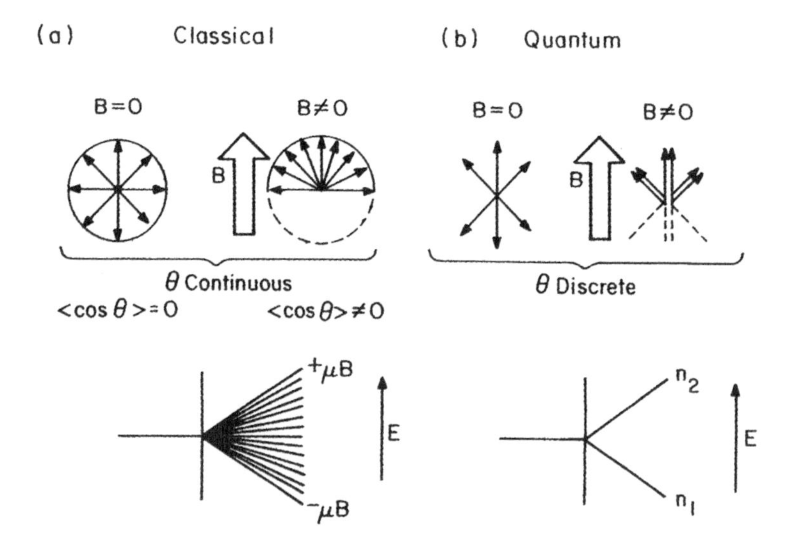

Figure 3.15 (*a*), Classical picture of a continuous distribution of spin orientations in zero field and the effect of a magnetic field that drives the spins to lower energy states in the field-split manifold; (*b*) the quantum picture in which only certain spin orientations are allowed.

(Compare the values of the bare Boltzmann factors with the normalized ones in the limit $T = 0$ and $B = 0$ for the two states.)

The magnetization M is proportional to the net spin imbalance per unit volume V, $M = \mu_m(N_1 - N_2)/V$:

$$M = N_v \mu_m \frac{e^x - e^{-x}}{e^x + e^{-x}}$$

where $x = \mu_m B/kT$ and $N_v = N/V$. Equivalently

$$M = N_v \mu_m \tanh(x) \tag{3.33}$$

This function (Fig. 3.16) is different from the Langevin form derived classically. The classical magnet is more difficult to saturate and is more easily demagnetized. This is because the classical magnetization can orient away from the applied field direction continuously but the quantized magnetization can do so only at discrete angles that depend on the size of the spin ($s = \frac{1}{2}$, $m_s = \pm\hbar/2$; $s = 2$, $m_s = \pm 2\hbar, \pm\hbar, 0$, etc.).

The general quantum mechanical case is not limited to $j = \pm\frac{1}{2}$, as derived above. More generally $m_J = -j, -(j-1), \ldots, 0, \ldots, (j-1), j$. If the full derivation is followed through, the summation above for $N_1 - N_2$ is a much more complicated function of j and x. It is called the *Brillouin function*:

$$B_j(x) = \frac{2j+1}{2j} \coth\left(\frac{2j+1}{2j} x\right) - \frac{1}{2j} \coth\left(\frac{x}{2j}\right) \tag{3.34}$$

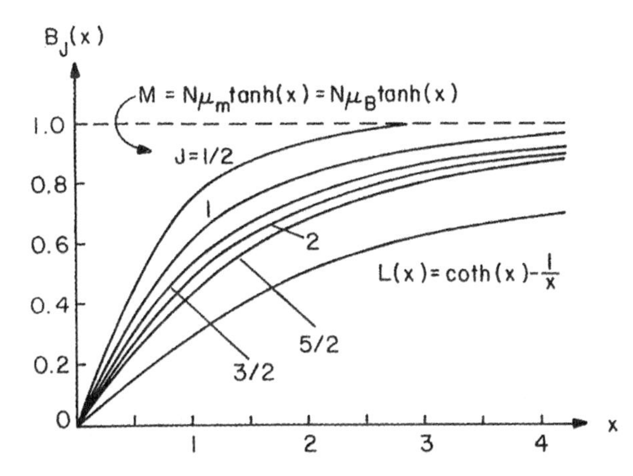

Figure 3.16 Brillouin function versus $x = \mu_m B/k_B T$ for various values of j. The spin $\frac{1}{2}$ limit is given by Eq. (3.33), and the infinite spin limit by the classical Langevin function is derived in Eq. (3.11).

Here μ_m is no longer the spin of a single electron μ_B but can now be $g\mu_B m_j$ or $g\mu_B[J(J+1)]^{1/2}$, where g is the Landé g factor [Eq. (3.24)], which indicates how the total magnetic moment, $\mu_l + \mu_s$, is related to the total angular momentum $\boldsymbol{J} = \boldsymbol{L} + \boldsymbol{S}$ to which μ_J is not necessarily parallel. It is important to note that when discussing paramagnetism, it is the *magnitude* of the total moment $|\mu_m| = g\mu_B[j(j+1)]^{1/2}$, called the *effective* (*paramagnetic*) *moment*, that determines the energy of the system in the field. When treating magnetization in ferromagnetic systems, the maximum value of the moment that can be aligned in a given direction is of importance. In that case $\mu_z = g\mu_B m_j$ is the relevant quantity. For the sake of generality, the magnetization can be expressed as

$$M = N_v g\mu_B J B_j(x) \tag{3.35}$$

where J is understood to be m_j for ferromagnetism and $[j(j+1)]^{1/2}$ for paramagnetism or exchange interactions (where it is the total magnetic moment that govern behavior).

Figure 3.16 shows the shape of $B_j(x)$ for various j values, and Figure 3.17 shows the remarkable fit of this theory to magnetization data in three materials whose paramagnetism is due to the presence of various $3d$ ions having different j values. It is important to note that the data scales as the ratio of the field to absolute temperature; measurements at lower temperatures are equivalent to higher fields as far as this kind of paramagnetism is concerned. In the limit H/T approaches infinity, the saturation magnetic moment of the ion results.

The Brillouin function must be examined in four limits: (1) small x, (2) large x, (3) large J, and (4) $J = \frac{1}{2}$.

1. In the limit of small x, the Brillouin function can be simplified using $\coth x = 1/x + x/3 \cdots$ to give [from Eq. (3.34)]:

$$B_J(x) \xrightarrow{x \ll 1} \frac{J(J+1)}{3J^2} x$$

   Thus the susceptibility is given by

$$\frac{M}{H} = \chi = \mu_0 \frac{N_v g^2 \mu_B^2 J(J+1)}{3J^2 k_B T} \tag{3.36}$$

   If $J = \frac{1}{2}$, this reduces to the result derived in Eq. (3.33). For larger J it approaches the Langevin susceptibility, Eq. (3.13), as it should.
2. In the high-field, low-temperature limit, that is, the limit of large x, $\coth(x)$ is unity and the Brillouin function becomes 1. The magnetization is saturated at $g\mu_B J$.
3. In the limit of large J, $B_J(x)$ becomes $\coth(x) - 1/x$, which is the

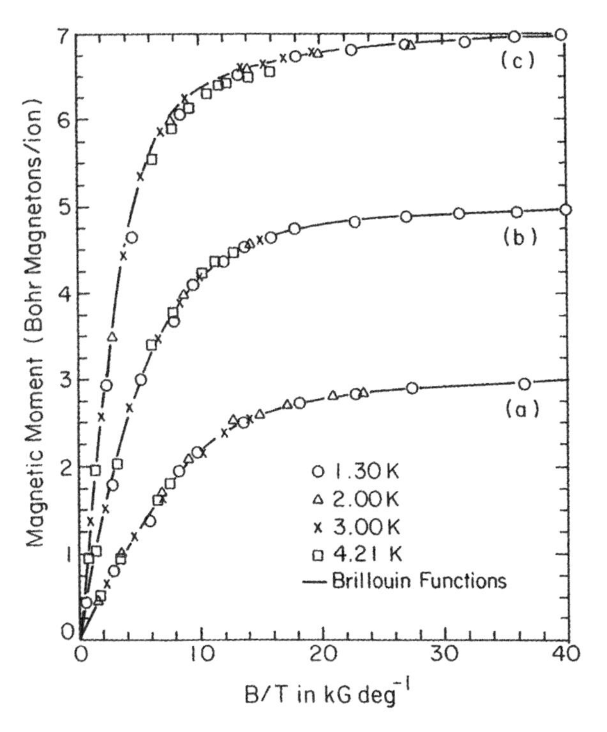

Figure 3.17 Magnetic moment versus B/T for (*a*) potassium chromium alum (Cr^{3+}, $s = \frac{3}{2}$), (*b*) ferric ammonium alum (Fe^{3+}, $s = \frac{5}{2}$) and (*c*) gadolinium sulfate octahydrate ($Gd^{3+}, s = \frac{7}{2}$), (After Henry, 1952).

Langevin function. In this case the behavior of $M(H, T)$ should be identical to that in Eq. (3.12).

4. For $J = \frac{1}{2}$, $B_{1/2}(x)$ reduces to $\tanh(x)$ as was derived for the spin-$\frac{1}{2}$ case[(Eq. (3.33)]. Here $B_{1/2}(x)$ has an initial slope that is 3 times greater than that for the Langevin function; thus the *quantum* paramagnetic susceptibility (small x) for $j = \frac{1}{2}$ is

$$\frac{M}{H} = \chi = \mu_0 \frac{N_v \mu_m^2}{k_B T}$$

Compare with Eq. (3.13) for the classical paramagnetic susceptibility.

3.6.1 Pauli Paramagnetism

The paramagnetism of isolated, distinguishable magnetic ions has been discussed, but not that of indistinguishable, free electron spins in a metal. Because of the Fermi distribution of electrons (which leaves most electrons far removed in energy from an empty state of opposite spin; see Fig. 3.18), it turns out that conduction electron paramagnetism is reduced from that of an equivalent distribution of moment-bearing ions by the factor $k_B T/E_F = T/T_F$ just as the conduction electron specific heat is so reduced. This coefficient reflects the fact that the only Fermi particles able to change their energy in response to an applied field are those within $\pm k_B T/2$ of the Fermi energy. It also accounts for the fact that as temperature increases, more carriers are excited above the Fermi level where they can be aligned by the field. Their magnetic response is called *Pauli paramagnetism* because it is the Pauli exclusion principle that limits two electrons to each orbital. Thus, from Eq. (3.36), for free electrons in weak fields, $x \ll 1$, the susceptibility becomes

$$\chi_{\text{Pauli}} \approx \frac{\chi_m kT}{E_F} = \frac{N_v \mu_m^2 \mu_0}{k_B T_F} \tag{3.37}$$

which is independent of temperature.

The Pauli susceptibility can be derived more precisely by considering the spin imbalance in two free electron bands subject to a weak Zeeman splitting ($\mu_m B \ll E_F$):

$$N_+ = \tfrac{1}{2}\int_{-\mu_m B}^{E_F} f(E)Z(E - \mu_m B)dE \approx \int_0^{E_F} f(E)Z(E)dE + \tfrac{1}{2}\mu_m BZ(E_F)$$

$$N_- = \tfrac{1}{2}\int_{+\mu_m B}^{E_F} f(E)Z(E + \mu_m B)dE \approx \int_0^{E_F} f(E)Z(E)dE - \tfrac{1}{2}\mu_m BZ(E_F)$$

Here $Z(E)$ is the zero-field free electron state density (states per eV per atom) and $f(E) = \{\exp[-(E - E_F)/kT] + 1\}^{-1}$, is the Fermi–Dirac distribution of

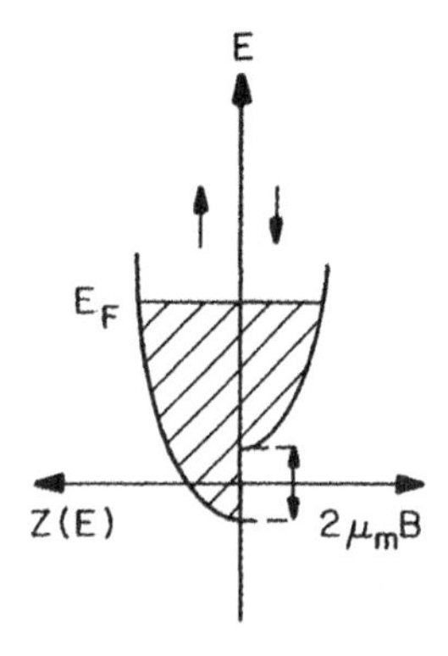

Figure 3.18 State density for metal in an applied B field where the electrons have magnetic moment $\mu_m = \mu_B$.

electrons over the states available. Thus $M = \mu_m(N_+ - N_-)$ gives

$$\chi_{\text{Pauli}} = \mu_0 \mu_m^2 Z(E_F)$$

Using the free electron density of states at E_F, $Z(E_F) = 3N_v/2k_B T_F$, the Pauli susceptibility can be written as follows:

$$\chi_{\text{Pauli}} = \frac{3\mu_0 \mu_m^2 N_v}{2k_B T_F} \tag{3.38}$$

The absence of temperature dependence in the susceptibility of free electrons thus reflects the cancellations of two opposing factors: increased promotion of electron spins at higher temperatures to states in which they can align with an external field and increased thermal disordering of those aligned spins with increasing temperature.

Figure 3.5 shows the temperature dependence of the magnetic susceptibility for a number of different materials. You should be able to distinguish free electron (Pauli) paramagnet(s) from Curie paramagnets and from diamagnets. Why is the sodium ion diamagnetic whereas the metal is paramagnetic? Why then is copper diamagnetic? Examples of the dependence of Pauli paramagnetism on density of states can be found in White (1970).

3.6.2 Quantum Diamagnetism

Section 3.2 presented a classical rationale for the phenomenon of diamagnetism. However, classical systems in equilibrium cannot show diamagnetism (or any magnetism for that matter). A quantum mechanical explanation of diamagnetism is therefore given.

The Hamiltonian operator for the kinetic energy of a quantum mechanical system in the presence of a magnetic field is

$$H = \frac{(\boldsymbol{p} + e\boldsymbol{A})^2}{2m}$$

The vector potential $\boldsymbol{A}$ is the momentum per unit charge stored in a magnetic field (Semon and Taylor 1996). Using the linear momentum operator $\boldsymbol{p} = -i\hbar\nabla$, leads to

$$H = -\frac{\hbar^2}{2m}\nabla^2 + \frac{ie\hbar}{2m}(\nabla \cdot A + A \cdot \nabla) + \frac{e^2}{2m}A^2 \tag{3.39}$$

If the field is uniform in the z direction, the vector potential may be represented as

$$A = \left(\frac{-yB}{2}, \frac{xB}{2}, 0\right) \tag{3.40}$$

This form satisfies the Coulomb gauge, where $\nabla \cdot A = 0$. The Hamiltonian then becomes

$$H = -\frac{\hbar^2}{2m}\nabla^2 + \frac{ie\hbar}{2m}\left(x\frac{\partial}{\partial y} - y\frac{\partial}{\partial x}\right)\frac{B}{2} + \frac{e^2B^2}{8m}(x^2 + y^2) \tag{3.41}$$

The first term gives the kinetic energy (independent of B); the next gives the orbital angular momentum L_z and it is responsible for orbital paramagnetism. The expectation value of the last energy term is recognized as

$$E = \frac{e^2B^2}{12m}\langle r^2 \rangle \tag{3.42}$$

and the magnetic moment response to the B field $\mu_m = -\partial E/\partial B$ is

$$\mu_m = -\frac{e^2\langle r^2 \rangle}{6m}B$$

or

$$\chi_d = \frac{\mu_m}{H} = -\frac{\mu_0 e^2\langle r^2 \rangle}{6m} \tag{3.43}$$

which is identical to Eq. (3.5) derived classically for a system of N_v particles per unit volume. (But $\langle r^2 \rangle$ must be calculated according to the rules of quantum mechanics.)

3.7 FERROMAGNETISM

You will notice that the data for Ni in Figure 3.5 show a Curie-like paramagnetic susceptibility above 360°C. What happens at $T = 358$°C where χ diverges? Nickel becomes ferromagnetic below 358°C, which is called its *Curie temperature* T_C. Weiss realized that this behavior could be modeled by assuming that a very strong internal field takes over below that temperature and allows the atomic moments to couple or act cooperatively despite the strong disordering effects of temperature (Weiss and Forrer, 1926). He also realized that you could extend the Langevin theory of paramagnetism by assuming that the applied field H_A can be replaced by $H_A + H_E$ where the effective field is given by

$$H_E = \lambda M \tag{3.44}$$

with $\lambda \approx 2000$. (Remember, there was no quantum mechanics in 1907 although the quantum paramagnet just described is also amenable to this generalization.)

When the treatment is carried through, the first result [cf. Eq. (3.13)] is that $M = \chi H = (C/T)H$ becomes $M = (C/T)(H + \lambda M)$, so that

$$\chi = \frac{M}{H} = \frac{C}{T - T_C} \tag{3.45}$$

with $T_C = \lambda C$. This explains the divergence of χ at $T = T_C$ rather than $T = 0$. T_C is the temperature that separates the ordered state ($T < T_C$), where the internal field dominates the thermal effect, from the disordered state ($T > T_C$), where thermal disorder reigns. Unfortunately, λ must be of order 10^3 to give T_C correctly, and such a strong internal field was difficult to justify classically.

The utility of this model goes further. The concept of an effective field may be extended to the argument of the Brillouin function, Eq. (3.35). Examine only the $J = \frac{1}{2}$ function, $B_{1/2}(x)$:

$$M = N_v \mu_m \tanh\left[\frac{\mu_m \mu_0 (H + H_g)}{k_B T}\right] \tag{3.46}$$

Because H_E contains M, this transcendental equation can be solved graphically. The applied field can be neglected relative to H_E. The reduced magnetization, $\sigma = M/N_v\mu_m = \tanh \xi$ (with $\xi \approx \mu_m \mu_0 H_E/k_B T$), is then plotted against ξ and $\sigma = k_B T\xi/\mu_0 \mu_m^2 \lambda N_v$ is plotted against ξ to look for their intersection. The process is illustrated in Figure 3.19. Note how the slope of the straight line

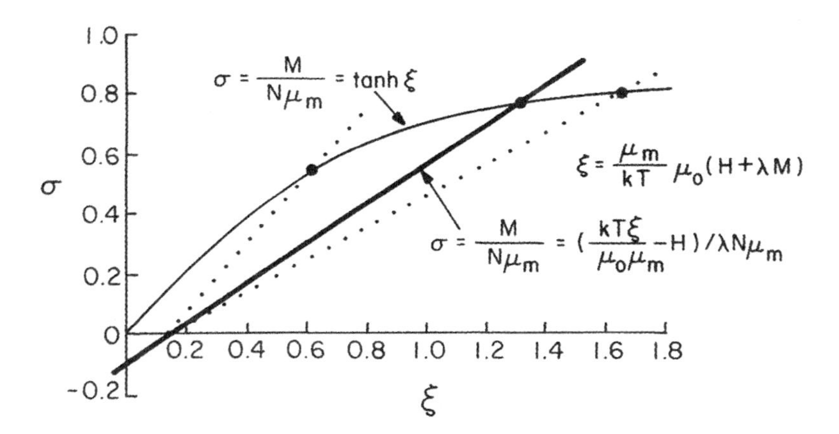

Figure 3.19 Construction for the solution, $\sigma(\xi)$, of the transcendental equation (4.25). The linear curve is shown for three different temperatures. The intersection of the linear and tanh curves is the solution for the temperature dependence of magnetization. Here the offset of the linear curve has been exaggerated to show the effect of an external field.

increases with increasing T so that above a certain temperature that turns out to be T_C, no solution exists (spontaneous magnetization vanishes). If the applied field is significant relative to λM, then the straight lines are shifted as indicated in the figure.

Figure 3.20 shows the intersection of the two theoretical curves in Figure 3.19 for the special case of $B_j(x)$ with $j = \frac{1}{2}$ appropriate to Ni. Data for Ni are also plotted. Such a plot of reduced magnetization versus reduced temperature is a useful way of comparing the magnetic behavior of different materials. When Brillouin functions corresponding to $J > \frac{1}{2}$ are used in Eq. (3.46), the solutions show less curvature than that for $J = \frac{1}{2}$. The solution for $J = \infty$ is shown as a dotted line. Because the spontaneous magnetization (the magnetization that appears in the absence of an external field) is a magnetic order parameter, it is not surprising that Figure 3.20 resembles an order–disorder curve for a cooperative process. If the graph were to extend above the Curie temperature, one could plot the inverse susceptibility, Eq. (3.35), which would be a straight line of positive slope originating at T_C. In actual practice, the magnetization curve can have a small tail of weak spontaneous magnetization extending into the paramagnetic region, and the inverse susceptibility can show

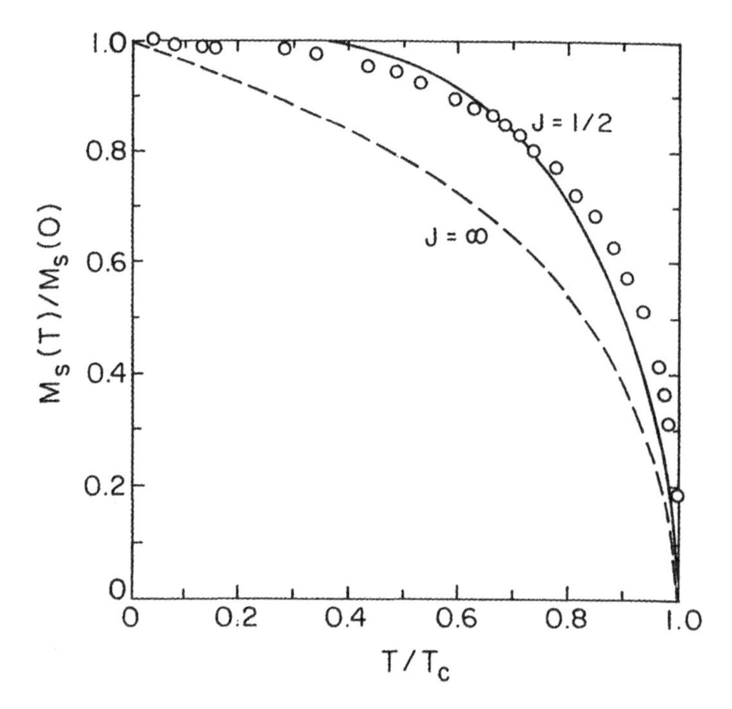

Figure 3.20 Reduced magnetization versus reduced temperature for nickel [open data points, from Weiss and Forrer (1926) and Brillouin function, $B_{1/2}(x)$ (solid line]. The dashed line is the classical solution for $J = \infty$.

positive curvature before going to zero. These effects can be due to short-range magnetic order (Smart 1970).

In the special case of $J = \frac{1}{2}$, it is not necessary to solve Eq. (3.46) graphically or with a computer. It can be done analytically (Whitaker 1989).

Table 3.2 lists the observed saturation magnetizations M_s, low temperature moments (unpaired spins per atom), and Curie temperatures for several magnetic materials. While the mean-field theory presented above gives the shape of the $M(H, T)$ curves, it is phenomenological and, therefore, cannot predict the values of M_s and it is orders of magnitude off in calculating the ordering temperature, $T_C = \lambda C$. Only a microscopic theory can give accurately M_s or T_C.

Note from the data for Mn, Fe, Co, Ni, and Cu ferrites that the moment per formula unit n_B *decreases* in nearly integral steps with *increasing* atomic number of the divalent species and the magneton numbers take on nearly integral values (mean deviation from nearest integer is 0.2). This might be expected for an integral number of unpaired spins per atom or per formula unit as the d states fill ($\mu_m = 0$ for d^{10}) with increasing atomic number. For metallic Fe, Co, and Ni, the magneton numbers also decrease with increasing number of valence electrons but not by near-integral steps. Further, n_B does not have near-integral values (mean deviation from $\mu_m/\mu_B = 2$, 1, 0 is 0.5 and from nearest integer is 0.3). Metals in fact are a little more difficult to understand than insulators. Therefore, magnetism in the transition metal oxides will be examined in Chapter 4; further treatment of metals will be deferred until Chapter 5, after introduction of more ideas about magnetism and bonding in

TABLE 3.2 Fundamental Magnetic Data for Various Crystalline Ferromagnets

Substance	Structure	M_s (290 K) (emu/cm^3)	M_s (0 K) (emu/cm^3)	$n_B{}^a = M_s/\mu_B N_v$ (μ_B)	$T_C(T_N)$ (K)
Fe	BCC	1707	1740	2.22	1043
Co	HCP, FCC	1440	1446	1.72	1388
Ni	FCC	485	510	0.606	627
$Ni_{80}Fe_{20}$	FCC	800	930	1.0	—
Gd	HCP	—	2060	7.63	292
Dy	HCP	—	2920	10.2	88
MnBi	NiAs(hex)	620	680	3.52	630
Ni_2MnGa	Heusler	480	—	—	373
CrO_2	—	515	—	2.03	386
$MnOFe_2O_3$	Spinel	410	—	5.0	573
$FeOFe_2O_3$	Spinel	480	—	4.1	858
$CoOFe_2O_3$	Spinel	—	—	3.2	—
$NiOFe_2O_3$	Spinel	270	—	2.4	858
$CuOFe_2O_3$	Spinel	135	—	1.3	728

[a]The Quantity n_B is called the magneton number, the number of bohr magnetons per atom or per formula unit in a material

nonlocalized-electron systems. Before treating these materials it is important to describe the quantum mechanics behind the Weiss molecular field, namely, the magnetic exchange interaction.

3.7.1 Magnetization at Low Temperature: Spin Waves

It appears from Figure 3.20 that $M_s(T)$ for Ni decreases from its $M_s(0)$ value with increasing temperature at a rate that is significantly greater than that indicated by the Brillouin function for $S = \frac{1}{2}$. The experimental thermal demagnetization is well fit by

$$\frac{\Delta M(T)}{M_s(0)} = -AT^{3/2} \tag{3.47}$$

At $T \approx 0.1\ T_C$ the data show $\Delta M/M \approx -2 \times 10^{-3}$. On the other hand, the Brillouin function $B_{1/2}(x)$ predicts $\Delta M/M = -2\exp\{-2 \times [(N_v \lambda \mu_m^2)/k_B T]\}$ at $T = 0.1\ T_C$, much smaller than observed.

The Brillouin function describes the thermal demagnetization process by random thermal fluctuations over the quantum states of the spin system in the face of a mean-field expression for exchange (plus a relatively small applied field) that tends to maintain the full saturation magnetization. But random, uncorrelated thermal spin fluctuations cost appreciable exchange energy because of the high degree of local spin misalignment. An exchange-coupled spin system, on the other hand, can also reduce its saturation magnetization by the formation of spatially correlated collective modes of demagnetization in which adjacent spins maintain a greater degree of alignment (Figure 3.21). These collective modes of demagnetization are called *spin waves* or *magnons*. They are for the spin system what phonons are to a crystal lattice: quantized modes of thermal excitation that are correlated by the wave nature of the displacement variable, ΔS_i^z for spins and Δx_i, for atomic positions (i is an atomic site).

The mathematical treatment of spin waves [(see Kittel (1986) or Barabara et al. (1988), for examples] indicates that the energy of excitation of a magnon of wavenumber $k = 2\pi/\lambda$ on an array of spins separated by the distance, a, goes as

$$\hbar\omega_k = 4JS[1 - \cos(ka)]$$

The long wavelength limit is

$$\hbar\omega_k \approx 4JSa^2k^2 \equiv Dk^2 \tag{3.48}$$

The parabolic energy in Eq. (3.48) indicates quantized energy levels that are equally spaced in energy as in the harmonic oscillator problem (which applies to phonons and photons as well):

$$\varepsilon_K = (n_K + \tfrac{1}{2})\hbar\omega_k = (n_F + \tfrac{1}{2})Dk^2$$

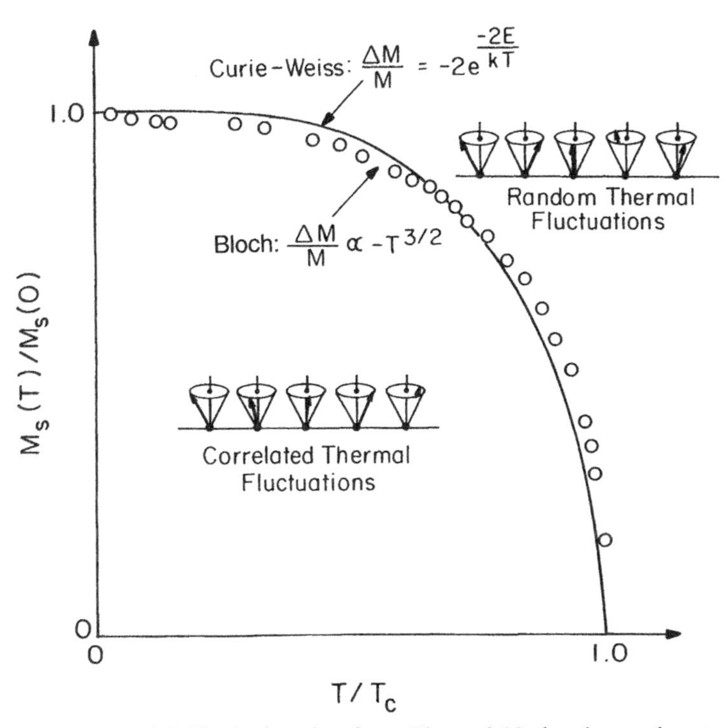

Figure 3.21 Data and Brillouin function from Figure 3.20 showing, at low temperature, the reduced magnetization falling off according to Bloch's $T^{3/2}$ law while the Curie–Weiss function drops more slowly. Inserts depict spins subject to random thermal fluctuations (Curie–Weiss) and spatially correlated thermal fluctuations or magnons (Bloch).

where n_k is the number of magnons of wavenumber k. Each magnon gives the same magnetization decrement that would result from reversal of one spin, but for these collective modes the decrement is distributed over the spin system as a wave with a small spin deflection at each site. The number of magnons, n_k, for the harmonic oscillator at a given temperature follows a Planck distribution

$$\langle n_k \rangle = \frac{1}{\exp(\hbar\omega_k / k_B T) - 1}$$

The net demagnetization due to all the spin waves is then

$$\Delta M = M_s \sum_k n_k$$

which can be shown to give

$$\frac{\Delta M}{M_s} \propto \left(\frac{k_B T}{2Js} \right)^{3/2} \tag{3.49}$$

This is the Bloch $T^{3/2}$ law that describes the low-temperature thermal demagnetization data of most systems. This derivation shows that an exchange coupled spin system can undergo thermal demagnetization more efficiently in a correlated, wave-like manner by conserving exchange energy compared to the uncorrelated demagnetization process described by the Brillouin function.

3.7.2 Curie Temperature Determination

In practice, the *saturation* magnetization does not vanish just above T_C as implied by Figure 3.20. The reason for this is that the susceptibility is very large there and short-range ordering of the moments can render the $M(H)$ curves nonlinear in H in this temperature range. The spontaneous magnetization, $M(H = 0, T)$, does vanish for $T \geqslant T_C$. It is often easier, therefore, to identify T_C by inspection of $M(H, T)$ curves taken in very weak fields; for smaller values of H, $M(T)$ show a sharper drop at T_C.

However, a more quantitative method, described by Belov and Goriaga (1956) and by Arrott (1957) is widely used when an accurate value of T_C is required. Near T_C, the magnetization is small, so a Landau expansion of the free energy is appropriate:

$$f_m = \frac{\alpha}{2} M^2 + \frac{\beta}{4} M^4 \cdots - \mu_0 HM \tag{3.50}$$

Here, α and β are positive functions of temperature. It costs energy to increase the magnetization in the face of thermal energy; hence the first two magnetization product terms in Eq. (3.50) are positive. The tendency of the magnetization to increase in an external field accounts for the negative sign of the Zeeman energy term, $-\mu_0 HM$. Terms of odd order in M and H are not present because they do not leave the energy invariant under time reversal. The form of energy in Eq. (3.50) can also be derived from the Weiss-Brillouin function [Eq. (3.46)] (Arrott 1957, Kouvel and Fisher 1964, Belov and Goriaga 1956). The equilibrium magnetization is the value that minimizes f_m with respect to M:

$$\alpha M + \beta M^3 = \mu_0 H \tag{3.51}$$

The coefficient α can be determined near T_C (small M) for either local moment systems or for itinerant magnets. The magnetization is small in weak fields near T_C, so from Eq. (3.51)

$$\mu_0 H = \alpha \chi H \tag{3.52}$$

For local moments, the Curie–Weiss law [Eq. (3.45)] combines with Eq. (3.52) to give

$$\alpha = \frac{\mu_0}{C}(T - T_C) \tag{3.53}$$

For itinerant magnets, it can be shown (Wohlfarth 1976) that

$$a \propto T^2 - T_C^2$$

Thus Eq. (3.51) may be written

$$M^2 = -C(T^n - T_C^n) + \frac{\mu_0}{\beta}\frac{H}{M} \tag{3.54}$$

where C is a constant while $n = 1$ for local moments and 2 for itinerant magnets. A plot of M^2 versus H/M should be a series of straight lines with an M^2 intercept of zero for $T = T_C$. Figure 3.22 shows a classic set of data on Ni from Weiss and Forrer (1926). The magnetization isotherms in (a) are represented in terms of the specific magnetization σ (emu/g) $= M/\rho$ where ρ is the mass density. Each line corresponds in panel (b) to σ^2–H/σ data at a different temperature. The linear parts of the σ^2–H/σ curves have zero intercept for temperatures near 360°C. T_C for Ni is determined from these plots to be 358°C. Such plots are called Arrott plots or Arrott–Belov–Goriaga plots.

The simple model presented here suggesting straight lines is seldom observed. Various factors can account for deviations from linearity, usually in weak fields. Nevertheless, extrapolation of higher magnetization data generally gives a reliable value for T_C.

Kouvel and Fisher (1964) show that the $M(H, T)$ data for Ni just below T_C are best described not by the Curie–Weiss law, but rather by χ proportional to $(T - T_C)^{-\gamma}$, where $\gamma = 1.35$. This so-called critical exponent γ describing the temperature dependence of the susceptibility near the critical point T_C touches on the topic of critical phenomena in magnetic systems. This subject [well reviewed in Domb (1965) and Stanley (1971)] is beyond the scope of this text.

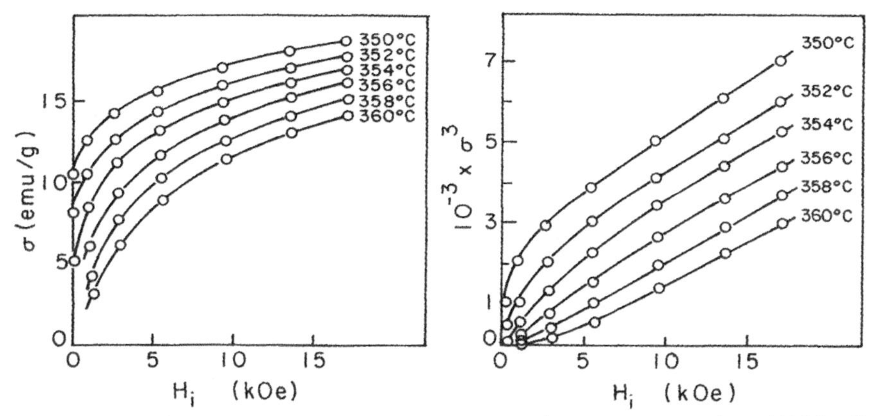

Figure 3.22 (*a*) σ versus H for Ni based on Weiss and Forrer data (1924); (*b*) Arrott–Belov–Goriaga plot of σ^2 versus H/σ using data in (*a*). [After Arrot (1957).]

3.8 SUMMARY

It was found that the orbital magnetic moment of a charged particle is proportional to the angular momentum of its orbit. The constant of proportionality is the gyromagnetic ratio, $\gamma = e/2m$. It was demonstrated that an orbital magnetic moment tends to precess about an applied magnetic field, where the Larmor precession frequency is given by γB. This precession contributes a field-induced change in the motion of the charged particles. As such, it represents a classical explanation of the phenomenon of diamagnetism.

Paramagnetism is the response to an applied field of noninteracting magnetic moments. A classical derivation based on a Boltzmann occupation of a continuum of energy levels for the moment in a $\boldsymbol{B}$ field gives the Curie law, $\chi = C/T$.

Table 3.3 outlines some similarities and differences between classical diamagnetism and paramagnetism.

A free electron model with energy levels based on angular momentum quantization was seen to provide a good explanation of the normal Zeeman effect. Explanation of the fine structure in optical spectra of materials in magnetic fields (the anomalous Zeeman effect) requires introduction of the concept of spin.

Spectroscopic data demanded a better explanation than could be afforded by the classical model; quantum mechanics had to be introduced. Quantum mechanics affords an explanation of spectroscopic fine structure, reveals the need to complement orbital angular momentum with an intrinsic angular

TABLE 3.3 Comparison of Various Aspects of Diamagnetism and Paramagnetism

	Diamagnetism	Paramagnetism
Classical form	$\chi_{\text{dia}} = -N_v e^2 \mu_0 \sum \langle r^2 \rangle / 6m$	$\chi_{\text{para}} = \mu_m^2 N_v \mu_0 / 3k_B T$
Source	B field acting on any charged particle including those with $L = 0(s)$.	Effect of B field on existing magnetic moments; hence core electrons do not contribute.
Temperature dependence	χ_{dia} is essentially independent of temperature.	$\chi_{\text{para}} = C/T$, Curie law.
Timescale	Occurs instantaneously; does not involve a relaxation process but only Larmor precession	Requires a change in projection of magnetic moment on field direction, therefore requires energy loss process; paramagnetism is observed only on time scales longer than the relaxation time.

TABLE 3A.1 Definition and Comparison of Parameters Describing Linear and Rotational Motion

	Linear	Rotational	
Displacement	x	θ	Angular displacement
Velocity	$v = dx/dt$	$\omega = d\theta/\mathrm{d}t$	Angular velocity
		$v = \omega \times r$	Linear velocity in rotation
Acceleration	$a = d^2x/dt^2$	$\theta = d^2\theta/\mathrm{d}t^2$	Angular acceleration
Mass	m	I	Moment of inertia
Force	$F = ma$	$\cdots T = r \times F \rightarrow T = Id^2\theta/\mathrm{d}t^2$	Torque
	$F = dp/dt$	$T = dL/dt$	
Momentum	$p = mv$	$\cdots L = r \times p \rightarrow L = I\omega$	Angular momentum

momentum—spin, and explains the very important spin–orbit coupling between the two.

Quantum mechanics gives a much better model for paramagnetism and diamagnetism and allows for incorporation of a strong internal molecular field to describe ferromagnetic behavior.

APPENDIX: LINEAR AND ANGULAR MOTION

Linear dynamics gives relations between displacement x, force F, mass m, velocity v, momentum $p = mv$, and acceleration, a. In rotational dynamics there are analogous quantities defined in Table 3A.1.

While the velocity vector $\boldsymbol{v}$ is oriented along the direction of motion, ω is a vector normal to the plane of rotation because it is related to $\boldsymbol{L} = \boldsymbol{r} \times \boldsymbol{p}$ by the moment of inertia tensor, $\boldsymbol{I}$. Care must be taken in rotational dynamics not to assume that motion is in the direction of torque. Rather, torque induces a change in angular momentum in the direction of $\boldsymbol{T}$. The $\boldsymbol{B}$ field in Figure 3.2 causes a new motion: precession of the original $\boldsymbol{L}$ about $\boldsymbol{B}$.

PROBLEMS

3.1 Use the Lorentz force to describe how the angular frequency ω changes for a circular electron orbit (e.g., a classical Bohr orbit) in the xy plane when a magnetic field is applied along the z axis.

3.2 Derive the expressions for total energy E_n and radius of circular electron orbits r_n in Bohr's model. Evaluate the constants E_1 and r_1 in SI units.

3.3 Calculate the classical angular momentum and the magnetic moment for a uniform shell of charge-$|e|$ and radius $r = e^2/(4\pi\varepsilon_0 mc^2)$ rotating with angular velocity ω. What is the gyromagnetic ratio for this particle? What is the surface velocity? (This problem was considered by Abraham in 1903, more than 20 years before Uhlenbeck and Goudsmit hypothesized that electrons spin with a gyromagnetic ratio twice that for orbital motion).

3.4 Compare essential characteristics of Lenz' law (macroscopic) and diamagnetism. How are they similar or different?

3.5 Calculate the dimension at which diamagnetism crosses over to paramagnetism for a metal with $\sigma = ne^2\tau/m = (ne^2/m)\lambda/\langle v\rangle = 10^7\ (\Omega\cdot\text{m})^{-1}$ (τ = relaxation time, λ = mean free path, $\langle v\rangle$ = mean drift velocity of change carriers).

3.6 **(a)** Evaluate the ratio of the integrals in the classical expression for $\langle\cos\theta\rangle$

$$\frac{\int_0^\pi \exp\left(\frac{\mu H}{k_B T}\cos\theta\right)\cos\theta\ \sin\theta d\theta}{\int_0^\pi \exp\left(\frac{\mu H}{k_B T}\cos\theta\right)\ \sin\theta d\theta}$$

(b) Show that $L(x) = \coth(x) - 1/x$ goes to $x/3$ in the limit x approaches zero. Discuss

(c) Show that $L(x) \to 1$ for $x = \infty$; discuss.

3.7 Show that the units of $\chi = N\mu_0\mu_m^2/k_BT$ and $\chi = N\mu_0 e^2/6m\,\Sigma r^2$ are $m^3 \times N$ so that if N is number of atoms per unit volume, χ is dimensionless. However, if N is Avogadro's number (in which case χ is the molar susceptibility χ_{mol}), it must be multiplied by $10^6\chi m^3/m^3$ to compare with $4\pi\chi_{\text{mol}} = 4\pi N\mu_m^2/k_BT$, or $4\pi Ne^2/6mc^2\,\Sigma r^2$ calculated in cgs units.

3.8 Calculate the paramagnetic susceptibility of diatomic oxygen at room temperature and compare it with the experimental, room temperature (cgs) value $\chi_{\text{mol}} = 3.4 \times 10^{-3}$.

3.9 Calculate the diamagnetic susceptibility of atomic He assuming $r^2 \approx a_0^2$ and compare with the room temperature cgs value $\chi_{\text{mol}} = -1.88 \times 10^{-6}$.

3.10 Derive a classical expression for the diamagnetic susceptibility of an electron in a circular orbit by considering the change in its angular momentum due to the electric field induced as a B field is slowly turned on normal to the orbit plane in time dt.

3.11 Analyze the condition for which the paramagnetic and diamagnetic susceptibilities are equal and opposite for a classical Bohr atom with orbital but not spin magnetic moment. Discuss.

3.12 Calculate the orbital magnetic moment of an electron in a circular Bohr orbit. Use $\omega = E/\hbar = 1.36\,\text{eV}/\hbar$ and $\omega = v/r_0 = \sqrt{(2E/m)}/r_0$ and compare the results. Which one is correct? Why?

3.13 A very thin film ($t <$ mean free path) may exhibit diamagnetism in an external field perpendicular to the film plane. Explain how application of the field in the plane of the film could change the sign of this effect.

3.14 Write the electronic configuration (e.g. $3d^4$), the spectroscopic notation ($^{2s+1}L_J$ e.g. $^3D_{5/2}$) and effective magneton number $n_{\text{eff}} = g\mu_B[J(J+1)]^{1/2}$, where $g = 1 + [J(J+1) + S(S+1) - L(L+1)]/2J(J+1)$ for Cr^{3+}, Fe^{3+} and Co^{2+}.

3.15 **(a)** Show that $B_j(x)$ reduces to the Langevin function $L(x)$, with $x = \mu_m B/k_B T$, in the limit J approaches infinity.

(b) Show that $B_{1/2}(x) = \tanh(x)$

(c) Show that for $x \ll 1$, $B_j(x)$ becomes

$$B_J(x) \xrightarrow{x \ll 1} \frac{J(J+1)}{3J^2}\, x$$

and thus in this limit

$$\chi = \mu_0 \frac{N_v g^2 \mu_B^2 J(J+1)}{3k_B T}$$

(d) Show that as x approaches infinity, $B_j(x)$ approaches 1, specifically, $M = N_v g \mu_B m_J$. Describe the physical significance of each case.

3.16 The Landé g factor is used to account for the fact that $\mu_J = \mu_L + \mu_S$ is not collinear with $\boldsymbol{J} = \boldsymbol{L} + \boldsymbol{S}$ because $\mu_L = \mu_B m_l = (eh/2m)m_l$ whereas $\mu_s = 2\mu_B m_s$. Derive the expression for g in terms of the quantum numbers l, s, and j. Make use of the facts that since $\boldsymbol{L}$ and $\boldsymbol{S}$ precess around $\boldsymbol{J}$, μ_L, μ_S, and μ_J also precess around $\boldsymbol{J}$ and it is the projection of μ_J on $\boldsymbol{J}$ that is measured.

3.17 What type(s) of magnetism (ferro-, para-, dia-) would you expect to find and why in **(a)** NaCl, **(b)** $MnSO_4 \cdot 4H_2O$, **(c)** Fe_3O_4, **(d)** H_2O (or Ne), and **(e)** metallic Cu?

BIBLIOGRAPHY

Ashcroft, N. W., and N. D. Mermin, *Solid State Physics* (W. B. Saunders Co., Philadelphia, PA, 1976).

Eisberg, R., and R. Resnick, *Quantum Physics of Atoms, Molecules, Solids, Nuclei and Particles*, Wiley, New York, 1974).

REFERENCES

Arrot, A., *Phys. Rev.* **108**, 1394 (1957).

Barbara, B., D. Gignoux, and C. Vettier, *Lectures in Modern Magnetism*, Springer Verlag, Berlin, (1988), p. 211.

Belov, K. P., and A. N. Goriaga, *Fiz. Met. Metallov.* **2**, 3 (1956).

Bozorth, R. M., *Ferromagnetic Materials* IEEE Press, New York, 1993.

Domb, C., in *Magnetism*, Vol. IIa, G. T. Rado and H. Suhl, eds., Academic Press, New York, 1965, p. 1.

Feynman, R. P., R. B. Leighton, and M. Suds, *Lectures on Physics*, Vol. II, Addison Wesley, Reading MA, 1964, p. 34–35.

Henry, W. E., *Phys. Rev.* **88**, 559 (1952).

Kittel, C., *Introduction to Solid State Physics*, 6th ed., Wiley, New York, 1986.

Kouve, J. S., and M. E. Fisher, *Phys. Rev.* **136**, A1626 (1964).

Leighton, R. B., *Principles of Modern Physics*, McGraw Hill, New York, 1959.

Semon, M. D., and J. R. Taylor, *Am. J. Phys.* **64**, 1361 (1996).

Smart, J. S., *Effective Field Theories of Magnetism*, Saunders, New York, 1970).

Stanley, H. E., *Phase Transitions and Critical Phenomena*, Oxford Univ. Press, New York, 1971, p. 79.

Weiss, P., and R. Forrer, *Ann. Phys.* **5**, 153 (1926).

Whitaker, M. A. B., *Am. J. Phys.* **57**, 45 (1989).

White, R. M., *Quantum Theory of Magnetism*, (McGraw-Hill, New York, 1970, p. 80.

Wohlfarth, E. P., in *Magnetism—Selected Topics*, S. Foner, ed., Gordon & Breach, New York, 1976, p. 74.

CHAPTER 4

QUANTUM MECHANICS, MAGNETISM, AND EXCHANGE IN ATOMS AND OXIDES

The chemical interactions that affect the electronic states and determine the physical properties of a material can be described by two limiting bonding types: polar bonds and covalent bonds. *Polar bonds* are formed between orbitals that differ significantly in their electronegativity ($E_A \neq E_C$), such as in compounds A_xC_{1-x} (where A and C represent the anion and cation, respectively). The orbitals must also satisfy symmetry and overlap conditions in order to form a bond. In the formation of a polar bond, charge is transferred from the orbital of higher energy (lower electronegativity, the cation) to that of lower energy (the anion). As a result of this charge transfer, the charge in the bond is biased toward the more electronegative species. The most familiar examples of polar bonding occur in oxides. In these cases, the interacting species can be treated as charged atoms and their electronic states are generally highly localized, atomlike, and readily described by quantum chemistry. Magnetism in polar bonded materials—insulators and oxides—are treated in this chapter.

Covalent bonds are formed between orbitals on two atoms that have similar electronegativities (i.e., similar electronic energies, $E_A \approx E_C$) as well as satisfying symmetry and overlap conditions. In a covalent bond, charge is delocalized from each of the atomic sites and builds up between the atoms. Bonding and antibonding hybrid orbitals are created. In alloys containing transition metal species, covalent bonds formed between partially occupied valence orbitals (i.e., near E_F) will often involve magnetic states. If $3d$ states are involved, they become more delocalized as a result of covalent bonding. Covalently bonded metal, especially those involving transition elements, are more difficult to treat than are polar bonded materials and hence will be treated in Chapter 5.

4.1 EXCHANGE INTERACTIONS

It was shown that for isolated atoms, Hund's rule dictates the formation of an atomic magnetic moment for incompletely filled orbitals. This is a reflection of an *intraatomic* exchange interaction to distinguish it from the interaction *between* atoms in a solid, *interatomic* exchange, that describes when and how strongly these atomic moments couple parallel or antiparallel with each other. To account for alignment between atomic magnetic moments in solids despite strong thermal disordering effects, Pierre Weiss postulated an internal or molecular field $H_{mol} = \lambda M$. But to explain the strength of the alignment, the molecular field had to be assumed to be much larger than Weiss was comfortable with. It was estimated to be of order 10^9 A/m (10^7 Oe or $B = 10^3$ T), which is larger than any human-made field. The mechanism of this strong exchange interaction will be seen to be an *electronic* interaction, not simply a *magnetic* interaction.

This chapter provides a qualitative description of interatomic exchange, with some reference to the quantum mechanical expressions that support these new concepts. The goal is to understand exchange in some simple systems such as magnetic oxides, where the electronic states of the ions may be treated as atomic states. The more difficult cases of magnetic exchange in metallic solids is deferred to Chapter 5.

4.1.1 Interacting Electrons

In order to understand magnetism in solids, it is important to know how electrons interact with each other and how those interactions affect spin. The two principles that must be understood are the Coulomb repulsion between electrons and the constraints imposed by the Pauli exclusion principle. These form the foundation for understanding magnetism in solids.

Consider the effects of the Coulomb interaction e^2/r_{ij} on the states of two electrons moving in similar potentials $V(r)$ (e.g., an H_2 molecule). The Hamiltonian operator for the electron pair is

$$
\begin{aligned}
H &= -\frac{\hbar^2}{2m}\nabla_1^2 - \frac{\hbar^2}{2m}\nabla_2^2 + V(1) + V(2) + \frac{e^2}{4\pi\varepsilon_0 r_{12}}, \\
&\equiv H_0 + \frac{e^2}{4\pi\varepsilon_0 r_{12}} \qquad (4.1)
\end{aligned}
$$

where the numbers 1 and 2 refer to the spatial coordinates of the two electrons with respect to a common origin O and r_{12} is their separation; $V(1)$ includes $-e^2/|\boldsymbol{r}_1 - \boldsymbol{R}_1|$ and $-e^2/|\boldsymbol{r}_1 - \boldsymbol{R}_2|$, the interaction of electron 1 with the nuclei at $\boldsymbol{R}_1$ and $\boldsymbol{R}_2$ (see Fig. 4.1). The constant $(4\pi\varepsilon_0)^{-1}$ has been ommitted from the electrostatic interactions. If the interaction between the two electrons were negligible, Eq. (4.1) could be separated into two independent equations, each

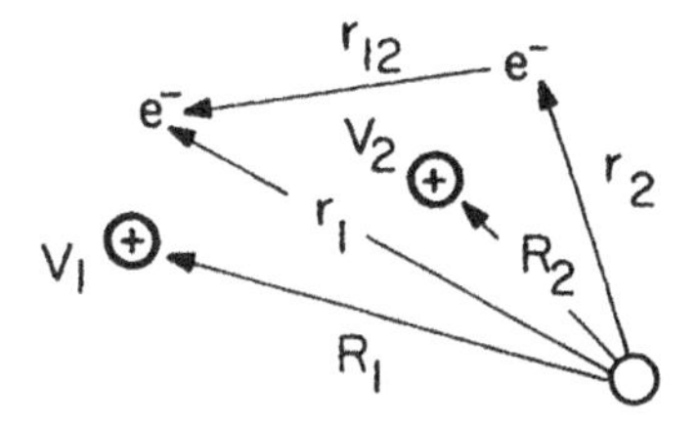

Figure 4.1 Coordinate system for two interacting electrons in double potentials, V_1 and V_2 as in Eq. (4.1).

involving the coordinates of a different electrons, $H^0\psi^0 = E^0\psi^0$, with

$$\psi^0 = \phi_i(1)\phi_j(2) \tag{4.2}$$

$$E^0 = E^{(i)} + E^{(j)} \tag{4.3}$$

Here ϕ_i and ϕ_j are the solutions for a single electron moving in potential, V_i and V_j. If the potentials are the same and if $|R_i - r_j| \gg |r_i - R_i|$, then the energy levels for the two atoms are the same: $E^{(i)} = E^{(j)}$. This situation for two non-interacting electrons in separate but identical potentials is depicted in Figure 4.2*a*.

The electrons are now allowed to interact with each other and we consider the interaction energy to be small compared to E^0. The effect of the interaction is calculated by first-order perturbation theory. The result is

$$\begin{aligned} E &= E^0 + \int \phi_i^*(1)\phi_j^*(2)\left[\frac{e^2}{r_{12}}\right]\phi_i(1)\phi_j(2)\,dv \\ &\equiv E^0 + C_{ij} \end{aligned} \tag{4.4}$$

where C_{ij} is an average of the Coulomb interaction of the two electrons in states i and j. The interaction energy raises the energy of the states relative to their unperturbed values (Fig. 4.2*b*).

4.1.2 Pauli Exclusion Principle

The symmetry of the potential in a quantum mechanical problem determines the symmetry of the wavefunction. For example, a Hamiltonian with a central potential dictates that the wavefunctions have the symmetry of spherical harmonic functions. A symmetric one-dimensional potential necessarily has solutions that are either symmetric or antisymmetric functions of the spatial coordinate. While this fact is based on group theory, it may be understood more concretely by considering that the observables of the system always involve even powers of ψ (e.g., the probability density is $\int \psi^*\psi d\tau$). Whether $\psi(x) = \pm\psi(-x)$, the even products of ψ have inversions symmetry like the potential.

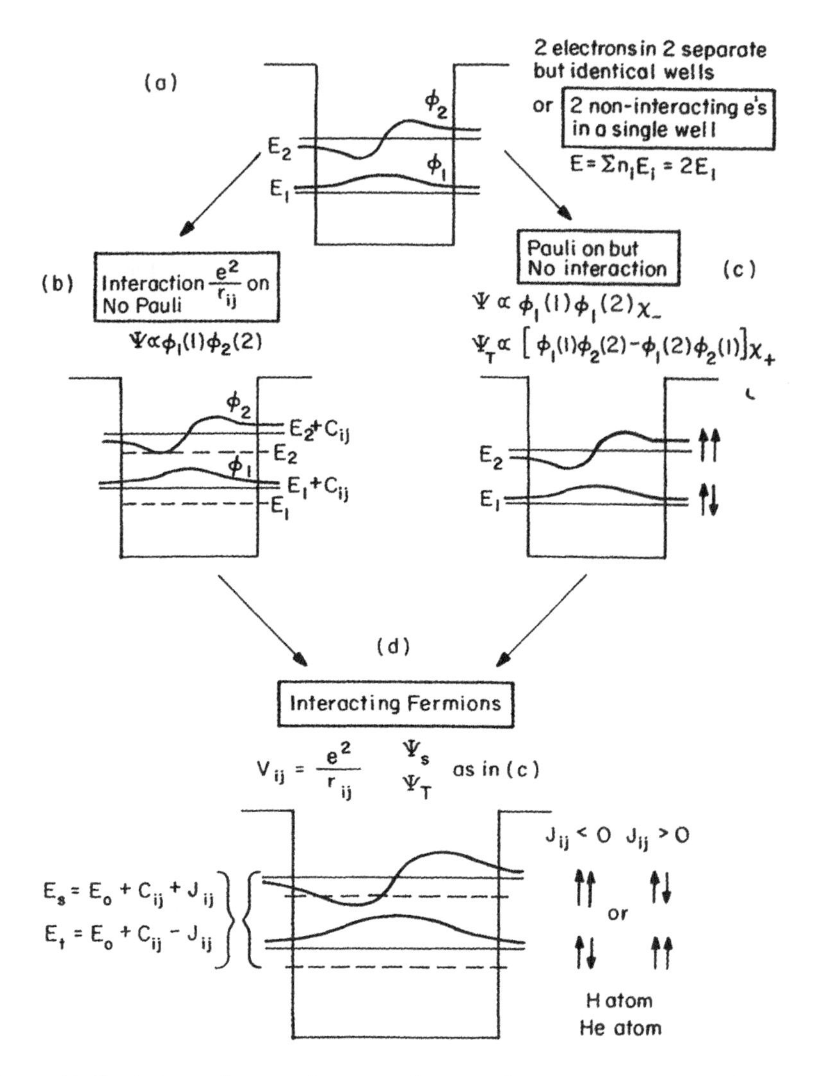

Figure 4.2 Evolution of states for two electrons in a common potential for (*a*) noninteracting, distinguishable, electrons; (*b*) interacting, distinguishable electrons; and (*c*) noninteracting, indistinguishable electrons, and (*d*) interacting, indistinguishable electrons. In cases (*c*) and (*d*), the triplet wavefunctions are superpositions of both states 1 and 2.

In a much more fundamental sense, a general multiparticle Hamiltonian exhibits a very simple but important form of symmetry. The Hamiltonian is unchanged on interchange of the space and spin coordinates of any two particles; that is, the Hamiltonian operator cannot distinguish particles. The probability densities resulting from two states that differ only by the interchange of particles (space and spin coordinates) must be indistinguishable if the particles are indistinguishable. This demands that the wavefunctions be either *even* or *odd* under that exchange.

The wavefunctions describing different types of identical particles (photons, electrons, neutrons, etc.) have one or the other of these symmetries (i.e., they are either *even* or *odd*) when you interchange particles between different states. Particles with integral intrinsic spin are called *bosons* and their wavefunctions are *symmetric* on interchange of space and spin coordinates of any two particles

$$\Psi(r_1, r_2, r_3, \ldots, s_1, s_2, s_3, \ldots) = +\Psi(r_1, r_3, r_2, \ldots, s_1, s_3, s_2, \ldots) \quad (4.5)$$

Such particles obey Bose–Einstein statistics, and any number of them may occupy the same quantum state simultaneously. Examples are photons, helium atoms, and certain coupled electron pairs (called *Cooper pairs*) which are important in superconductivity.

Particles with half-integral intrinsic spin are called *fermions* and they can be described only by wavefunctions that are *antisymmetric* on interchange of the space and spin coordinates of any two particles

$$\Psi(r_1, r_2, r_3, \ldots, s_1, s_2, s_3, \ldots) = -\Psi(r_1, r_3, r_2, \ldots, s_1, s_3, s_2, \ldots) \quad (4.6)$$

Fermions obey Fermi–Dirac statistics, that is, no two can occupy the same quantum state simultaneously. (Examples are electrons, protons, and neutrons.) This is a statement of the Pauli exclusion principle.

The "exchange" symmetry of a Hamiltonian operator may be examined by considering what is called a permutation or exchange operator:

$$P_{12}\Psi(1, 2, 3\ldots) = \Psi(2, 1, 3\ldots)$$

It should be noted that P commutes with H (i.e., $HP\Psi = PH\Psi$); therefore eigenfunctions of P are constants of the motion. Double application of P_{ij} to a wavefunction gives the same wavefunction:

$$P^2\Psi = P_{ij}P_{ij}\Psi(1, 2, \ldots, ij\ldots) = \Psi(1, 2, \ldots, ij, \ldots)$$

so the eigenvalues of P are ± 1:

$$P\Psi = \pm\Psi$$

These exchange symmetries have several important consequences especially for what are called "cooperative phenomena" such as magnetism or superconductivity. Most important are the following:

1. Bosons ($P\Psi = +\Psi$), exhibit an extra attractive force on one another and hence are found closer together, and fermions ($P\Psi = -\Psi$) exhibit a repulsive force and are further apart, relative to equivalent distinguishable particles.

2. The energy and pressure of a system of bosons is less, and fermions, greater than a comparable system of distinguishable particles.
3. If the particles of a system experience an external force but remain independent of each other, the quantum states of the system correspond to the single-particle states of the various particles. Bosons can all occupy the same state and fermions cannot.

These permutation symmetry properties of identical particles are responsible for ferromagnetism, Bose condensation (superfluidity of He and superconductivity), the *ortho*- and *para* states of hydrogen, and the saturation of covalent chemical bonds.

4.1.3 Noninteracting, Indistinguishable Electrons

The simplest case of two identical but noninteracting electrons (Fig. 4.2*a*) will be considered. When particles do not interact, the Hamiltonian operator describing their behavior can be separated into different parts for the coordinates of the distinct particles, and Eqs. (4.3) and (4.4) give the solutions.

One way to guarantee that such product wavefunctions satisfy either Fermi–Dirac or Bose–Einstein statistics is to write then as symmetric or antisymmetric combinations of the individual wavefunctions

$$\psi^{\pm} = \phi_1(1)\phi_2(2) \pm \phi_1(2)\phi_2(1) \tag{4.7}$$

and two additional symmetric forms

$$\psi_a = \phi_1(1)\phi_1(2) \quad \text{and} \quad \psi_b = \phi_2(1)\phi_2(2)$$

In Eq. (4.7) the plus sign represents bosons and the minus sign represents fermions if only the spatial part of the wavefunction is considered. Also, $\phi_2(1)$ indicates that electron 1 is in state 2 and so forth. Note that two fermions in the same state $[\phi_1(2)\phi_1(1)]$ or the same electron in both states $[\phi_1(1)\phi_2(1)]$ results in $\psi^- = 0$ in Eq. (4.7).

But spin still has to be included. In the simplest case, the spin and spatial coordinates are noninteracting (separable) variables, so the full wavefunction again is a product of space ψ and spin χ functions

$$\Psi = \psi(r_i)\chi(S_i) \tag{4.8}$$

Neglecting spin–spin interactions $\chi(s)$ can be written by analogy with Eq. (4.7) as

$$\chi = \chi_+(1)\chi_-(2) \pm \chi_+(2)\chi_-(1) \tag{4.9}$$

where $\chi_-(2)$ stands for electron 2 in a spindown state and so on. The antisymmetric $(-)$ form of the spin function describes a spin singlet state $(S = 0)$ $2S + 1 = 1$; the symmetric form, a triplet state $(S = 1)$, $2S + 1 = 3$. But there are two other obvious forms for a triplet spin function, namely, $\chi_+(1)\chi_+(2)$ and $\chi_-(1)\chi_-(2)$ by analogy with the additional spatially symmetric functions. These three triplet functions describe the three states $m_s = 0, \pm 1$. Equations (4.7)–(4.9) contain all the assumptions needed in order to show the most fundamental and far-reaching connections between bonding and magnetism.

It was indicated that for fermions, the total wavefunction $\Psi(r, s)$ must be antisymmetric on interchange of both space and spin coordinates of any two particles. Clearly, if the spatial function is symmetric, then the spin function must be antisymmetric and vice versa. What does this mean? Figure 4.2*c* shows that even for otherwise noninteracting electrons, the imposition of the Pauli exclusion principle requires the lower energy state, which is spatially symmetric like an *s* orbital, to be a spin singlet state $(S = 0)$. Similarly, the higher energy state in Figure 4.2*c*, which is spatially antisymmetric, must be a spin triplet $(S = 1)$.

As argued above, the appropriate singlet or triplet wavefunctions for the system of fermions are given by

$$\begin{aligned} \Psi_S &= \left(\tfrac{1}{2}\right)^{1/2}[\phi_1(1)\phi_2(2) + \phi_1(2)\phi_2(1)]\chi_S, \qquad & E_S^0 &= E_1 + E_2 \\ \Psi_T &= \left(\tfrac{1}{2}\right)^{1/2}[\phi_1(1)\phi_2(2) - \phi_1(2)\phi_2(1)]\chi_T, \qquad & E_T^0 &= E_1 - E_2 \end{aligned} \tag{4.10}$$

and the χ values are given by Eq. (4.9).

4.1.4 Electron–Electron Interactions

If the electron–electron interactions are weak, their effects on the situation in Figure 4.2*c* can be calculated using perturbation theory. The unperturbed fermion functions [Eq. (4.10)], lead, for weak electron interactions, to

$$\begin{aligned} E_S &= E^0 + C_{ij} + \mathcal{J}_{ij} \\ E_T &= E^0 + C_{ij} - \mathcal{J}_{ij} \end{aligned}$$

where

$$\mathcal{J}_{ij} = \int \phi_i^*(1)\phi_j^*(2)\,\frac{e^2}{4\pi\varepsilon_0 r_{12}}\,\phi_i(2)\phi_j(1)\,dv \tag{4.11}$$

is the exchange energy of two electrons in states i and j; $\mathcal{J}_{ij}$ has no classical interpretation. Quantum mechanically, it is the strength of the interaction between electrons in the two states $\phi_i(1)\phi_j(2)$ and $\phi_i(2)\phi_j(1)$. Either the singlet or triplet can be lower in energy depending on the sign of $\mathcal{J}_{ij}$. Previously, knowledge of the relative energies of bonding and antibonding states in molecules was used to assign the triplet spin function to the antisymmetric, antibonding state. This type of calculation was first carried out by Pauli for the helium atom and by Heitler and London for the hydrogen molecule. In both of those cases, $\mathcal{J}_{ij}$ is negative and the singlet state lies lower.

When the Coulomb interaction between two electrons is strong, the integrals C_{ij} [Eq. (4.4)] and $\mathcal{J}_{ij}$ [Eq. 4.11)] are no longer perturbations on the energy E_0 and the ground state of the system can be one of either parallel or antiparallel spin. This is represented in Figure 4.2*d*.

The effects on magnetism of bonding are now considered in a simple model system: a molecule with one electron in the field of two protons at an arbitrary separation, R, ($H_2^+ \rightarrow He^+$ at $R = 0$). The transition from atomic moments due to Hund's rule to magnetic moments in molecules showing interatomic exchange becomes evident.

4.1.5 Hydrogen Molecule Ion, H_2^+

A good model system for illustrating the effects of interatomic distance on magnetism, originally studied by Edward Teller, is the hydrogen molecule ion H_2^+ with variable nuclear separation. The Hamiltonian for this system, with a single electron shared between two protons, is

$$H = \frac{P_1^2}{2M} + \frac{P_2^2}{2M} + \frac{p^2}{2m} + \frac{e^2}{4\pi\varepsilon_0}\left(\frac{1}{R} - \frac{1}{r_1} - \frac{1}{r_2}\right) \tag{4.12}$$

Here M and m indicate the mass of each proton and of the electron, respectively; P and p are their respective momentum operators, and R is the nuclear separation and r_1 and r_2 are the distances of the electron from the two nuclei. The kinetic energy of the nuclei can be safely neglected relative to that of the electrons. This is the Born–Oppenheimer approximation (1927). The relative potential energy of the nuclei ($+e^2/R$) will be treated as a constant (it involves nuclear vibrations that correspond to energies $\ll 1$ eV) to be added after solution of the electronic problem:

$$H_{\text{el}} = \frac{p^2}{2m} - \frac{e^2}{4\pi\varepsilon_0}\left(\frac{1}{r_1} + \frac{1}{r_2}\right) \tag{4.13}$$

Taking the protons at $\pm R/2$ along the x-axis, Schrödinger's equation becomes

$$-\frac{\hbar^2}{2m}\nabla^2\psi_n - \frac{e^2}{4\pi\varepsilon_0}\left(\frac{1}{\sqrt{(x+R/2)^2+y^2+z^2}} + \frac{1}{\sqrt{(x-R/2)^2+y^2+z^2}}\right)\psi_n = E_n\psi_n \tag{4.14}$$

Because this Hamiltonian has even parity [i.e., $H(x, y, z) = H(\pm x, \pm y, \pm z)$], its wavefunctions must be odd or even functions of the coordinates, so the form of the hydrogenic eigenfunctions at the two sites $\pm R/2$, is taken to be

$$\psi_n^s = \phi_n\left(x + \frac{R}{2}, y, z\right) + \phi_n\left(x - \frac{R}{2}, y, z\right)$$

$$\psi_n^a = \phi_n\left(x + \frac{R}{2}, y, z\right) + \phi_n\left(x - \frac{R}{2}, y, z\right) \tag{4.15}$$

Because there is one electron to distribute between these two states, a normalization factor of $1/\sqrt{2}$ is required for each function in addition to the normalization factor that already applies to the hydrogenic functions ϕ_n.

At large separations the symmetric function ψ_n^s corresponds to the atomic 1*s* function (see Appendix at end of this chapter), $\psi_{100} = (\pi a_0^3)^{-1/2} e_0^{-r/a}$. Figure 4.3*a* shows the evolution of ψ^s from two atomic hydrogen ψ_{100} states at $r = \infty$, through a bonding state, to a helium ion He^+ 1*s* wavefunction at $r = 0$. The antisymmetric function ψ_n^a evolves also from a hydrogenic form at $r = \infty$, through an antibonding function at intermediate *r*, to the ψ_{210} excited state of He^+. (ψ_n^a does *not* evolve to ψ_{200} because it has a node at the origin, characteristic of the symmetry of a 2*p* orbital.)

Note that for the bonding orbital ψ_0^s, where the electrons simultaneously occupy some common ground between the two nuclei (large overlap), the spatial wavefunction is *symmetric* $\psi_s = \phi_1(1)\phi_2(2) + \phi_1(2)\phi_2(1)$, so the *spin factor must be antisymmetric* $\chi_a = \chi_+(1)\chi_-(2) - \chi_+(2)\chi_-(1)$; that is, the spins are paired off, $s = \Sigma s_i = 0$, for a bonding orbital (singlet state). For the antibonding orbital, the two lobes of the wavefunction are separated by a node, the spatial wavefunction is antisymmetric (ψ_a), and therefore the spins in the two lobes are parallel (a triplet state).

The preceding simple assumptions about the form of $\Psi(r, s) = \psi(r)\chi(s)$ have forced obedience to the Pauli exclusion principle—as two or more electrons are confined to the same volume, the more strictly must their spins be paired ($s_1 + s_2 = 0$). This is a remarkable result. A Hamiltonian and a form of the wavefunction have been assumed that describe a situation where there is no explicit interaction between space and spin coordinates; spin does not enter the Hamiltonian at all. The result tells us that a necessary and subtle interaction between space and spin coordinates is implied by antisymmetrization. Spatially symmetric (bonding) orbitals do not support a magnetic moment whereas spatially antisymmetric (antibonding) orbitals do! The Pauli exclusion principle, Fermi–Dirac statistics, and magnetism are intimately connected.

Edward Teller (1930) calculated the energies of these states at different radii. His results are shown in Figure 4.3*b*. Molecular orbital (MO) notation is used: σ is an MO having axial symmetry about the bond direction while π and *d* denote MOs having one and two nodal planes, respectively, about the bond direction. First note that the energy of the symmetric state decreases monotonically as the two nuclei move together. This provides an attractive force to

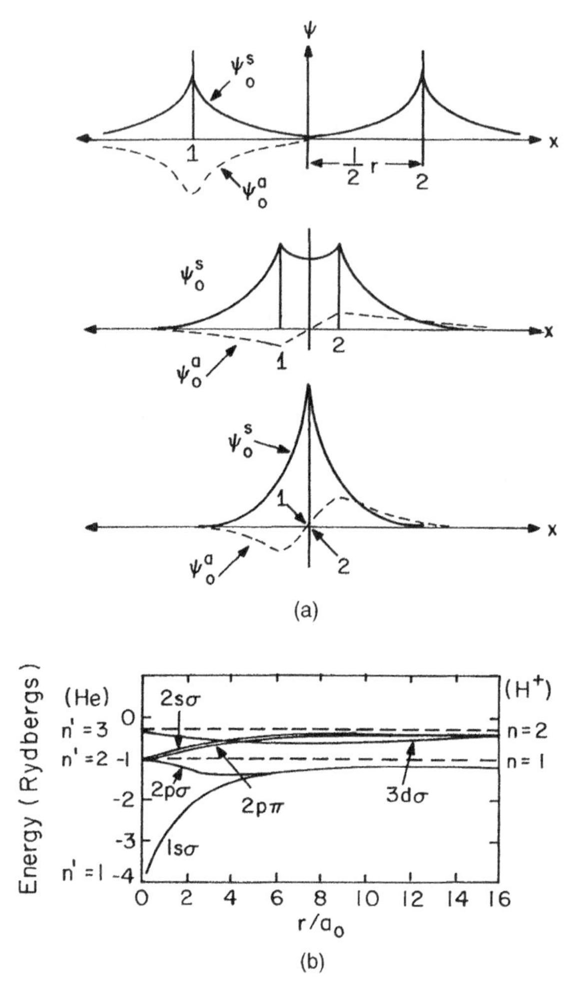

Figure 4.3 (*a*) Form of low-energy wavefunctions for H_2^+ molecule at different internuclear spacings normalized to the Bohr radius and, at $r = 0$, a He^+ atom; (*b*) electronic energy of symmetric and antisymmetric states calculated above as a function of nuclear separation r in units of Bohr radii. [After Teller (1930)].

stabilize the formation of the molecule. On the other hand, the energy of the antisymmetric state is the same at $r = \infty$ and $r = 0$; it dips slightly when the nuclei are closer to the positions of maximum antisymmetric charge density.

Note how the 1*s* hydrogenic eigenstates evolve with decreasing r: (1) the symmetric $1s(H)_s$ state evolves to a 1σ bonding orbital then to a 1*s* He function and (2) the antisymmetric $1s(H)_a$ state evolves to a σ^* antibonding orbital then to a 2 a *p* He function. The $n = 2$ excited states of hydrogen (2*s*, 2*p*) evolve as

1. The symmetric $2s(H)_s$ state evolves to a σ bonding orbital then to a $2s$(He).

2. The antisymmetric $2s(\mathrm{H})_a$ state evolves to a σ^* antibonding MO, then to $3d$ (He).
3. The $2p_z(\mathrm{H})$ state evolves to a σ^* antibonding MO, then to $2p$(He).
4. The $2p_{xy}(\mathrm{H})_{s,a}$ states evolve to π-bonding MOs.
5. The antisymmetric $2p_{xy}(\mathrm{H})_a$ states evolve to π^* antibonding MOs to $3d$(He) Symmetry dictates which functions evolve to which.

Finally, when the repulsive nuclear potential energy e^2/R is added to the electronic energies, the stability of the H_2^+ molecular ion is seen clearly in an energy minimum at $R = 2a_0$ for the $1s$ bonding state.

This sampling of concepts from quantum mechanics and bonding should make it clear that when atoms are strongly bonded (charge density concentrated between the atomic centers), antiferromagnetism is most likely. When the bonding functions have nodes between the atom centers (antibonding states), ferromagnetism becomes possible.

4.1.6 Heisenberg Hamiltonian

In 1923 Dirac showed that for the special case of *localized electrons* in orthogonal orbitals the effect of the Pauli principle was equivalent to the introduction of a term in the Hamiltonian of the form

$$-\sum_{i \neq j} \mathscr{J}_{ij} [\tfrac{1}{2} + 2\mathbf{S}_i \cdot \mathbf{S}_j] \tag{4.16}$$

The subscripts label spins on different atomic sites. This result suggested that the spin-dependent energy arising from the Pauli exclusion principle could be modeled as a spin–spin interaction in a vector model:

$$H_{\mathrm{Heis}} = -2 \sum_{i<j} \mathscr{J}_{ij} \mathbf{S}_i \cdot \mathbf{S}_j \tag{4.17}$$

This expression is universally known as the *Heisenberg Hamiltonian* even though it was first deduced by Dirac and first used extensively by Van Vleck. Clearly the Heisenberg Hamiltonian favors parallel spins if $J > 0$ and antiparallel spins if $\mathscr{J}_{ij} < 0$. The exchange interaction as described here is isotropic relative to any externally fixed spatial direction.

Slater (1953) discusses the disadvantages of this Hamiltonian in the proceedings of the first Conference on Magnetism and Magnetic Materials, which make quite interesting reading. The form of the Heisenberg Hamiltonian is now accepted to be a valid starting point for theories of magnetism in *insulators* where the magnetic electrons are quite well localized. For more extensive discussion of the complexities of exchange integrals, see the articles by Anderson (1963) and Herring (1966).

The Heisenberg exchange energy has a value of order 0.05 eV/atom (10^{-20} J/atom) to account for the fact that exchange breaks down at $T_C \approx 600$ K. Note that this interaction is spatially isotropic, i.e., the important variable is the *relative angle* between spins, not the angle of S_i with respect to some fixed direction in the material. Further, the exchange interaction has a symmetry that is clearly different from that of the dipole–dipole interaction, specifically, the potential energy of one magnetic moment in the field [Eq. (1.20)] of another.

4.2 MOLECULAR FIELD THEORY

Because of the small spatial extent of atomic wavefunctions, especially in insulators, it is usually sufficient to consider only nearest-neighbor interactions in Eq. (4.17), thus reducing the complexity of the problem significantly. Also, in this case it can sometimes be safely assumed that the exchange interaction is the same for each nearest-neighbor pair:

$$E_{\text{exch}} = -2\mathcal{J}_{ij} \sum_{i<j}^{nn} \mathbf{S}_i \cdot \mathbf{S}_j$$

This is still a difficult Hamiltonian to deal with. It can be simplified by considering the energy of a particular atom, i, interacting with its j nearest neighbors:

$$E^i_{\text{exch}} = -2JS_i \cdot \sum_j S_j \tag{4.18}$$

while for the entire material

$$E_{\text{exch}} = -\tfrac{1}{2} \sum_i E^i_{\text{exch}}$$

The form of E^i_{exch} suggests that it may be related to the energy of a magnetic moment μ_m proportional to S_i, in an effective field, H_{eff} proportional to ΣS_j. Thus, the discrete, pairwise interactions can be replaced by assuming that the magnetic moment, $\mu^i_m = g\mu_B S_i$, at site i interacts with a molecular field H_{eff} given by the net effect of the z nearest-neighbor spins:

$$E^i_{\text{exch}} = -\mu_0 \mu^i_m \cdot H_{\text{eff}} = \mu_0 g \mu_B S_i H_{\text{eff}}$$

where S_i is understood to be an eigenvalue. Comparison with Eq. (4.18) gives for the effective field:

$$H_{\text{eff}} = \frac{2\mathcal{J}}{\mu_0 g \mu_B} \sum_j S_j \cong \frac{2z\mathcal{J}}{\mu_0 g \mu_B} S_j\rangle \tag{4.19}$$

Here, the sum over z neighboring spins has been replaced by z times the average spin value $\langle S_i \rangle$. Using $M = N_v g \mu_B \langle S_j \rangle$ [Eq. (4.19)] gives

$$H_{\text{eff}} = \frac{2z\mathcal{J}}{\mu_0 N_v g^2 \mu_B^2} M$$

It can be seen that H_{eff} is the Weiss molecular field $H_{\text{mol}} = \lambda M$ described earlier if

$$\lambda = \frac{2z\mathcal{J}}{\mu_0 N_v g^2 \mu_B^2} \tag{4.20}$$

Using the value estimated earlier for the molecular field coefficient, $\lambda \approx 10^3$, $\mathcal{J}$ is calculated to be of order 2×10^{-21} J, or 0.01 eV/atom. That is, exchange interactions are weaker than the Coulomb interactions that distinguish levels of different principal and orbital quantum number (1 – 10 eV), but they are stronger that the spin-orbit interaction (10^{-4} eV/atom). Recall from the treatment of the Weiss molecular field [Eqs. (3.13) and (3.36)] that $T_C = \lambda C$, where the Curie constant $C = N_v \mu_m^2 \mu_0 / 3k_B$. Therefore, from the expression above for λ, the important relation follows:

$$T_C = \frac{2z\mathcal{J}s(s+1)}{3k_B} \tag{4.21}$$

Here, $\mu_m^2 = g^2 \mu_B^2 s(s+1)$ has been used. The values of T_C calculated from (4.21) using $\mathcal{J} = 2 \times 10^{-20}$ J are of order 10^4–10^5 K, much too large to consider this model to be quantitatively correct. Equation (4.21) is useful for the dependence it assigns to the Curie temperature and for the insight it provides to the Weiss molecular field. Mean-field models generally overestimate ordering temperatures.

The effect of the Heisenberg (or the mean-field) interaction is to align spins parallel with each other if $\mathcal{J}$ (or λ) is positive and to give an antiferromagnetic arrangement if $\mathcal{J}$ (or λ) is negative. A few simple sketches should convince the reader that the existence of true antiferromagnetism depends critically on the topology and dimension of the structure on which the spins are located.

For the purposes of macroscopic calculations it is useful to write the average value of Eq. (4.17) as

$$E_{\text{ex}} = -2 \sum_{i>j} \mathcal{J}_{ij} S^2 \cos\theta_{ij} \tag{4.22}$$

where θ_{ij} is the angle between the directions of adjacent spins $\boldsymbol{S}_i$ and $\boldsymbol{S}_j$. This semiclassical approach is valid only when θ_{ij} is small. In that case $\cos\theta_{ij}$ may be expanded as $1 - \theta_{ij}^2/2!\cdots$ so that

$$E_{ex} = \mathcal{J}S^2 \sum_{i>j} \theta_{ij}^2 + \text{constant} \tag{4.23}$$

This equation indicates that the exchange energy between two adjacent spins is proportional to the square of the angle between them. In terms of continuous variables, this angular change θ_{ij} may be expressed as $a\,\partial\theta/\partial x$, where a is the distance between spins. Equation (4.23) may then be written as

$$\frac{E_{ex}^{ij}}{V} = A\left(\frac{\partial\theta_{ij}}{\partial x_{ij}}\right)^2 = A\left(\frac{\nabla M}{M}\right)^2 \tag{4.24}$$

where $A = s^2a^2\mathcal{J}N_v'/2$ is called the *exchange stiffness constant* and N_v' is the number of nearest-neighbor atoms per unit volume; A is a macroscopic measure of the stiffness of coupling in the spin system. The units of $\mathcal{J}$ are those of energy while A is measured in energy/length. A value $A = 1$ to 2×10^{-11} J/m (1–2×10^{-6} erg/cm) is typical for most ferromagnets.

Equation (4.24) describes only the spatially varying part of the exchange energy, which is assumed small compared to the constant term. The coefficient A describes the low-temperature spin stiffness and hence can be used to characterize long wavelength spin waves and low-temperature demagnetization. The stiffer the spin system, the more slowly it demagnetizes as temperature increases. Conversely, a small value of A corresponds to easy thermal demagnetization. Demagnetization by spin waves is characterized by a magnetization temperature dependence of $T^{3/2}$. The constant term as well as the spatially dependent part of E_{ex} determine the Curie temperature at which long-range spin order vanishes.

Calculations for specific lattices using Eq. (4.21) give

$$\begin{aligned} \mathcal{J} &= 0.54\,kT_C \quad && \text{for simple cubic, } s = 1/2 \\ \mathcal{J} &= 0.34\,kT_C \quad && \text{BCC, } s = \tfrac{1}{2} \\ \mathcal{J} &= 0.15\,kT_C \quad && \text{BCC, } s = 1 \end{aligned} \tag{4.25}$$

Note that for Fe, $T_C = 1044$ K, so from the BCC value above $s = 1$, $\mathcal{J} \approx 2 \times 10^{-21}$ J gives $A = 7 \times 10^{-12}$ J/m.

It is worth distinguishing again the situations in which an experiment measures $|\mu_m| = g\mu_B[s(s+1)]^{1/2}$ from those in which it measures $\mu_m|_z = g\mu_B m_s$. Paramagnetic susceptibility, the exchange interaction, $\mathcal{J}\mathbf{S}_i \cdot \mathbf{S}_j$ (which gives rise to the Curie temperature), and magnetic contributions to entropy all depend on the *magnitude* of the magnetic moment, called the effective paramagnetic moment, $p_{eff} = g\mu_B[s(s+1)]^{1/2}$ or $g\mu_B[j(j+1)]^{1/2}$. On the other hand, the saturation magnetization achieved in a ferromagnet is a measure of the projection of that moment on a specific direction. In this case, it is the z

component of $\langle \boldsymbol{J} \rangle$ that is important, so the ferromagnetic moment is $\mu_m = g\mu_B m_j$.

4.3 MAGNETISM IN OXIDES: SUPEREXCHANGE

Magnetism in oxides is treated before magnetism in metals because the electronic states in oxides are more atomlike. The electrons in oxides can be described by the same quantum numbers that apply to isolated atoms. It will be seen that this is not the case in metals. Hence it is appropriate to discuss ferromagnetism in oxide solids from an atomic point of view.

The Heisenberg exchange interaction, $H = -2\mathscr{J}\Sigma \boldsymbol{S}_i \cdot \boldsymbol{S}_j$, is used to describe the tendency of neighboring localized magnetic moments to align parallel ($\mathscr{J} > 0$) or antiparallel ($\mathscr{J} < 0$) to each other in a magnetic material. The strength of the exchange interaction depends on orbital overlap, as will be seen shortly. The Heisenberg form of exchange applies to some localized systems, but fails even for ionic, oxide magnets such as $FeO \cdot Fe_2O_3$. The magnetic behavior of transition metal oxides is governed by a less familiar form of exchange, but one that is easier to explain. This case is treated here.

Consider MnO, whose rocksalt crystal structure has the antiferromagnetic structure shown in Figure 4.4. The spins on some nearest-neighbor transition metal ions are parallel, (*a*) and (*b*); those on other nearest neighbors are antiparallel, (*b*) and (*c*). But next nearest neighbors always have antiparallel spins: (*a*) and (*c*) or (*c*) and (*d*). Thus the spins within a given (111) plane are parallel to each other and antiparallel to those on the two adjacent (111) planes. The preferred direction of magnetization of a given sublattice in MnO is along the [100] crystal direction; other preferred directions and even more complex spin structures are possible in different antiferromagnetic materials.

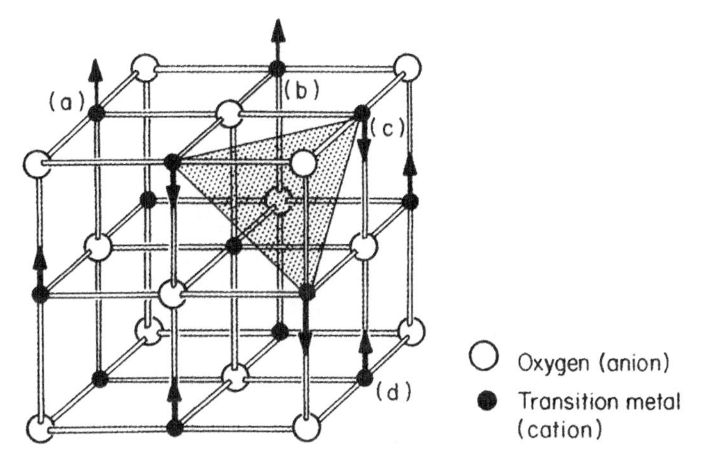

Figure 4.4 Rocksalt structure common to transition metal monoxides. The spin structure of MnO is indicated by arrows on the transition metal ions.

There must be a magnetic interaction that couples second nearest neighbors so that their spins have opposite orientations. Oxygen (or F, Cl, Te, etc.) atoms always lie midway on the line between next nearest neigbors. Next-nearest neighbor sites, two lattice constants apart, are too far apart to be involved in Heisenberg exchange; there is just not enough wavefunction overlap. Therefore some form of mediated or "indirect" exchange interaction must be operating. A model proposed by Néel, called superexchange, and formulated in detail by Anderson (1963), describes this effect quite well. Superexchange operates in many ionic oxides and couples localized, usually 3*d*, moments.

4.3.1 Superexchange

A simple picture that adequately describes superexchange is shown in Figure 4.5. Two transition (T) metal ions (e.g., *a* and *c* in Figure 4.4) are separated by a *p* ion (represented here by oxygen). The *p* orbital, which is filled in the ground state, can exchange an electron with each of the adjacent 3*d* orbitals ($d_{x^2-y^2}$ MOs are shown here). Thus, the bonding is mostly ionic, T^{2+} and O^{2-}, but some hopping is allowed. The positive and negative phase parts of the wavefunction for each orbital are represented by solid and dashed lines, respectively. The doubly occupied p_x orbital has two electrons of *opposite spin.*

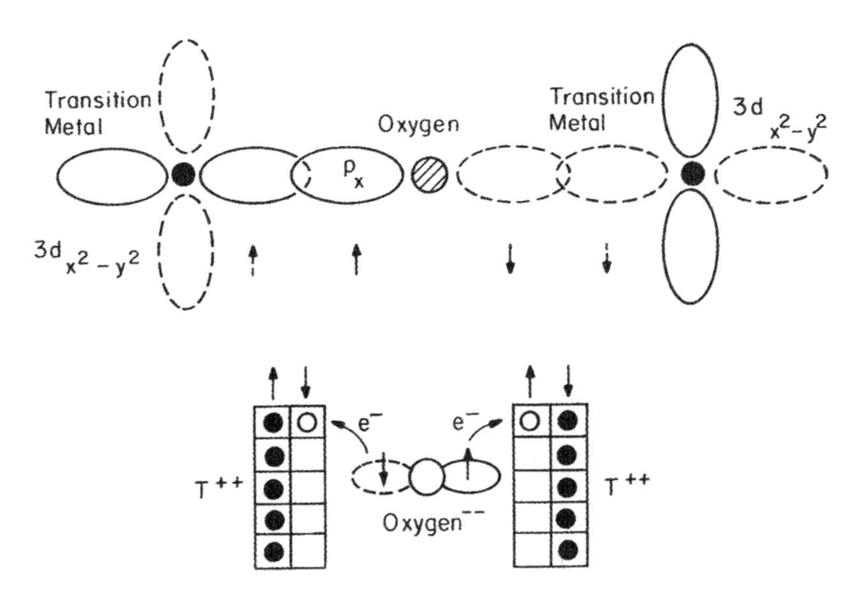

Figure 4.5 Schematic of *d* and *p* orbitals important to the superexchange interaction. When the oxygen *p* orbital is in its ground state (two electrons), the electrons there must be of opposite spin. When an electron is exchanged or shared by the overlapping orbitals between two sites, the spin is conserved as the electron hops from the *p*-like to the *d*-like orbital. The result is that the two *T* ions must have opposite spin directions.

It is not enough for this to be the ground state; the $3d$ and p orbitals must exchange electrons; that is, some of the excited states of the system must be partially occupied. A simpler electronic state picture is shown in the lower part of Figure 4.5. When one of the p_x electrons is excited into an empty d state to form a $(p-d)$ σ or σ^* bond, it leaves behind an electron of *opposite* spin, which may be exchanged with the d states of the other T species coupled to this p_x orbital. If the two d ions have the same ground-state electronic configuration, the net effect is an effective antiferromagnetic coupling between the two T ions.

At its root, the superexchange interaction is a consequence of the different symmetry of p and d states. Notice that one of the three p orbitals represents two electron states (spinup and spindown) whose spatial wavefunctions have 360° (single-fold) rotational symmetry. One of the five d orbitals represents two states that have 180° rotational symmetry. Thus a p state has spinup and spindown electrons concentrated in lobes 180° apart while a d state has opposite spin concentrations every 90° rotation about the ion. These symmetry considerations, plus the fact that electron hopping conserves spin orientation, produce the antiferromagnetic spin coupling between nearest neighbor transition metal ions.

Superexchange applies to transition metal monoxides of the rocksalt structure as well as to many of the more complex spinel, garnet, and perovskite oxides.

4.3.2 Iron Oxides

The physical properties of the most frequently encountered iron oxides are listed in order of increasing oxidation state in Table 4.1.

The electronic states responsible for magnetism in transition metal oxides are well-defined states, little affected by their environment once their valence is established. A fixed number of valence electrons per atom is transferred to the

TABLE 4.1 Structure and Magnetic Properties of Iron Oxides

Iron Oxide	Structure	Magnetic Structure	$\mu_m(\mu_B)$	σ(emu/g)	T_N(K)
FeO	Rocksalt	Antiferro-magnetic	4	—	—
γ-Fe_2O_3 maghemite	Metastable defect spinel	Ferrimag-netic	5.0	74	863–945
$FeO \cdot Fe_2O_3$ magnetite	Spinel	Ferrimag-netic	4.1	84	850
α-Fe_2O_3 hematite	Corundum (hexagonal)	Antiferro-magnetic	5	0	—

anion lattice leaving cations with atom-like electron configurations whose moments can be estimated from Hund's rules. For example, the iron ions in Fe_2O_3 have a valence of $+3$ and an electronic configuration $3d^5$. The Fe^{3+} moment is $\mu_m \approx 5\mu_B$. The valence electron configuration of Fe^{2+} in FeO is $3d^6$ and its moment is $\mu_m \approx 4\mu_B$ because

$$\mu_m = (n^+ - n^-)\mu_B = (5 - 1)\mu_B$$

The most stable phase of Fe_2O_3, hematite designated α-Fe_2O_3, has the corundum (hexagonal) structure with linear Fe^{3+}–O^{2-}–Fe^{3+} ligands. It is antiferromagnetic with the moments of the two Fe ions opposing each other. Magnetite, Fe_3O_4, more accurately written as $FeO \cdot Fe_2O_3$, assumes the more complex spinel structure, and it contains one Fe^{2+} ion in addition to the two trivalent Fe ions per formula unit. Therefore, it cannot be antiferromagnetic. Maghemite is a defect spinel and is the magnetic oxide used often in recording tapes.

4.3.3 Spinel Structure

Spinel ferrites such as Fe_3O_4 have a complex cubic unit cell, subunits of which are shown in Figure 4.6. Oxygen atoms occupy four of the vertices of each of the eight small cubes which make up the unit cell. There are two distinct types of sites which the transition metal ions can occupy, tetrahedral (A) sites and octahedral (B) sites. Of the 64 possible tetrahedral sites in the spinel unit cell, only eight are occupied. Of the 32 possible octahedral sites, only 16 are occupied. Therefore, in each formula unit, Fe_3O_4, one A site and two B sites are occupied. The A-occupied sub-cells are tetrahedrally arranged within the unit cell and the B-occupied cells are arranged in a complementary tetrahedron (Fig. 4.6).

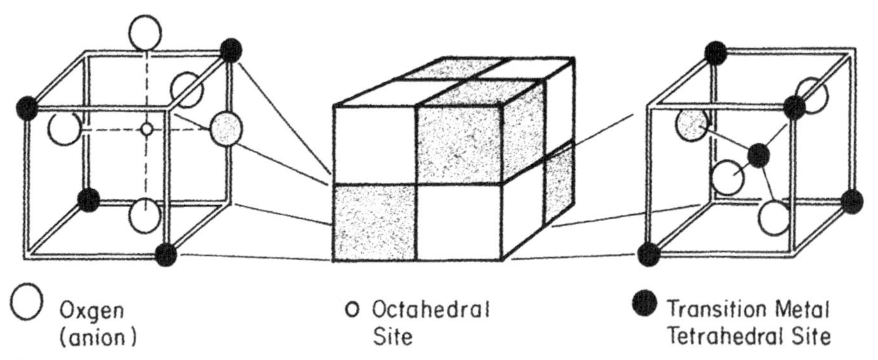

Figure 4.6 Spinel structure showing transition metal sites that are octahedrally and tetrahedrally coordinated by oxygen anions. The shaded oxygen atom is shown in two subcells; it links A and B sites.

It is observed that the moment of Fe_3O_4 is approximately $4\mu_B$ per formula unit. Exactly how the three moments interact to give this may not be immediately clear, but one might explain it by assuming the two trivalent moments cancel each other out as in antiferromagnetic Fe_2O_3 and the net moment is then that of the divalent iron in FeO, namely, $4\mu_B$. This turns out to be correct; let us see why.

Exchange Interactions Recall that the magnetic exchange interactions between $3d$ cations in ionic solids are antiferromagnetic superexchange couplings. The exchange is mediated by an intervening anion p orbital and is therefore strongest in structures for which the three ions are collinear. In the spinel structure, the A—O—B bond angle is 125° (Fig. 4.6), the B—O—B angle is 90° and there are no A—O—A bonds. Hence, the strongest superexchange interaction is the 125° antiferromagnetic A—B coupling. B sites exhibit a weak 90° antiferromagnetic coupling among themselves. These interactions are most economically accommodated if all A moments are parallel to each other and antiparallel to all B moments. It so happens that in Fe_3O_4 the divalent iron occupies one of the B sites with a trivalent Fe and that the other trivalent iron occupies an A site (see Table 4.2). Thus the trivalent moments are on different sublattices and cancel each other out, leaving the divalent Fe moment, $4\,\mu_B$ per formula unit.

The preference of the Fe^{3+} ion rather than Fe^{2+} for the smaller A site seems natural based on the smaller size of the Fe^{3+} ion. However, site preference was first thought to be determined purely by valence: T^{2+} on a fourfold coordinated site (A) and T^{3+} on a sixfold coordinated site (B). Hence Fe_3O_4 was called an "inverse" ferrite, because the divalent ion was not where it was considered "normal" to be, namely, on a sixfold coordinated site.

4.3.4 Ferrite Moments

Let us consider the magnetic moment variation in a series of ferrites based on magnetite, Fe_3O_4, which have the spinel structure.

When transition metal species T are substituted for iron in the spinel ferrite, magnetite, technologically useful magnetic ferrites result, such as Ni ferrite and

TABLE 4.2 Possible Distributions of Two Fe^{3+} Ions and One Fe^{2+} Ion in Fe_3O_4 among the One A Site and Two B Sites that Can Be Occupied in the Spinel Structure

"Inverse"		"Normal"	
A (tetrahedral)	B (octahedral)	A (tetrahedral)	B (octahedral)
$Fe^{3+}+5\mu_B$	$Fe^{2+}Fe^{3+}-(4+5)\mu_B$	$Fe^{2+}+4\mu_B$	$2Fe^{3+}2(-5)\mu_B$

[a] The observed magnetic moment of $4\mu_B$/FU suggests the "inverse" occupation applies.

Mn ferrite. The valence of T = Cu, Ni, Co, Mn substituents for Fe in Fe_3O_4 is generally found to be close to 2^+. The moment on Ni^{2+} ($3d^8$) is $2\mu_B$ and on Mn^{2+} ($3d^5$), $5\mu_B$, and so on. With three different moment-bearing magnetic ions in Ni ferrites or Mn ferrites, the variation of magnetization with composition should provide an interesting test of an atomic model of ferrite moments. It turns out that for full T substitution for Fe^{2+}, $TO \cdot Fe_2O_3$, the net moment per formula unit is approximately 1.3, 2.2, 3.3, 4.2, and $5.0\mu_B$ for T = Cu, Ni, Co, Fe, and Mn, respectively. So in all of these compounds, the moments of the two remaining Fe^{3+} ions still appear to be canceling each other out. The net moment increases roughly as the moment of the divalent T metal (Fig. 4.7 ordinate). It would be of interest if there were a way to unlock some of the potential ten Bohr magnetons of the two antiferromagnetically coupled Fe^{3+} ions.

Attention is now given to the thought-provoking ferrite magnetic moment data first published in 1951 by Guillaud, and added to by Gorter in 1954, in which Zn^{2+} is substituted for T^{2+} (Figure 4.7). The moments of the $TO \cdot Fe_2O_3$ compounds just discussed are shown on the ordinate.

Note that all the Zn-substituted ferrites show an initial *increase* in magnetic moment as the magnetic T species is replaced by zinc (Zn^{2+} has no magnetic moment, Zn^{3+} has one Bohr magneton)! How can this apparent creation of magnetism by zinc substitution be understood?

Before these questions can be answered, a better understanding is needed of the site preferences of transition metal ions of different valences.

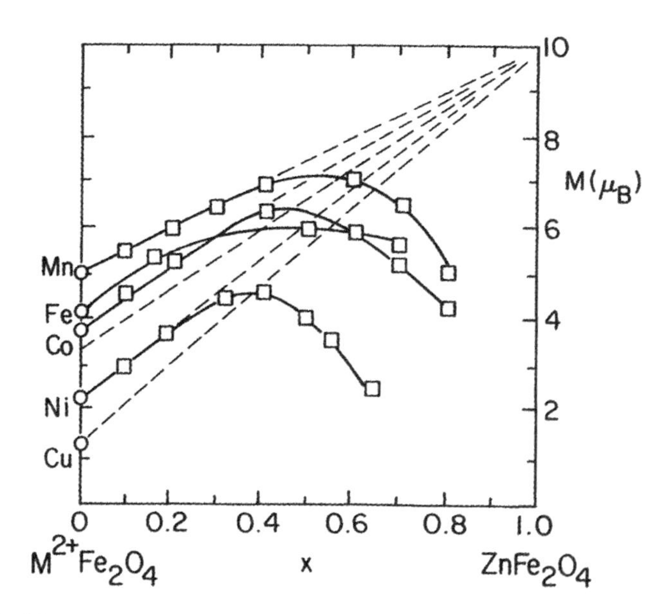

Figure 4.7 Magnetic moments in transition metal–zinc ferrites as T = Cu, Ni, Co, and so on are substituted for divalent iron (Guillaud 1951, Gorter, 1954).

4.3.5 Site Preference

The original expectation of "normal" and "inverse" site selection based on ionic valence is outlined in Table 4.2. It is now realized that other factors such as cation size, crystal field (see Chapter 6), and valence play a role in site selection. Figure 4.8 summarizes the calculated and observed site preference energies for various cations in spinel ferrite. The sequence of cations on the abscissa is chosen simply in terms of increasing observed or calculated octahedral site preference. Consider magnetite which has two Fe^{3+} ions and one Fe^{2+} ion. From Figure 4.8, it can be seen that the Fe^{3+} ion has a much stronger tetrahedral (inverse) site preference than Fe^{2+}. One of the Fe^{3+} ions occupies the A site and relegates to the other two ions (Fe^{3+} and Fe^{2+}) to the two B sites per formula unit. Hence magnetite has the "inverse" site occupations shown in Table 4.2.

It is important to notice that Zn^{2+} has by far the strongest preference for A site occupation of any of the ions studied. The data in Figure 4.8 can be used to determine the distribution of a given set of transition metal ions among the A site (one ion) and the two B sites (two ions).

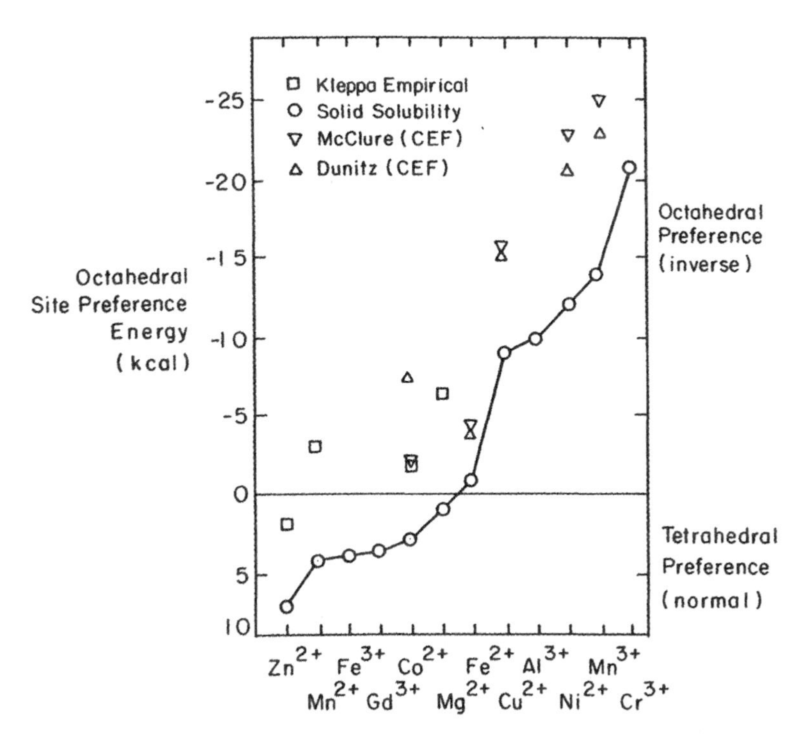

Figure 4.8 Cation site preference energy for various transition metal ions in A or B sites. CEF stands for crystalline electric field (Navrotsky and Kleppa 1968).

Consider divalent transition metal substitutions for iron:

$$Fe_3O_4 = Fe^{2+}O^{2-} \cdot (Fe^{3+})_2(O^{2-})_3 \rightarrow T^{2+}O^{2-} \cdot (Fe^{3+})_2(O^{2-})_3$$

Figure 4.8 indicates that only Mn^{2+} and Zn^{2+} prefer the A site more than Fe^{3+} does. This explains the site occupations for TO–Fe_2O_3, MnO-Fe_2O_3, and ZnO–Fe_2O_3, shown in Table 4.3.

Considering that the A and B sublattices are antiferromagnetically coupled, the site occupations shown in Table 4.3 explain the simple transition metal ferrite moments shown along the ordinate in Figure 4.7. The Fe^{3+} moments cancel for the compounds shown in the first line leaving only the T^{2+} moments. In the case of Mn ferrite, the iron moments produce a B sublattice moment of $10\mu_B$/FU but the Mn^{2+} moment of $5\mu_B$/FU on the oppositely magnetized A sublattice reduces the net moment to $5\mu_B$/FU. This is also consistent with Figure 4.7.

Note the initial increase in moment observed with zinc substitution in TO·Fe_2O_3, specifically, $(T^{2+})_{1-x}(Zn^{2+})_x(Fe^{3+})_2O_4$ (Fig. 4.7). Zn^{2+} ions show a stronger normal tendency than any of the ions shown, particularly stronger than Fe^{3+}. Thus although Zn^{2+} substitutes for T^{2+}, it preferentially occupies A sites, driving the A site Fe^{3+} ions to B sites (see Table 4.4).

Note that the total Fe^{3+} content remains as two ions per FU but the iron ions are forced to the B sites as x increases. Taking the sign of the B site moments to be positive, the net moment per FU becomes

$$\frac{\mu_m}{\text{FU}} = 5(1 + x) + \mu_T(1 - x) - 5(1 - x)$$

$$= 10x + \mu_T(1 - x) \tag{4.26}$$

where μ_T is the moment of the divalent transition metal ion for which zinc is substituted. At $x = 0$ Eq. (4.26) describes the moment of the divalent transition metal species as observed in Figure 4.7. With increasing x, the net moment is predicted to increase linearly toward $10\,\mu_B$/FU. These lines describe the data fairly well until $x \approx 0.3$–0.4. At that point most of the magnetism is on the B

TABLE 4.3 Site Distribution in $T^{2+}Zn^{2+}2Fe^{3+}O_4$ Based on Data in Figure 4.8

A Site	B Site[a]
Fe^{3+}	$T^{2+}Fe^{3+}$
Zn^{2+}	$2Fe^{3+}$
Mn^{2+}	$2Fe^{3+}$

[a] T = Cr, Fe, Co, Ni, Cu.

TABLE 4.4 Site Distribution of Ions and Consequent Moment Distribution in Zinc-Substituted Transition Metal Ferrites

	A Site	B Site
Ion	$(Zn^{2+})_x(Fe^{3+})_{1-x}$	$(T^{2+})_{1-x}(Fe^{3+})_{1+x}$
Moment	$5(1-x)$	$\mu_T(1-x) + 5(1+x)$

sites and relatively few A sites have magnetic moments. At large Zn concentrations the antiferromagnetic B–B interactions start flipping those B site Fe^{3+} moments that are farthest from an A moment. The net moment starts to decrease. It is also a factor in the data that the A–B exchange interaction is weakened as Zn content increases.

To summarize, it has been shown that for oxides or ionic solids, the valence electronic configuration is a good measure of the magnetic moment localized on a particular species. These local moments can be added vectorially to get the net magnetic moment per FU provided something is known about site preference and about the exchange interactions between different sites. The temperature dependence of magnetization in such a two-sublattice ferrimagnetic material is now considered.

4.4 TEMPERATURE DEPENDENCE OF MAGNETIZATION IN FERRIMAGNETS

If one measures the temperature dependence of magnetization in a ferrimagnetic material, the behavior is seldom like that of a Brillouin function that describes $M_s(T)$ quite well in many ferromagnetic systems. The behavior shown in Figure 4.9 for ferrimagnetic nickel–iron vanadates shows compensation temperatures where the net magnetization vanishes at a temperature below the Néel temperature where the individual sublattice magnetizations vanish. This suggests that perhaps the two spin sublattices in a ferrimagnetic lose their magnetism at different rates as temperature increases, thus complicating the net $M(T)$ behavior. A simple adaptation of the mean field theory outlined in Sections 3.8 and 4.2 can be used to describe and predict the temperature dependence in magnetic systems comprised of two interacting magnetic sublattices.

Néel (1948) assumed that the molecular field [Eq. (3.44)] acting on the spins in one sublattice, A or B, could be written as the sum of magnetic interactions with other spins on the same sublattice plus that with the spins on the other sublattice:

$$H_{\text{eff}}^A = \lambda_{AA}\mathbf{M}_A + \lambda_{AB}\mathbf{M}_B = \lambda_{AA}M_A - \lambda_{AB}M_B \tag{4.27}$$

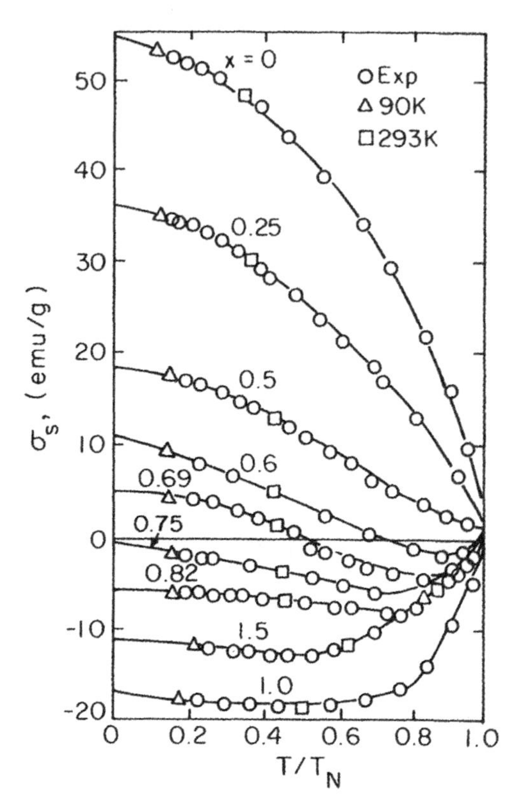

Figure 4.9 Temperature dependence of the magnetization in $NiFe_{2-x}V_xO_4$. [After Blasse and Gorter (1962)].

and

$$H_{\text{eff}}^{B} = \lambda_{BB} M_B - \lambda_{AB} M_A \tag{4.27$'$}$$

These definitions can be made to reflect a particular crystal structure (Smart 1966), but here, they are kept structure neutral. Each of the three molecular field coefficients could be related by Eq. (4.20) to exchange interactions within each sublattice, $\mathcal{J}_{AA}$ and $\mathcal{J}_{BB}$, and between the two sublattices, $\mathcal{J}_{AB} = \mathcal{J}_{BA}$.

The field-dependent and temperature-dependent magnetization in each sublattice is then described by a Brillouin function with an argument appropriate to each sublattice:

$$M^A(H, T) = M_s^A B_J^A \left[\frac{\mu_0 \mu_m^A (H + H_{\text{eff}}^A)}{k_B T} \right] \tag{4.28}$$

$$M^B(H, T) = M_s^B B_J^B \left[\frac{\mu_0 \mu_m^B (H + H_{\text{eff}}^B)}{k_B T} \right] \tag{4.28$'$}$$

These two equations are solved simultaneously for the temperature dependence of the magnetizations of the two sublattices given (or assuming) values for the three exchange interactions. The temperature dependence of the net magnetization, $M^A + M^B$, then takes on characteristic forms depending on the values of the exchange constants, $\mathcal{J}_{AA}$ and $\mathcal{J}_{BB}$.

The paramagnetic case is illustrated first. The results are more clearly expressed in terms of Curie constants for the two sublattices

$$C_A = \frac{N_v^A \mu_0 \mu_{m^2}^A}{k_B} \quad \text{and} \quad C_B = \frac{N_v^B \mu_0 \mu_{m^2}^B}{k_B}$$

where N_v^i are the volume densities of the number of spins on the ith sublattice and $\mu_m^i = g^i \mu_B \sqrt{j(j+1)}$ is the magnetic moment per transition metal in the ith sublattice.

Instead of the one-sublattice, ferromagnetic form of the Curie–Weiss law, $\chi = C/(T - \theta)$, the ferrimagnetic result is

$$\frac{1}{\chi} = \frac{T - \theta}{C_A + C_B} - \frac{\zeta}{T - \theta'} \tag{4.29}$$

where

$$\zeta = \frac{C_A C_B}{C_A + C_B} \lambda_{AB}^2 [C_A(1 + \lambda_{AA}) - C_B(1 + \lambda_{BB})]^2,$$

$$\theta = \frac{C_A C_B}{C_A + C_B} \lambda_{AB} \left(\lambda_{AA} \frac{C_A}{C_B} - \lambda_{BB} \frac{C_B}{C_A} - 2 \right)$$

and

$$\theta' = \frac{C_A C_B}{C_A + C_B} \lambda_{AB} (\lambda_{AA} + \lambda_{BB} + 2)$$

Equation (4.29) describes Curie-Weiss-like behavior with a temperature intercept that is shifted by the new factor θ', which can be positive or negative depending on the signs and magnitudes of λ_{AA} and λ_{BB}. This result is shown in Figure 4.10. Unlike the ferromagnetic case, where $1/\chi$ is linear in T, for the ferrimagnet $1/\chi$ is not linear.

The Néel temperature is defined from the positive root of the inverse susceptibility in Eq. (4.28) as

$$T_N = \frac{\lambda_{AB}}{2} \{C_A \lambda_{AA} + C_B \lambda_{BB} + [(C_A \lambda_{AA} - C_B \lambda_{BB})^2 - 4 C_A C_B]^{1/2}\} \tag{4.30}$$

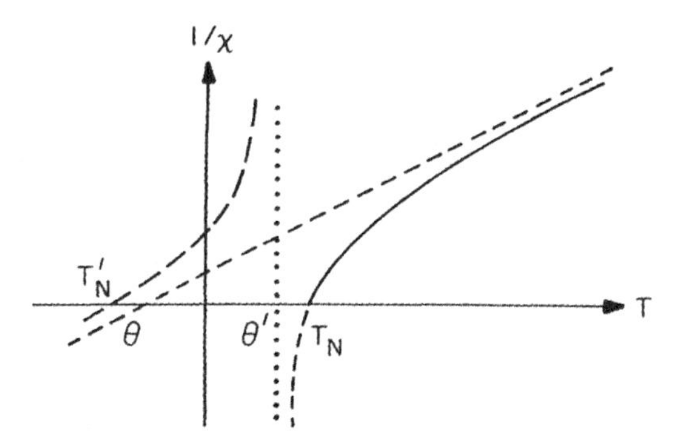

Figure 4.10 Graph of Eq. (4.29) showing the curvature in the inverse susceptibility and the Néel temperature as well as the other temperatures defined for the ferrimagnetic susceptibility.

Figure 4.11 shows the various regions in the space of the two intrasublattice molecular field coefficients in which the paramagnetic, ferrimagnetic, and ferromagnetic solutions are found.

In the strongly magnetic regime, the net magnetization can show the unusual shapes depicted in Figure 4.11 for different combinations of the

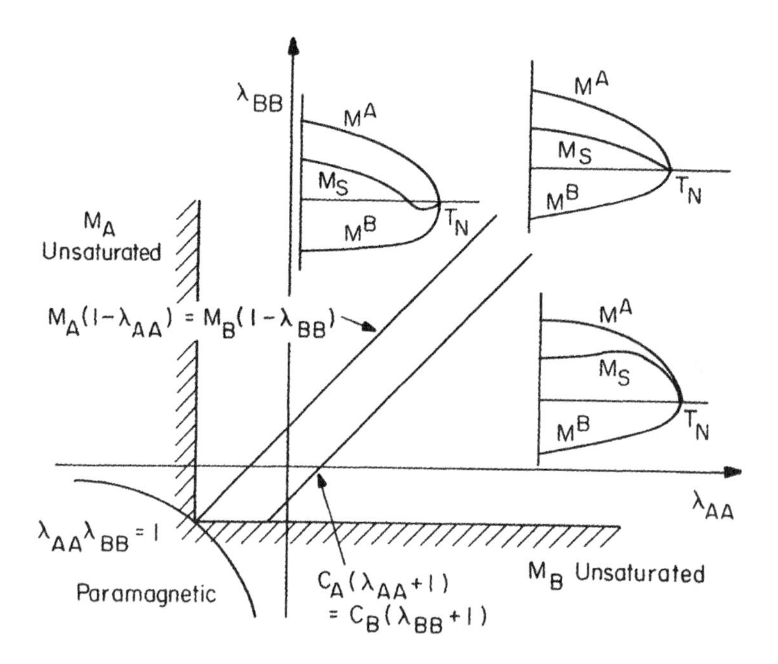

Figure 4.11 Diagram in λ_{AA}–λ_{BB} space where different forms of net magnetization temperature dependence are observed in ferrimagnets.

exchange parameters λ_{AA} and λ_{BB}. A particular value of λ_{AB} and $M_s^A > |M_s^B| > 0$ have been assumed in Figure 4.10. Clearly when $|\lambda_{AA}| > |\lambda_{BB}|$, the Brillouin function for M_A is more square than that for M_B and vice versa for $|\lambda_{AA}| < |\lambda_{BB}|$. Consequently, if $|\lambda_{AA}| > |\lambda_{BB}|$ the net magnetization, $M^A + M^B$, can *increase* with increasing temperature before vanishing at T_N or if $|\lambda_{AA}| < |\lambda_{BB}|$, it can show compensation points at which the net magnetization vanishes below T_N even though each sublattice remains magnetically ordered. Beyond the dashed lines at $\lambda_{AA} < -2$ or $\lambda_{BB} < -1$, noncollinear moment arrangements are possible (Yafet and Kittel 1952, Lyons et. al. 1961).

4.5 SUMMARY

In this chapter, the quantum mechanical origin of magnetism in insulator and ionic solids has been discussed. It began with a description of the factors affecting the energies, wavefunctions, and spin configuration in a two-electron system, namely, electron–electron interactions and the Pauli exclusion principle. The latter was seen to impose a spin structure on the states of two indistinguishable particles, that is, ferromagnetic (symmetric in spin) for antisymmetric spatial wavefunctions and antiferromagnetic (spin antisymmetric) for symmetric spatial wavefunctions. These principles were illustrated for the hydrogen molecule ion, H_2^+.

The Heisenberg Hamiltonian was defined and expressed as a mean-field exchange interaction to give substance to the concept of the Weiss molecular field.

The chapter ended with a treatment of magnetism in insulators, explaining superexchange, site selection, and moment variations with composition and temperature in spinel ferrites.

APPENDIX: BRIEF REVIEW OF QUANTUM MECHANICS AND SCHRÖDINGER'S EQUATION

4A.1 Eigenvalue Equations

In Newtonian mechanics, algebraic or differential equations can be solved for the position or momentum of a body subjected to known forces. With enough information, a definite answer can be obtained. Witness the accuracy with which a satellite can be put in orbit or a *Voyager* spacecraft can be sent to a pinpoint rendezvous with a planet millions of miles away.

Schrödinger's equation is not a simple algebraic equation but a differential equation of a particular kind. It is an eigenvalue equation, that is, one in which an operation on a function, $\phi(x)$, returns that function times a simple numerical factor:

$$N\phi(x) = n\phi(x). \tag{4A.1}$$

The number n is the eigenvalue, or quantum number, corresponding to the operator N. If the operator is the energy operator, the eigenvalue is the energy of the allowed state. If the operator is the momentum operator, the eigenvalue is the momentum of the state described by the eigenfunction, $\phi(x)$. For example, the momentum operator, $p_x = -i\hbar(\partial/\partial x)$, applied to a free electron wavefunction, $\phi = A\exp[i\boldsymbol{k}\cdot\boldsymbol{x}]$, gives the eigenvalue for momentum of a plane wave:

$$p_x\phi = -i\hbar\frac{\partial}{\partial x}\phi = \hbar k_x\phi = \frac{h}{\lambda}\phi \tag{4A.2}$$

In other words, the value of the momentum for a plane wave is $p_x = \hbar k_x = h/\lambda$. This may be recognized as the de Broglie relation, which expresses the wave-particle duality of nature and whose effects are observed on the quantum level (waves have momentum and particles are characterized by a wavelength). Most eigenvalues in quantum mechanics contain Planck's constant, the quantum of action (energy times time), $h = 6.6 \times 10^{-34}$ J/s or $\hbar = 1.06 \times 10^{-34}$ J/s. This indicates that measurable quantities may take on discrete values, albeit values that on human scales are so close together that they appear continuous.

An operator equation can be considered as a statement of the results of a particular quantum mechanical measurement. $\mathcal{H}\psi = E\psi$ says that if you measure the energy of a system characterized by ψ, you get the result, E. ($\mathcal{H}$ stands for the Hamiltonian operator, which is the operator for the total energy: kinetic energy, T plus potential energy V).

In quantum mechanics, it is not possible to speak with certainty about the result of a measurement; only the *probability* of a given result can be given. The probabilistic weighting of various outcomes of a measurement is called the *expectation value*. The expected value of a given measurement or variable is obtained by operating with the appropriate operator on a suitable wavefunction for the system [e.g., Eq. (4A.2)]. The wavefunction is then removed from the expectation value by making use of its normalization condition. For the plane wave considered above, the expectation value of the momentum $\langle p\rangle$ is

$$\langle p_x\rangle = -i\hbar\int\phi^*\frac{\partial}{\partial x}\phi dx = \frac{h}{\lambda}\int\phi^*\phi dx = \frac{h}{\lambda} \tag{4A.3}$$

The operator in the integral operates to the right, on ϕ; if ϕ is an eigenfunction of the operator, this process gives the number h/λ times ϕ. Because h/λ is a number, it can be removed from the integral which is then evaluated. The integral $\int\phi^*\phi dx$ is the probability density of finding the particle in the region over which the integral was done. If ϕ has been normalized, the integral is unity, that is, there is unit probability of finding the particle somewhere. (The probability of finding the plane wave particle in a small volume of space is small because the particle is uniformly spread over all space.) More compact ways of writing Eq. (4A.3) include $\langle \boldsymbol{p} \rangle = (\phi^*, \boldsymbol{p}, \phi) = \langle \phi^* | \boldsymbol{p} | \phi \rangle = h/\lambda$. The brackets around the momentum operator symbolize both the average value in the conventional sense but, more importantly, imply the integration in Eq. (4A.3).

Much of the toil of a quantum mechanism may be outlined as follows: (1) find the appropriate operator, the Hamiltonian, for the system of interest; then (2), pick a complete, orthogonal set of trial wavefunctions (the basis set) that is expected to represent the solution, and use only a few of them. The operator equation is then solved, preferably by reducing it to a matrix equation. Solution of the matrix equation consists of diagonalizing the matrix (which is easier for sparse matrices). The solution may be improved by including more wavefunctions. The method is anything but mechanical; more art and intuition is involved than might be expected. It is important to understand some frequently encountered results.

4A.2 Uncertainty and Wavepackets

It is often sufficient to begin solving a problem by considering plane wave solutions (which have equal probability everywhere). This method is used occasionally in the present text. At other times it is more important to consider a wavepacket that might describe an electron confined to a specific volume. By confining the electron to a region of, say, $\Delta x = a$, it is not possible to specify its position any better than Δx. The possible values of its wavelength are limited to $2a$ and smaller: $\lambda < 2a$. This limits its possible values of $k = 2\pi/\lambda$, to $k > \pi/a$. This linear dependence of Δx on a and the inverse depedence of Δk on a, requires that their product satisfy $\Delta x \Delta p_x \geqslant \hbar \Delta x \Delta k_x \approx h/2$. Heisenberg recognized the implications of this relation and formulated the uncertainty principle that bears his name. In its more exact form, it states that the product of the uncertainty of two conjugate variables (x and p are conjugates and E and t also are) must be greater than or equal to $h/2$:

$$\Delta x \Delta p_x \geqslant \frac{h}{2} \qquad (\text{or } \Delta x \Delta k \geqslant \pi)$$

$$\Delta E \Delta t \geqslant h \qquad (\text{or } \Delta \nu \Delta t \geqslant 1) \tag{4A.4}$$

These relations are often useful in placing bounds on the values of the conjugate variables of a particle. For example, if you can do a measurement that tells you the position of a particle with an uncertainty $\Delta x = a$, then you cannot know the momentum of that particle with an uncertainty less than $\Delta p = h/2a$.

4A.3 Schrödinger Equation

One way to confine a wave to a location of width Δx is to force the wavefunction to have zero amplitude at $x = \pm a$. Combination of two waves having values of λ differing by $2\Delta x$ produces beats in the total wave with nodes separated by Δx. In one dimension, such a wave may be written as a sum of two plane waves:

$$\phi_1 = A \exp\{i[(k + \Delta k)x - \omega t]\} \quad \text{and} \quad \phi_2 = A \exp\{i[(k - \Delta k)x - \omega t]\}$$

$$\begin{aligned} \phi_{\text{tot}} &= A \exp\{i(kx - \omega t)\}\{\exp[i\Delta kx] + \exp[-i\Delta kx]\} \\ &= A \exp\{i(kx - \omega t)\} 2 \cos(\Delta kx) \end{aligned} \tag{4A.5}$$

This is a traveling wave with nodes at $\Delta kx = n\pi/2$ ($n = \pm 1, \pm 3, \ldots$). Unfortunately this expression does not confine our electron to one region, but rather to many. A *continuous* distribution of wavelengths in the range $\lambda_0 - \Delta x < \lambda < \lambda_0 + \Delta x$ results in cancellation of all beats except the one centered at x_0.

The total energy, $E = T + V$ (kinetic plus potential energies) of a quantum mechanical system can be determined from application of the Hamiltonian operator:

$$\mathscr{H} = T + V = \frac{p^2}{2m} + V(r) \tag{4A.6}$$

(Note that the kinetic energy of our plane wave is

$$\langle T \rangle = \frac{\hbar^2 k^2}{2m} \tag{4A.7}$$

as it should be.)

Schrödinger's equation is a statement that the total energy operator applied to a wavefunction returns the total energy eigenvalue(s) for the system times the wavefunction:

$$\mathscr{H}\psi = E\psi \tag{4A.8}$$

The Hamiltonian operator for the hydrogen atom is

$$\mathscr{H} = -\frac{\hbar^2}{2m}\nabla^2 - \frac{e^2}{4\pi\varepsilon_0 r_{12}} \tag{4A.9}$$

with the del operator expressed in spherical coordinates. Equations (4A.8) and (4A.9) give

$$\nabla^2\psi + \frac{2m}{\hbar^2}[E - V(r)]\psi = 0 \tag{4A.10}$$

With $V(r) = -e^2/(4\pi\varepsilon_0 r)$, this becomes the equation for the electronic states in a hydrogen atom, the only atom whose electronic structure can be solved exactly. In spherical symmetry, the full Schrödinger equation depends on all three spherical variables r, θ, and ϕ:

$$-\left(\frac{\hbar^2}{2m}\right)\left\{\frac{1}{r^2}\frac{\partial}{\partial r}\left(r^2\frac{\partial}{\partial r}\right) + \frac{1}{r^2\sin\theta}\left(\sin\theta\frac{\partial}{\partial\theta}\right) + \frac{1}{r^2\sin^2\theta}\frac{\partial^2}{\partial\phi^2}\right\}\psi = (E - V)\psi \tag{4A.11}$$

where the radial r and the angular coordinates θ and ϕ are independent variables; writing the solution as a product $\psi = R(r)Y_l^m(\theta, \phi) = R(r)\Theta(\theta)\Phi(\phi)$ allows us to solve Eq. (4A.11) by separation of variables. First, isolate the ϕ dependence:

$$\begin{aligned}\frac{1}{\Phi}\frac{\partial^2\Phi}{\partial\phi^2} &= -\frac{\sin\theta}{\Theta}\frac{\partial}{\partial\theta}\left(\sin\theta\frac{\partial\Theta}{\partial\theta}\right) - \frac{\sin^2\theta}{R}\frac{\partial}{\partial r}\left(r^2\frac{\partial R}{\partial r}\right) - \frac{2m}{\hbar^2}r^2\sin^2\theta(E - V)\\ &= f(r, \theta) = \text{constant}\end{aligned} \tag{4A.12}$$

The constant is conveniently chosen to be $-m_l^2$, so that the ϕ-dependent factor in the solution is:

$$\Phi(\phi) = \exp(im_l\phi) \tag{4A.13}$$

The solution works only for certain values of m_l, called *eigenvalues*. Equation 4A.11 then becomes

$$\begin{aligned}\frac{m_l^2}{\sin^2\theta} - \frac{1}{\Theta\sin\theta}\frac{\partial}{\partial\theta}\left(\sin\theta\frac{\partial\Theta}{\partial\theta}\right) &= \frac{1}{R}\frac{\partial}{\partial r}\left(r^2\frac{\partial R}{\partial r}\right) + \frac{2m}{\hbar^2}r^2(E - V(r))\\ &= g(r) = \text{constant}\end{aligned} \tag{4A.14}$$

Here it is convenient to choose the constant to be $l(l + 1)$, giving for the θ-dependent solution

$$\Theta(\theta) = \sin^{|m_l|}(\theta)P_l^{|m_l|}(\cos\theta). \tag{4A.15}$$

Again, this solution is acceptable only for certain eigenvalues of l. The right-hand side (RHS) of Eq. (4A.14) then gives

$$\frac{1}{r^2}\frac{\partial}{\partial r}\left(r^2\frac{\partial R}{\partial r}\right)-\frac{\ell(\ell+1)}{r^2}R+\frac{2m}{\hbar^2}\left(E+\frac{Ze^2}{4\pi\varepsilon_0 r}\right)R=0 \qquad (4A.16)$$

with the solution

$$R(r)=\exp(-ar)r^l F(r) \qquad (4A.17)$$

where

$$a=\sqrt{-\frac{2mE}{\hbar^2}}$$

and $F(r)$ is related to the $(2l+1)$th derivative of a Laguerre polynomial.

With the acceptable values of m_l and l, Eq. (4A.16) has solutions only for certain values of E, often labeled E_n. Thus, a set of three eigenvalues has been identified as necessary to specify a solution. The solutions to the Schrödinger equation for a hydrogen atom are then labeled ψ_{nlm} according to the set of eigenvalues to which the solutions belong:

$$\psi_{100}=\frac{1}{\sqrt{\pi}}\left(\frac{Z}{a_0}\right)^{3/2}e^{-Zr/a_0}$$

$$\psi_{200}=\frac{1}{4\sqrt{2\pi}}\left(\frac{Z}{a_0}\right)^{3/2}\left(2-\frac{Zr}{a_0}\right)e^{-Zr/2a_0}$$

$$\psi_{210}=\frac{1}{4\sqrt{2\pi}}\left(\frac{Z}{a_0}\right)^{3/2}\frac{Zr}{a_0}e^{-Zr/2a_0}\cos\theta$$

$$\psi_{211}=\frac{1}{8\sqrt{\pi}}\left(\frac{Z}{a_0}\right)^{3/2}\frac{Zr}{a_0}e^{-Zr/2a_0}\sin\theta e^{i\varphi}$$

$$\cdots$$

where $a_0=0.53$ Å, is the Bohr radius.

The total energy of the 1s electron is the expectation value of the energy operator H evaluated for the state in question: $E_n=\langle\psi_n^* H\psi_n\rangle=\psi_{1s}^* H\psi_{1s}$ $=13.6\,\text{eV}-27.2\,\text{eV}=-13.6\,\text{eV}$.

The angular part of the Hamiltonian in Eq. (4A.11) and its solution were given in Eq. 3.15:

$$\left\{\frac{1}{\sin\theta}\frac{\partial}{\partial\theta}\left(\sin\theta\frac{\partial}{\partial\theta}\right)+\frac{1}{\sin^2\theta}\frac{\partial^2}{\partial\varphi^2}\right\}Y_l^m=-l(l+1)Y_l^m \qquad (4A.18)$$

which is really the equation for the angular momentum operators

$$L^2Y_l^{m_l}=\hbar^2 l(l+1)Y_l^{m_l} \quad \text{and} \quad L_z\Phi=\hbar m_l\Phi \qquad (4A.19)$$

Only certain values of l and m_l allow solutions of Eq. (4A.18); they are the quantized values of the magnitude and the z component of the angular momentum in units of $\hbar$. Thus the magnitude of the total angular momentum has the values $\langle L \rangle = l = \hbar[l(l+1)]^{1/2}$ and its z component $\langle L_z \rangle = l_z = \hbar m_l$. These results are of fundamental importance to modern magnetism.

From the form of the momentum operator in Eq. (4A.2), the Cartesian components of angular momentum can be written

$$\boldsymbol{L} = \boldsymbol{r} \times \boldsymbol{p}:$$

$$L_x = yp_z - zp_y = -i\hbar\left(y\frac{\partial}{\partial z} - z\frac{\partial}{\partial y}\right)$$

$$L_y = -i\hbar\left(z\frac{\partial}{\partial x} - x\frac{\partial}{\partial x}\right)$$

$$L_z = -i\hbar\left(x\frac{\partial}{\partial y} - y\frac{\partial}{\partial x}\right)$$

which become, in spherical coordinates

$$L_x = i\hbar\left(\sin\phi\frac{\partial}{\partial\theta} + \cos\theta\cos\phi\frac{\partial}{\partial\phi}\right)$$

$$L_y = i\hbar\left(-\cos\phi\frac{\partial}{\partial\theta} + \cos\theta\sin\phi\frac{\partial}{\partial\phi}\right)$$

$$L_z = -i\hbar\frac{\partial}{\partial\phi}$$

From these, Eqs. (4A.19) follow.

PROBLEMS

4.1 Magnetic moments in ferrites.

(a) Give the outer electron configurations and magnetic moments for the following ions: Fe^{2+}, Fe^{3+}, Mn^{2+}, and Zn^{2+}.

(b) Using the site selection from Figure 4.8, determine the preferred site occupations in $MnFe_2O_4$ (assume Mn^{2+}) and give its magnetic moment per formula unit.

(c) Describe what happens to the site occupation and net moment as Zn^{2+} is first substituted for Mn^{2+}.

4.2 Explain what happens as Zn substitutes for Ni in nickel ferrite, $NiFe_2O_4$. Again, describe the valence electronic structure of the ions and magnetic moment variation per formula unit. Be quantitative where possible.

4.3 Use the data for paramagnetic Ni in Figure 3.5 to calculate the molecular field coefficient λ in $H^{eff} = \lambda M$ in the following two ways:

(a) Calculate the Curie constant from Eq. (3.45) using a lattice constant of 3.6 Å for FCC Ni.

(b) Determine the Curie constant then λ from two data points on $\chi(T)$ in Figure 3.5.

BIBLIOGRAPHY

Becker R., and W. Doring, *Ferromagnetisms*, Springer-Verlag, Berlin, 1939.

Eisberg, R., and R. Resnick, *Quantum Physics of Atoms, Molecules, Solids, Nuclei and Particles*, Wiley, New York, 1974.

Goodenough, J. B., *Magnetism and the Chemical Bond*, Wiley, New York, 1963.

Krupicka, S., and P. Novak, "Oxide spinels," in *Ferromagnetic Materials*, Vol. 3, E. P. Wohlfarth, ed., North Holland, Amsterdam, 1989, p. 189.

Leighton, R. B., *Principles of Modern Physics*, McGraw-Hill, New York, 1959.

White, R. M., *Quantum Theory of Magnetism*, McGraw-Hill, New York, 1970, p. 86.

REFERENCES

Anderson, P. W., "Exchange in insulators, superexchange, direct exchange and double exchange," in *Magnetism*, Vol. I, G. T. Rado and H. Suhl, eds., Academic Press, New York, 1963, p. 25.

Blasse, G., and E. W. Gorter, *J. Phys. Soc. Jpn.* **17**, Suppl. B-1, 176 (1962).

Born, M., and R. Oppenheimer, *Ann. Physik* **84**, 457 (1927).

Gorner, E.W., *Phil. Res. Rept.* **9**, 295 (1954).

Guillaud, P. C., *J. Phys. Rad.* **12**, 239 (1951).

Herring, C., "Direct exchange between well separated atoms," in *Magnetism*, Vol. IIb, G. T. Rado and H. Suhl, eds., Academic Press, New York, 1966, p. 11.

Leighton, R. B., *Principles of Modern Physics*, McGraw-Hill, New York, 1959.

Lyons, D. H., T. Kaplan, K. Dwight, and N. Menyuk, *Phys. Rev.* **126**, 540 (1962).

Navrotsky, A., and O. J. Klepa, *J. Inorg. Nucl. Chem.* **30**, 479 (1968).

Néel, L., *Ann. Phys.* (France) **3**, 137 (1948).

Slater, J. C., *Rev. Mod. Phys.* **25**, 199 (1953).

Smart, J. S., *Effective Field Theories of Magnetism*, Saunders, Philadelphia, 1966.

Teller, E., *Z. Physik* **61**, 458 (1930).

Yafet, Y., and C. Kittel, *Phys. Rev.* **87**, 290 (1952).

CHAPTER 5

QUANTUM MECHANICS, MAGNETISM, AND BONDING IN METALS

The last chapter considered magnetism in atoms as well as in oxides and other materials for which the electrons remain fairly well localized on the atomic or ionic sites; that is, there is little hopping or itinerant character in the valence electrons. There, two mechanisms were considered, superexchange and Heisenberg exchange, to account for the nearly ubiquitous antiferromagnetic coupling in such systems. Both of these exchange mechanisms are consequences of the Pauli exclusion principle, specifically, of the antisymmetric nature of the electronic wavefunction. The electron–electron interaction between different sites, although present in oxides, is generally weak.

A foundation is now laid for understanding magnetism in metals; systems are considered in which electrons from two or more sites occupy a common bond, covalent or metallic. The difficulty in treating transition metals in terms of their electronic structure stems from the fact that d electrons in metals are neither free electrons nor are they atomic-like. Instead, they have attributes that in some cases are more free-electron-like and in some cases, more atomic-like. While the Heisenberg exchange interaction is defined for localized electrons that interact only weakly, it is often used formally to describe ferromagnetism ($J_{ij} < 0$) or antiferromagnetism ($J_{ij} > 0$) in metals. In reality, exchange in metals is much more complex than this. A number of examples are given to show the magnetic consequences of electron–electron interactions in simple systems. Exchange in metals is usually introduced into band structure calculations by means of an exchange/correlation factor. While the aspects of band calculations of importance to theorists are not covered, some important results for BCC iron and FCC nickel are described. Further, the effects of various impurities on magnetic moments are covered. Also discussed in this chapter is an indirect exchange interaction in which the conduction electrons couple spins that are localized on separate atoms.

As usual, this chapter begins with an experimental observation that epitomizes the subject of the chapter and demands an explanation. It is the Slater–Pauling curve showing the variation of saturation magnetic moment with composition in many magnetic alloys.

5.1 SLATER–PAULING CURVE

The magnetic moment variations in oxides were treated for the simple but instructive case of spinel ferrites in Chapter 4. For oxides the electronic states of the ions are atomic-like and Hund's rules provide a good starting point for determining ionic moments. The magnetic moments of oxide compounds can then be determined from a knowledge of ionic valence, site selection, and simple addition of ionic moments. The concepts needed to understand moment formation in magnetic metals, while more complex, still build on the material covered in Chapter 4. It is important to know the relation between the electronic structure of a metal such as Fe or Ni and its magnetic properties. This will form a basis for understanding the consequences for magnetism of alloying Fe, Co, and Ni with each other as well as other species with these metals.

One of the most evocative sets of data for physicists of the 1930s was the quite regular variation of magnetic moment with composition in $3d$ metals and alloys (Fig. 5.1). This is called the *Slater–Pauling curve* because of the contributions of these two scientists, John Slater and Linus Pauling, to its interpretation. Note first that the average moment μ_m per transition (T) metal atom, expressed in Bohr magnetons, $n_B = \mu_m/\mu_B$, is $2.2\mu_B$, $1.7\mu_B$, and $0.6\mu_B$ for Fe, Co, and Ni, respectively. Further, n_B for $Fe_{50}Ni_{50}$ is very close to that for Co, and both have the same average atomic number, $Z = 27$ or average number of valence electrons, $n_v \approx 9$. Looking at lower electron concentrations from copper ($Z = 29$), it can be seen that n_B increases nearly linearly from zero at 60% Cu in Ni: $\partial n_B/\partial Z \approx -1$ for the data to the right of the maximum. This is also true for FeNi, FeCo, and CoNi alloys over most of the FCC and HCP range ($n_v > 8.6$). The magneton number n_B reaches a maximum value of about $2.5\mu_B$ near the average electron concentration $Z = 26.5$ electrons/atom for BCC metals and alloys ($n_d = 7.5$ assuming one $4s$ electron per atom). A simple explanation of this three-fourths filling of the d states and the uniform slope to the right of the peak are possible in terms of the *rigid-band model* of transition metal magnetism.

The magnetic moments of the elements and alloys displayed in Figure 5.1 can be determined accurately from band structure calculations (Dederichs et al. 1991). (The electronic structure of iron and nickel are reviewed in Section 5.4). What is physically occurring with the electronic states to produce such regular behavior can be appreciated by considering simpler band models that

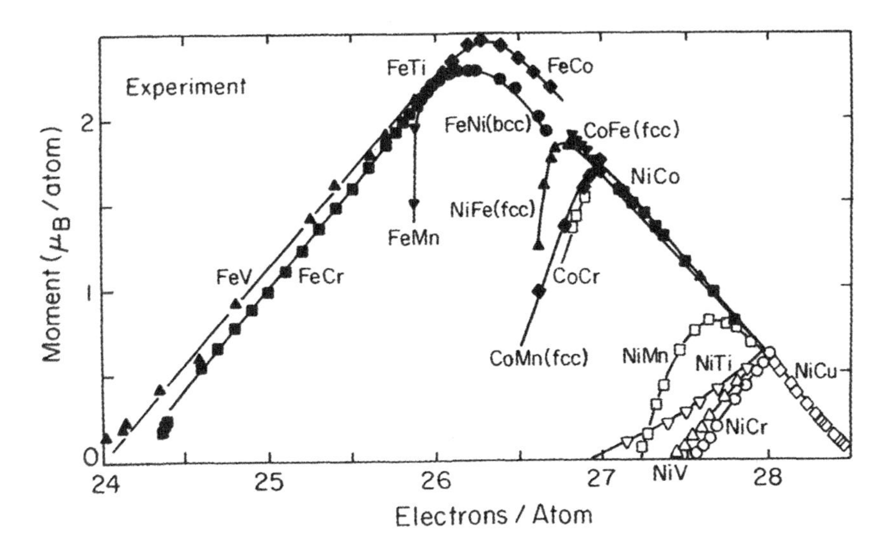

Figure 5.1 The Slater–Pauling curve showing moment per atom (in Bohr magnetons) for metallic alloys as a function of valence electron concentration or alloy composition. [After Dederichs et al. (1991).]

represent the density of valence electron states. In order to gain insight into the physics behind the Slater–Pauling curve, three concepts are needed:

1. Figure 5.2*a* illustrates the familiar broadening of atomic levels (in this case focusing on the 4*s* and 3*d* states) into bands when atoms are brought together to form solids. Note that the $l = 0$ (4*s*) atomic states that get closer

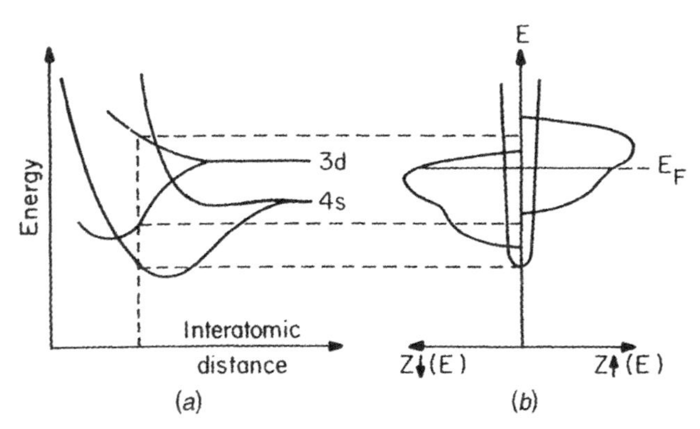

Figure 5.2 (*a*) Evolution of atomic 4*s* and 3*d* states at large interatomic spacing to bands at smaller spacing (r_0 occurs when the net repulsive force $-\partial E/\partial r$ from 4*s* electrons exactly balances the net attractive force from 3*d* electrons); (*b*) density of states of 4*s* and 3*d* states split to reflect exchange preference for spins of one direction.

to the nucleus, have lower energy than $l \neq 0$ ($3d$) states. But the $l = 0$ states also stray much farther from the nucleus than the $l \neq 0$ states, so they also interact (bond) more with neighboring atoms as interatomic distance decreases. For this reason the low lying $4s$ levels begin broadening into bands at larger interatomic distances than do the $3d$ levels. The states in the lower half of a given band are predominantly bonding states whose wavefunctions have large amplitude between atomic sites. Those in the upper half of a band are predominantly antibonding in character with a node between the atomic sites; thus the wavefunctions are pushed back onto the atomic sites and are more localized. The density of states $Z(E)$ [states/(eV)(atom)] can be represented at the equilibrium spacing as in Figure 5.2*b*. This figure describes a broad free-electron-like *s* band (Z goes as $E^{1/2}$) and a narrow, higher-density-of-states *d* band. The areas under these curves, $\int Z(E)dE$ (with units states/atom), must be in the ratio of 2 to 10. As E_F assumes different positions in such a diagram (perhaps simulating a change in *d* electron concentration with alloying), there is a much larger change in the number of *d* electrons than *s* electrons. This has important consequences for magnetism because of the central role played by the *d* states in magnetism.

2. Next, recall that Weiss postulated a strong internal, effective field $H_E = \lambda M$ to account for ferromagnetism. Quantum mechanics tells us that this is really a Coulomb interaction that occurs in systems whose wavefunctions obey the Pauli exclusion principle. [It is not the Coulomb integral C_{ij} but the exchange integral $\mathcal{J}_{ij}$ of Eq. (4.38).] This interatomic exchange interaction requires a shift in the spinup and spindown parts of the *d* band relative to each other as in Figure 5.2*b*. Because the conduction electrons are essentially free, there is very little exchange splitting and the population of the spinup and spindown *conduction* bands is nearly equal. One of the simplest ways to begin to understand interatomic magnetic exchange in metals is to consider it as a vestige of Hund's first rule in atoms: electrons fill otherwise degenerate states with parallel spins first. This minimizes their Coulomb repulsion because they then occupy different orbital states that have minimal spatial overlap. But in a band the different states are clearly not degenerate; there is a significant energy cost to putting all electrons in the spinup band to satisfy Hund's rule. That energy cost is greater if the states in the band are spread over a broader energy range, that is, if $Z(E)$ is small. This competition between exchange energy savings, $\mathcal{J}$ (shifting the spin subbands relative to each other thus favoring parallel spins) and kinetic energy cost proportional to $1/Z(E)$ (favoring paired spins) is expressed quantitatively as the Stoner criterion for the occurrence of magnetism in systems where a band picture is appropriate:

$$\mathcal{J}(E_F)Z(E_F) > 1 \tag{5.1}$$

Ferromagnetism is favored in systems with strong exchange integrals and large state densities at E_F.

3. Finally, this picture must reflect the fact (from Chapter 4) that bonding states between two atoms favor paired, antiparallel spins (presumably one from each atom) and antibonding states favor parallel spins. It is now possible to appreciate one of the reasons why antiferromagnetism is observed in the first half of the 3*d* series—V, Cr, Mn (E_F lies in the bonding part of the *d* band)—and why ferromagnetism occurs in the second half of the 3*d* series—Fe, Co, Ni—(E_F lies in the antibonding part of the *d* band). Narrow *d* bands [large $Z(E_F)$, more atomic-like states] favor magnetism; bonding weakens magnetism; ferromagnetism is unlikely, although not impossible, in a truly free-electron-like [low-$Z(E_F$, *s*-band] metal. (A homogeneous electron gas does show a tendency toward ferromagnetism at low density. The reason for this is that the kinetic energy decreases more rapidly with decreasing density than does the exchange interaction. This sort of ferromagnetism is not seen in metals because their electron densities are too large.)

In alloys containing transition metal species, the covalent bonds formed between partially occupied valence orbitals (i.e., near E_F) often involve magnetic states. If 3*d* states are involved, they become more delocalized as a result of covalent bonding. This delocalization results in a loss of *d* character and hence weaker intraatomic exchange $I(E_F)$, and in a suppression of $D(E_F)$; both of these effects weaken magnetic moment formation [Eq. (5.1)]. If a polar bond is formed in a metal (and it may form between *d* orbitals), the conduction electrons (*s*) will redistribute themselves to screen the bond charge transfer and maintain some degree of local charge neutrality. The screened polar bond still contributes to the chemical stability of the alloy. However, it will affect magnetic properties only if one of the orbitals involved contributes to the magnetism, such as a 3*d* orbital.

Example 5.2 A quantitative example is given to illustrate the consequences of antisymmetrized wave functions on magnetic exchange in order to illustrate for interacting electrons the effects described in paragraphs 2 and 3 above. The average value (expectation value) must be evaluated for the square of the distance between two particles $\langle (r_1 - r_2)^2 \rangle$, which reflects the repulsive force between two electrons. The spatial wavefunction for the electron pair is given by

$$\psi = \phi_a^{(1)} \phi_b^{(2)} \pm \phi_a^{(2)} \phi_b^{(1)} \tag{5.2}$$

The + and − apply for symmetric and antisymmetric spatial states, respectively. The orthogonality of the single particle wavefunctions, $\int \phi_1 \phi_2 \, dv = 0$, implies

$$\langle (r_1 - r_2)^2 \rangle = \langle r^2 \rangle_a + \langle r^2 \rangle_b - 2\langle r \rangle_a \cdot \langle r \rangle_b \pm |\langle a|r|b \rangle|^2 \tag{5.3}$$

where

$$\langle r^2 \rangle_n = \int \phi_n^* r^2 \phi_n \, dv \tag{5.4a}$$

$$\langle r \rangle_n = \int \phi_n^* r \phi_n \, dv \tag{5.4b}$$

$$(a|r|b) = \int \phi_a^* r \phi_b \, dv \tag{5.4c}$$

The first three terms on the RHS of Eq. (5.3) are Coulomb integrals [Eq. (4.4)]. The fourth term involves the hopping or exchange integral between states a and b [Eq. (5.4c), cf. Eq. 4.11] and it changes sign depending on whether the wavefunction is symmetric [giving *smaller* $\langle (r_1 - r_2) \rangle^2$, i.e., the two particles attract each other] or antisymmetric [giving *larger* $\langle (r_1 - r_2)^2 \rangle$, a repulsion for ψ_a particles]. In other words, on average the two electrons in a spatially symmetric orbital ψ_s are closer to each other (and have an antisymmetric spin function) than are those in an antisymmetric orbital ψ_a (which will have a symmetric spin function). These attractive and repulsive exchange forces are real and add to the classical repulsion between like charged particles. They come from Pauli's exclusion principle and Fermi–Dirac statistics and are therefore unique to quantum mechanics.

The implications of this example for metallic magnetism are that an exchange interaction exists for interacting electrons (paragraph 2, p. 146) and that antibonding states are more conducive to moment formation (paragraph 3, p. 147). For spatially symmetric, bonding wavefunctions, Eq. (5.3) shows that the electrons are more likely to be found near each other, so the Pauli exclusion principle forces them into opposite spin states. For spatially antisymmetric wavefunctions the electrons are, on average, farther apart and parallel spins are less costly in energy. These concepts provide a good foundation on which to build an understanding of magnetism in metals and covalent solids.

Figure 5.3 schematically summarizes the dependence of the exchange interaction on interatomic distance for interacting electrons. It is often called a *Bethe–Slater curve.* Pauli's exclusion principle demands that electrons that get too close to each other (same spatial coordinates) have opposite spin (antiferromagnetism). Parallel spin is favored in materials when electrons share the same wavefunction but are confined to separate regions of space; that is, the wavefunction has a node between the respective electron positions. This describes an antibonding wavefunction in a solid. In fact, ferromagnetism is generally observed in $>\frac{1}{2}$-filled bands (antibonding states) as opposed to nearly-empty bands even though Hund's rules favor parallel spins within atoms equally in the two cases. Finally, at large separations, electrostatic interactions become negligible and the difference in energy between parallel and antiparallel spin arrangements vanishes, so there is no exchange interaction.

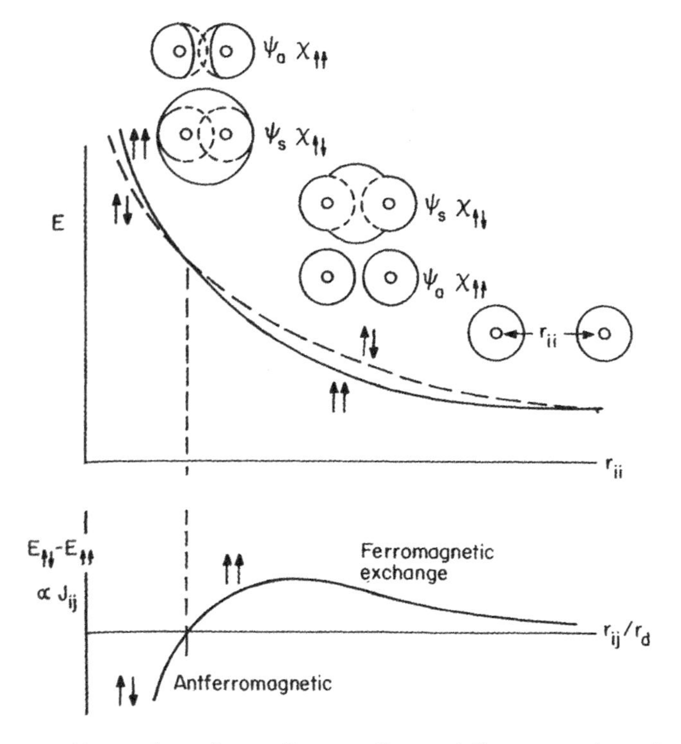

Figure 5.3 Above, dependence of energy for spatially symmetric and antisymmetric wavefunctions versus atomic separation r_{ij}. Below, energy difference versus r_{ij} normalized to d-electron radius showing regions favoring ferromagnetic and antiferromagnetic magnetic coupling.

Thus, a simple band model, with exchange splitting and recognition of the tendency toward ferromagnetism in antibonding states, affords a simple basis for the ferromagnetism of Fe, Co, and Ni. This picture is now extended to magnetic alloys.

The simple band model shown in Figure 5.4 illustrates some of the important differences between the magnetism of metallic Fe and that of metallic Ni. Fe has 8 valence electrons to spread over the $3d$ and $4s$ states; Ni has 10. Transport and other measurements indicate that Fe has slightly less than one electron that can legitimately be called free or itinerant ($4s^{0.95}$); the remaining 7.05 electrons occupy the more localized $3d$ band. The number of d electrons per atom in each spin subband is therefore

$$N_d^+ + N_d^- = 7.05 \tag{5.5}$$

The observed magnetic moment $2.2\mu_B/\text{Fe}$ tells us that

$$N_d^+ - N_d^- = 2.2 \tag{5.6}$$

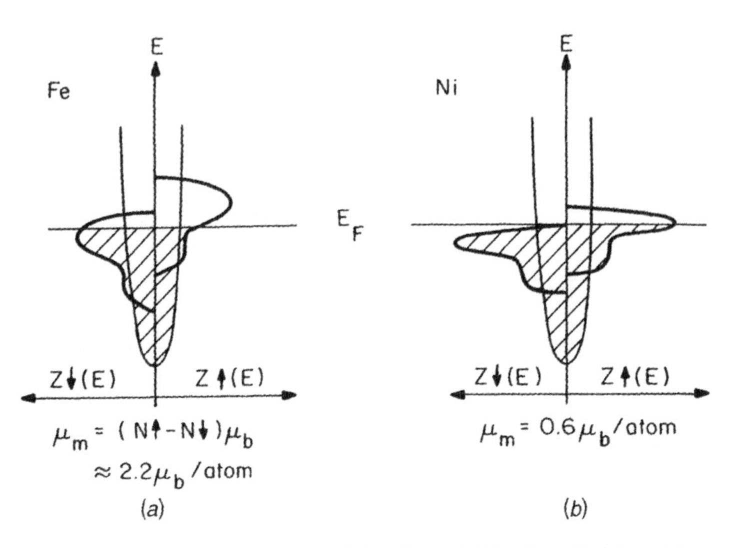

Figure 5.4 Density of states compared for Fe and Ni. The Ni *d* band is narrower in energy and the Fermi level is closer to the top of the Ni *d* band.

Solving for N_d^+ and N_d^- indicates that 4.62 of these 7.05 3*d* electrons are in 3*d* up and 2.42 in 3*d* down. That is, both 3*d* subbands are partially occupied (Fig. 5.4*a*).

Ni on the other hand, has 0.6 free electrons ($4s^{0.6}$) and $3d^{9.4}$. Its magnetic moment of $0.6\mu_B$/atom tells us that 3*d* up is full and 3*d* down has 4.4 electrons or 0.6 hole to account for the net spin imbalance (Fig. 5.4*b*).

Metals that, like iron, exhibit an exchange splitting that is less than the energy difference between E_F and the top of the *d* band E_d are called *weak* ferromagnets. They have, by definition, holes in both the minority and majority spin bands. They are found to the left of the peak in the Slater–Pauling curve. Metals that, like nickel, have an exchange splitting greater than $E_d - E_F$ are called *strong* ferromagnets. By definition they have holes in only the minority, 3*d* band. They are found to the right of the peak of the Slater–Pauling curve.

5.2 RIGID-BAND MODEL

The earliest attempt to explain the curious behavior of the Slater–Pauling curve, and indeed many other physical properties of alloys, was the *rigid-band model*. It assumes that the *s* and *d* bands are rigid in shape as atomic number changes. This simplifies modeling the behavior of different alloys by simply moving E_F up or down through the bands in Figure 5.4 according to the number of electrons present. This premise of the rigid-band model is not correct. Nevertheless, this model can account for some trends in physical properties observed on alloying and thus serves as a suitable starting point for qualitative discussion of alloying effects.

The rigid-band model makes several simplifying assumptions about the electronic structure of alloys. The rigid-band assumption referred to above has many corollaries. It implies that the species involved in alloying (e.g., A and B) are sufficiently similar that the periodic potential of metal A or B is only weakly perturbed by the presence of the other species. Further, all valence electron states, regardless of their orbital angular momentum, are assumed to keep the same energy relative to each other as alloy composition varies. The only effect then of the different number of electrons of A and B would be to shift the Fermi level relative to the rigid-band density of states. It will be seen later that the shape of the state densities for BCC Fe and FCC Ni are quite different from each other and are often characterized as illustrated in Figure 5.5. Hence, it may seem surprising that any observations can be accounted for by a model that assumes that a variety of transition metals and alloys can be represented by a common, simple density of states as depicted at the right in Figure 5.5. In fact, magnetic moment depends only on the number of occupied or unoccupied states in the two spin d bands, not on the shape of $Z(E)$ or on the state density at E_F. Other metallic properties such as susceptibility, specific heat, and phenomena arising from spin–orbit interactions (proportional to L_z at E_F) are sensitive to the shape of the state density curve.

As an example of the extension of the rigid-band model to alloys, consider Ni substituted in Fe: $Fe_{1-x}Ni_x$ (ignore for now the fact that the former is FCC and the latter, BCC in the pure metallic state). The valence electron charge density of the alloy is $n_v = 8(1 - x) + 10x$, so as Ni content increases, the Fermi level is assumed to move up through the rigid, unshifted band structure as shown in Figure 5.6.

The magnetic moment μ_m per atom of an alloy is given by the spin imbalance, $\mu_m = (n^{\uparrow} - n^{\downarrow})\mu_B$, where $n^{\uparrow}$ and $n^{\downarrow}$ are the spinup and spindown

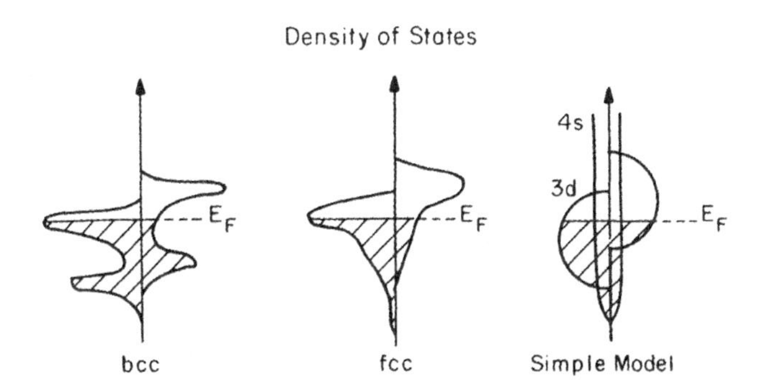

Figure 5.5 Representation of spin-resolved state densities in BCC and FCC metals with Fermi levels located at energies appropriate for Fe and Ni, respectively, and right, a simplified transition metal density of states model.

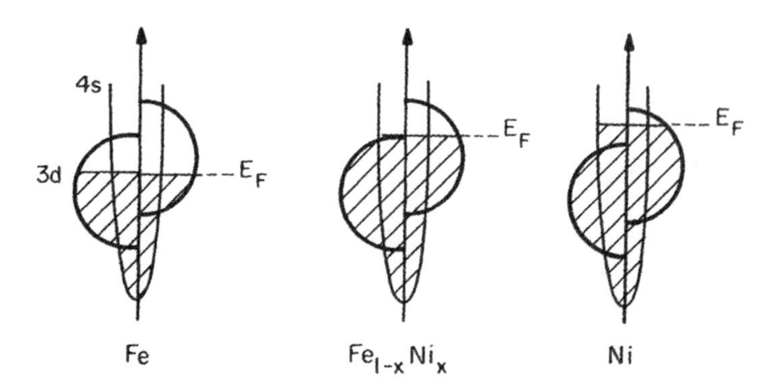

Figure 5.6 Simple valence band pictures appropriate to iron (left), nickel (right), and iron–nickel (center) alloys. See text for shortcomings of such a model.

band populations. In transition metal alloys, the net spin imbalance is due mostly to the d electrons (core electrons and s electrons are not as strongly polarized as the d electrons). Thus

$$\mu_m \approx (n_d^{\uparrow} - n_d^{\downarrow})\mu_B \tag{5.7}$$

where the $n_d^{\uparrow\downarrow}$ are the d subband populations. Generally, both $n_d^{\uparrow}$ and $n_d^{\downarrow}$ can vary on alloying. However, when the Fermi energy lies above the top of the spinup band (such alloys are called *strong ferromagnets*; Fig. 5.6, right), the magnetic moment per atom may be simply calculated. For such strong ferromagnets, $n_d^{\uparrow} = 5$, so from Eq. (5.7), $\mu_m = (5 - n_d^{\downarrow})\mu_B$. Because $n_d^{\downarrow} = n_d - 5$, where $n_d = n_d^{\uparrow} + n_d^{\downarrow}$, it follows that

$$\mu_m = (10 - n_d)\mu_B \tag{5.8}$$

For strong ferromagnets, this equation is a straight line with slope -1, adequately describing the data on the RHS of the Slater–Pauling curve. (It has been assumed that the number of valence electrons occupying the nonmagnetic 4s states remains constant with composition.) Equation (5.8) explains why the average moment of cobalt should be so close to that of NiFe; both have the same valence electron concentration and thus about the same value of n_d. Further, this model explains the observation of nonintegral average magnetic moments in 3d alloys. Also, it could be argued that the Slater–Pauling curve peaks at $2.5\mu_B$ for BCC alloys because the Fermi level is stable when it coincides with the density of states minimum near the center of the BCC minority spin band (Fig. 5.5) and the majority 3d band is full ($n_d = 7.5$). For lower d electron concentration than 7.5, majority as well as minority states are empty and μ_m decreases with decreasing n_d. When both spin subbands have empty states, the alloy is referred to as a *weak ferromagnet*.

An alternate form of Eq. (5.8) often found in the literature can be simply derived. The number of d electrons in the alloy $A_{1-x}B_x$ is given by

$$n_d^{\text{alloy}} = (1 - x)n_d^A + xn_d^B$$

If the atomic number of species B differs from that of A by ΔZ, then $n_d^B \approx n_d^A + \Delta Z$. Thus

$$n_d^{\text{alloy}} \approx (1 - x)n_d^A + x(n_d^A + \Delta Z) = n_d^A + x\,\Delta Z$$

When this result is used in place of n_d in Eq. (5.8), the alloy moment is given by

$$\mu_m^{\text{alloy}} = (10 - n_d^A - x\,\Delta Z)\mu_B$$

or, noting that $10 - n_d^A = \mu_{\text{host}}/\mu_B$

$$\mu_m^{\text{alloy}} = \mu_{\text{host}} - x\,\Delta Z\,\mu_B \tag{5.9}$$

This equation says that the magnetic moment per average atom in a strong ferromagnetic alloy, $A_{1-x}B_x$, differs from that of the host by an amount proportional to the atomic number difference of B relative to A. The rigid-band model does not apply to weak ferromagnetic alloys because then it is not simple to determine into which band the impurity electrons go.

The rigid-band approximation should obtain for only small concentrations ($c \ll 1$) of impurities that weakly perturb the periodic potential of the matrix ($|\Delta Z| \approx 1$). Nevertheless, this model affords a simple explanation for the compositional dependence of μ_m in strong ferromagnetic alloys over broad concentration ranges. For example, in Ni, dilute Fe, or Co additions, $\Delta Z = -2$ and -1, respectively, increase the magnetic moment (Fig. 5.1). The rigid-band approximation should fail for $|\Delta Z| \geqslant 2$ as is the case for Mn, Cr, V, and in Co or in Ni. Figure 5.1 shows that many of these substitutions decrease the average moment even though $\Delta Z < 0$.

The rigid-band model is naive because the band structure and the shape of the state density curve of most alloys does change with alloy composition as suggested in Figure 5.5. Another problem with the rigid-band model is that the atomic magnetic moments are known from neutron scattering data to be different on different species in a given alloy, and to be independent of composition over fairly wide composition ranges in some cases. Figure 5.7 is a collection of site-resolved moments in $3d$ alloys collected from various sources. Note that upon alloying Co in BCC Fe, the Fe moment increases sharply while the Co moment remains constant at about $1.7\mu_B$. The *average* moments of these alloys, measured, for example, by magnetometry, are simply linear combinations of the distinct elemental moments.

Another shortcoming of the rigid-band model is the fact that the number of conduction electrons per atom is experimentally observed to vary across the Slater–Pauling curve. For example, the number of $4s$ electrons is about 0.6 for

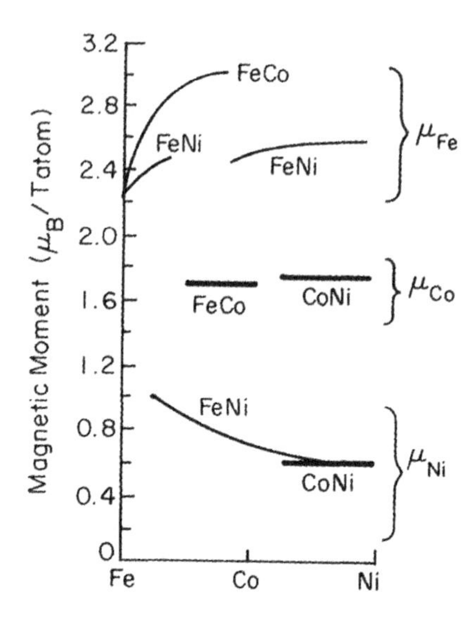

Figure 5.7 Local moments on Fe, Co, and Ni sites for FeNi, FeCo, and CoNi alloys as determined by neutron scattering measurements compiled from Collins and Wheeler (1963) and Collins and Forsyth (1963).

Ni, about 0.95 for iron, and about 1.0 for elements to the left of Fe in the 3*d* period. Thus, the 3*d* states must be more stable (lower energy) relative to the 4*s* conduction states toward the right of the transition series. (The Ni core potential is stronger than that of Fe). The resulting stabilization of the 3*d* band of late T species makes them more atomic-like and more localized, and hence they exhibit a higher density of states compared to early T species. This relative shift of different features in the band structure is not explained by the rigid-band model; if it were accounted for, it should lead to a better model. Band shifting effects should be most significant when the atomic number of the species involved differ significantly, that is, when $|\Delta Z| > 2$. The relative energy of band features associated with different species in an alloy is described in the dilute alloy limit by the virtual bound-state model and in the concentrated alloy limit by the split-band model. These models are now discussed.

5.3 VIRTUAL BOUND STATES AND THE SPLIT-BAND MODEL

It may be asked why the 3*d* states are found at higher energies than the bottom of the 4*s* conduction band when the 3*d* wavefunctions are more localized, more atomic-like than the 4*s* states? The answer lies in the form of the potential in the radial part of Schrödinger's equation [Eq. (4A.16)]. The orbital contribution, $+h^2 l(l+1)/2mr^2$, to the atomic potential energy, $-Ze/(4\pi\varepsilon_0 r)$, is a rotational kinetic energy (associated with a centrifugal force) that may raise the energy of an otherwise bound state above the muffin tin potential of the solid. When this happens, the state is called a virtual bound state (VBS). A $3d$ state in a pure transition metal is virtually bound in the sense that it lies above the muffin tin zero, in the continuum of conduction electron states, but its amplitude is enhanced over the atomic sites; it may be thought of as localized periodically, although, strictly speaking, it is a scattering resonance. It would be a bound state were it to have $\langle L^2 \rangle = 0$. This effect can be rationalized also in terms of the form of the $3d$ and $4s$ wavefunctions. The former are concentrated radially in a relatively narrow shell about the nucleus and have, therefore, a fairly well defined range of energies over which they can be found; the latter can be found over a greater range of r values, causing them to have energies that range from below the $3d$ energy to above it.

A VBS in the impurity sense is similar to a $3d$ state but has peak amplitude only *at the impurity site*; that is, it breaks the periodicity of the host potential. Figure 5.8*a* shows the potential and schematic VBS, and Figure 5.8*b* depicts the density of states for a true bound impurity state and a VBS (the VBS may lie above or below E_F). Friedel (1958) developed these concepts to explain many of the physical properties of magnetic alloys containing dilute magnetic species.

Friedel's VBS model is useful in accounting for the sharply decreasing moments of Co- or Ni-base alloys containing light ($\Delta Z < -2$) transition metal solutes. According to this model, a fivefold-degenerate $3d$ virtual bound state located near the impurity is lifted out of the $3d$ band because of its repulsive potential. This VBS is exchange split to spinup and spindown states. If the majority-spin VBS remains below the Fermi level, then the magnetic solute affects the electronic properties only because of the difference of its $3d\downarrow$ population relative to that of the host [as expressed in Eqs. (5.7) and (5.9)]; as long as $n_d^\downarrow$ decreases as a result of the impurity, the moment increases. If the impurity potential is sufficiently repulsive to move also the majority-spin VBS above the Fermi level, then Eq. (5.9) must be modified. Five $3d$ electrons per impurity atom will be transferred from the $3d\uparrow$ (moment reduced by $5x\mu_B$) to

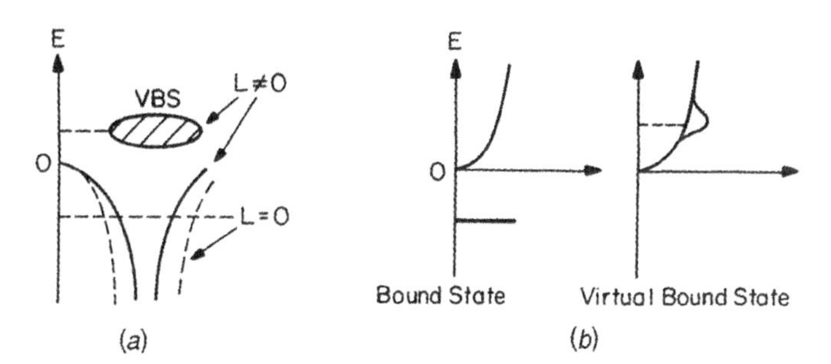

Figure 5.8 (*a*) Schematic representation of a state above the "muffin tin" potential that is virtually bound by its angular momentum; (*b*) the effects of a VBS on an otherwise free electron DOS.

the host $3d\downarrow$ states (moment reduced by another $5x\mu_B$): the alloy magnetic moment is reduced by $10\mu_B$ in addition to the change due to the valence electron difference [$-\Delta Z$ in Eq. (5.9)]. Thus, for sufficiently light magnetic metal solutes, in the VBS approximation, Eq. (5.9) becomes

$$\mu_{\text{av}} = \mu_{\text{host}} - (\Delta Z + 10)x\mu_B \tag{5.10}$$

This equation suggests that the lighter the transition metal substituent (more negative ΔZ) in a strong ferromagnetic host, the less should be the moment suppression. This behavior is followed more for Ni-base alloys than for Co-base alloys (Fig. 5.1).

The potential that gives rise to the virtual bound state is clearly localized at the impurity atom. The wavefunctions of the electrons associated with the VBS extend into the matrix with a range that decreases with increasing density of states at E_F. Holes in the virtual bound state, therefore, give rise to a magnetic moment identified with the impurity site, but not necessarily localized there to the same extent that the host moment is localized. Charge displaced from the VBS may enhance the impurity moment at the expense of the moments in the surrounding matrix.

As an illustration of virtual bound states, Figure 5.9 depicts that part of the electronic state density that is localized near the impurity for (*a*) V impurities in Ni, (*b*) Fe impurities in Ni, and (*c*) Ni impurities in Fe (Dederichs et al. 1991). In the first case most of the density of both spinup and spindown states of the V impurity lie above the Fermi level while some amplitude (resonance with the lower lying Ni states) occurs below E_F. For dilute Fe in Ni, mostly the minority-spin states of Fe appear above E_F while the majority-spin states appear below E_F. For Ni impurities in Fe, most of the Ni states lie below E_F while there is a small resonance of the Ni spindown states with the Fe spindown states above E_F.

The impurity atomic number and the energy of the impurity states relative to those of the host determine the impurity moment and how it affects that of the host. For example, *V*, $Z = 23$ (approximately $3d^4 4s^1$) gives most of its 5 valence electrons to the lower-lying host *d* states in its vicinity. From Figure 5.9*a* it is clear that at the V site, there are more minority-spin states below E_F than there are majority-spin states. This is because the vanadium VBSs are closer in energy to the Ni $3d\downarrow$ states and hence show more of a resonance with them. Thus, at the V site, there exists a negative magnetic moment. Typically, early transition metal impurities exhibit a moment that is opposite to that of an iron or nickel host; the converse applies for a late transition impurities. Figure 5.10, from Dederichs et al. (1991), shows calculated and measured *impurity moments* for dilute 3*d* elements in iron. Further, the charge displaced from the impurity *d* states also perturbs the *host moments* in the vicinity of the impurity. The net change in iron magnetic moment per impurity atom is shown in Figure 5.11 (Dederichs et al. 1991). The stronger suppression of the iron moment (Fe is a weak ferromagnet) by earlier 3*d* impurities, is opposite the

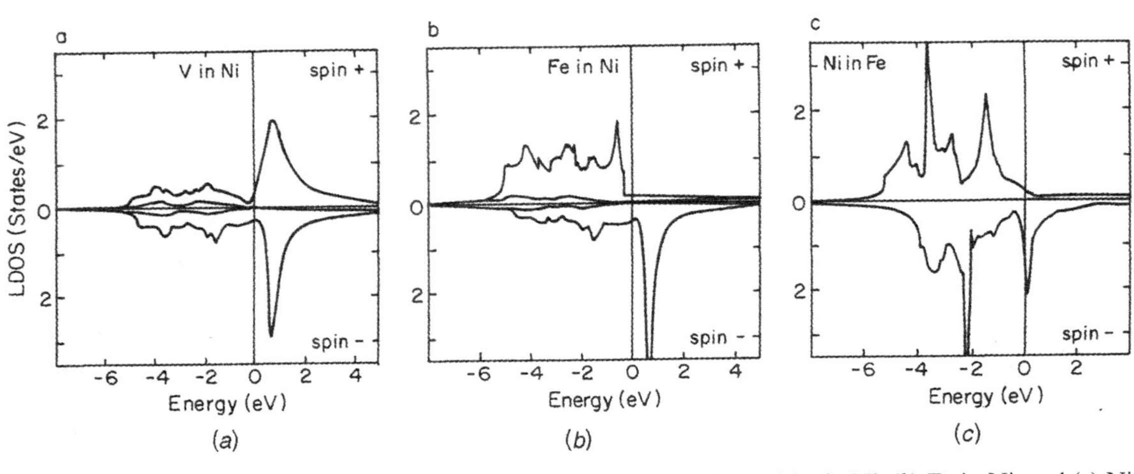

Figure 5.9 Spin-resolved, calculated state densities for (*a*) vanadium impurities in Ni, (*b*) Fe in Ni, and (*c*) Ni in Fe (Dederichs et al. 1991).

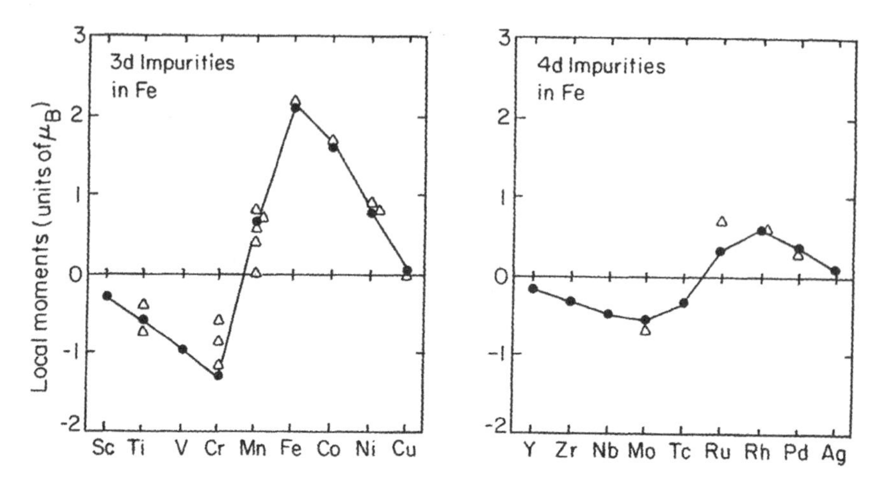

Figure 5.10 Calculated local moments for 3*d* and 4*d* impurities in Fe [full circles, Dederichs et al. (1991)] and experimental values [open triangles, Dritter et al. (1989)].

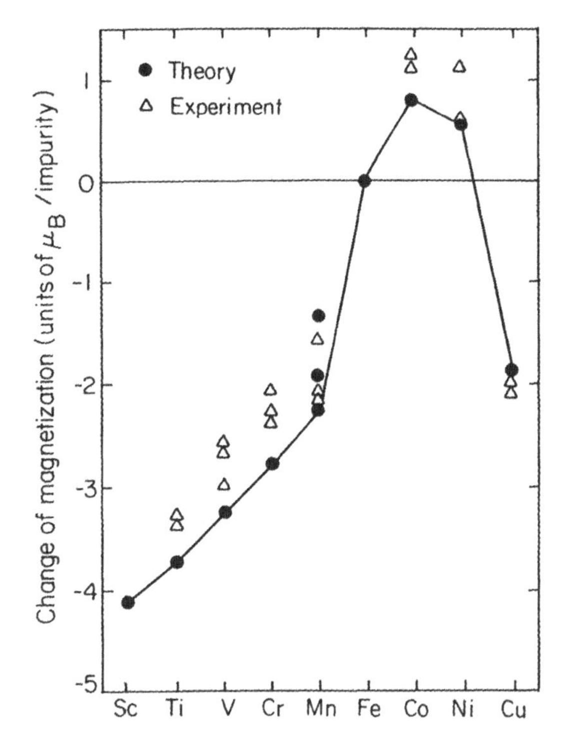

Figure 5.11 Change of magnetization ΔM per dilute 3*d* impurity atom in Fe (Dederichs et al. 1991). The triangles are experimental values from Dritter et al. (1989).

effect described for strong ferromagnets in Eq. (5.10). It will be seen in Chapter 15, on transport properties, that these magnetic impurity states also have an effect on electrical resistivity that reflects the energy of the VBS relative to E_F.

Some of the concepts of the VBS model are now extended to more concentrated alloys, $A_{1-x}B_x$ ($Z_B > Z_A$). If the energy difference between the two sets of states is greater than their average bandwidth [i.e., $E_A - E_B > (w_A + w_B)/2$], then the states associated with the A and B sublattices will remain distinct in energy (Fig. 5.12). This condition is generally met when the atomic numbers of the two species differ by at least 2: $|\Delta Z| > 2$.

The significance of this band splitting can be seen if by returning to the Fe-Ni example used earlier. Figure 5.13 depicts the densities of states for Fe (left), Ni (right) and a concentrated Fe-Ni alloy (center). On increasing Ni content, the Fermi energy moves *down* through the upper split band and into the band of Ni origin. This is exactly the opposite of what was illustrated in the rigid-band model for this system (Fig. 5.6). Because the magnetic moment depends on the number of electrons or holes in each band, the consequences of the split-band model for the net moment are the same as those for the rigid-band model because the number of electrons is the same in either case. However, the split-band model provides a qualitative rationale for the different moments on each species as observed in Figure 5.7. The lower-lying *d* states in the center panel of Figure 5.13 have lower energy because they tend to be more localized at the Ni sites and conversely the higher-energy subband represents states more strongly localized at the iron sites. Clearly, then, the split-band model for Fe-Ni suggests a much smaller moment at the Ni sites and a larger one on at the Fe sites.

Further, the split-band model contains information about angular momentum that is missing from the rigid-band model. It is of interest to know alloy compositions for which certain properties that depend on spin–orbit interac-

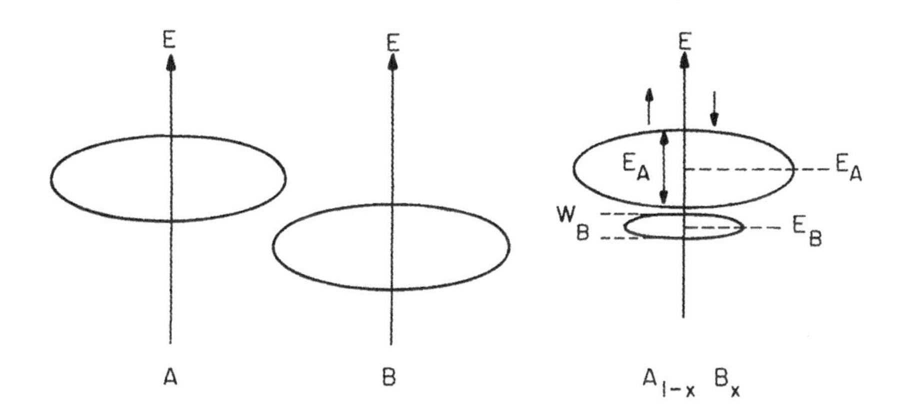

Figure 5.12 Split-band model for alloy $A_{1-x}B_x$, where $Z_B > Z_A$. The components of the composite DOS can be resolved if $\Delta Z \geqslant 2$.

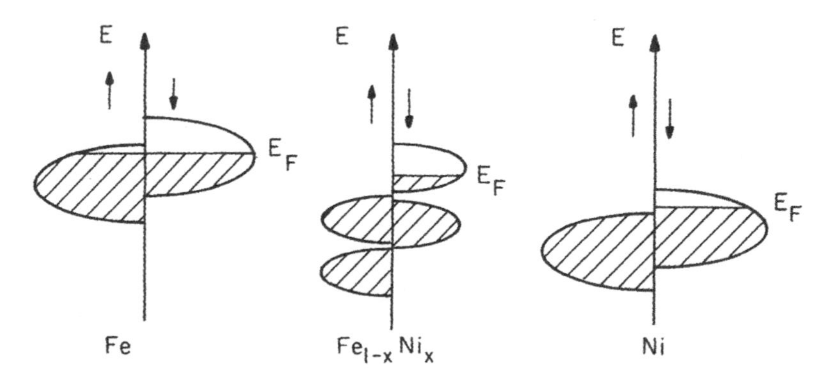

Figure 5.13 Left and right panels show schematics of the band structure of Fe and Ni, respectively. Center, split-band model for FeNi alloys (cf. Fig. 5.8).

tion vanish while a material retains a magnetic moment. (The trivial case is that $\langle L_z \rangle = 0$ when the Fermi level is outside the *d* band, but then there is no magnetic moment.) Does $\langle L_z \rangle$ ever vanish when a material retains a magnetic moment? It is known from quantum mechanics of atomic spectra that when the degeneracy of a set of levels is lifted, the new manifolds take on new quantum numbers and $\langle L_z \rangle = 0$ for a filled manifold (e.g., the fivefold degenerate *d* levels split into e_g and t_{2g} levels in octahedral symmetry; the t_{2g} levels are characterized by $m_L = 0, \pm 1$, so when this triplet is fully occupied, $\langle L_z \rangle = 0$). The split-band model shows us just where to find $\langle L_z \rangle = 0$ in a *d* band even though there may be some empty *d* states.

It is possible to calculate the compositions for which the Fermi level coincides with the gap or minimum density of states between two split bands for strong ferromagnets (majority bands full). For $A_{1-x}B_x$ and $Z_B > Z_A$, the number of *d* states in the higher-energy minority band is $5(1-x)$ and in the lower-energy minority band, $5x$. The Fermi level will lie between the two minority split bands when the number of holes in the alloy $(1-x)h_A + xh_B$ is equal to the number of states in the higher-energy split band $5(1-x)$. The number of holes for Ni, Co, and Fe is roughly 0.6, 1.6, and 2.6, so the condition for $\langle L_z \rangle = 0$ in $Fe_{1-x}Ni_x$ is

$$5(1-x) = 2.6(1-x) + 0.6x$$

or $x = 0.8$. This is the approximate Ni concentration for the famous zero-magnetostriction permalloys, which will be covered in Chapter 10. (A more accurate counting of *s* and *d* electrons shows that $n_h = 2.8$ for Fe, which gives $x = 0.786$, much closer to the zero magnetostriction composition in permalloy.)

The first direct evidence for split bands came from photoemission data on a series of NiCn alloys (Seib and Spicer, 1968). An even sharper resolution of split band features is seen in amorphous PdZr and CuZr early/late transition metal (TE-TL) alloys studied by Güntherodt's group in Basel using ultraviolet

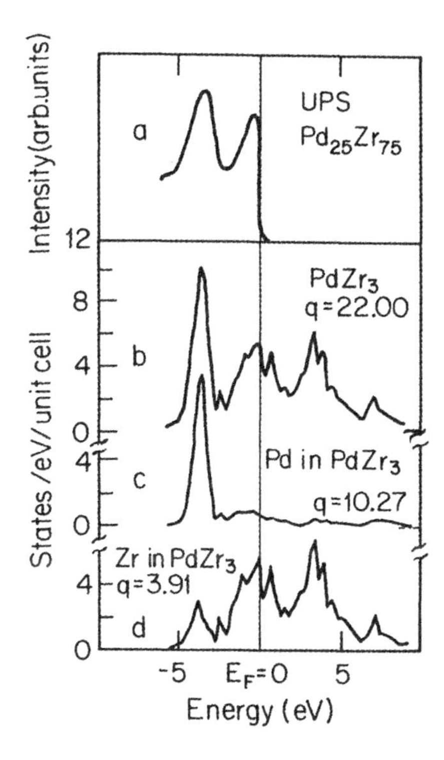

Figure 5.14 Above, density of occupied states in amorphous PdZr determined by ultraviolet photoelectron spectroscopy (UPS). [From Moruzzi et al. (1983)]; below, calculated state densities for $PdZr_3$ (Cu_3Au structure) and partial state densities for Pd and Zr from this calculation. [From Oelhafen et al. (1979).]

photoelectron spectroscopy (UPS) (Oelhafen et al. 1979). Figure 5.14 (top panel) shows the experimental data for amorphous PdZr (Moruzzi et al. 1983). The lower-energy (greater binding energy) feature reflects the chemical stabilization due to the more attractive core potential at the Pd site compared with that at the Zr site.

This approach forms a conceptual bridge from the virtual bound states of transition metal impurities in metallic hosts to a concentrated alloy version of this model called the *split-band model.* Many shortcomings of the rigid-band model are remedied by the split-band model which is an extension to concentrated alloys of the concepts of virtual bound states used to explain many features of dilute alloys.

5.4 ELECTRONIC STRUCTURE OF TRANSITION METALS

While it is not necessary here to go into the details of band structure calculations for magnetic materials, it is useful to show some representative

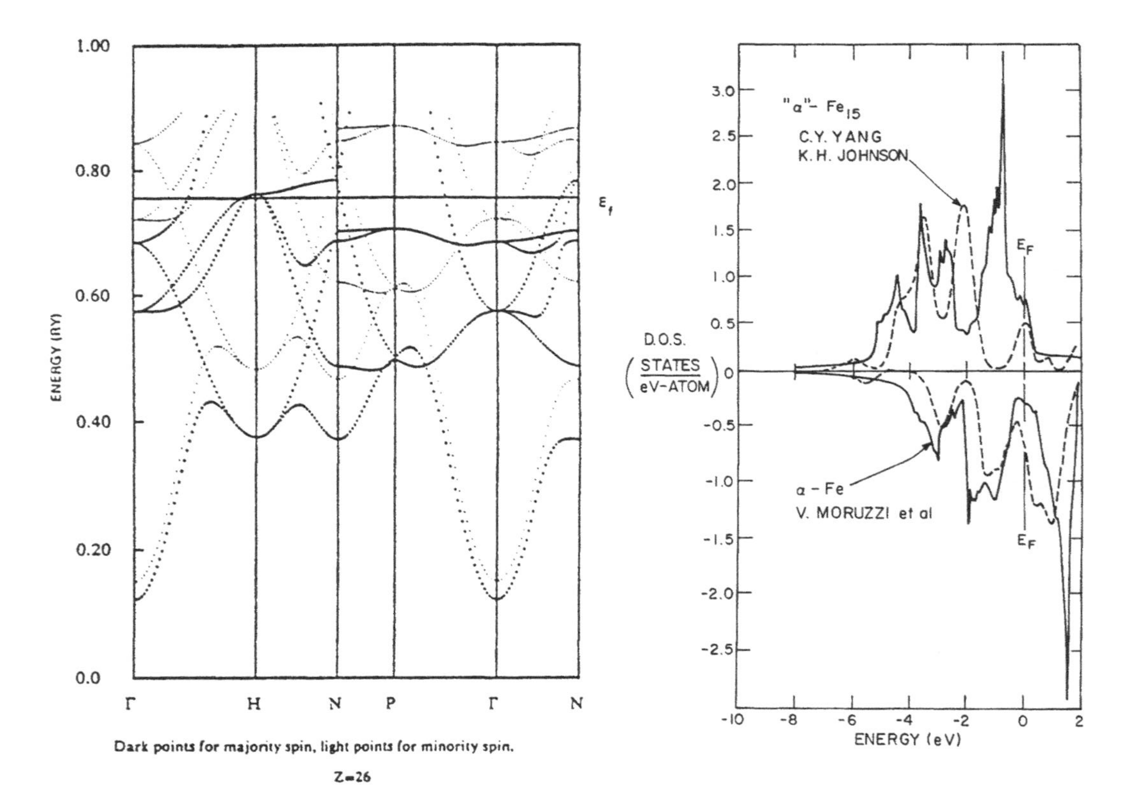

Figure 5.15 Left, calculated, spin-resolved band structure for BCC Fe (Moruzzi et al. (1979); solid lines are spinup bands, dotted lines, spindown. Right, spin-resolved density of states for α-Fe: solid line from Moruzzi et al. (1978) dashed line from cluster calculations by Yang et al. (1981).

results and place them in a materials and property context. It should be mentioned that present density functional methods of calculating electronic structure are vastly superior to the older linear combination of atomic orbitals (LCAO) and tight binding methods in terms of accuracy. Nevertheless, considerable physical insight can still be gleaned from the simpler methods.

Iron and nickel represent two different types of magnetic material in as much as the Fermi level cuts across *both* spin bands in Fe (Fig. 5.15) but lies exclusively in the minority-spin band in Ni (Fig. 5.16). Thus while alloying additions in Ni typically cause moment changes proportional to concentration (simple changes in the number of $3d\downarrow$ holes), the situation in Fe is more complicated. Because the exchange splitting in Fe-like materials is less than the energy difference between the top of the d band and E_F, they are referred to as weak ferromagnets. Conversely in Ni-like materials, $E_{ex} > E_{top,d} - E_F$ and they are referred to as strong ferromagnets.

Figure 5.15 shows the band structure of α-Fe calculated by Moruzzi et al. (1978). They used a nonrelativistic, density functional within the local-density approximation to account for electronic exchange and correlation and a muffin-tin approximation for the form of the potential (electronic density is spherically symmetric inside the muffin tin radius and constant between these spherical regions). The relatively flat (dispersionless) states near $E_F(E > -6\,\text{eV})$ are the d states. Note the difference in energy of the spinup and spindown d states but the nearly similar energy of the spinup and spindown free-electron-like states ($E \propto k^2$).

The density of states (DOS) is represented next to the band structure for the two spin states. This shows the nature of the band structure in terms that can be compared with the simple band models described earlier in the chapter. The Fermi level lies near the top of the $3d\uparrow$ band and near the center of the $3d\downarrow$ band. The dashed line in the background shows the DOS determined from self-consistent-field (SCF) calculations of molecular orbitals (MOs) on small BCC-like Fe clusters (Yang et al. 1981). The sharp energy states calculated for clusters have been broadened to reflect what would happen in an extended solid. The agreement between these two very different methods of calculating electronic structure in metals is reasonable and instructive; the MO calculations allow identification of various features in the d-band.

Figure 5.16 shows the band structure and state densities for FCC Ni. Note the difference in shape between the Ni $3d$ state density and that of Fe which shows a clearer separation between the lower energy d states (related to t_{2g} MOs) and the higher-energy d states (related to e_g MOs).

Figure 5.17 shows the variation of the valence band parameters across much of the $3d$ series measured relative to the top of the muffin-tin potential. E_c is the bottom of the conduction band, E_{d1} and E_{d2} are the low and high-energy limits of the $3d$ band, and E_F is the Fermi energy. Both the $4s$ and $3d$ bands become more stable, the d band shows significant narrowing and the Fermi level generally appears at lower energy, with increasing atomic number. The d band splits into spin subbands (dashed lines) for Fe, Co, and Ni. Note the

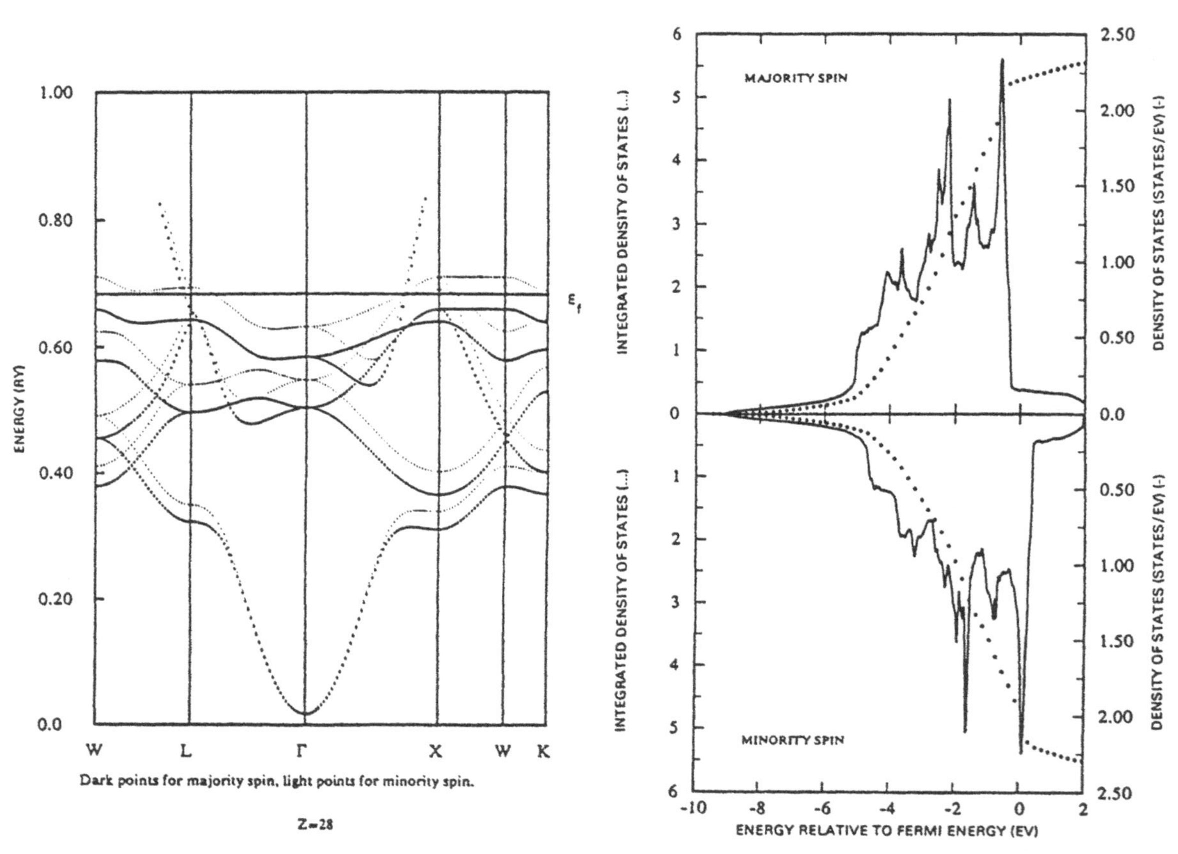

Figure 5.16 Calculated, spin-resolved band structure for FCC Ni. Solid lines are spinup bands; dotted lines, spindown (Moruzzi et al., 1978).

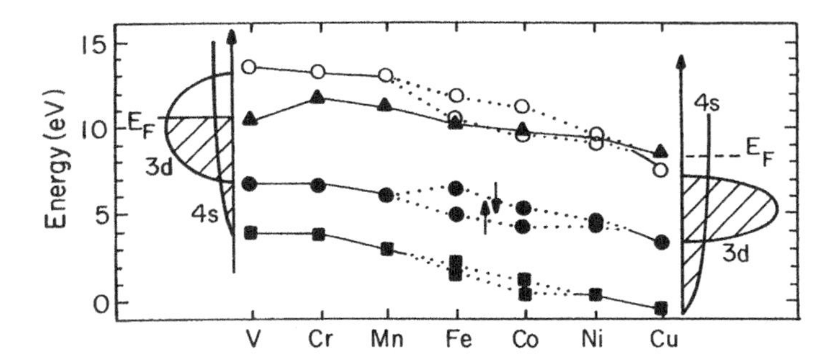

Figure 5.17 Variation of valence band structure across the 3*d* series from vanadium to copper. Energies are measured from the top of the muffin tin potential. Shown are the bottom of the 4*s* conduction band, the bottom and top of the 3*d* band, and the Fermi level. The band limits show an exchange splitting for Fe, Co, and Ni. [Data adopted from calculations by Moruzzi et al. (1979) and MacAlister et al. (1975).]

crossover from weak magnetism ($E_F < E_{d2}^{\uparrow} < E_{d2}^{\downarrow}$) to strong magnetism ($E_{d2}^{\uparrow} < E_F < E_{d2}^{\downarrow}$) between Fe and Co.

It has long been appreciated that magnetism—especially in 3*d* alloys—is predominantly a local phenomenon determined by the immediate environment about potentially magnetic atoms. Hence the importance and success of local environment models as reviewed by Kouvel (1969) and used by Niculescu et al. (1976) and by Stearns (1981). Although many properties of ferromagnets (saturation moment, magnetic anisotropy) can be calculated in the context of band theory, it is the dependence of the electronic energy levels on local structure, rather than the Bloch nature of the wavefunctions, that is critical.

Four aspects of short-range order can be identified as important to magnetism: the *number*, *type*, *distance*, and *symmetry* of the nearest neighbors about a given site. The dependence of magnetism on these measures of short-range order is illustrated in simple terms by the Stoner criterion for the existence of a local moment [Eq. (5.1)] or by the molecular field expression for the Curie temperature [Eq. 4.21)]: $T_C = 2J(r)Z_T S(S+1)/3k_B$. Here $J(r)$ is the *distance*-dependent interatomic exchange integral, Z_T is the coordination *number* (presumably by strongly magnetic species) about the transition metal T site, S is the atomic spin quantum number, $I(E_F)$ is the Stoner integral evaluated at E_F reflecting what remains in the solid of intraatomic (Hund's rule) exchange, and $D(E_F)$ is the electronic density of states at the Fermi energy. The *number*, *type*, and *distance* of nearest neighbors enter Eqs. (5.1) and (4.21) explicitly through Z_T and $J(r)$ and implicitly through $D(E_F)$ and $I(E_F)$. The *symmetry* of the nearest-neighbor arrangement affects $D(E_F)$ and $I(E_F)$ by changing the degeneracy and hence the distribution of the electronic states. Thus local magnetic moment formation is determined by the extent to which intraatomic exchange is free to operate. A full moment equal to half the population of an orbital can be achieved in some isolated atoms or ions (e.g.,

$\mu_{Fe^{3+}} = 5\mu_B$ per Fe^{3+}). However, when the energy levels are broadened by bond formation to the extent that they overlap, the local magnetic moment is reduced (e.g., $\mu_{Fe} = 2.2\mu_B$ in BCC Fe). The Curie temperature reflects more cooperative, interatomic effects [through $J(r)Z_T$], as well as depending on the magnitude of the local moment.

5.5 MAGNETIC PRESSURE

This section introduces a concept that illustrates powerfully the implications of band filling for physical properties, particularly magnetic moment formation.

Consider first the p series of diatomic molecules B_2, C_2, N_2 O_2, and F_2 shown in Figure 5.18. Their molecular orbital ladder of allowed states is shown with the π or σ symmetry of the orbitals indicated. B_2 is paramagnetic; the two electrons in its $2p$–π bonding orbital have parallel spins. Does that not violate what was just proved? No! It was proved that two electrons in a common bonding orbital have antiparallel spins. But the $2p$–π bond is doubly degenerate; it arose from p_x and p_y orbitals and has a capacity of 4, not 2, electrons. So the two electrons in the π bond of B_2 are not forced into the same region of space and the Pauli exclusion principle does not come into play. They can have parallel spins, and they do.

The C_2 molecule completes the filling of the π bonding state so its net spin is zero. N_2 has a filled $3\sigma^2$ molecular orbital, so it also is not magnetic. However, O_2 has a half-filled, doubly degenerate π^* orbital (2 electrons in 4 states) so the two spins are parallel as for B_2. Next is F_2 with 4 electrons in a degenerate π^* orbital. Clearly, F_2 should not have a paramagnetic moment. Finally, Ne completes the series. It is not stabilized by forming a diatomic molecule because just as many antibonding (higher-energy) states as bonding (lower-energy) states are occupied. That is, it is an inert gas and does not form stable molecules, even with itself.

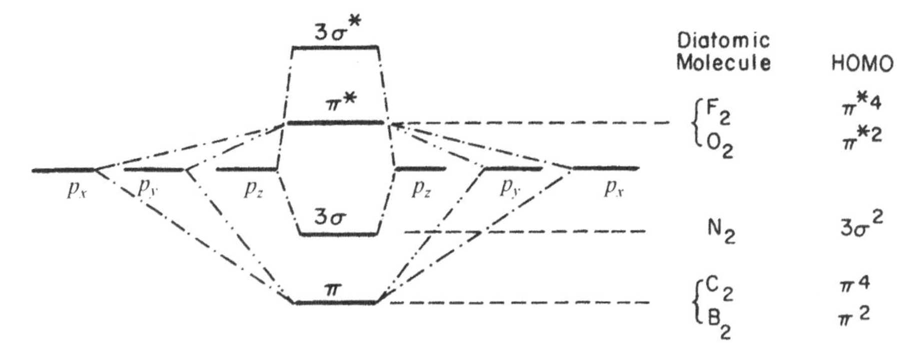

Figure 5.18 Molecular orbital structure (left) for homonuclear $2p$ diatomic molecules; right, symmetry of highest occupied molecular orbital (HOMO) in each molecule.

The implications of the electronic structure of this series of molecules for physical properties (Fig. 5.19) is now considered. Note that as the six hybrid orbitals fill, the molecules first become more stable (valence electrons are occupying states of energy lower than the average for the manifold) then less stable (higher-energy states are now occupied). This is reflected in the bond length and in the stiffness of these molecules.

To understand these trends, one must consider the energies of the states occupied by the electrons in the molecules relative to their energies in the noninteracting atoms (Fig. 5.18). For early members of this *p* series, bonding states near the bottom of the MO manifold fill first; this constitutes a decrease in energy for these electrons relative to their energies in the free atoms. This accounts for the dramatic decrease in interatomic volume and a concomitant increase in stability and stiffness in the first half of the series. The occupation of antibonding states in the second half of the series cancels the energy stabilization gained in the first half, and the trends reverse.

A similar but much more impressive trend occurs in filling the *d* bands in the metallic 3*d*, 4*d*, and 5*d* transition series as shown in Figure 5.20. The 4*d*

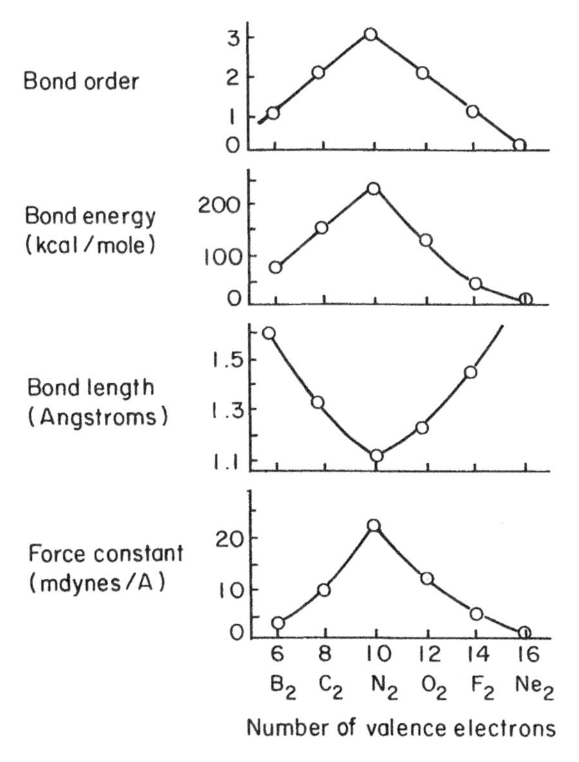

Figure 5.19 Trends in properties of 2*p* diatomic molecules with atomic number of bond occupation. Note that stronger bonds are shorter and stiffer (110 kcal/mole = 4.34 eV, 10 mdnes/Å = 10^3 N/m). [After Pemental and Sprotley (1969).]

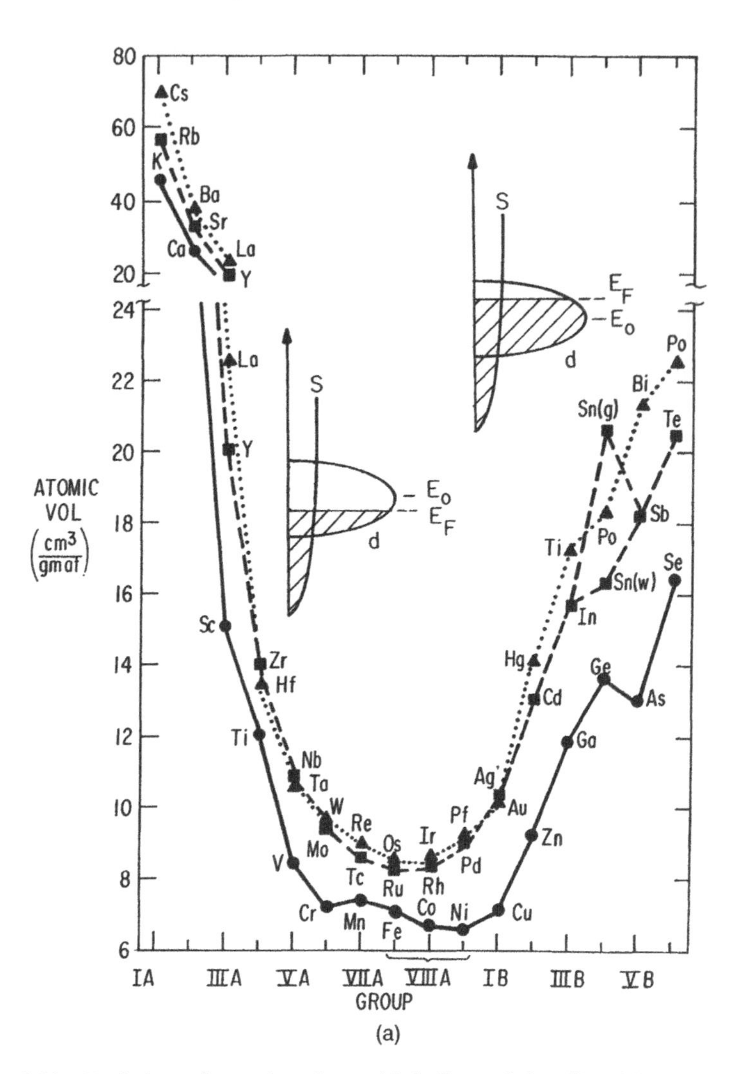

Figure 5.20 Variation of atomic volume (*a*) bulk modulus (b) with atomic number across 3*d*, 4*d* and 5*d* series. Stiffest and shortest bonds (strongest bonding) are observed near half-filled bands. Two valence given for tin are for the gray and white forms (Gschneider 1964). See Janak and Williams (1976) for an explanation of the anomaly in moment-bearing 3d metals.

and 5*d* series show a nearly perfect parabolic trend as expected for filling a band (see Problem 5.5). This trend can be understood from the state densities inserted on Figure 5.20*a* and arguments similar to those used to describe the results of Figure 5.19. A more quantitative explanation of the parabolic trends in Figure 5.20 comes from calculations of the bonding in bulk transition metals [Fig. 5.21, after Gelatt et al. (1977)], which show that the variation of 3*d*

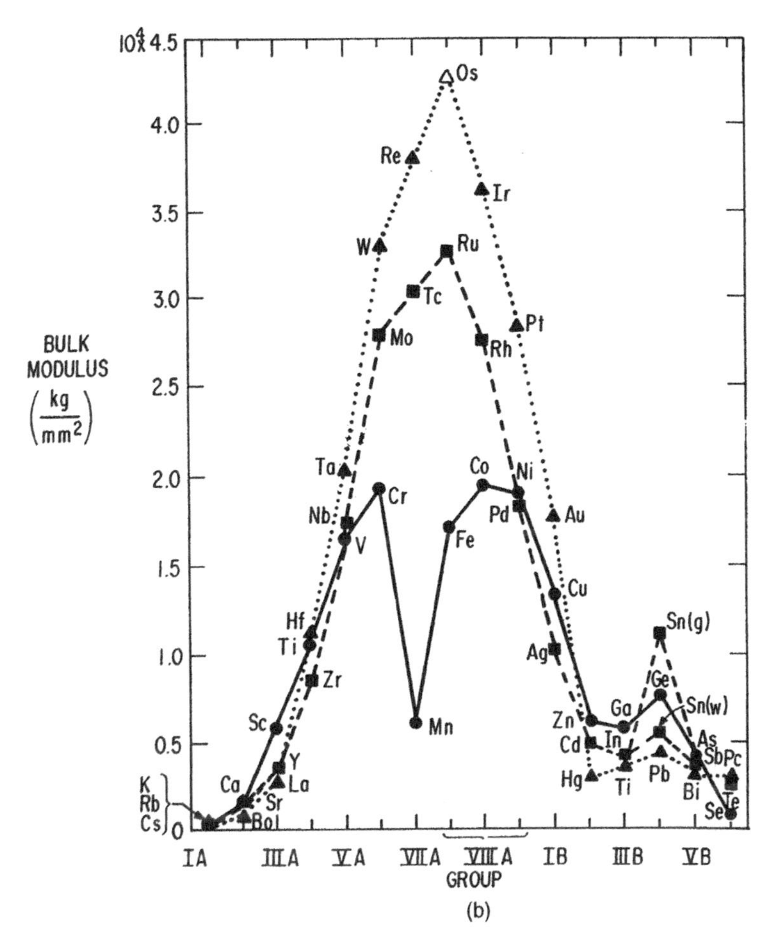

Figure 5.20 (*Continued*)

electron energy with interatomic distance is such that d electrons provide an *attractive* interatomic force $F = -\partial E/\partial r$. The strongly attractive bonding of $3d$, $4d$, and $5d$ electrons is nowhere more evident than in the data of Figure 5.20. The conduction electrons, on the other hand, are under compression and provide a repulsive interatomic force; this is evident from their potential gradient at the equilibrium interatomic spacing in Figure 5.21. These two factors balance each other at the equilibrium bulk atomic spacing so that the total energy is a minimum at r_0.

There is another important feature of Figure 5.20 that bears directly on magnetic moment formation. Anomalies are observed in both interatomic distance and elastic stiffness near the middle of the $3d$ series for the moment-bearing elements: Cr, Mn, Fe, Co, and Ni. These anomalies give an important

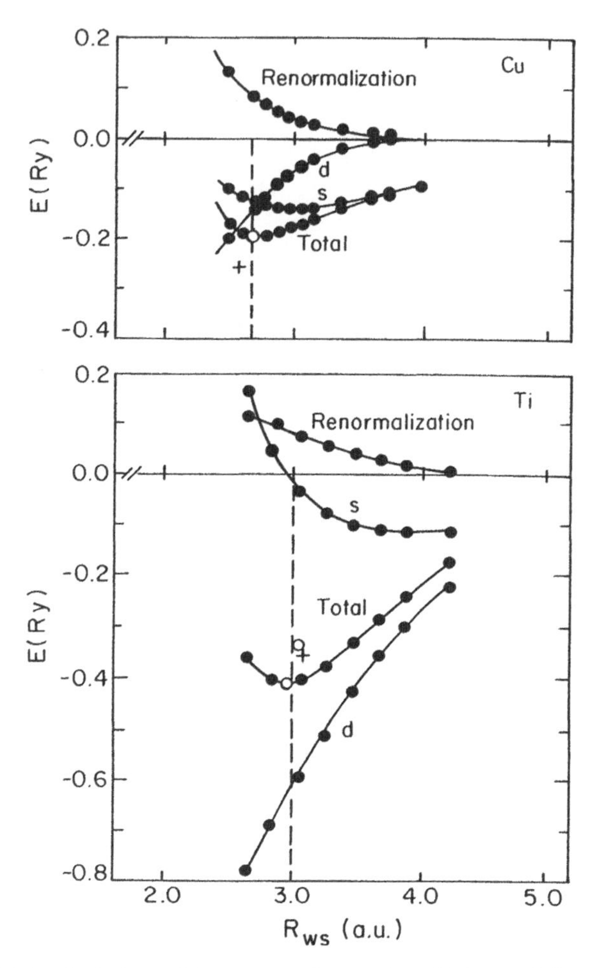

Figure 5.21 Variation with interatomic spacing (Weigner–Seitz radius, R_{WS}) of the energy s and d electrons and of total, cohesive energy for Cu and Ti. The open circles are the calculated energy minima and the crosses mark the experimental values. [Adapted from Gelatt et al. (1977).]

insight into the kinetic-energy cost of forming a magnetic moment in a solid. Magnetic moments are known to form on atoms in the 3d series metals Cr, Mn, Fe, Co, and Ni. [In the first two, these moments order antiferromagnetically and in the latter three species, the moments order ferromagnetically.] The average kinetic energy of the d electrons is given by

$$\langle E_K \rangle \approx \int_0^{E_F} EZ(E)\, f(E)\, dE \tag{5.11}$$

where $f(E)$ is the Fermi–Dirac distribution function and $Z(E)$ is the density of states. The value of $\langle E_K \rangle$ increases when there is a spin imbalance (to partially satisfy Hund's rule) because more electrons now find themselves in states farther from the bottom of the spinup band. Just as increased kinetic energy in a gas causes expansion, so, too, in metals, the kinetic-energy increase due to moment formation causes an increase in atomic volume. In terms of the results in Figure 5.21, the increased kinetic energy of the *d* electrons when they are exchange split, raises the *d*-electron energy curve near r_0. This increase in $E(r_0)$ decreases its slope, which is responsible for the attractive part of the potential. As a result, the material expands. The forces accompanying magnetic moment formation are tremendous as might be expected for Coulomb interactions. Moment formation leads to a giant internal pressure equivalent to several megabars! Note the concomitant softening of the interatomic pair potential (bulk modulus Fig. 5.20b) as atomic volume increases for these metals relative to the parabolic trend of the 4*d* and 5*d* series where magnetic moments do not form. From Figure 5.20, it can be seen that Mn shows an anomalous magnetic expansion of order 10%! The need to accommodate such an expansion drives magnetic iron from what would ordinarily be its stable low-temperature phase, FCC, to the BCC structure that it is observed to assume below 912°C. Above this temperature iron reverts to its nonmagnetic, stable low-temperature phase, FCC. (In most materials, the open BCC structure normally appears only at elevated temperatures where entropy demands the excess volume the BCC structure provides. In fact, iron naturally reverts to the BCC structure above its FCC temperature range.)

It is important to ask whether these volume and stiffness anomalies vanish at the Curie temperature where long-range magnetic order vanishes. Or you might ask why Fe assumes the BCC structure up to 1600°C (presumably stabilized by its magnetic moment) when its Curie temperature is 770°C. It is known that thermal expansion and elastic constants show anomalies at T_C (see Chapter 7); are these Curie temperatures anomalies related to the magnetic expansion in Figure 5.20? Thermal expansion anomalies at T_C amount to strain changes of about 10^{-4}. This is much smaller than the expansions of several percent observed in Figure 5.20 for the magnetic transition metals at zero Kelvin. Thermal expansion anomalies near T_C erase only a small fraction of the excess volume that is a consequence of moment formation. This suggests that while long-range order vanishes at T_C, a local moment may still persist to temperatures well above T_C. There is, in fact, direct evidence (from neutron scattering and from photoemission) that local magnetic moments persist above T_C.

Photoemission data indicate that the exchange splitting between spinup and spindown bands in a magnetic metal does not go to zero at T_C. T_C is the temperature above which *long-range* order vanishes; local magnetic moments still exist above T_C. Electronic band structure reflects relatively local structural aspects of a metal, particularly for *d* states. (Because of the short screening lengths in metals, a given atom is little affected by what goes on more than a few lattice constants away.) The existence of local moments above T_C is also demonstrated directly by neutron scattering studies. The persistence of a local moment above T_C implies a different energy for the spinup and spindown states there. Thus, *intra*atomic exchange persists above temperatures for which *inter*atomic exchange is weakening. It is this intraatomic exchange energy above the Curie temperature that is responsible for the local moment on the α phase of Fe, the phase with more volume per Fe atom. This is the energy that stabilizes BCC Fe relative to FCC iron.

5.6 INDIRECT EXCHANGE

In rare-earth metals and alloys, the magnetic moments are determined by the partial filling of the highly localized $4f$ transition states. The valence states are $5d^1 6s^2$. Many rare-earth metals and alloys order magnetically (ferromagnetism, antiferromagnetism, and/or helimagnetism) up to temperatures of several hundred degrees Kelvin. How is the magnetic moment on one rare-earth metal site coupled to that on another site? Even though Heisenberg exchange is defined for localized wavefunctions, it requires some electron hopping, or exchange from site to site. The probability for an electron to hop from a $4f$ state at one site to that at another site is too small for Heisenberg exchange to operate here. (The radius of the $4f$ wavefunction is only about 10% of an interatomic spacing.)

It turns out that the conduction electrons mediate the exchange interaction. The $4f^n$ moment at one site polarizes the conduction electrons there ($6s$ wavefunctions have appreciable amplitude at small r). These weakly polarized conduction electrons then communicate the spin information to other sites because their wavefunctions are extended. This form of indirect exchange is usually described by a model developed independently by Ruderman and Kittel and by Kasuya and Yosida (Kittel, 1963, 1968) to describe the coupling of nuclear spins to s-election spins by a contact interaction. This model is therefore often referred to as the RKKY model.

When an impurity is placed in an electron gas, the background charge is redistributed to neutralize the impurity charge. The electron gas effectively forms a Fourier series of charge oscillations that sum to cancel the impurity charge. However, the electron gas has a Fermi energy, $h^2 k_F^2/2m$, which places a lower limit on the wavelength of the oscillations that can be called on to cancel the impurity charge. Thus, the Fourier series is truncated and the electron gas is left with oscillations characterized by $k \geqslant k_F$ that are uncompensated. These are called *Friedel oscillations.* Similarly, oscillations in the spin density remain near a magnetic impurity if $k_F^\uparrow \neq k_F^\downarrow$ for the electron gas.

It will be clear after studying the electronic charge distribution about an impurity in a metal (Friedel oscillations; Kittel, 1963) that the spin density about an impurity can be expressed as

$$\rho\uparrow\downarrow(x) = n\uparrow\downarrow\left[1 \pm \frac{9n\uparrow\downarrow}{E_F}\pi J S^z F(2k_F^{\uparrow\downarrow} r)\right] \tag{5.12}$$

where

$$F(x) = \frac{\sin(x) - x\cos(x)}{x^4}$$

and n^i is the concentration of electrons of one spin. This equation describes a spin density that shows damped oscillations with distance from its source. The period of the oscillations lengthens for smaller conduction electron density because $k_F \propto (3\pi^2 N/V)^{1/3}$. The spin polarization $\rho\uparrow - \rho\downarrow$ is proportional to $F(x)$, and at large x it may be expressed as

$$\rho^{\uparrow} - \rho^{\downarrow} \approx \frac{n^2}{E_F} J\langle S^2\rangle \frac{\cos(2k_F r)}{(k_F r)^3} \tag{5.13}$$

The impurity potential that gives rise to this oscillating spin polarization is a localized moment s_i whose spin-dependent exchange interaction with the conduction electrons J affects spinup and spindown electrons differently. Because these oscillations (Fig. 5.22) carry spin information away from the local moment, they allow it to interact with other moments beyond the range of direct exchange interactions. The exchange Hamiltonian for coupling of this spin polarization with a second spin S_j is

$$H_{\mathrm{RKKY}}(x) = S_i \cdot S_j \frac{4J^2 m^* k_F^r}{(2\pi)^3} F(2k_F r) \tag{5.14}$$

This model describes conduction-electron-mediated, indirect exchange interactions. (It was originally formulated to describe the nuclear hyperfine interac-

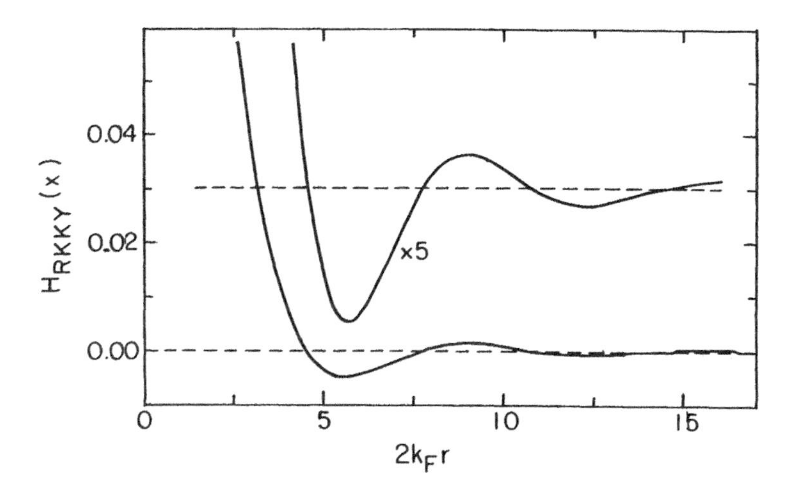

Figure 5.22 Variation of the RKKY oscillatory function $F(x)$ showing how the sign of the indirect exchange interaction can change for atoms at different distances from a given site in a material.

tion between an *s* electron and a nuclear spin). It is very well suited to describing the magnetic properties of rare-earth metals and alloys; the highly localized nature of the $4f$ orbitals precludes an explanation of their Curie temperatures on the basis of direct exchange. Rare-earth metals have higher Curie temperatures than do rare-earth oxides. Because of the oscillations in spin density with distance from a local moment, the RKKY interaction can give rise to ferromagnetic, antiferromagnetic, or helimagnetic spin orderings. The conduction electrons are spin-polarized with an amplitude that oscillates with different periods in different crystallographic directions. Neighbors at different distances may experience different signs of exchange coupling. This coupling is weaker than the interatomic exchange coupling in $3d$ metals and weaker than the superexchange, in $3d$ magnetic oxides.

Exchange Mechanisms in Iron Heisenberg-like exchange, $-2J(S_i \cdot S_j)$, is *not* responsible for interatomic coupling of magnetic moments in iron. The $3d$ electrons have a small spatial extension (about half of the interatomic distance). The free electron specific heat shows only a small magnetic anomaly. Neutron scattering shows the iron moment to be highly localized and to persist above T_C. Despite this, the $3d$ electrons do hybridize partially with the conduction bands. The question is then open as to the mechanism of interatomic exchange coupling in $3d$ metals.

Stearns (1971) noticed that the magnetic moment on iron was unperturbed by silicon additions up to 6 wt%. The observed alloy moment decrease was due to simple dilution alone. If the silicon atoms were randomly distributed over the sites in the BCC iron lattice, there would be a known probability of any given iron atom having a silicon neighbor in its first, second, third, and so on coordination shells (having 6, 8, 12, 24,... sites per shell). Some of the silicon valence electrons would be expected to enter the conduction ($4s$) band where they would dilute the *s*-electron spin polarization (coming from iron *s* electrons). Basically a silicon atom behaves like a hole in the iron lattice, specifically, a deficit in the *s*-electron spon polarization. A sensitive measure of the *s*-electron polarization is the hyperfine field at the nucleus which can be measured by ^{57}Fe Mössbauer effect spectroscopy.

Stearns chose a series of disordered $Fe_{1-x}Si_x$ alloys for investigation. The six peaks of the Mössbauer absorption spectra showed shifts to weaker H_{hf} with increasing silicon content. Analysis of the lineshapes of the outermost absorption peaks allowed an assignment of a hyperfine field contribution to each coordination shell. The strongest contribution to conduction electron polarization at a Fe site was found to arise from the first two coordination shells (14 sites), *and it is negative*. Thus the conduction electrons are spin-polarized opposite to the spin of a central atom out to just beyond the second neighbors. They cannot account for the ferromagnetism of iron. This negative conduction electron spin polarization can be understood to be a consequence of s–d scattering. Conduction electrons may be scattered into vacant $3d$ states of the same spin and energy. There are more $3d$ spindown holes so conduction electrons tend to have negative polarization.

Stearns (1976) proposed an explanation for the strong exchange coupling in terms of indirect coupling mediated by the small fraction of the d electrons that are itinerant (due to their mixing with the s band). Such electrons would be more polarizable and, as they are fewer in number, give rise to longer range oscillations than for conduction electrons. This might leave the nearest neighbors *inside* the first crossover in the RKKY function. Stearns used this argument and the calculated band structure of iron to argue that 5–8% of the seven d electrons in metallic iron (i.e., half a $3d$ electron per atom) are sufficiently itinerant to mediate the interatomic exchange by an RKKY-like mechanism. This model, although controversial, remains a viable candidate for explaining the mystery of long-range magnetic ordering in iron.

5.7 SUMMARY

The magnetic moment variation with valence electron concentration in transition metal alloys shows a striking regularity. Early explanations based on the rigid-band model are deceptive in their simplicity. This model overlooks changes in electronic structure with composition and even crystal structure and is unable to account for the difference in local moment at different sites in an alloy.

A partial improvement to the rigid-band model begins with a treatment of dilute impurities in a transition metal matrix, the VBS model. Extension of the dilute alloy concepts to more concentrated alloys leads to the split-band model. This model accounts for moment differences from site in alloys and for the anomalous branches on the Slater–Pauling curve. Some representative band structures of magnetic materials were interpreted.

The concepts of the band theory of magnetism afford an understanding of elegant sets of data showing variations in modulus and atomic volume in each of the d-transition series. The anomalous compliance and increased atomic volume of the species which bear a local magnetic moment was explained as a consequence of a giant internal pressure of the d-electron gas. This pressure is due to the promotion of minority spin electrons to higher-kinetic energy, spinup states.

The mechanism of conduction-electron-mediated exchange was described in terms of the RKKY model of spin oscillations about a perturbation.

PROBLEMS

5.1 Calculate the electron density needed to produce the first crossover in Figure 5.22. What would the charge density have to be for the first crossover to occur between the second and third nearest neighbors of iron ($a_{Fe} = 2.86$ Å)^2.

5.2 Describe and contrast the valence electronic structure and magnetic characteristics of (**a**) free iron atoms, (**b**) atoms in metallic iron, and (**c**) iron ions in magnetic Fe_3O_4.

5.3 Explain what happens as copper is alloyed with nickel, $Ni_{1-x}Cu_x$. Describe the valence electronic structure and the magnetic moment in terms of a simple band model. Be quantitative where possible.

5.4 Calculate the effect on the total energy of an alloy as states in a flat *d* band are filled.

5.5 Verify Eqs. (5.7) and (5.9).

5.6 (**a**) Explain why ferromagnetism is observed in nickel $(4s\,3d)^{10}$ but not in titanium $(4s\,3d)^4$.

(**b**) Explain why ferromagnetism is observed in the metals of the 3*d* transition series and not in the metals of the 3*p* series (e.g., Al).

5.7 You are running a specialty steel operation. You want to improve the corrosion resistance of a magnetic iron alloy but do not want the saturation magnetization to decrease too much. While talking in your office to an old classmate, you receive a phone call that you can buy several tons of either Mg or Ti at a bargain price. All you know is that Mg has a positive heat of formation with iron (+20 kJ/mol) and Ti has a negative heat of formation (−26 kJ/mol).

(**a**) If both are equally effective in improving corrosion resistance, which would you buy?

(**b**) Justify your choice.

5.8 (**a**) Use your understanding of the electronic structure of metals to explain the general parabolic trend and the magnetic exceptions to that trend shown in Figures. 5.20*a*,*b*. While answering this question, be sure to sketch representative electronic state densities, explain the relation between atomic volume and bulk modulus, and discuss the relation between electronic energy and atomic volume,

(**b**) From Fig. 5.20, estimate the fractional volume change in Fe due to its magnetic moment and give a numerical value for the pressure needed to cause such a volume change. Discuss the possible relation of this effect to the structure Fe assumes at room temperature.

5.9 (**a**) Calculate the dipole energy of a spin $\frac{1}{2}$ particle in the field of another spin $\frac{1}{2}$ particle 2 Å away. Assuming they are free to rotate, what relative alignment will they assume?

(**b**) Calculate the Coulomb energy of two electrons 2 Å apart.

(c) Estimate the Coulomb integral

$$C_{ij} = +\int \phi_a^*(1)\phi_b^*(2) \frac{e^2}{4\pi\varepsilon_0 r_{ij}} \phi_a(1)\phi_b(2)\, d\tau$$

$$= +\int \frac{\rho_a(1)\rho_b(2)}{4\pi\varepsilon_0 r_{ij}}\, d\tau$$

and the exchange integral

$$\mathcal{J}_{ij} = \pm\int \phi_a^*(1)\phi_b^*(2) \frac{e^2}{4\pi\varepsilon_0 r_{ij}} \phi_a(2)\phi_b(1)\, d\tau$$

5.10 Ferromagnetic metals:

(a) Explain why ferromagnetism is observed in nickel and not in aluminum.

(b) Explain why ferromagnetism is observed in the second half of the 3*d* transition series and not in the first half.

BIBLIOGRAPHY

Anderson, P. A., in *Magnetism*, Vol. I, G. Rado and H. Suhl, eds., Academic Press, New York, 1963.

Friedel, J. *Nuoro Cim. Suppl.* **7**, 287 (1951).

Herring, C., in *Magnetism*, Vol. IV, G. Rado and H. Suhl, eds., Academic Press, New York, 1966.

Kouvel, U. S. in *Magnetism and Metallurgy*, Vol 2, A. E. Berkowitz and E. Kueller, eds., Academic Press, New York, 1969, p.523.

Moruzzi, V. L. and P. M. Marcus, in *Ferromagnetic Materials* **7**, 97 (1993).

Niculescu, V., K. Raj, J. I. Budnick, T. J. Burch, W. A. Hines, and A. W. Menotti, *Phys. Rev.* **B14**, 4160 (1976).

Seib, D. H., and W. E. Spicer, *Phys. Rev. Lett.* **20**, 1441 (1968).

Slater, J. C., *Rev. Mod. Phys.* **25**(1) (1953).

Stearns, M. B. *Phys. Rev.* **B4**, 4069 (1971).

Stearns, M. B., *Physics Today* (April 1978).

REFERENCES

Collins, M. F., and D. A. Wheeler, *Proc. Roy. Soc.* **82**, 633 (1963); M. F. Collins and J. B. Forsyth, *Phil. Mag.* **8**, 401 (1963).

Dederichs, P. H., R. Zeller, H. Akai, and H. Ebert, in *Magnetism in the Nineties*, A. J. Freeman and K. Gschneider, Jr., eds., North Holland, Amsterdam, 1991, p. 241.

Dritter, B., N. Stefanou, S. Blügel, R. Zeller, and P. H. Dederichs, *Phys. Rev.* **B40**, 8203 (1989).

Gelatt, D. C., H. Ehrenreich, and R. E. Watson, *Phys. Rev. B* **15**, 1613 (1977).

Gschneider, K. in *Solid State Physics*, Vol. 16, F. Seitz and D. Turnbull, eds., Academic Press, New York, 1964.

Janak, J., and A. R. Williams, *Phys. Rev.* **B14**, 4199 (1976).

Kittel, C., *Quantum Theory of Solids*, Wiley, New York, 1963.

Kittel, C., in *Solid State Physics*, Vol. 22., F. Seitz, and D. Turnbull, eds., Academic Press, New York, 1968.

MacAlister, A. J., J R. Cuthill, R. C. Dobbyn, M. L. Williams, and R. E. Watson, *Phys. Rev.* **B12**, 2973 (1975).

Moruzzi, V. L., J. Janak, and A. R. Williams, *Calculated Electronic Structure of Metals*, Pergamon Press, Elmsford, NY, 1978.

Moruzzi, V. L., P. Oelhafen, A. R. Williams, R. Lapka, and H.-J. Güntherodt, *Phys. Rev.* **B27**, 2049 (1983).

Oelhafen, P., E. Hauser, H.-J. Güntherodt, and K. Benneman, *Phys. Rev. Lett.* **43**, 1134 (1979).

Pimental, G. B., and R. D. Sprotley, *Chemical Bonding Clarified through Quantum Mechanics*, Holden-Day, San Francisco, 1969, p. 115.

Yang, C. Y., K. H. Johnson, D. R. Salahub, J. Kaspar, and R. P. Messmer, *Phys. Rev.* **B24**, 5673 (1981).

CHAPTER 6

MAGNETIC ANISOTROPY

When a physical property of a material is a function of direction, that property is said to exhibit anisotropy. The preference for the magnetization to lie in a particular direction in a sample is called *magnetic anisotropy*. A parameter that describes the magnetization process could be the permeability or the susceptibility. Thus, in general, these parameters are functions of the direction in which the field is applied to the material: $\mu = \mu(\phi, \theta)$ or $\chi = \chi(\theta, \phi)$. Magnetic anisotropy can have its origin in sample shape, crystal symmetry, stress, or directed atomic pair ordering. Shape anisotropy was covered in Chapter 2. Attention is focused here on magnetocrystalline anisotropy. The conventional observations that define magnetocrystalline anisotropy in Fe, Co, and Ni are first described. The theoretical treatment of magnetic anisotropy is then described in three stages, moving from a purely phenomenological, macroscopic picture, to microscopic phenomenology, and finally, to a discussion of the physical origin of magnetocrystalline anisotropy. Throughout, experimental results are included to illustrate the phenomena. We conclude the chapter with a survey of experimental techniques for measuring anisotropy. An appendix is included to outline a simple implementation of group theory in this context.

(Other properties of magnetic materials can also show anisotropy. These include the electrical resistivity (Chapter 16), elastic properties (Chapter 7) and, in certain thin films, the magnetic ordering temperature).

6.1 OBSERVATIONS

Figure 6.1 shows the most obvious experimental manifestations of magnetic anisotropy in single crystals of Fe, Ni, and Co. In BCC Fe, the magnetization

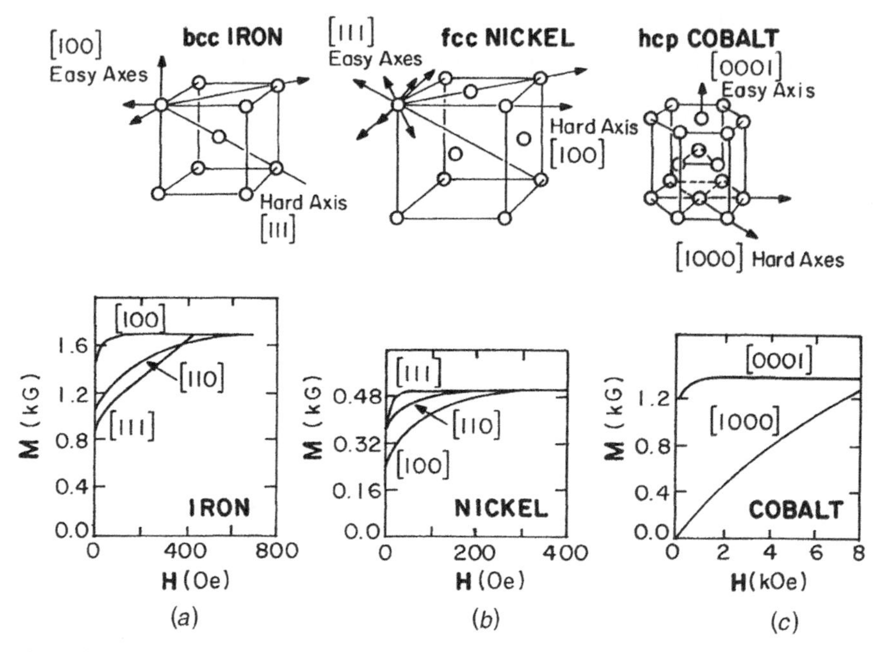

Figure 6.1 Crystal structure showing easy and hard magnetization directions for Fe(*a*), Ni(*b*), and Co(*c*), above. Respective magnetization curves, below.

process is said to be *easy* in the ⟨100⟩ directions and *hard* in the ⟨111⟩ directions; that is, the field needed to magnetize iron to saturation is smaller in the ⟨100⟩ directions than in any others. In FCC Ni, the case is just the opposite: ⟨111⟩ directions are *easy*, ⟨100⟩ *hard*, and the fields required for saturation in the hard directions are smaller for Ni than for Fe. Cobalt is hexagonal, and its easy direction of magnetization is the *c* axis; saturating the sample in the basal plane is very difficult, more than an order of magnitude harder than in the ⟨111⟩ directions in Fe. Note the different scale for the cobalt field axis.

After saturation, reduction of the field to zero leaves more of the magnetization remaining in the direction in which the field had been applied if it is an easy as opposed to a hard direction. In the absence of an external field, the magnetization prefers to lie along the easy directions. Also note that the magnetization remaining at $H = 0$, called the *remanence*, is nonzero for Fe and Ni magnetized in hard directions whereas it is zero for Co magnetized in a hard direction.

A quantitative measure of the strength of the magnetocrystalline (or any other) anisotropy is the field, H_a, needed to saturate the magnetization in the hard direction. This field is called the *anisotropy field*. (It will be seen in Chapter 9 that H_a is also the parameter that governs small rotations of the magnetization from a uniaxial easy direction.) The energy per unit volume needed to saturate a material in a particular direction is given by a generalization of Eq. (2.33):

$$u_a = \mu_0 \int_0^{M_s} H(M)\, dM \xrightarrow[\text{1st order}]{} \mu_0 \frac{M_s H_a}{2} \quad (J/m^3)$$

$$\left[u_a = \int_0^{M_s} H(M)\, dM \xrightarrow[\text{1st order}]{} \frac{M_s H_a}{2} \quad (\text{erg/cm}^3) \right] \tag{6.1}$$

The first-order expressions above apply to magnetization curves linear in the field. The sloped portions of the magnetization curves in Figure 6.1 suggest a process of rotating the magnetization from an easier direction into a harder direction with increasing field in the hard direction. Magnetic anisotropy energy density is the area between the magnetization curves in different crystallographic directions.

The simplest case to consider is that of uniaxial anisotropy, as, for example, in cobalt (Fig. 6.1*c*). The uniaxial anisotropy energy density is the energy needed to saturate the magnetization in the basal plane minus that needed to saturate along the *c* axis (which is essentially zero, by comparison). From Figure 6.1*c*, K_u is about $1400 \times 7000/2$ Oe·emu/cm^3 = 4.9×10^6 erg/cm^3 (or 4.9×10^5 J/m^3), using an average straight line through the curved magnetization–field behavior. This energy density is 20% greater than the measured value of $K_u = 4.1 \times 10^5$ J/m^3. (It should be clear eventually that this 20% difference is not an error.) To calculate the anisotropy in SI units from the data in Figure 6.1, the magnetization must be converted to flux density in tesla ($\mu_0 M = 4\pi M \times 10^{-4}$) and the field axis converted to A/m (H = Oe × 80). Thus, $K_u \approx 1.76 \times 5.6 \times 10^5/2 = 4.8 \times 10^5$ J/m^3. When the strength of the applied field is decreased, the magnetic moment reverts to *c*-axis orientation and its component in the hard direction is zero.

For the anisotropy energy of iron and nickel, more care must be taken in identifying two directions between which the anisotropy energy is defined. If a {110} disk-shaped sample were cut from a cubic material as shown in Figure 6.2, all three principal cubic directions would be found in one plane.

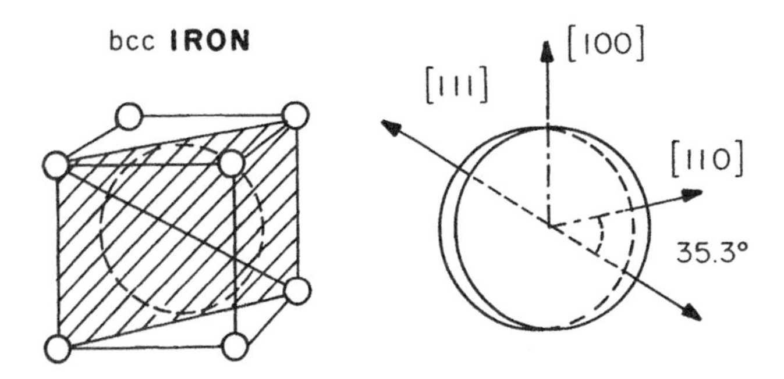

Figure 6.2 A (110) sample cut from a cubic crystal contains all three major symmetry directions: {100}, {110}, and {111}.

By measuring the magnetization in the plane of such a (110) iron disk as a function of angle at constant field, data similar to those depicted in Figure 6.3 would be collected. It clearly shows that the magnetization *below saturation* is anisotropic. This curve is really a representation of the negative of the magnetic potential energy of the system $U/V = -\boldsymbol{M} \cdot \boldsymbol{B}$. Multiplication of the $M(H)$ values by $-B$ would give absolute minima at $\theta = 0°$ and $180°$ ($\langle 100 \rangle$ directions) and local minima at $90°$ and $270°$ ($\langle 110 \rangle$ directions). These energy minima are easy directions of magnetization.

Consider the three following consequences of magnetic anisotropy.

1. Square or rectangular samples of iron cut to explore (001) faces have domain patterns (when demagnetized) as shown in Figure 6.4. The net magnetization is zero but within each domain $\boldsymbol{M}$ points along a crystallographic easy direction.

2. Consider a spherical sample of Ni saturated along a hard [100] direction. As the field decreases below $H_a \approx 230$ Oe (Fig. 6.1), some of the magnetization rotates away from the [100] direction. It is not immediately clear whether it goes toward the four adjacent $\langle 110 \rangle$ directions or the four nearest $\langle 111 \rangle$ directions. (See Problem 6.3.) As H decreases below about 150 Oe, the magnetization rotates toward the four easy $\langle 111 \rangle$ directions that have positive components along [100] because that orientation maximizes M_{100} and thus minimizes the energy. When $H = 0$, all the moments lie in $\langle 111 \rangle$ directions and the remanence is reduced to about $1/\sqrt{3} = 0.577$ of M_s. Contrast this with the behavior of cobalt where the hard axis remanence is zero.

3. Finally, a *polycrystalline* sample of iron may show an M–H curve like that depicted in Figure 6.5 for the *field applied in any direction* (shape effects

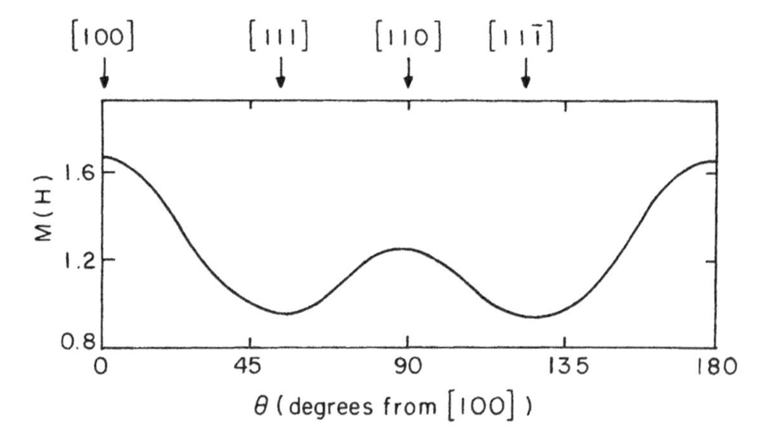

Figure 6.3 Schematic of magnetization for iron measured in a field of about 80 Oe as a function of angle from the [001] direction of a sample such as that shown in Figure 6.2.

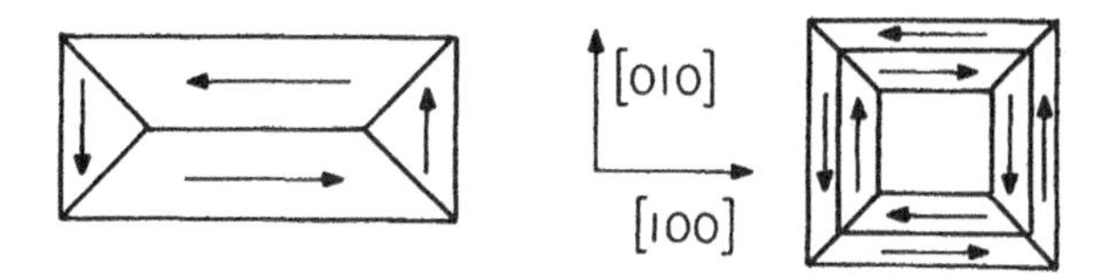

Figure 6.4 Schematic of domain magnetization directions for iron crystals cut along the $\langle 100 \rangle$ directions. The picture frame geometry shown at right was used for early studies of 180° domain wall motion.

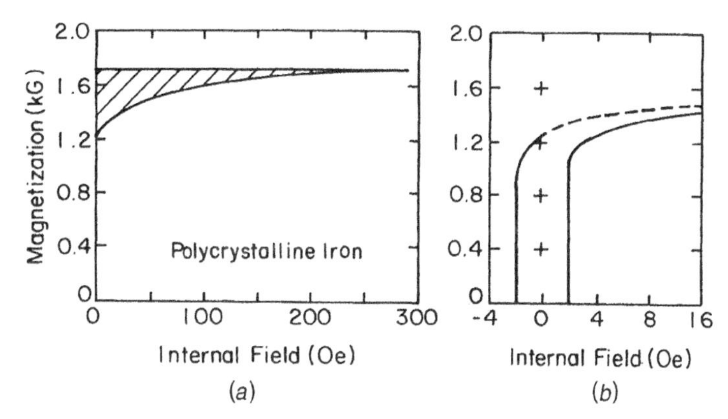

Figure 6.5 Magnetization versus internal field (shape effects removed) for polycrystalline iron: (*a*) in moderate fields; (*b*) in low fields.

removed). Even though the magnetization process is isotropic in the material, saturation is not achieved easily. Note that the low-field loop (right) appears to approach a "technical saturation" well below the true value of about 1740 emu/cm^3. In polycrystalline materials the ease or difficulty in saturating the magnetization in any direction is an average of the single crystal anisotropies as described later in this chapter and in Chapter 9. The area between the M–H curve and the $H = 0$ and $M = M_s$ axes gives $\langle K \rangle = \int w_i K_i \, d\Omega$ with the crystal anisotropies weighted by the number and orientation of the symmetry axes of the crystallites.

6.2 PHENOMENOLOGY OF ANISOTROPY

6.2.1 Uniaxial Anisotropy

Measurements such as those shown in Figure 6.3 on suitably cut crystals of cobalt show that there is negligible anisotropy in the basal plane. So, on the basis of these data and those in Figure 6.1*c*, cobalt is described as uniaxial with a preference for magnetization along the *c* axis. Thus, without any consideration of microscopic mechanisms, it is clear that the free energy of a uniaxial magnetic material must depend on the orientation of the magnetization vector in a way that minimizes the energy when $\boldsymbol{M}$ lies along the easy axis (absent an external field). This contribution to the free energy can be visualized as a three-dimensional energy surface for the orientation of the magnetization vector. The distance from the origin a point on to the surface gives the anisotropy energy for magnetization in that direction in the crystal. In the case of cobalt or other uniaxial crystals (e.g., hexagonal rare earths, barium ferrite, or $Fe_{14}Nd_2B_1$), the energy surface has a minimum in the c direction (Fig. 6.6c) or in the plane normal to the c axis. Figures 6.6a and 6.6b show cubic magnetic anisotropy energy surfaces that will be discussed later. For now, attention will be focused on uniaxial materials.

A magnetic material has an energy function that describes exactly what the equilibrium orientation of magnetization should be under various conditions of field, stress, temperature, and other parameters. There are many different ways of approximating this energy function or surface; most of them involve series expansions in an appropriately chosen set of basis functions. The uniaxial crystal anisotropy energy density is often expressed as a power series of the form

$$u_a = \frac{U_0}{V_0} = \sum_n K_{un} \sin^{2n}\theta \tag{6.2}$$

Careful analysis of the magnetization–orientation curves indicates that for most purposes it is sufficient to keep only the first three terms:

$$u_a = K_{u0} + K_{u1}\sin^2\theta + K_{u2}\sin^4\theta + \cdots \tag{6.3}$$

where K_{u0} has no meaning for anisotropic properties because it is independent of the orientation of $\boldsymbol{M}$. In the convention of Eq. (6.3), $K_{u1} > 0$ implies an easy axis.

For *cobalt* at room temperature, experiments indicate

$$K_{u1} = 4.1 \times 10^5\ \text{J/m}^3 \quad \text{and} \quad K_{u2} = 1.5 \times 10^5\ \text{J/m}^3 \qquad \textit{cobalt}$$

The shape of the energy surface for the first two terms in Eq. (6.2), when $K_{v1} > 0$, is an *oblate* spheroid as shown in Figure 6.6c. When there is no field applied, the magnetization seeks the lowest energy orientation on this surface, namely, along the $\pm z$ axis. The energy surface for a uniaxial material with $K_{u1} < 0$ is a *prolate* spheroid extended along the z axis and having minimum energy in the x–y plane. In zero field, magnetization along any direction in the x–y plane is preferred to any magnetization orientation having a component out of this plane.

As suggested above, the simple expansion in Eq. (6.3) is not unique, although it is used widely for uniaxial materials. It was pointed out by Callen

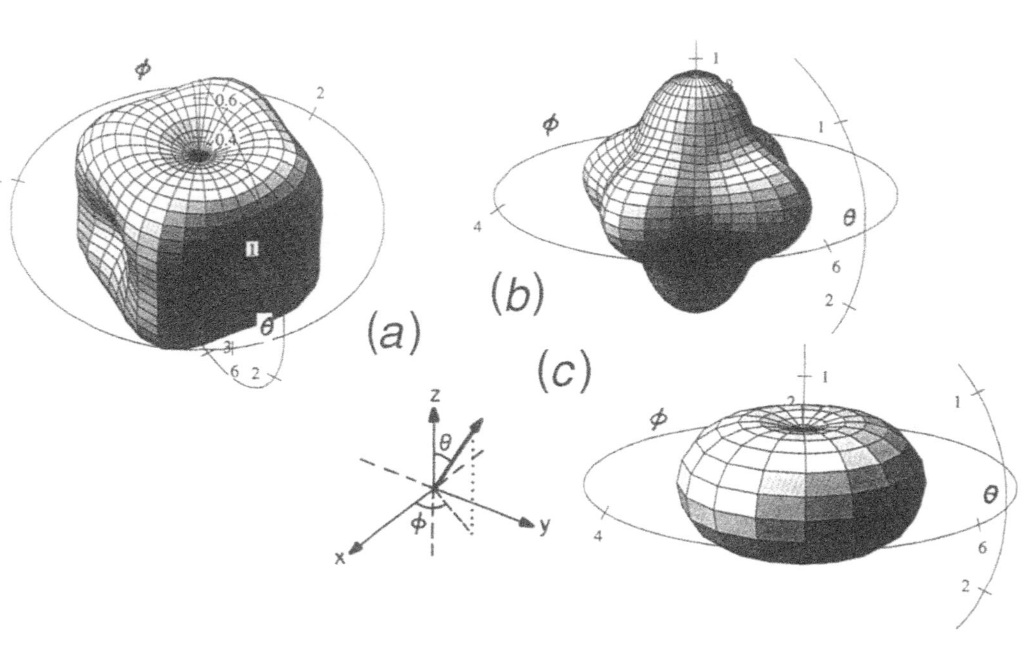

Figure 6.6 First-order anisotropy energy surfaces for iron (*a*), nickel (*b*), and cobalt (*c*). Insert shows the coordinate system. The length of the radius vector to any point on the surface defines the anisotropy energy density in that direction, (θ, ϕ).

and Callen (1960) that expanding the magnetic anisotropy energy in orthogonal, normalized functions, based on Legendre polynomials, is most appropriate:

$$u_a^{\text{hex}} = \sum_{l=0} k_l g_l(\alpha) = k_0 + k_2(\alpha^2 - \tfrac{1}{3}) + k_4(\alpha^4 - \tfrac{6}{7}\alpha^2 + \tfrac{3}{35}) + \cdots \qquad (6.4)$$

where

$$g_l(\alpha) = \sum_{m=-l}^{l} A_l^m Y_l^m(\alpha)$$

The argument α represents the direction cosines of the magnetization. The coefficients A_l^m of the spherical harmonics, Y_l^m are defined so that the polynomials $g_l(\alpha)$ belong to the fully-symmetric representation of the crystal point group. The anisotropy coefficients k_l are found more often in the literature on rare-earth materials. When these orthonormal polynomials are used it should be noted that the anisotropy coefficients take on different values:

$$k_2 = -K_{u1} - \tfrac{8}{7}K_{u2} \quad \text{and} \quad k_4 = K_{u2}$$

so while $K_{u1} > 0$ implies an easy axis, $k_2 > 0$ implies an easy plane. We can now look at the anisotropy of some rare-earth metals.

For the second half of the rare-earth series where data are available, the strength of the anisotropy increases from a very small value at Gd and peaks for Tb and Dy (Fig. 6.7). Ho and Er are both hexagonal and show uniaxial anisotropy of opposite sign: $k_2 = +2.0 \times 10^7$ (easy plane) and -1.2×10^7 J/m^3 (easy axis) at 4.2 K, respectively. Alloys of these two metals show

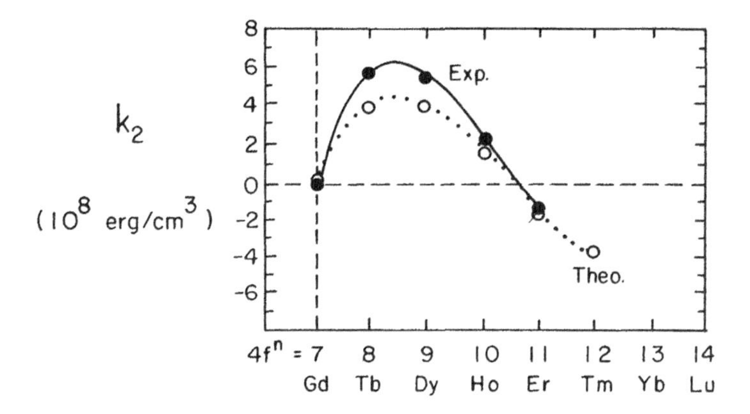

Figure 6.7 First anisotropy constant, k_2 across the second half of the rare-earth series of metals, $k_2 > 0$ implies an easy plane. [Values after Rhyne (1972).]

intermediate anisotropy with $k_2 \approx 0$ for Ho_xEr_{1-x} with $x \approx 0.4$. At this composition the energy surface is nearly spherical.

In order to describe quantitatively the *hard-axis* magnetization process in uniaxial materials, the Zeeman energy, $-\mu_0 M_s \cdot H = -\mu_0 M_s H \sin\theta$ (θ defined as in Figure 6.6), is added to the second-order, uniaxial anisotropy energy of Eq. (6.3). The zero-torque condition, $-\partial u/\partial\theta = 0$, gives

$$2K_{u1}\sin\theta\cos\theta + 4K_{u2}\sin^3\theta\cos\theta = \mu_0 M_s H\cos\theta$$

Below saturation ($\cos\theta > 0$) both sides may be divided by $\cos\theta$ because it is not zero. The result, using $m = M/M_s = \sin\theta$ for the component of magnetization in the field direction, is

$$\mu_0 H M_s = 2K_{u1}m + 4K_{u2}m^3 \tag{6.5}$$

This equation describes a magnetization process that includes a higher-order term; when $K_2^u = 0$, $M(H)$ saturates, $m = 1$, at a field $H_a = 2K_{u1}/M_s$. This equation can be solved analytically (*Standard Mathematical Tables*, 12th ed., CRC Press, Cleveland, OH, 1959) or it is easily graphed on a computer as $HM_s = f(m, m^3)$; see Figure 6.8. The result shows the effect of the K_{u2} term on the magnetization process. The linear anisotropy energy density, K_{u1}, is seen to be the area inside the triangle; K_{u2} is the added area defined by the curved part of the M–H curve (Fig. 6.8). This can be verified quantitatively by integrating Eq. (6.5) as shown in Eq. (6.1) to get the total anisotropy energy, $\int H(M)\,dM = u_a = K_{u1} + K_{u2}$.

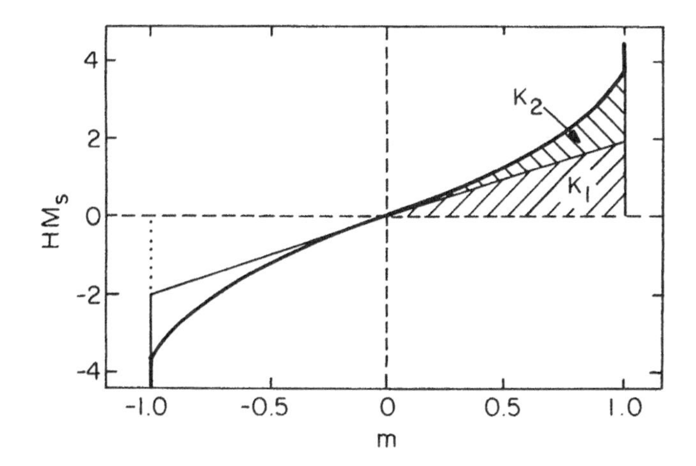

Figure 6.8 Shaded areas indicate K_{u1} and K_{u2} values. The fine line represents the linear magnetization process and the bold line is the result of Eq. (6.5) to second order with $K_{u1} = 1$ and $K_{u2} = 0.38$ (approximate ratio for Co).

Earlier in this chapter, the anisotropy energy of cobalt was estimated from Figure 6.1. The result included the effects of $M(H)$ curvature and so was closer to $K_{u1} + K_{u2}$ than to K_{u1}. If the area had been taken for the triangle extrapolated from the initial linear portion of the M–H curve, a value of K_{u1} very close to 4.1×10^5 J/m^3 would have been determined. The remaining area enclosed by the $M(H)$ curve is due to $K_{u2'}$. What describes the energy needed to saturate cobalt, K_{u1} or K_{u1} plus K_{u2}?

6.2.2 Cubic Anisotropy

The energy surfaces for cubic crystals are not as easy to imagine or construct as uniaxial ones. One could conceive an energy surface for Fe with $\langle 100 \rangle$ easy axes, similar to that shown in Figure 6.6a and try to write a trigonometric function for it. Fortunately, there are more rigorous and elegant ways to get the correct result. An instructive way is to expand the free energy in powers of the direction cosines, α_1, α_2, and α_3, of the magnetization along the three coordinate axes and apply symmetry operations to reduce the number of independent terms. (Note that $\alpha_i = m_i = M_i/M_s$.) Of all the possible terms in the expansion, only those that leave the energy invariant under the symmetry operations of the crystal can be kept. Details can be found in the end Appendix or in the references by Nye (1957) or Juretschke (1974) listed at the end of this chapter.

The anisotropy of a cubic system may then be written as

$$u_a = K_0 + K_1(\alpha_1^2\alpha_2^2 + \alpha_2^2\alpha_3^2 + \alpha_3^2\alpha_1^2) + K_2(\alpha_1^2\alpha_2^2\alpha_3^2) + \cdots \tag{6.6}$$

Using the trigonometric functions of spherical coordinates (θ, ϕ), the coefficient of K_1 in Eq. (6.6) reduces to

$$\sin^4\theta \cos^2\phi \sin^2\phi + \cos^2\theta \sin^2\theta = \frac{\sin^4\theta \sin^2 2\phi + \sin^2 2\theta}{4}$$

Examination of this function in the x–y plane ($\theta = 90°$) shows that it does, indeed, resemble the appropriate cut through the cubic energy surface in Figure 6.6*a*. The second-order term improves the quality of the fit between the model and reality.

The correct cubic expansion in the orthonormal polynomials of Eq. (6.4) is

$$\begin{aligned} u_a^{\text{cubic}} &= k_0 + k_4(\alpha_1^2\alpha_2^2 + \text{cycl.} - \tfrac{1}{5}) \\ &\quad + k_6[\alpha_1^2\alpha_2^2\alpha_3^2 - \tfrac{1}{11}(\alpha_1^2\alpha_2^2 + \text{cycl.} - \tfrac{1}{5}) - \tfrac{1}{105} + \cdots \end{aligned} \tag{6.7}$$

For most purposes, it is not necessary to distinguish the cubic constants k_4 and K_1

$$k_4 = K_1 + \tfrac{1}{11} K_2 \quad \text{and} \quad k_6 = K_2$$

whereas it is always necessary to distinguish k_2, k_4 from K_1, K_2 in uniaxial materials.

The experimental anisotropy constants for *iron* are

$$K_1 = 4.8 \times 10^4 \quad \text{and} \quad K_2 = -1.0 \times 10^4 \text{ J/m}^3 \qquad \textit{iron}$$

and for *nickel*

$$K_1 = -4.5 \times 10^3 \quad \text{and} \quad K_2 = -2.3 \times 10^3 \text{ J/m}^3 \qquad \textit{nickel}$$

at room temperature.

Figure 6.9 shows the composition dependence of the first cubic anisotropy constant for FCC FeNi alloys (*a*) and for BCC FeCo alloys (*b*). The change in sign of K_1 signals a change in the first-order energy surface from one like panel (*a*) to panel (*b*) in Figure 6.6. Thus, the easy axes change from $\langle 100 \rangle$ to $\langle 111 \rangle$ for more than about 75% Ni in Fe or more than about 45 at% Co. Clearly at the point where K_1 changes sign, the energy surface is close to spherical and the anisotropy vanishes to first order. At these compositions, it is very easy to change the direction of magnetization with an applied field, $H_a = 2K_1/M_s \approx 0$. Both of these near-zero anisotropy compositions are technologically important. Iron–nickel alloys, known generically as *permalloys*, show exceptionally soft magnetic properties near the $K_1 = 0$ composition. Equiatomic FeCo alloys, known as *permendurs*, show large saturation magnetization and relatively easy magnetization. See Chapter 13.

Note that near the Ni_3Fe composition the anisotropy depends strongly on chemical ordering of the species on the FCC lattice (see Chapter 14). The order–disorder transformation at Ni_3Fe is from the FCC structure, *Fm*3*m*, above, to the Cu_3Au, *Pm*-3*m*, structure below 517°C. In the ordered state (Ni atoms on the face-centered positions, Fe on cube corners; simple cubic structure), the first cubic anisotropy constant is more negative than it is in the disordered state (where every site in the FCC lattice has a 75% chance of being occupied by a Ni atom and 25% by Fe). Thus, chemical ordering in this system favors $\langle 111 \rangle$ magnetization. FeCo also shows an order–disorder transformation with the BCC structure, *B*1 (*Im*3*m*), stable above, and *B*2 (*Im*-3*m*) stable below 730°C. In the FeCo case, ordering favors $\langle 100 \rangle$ magnetization.

Figures 6.10*a* and 6.10*b* show the phase diagrams for the FeNi and FeCo systems, respectively. In the FeNi system, besides the stable ordered region, Ni_3Fe, there is a two-phase region below 347°C in which α-Fe can coexist with Ni_3Fe. Note also the Curie temperatures above this two-phase region. T_C for the α-Fe and Ni_3Fe components are constant over this range but the volume fractions of the two phases change across this field. For FeCo, the α' field defines the temperature-composition range over which the ordered phase is thermodynamically stable. The cobalt-rich stable phase is HCP up to 410°C

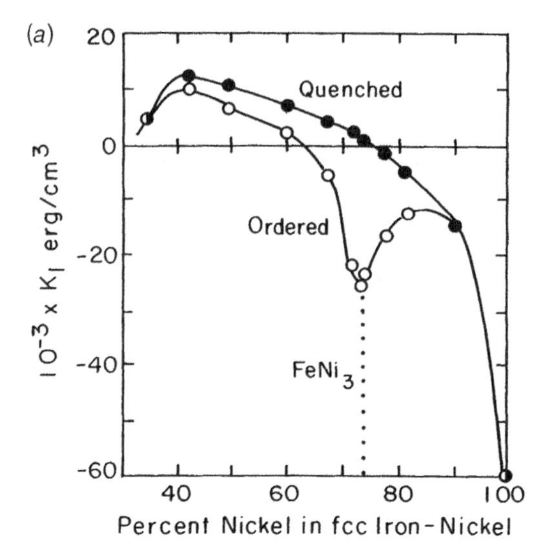

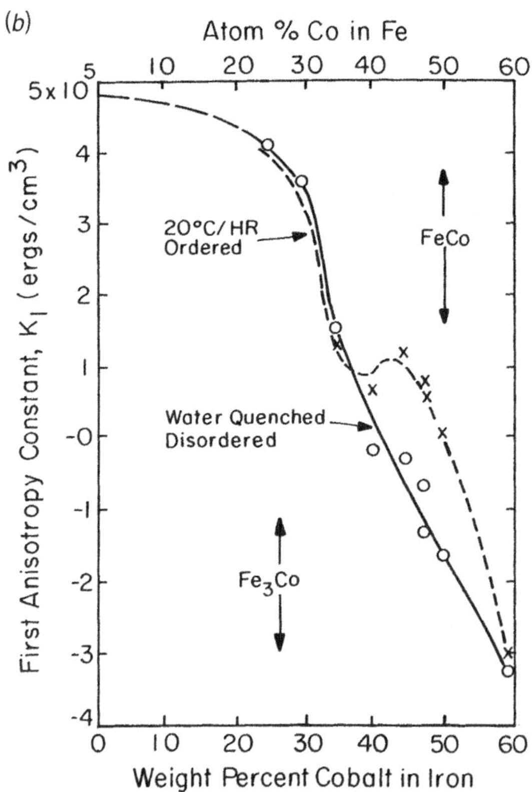

Figure 6.9 First-order anisotropy constant for (*a*) FCC NiFe alloys (Bozorth 1993) and (*b*) BCC FeCo alloys (Hall 1959) at room temperature.

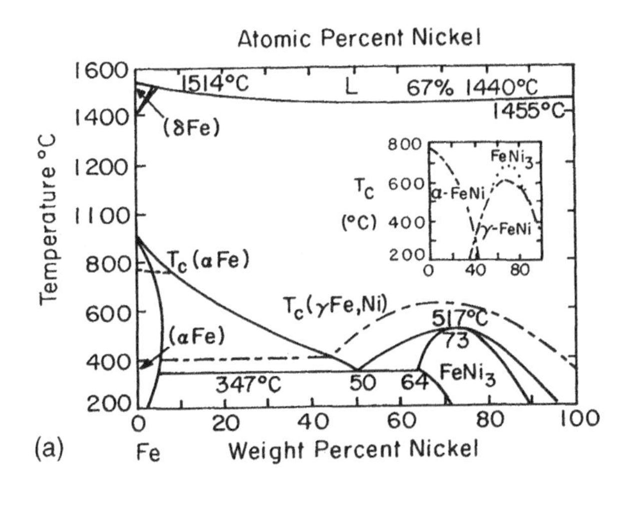

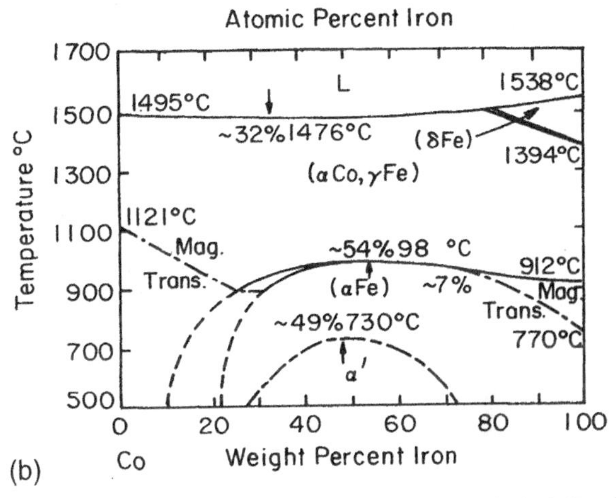

Figure 6.10 Phase diagrams for FeNi (*a*) and FeCo (*b*). (*ASM Handbook* 1994).

and FCC above that. Order–disorder transformations also effect magnetostriction (Chapter 7) and field-induced anisotropy (Chapter 14) in these alloys.

Table 6.1 lists the anisotropy constants for a variety of magnetic materials. Note the sign change in K_2 for Ni from low temperature to room temperature. When reading or quoting anisotropy constants, the form of the free energy should always be clearly understood. For uniaxial materials, $K_1 < 0$ ($k_2 > 0$) implies an easy plane rather than an easy axis. Easy plane uniaxial magnetic materials do not make good permanent magnets, easy axis materials may, provided K_1 is very large (see Chapter 13). The hexagonal compound $SmCo_5$ is the basis of many permanent magnets. It shows preferred c-axis magnetization and a very linear hard-axis M–H loop that saturates at room temperature at $H_a \approx 30$ kOe.

TABLE 6.1 Magnetic Anisotropy Constants K_1 and K_2 (in erg/cm³) for Selected Materials[a]

	($T = 4.2$ K)		(RT)	
	K_1	K_2	K_1	K_2
		3d Metals		
Fe	5.2×10^5	-1.8×10^5	4.8×10^5	-1.0×10^5
Co^u	7.0×10^6	1.8×10^6	4.1×10^6	1.5×10^6
Ni	-12×10^5	3.0×10^5	-4.5×10^4	-2.3×10^4
$Ni_{80}Fe_{20}$	—	—	-3×10^3	—
$Fe_{50}Co_{50}$			-1.5×10^5[b]	
		4f Metals		
Gd^u	-1.2×10^6	$+8.0 \times 10^5$	$+1.3 \times 10^5$	—
Tb^u	-5.65×10^8	-4.6×10^7	—	—
Dy^u	-5.5×10^8	-5.4×10^7	—	—
Er^u	$+1.2 \times 10^8$	-3.9×10^7	—	—
		Spinel Ferrites		
Fe_3O_4	-2×10^5	—	-0.9×10^5	—
$NiFe_2O_4$	-1.2×10^5	—	-0.7×10^5	—
$MnFe_2O_4$	$\approx -4 \times 10^5$	$\approx -3 \times 10^5$	-3×10^4	—
$CoFe_2O_4$	$+ 10^7$	—	2.6×10^6	—
		Garnets		
YIG	-2.5×10^4	—	1×10^4	—
GdIG	-2.3×10^5	—	—	—
		Hard Magnets		
$BaO{\cdot}6Fe_2O_3^u$	4.4×10^6	—	3.2×10^6	—
Sm^uCo_5	7×10^7	—	$1.1 - 2.0 \times 10^8$	—
Nd^uCo_5	-4.0×10^8	—	1.5×10^8	—
$Fe_{14}Nd_2B^u$	-1.25×10^8[c]	—	5×10^7	—
$Sm_2Co_{17}^u$	—	—	3.2×10^7	—
$TbFe_2$	—	—	-7.6×10^7	—

[a]Uniaxial materials are designated with a superscript u and their values K_{u1} and K_{u2} are listed under K_1 and K_2 respectively. The sign convention for the uniaxial materials is based on the $\sin^2\theta$ notation of Eq. (6.2): $K_1 > 0$ implies an easy axis. Units are erg/cm³; divide these values by 10 to get J/m³.

[b]Disordered; $K_1 \approx 0$ for ordered phase.

[c]Net moment canted about 30° from [001] toward [110].

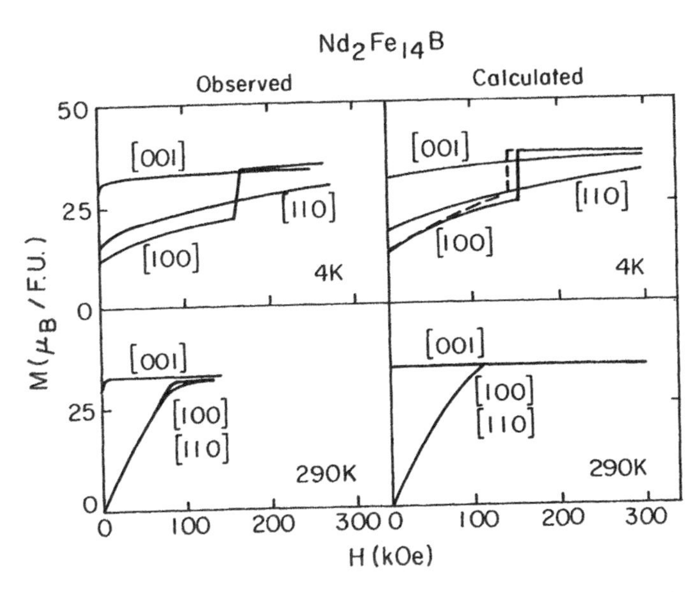

Figure 6.11 Magnetization curves observed and calculated at 4 K and 290 K for $Nd_2Fe_{14}B$. [Adapted from Yamada et al. (1988).]

Tetragonal $Fe_{14}Nd_2B$ also is an important phase for permanent magnets. It shows easy axis anisotropy at room temperature (***M*** along *c* axis), but below 135 K a spin reorientation transition leaves the moments of the various sites canted about 30° toward the [110] direction. Figure 6.11 shows easy- and hard-axis magnetization curves at 4.2 K and at room temperature (Yamada et al. 1988). The behavior at room temperature is what would be expected for a [001] easy-axis magnet: for field applied along [100] or [110], a linear approach to saturation, $\mu_0 M_s = 1.6\,\text{T}$ ($4\pi M_s = 16\,\text{kG}$), is observed with $H_a \approx 90\,\text{kOe}$ indicating $K_1 = -(\frac{3}{2})k_2 \approx 4.8 \times 10^7\,\text{erg/cm}^3$. The hard-axis magnetization behavior at 4.2 K is different. The discontinuity in the [100] magnetization curve suggests a first-order magnetization reorientation transition from the canted orientation to the [100] direction. This is a field-induced manifestation of the transformations observed in zero field at 135 K. Note that the magnetization process along [001] causes a gradual rotation of the net moment toward the *c* axis while the high value of remanence in this direction indicates that the canting leaves the zero-field, net magnetization closer to [001] than to [100]. Application of a field along [110] gives a gradual (second order) magnetization process. These processes can be used to determine the shape of the zero-field energy surface.

The use of mathematical analysis software affords an instructive view of these anisotropy surfaces. The Mathematica® programs used to generate the surfaces in Figure 6.6 are listed below. This program uses *a*, *b*, and *c* for the direction cosines and $t = \theta$, $f = \phi$ in the coordinate system of Figure 6.6. The first-order anisotropy energy surface for iron (or any cubic material with $K_1 > 0$) can be generated in Mathematica by the following statements:

```
a = Sin[t] Cos[f]; b = Sin[t] Sin[f]; c = Cos[t]
r = k0 + k1 (a^2 b^2 + b^2 c^2 + c^2 a^2)
k1 = 0.7; k0 = 0.3
ParametricPlot3D[r a,r b,r c}, {t, 0, Pi},
        {p, 0, 2 Pi}]
```

For nickel, (or any cubic material with $K_1 < 0$) substitute the following step for the third one above:

```
k1 = −0.6; k0 = 1.0
```

For cobalt (or any uniaxial material with $K_u > 0$) substitute these two steps for the middle two above:

```
r = k0 + k1 c^2
k1 = 0.7; k0 = 0.3
```

The nonzero values used for K_0 are arbitrarily chosen to make the shape of the energy surface easier to follow. As an exercise, the student might generate the energy surface for a uniaxial material with $K_u < 0$, look at the shape of $K_2(\theta, \phi)$ for the cubic case, or add a K_4 term to the first-order cubic energy surface generated above. It is possible to create a uniaxial energy surface that leads to the low temperature, [100] magnetization curve of $Fe_{14}Nd_2B$ in Figure 6.11.

6.3 PHYSICAL ORIGIN OF ANISOTROPY

The phenomenological approach of the preceding section told us nothing about the causes of magnetic anisotropy. Macroscopic shape anisotropy has its origins in dipole interactions arising from free poles at surfaces. (As shown earlier, it can be represented in terms of the fields created external to the sample.) There is no mystery to understanding its origin, and only careful effort is required to quantify its effects (Chapter 2). Magnetocrystalline anisotropy on the other hand is a little more subtle in its microscopic origin.

6.3.1 Crystal Field Symmetry and Spin–Orbit Interaction

Magnetocrystalline anisotropy is *not* due simply to the anisotropy of the dipole interaction, although in some materials dipolar interactions are important. Note that for spins constrained to be ferromagnetically aligned by a strong exchange interaction, dipole interactions lower the energy for collinear spins and raise the energy for side-by-side spins. Thus a simple dipole mechanism would fail to explain the preference for FCC Ni to be magnetized perpendicular to its densest atomic planes, {111}. In addition, any interaction of dipolar symmetry vanishes when summed over atoms on a cubic lattice.

How, then, does the local magnetic moment distinguish between different crystallographic directions? In other words, how is $\boldsymbol{\mu}_J$ coupled to the lattice?

The answer lies in the coupling of the spin part of the magnetic moment to the electronic orbital shape and orientation (spin–orbit coupling) as well as in the chemical bonding of the orbitals on a given atom with their local environment (crystalline electric field). If the local crystal field seen by an atom is of low symmetry *and* if the bonding electrons of that atom have an asymmetric charge distribution ($L_z \neq 0$), then the atomic orbitals interact anisotropically with the crystal field. In other words, certain orientations for the molecular orbitals or bonding electron charge distributions are energetically preferred. It is important for magnetocrystalline anisotropy that there be a significant directional character to the bonding.

The roles of crystal field anisotropy and orbital anisotropy are simply illustrated by the analogy shown in Figure 6.12. The three shapes at the top of the figure represent in two dimensions the symmetry of atomic or ionic wavefunctions; the three holes in the soft structure at the bottom represent in two dimensions the symmetry of the crystalline electric field. If the atomic orbital has zero angular moment (spherical charge distribution), it does not matter what the symmetry of the crystal field is; the orbital can take on any orientation with respect to the crystal. Further, since there is no coupling between the direction of spin and orbital angular momentum on the atom in question (i.e., if $\xi \boldsymbol{L} \cdot \boldsymbol{S} = 0$), the spin magnetic moment is free to assume any direction in space dictated by other factors such as applied field.

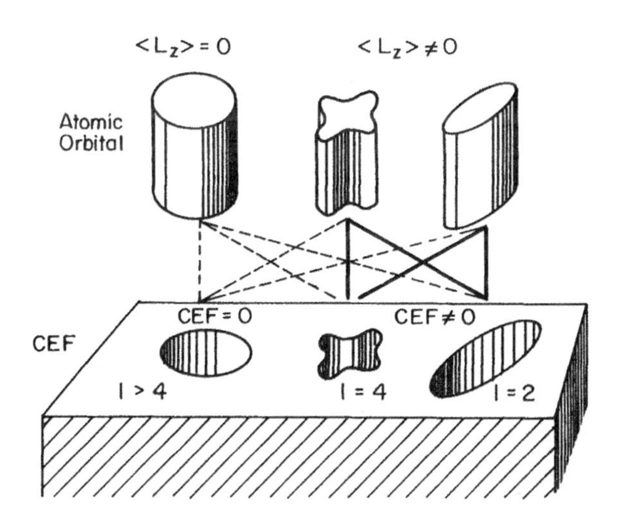

Figure 6.12 Simple representation of the role of orbital angular momentum $\langle L_z \rangle$ and crystalline electric field (CEF) in determining the strength of magnetic anisotropy. Both must have less than spherical symmetry for the orbitals to prefer a particular orientation. Further, if $\boldsymbol{L} \cdot \boldsymbol{S}$ is appreciable, the spin also will prefer particular crystallographic directions. Solid lines show combinations favorable to strong anisotropy; dashed lines, weak.

If the orbitals have nonzero $\langle L_z \rangle$, they may assume any orientation in a spherically symmetric crystal field, but only certain orientations will be preferred in crystal fields of lower symmetry. Further, if $\xi \mathbf{L} \cdot \mathbf{S}$ is not zero, the spin will prefer a specific orientation relative to $\mathbf{L}$.

When there is magnetic anisotropy, application of a field will cause a torque on $\boldsymbol{\mu}_S$, but $\mathbf{S}$ may be coupled to $\mathbf{L}$. Two limiting responses can be distinguished depending on which is stronger: (1) the crystal field energy D (coupling $\mathbf{L}$ to the lattice) or (2) the spin–orbit interaction (coupling $\mathbf{L}$ to $\mathbf{S}$). In case 1, $D > \xi \mathbf{L} \cdot \mathbf{S}$ (as is generally the case in $3d$ transition metals and alloys) μ_S will show a weakly anisotropic response to H_{ext} while μ_L is mostly quenched, that is, locked to a fixed direction by the crystal field energy D. In case 2, $\xi \mathbf{L} \cdot \mathbf{S} > D$ (as is typically the case in rare-earth systems) $\mu_J = \mu_L + \mu_S$ will respond to H_{ext}, but the pull of the crystal field in certain directions may be strong if it is of low symmetry. In the latter case, very large magnetocrystalline anisotropy results and, as the moment is rotated relative to the crystal (necessarily by large fields), the rotation of the anisotropic orbital charge distribution causes large anisotropic strains (magnetostriction). Note from Table 6.1 that the adjacent rare-earth elements ($4f$) Gd and Tb both have uniaxial structures (hence low-symmetry crystal fields) but their first anisotropy constants differ by more than two orders of magnitude. The reason for this difference is that the magnetic state of Gd, $4f^7$, is $\langle L_z \rangle = 0$ while that of Tb is $4f^8$, $\langle L_z \rangle = 3$. Thus the spin–orbit interaction essentially vanishes in Gd but is strong in Tb.

6.3.2 Pair Interaction Model

When the spin is coupled to the orientation of the orbitals, the material shows a magnetic anisotropy that has the symmetry of the crystal field, specifically, of the local atomic environment. Thus, the anisotropic energy of interaction between two atoms can be expressed as an expansion in Legendre polynomials in the angle ψ between the magnetization and the vector to a particular neighbor:

$$w_{ij}(r, \psi) = g(r) + l(r)\left(\cos^2\psi - \tfrac{1}{3}\right) + q(r)\left(\cos^4\psi - \tfrac{6}{7}\cos^2\psi + \tfrac{3}{35}\right) + \cdots \qquad (6.8)$$

Note that the energy dependence on ψ is separated from that on the distance between the pair of atoms r (see Fig. 6.13). The form of the angular terms in the microscopic energy, Eq. (6.8), is the same as those in the orthonormal, macroscopic uniaxial energy expansion, Eq. (6.4). The first term, which is independent of ψ, formally includes spatially isotropic effects such as the exchange interaction, $-J_{ij}S_i \cdot S_j$. It does not contribute to magnetic anisotropy. The second, dipolar term describes anisotropies of uniaxial symmetry. The third, quadrupolar term becomes important in cubic symmetry. The coefficients, $g(r)$, $l(r)$, and $q(r)$, of the first three Legendre polynomials describe how the strength of each component varies with the distance between two atoms. The magnetic anisotropy energy of a crystal can be calculated by

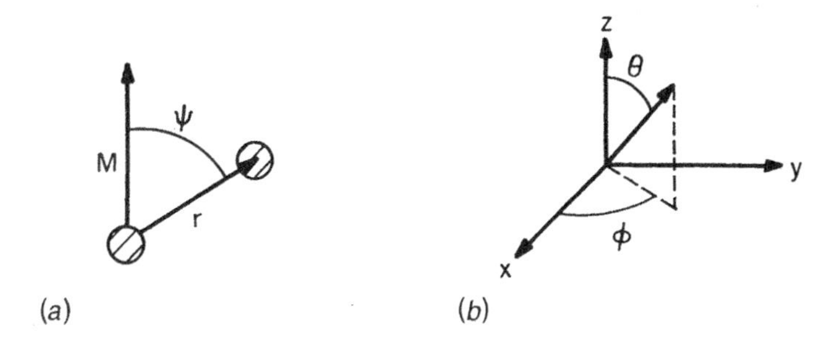

Figure 6.13 Definitions of (*a*) the angle ψ between $\boldsymbol{M_S}$ and the bond axis $\boldsymbol{r}$ and (*b*) θ and ϕ, the angles used to specify the direction of a vector (e.g., $\boldsymbol{M_S}$) with respect to the coordinate axes.

summing this atomic interaction energy $w(r, \psi)$ over all pairs of atoms. When this energy is summed over the nearest neighbors in various cubic structures, the results at $T = 0\,\text{K}$ are

$$W = \frac{-q}{2[\sin^2 2\theta + \sin^2 2\phi \sin^4\theta]} \qquad SC$$

$$W = \frac{4q}{9[\sin^2 2\theta + \sin^2 2\phi + \sin^4\theta]} \qquad BCC$$

$$W = \frac{q}{4[\sin^2 2\theta + \sin^2 2\phi + \sin^4\theta]} \qquad FCC$$

Note that the dipole terms vanish when summed over a cubic lattice. Comparison with Eqs. (6.6) and (6.7) gives values for a *microscopic* parameter, the quadrupole interaction energy, if the macroscopic cubic anisotropy is measured.

$$q = -2K_1 V = \frac{3k_2}{4} \qquad SC$$

$$q = \frac{9K_1}{16} = \frac{-27k_2}{64} \qquad BCC$$

$$q = K_1 = \frac{-3k_2}{2} \qquad FCC$$

Thus for iron, $q = 2.6 \times 10^4\,\text{J/m}^3$ and for nickel $q = -4.5 \times 10^3\,\text{J/m}^3$. These interaction energies should be regarded with cautious interest. The pair interaction model assumes localized spins and this is not completely accurate for transition metals. Nevertheless, it should be noted that the anisotropy calculated using the pair interaction model on Co–Pd multilayers agrees well with both electronic structures calculations and experiment (Victoria and MacLaren 1993).

The pair interaction model thus adds a microscopic dimension to the macroscopic phenomenology developed from group theory.

6.3.3 Crystal Field Splitting

The strength of the anisotropy of the crystalline electric field determines the strength of the magnetic anisotropy that an ion in that field can exhibit. The term "single-ion anisotropy" is given to this type of interaction because a single magnetic species can exhibit anisotropy in a nonmagnetic environment. (This is to be distinguished from anisotropic exchange between two magnetic ions which will be described below.) The single-ion anisotropy coefficient, D is not easily determined for $3d$ metals because their extended wavefunctions give rise to a band of energies rather than a few discrete levels. In insulators and $4f$ metals, where the electronic states important for magnetism are well defined functions with good quantum numbers, it is easier to relate D to fundamental quantities. It has been shown that what is needed for the existence of magnetic anisotropy is (1) an asymmetric crystal field, (2) a nonzero *orbital* angular momentum for the highest occupied electronic state so that the orbital senses the crystal field symmetry, and (3) a nonzero spin–orbit interaction for the *spin* to couple to the crystal field.

For transition metal ions, magnetic anisotropy is usually treated by examining the crystal field splitting of the valence states of the magnetic ion of interest and adding spin–orbit coupling as a perturbation.

The $3d$ electronic wavefunctions ψ_{ml} can be written in spherical coordinates as

$$\begin{aligned}\psi_0 &\propto 3(\cos^2\theta - \tfrac{1}{3})\\ \psi_{\pm 1} &\propto \sin\theta\cos\theta\exp(\pm i\phi)\\ \psi_{\pm 2} &\propto \sin^2\theta\exp(\pm 2i\phi)\end{aligned} \tag{6.9}$$

The commonly used linear combinations of these functions are as follows [see also Goodenough (1963, pp. 51, 10]:

Orbital	Notation: Bethe	Notation: Mulliken
$d_{z^2} \propto \psi_0$ $d_{x^2-y^2} \propto [\psi_2 + \psi_{-2}]$	$d\gamma, \Gamma_3$	e_g
$d_{yz} \propto [\psi_1 - \psi_{-1}]$ $d_{zx} \propto [\psi_1 + \psi_{-1}]$ $d_{xy} \propto [\psi_2 - \psi_{-2}]$	$d\varepsilon, \Gamma_5$	t_{2g}

(6.10)

These functions are sketched in Figure 6.14. The *solid lines* depict contours of constant *positive* wavefunction amplitude and the *dotted* lines, *negative* amplitude. For a free atom, the five d orbitals of a given spin are all of the same energy; that is, they are degenerate. The exchange splitting between spinup and spindown states is generally greater than the crystal field splitting for $3d$ ions. Thus, majority spin states of both t_{2g} and e_g symmetry are generally occupied first, then minority spin states are occupied. In a cubic crystal field of octahedral symmetry (Fig. 6.14), the d_z^2 and $d_{x^2-y^2}$ orbitals, whose electronic wavefunctions point *toward* the six neighboring sites, take on a different energy relative to the three d_{xy}, d_{yz}, d_{zx} orbitals, which are directed *between* neighboring sites. This energy shift is due to the Coulomb interaction between the electronic charge distributions on the various orbitals and that on the neighboring ions, which are generally considered to be point charges. As a result, the $3d$ levels at octahedral or tetrahedral sites split into a triply degenerate manifold and a doubly degenerate manifold of states. If the d states are singly occupied, $3d^1$ or $3d^6$ (i.e., interatomic orbital interactions are bonding in nature), then e_g levels are lower in energy than t_{2g} in octahedral symmetry. If

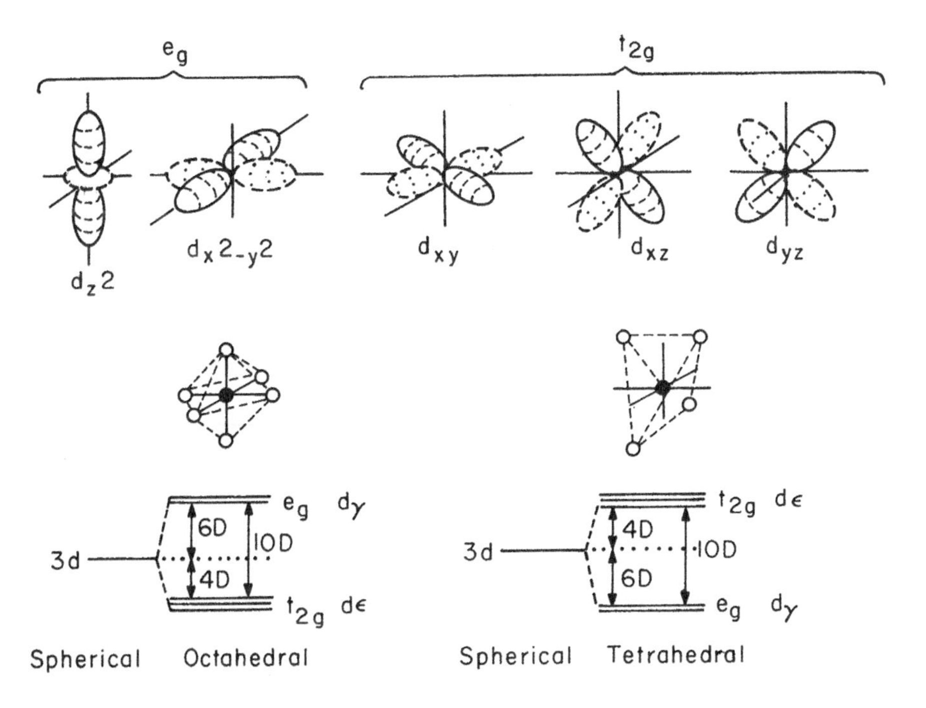

Figure 6.14 Topologies of the five d-orbital wavefunctions are illustrated at top. Center, we show the symmetry of octahedral and tetrahedral sites. Below we show the crystal field splitting for $3d^4$ or $3d^9$ levels in octahedral and tetrahedral site symmetry. The spinup and spindown manifolds are assumed to be separated by energy greater than $10D$.

the $3d$ states are characterized by a single hole, $3d^4$ or $3d^9$ (interatomic interaction is antibonding, which is more often the case in magnetic materials), then t_{2g} levels are stabilized relative to e_g in octahedral symmetry. The antibonding case is illustrated in Figure 6.14.

This crystal field splitting renders invalid the original orbital quantum numbers ($L_z = 0, \pm 1, \pm 2$ defined for a symmetric central field); each manifold now takes on new orbital quantum numbers that sum to zero over the manifold, such as $L_z = 0, \pm 1$ for the triply degenerate manifold. Further, the energetic center of gravity is conserved so that the crystal field energy shifts are in the ratio $\Delta E(e_g)/\Delta E(t_{2g}) = \frac{3}{2}$. Thus $\Delta E(e_g) = +6D$ and $\Delta E(t_{2g}) = -4D$ for a total splitting of $\Delta E = 10D$.

Table 6.2 lists some parameters related to divalent transition metal ions in octahedral and tetrahedral sites.

Magnetic anisotropy is expected to vanish when a manifold is completely filled (i.e., for a singlet state, $\langle L_z \rangle = 0$); the existence of anisotropy should, therefore, be associated with a manifold that is partially filled. Inspection of the anisotropy of the spinel ferrites in Table 6.1 reveals cobalt (in the octahedral site) to have the potential for creating strong anisotropy. Small concentrations of Co^{2+} in magnetic cause the easy axis of Fe_3O_4 to change from $\langle 111 \rangle$ to $\langle 100 \rangle$. From Figure 4.8 it is known that Co^{2+} occupies the octahedral site in the spinel structure (Table 4.4). This leads to a triply degenerate ground state with nonzero orbital angular momentum. This case will be examined in more detail. Whatever a degenerate manifold is partially filled, the system can lower its energy by a further distortion that breaks the symmetry responsible for the degeneracy. This is a statement of the Jahn–Teller theorem [see, e.g., Aschroft

TABLE 6.2 Level Fillings and Degeneracies for Selected Transition Metal Ions[a]

		Octahedral			Tetrahedral		
Ion	Electron Configuration	Orbital Configuration	$\langle L_z \rangle$	Degeneracy	Orbital Configuration	$\langle L_z \rangle$	Degeneracy
Cr^{2+}	$3d^4$	$t_{2g}^3 e_g^1$	$\neq 0$	2	$e_g^2 t_{2g}^2$	$\neq 0$	3
Mn^{2+}	$3d^5$	$t_{2g}^3 e_g^2$	0	0		0	0
Fe^{3+}							
Fe^{2+}	$3d^6$	t_{2g}^1	$\neq 0$	3	e_g^1	$\neq 0$	2
Co^{2+}	$3d^7$	t_{2g}^2	$\neq 0$	3	e_g^2	0	1
Ni^{2+}	$3d^8$	t_{2g}^3	0	0	$e_g^2 t_{2g}^1$	$\neq 0$	3
Cu^{2+}	$3d^9$	$t_{2g}^3 e_g^1$	$\neq 0$	2	$e_g^2 t_{2g}^2$	$\neq 0$	3

[a]After the second entry, $Mn^{2+}(Fe^{3+})$, the completed majority spin filling $(t_{2g})^3(e_g)^2$ is not listed explicitly, only the partial minority-spin state filling is given.

and Mermin (1976)]. Such a distortion could explain the exchange-striction or magnetostriction observed for certain cation configurations. But how can the degeneracy be linked to magnetic anisotropy?

To understand magnetic anisotropy in these systems, it is necessary to look beyond the first neighbor environment. Figure 6.15 shows the environment about the octahedral site in the spinel structure. The six anion (oxygen) neighbors account for the octahedral symmetry; this cubic symmetry alone would not allow for the strong anisotropy exhibited by cobalt ferrites. The next nearest neighbors, six cations, have a trigonal arrangement about the $\langle 111 \rangle$ axis. These second neighbors lower the crystal field symmetry enough to break the degeneracy of the t_{2g} triplet state.

The three d_{xy}-like states recombine to form three new orbitals compatible with trigonal symmetry (much like the $3p$ states and s states combine in tetrahedral environments to form sp^3 hybrid orbitals). The result is a singlet state with charge distribution concentrated along the trigonal axis and a doublet with charge distribution in the plane perpendicular to the trigonal axis. Again, depending on the nature of the cation–cation interactions, bonding or antibonding, the singlet or doublet will be stabilized. *Bonding* second neighbor, cation–cation interactions prevail along the trigonal axis in cobalt ferrite and the singlet state is stabilized (Fig. 6.16). For Fe^{2+} in a trigonal environment as shown in Figure 6.15, the ground state is nondegenerate and $\langle L_z \rangle = 0$. Thus, for Fe^{2+} there is no significant spin–orbit coupling to link the spin direction to a particular crystallographic direction. For cobalt, the highest occupied molecular orbital is the doubly degenerate state for which $\langle L_z \rangle \neq 0$. In fact $\langle L_z \rangle$ is quantized along the trigonal axis for Co^{2+}. Thus spin–orbit coupling can link the spin to the orbital angular momentum that is quantized along $\langle 111 \rangle$.

This model, developed in detail by Slonczewski (1963) explains the stronger anisotropy of Co^{2+} ions relative to the Fe^{2+} ions in spinel ferrites. Further, it

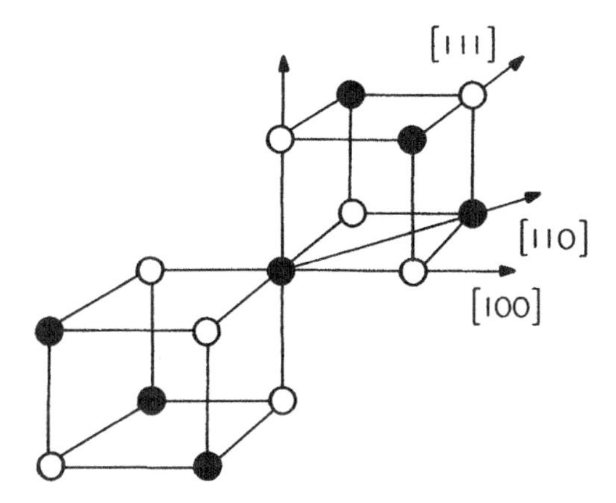

Figure 6.15 Cation in spinel structure is octahedrally coordinated by nearest-neighbor anions and trigonally coordinated by next-nearest-neighbor cations.

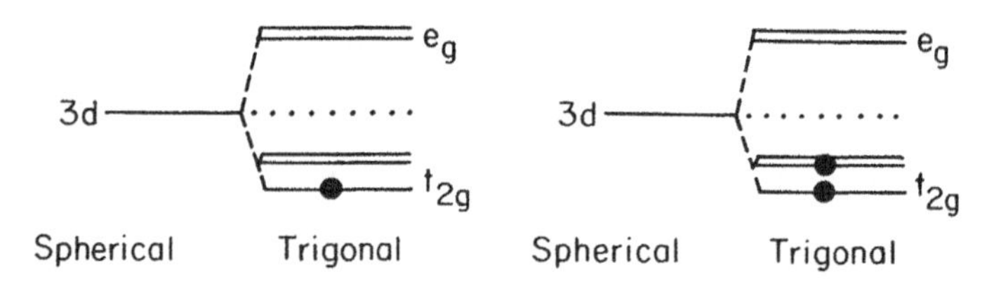

Figure 6.16 Crystal field splitting of antibonding $3d$ levels in trigonal site symmetry. At left the filling of the minority spin states is shown for Fe^{2+} and at right, for Co^{2+}.

accounts for the temperature dependence of the anisotropy and explains the mechanism for magnetic annealing in the technologically important ferrites.

For rare-earth (R) metals and R-intermetallic compounds, the crystal field splittings are weak compared to the spin–orbit energy because the $4f$ states are screened from the crystal field by the $5d$ and $6s$ electrons. Crystal field effects are therefore calculated as perturbations on the atomic energies and thus involve the matrix elements of the crystal field energy evaluated between the $4f$ states that are characterized by their total angular momentum, J:

$$\langle \psi_J^* | E_{\text{cryst}} | \psi_J \rangle \propto \alpha_l \langle r^l \rangle O_1^{|m|}(\boldsymbol{J}) \tag{6.11}$$

The α_l terms, called *Stevens factors* (Stevens 1952), are the lth moments of the $4f$ wavefunctions: $\alpha_l > 0$ indicates prolate orbitals, $\alpha_l < 0$, oblate orbitals. The operators O_l are related to the Legendre polynomials which describe the symmetry of the crystal field (Elliott 1972).

It is known that the shape of the $4f$ orbitals is spherical at the beginning, middle, and end of each half-period. Thus, for rare-earth ions, the Stevens factor and the crystal field anisotropy are expected to change sign between $4f^3$ and $4f^4$, at $4f^7$, and between $4f^{10}$ and $4f^{11}$. That is indeed the case as shown in Figure 6.17 for the calculated (solid line) and observed (open data points) crystal field splitting for rare-earth impurities in Gd. The Stevens factors are shown by the square data points. Compare the variation of these parameters with the metallic anisotropies in Figure 6.7.

6.4 TEMPERATURE DEPENDENCE

Figure 6.18 shows the temperature dependences of the principle anisotropy constants for Fe (Gengnagel and Hofmann, 1968), Ni (Franse 1971) and Co (Pauthenet et al. 1962). For iron and nickel the figures show that $|K_1| > |K_2|$ over the entire temperature range, so their easy axes, $\langle 100 \rangle$ and $\langle 111 \rangle$ respectively, are the same at all temperatures. In cobalt, on the other hand, the easy direction of magnetization is the *c axis* for $K_1 > 0$ and $K_2 > 0$ ($T <$ about 500 K) while the *c plane* is easy for $K_1 < 0$ and $K_1 + 2K_2 < 0$ ($T >$ about

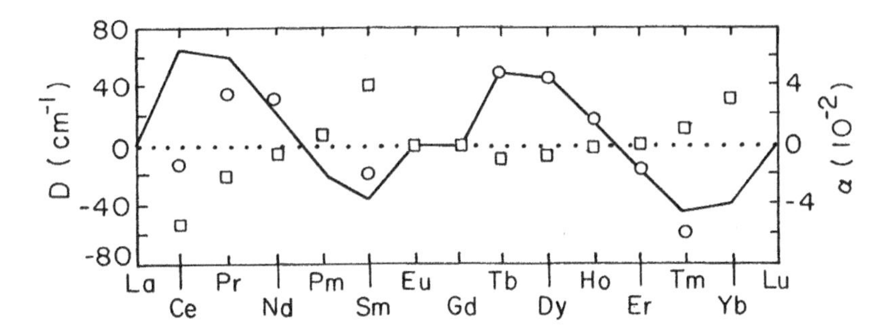

Figure 6.17 Crystal field strength D for rare-earth impurities in Gd: calculated (solid line) and measured (round data points) (Chikazumi et al. 1971). Stevens factors are given by square data points.

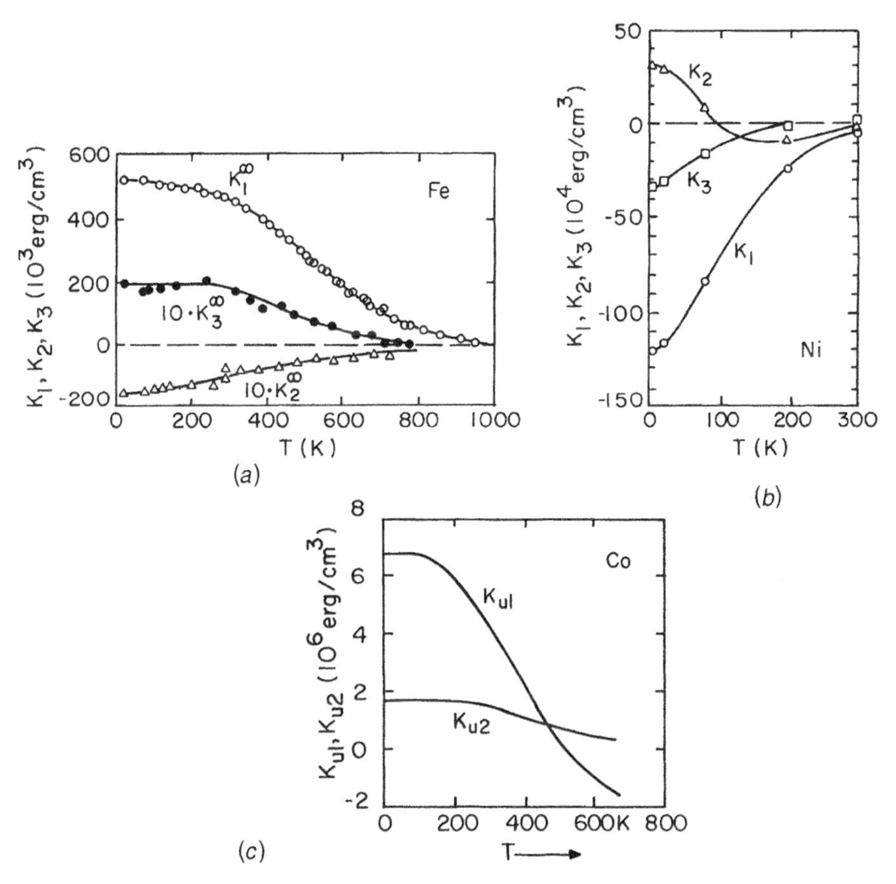

Figure 6.18 Temperature dependence of the principal anisotropy constants of (*a*) Fe (Gengnagel and Hofmann 1968), (*b*) Ni (Franse 1971), and (*c*) Co (Pauthenet et al. 1962).

600 K). In between these limits, Figure 6.18*c* shows $K_1 < 0$ and $K_1 + 2K_2 > 0$, and the magnetization lies on a cone of angle $\theta_0 = (-K_1/2K_2)^{1/2}$. Figure 6.19 shows $-K_1(T)$ in YIG (Rodrigue et al. 1960). The reader is referred to the article by Callen and Callen (1966) for details. Figures 6.20*a* and 6.20*b* show $K(T)$ for Tb and for Co_5Sm and Co_5Nd, respectively; the behavior of $K_1(T)$ for Co_5Sm, increasing in magnitude with increasing temperature, is anomalous.

Naturally, the magnetocrystalline anisotropy vanishes above the Curie temperature as does the long-range magnetic order. However, it is clear from a cursory comparison of these data with the Brillouin function or the $M(T)$ data in Figures 4.8, that the anisotropy generally decreases much more sharply with increasing temperature than does the magnetization. This is of fundamental importance.

The strong temperature dependence of anisotropy energy may be understood from the phenomenology developed earlier in this chapter without yet knowing its physical origin. Consider a cubic energy surface cut as shown in Figure 6.21*a*. The temperature-dependent anisotropy can be defined from the energy surface and Eq. (6.6) as the difference between the hard-axis and easy-axis free energies:

$$u_k^{110} - u_k^{100} = \frac{K_1(T)}{4} \tag{6.12}$$

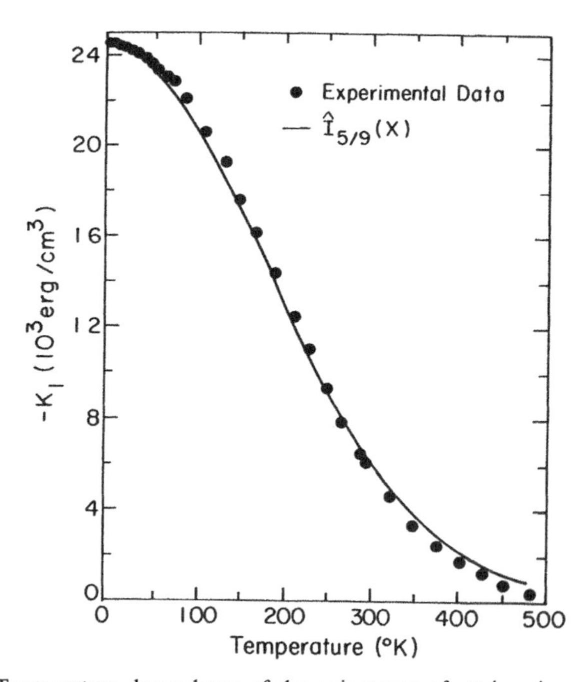

Figure 6.19 Temperature dependence of the anisotropy of yttrium iron garnet (YIG). (After Rodrigue *et al.* 1960.)

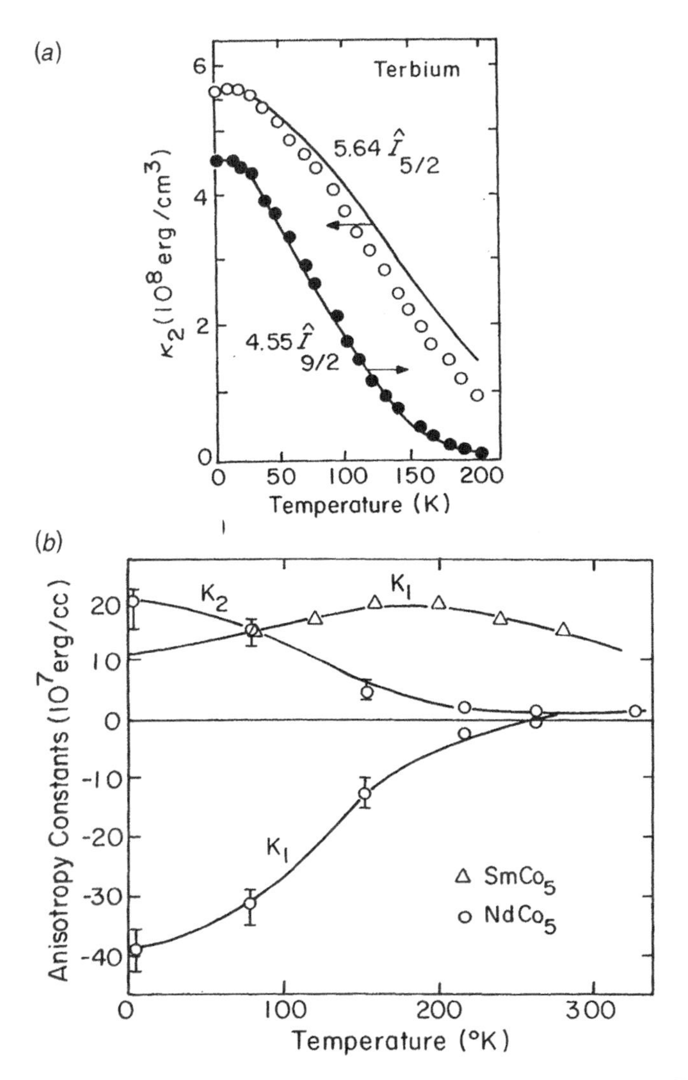

Figure 6.20 (*a*) First two anisotropy constants for terbium after Rhyne (1972)—open circles, K_2, closed circles, K_4 ($\hat{I}_{l+1/2}$ is the modified Bessel function of order l). (*b*) anisotropy constants for $SmCo_5$ and $NdCo_5$ (Tatsumoto et al. 1971).

The temperature dependence of the physical mechanisms responsible for anisotropy should be reflected as a temperature dependence of the energy surface. In 1936, Akulov turned the problem around to give a classical explanation of $K_1(T)$.

At elevated temperatures, thermal energy causes the magnetization to "sample" the energy surface over a small angular range about the minimum energy orientation (Fig. 6.21*b*). Akulov therefore asumed that Eq. (6.12), where

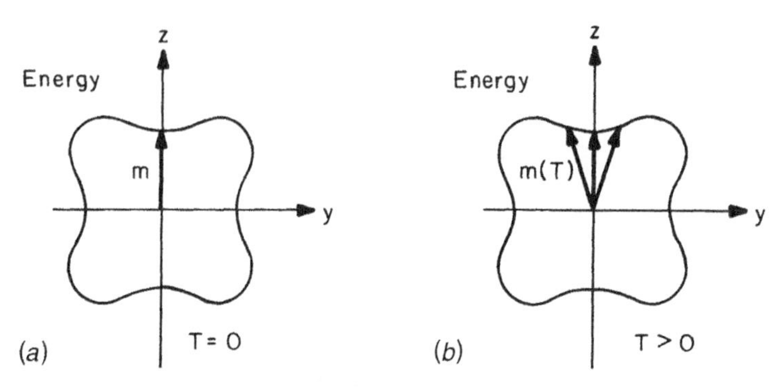

Figure 6.21 A cut through the cubic energy surface [Eq. (6.5), $\phi = 0$] showing that anisotropy energy can be calculated from the difference of the energy between easy and hard directions. When the temperature dependence is treated as a dispersed sampling of this surface, the anisotropy energy decreases more sharply with increasing temperature the sharper the curvature of the energy surface.

$K_1 = K_1(T)$ and the direction of magnetization is independent of temperature, can be approximated by putting the temperature dependence into the orientation of $\boldsymbol{M}$. The energy surface is now assumed to be independent of temperature, and the direction cosines of $\boldsymbol{M}$ become functions of temperature to describe the assumption that the magnetization is thermally distributed over a *small* angular range about some direction on the energy surface (Fig. 6.21*b*). It can be shown by expanding the direction cosines about the [100] and [110] directions that

$$u_k^{100} = K_0 + K_1 \langle \delta\theta^2 \rangle$$

and

$$u_k^{110} = K_0 + \frac{K_1(0)}{4} \langle 1 - \delta\theta^2 \rangle$$

where $\delta\theta \ll 1$ is measured from the [100] or the [110] direction for the two cases, respectively. The anisotropy energy difference between these two directions can then be expressed

$$u_k^{110} - u_k^{100} + \frac{K_1(0)}{4} \langle 1 - 5\delta\theta^2 \rangle \tag{6.13}$$

where $K_1(0)$ has been written to emphasize that the shape of the energy surface is now independent of temperature. We can relate $\delta\theta$ to the temperature dependence of reduced magnetization at low temperatures:

$$m(T) = \frac{M(T)}{M(0)} = \langle \cos\theta \rangle \approx \left\langle 1 - \frac{\delta\theta^2}{2} \cdots \right\rangle$$

Thus in terms of the magnetization, Eq. (6.13) becomes

$$u_k^{110} - u_k^{100} + \frac{K_1(0)}{4} \frac{\langle 1 - 10\delta\theta^2 \rangle}{2} \cdots = \frac{K_1(0)}{4} [m(T)]^{10}$$

and, using Eq. (6.12), cubic anisotropy may be written

$$\frac{K_1(T)}{K_1(0)} = [m(T)]^{10} \tag{6.14}$$

What this says is that the magnetic anisotropy of a cubic system drops off much faster with increasing temperature than does the magnetization itself.

If the same arguments are applied to uniaxial systems one obtains the following at low temperatures:

$$\frac{K_1(T)}{K_1(0)} = [m(T)]^3 \tag{6.15}$$

These steep power dependences turn out to be very important in understanding magnetic anisotropy. The sharper the curvature in the energy surface, the more sharply does the anisotropy drop with increasing temperature.

It can be shown quite generally that if the anisotropy coefficients are regrouped and expressed as coefficients of orthonormal harmonics [Eq. (6.4)], then at low temperature

$$\frac{k_l(T)}{k_l(0)} = m^{l(l+1)/2} \tag{6.16}$$

Here l is the order of the spherical harmonics describing the angular dependence of the local anisotropy ($l = 2$, for uniaxial, $l = 4$ for cubic, and so on). This was derived classically in 1954 by Zener and from quantum mechanics by Callen and Callen. For a review, see, Callen and Callen (1966). Essentially, the magnetization power law for the temperature dependence of the anisotropy is related to the symmetry of the anisotropy energy surface; the sharper the curvature of u_a with θ, the steeper is the temperature dependence of $k_l(T)$. This appears to be a unique case in which the temperature dependence of an effect is determined by symmetry alone. The data in Figures 6.18–6.20 are fit with the general form of the result in Eq. (6.16). The theory is quite successful in accounting for the temperature dependence of anisotropy especially in oxide and rare-earth magnetic materials because in these cases the magnetic moment is well localized. The theory meets with partial success in transition metals where the 3*d* electrons and the magnetic moments are less well localized (Figs. 6.18*a*–6.18*c*). Some anomalies in the temperature dependence of anisotropy constants still occur (e.g., K_2 for nickel changes sign). These appear to be related to contributions from occupation of excited states of different anisotropy which play an increasingly important role as temperature increases.

Callen and Callen (1966) put the theory of magnetic anisotropy (and magnetostriction, as we will see in the next chapter) on solid quantum mechanical footing. A general magnetic Hamiltonian with exchange and crystal field terms can be formulated as outlined below.

Exchange		Crystal field
$S_i \cdot J_{ij} \cdot S_j$	$+$	$S_i \cdot D \cdot S_i$

Note that the exchange terms describe an interaction between spins at two sites while the crystal field term involves the spin at only one site. Figure 6.22 schematically represents these two interactions. The parameter J_{ij} and the third-rank tensor D describe the strength of the *two-ion* exchange and *single-ion* spin–spin interaction, respectively. The diadic terms can be separated into isotropic and anisotropic components:

$S_i J_{ij} S_j$	$+$	$S_i D S_i$	isotropic
$S_i^z J_{ij} S_j^z$	$+$	$D(S_i^z)^2$	anisotropic

Of the two exchange terms, the first is of the form of the usual isotropic *Heisenberg exchange* interaction. The second expresses an *anisotropic exchange* interaction: one that couples spins more strongly when they are aligned along certain directions rather than others. In the case of the crystal field terms, the first term expresses the magnetic energy due to the isotropic parts of the crystal field energy; it contributes to the Madelung energy and has no consequences

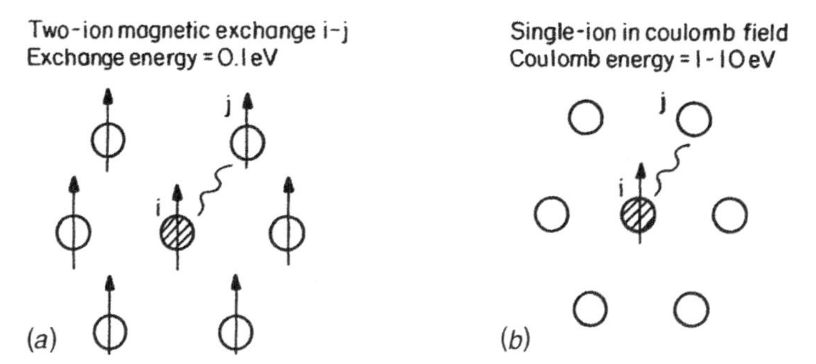

Figure 6.22 Schematic representation of two-ion (*a*) and single-ion (*b*) interactions. The two-ion interaction is magnetic in the sense that it directly couples two spins to different extents depending on the magnitude of their z components. The single-ion interaction does not require spins on the neighboring ions. The site-to-site crystal field interaction is purely Coulombic; the spin enters only through the spin–orbit interaction at one site.

for magnetic anisotropy. The second *crystal field* term describes the anisotropic parts of the magnetic energy due to the nonspherical part of the crystal field. This single-ion, crystal field term is the one that has been focused on here until now.

The exchange interactions are called *two-ion*, or *two-spin*, interactions, while the crystal field terms are called *single-ion* terms because they do not represent a magnetic interaction between two different spins, but rather describe the effect of the Coulomb interaction with the crystal field on the orientation of a moment at a given site. Thus the concept of single-ion, crystal field anisotropy must be added to the anisotropic part of the exchange interaction between two spins.

The anisotropy energy is given by the expectation value of the Hamiltonian $E = \langle H \rangle$ which contains expectation values of spin operators, called *spin correlation functions*, such as $\langle S_i \cdot S_j \rangle$, $\langle S_i^z S_j^z \rangle$, and $\langle (S_i^z)^2 \rangle$. These correlation functions play a central role in understanding the temperature dependence of the anisotropy. Essentially, if the spins at two sites are not well correlated spatially or temporally as T increases, the correlation function falls off more quickly. Recalling Akulov's macroscopic picture, in low crystal field symmetry, site-to-site correlation is not sharply reduced as the magnetic moment samples different directions with increasing temperature. In high crystal field symmetry, spin correlation drops sharply with increasing temperature because the anisotropy energy changes over smaller angular intervals. The coefficients of these correlation functions, J_{ij} and D, contain the physics (exchange interactions, spin–orbit interactions) and determine the strength of the effect. This formalism extends also to magnetoelastic effects (Chapter 7).

The symmetry and insights afforded by Callen and Callen's results can be summarized as follows:

$[\mu_m(T)]^2 \propto \langle S_i \cdot S_j \rangle$ is the local moment *magnitude* two-site correlation; it shows very weak temperature dependence.

$\mu_z^2 \propto \langle (S_i^z)^2 \rangle$ is the single-site local magnetization; it shows weak temperature dependence.

$m_z^2 \propto \langle S_i^z S_j^z \rangle$ is the two-site magnetization longer-range order; it shows stronger temperature dependence than the other correlation functions.

For the reduced, single-ion anisotropy, the theory gives

$$\frac{k_l(T)}{k_l(0)} = \kappa(T) = \frac{\langle S_i^z(T)^2 \rangle}{\langle S_i^z(0)^2 \rangle} \Rightarrow \frac{I_{l+1/2}(X)}{I_{1/2}(X)} \equiv \hat{I}_{l+1/2}(X)$$

where $I(X)$ is a Bessel function and $\hat{I}(X)$ is a reduced, modified Bessel function. The argument X is defined by the temperature dependence of the reduced magnetization $m(T) = I_{3/2}(X)$ so that X can be formally written as $X = I_{3/2}^{-1}(m)$. The temperature dependence of the magnetization is given by the expression

$$\frac{\langle S_i^z(T)S_j^z(T)\rangle}{\langle S_i^z(0)S_j^z(0)\rangle} \equiv m^2$$

With these definitions the first uniaxial and cubic anisotropies vary with temperature as

$$l = 2: \qquad \kappa_2(T) \sim I_{5/2}(X)$$

$$l = 4: \qquad \kappa_4(T) \sim I_{9/2}(X)$$

It is common to plot $\kappa(T)$ data as a function of the parameter m which ranges from unity at $T = 0$ to zero at $T = T_C$. The form of these reduced, Bessel functions is shown in Figures 6.19 and 6.20*a*. The first anisotropy constant of YIG (Fig. 6.19) is well described over the entire temperature range by $I_{5/2}(X)$. While the first anisotropy constant of Tb is not well described by $I_{5/2}(X)$, the second constant agrees well with $I_{9/2}(X)$ right up to T_C (Fig. 6.20).

Approximate temperature dependences of the modified Bessel functions have been calculated by Callen and Callen; they take on different forms above and below approximately $0.6T_C$. Table 6.3 shows the low-T and high-T approximations for single-ion anisotropy for $l = 2, 4$.

Anisotropy originating from shape has a temperature dependence proportional to $[m(T)]^2$ so, except for $l = 2$ above $0.6\,T_C$, it can be distinguished readily from crystal field anisotropy by its temperature dependence.

6.5 MEASUREMENT OF ANISOTROPY

Figures 6.3, 6.5, and 6.8 illustrate simple ways of determining magnetic anisotropy using a conventional magnetometer. Essentially, the anisotropy energy is the energy needed to magnetize a sample to saturation in a specific direction.

More accurate anisotropy measurements are often made using a torque magnetometer. A torque magnetometer works on the principle of balancing a known mechanical torque from a suspension wire against an unknown torque

TABLE 6.3 Approximate Single-Ion Magnetization Power Law Dependences for Uniaxial ($l = 2$) and Cubic ($l = 4$) Systems at Low Temperature and High Temperature

Approximations to $I_{l+1/2}(X)$		$l = 2$	$l = 4$
Low T	$\sim m^{l(l+1)/2}$	m^3	m^{10}
High T	$\sim m^l$	m^2	m^4

associated with the magnetization being rotated away from its easy axis. In this instrument (Fig. 6.23*a*), a sample, such as that shown in Figure 6.2, is suspended on a wire (beryllium copper or platinum are good choices) in a magnetic field. The wire is fixed above to a dial that rotates on a fixed, angle-calibrated scale (two protractors will do). Below the sample holder, a rigid rod hangs suspended through a frictionless bearing in an angle-calibrated scale. Affixed to the rod is an adjustable dial. The sample holder and rod provide sufficient mass to keep uniform tension in the suspension wire.

In zero field, the sample assembly rotates freely to establish zero torque from the suspension wire. The sample orientation can be chosen by setting the orientation of the upper dial; the lower dial can be set to read the same angle, indicating no torque from the suspension. In $B = 0$ the magnetization of the sample will lie along one of the easy directions in the sample where it experiences no anisotropy torque of the form $\partial u_a/\partial\theta$. This easy direction may have an arbitrary orientation θ_0 relative to the direction of a future applied field (Figs. 6.23*b*, and 6.23*c* above).

In a nonzero field whose direction defines $\theta = 0$ (Fig. 6.23*c*, lower), the *sample* is in equilibrium and the *magnetization vector* is in equilibrium. For the

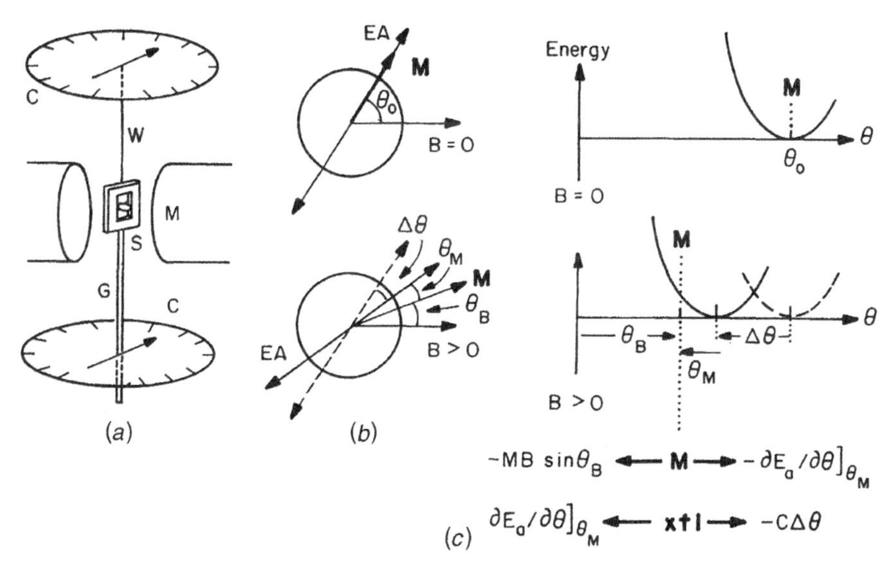

Figure 6.23 Schematic of torque magnetometer, showing (*a*) suspension wire W, guide rod G, sample holder S, calibrated circles C, and magnet pole pieces, M; (*b*) the sample viewed from above (the angles are positive when increasing counterclockwise from the field direction) for $B = 0$ and for $B > 0$; and (*c*) energy diagrams, $E(\theta)$ for $B = 0$, above and $B > 0$ below. For $B = 0$, $\boldsymbol{M}$ coincides with the EA at θ_0. For $B > 0$, the EA is rotated by $\Delta\theta$ and $\boldsymbol{M}$ is rotated by $\Delta\theta + \theta_M$ so that it experiences a counterclockwise torque from the crystal anisotropy energy, which balances the clockwise torque it experiences from the field. Below the $B > 0$ energy surface we schematically show the torque balance conditions on the magnetization vector $\boldsymbol{M}$ and on the crystal (xtl).

magnetization vector, the field torque on $\boldsymbol{M}$, $-MB_0 \sin\theta_B$, is balanced by the crystal anisotropy torque on $\boldsymbol{M}$: $-\partial u_a/\partial\theta]_{\theta M}$. For the *sample*, the suspension torque $-C\Delta\theta$ (counterclockwise in Figure 6.23c, below) is balanced by the reaction of the crystal to the torque it exerts on M at θ_M, namely, $\partial u_a/\partial\theta]_{\theta M}$. The angles θ_0 and $\Delta\theta$ are the measured quantities from the top circle and top-to-bottom circle difference, respectively. The values of C, M_S and B are known. From the crystal equilibrium equations depicted in Figure 6.23 we get $C\Delta\theta = -\partial u_a/\partial\theta$. When $\Delta\theta = 0$, it is required that $\theta_M = 0$ and $\partial u_a/\partial\theta]_{qM} = 0$. As $|\Delta\theta|$ increases, $-C\Delta\theta$ changes from zero. Thus, the measured torque, $-C\Delta\theta$, is numerically equal to $\partial u_a/\partial\theta]_{\theta M}$. At saturation, θ_M is given by $\theta_0 - \Delta\theta$, which is measured.

Measurements of torque versus θ_M for an iron sample like that shown in Figure 6.2, would give a torque curve related to Figure 6.3 by $T = -\partial u_a/\partial\theta$. Conversely, integration of a torque curve

$$\int_0^\theta T(\theta)\, d\theta = -u_a + \text{constant}$$

gives the shape of the anisotropy energy surface $U(\theta)$ (see Fig. 6.24). Torque magnetometers are useful for studying single-crystal anisotropy and thin-film magnetization and anisotropy.

Magnetic anisotropy can also be determined by ferromagnetic resonance and qualitatively by Mössbauer spectroscopy. These methods are not covered here.

6.6 SUMMARY

Several examples of magnetocrystalline anisotropy have been given to define the problem. A phenomenological model, independent of microscopic mechanisms, is able to describe the form of uniaxial, cubic and other anisotropy energy functions (surfaces) by various expansions. These functions, combined with a Zeeman energy, allow calculation of simple magnetization curves. A hard-axis M–H loop in a uniaxial material was calculated to second order in the anisotropy expansion and found to approach saturation more slowly than for the linear (first-order) solution.

The physical origin of anisotropy is based on two necessary effects, a low-symmetry crystal field and a nonzero spin–orbit interaction. A microscopic model of anisotropy was described in which the crystal field symmetry came from summing a magnetic pair interaction over the nearest neighbors in a given lattice. The results were consistent with the macroscopic phenomenology, and comparison of the results of the two methods provides a means of calculating the anisotropy constants K_i or measuring the dipole or quadrupole interactions, $l(r)$ or $q(r)$, respectively. A consideration of crystal field effects in

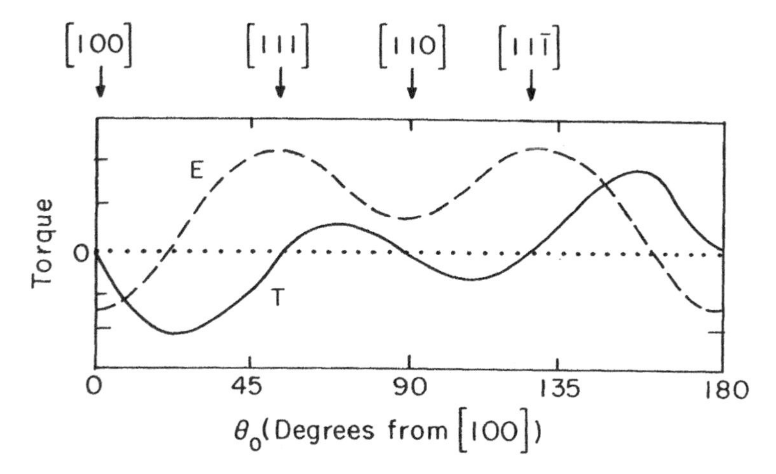

Figure 6.24 Torque curve for iron single crystal (solid line) and orientation dependence of energy (dashed line) from Figure 6.3.

3*d* oxides and rare-earth metals revealed the role of orbital topology in magnetic anisotropy, in particular explaining the strong anisotropy of spinel cobalt ferrites relative to that of magnetite, Fe_3O_4.

The strong temperature dependence of magnetic anisotropy constants was shown to be intimately related to the symmetry of the anisotropy energy function. Uniaxial ($l = 2$) and cubic ($l = 4$) anisotropies drop with increasing temperature as powers of the reduced magnetization, $m^{l(l+1)/2} = m^3$ and m^{10}, respectively. Agreement with this theory is particularly good for local moment materials, namely, oxides and rare earths.

The measurement of magnetic anisotropy by torque magnetometry was described.

APPENDIX: SYMMETRY

It is of interest to express the angular dependence of the free energy for a crystal of a given symmetry. First, the free energy is expanded in spherical harmonics or in powers of the direction cosines α_1, α_2, and α_3 of the magnetization along the three coordinate axes. The latter is more common in the magnetics literature. Symmetry operations are then applied to reduce the number of independent terms in the expansion. [See Nye (1957), or Juretschke (1974), for details.] *Of all the possible terms in the expansion, only those are kept that leave the energy invariant under the symmetry operations of the crystal.* This can be achieved by changing the indices on the direction cosines in the energy terms according to rules defined by various matrices which belong to the symmetry group of the crystal. If the form of the term changes, it is not an allowed combination for the symmetry group of the system. For example, the operations giving cyclic permutation, mirror reflection, and rotation about x by 90° may be represented by the following matrices:

$$P = \begin{pmatrix} 0 & 0 & 1 \\ 1 & 0 & 0 \\ 0 & 1 & 0 \end{pmatrix}, \quad M = \begin{pmatrix} 1 & 0 & 0 \\ 0 & 1 & 0 \\ 0 & 0 & -1 \end{pmatrix}, \quad R_x = \begin{pmatrix} 1 & 0 & 0 \\ 0 & 0 & 1 \\ 0 & -1 & 0 \end{pmatrix} \tag{6A.1}$$

We represent the indices on the direction cosines as a column vector (1, 2, 3) or (x, y, z) and operate on them with the appropriate matrices. The matrix defines the transformation of indices and these transformation rules are applied to the energy expression. Under the operation R_x

$$\begin{pmatrix} 1 & 0 & 0 \\ 0 & 0 & 1 \\ 0 & -1 & 0 \end{pmatrix} \begin{pmatrix} x \\ y \\ z \end{pmatrix} = \begin{pmatrix} x \\ z \\ -y \end{pmatrix} \tag{6A.2}$$

a term such as $\alpha_1\alpha_2^2\alpha_3$ becomes $\alpha_1\alpha_3^2(-\alpha_2)$, which fails the invariance test, so this form is not allowed. Clearly, a term such as $\alpha_1^2\alpha_2^2\alpha_3^3$ would be invariant under R_x.

In magnetism, one other symmetry operation—time reversal invariance—must be considered. Time-reversal changes the direction of electron spin and orbital rotation and hence changes the direction of $\boldsymbol{M}$. Time reversal is equivalent to the inversion operation

$$I = \begin{pmatrix} -1 & 0 & 0 \\ 0 & -1 & 0 \\ 0 & 0 & -1 \end{pmatrix} \tag{6A.3}$$

If the energy is to be invariant under reversal of $\boldsymbol{M}$ [i.e., $u(M) = u(-M)$], then each term in an anisotropy energy expression must have only *even powers* of any direction cosine. Care must be taken with energy terms describing application of a magnetic field; both H and M are inverted on time reversal, so the direction cosines of either H or M can appear in odd powers as long as they are multiplied by odd powers of the direction cosines of the other. The direction cosines of H seldom appear explicitly; its direction is usually chosen as the z axis and $\boldsymbol{M}$ is referred to this system.

Some examples are given for concreteness. An expression such as $K_1\alpha_1\alpha_2$ survives unchanged on inversion but not cyclic permutation, mirror reflection, or rotation operations. $K_2\alpha_1^2\alpha_2^2\alpha_3^2$ survives all four operations. Show that $K_1(\alpha_1\alpha_2 + \alpha_2\alpha_3 + \alpha_3\alpha_1)$ fails to satisfy all of the cubic symmetry operations. This process leads us to what is called the *irreducible representation* for the symmetry at hand. The equivalence of x, y, and z directions in a cubic system demands that α_1^2, α_2^2, and α_3^2 must appear in the anisotropy energy density in such a form that $u_a(\alpha_i, \alpha_j, \alpha_k) = u_a(\alpha_i, \alpha_k, \alpha_j) = u_a(\alpha_k, \alpha_i, \alpha_j)$ and so on. Thus the lowest order term should be $\alpha_1^2 + \alpha_2^2 + \alpha_3^2$, which is equal to 1. Next we consider the cubic terms $\alpha_1^2\alpha_2^2 + \alpha_2^2\alpha_3^2 + \alpha_3^2\alpha_1^2$ and $\alpha_1^4 + \alpha_2^4 + \alpha_3^4$. But

$$1 = (\alpha_1^2 + \alpha_2^2 + \alpha_3^2)^2 = \alpha_1^4 + \alpha_2^4 + \alpha_3^4 + 2(\alpha_1^2\alpha_2^2 + \alpha_2^2\alpha_3^2 + \alpha_3^2\alpha_1^2) \tag{6A.4}$$

so

$$\alpha_1^2\alpha_2^2 + \alpha_2^2\alpha_3^2 + \alpha_3^2\alpha_1^2 = \frac{1 - (\alpha_1^4 + \alpha_2^4 + \alpha_3^4)}{2} \tag{6A.5}$$

and these forms are equivalent within a constant.

So the anisotropy of a cubic system may be written as

$$u_a = K_0 + K_1(\alpha_1^2\alpha_2^2 + \alpha_2^2\alpha_3^2 + \alpha_3^2\alpha_1^2) + K_2(\alpha_1^2\alpha_2^2\alpha_3^2) + \cdots \tag{6A.6}$$

This is Eq. (6.6) in the text.

A similar process can be used to generate uniaxial, tetragonal, and other energy functions.

PROBLEMS

6.1 Show in Cartesian coordinates that **(a)** $\alpha_1^2 + \alpha_2^2 + \alpha_3^2 = 1$ and **(b)** the coefficient of K_1 in Eq. (6.6) is given by (6.7).

6.2 Estimate the hard-axis anisotropy energies of Fe and Ni from Eq. (6.1) and Figure 6.1. Compare your results with those in the text.

6.3 A Ni sphere is saturated in the [100] direction. As the field is decreased the magnetization starts to rotate away from [100] below an internal field of about 230 Oe. Determine from Eqs. (6.6) and the data in Table 6.1 whether it rotates first toward the four nearest $\langle 110 \rangle$ directions or the four nearest $\langle 111 \rangle$ directions.

6.4 Work out the energy difference in Eq. (6.12). [See *J. Phys. Chem. Sol.* **27**, 1271 (1966)] on the two-dimensional cubic energy surface at $\phi = 0$.

6.5 The uniaxial energy surface in Figure 6.6*c* can be represented by Eq. (6.4).

(a) Write the appropriate expression for the case where the z axis is along a hard direction.

(b) Following Akulov's method described in Section 6.4, evaluate $u_K^{\text{hard}} - u_K^{\text{easy}}$ for small angles to show that for a uniaxial magnet $K_1(T)/K_1(0) = [m(T)]^3$ at low temperature.

6.6 Write the first- and second-order anisotropy expressions for tetragonal symmetry analogous to Eq. (6.6) (cubic symmetry). Give the expressions in terms of the direction cosines, and also express them in terms of the spherical angles θ and ϕ in order to combine similar terms.

6.7 Give a quantitative explanation of why the remanence after magnetization in the hard direction is nonzero for Fe and Ni crystals but is zero for Co (see Fig. 6.1).

6.8 Solve for and plot the field dependence of magnetization in a cubic system, $K_1 > 0$, $K_2 = 0$, for a field applied along [110] and along [111].

6.9 What are the conditions on the magnitude of M_s and out-of-plane crystal anisotropy for a thin film to have no in-plane magnetization?

BIBLIOGRAPHY

Chuang, D. S., C. A. Ballentine, and R. C. O'Handley, *Phys. Rev.* **B49**, 15084 (1994).

Hutchings, M. T. in *Solid State Physics*, Vol. 16, H. Ehrenreich and D. Turnbull, eds., Academic Press, New York, 1964, p. 227.

Kanamori, J., in *Magnetism*, Vol. 3, G. Rado and H. Suhl, eds., Academic Press, New York, 1963, p. ??.

Rhyne, J. J., and T. R. McGuire, *IEEE Trans.* **MAG8**, 105 (1972).

Wakayama, T., in *Physics and Engineering Applications of Magnetism*, Y. Ishikawa and N. Miura, eds., Springer Series in Solid-State Science, Berlin, 1991, p. 133.

REFERENCES

Akulov, N., *Z. Phys.* **100**, 197 (1936).

Aschcroft, N. W., and N. D. Mermin, *Solid State Physics*, W. B. Saunders, Philadelphia, PA, 1976.

Bozorth, R. M., *Ferromagnetic Materials*, IEEE Press, New York, 1993.

Callen, E. R., and H. B. Callen, *J. Phys. Chem. Solids* **16**, 310 (1960).

Callen, H. B., and E. Callen, *J. Phys. Chem. Sol.* **27**, 1271 (1966).

Chikazumi, S., K. Tajima, and K. Toyama, *J. Phys. Coll. C1* **32**, 179 (1971).

Elliott, R. J., in *Magnetic Properties of Rare Earth Metals*, R. J. Elliotet, ed., Plenum Press, New York, 1972, p. 1.

Franse, J. J. M., *J. Phys. Coll C1* **32**, 186 (1971).

Gengnagel, H., and U. Hoffmann, *Phys. stat. solidi* **29**, 91 (1968).

Goodenough, J. B., *Magnetism and the Chemical Bond*, Wiley, New York, 1963.

Hall, R. C., *J. Appl. Phys.* **30**, 816 (1959).

Juretschke, H. J., *Crystal Physics*, Benjamin, New York, 1974.

Nye, J. F., *Physical Properties of Crystals*, Oxford Univ. Press, Oxford, UK, 1957.

Pauthenet, R., Y. Barnier, and G. Rimet, *J. Phys. Soc. Jpn.* **17**, Suppl. B-1, 309 (1962).

Rhyne, J. J., in *Magnetic Properties of Rare Earth Metals*, R. J. Elliott, ed., Plenum Press, New York, 1972, p. 105.

Rodrigue, G. P., H. Meyer, and R. V. James, *J. Appl. Phys.* **31**, 376S (1960).

Slonczewski, J. C., in *Magnetism*, Vol. 3, G. Rado and H. Suhl, eds., Academic Press, New York, 1963.

Stevens, K. W. H., *Proc. Phys. Soc.* (London) **A65**, 209 (1952).

Tatsumoto, E., T. Okamoto, H. Fuji, and C. Inoue, *J. Phys. Suppl.* **32**, C1-550 (1971).

Van Vleck, J. H., *Phys. Rev.* **52**, 1178 (1937).

Victora, R. H., and J. H. MacLaren, *Phys. Rev.* **B47**, 11583 (1993).

Yamada, M., H. Kato, H. Yamamoto, and Y. Nakagawa, *Phys. Rev.* **B38**, 620 (1988).

CHAPTER 7

MAGNETOELASTIC EFFECTS

7.1 OBSERVATIONS

The various contributions to the thermal expansion of a magnetic material are depicted in Figure 7.1*a*. The dot/dashed line (marked "no local moment") indicates normal linear thermal expansion for a solid (Grüneisen behavior). The dashed line labeled "local moment" is displaced from the first to indicate the volume expansion that accompanies the formation of a local magnetic moment. Because the *local* magnetic moment does not vanish immediately above T_C but merely loses its long-range ordering, the internal pressure associated with it does not vanish completely above T_C (Fig. 7.1*a*, solid line). The solid lines show the form of the thermal expansion for a ferromagnet above and below T_C. Below T_C, additional magnetovolume effects due to *long-range* magnetic ordering are turned on; they may add to, or subtract from the volume expansion due to the presence of a local moment. The slope of these solid lines (the thermal expansion coefficient α) can be of either sign just below the Curie temperature. All of these effects are isotropic, involving the bulk modulus. Figure 7.1*b* shows the volume expansion, $\omega = \Delta V/V$, and the linear coefficient of thermal expansion, $\alpha = \Delta l/l\Delta T$, measured for Ni by Kollie (1977). The volume expansion is referenced to the paramagnetic state so $\omega < 0$. Note that for Ni, ω follows the type of behavior indicated by the lower solid curve in Figure 7.1*a*. The fractional volume deficit in Ni at 4.2 K is of order $\Delta V/V \approx -0.12\%$ relative to the extrapolated high-temperature ($T \gg T_C$) volume. (This is a small effect compared to the 4% local-moment expansion in Ni indicated in Fig. 5.20). In some alloys where T_C is just above room temperature and the magnetovolume effects associated with long-range order turn on a positive

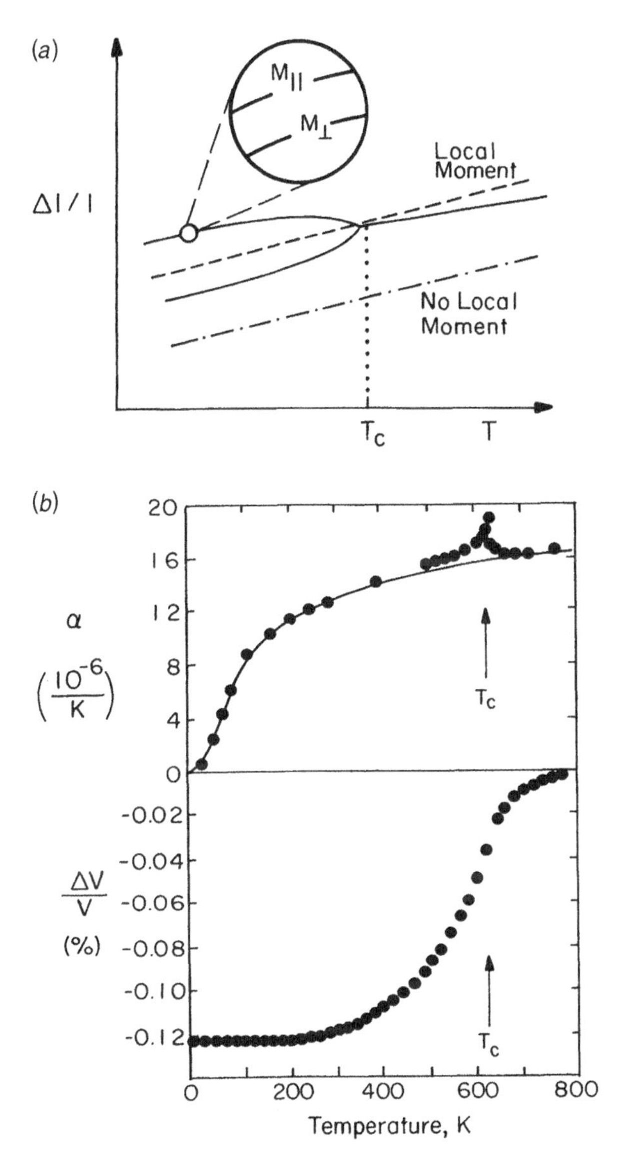

Figure 7.1 (*a*) Schematic of the thermal expansion of a magnetic material as a function of temperature illustrating the increased volume due the presence of a local magnetic moment and the onset of magnetic anomalies below the Curie temperature; (*b*) volume expansion and thermal expansion coefficient of Ni (Kollie 1977). A small anisotropic strain, depending on the direction of magnetization (circled inset, left), is also observed below T_C. The latter is usually referred to as anisotropic magnetostriction.

strain below T_C, the thermal expansion can be very close to zero over several tens of degrees. These alloys are called *invar alloys* because their dimensions are invariant with temperature. The classic example is $Fe_{70}Ni_{30}$.

These *isotropic* effects are called *volume magnetostriction* or, when the magnetic ordering is produced by an applied field, they are called forced *magnetostriction*. Such isotropic strains are not the main focus of this chapter.

On a smaller scale, the volume expansion can show an anisotropy for $T < T_C$, that is, the linear strain is different in different directions relative to the direction of magnetization (Fig. 7.1*a*, circled inset). Thus, the magnetization vector $\boldsymbol{M}$ has associated with it a stress which causes a mechanical deformation of the material. This anisotropic strain is the main subject of the present chapter.

The field dependence of this anisotropic strain is shown schematically in Figure 7.2 for an isotropic material ($e_{\parallel}$ for strain measured parallel to the field and $e_{\perp}$ for strain measured perpendicular to the field). This anisotropic strain associated with the direction of magnetization was first observed in iron in 1842 by Joule. These strains, $\Delta l/l = \lambda$, called Joule or anisotropic magnetostriction, can range from zero ($\lambda < 10^{-7}$) to nearly $\pm 10^{-4}$ in 3*d* metals and alloys and to over $\pm 10^{-3}$ in some 4*f* metals, intermetallic compounds, and alloys. The magnetic stress tensor, called the magnetoelastic coupling coefficient, with components B_{ij}, can be related to its magnetostrictive strains by a analogy with Hook's law: $B_{ij} \propto -c_{ijkl}\lambda_{kl}$. For Ni, $B_1 = 6.2\,\text{MPa}$ and, given its Young's

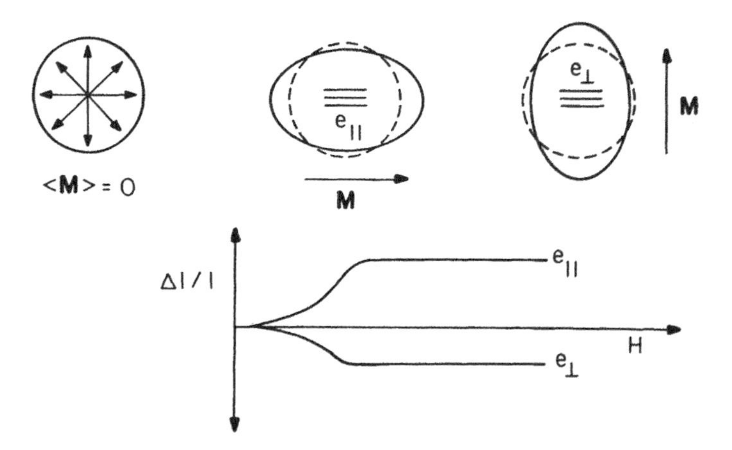

Figure 7.2 When a demagnetized sample has its magnetization aligned by an external field, the sample strains anisotropically. The direction in which the strain is measured in the samples above is indicated by the three parallel lines. The strain in the direction of magnetization will be opposite in sign to the strain perpendicular to the direction of magnetization. Notice that $e_{\parallel}$ and $e_{\perp}$ need not be related by Poisson's ratio because of ;the arbitrariness of the zero-field magnetization configuration, which defines the zero-field strain. The strains depicted above are those for a material with a positive magnetostriction constant: $(\Delta l/l)_{\parallel} > 0$.

modulus, $E = 200$ GPa, and the observed magnetostrictive strains are of order 30×10^{-6}. In describing anisotropic magnetostriction in a material one can refer to its magnetostrictive constants, λ_S, the strains produced at magnetic saturation, or to its magnetoelastic coupling coefficients, B_{ij}, the magnetic stresses causing λ_S.

The inverse effect is also important. Stressing or straining a magnetic material can produce a change in its preferred magnetization direction, which is manifested in its magnetization curve (Fig. 7.3). These phenomena are called *inverse Joule effects*, *Villari effects*, *piezomagnetism*, or, most often, *stress-induced anisotropy*. If λ_S is positive, it is easier to magnetize the material in the tensile stress ($\sigma > 0$) direction. It is harder to magnetize a material in a direction for which $\lambda_s < 0$ and $\sigma > 0$ or for which $\lambda_S > 0$ and $\sigma < 0$.

Torsional effects can also result from magnetostriction, but they are associated with specific magnetization distributions in a material. A current passing through a magnetic material in the direction of magnetization causes a twisting of the magnetization around the current axis, and, if $\lambda_S \neq 0$, a torsional motion of the sample occurs. This is the Wiedemann effect. In the inverse Wiedemann effect, named after Matteucci, a mechanical twisting of the sample causes a voltage to appear along the sample length, consistent with Faraday's law and the strain-induced magnetization change.

The existence of anisotropic magnetoelastic (ME) effects implies the existence of a coupling between the magnetization direction and mechanical strains. Thus, the magnetic anisotropy energy must contain ME terms that depend on both strain and the magnetization direction. The anisotropic strain effects arising from ME terms in the free energy are the focus of this chapter.

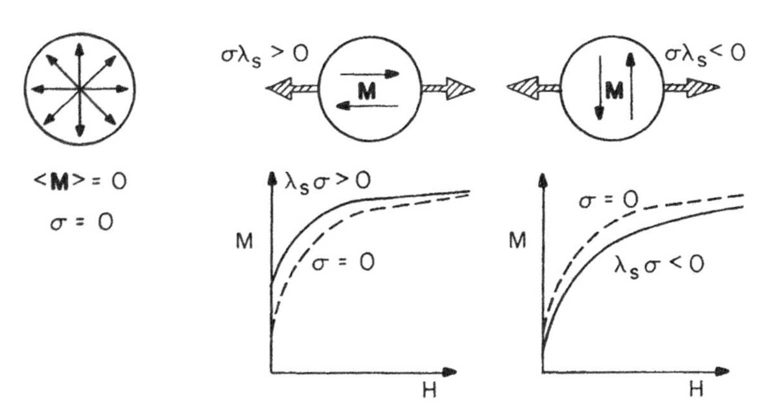

Figure 7.3 Imposing a strain on a magnetic material by a mechanical stress alters the preferred direction of magnetization (magnetic anisotropy) and thus changes the shape of the M–B_0 curve below saturation. The cases above illustrate the changes observed for an isotropic material. When the product $\lambda_s\sigma$ is positive, the magnetization is favored along that stress axis (shaded arrows).

Field Dependence of Joule Magnetostriction The *anisotropic magnetostrictive strain* relative to the direction of magnetization may be described for an *isotropic material* by the relation

$$e = \tfrac{3}{2}\lambda_s(\cos^2\theta - \tfrac{1}{3}) \tag{7.1}$$

Here $e = \Delta l/l$ is the strain measured at an angle θ relative to the saturation magnetization direction and λ_s, the saturation magnetostriction coefficient, is a measure of the magnitude of the strain on changing the direction of magnetization in the material. The strain e is sometimes called the *magnetostriction* λ. From Eq. (7.1), e is a function of the direction of $\boldsymbol{M}$ or of the applied field, so it should not be confused with the parameter λ_s which is a material constant. Figure 7.2 shows the variation of strain along a fixed direction in a material as an external field is increased either parallel or perpendicular to the strain measuring direction.

Consider the hard-axis magnetization process in a first-order uniaxial material, which from Eq. (6.5) is linear in the field: $M = M_s H/H_a$ or $m = h$. From Figure 7.4, left, and $m = \cos\theta$, Eq. (7.1) may be written as

$$e = \tfrac{3}{2}\lambda_s(m^2 - \tfrac{1}{3}) \tag{7.2}$$

In other words, the magnetostrictive strain in a hard uniaxial direction is quadratic in $m = h$, so e is proportional to H^2 (Fig. 7.4, right).

Above saturation, the two strain curves in Figure 7.2 are parallel to each other (if high field susceptibility and forced magnetostriction are neglected) and their difference is proportional to λ_s. To see this, Eq. (7.1) gives

$$e_{\|} = \frac{3}{2}\lambda_s\left(1 - \frac{1}{3}\right) = \lambda_s \qquad e_{\perp} = \frac{3}{2}\lambda_s\left(0 - \frac{1}{3}\right) = -\frac{\lambda_s}{2}$$

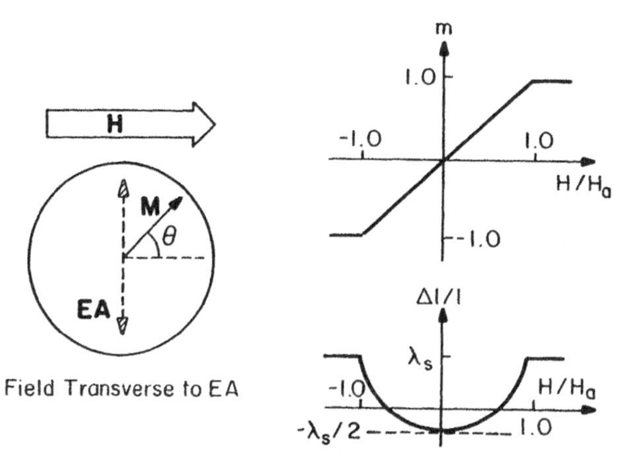

Figure 7.4 Application of a magnetic field perpendicular to the easy axis (EA) of a material causes a rotation of the magnetization direction and results in a linear M–H characteristic and a quadratic dependence of strain on field.

so for isotropic materials

$$e_{\|} - e_{\perp} = \tfrac{3}{2}\lambda_s \qquad \text{or} \qquad \lambda_s = \tfrac{2}{3}(e_{\|} - e_{\perp}) \tag{7.3}$$

The saturation magnetostriction λ_s can be measured in the direction of the applied field, starting with a sample in the randomly magnetized state ($\langle \cos^2\theta \rangle = \frac{1}{3}$). However, because a completely demagnetized state is not easily achieved with certainty, measurement of both $e_{\|}$ and $e_{\perp}$ in Eq. (7.3) is recommended to determine λ_s in isotropic materials. Measurement of strain in more than two directions is required to fully specify λ_{ij} for lower-symmetry materials.

The magnetostrictive strain is the same in each of two domains separated by a 180° domain wall (see Chapter 9) because $\cos^2\theta$ is the same in each domain. This is due to time-reversal invariance; the system is the same whether the microscopic currents rotate to give $\boldsymbol{M}$ in one direction or they reverse to give $\boldsymbol{M}$ in the opposite direction. Figure 7.5 shows two limiting initial domain states that differ in the orientation of their easy-axis directions relative to the strain sensing direction. The sample at left is in an initial state of contraction while that at the right is in extension, when measured as indicated. Application of a field either parallel or perpendicular to the strain sensing direction has different results in the two cases as indicated below the idealized domain structures. There is no magnetostrictive shape change associated with a magnetization process involving only 180° domain walls ($e_{\perp}$ at left or $e_{\|}$ at right). (The motion of any other kind of domain wall will result in a net shape change provided the magnetostriction constant is not zero.) Conversely, a

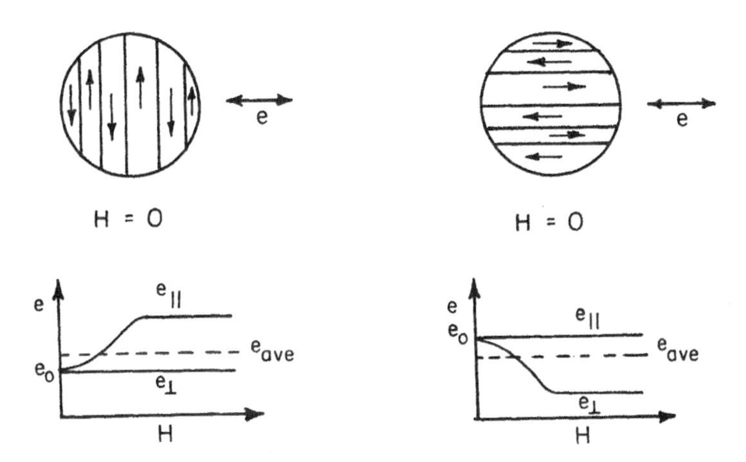

Figure 7.5 Above, schematic illustration of two demagnetized states that have different orientations relative to the strain sensing direction (indicated by e); below, field dependence of strain for each case in the presence of fields applied either parallel ($e_{\|}$) or perpendicular ($e_{\perp}$) to the strain-sensing direction.

strain of $\frac{3}{2}\lambda_s$ is measured in magnetizing a sample with initially transverse magnetization ($e_{\|}$ at left or $e_{\perp}$ at right). This illustrates the importance of the initial state when strain is being measured or used in a device.

Some Data So far, magnetostriction has been described for isotropic materials. For materials that are not isotropic, the magnetostriction constant can be different in different directions. For example, for *iron*

$$\lambda_{100} = 20.5 \times 10^{-6} \quad \text{and} \quad \lambda_{111} = -21.5 \times 10^{-6} \qquad \textit{iron}$$

and for *nickel*

$$\lambda_{100} = -46 \times 10^{-6} \quad \text{and} \quad \lambda_{111} = -25 \times 10^{-6} \qquad \textit{nickel}$$

Thus, magnetizing an iron crystal in its $\langle 100 \rangle$ direction causes an elongation along $\langle 100 \rangle$ but magnetizing it in the $\langle 111 \rangle$ direction causes a contraction along the $\langle 111 \rangle$ direction. Nickel contracts in the direction of magnetization for any crystal orientation. Recall that the $\langle 100 \rangle$ directions are easy for BCC Fe and the $\langle 111 \rangle$ directions are easy for FCC Ni. It is left as an exercise for the reader to show that the signs of these magnetostrictions cannot be explained by considering dipole field interactions between magnets (Problem 7.1).

In addition, distortions of higher order than the dipolar strain in Eq. (7.1) may also be important in anisotropic materials. For example, magnetizing a uniaxial material perpendicular to its symmetry axis can result in a magnetostrictive strain along the c axis that is different from that perpendicular to both c and $\boldsymbol{M}$.

Values of magnetostriction for selected materials having various symmetries are listed in Table 7.1. The experimental notation (λ_{100}, λ_{111}) is used because of its prevalence in the literature. A more universal definition of $\lambda[\lambda^{\gamma,2} \approx 3\lambda_{100}/2$ and $\lambda^{\varepsilon,2} \approx 3\lambda_{111}/2$; Lacheisserie (1994)], will be introduced later. Where the literature indicates $\lambda^{\gamma,2}$ or $\lambda^{\varepsilon,2}$ was measured, those values are listed in parentheses in the table.

We now consider the compositional variation of magnetostriction in some technically important series of $3d$ alloys. Figure 7.6 shows the compositional variation of λ_{100} and λ_{111} for FCC NiFe alloys. (For alloys of less then 30% Ni, the structure is BCC. At the FCC–BCC transformation near 30% Ni, T_C drops to zero and the room temperature magnetostriction coefficients vanish.) Both magnetostriction coefficients are positive for the FCC FeNi alloys up to about 80% Ni, where they change sign as they approach the negative values of pure Ni. K_1 also changes sign in this composition range (Fig. 6.9). The simultaneous vanishing or minimization of magnetocrystalline anisotropy and magnetostriction leads to very easy magnetization and therefore high permeability. Actually, K_1 is negative where λ_{100} and λ_{111} vanish. Because the $\langle 111 \rangle$ directions are the easy magnetization axes for $K_1 < 0$, it is more important for soft magnetic properties that λ_{111} vanishes than it is that λ_{100} vanishes. The

TABLE 7.1 Magnetostriction Constants λ_{100} and λ_{111} ($\times 10^6$) at 4.2 K and Room Temperature for Several Materials[a]

	$T = 4.2$ K		Room Temperature		
	$\lambda_{100}(\lambda^{\gamma,2})$	$\lambda_{111}(\lambda^{\varepsilon,2})$	$\lambda_{100}(\lambda^{\gamma,2})$	$\lambda_{111}(\lambda^{\varepsilon,2})$	Polycrystal λ_s
		3d Metals			
BCC-Fe	26	−30	21	−21	−7
HCP-Co[u]	(−150)	(45)	(−140)	(50)	(−62)
FCC-Ni	−60	−35	−46	−24	−34
BCC-FeCo	—	—	140	30	—
a-$Fe_{80}B_{20}$	48 (isotropic)	—	—	—	+32
a-$Fe_{40}Ni_{40}B_{20}$	+20	—	—	—	+14
a-$Cos_{80}B_{20}$	−4	—	—	—	−4
		4f Metals/Alloys			
Gd[u]	(−175)	(105)	(−10)	0	—
Tb[u]	—	(8700)	—	(30)	—
$TbFe_2$	—	4400	—	2600	1753
$Tb_{0.3}Dy_{0.7}Fe_2$	—	—	—	1600	1200
		Spinel Ferrites			
Fe_3O_4	0	50	−15	56	+40
$MnFe_2O_4$[u]	—	—	(−54)	(10)	—
$CoFe_2O_4$	—	—	−670	120	−110
		Garnets			
YIG	−0.6	−2.5	−1.4	−1.6	−2
		Hard Magnets			
$Fe_{14}Nd_2B$[u]	—	—	—	—	—
$BaO{\cdot}6Fe_3O_4$[u]	—	—	(13)	—	—

[a]Some polycrystalline room-temperature values are also listed. The prefix a- designates an amorphous material. For uniaxial materials (superscript u) where $\lambda^{\gamma,2}$ or $\lambda^{\varepsilon,2}$ was reported, their values are given in parentheses in the λ_{100} and λ_{111} columns, respectively

soft Ni-rich alloys near these λ_s and K_1 zeros are called *permalloys*, a generic use of the trade name originally registered in 1935 by Western Electric. More will be said about these alloys in Chapter 10.

Unlike the anisotropy constants of NiFe (Fig. 6.9), the magnetostriction coefficients show a relatively weak dependence on chemical ordering near the $FeNi_3$ composition, and here, only λ_{100} is significantly affected.

Figure 7.7 shows the variation of the polycrystalline magnetostriction coefficient of NiFe alloys over the full composition range. Data for amorphous $(FeNi)B_{20}$ alloys are also shown. While the magnetostriction coefficient of the crystalline alloys vanishes near 80% Ni, a similar effect cannot

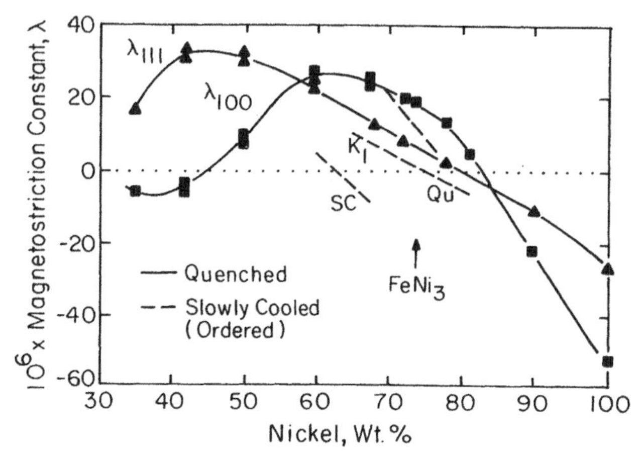

Figure 7.6 Room temperature magnetostriction constants for FCC FeNi alloys (Bozorth 1962; Hall 1960). Crossover behavior of anisotropy constant K_1 for slow-cooled (SC) and quenched (Qu) FeNi alloys are also shown for reference.

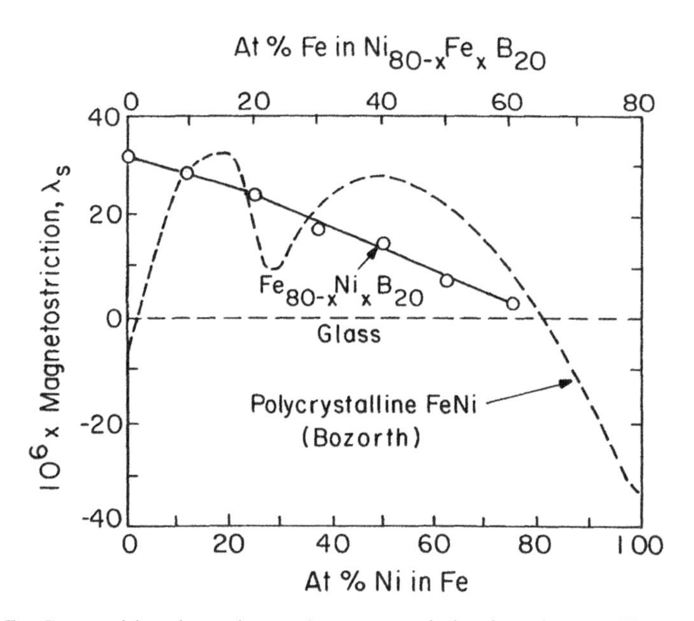

Figure 7.7 Composition dependence of magnetostriction in polycrystalline FeNi alloys (dotted line, after Bozorth 1993) and in amorphous Fe-Nibase alloys (after O'Handley 1978). The dip in λ for the polycrystalline alloys near 25% Ni is due to the BCC-FCC transformation, which results in invar alloys.

be confirmed in the amorphous alloys based on boron because the magnetization goes toward zero and the alloys are unstable as iron content is reduced below 20 at%. (Chapters 10 and 11 contain more information on amorphous magnetic alloys).

Figure 7.8 shows the composition dependence of the magnetostriction coefficients in polycrystalline FeCo and amorphous $(FeCo)_{80}B_{20}$ alloys. Note that for the crystalline alloys, magnetostriction is negative near α-Fe and in the HCP ε-Co-rich phase whereas positive magnetostriction is observed over the FCC range and most of the BCC range. Recall that the anisotropy of BCC FeCo alloys goes from positive to negative as Co content increases beyond about 50% (Fig. 6.9*b*). While the amorphous FeCo-based alloys show a zero-magnetostriction composition near the cobalt end of the series, the magnetostriction at the iron end remains strong and positive, unlike that of α-Fe. The magnetostriction value shown in Figure 6.7 for cobalt is considerably smaller than that shown in Table 7.1. The reason for the smaller value in the figure is that the field applied in that case was probably not sufficient to saturate the polycrystalline sample, given the very large magnetocrystalline anisotropy of cobalt.

Note that in either alloy series (Fig. 7.7 or 7.8) the magnetostriction constants vanish in the dilute-iron compositions at nearly the same transition metal ratio for the amorphous alloys as for the polycrystalline alloys: Fe:Ni ≈ 4:1 (in Fig. 7.7) and Fe:Co ≈ 9:1 (in Fig. 7.8). Comparison of these

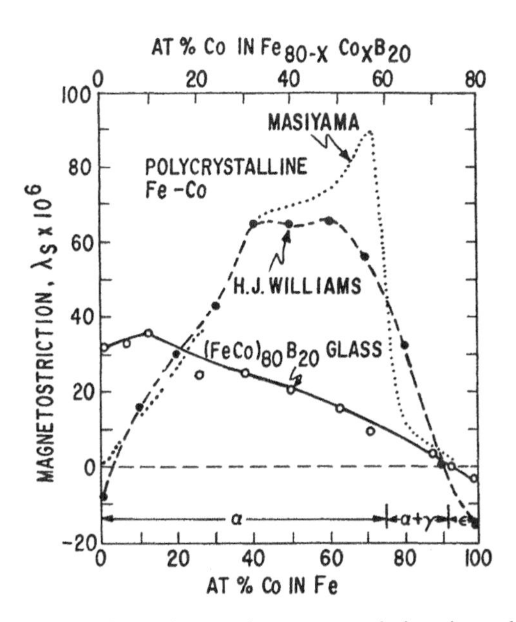

Figure 7.8 Composition dependence of magnetostriction in polycrystalline FeCo alloys (after Bozorth 1993 and Hall 1960) and in FeCo-base amorphous alloys (after O'Handley 1978).

figures helps to distinguish features that may be associated with long-range order (and hence different in amorphous and crystalline alloys) from those associated with short-range order (and hence possibly similar in amorphous and crystalline alloys).

Figure 7.9 gives a summary perspective of magnetostriction data in the form of a ternary diagram showing the composition dependence of $\lambda_s = 0$ lines in crystalline FeNiX alloys (X = Co, Cr, Mo,...). The BCC–FCC and FCC–HCP phase boundaries are indicated by the dashed lines for reference. There are several noteworthy features of this diagram. Starting at the NiFe side of the diagram, $\lambda_s = 0$ lines extend from the $Ni_{81}Fe_{19}$ permalloy composition with different slopes depending on the valence electronic structure of the third element, X, and on the method of preparation: the solid lines correspond to bulk materials, the dotted line to thin films. Ni-rich FeNiCo thin films show a larger field of negative magnetostriction than do bulk crystalline materials. Alloying FeNi with earlier transition elements such as Cr or Mo (dot/dashed line) constricts the negative field even more. From the Co-rich corner of the diagram, there is a significant difference between the $\lambda_s = 0$ lines for bulk and thin films: thin films show a pocket of negative magnetostriction about the Co corner; bulk, FCC phases (Miyazaki et al. 1994) show a swath of negative

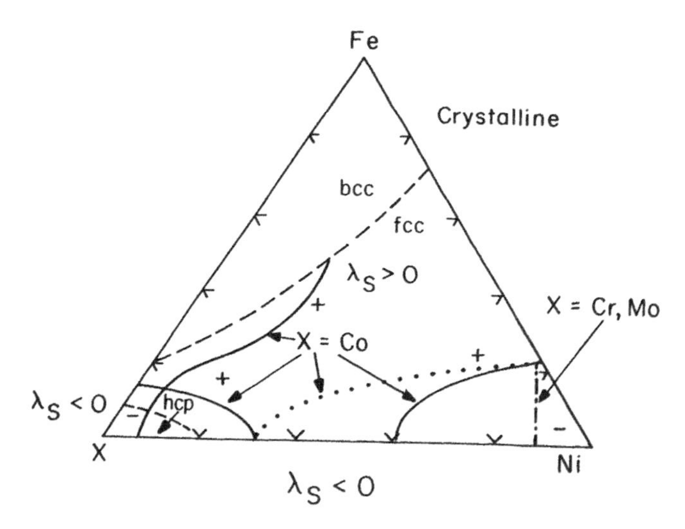

Figure 7.9 Bold lines are experimentally observed lines of zero magnetostriction for bulk crystalline alloys [Miyazaki et al. (1994)]. Dot/dashed line is an extension of the zero-magnetostriction line identified by Sirota et al. (1969, 1972). The sign of λ_s in the different fields is indicated by the plus and minus signs. Fine, solid line shows the observed line for $\lambda_s = 0$ for polycrystalline thin films deposited at 100°C. [Tolman (1967) and Lampert et al. (1968)]. If the substrates are held at a higher temperature, the curve moves toward the bulk line, almost reaching it for $T_{sub} = 300°C$ (dotted line). The dashed line is a prediction of zero magnetostriction from split-band model for FeNi(Mo or Cr) alloys.

magnetostriction compositions between the BCC–FCC boundary and the solid line to its right. This latter $\lambda_s = 0$ line is not closed; the positive magnetostriction indicated in Figure 7.8 for bulk FCC FeCo alloys demands that it loop around and intersect the FeCo side of the diagram near 80% Co. Indeed, earlier, little-known data from Sirota et al. (1969, 1972) indicate that the $\lambda_S < 0$ field is closed in this manner. Their measurements show the field of negative magnetostriction extending farther along the phase boundary toward the FeNi side than the more recent data shown here.

The rare-earth metals and many rare-earth intermetallic compounds are characterized by strong magnetic anisotropy, which makes it difficult to get an accurate measurement of saturation magnetostriction values. Some data are given in Table 7.1, and temperature dependence of magnetostriction in Tb and Gd will be displayed below. Further data on these systems can be found in the reviews by Clark (1980), Morin and Schmitt (1990), and Cullen et al. (1995). The treatment of magnetostriction in amorphous rare-earth intermetallics involves an understanding of the concept of random anisotropy (Chapter 11) as reviewed by del Moral (1993).

Surface Magnetostriction It has recently been reported that the magnetostriction or magnetoelastic coupling coefficient of a thin film can depart sharply from the value observed in thicker films and in bulk. These effects, which may have a fundamental origin associated with reduced symmetry or be due to microstructural or strain changes with film thickness will be treated in Chapter 16.

7.2 PHENOMENOLOGY

What these data indicate collectively is that there is a coupling between the magnetization direction and mechanical deformations in a material, and this effect depends on the direction of magnetization relative to the crystal axes. Therefore the magnetic free energy must contain terms that couple the direction of magnetization, specified by the direction cosines α_i, to the components of the strain tensor, e_{ij}. The magnetic energy, $u(\alpha, e)$, is expanded in a Maclaurin series as follows:

$$u_m = \frac{U_m}{V_o} = f_o + K_1\alpha_i^2\alpha_j^2 + K_2\alpha_i^2\alpha_j^2\alpha_k^2 + \cdots + c_{ijkl}e_{ij}e_{kl} + H_{ijklmn}e_{ij}e_{kl}e_{mn}$$
$$+ \cdots + B_{ij}e_{ij}\alpha_i\alpha_j + \cdots + D_{ijkl}e_{ij}e_{kl}\alpha_i\alpha_j\alpha_k\alpha_l + \cdots \tag{7.4}$$

In this formal expression, the two lowest-order terms of three types have been included. The first-type terms, independent of strain, depend only on the direction of magnetization and thus describe the magnetocrystalline anisotropy discussed in the Chapter 6. The second-type terms, quadratic and cubic in the strain components and independent of the α terms, describe the pure elastic energy (with the elastic stiffness constants as coefficients) and the strain dependence of the c terms, respectively (see Appendix 7A for a quick refresher). The third-type terms, linear and quadratic in the strain components *and dependent on the direction cosines of the magnetization*, describe the first-order and second-order magnetoelastic (ME) energy density. The B values are ME coefficients that express the coupling between the strains e_{ij} and the direction of magnetization given by θ or α_i. Later these coefficients will be related to the magnetostriction constants, λ_{100} and λ_{111}. Essentially, the B values are the stresses of magnetic origin that cause the magnetostrictive strains e_{ij} with elastic moduli as constants of proportionality. The exact form of the terms in Eq. (7.4) depends on symmetry as described in Chapter 6 and its Appendix and as expanded on in Section 7.4, below.

For a cubic material, the surviving first-order terms are

$$\begin{aligned} u = u_a + u_{me} + u_{el} &= \frac{U_o}{V_o} + K_1(\alpha_1^2\alpha_2^2 + \alpha_2^2\alpha_3^2 + \alpha_3^2\alpha_1^2) + K_2\alpha_1^2\alpha_2^2\alpha_3^2 \cdots \\ &+ B_1(\alpha_1^2 e_{xx} + \alpha_2^2 e_{yy} + \alpha_3^2 e_{zz}) + B_2(\alpha_1\alpha_2 e_{xy} + \alpha_2\alpha_3 e_{yz} + \alpha_3\alpha_1 e_{zx}) \\ &+ (\tfrac{1}{2})c_{11}(e_{xx}^2 + e_{yy}^2 + e_{zz}^2) + c_{12}(e_{xx}e_{yy} + e_{yy}e_{zz} + e_{zz}e_{xx}) \\ &+ (\tfrac{1}{2})c_{44}(e_{xy}^2 + e_{yz}^2 + e_{zx}^2) + \text{const} \times e_{ij} \end{aligned} \tag{7.5}$$

The reason for the last term, independent of direction cosines, will become clear below.

It is of interest to know the relative importance of the terms in Eq. (7.5). Table 7.2 gives the values of some relevant parameters for Fe, Co, and Ni. In order to compare the magnitude of the energies contributing to Eq. (7.5), the B values must be multiplied by a strain. Clearly, magnetoelastic contributions to the free energy are relatively more important in Ni than in Fe. In fact, while an imposed strain of order 2% would be required for the ME contribution to be comparable to K_1 in Fe, a strain of less than 0.1% results in an ME anisotropy comparable to K_1 in Ni. Strains exceeding about 0.1% are greater than the elastic limit of most metals; they can exist in thin films.

TABLE 7.2 Bulk Magnetic Properties of Fe, Co, and Ni at Room Temperature[a]

	BCC Fe	FCC Co	HCP Co	FCC Ni
$\mu_0 M_s^2/2$ (10^6)	1.9	1.3	1.3	0.14
K_1 (10^4)	4.8	−12	35	−0.45
K_2 (10^4)	−1.5	0	15	0.23
B_1 (10^6)	−2.9	−16	6	6.2
B_2 (10^6)	+2.9	26	13	4.3

[a]Units are J/m^3 for the first three rows and N/m^2 for the B rows.

The microscopic origin of these ME phenomena may be understood better by following Néel's treatment (1954) of the effects of strain on the anisotropic magnetic interaction between atoms on a lattice. The interaction energy between two atoms in a solid can be expanded in Legendre polynomials [as done in Chapter 6, Eq. (6.7)] where ψ is the angle between the bond direction $\boldsymbol{r}$ and the direction of magnetization.

In Eq. (6.8), $g(r)$ describes isotropic (possibly exchange) interactions because there is no reference to a crystallographic direction for the magnetization. On the other hand, the dipolar and quadrupolar terms, proportional to $l(r)$ and $q(r)$, respectively, are functions of the angle between $\boldsymbol{M}$ and the bond directions. The dipole coefficient may, in the simplest case, be given by $\mu_0 \mu_m^2/r_0^3$, a true dipole–dipole, *magnetostatic* interaction; it is rarely important in magnetostriction. However, $l(r)$ may also contain any other interactions of dipole symmetry, pseudo-dipole interactions, which couple $\boldsymbol{M}$ to the crystal structure. Dipole interactions become important when stronger effects vanish, as, for example, in s-state ions ($L = 0$) such as Gd^{3+}.

If the direction cosines of the magnetization $\boldsymbol{M}$ within a domain are taken to be α_1, α_2, and α_3 and those of the bond directions to be β_1, β_2, and β_3,

$$\boldsymbol{M} = M_s(\alpha_1, \alpha_2, \alpha_3) = M_s(m_x, m_y, m_z)$$
$$\boldsymbol{r} = r(\beta_1, \beta_2, \beta_3),$$

then $\cos\psi = \boldsymbol{M}\cdot\boldsymbol{r}/|M||r| = \alpha_1\beta_1 + \alpha_2\beta_2 + \alpha_3\beta_3$. Thus, the first terms in Eq. (6.7) may be written

$$u_a(r_0, \psi_0) = g(r_0) + l(r_0)[(\alpha_1\beta_1 + \alpha_2\beta_2 + \alpha_3\beta_3)^2 - \tfrac{1}{3}] + \cdots \tag{7.6}$$

If a crystal is strained, for example, with a bond in the x direction taking on the new length $r_0(1 + e_{ij})$, then it is possible that the free energy will have its minima shifted to define new preferred directions of magnetization. Conversely, if the direction of magnetization is changed by application of a field, the crystal may be able to lower its energy by changing its equilibrium bond lengths. The mathematics of this coupling between magnetic and elastic effects must be determined. For simplicity, consider the energy due to bonds in the x direction: $r_0 = r_x$ and $\beta = (1, 0, 0)$. Under a uniaxial strain, $r_x = r_0(1 + e_{xx})$ and, to first order in the engineering strain, $\beta = (1, e_{xy}/2, e_{xz}/2)$. Expanding $l(r)$, Eq. (7.6) becomes

$$u_a(r, \psi) = g\{r_0(1 + e_{xx})\} + \left[l(r_0) + \left(\frac{dl}{dr}\right)_{r0} e_{xx}\right]$$
$$\times\left[\alpha_1 + \frac{\alpha_2 e_{xy}}{2} + \frac{\alpha_3 e_{yz}}{2}\right)^2 - \frac{1}{3}\right] + \cdots \tag{7.7}$$

Subtracting Eq. (7.6) from (7.7) gives the change in magnetic energy due to strain, which is, by definition, the magnetoelastic energy density:

$$u_a(r, \psi) = u_a(r, \psi) - u_a(r_0, \psi_0) \equiv u_{me} = \left(\frac{dg}{dr}\right)_{r0} e_{xx} + \left(\frac{dl}{dr}\right)_{r0} e_{xx}\left(\alpha_1^2 - \frac{1}{3}\right)$$

$$+ l(r_0)[\alpha_1\alpha_2 e_{xy} + \alpha_1\alpha_3 e_{xz}] + \cdots + 0(e_{ij}^2)\cdots$$

Similar expressions result for the bonds in y and z directions. These pair interactions are summed over the nearest neighbors in various cubic lattices to give

$$u_{me} = B_1[e_{xx}(\alpha_1^2 - \tfrac{1}{3}) + e_{yy}(\alpha_2^2 - \tfrac{1}{3}) + e_{zz}(\alpha_3^2 - \tfrac{1}{3})]$$
$$+ B_2[e_{xy}\alpha_1\alpha_2 + e_{yz}\alpha_2\alpha_3 + e_{zx}\alpha_3\alpha_1] + \cdots \tag{7.8}$$

where the B_i values for different cubic lattices take on the values

$$B_1 = \left(\frac{N}{V}\right)\left(\frac{\partial l}{\partial r}\right)_{r0}, \qquad B_2 = 2\left(\frac{N}{V}\right) l(r_0) \qquad SC$$

$$B_1 = \left(\frac{8N}{3V}\right)\left(\frac{\partial l}{\partial r}\right)_{r0}, \qquad B_2 = \left(\frac{8N}{9V}\right)\left[l(r_0) + \left(\frac{\partial l}{\partial r}\right)_{r0}\right] \qquad BCC$$

$$B_1 = \left(\frac{N}{2V}\right)\left[6l(r_0) + \left(\frac{\partial l}{\partial r}\right)_{r0}\right], \qquad B_2 = \left(\frac{N}{V}\right)\left[2l(r_0) + \left(\frac{\partial l}{\partial r}\right)_{r0}\right] \qquad FCC$$

The ratio N/V is the number of atoms per unit volume, rendering the results of this atomic model in units of energy per unit volume. The term $-\frac{1}{3}B_1 e_{ii}$ in the ME free energy is not simply a constant with no physical consequences other than expressing the energy as an expansion in orthonormal polynomials [as is the term $-\frac{1}{3}k_l$ in the anisotropy, Eq. (6.4)]. Here it contains a variable, e_{ii}, so it cannot be neglected without serious consequences. It will be seen later that this term is important in defining the reference state from which the strain is measured. So Eq. (7.8) is the correct form for the ME energy density in a cubic crystal; the form in Eq. (7.5) had not defined the terms $-e_{ii}/3$.

The free energy can be minimized with respect to various strain components, $\partial u/\partial e_{ij} = 0$, to get the equilibrium Joule strain $e_{ij}^0 = e(M)$ or $e(H)$ as in Figure 7.2. Alternatively, the free energy could be minimized with respect to the orientation of $\boldsymbol{M}$, $\partial u/\partial\theta = 0$ to get the equilibrium orientation of magnetization $\theta(e, K)$, which gives $M(e)$ as in Figure 7.3 (inverse Joule effect).

The cubic magnetostrictive strains are now derived by minimization of Eq. (7.8). The equilibrium strain configuration is obtained by setting $\partial u/\partial e_{ij} = 0$:

$$\frac{\partial u}{\partial e_{ii}} = c_{11}e_{ii} + c_{12}(e_{jj} + e_{kk}) + B_1\left(\alpha_i^2 - \frac{1}{3}\right) = 0$$

$$\frac{\partial u}{\partial e_{ij}} = c_{44}e_{ij} + B_2\alpha_i\alpha_j = 0 \qquad (i \neq j) \tag{7.9}$$

The solutions to these six equations give the components of the magnetostrictive strain tensor:

$$(e_{ij}) = \begin{pmatrix} -\dfrac{B_1}{(c_{11}-c_{12})}\left(\alpha_1^2-\dfrac{1}{3}\right) & -\dfrac{B_2}{c_{44}}\alpha_1\alpha_2 & -\dfrac{B_2}{c_{44}}\alpha_1\alpha_3 \\ -\dfrac{B_2}{c_{44}}\alpha_1\alpha_2 & -\dfrac{B_1}{(c_{11}-c_{12})}\left(\alpha_2^2-\dfrac{1}{3}\right) & -\dfrac{B_2}{c_{44}}\alpha_3\alpha_2 \\ -\dfrac{B_2}{c_{44}}\alpha_1\alpha_3 & -\dfrac{B_2}{c_{44}}\alpha_3\alpha_2 & -\dfrac{B_1}{(c_{11}-c_{12})}\left(\alpha_3^2-\dfrac{1}{3}\right) \end{pmatrix} \tag{7.10}$$

Note that the diagonal strains are proportional to the ME stress B_1 and are inversely proportional to $c_{11}-c_{12}$, which is the stiffness resisting uniaxial distortions of the cube. The off-diagonal strains are proportional to the ME shear stress B_2 and inversely proportional to c_{44}, the elastic shear modulus G. Thus, what was asserted earlier has been demonstrated, namely that the ME stresses and strains are related by a magnetic analog of Hook's law, $B_{ij} = -c_{ijkl}\lambda_{kl}$. The minus sign reflects the convention that a positive magnetic stress (compressive) results in a negative strain.

The strain in any direction is specified by projecting e_{ij} on the vector $(\beta_1, \beta_2, \beta_3)$, defining the strain-measuring direction (see Appendix 7B):

$$\frac{r-r_0}{r_0} = \frac{\partial l}{l} = (\beta_1, \beta_2, \beta_3)(e_{ii})\begin{pmatrix}\beta_1\\ \beta_2\\ \beta_3\end{pmatrix} = \sum e_{ii}\beta_i^2 + \sum_{i<j} e_{ij}\beta_i\beta_j \tag{7.11}$$

Substituting in this expression the values of e_{ii} and e_{ij} given in Eq. (7.10), gives the strain in direction β for magnetization in direction α:

$$\frac{\partial l}{l} = -\frac{B_1}{c_{11}-c_{12}}\left(\alpha_1^2\beta_1^2+\alpha_2^2\beta_2^2+\alpha_3^2\beta_3^2-\frac{1}{3}\right) - \frac{B_2}{c_{44}}(\alpha_1\alpha_2\beta_1\beta_2+\alpha_2\alpha_3\beta_2\beta_3+\alpha_3\alpha_1\beta_3\beta_1) \tag{7.12}$$

This equation shows how the deformation of a magnetic sample known as Joule or anisotropic magnetostriction depends on the direction of magnetization through the α terms and on the strain-measuring direction through the β terms. Equation (7.12) also reflects the fact that a demagnetized crystal with a random distribution of domain magnetizations is taken, by convention, to have $dl/l = 0$. In this state the magnetization direction in each grain is along one of the easy axes so an average over all possible strain directions gives zero strain. Thus the term $-\frac{1}{3}B_1 e_{ij}$ in the ME energy sets the proper strain reference state to that of the unstrained cubic crystal.

For a sample magnetized in the [100] direction, $\alpha = (1, 0, 0)$, the strain measured in the same direction, $\beta = (1, 0, 0)$, is defined as λ_{100}. In that case Eq. (7.12) gives

$$\lambda_{100} = -\frac{2}{3}\frac{B_1}{c_{11} - c_{12}} \tag{7.13}$$

For λ_{111}, $\alpha_i = \beta_i = (1/\sqrt{3})$ and the result is

$$\lambda_{111} = -\frac{1}{3}\frac{B_2}{c_{44}} \tag{7.14}$$

Similarly for λ_{110}, $\alpha_1 = \alpha_2 = \beta_1 = \beta_2 = (1/\sqrt{2})$, $\alpha_3 = \beta_3 = 0$, giving

$$e_{110} = \lambda_{110} = -\frac{1}{6}\frac{B_1}{(c_{11} - c_{12})} - \frac{1}{4}\frac{B_2}{c_{44}} = \frac{1}{4}\lambda_{100} + \frac{3}{4}\lambda_{111} \tag{7.15}$$

In a cubic material such as iron, where $\lambda_{100} > 0$ and $\lambda_{111} < 0$, the dependence of strain on applied field can be more complicated than shown in Figure 7.2. The reason is that for certain orientations of the applied field, the magnetization is rotated through directions for which the strain has different signs. A demagnetized sample (***M*** equally distributed among $\pm\langle 100 \rangle$), in a [110]-directed field initially expands in the field direction as the magnetization in the $\pm$[001] directions rotates to [100] or [010]. (There is no elongation for the change in magnetization from $-$[100] to [100] and $-$[010] to [010].) Increasing the field further causes the magnetization along [010] and [100] to rotate to [110]. This process causes a contraction because $\lambda_{110} < 0$ [see Eq. (7.15)].

For an isotropic material, it is expected that $\lambda_{100} = \lambda_{111} = \lambda_s$. Then Eq. (7.12) reduces to Eq. (7.1), which was presented without justification. All reference to crystallographic direction is gone (see Problem 7.4). For a completely *random polycrystalline material*, in a state of zero net stress, Callen and Goldberg (1965) obtained

$$\lambda_s = \tfrac{2}{5}\lambda_{100} + \tfrac{3}{5}\lambda_{111} \tag{7.16}$$

Another form for the magnetostriction of a polycrystalline sample is also given by Callen and Goldberg for situations of zero net strain. Lacheisserie (1994) shows that most data fall between these two limits. The values usually assigned to λ_s for polycrystalline nickel and iron are -34×10^{-6} and -7×10^{-6}, respectively. Use of Eq. (7.16) is justified only for an untextured polycrystal and for amorphous alloys.

7.3 MAGNETOELASTIC CONTRIBUTION TO ANISOTROPY

Inspection of Eq. (7.8) shows that for a given strain, the magnetoelastic energy resembles a magnetic anisotropy energy inasmuch as it is a function of the magnetization orientation. However, the first two *cubic* ME terms are of lower order in direction cosines, that is, of lower symmetry, than K_1 and K_2 terms. Thus, an imposed strain may change a cubic material, having fourfold easy axes, to a uniaxial system.

Inverse magnetoelastic effects need to be considered in two situations: (1) an *imposed mechanical strain* alters the energy surface and hence changes the magnetization process below saturation and (2) a *magnetostrictive strain* (resulting from a change in magnetization direction) alters the energy surface and hence changes the approach to saturation. This is a second-order effect.

Each of these magnetoelastic effects on magnetic anisotropy is considered, the first-order effect, then the second-order effect.

First-Order Anisotropy Due to an External Strain In Chapter 6 the effects of magnetic anisotropy on the preferred direction of magnetization were illustrated through the use of energy surfaces. The energy surface is also used here to illustrate the effects of strain on the free energy function.

If the cubic magnetic anisotropy and magnetoelastic energy expressions in Eq. (7.5) are considered, it is clear that imposition of an external strain e_{ij}^0 alters the anisotropy energy density. For example, a tensile stress that produces an elongation in the x direction, e_{11}^0, will also cause lateral contractions given by $e_{22}^0 = e_{33}^0 = -\upsilon e_{11}^0$, where υ is Poisson's ratio (which has a value close to 1/3 for many metals). Hence, the anisotropy energy in this case is written:

$$u_a = K_1(\alpha_1^2\alpha_2^2 \cdots) + B_1 e_{11}^0[\alpha_1^2 - \upsilon(\alpha_2^2 + \alpha_3^2)]$$

Figure 7.10 shows the cubic energy surface cross section ($K_1 > 0$) in the xy

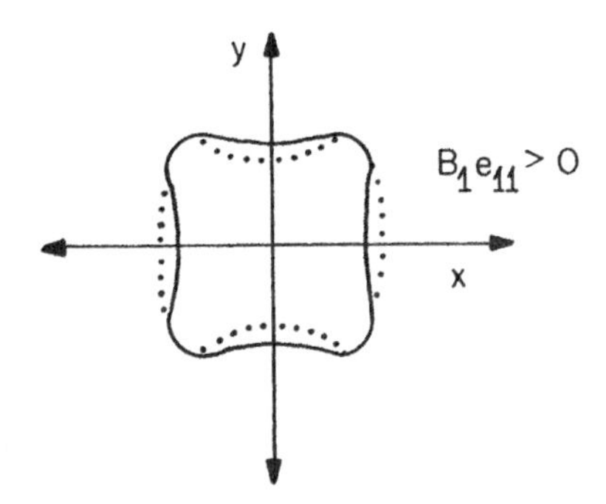

Figure 7.10 Cross section of cubic anisotropy energy surface for zero strain (solid line) and for $B_1 e_{11} > 0$ showing stabilization of magnetization in directions transverse to the strain axis.

plane for $e_{11}^0 = 0$ (solid) and for $B_1 e_{11}^0 > 0$ ($\lambda_{100} e_{11}^0 < 0$) (dashed). Magnetization in the y direction is stabilized relative to the x direction as a result of the strain.

To accommodate this strain in a cubic material in three dimensions, the y–z *plane* becomes a more favored plane of magnetization relative to the x axis. For $B_1 e_{11}^0 < 0$, the y–z plane becomes less stable and the x axis, more stable.

What is important for the impact of ME effects on the magnetization process is the ratio of $|Be^0|$ to the other energies influencing the direction of magnetization. Table 7.3 shows the critical strain values above which magnetoelastic energy $|Be^0|$ exceeds magnetocrystalline anisotropy

$$e_c^K = \frac{|K_1|}{|B_1|}$$

or above which it exceeds the magnetostatic energy

$$e_c^M = \frac{\mu_0 M_s^2}{2|B_1|}$$

for iron and nickel. It is assumed that equilibrium values of K_1, M_s, and B_1 apply in a high strain situation (and they are not expected always to apply). Nickel is especially sensitive to strain-induced changes in its magnetic response because of its strong magnetoelastic coupling, its relatively weak crystalline anisotropy, and its small magnetization. A strain of only 0.1% in nickel gives rise to an ME anisotropy comparable to K_1. For iron, a much larger strain (1.7%) is required to give an ME anisotropy comparable to the crystal anisotropy but would not change magnetization distributions governed by magnetostatic effects.

It is worth noting that through the ME interaction, a small lattice strain, e, can produce a much larger strain in the anisotropic energy surface. The

TABLE 7.3 Assuming Bulk Values for K_1, B_1, and M_s in Fe and Ni, the Critical Strains e_c^K and e_c^M Indicate the Strain above which Magnetoelastic Anisotropy Strength Exceeds Crystal Anisotropy or Magnetostatic Energy, Respectively

	Fe	Ni
K_1 (10^4 J/m^3)	+4.8	−0.45
B_1 (10^6 J/m^3)	−2.9	+6.2
$e_c^K = K_1/B_1$	1.7%	0.1%
$e_c^M = \mu_0 M_s^2/(2B_1)$	66%	2.4%

ME-induced strain in the energy surface is given by $B_1 e/K$. The ratio of this strain to the lattice strain is simply B_1/K. From Table 7.3, this quantity is equal to $(e_c^K)^{-1}$. Thus the energy surfaces of iron and nickel strain 60 times and 100 times the amount by which their lattices are strained, respectively. Small changes in sample shape can cause much larger changes in the magnetization process defined by the energy surface.

Example 7.1: Effects of Imposed Strain Consider a magnetic thin film with a uniaxial anisotropy $K_u \cos^2\theta$ in the plane of the film. (Magnetostatic energy is assumed to be great enough to keep $\boldsymbol{M}$ in plane.) A tensile stress and an external magnetic field are applied along the hard axis as shown in Figure 7.11.

It is important to identify clearly the effect of the stress on the M-H curve. For $e = 0$, the magnetization ideally follows $M = M_s H/H_a$. The problem is solved for nonzero stress assuming no out-of-plane magnetization ($\alpha_3 = 0$). A magnetoelastic term [Eq. (7.8)]

$$u_{me} = B_1(e_{xx}\alpha_x^2 + e_{yy}\alpha_y^2) + B_2 e_{xy}\alpha_x\alpha_y$$

is, therefore, added to the Zeeman and anisotropy energy densities. (The $-\frac{1}{3}$ terms have been dropped because an imposed strain is assumed; these terms must be retained when calculating magnetostrictive strains.) From elasticity theory, $e_{xx} = \sigma/E$, $e_{yy} = -\upsilon e_{xx}$, and $e_{ij} \approx 0$, and the ME energy density can then be written

$$u_{me} = B_1 e_{xx}(\cos^2\theta - \upsilon \sin^2\theta) \tag{7.17}$$

Note that for $B_1 e_{xx} > 0$ this term lowers the energy density near $\theta = \pi/2$ and raises the energy for θ near zero or π. When the torque from Eq. (7.17) is combined with the earlier torque expressions for the zero-strain, uniaxial problem, the magnetization response becomes (see Problem 7.9)

$$\frac{M}{M_s} = m = \frac{M_s H}{2[K_u + B_1 e_{xx}(1 + \upsilon)]} = \frac{M_s}{2K_u^{\text{eff}}} H \tag{7.18}$$

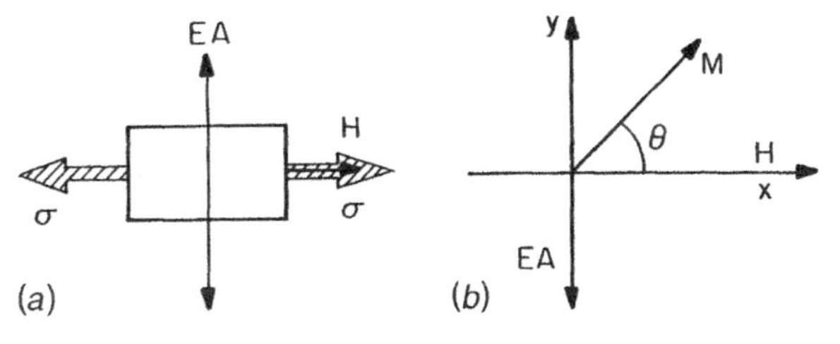

Figure 7.11 (*a*) A schematic representation of a sample subjected to collinear tensile stress and external field at right angles to the easy axis; (*b*) the appropriate coordinate system for the problem.

For $B_1 e_{xx} > 0$ ($\lambda_{100} e_{xx} < 0$), K_u^{eff} is increased by the magnetoelastic anisotropy and the slope $\partial M_x / \partial H$ decreases (saturation is harder to achieve as in Fig. 7.12) for the applied field H less than saturation. Equation (7.18) can also be written as

$$m_x = \frac{H}{H_a^{\text{eff}}} \qquad \text{where} \qquad H_a^{\text{eff}} = 2\frac{K_u^{\text{eff}}}{M_s} = H_a + B_1 e_{xx} \frac{1 + \upsilon}{M_s} \tag{7.19}$$

For $B_1 e_{xx} < 0$ ($\lambda_{100} e_{xx} > 0$), K^{eff} is decreased by the magnetoelastic contribution (Fig. 7.12). The magnetization still reverts to zero when $H = 0$, *regardless of the stress* as long as $|B_1 e_{xx}(1 + \upsilon)| < K_u$, that is, $K_u^{\text{eff}} > 0$, because m is linear in H. However, if the ME energy is so negative that $K_u^{\text{eff}} \leqslant 0$, m_x may be finite at $H = 0$ because the x axis becomes the easy axis.

Equations (7.18) and (7.19) are often written in terms of the magnetostriction constant for an isotropic material. In those cases, $B_1 e_{xx}(1 + \upsilon)$ is replaced by $-(3/2)\lambda_s \sigma$ where σ is the external stress imposing the strain:

$$K^{\text{eff}} = K_u - \left(\frac{3}{2}\right)\lambda_s \sigma \qquad \text{or} \qquad H_a^{\text{eff}} = H_a - \frac{3\lambda_s \sigma}{M_s} \tag{7.20}$$

Essentially, tension in a positive magnetostriction material ($B_1 e_{xx} < 0$) facilitates the magnetization process in the tensile direction. Contrast this behavior as summarized in Figure 7.12 with that of a cubic film ($K_1 > 0$, i.e., $\langle 100 \rangle$ easy axes) magnetized and stressed along [110] in Problem 9.4.

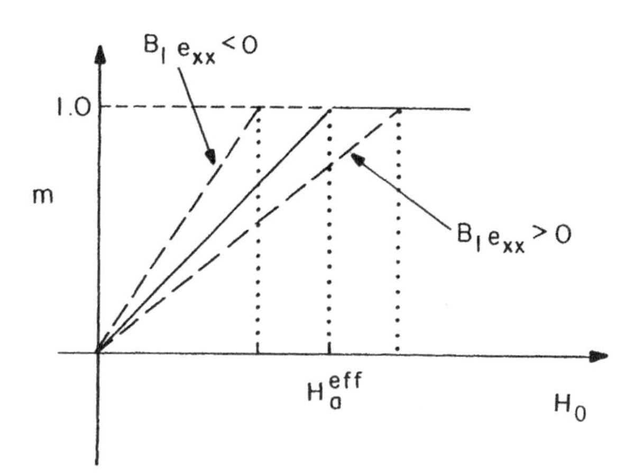

Figure 7.12 Solution to the problem represented in Figure 7.11. The solid line represents zero stress and is the same solution as in Figure 7.4. The dashed lines show how the magnetization is affected by stress for different values of the product of strain and magnetoelastic coupling coefficient B_1.

There are circumstances for which an externally imposed strain is so large that second-order effects, $D_{ijkl}e_{ij}e_{kl}\alpha_i\alpha_j\alpha_k\alpha_l$ [Eq. (7.4)], must be considered. In such cases, these effects show up as apparent changes in the first-order ME coefficients:

$$\begin{aligned} u_{me} &= B_{ij}e_{ij}\alpha_i\alpha_j + \cdots + D_{ijkl}e_{ij}e_{kl}\alpha_i\alpha_j\alpha_k\alpha_l + \cdots \\ &= B^{\text{eff}}e_{ij}\alpha_i\alpha_j \end{aligned}$$

where

$$B^{\text{eff}} = B_{ij} + D_{ijkl}e_{kl}\alpha_k\alpha_l$$

An example of second-order ME effects will be seen in epitaxial thin films (Chapter 16) where a lattice mismatch of a few percent can cause an apparent change in the ME coupling coefficient. Attention is now turned to the effect of a magnetostrictive strain (not an externally imposed strain) on the magnetic anisotropy.

Second-Order Anisotropy Due to Magnetostriction To consider the change in anisotropy that results from spontaneous magnetostriction during the magnetization process in an unconstrained sample, the strain in Eq. (7.8) must be expressed as a function of magnetization. By replacing the strains in Eq. (7.8) with the results in Eq. (7.10), the magnetostrictive contribution to crystal anisotropy ΔK_1 in a cubic system is identified. The result is

$$u = (K_1 + \Delta K)(\alpha_1^2\alpha_2^2 + \alpha_2^2\alpha_3^2 + \alpha_3^2\alpha_1^2) \tag{7.21}$$

with

$$\Delta K = \frac{B_1^2}{(c_{11} - c_{12})} - \frac{B_2^2}{2c_{44}}$$

or using Eqs. (7.13) and (7.14):

$$\Delta K = \tfrac{9}{4}[c_{11} - c_{12})\lambda_{100}^2 - c_{44}2\lambda_{111}^2] \tag{7.22}$$

Note first that this second-order spontaneous magnetostrictive contribution to magnetic anisotropy does not lower the symmetry of the energy function as does an imposed strain. Further, ΔK_1 is independent of the signs of λ_{100} and λ_{111} and depends essentially on the difference in their magnitudes. For that reason ΔK is very small in iron where $|\lambda_{100}| \approx |\lambda_{111}|$; ΔK contributes a tenth of a percent to K. For Ni, $\Delta K/K \approx 1\%$; $\Delta K/K$ is also large in rare-earth alloys where $|\lambda_{111}| \gg |\lambda_{100}|$. For $TbFe_2$, $\Delta K/K \approx 20\%$ (Clark 1980).

7.4 ΔE EFFECT AND THERMODYNAMICS OF MAGNETOMECHANICAL COUPLING

It has just been seen that the Joule magnetostriction induced during the magnetization process has a second-order effect making magnetization easier or harder depending on the relative magnitudes of $(c_{11} - c_{12})\lambda_{100}^2$ and $2c_{44}\lambda_{111}^2$. The magnetostrictive strain has another second-order effect; it makes a material appear mechanically softer in the small strain regime. This is illustrated in Figure 7.13. Application of a stress to a magnetostrictive material induces a strain which tends to rotate the magnetization (***M*** moves toward a tensile stress direction for $\lambda > 0$, $B < 0$). The stress-induced rotation of ***M*** brings with it the magnetorestrictive strain in addition to the mechanically induced strain. Thus, the material appears more compliant by the ratio of the magnetostrictive strain to the purely mechanical strain. The reader should demonstrate that the material appears mechanically softer regardless of the sign of the magnetostriction.

The added magnetic strain, $e_M \approx \lambda$, is generally insignificant compared to the elastic strain for large stresses. Thus, this ΔE effect, as it is called, is more important for acoustic weaves, vibrations, and damping, than it is for mechanical strength.

The total strain e_{tot} that the ferromagnetic sample experiences under stress σ can be expressed as

$$e_{\text{tot}} = \frac{\sigma}{E_M} + \frac{3}{2}\lambda_s\left[\cos^2\theta - \frac{1}{3}\right] \tag{7.23}$$

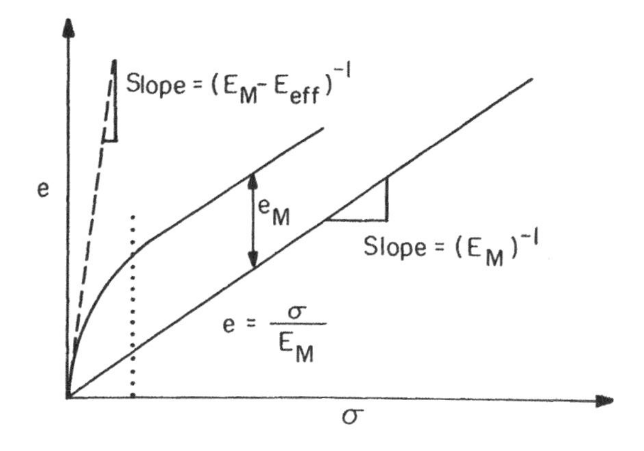

Figure 7.13 Strain versus stress in a magnetic material in the elastic regime. When the magnetization is fixed, the small strain behavior is purely linear. When the magnetization is free to respond to the applied stress, the material appears softer because of the additional magnetostrictive strain e_M. [Adapted from Lacheisserie (1994)].

Here E_M is the Young's modulus for fixed magnetization (no magnetostrictive contribution) that controls the purely elastic strain [first term in Eq. (7.23)]. The magnetostrictive contribution to total strain [second term in Eq. (7.23)] is taken from Eq. (7.1), where θ is the angle between the magnetization and the strain measuring direction. Because $\cos\theta = m = H/H_a^{\rm eff}$ for a hard-axis magnetization process, Eq. (7.19) can be substituted for $\cos\theta$. The apparent modulus in the face of this magnetostrictive strain is obtained from Eq. (7.23)

$$\frac{1}{E_{\rm eff}} \equiv \frac{1}{E_H} = \frac{\partial e_{\rm tot}}{\partial \sigma} = \frac{1}{E_M} - \frac{3\lambda_s H^2}{(H_a^{\rm eff})^3}\frac{\partial H_a^{\rm eff}}{\partial \sigma}$$

Using Eq. (7.20) for the strain dependence of $H_a^{\rm eff}$ gives

$$\frac{E_M - E_{\rm eff}}{E_M} \equiv \frac{\Delta E}{E} = \frac{9\lambda_s^2 H^2}{M_s H_a^3} E_{\rm eff} \tag{7.24}$$

This formula applies when the magnetization is initially orthogonal to the stress direction. Starting with $\boldsymbol{M}$ parallel to the stress would produce a magnetostrictive strain as M rotates away from the stress axis. But in this case the magnetic strain is only $\lambda_s/2$.

The ΔE effect is used in many magnetoacoustic devices where the resonance frequency or the sound velocity, both proportional to $E_H^{1/2}$, can be modulated by an applied field.

Thermodynamics of Magnetomechanical Coupling It is important in many cases to consider the change in internal energy, dU [Eq. (2.39)], of a magnetic material as it strains in response to being magnetized, $dl/l = \lambda$, where λ is the magnetostriction coefficient. In Chapter 2, the convention for the magnetoelastic contribution to the internal energy of a sample as it strains on being magnetized is given as $-V\sigma_M\lambda$. This term is analogous to the elastic energy $-p\,dV$ (a product of an intensive force variable and the change in an extensive measure of sample size). Thus the change in internal energy on magnetizing a magnetostrictive material is written as an extension of Eq. (2.39):

$$dU = T\,dS - p\,dV + (\mu_0 H\,dM - \sigma_M \lambda)V \tag{7.25}$$

The two most common thermodynamic functions, the Helmholtz free energy $F = U - TS$ and the Gibbs free energy $G = U + (p + \sigma e)V - (TS + \mu_0 HMV)$, have the differentials

$$\begin{aligned} dF &= -S\,dT - p\,dV + (\mu_0 H\,dM - \sigma_M de)V \quad \text{(independent variables: } T, V, M, e) \\ dG &= -S\,dT + V\,dp - (\mu_0 M\,dH - e\,d\sigma_M)V \quad \text{(independent variables: } T, p, H, \sigma) \end{aligned} \tag{7.26}$$

The function dF is appropriate in elastically free situations (e.g., magnetizing a sample that is free to strain). The function dG is useful in clamped situations (e.g., magnetizing a magnetostrictive film on a rigid substrate). Important relations follow from the partial derivatives of the free energies. For example

$$\left.\frac{\partial G}{\partial T}\right)_{p,H,\sigma} = -S \qquad \left.\frac{\partial G}{\partial p}\right)_{T,H,\sigma} = V$$

$$\left.\frac{\partial G}{\partial H}\right)_{T,p,\sigma} = -\mu_0 M V \qquad \left.\frac{\partial G}{\partial \sigma}\right)_{T,p,H} = Ve \tag{7.27}$$

(The μ_0 is omitted in cgs units.) Making use of the fact that dG is a perfect differential gives, among other things, the following useful relations:

$$\text{Volume thermal expansion} = \frac{\partial^2 G}{\partial T\,\partial p} = \frac{\partial V}{\partial T} = -\frac{\partial S}{\partial p} \equiv \alpha$$

and

$$\frac{1}{V}\frac{\partial^2 G}{\partial H\,\partial \sigma} = -\mu_0\frac{\partial M}{\partial \sigma} = \frac{\partial e}{\partial H} \equiv \mathbf{d} \tag{7.28}$$

The second relation in Eq. (7.28) defines the magnetostrictivity **d**, the strain produced per unit field or the change in magnetization per unit stress. These definitions form the basis of various methods of measuring the magnetostrictive strain or the magnetoelastic coupling coefficient (see Section 7.7). If the isotropic strain expression, Eq. (7.1), is used in Eq. (7.28), along with $m = \cos\theta = H/H_a$ for hard-axis magnetization, the following expression results for the magnetostrictivity:

$$\mathbf{d} = \frac{3\lambda_s H}{H_a^2} \tag{7.29}$$

This equation applies up to saturation and $\mathbf{d} = 0$ for $H > H_a$ (Livingston 1982). Integrating this form of **d** as suggested by the last equality in Eq. (7.28) yields

$$e(H) = \int_0^H \mathbf{d}(H)dH = \frac{3}{2}\lambda_s\left(\frac{H}{H_a}\right)^2$$

where $e(H)$ is the field-dependent magnetostrictive strain at H; that is, the magnetostriction is quadratic in the hard-axis field below saturation as noted at the beginning of this chapter (Fig. 7.14*a*). Figure 7.14*b* shows the variation of $\mathbf{d}(H)$ for a material with transverse anisotropy. Note that $\mathbf{d}(H)$ is the field derivative of $\lambda(H)$ from Eq. (7.2).

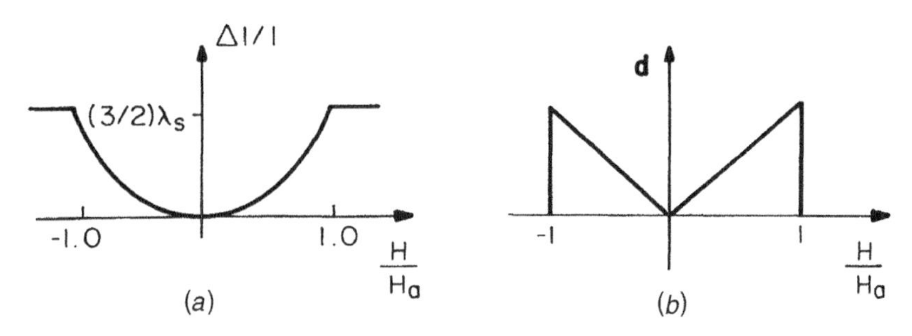

Figure 7.14 (*a*) Strain versus applied field transverse to easy axis; (*b*) magnetostrictivity for the same conditions.

The magnetostrictivity is important for transducer applications because it describes the stress sensitivity of magnetization (i.e., it describes a magneto-acoustic microphone) or the field sensitivity of strain (magnetoelastic speaker or sonar projector) [Eq. (7.28)]. What is needed for these applications is a material with a large value of λ_s and, more importantly, a small value of H_a so that saturation is achieved in a relatively weak field. Hence a useful figure of merit for transducer materials is the ratio λ_s/H_a^2 as suggested by Eq. (7.29). Terfenol-D, $(Tb_{0.27}Dy_{0,73})Fe_2$, is an example of a material developed to optimize this figure of merit. Keeping in mind that the area under $\mathbf{d}(H)$ (Fig. 7.14*b*) is fixed by λ_s [Eq. (7.29)], by decreasing the anisotropy field, the peak value of $\mathbf{d}(H)$ is increased.

The magnetic and elastic response of a system, $M = \chi H$, and $e = s^H\sigma$, are modified for a *coupled* magnetoelastic system by the addition of new terms from the integrals of Eq. (7.28):

$$M = \chi^\sigma H - \frac{1}{\mu_0}\int_0^\sigma \mathbf{d}d\sigma = \chi^\sigma H - \frac{1}{\mu_0}\mathbf{d}\sigma$$

$$e = \frac{\sigma}{E^H} + \int_0^H \mathbf{d}dH = S^H\sigma + \mathbf{d}H \tag{7.30}$$

Here, χ^σ is the magnetic susceptibility at constant stress and $S^H = 1/E^H$ is the elastic compliance at constant field. With Eq. (7.30), the magnetic and elastic contributions to the Gibbs free energy density [Eq. (7.26)] are

$$\frac{1}{V}dG = d\left\{\left(\frac{1}{2}\right)s^H\sigma^2 + \mathbf{d}H\sigma - \left(\frac{1}{2}\right)\chi^\sigma H^2\right\}. \tag{7.31}$$

Note how Eqs. (7.30) follow from dG by taking the derivatives described in Eq. (7.27). The form of the Gibbs free energy in Eq. (7.31) expresses the partition of energy between pure elastic modes, $\frac{1}{2}s^H\sigma^2$, pure magnetic modes $\frac{1}{2}\chi^\sigma H^2$, and coupled magnetoelastic modes, $\mathbf{d}H\sigma$. The coupling strength of a magnetoelastic device k is defined as the ratio of the energy in coupled magnetoelastic modes to the geometric mean of that in pure magnetic and elastic modes. Thus, from Eq. (7.31)

$$k = \frac{\mathbf{d}H\sigma}{\sqrt{\frac{1}{2}s^H\sigma^2\frac{1}{2}\chi H^2}} = \frac{2\mathbf{d}}{\sqrt{s^H\chi^\sigma}} = 2\mathbf{d}\sqrt{\frac{E^H}{\chi^\sigma}} \tag{7.32}$$

The magnetomechanical coupling factor k can take on values between zero and unity; the latter is the condition of complete coupling. In a crystal, k is a tensor with k_{33} pertaining to longitudinal mode coupling. Similar expressions apply for shear-mode coupling constants.

In terms of the ΔE effect, it can be shown, using Eqs. (7.24), (7.29), and (7.32), that

$$k^2 = \frac{E^M - E^H}{E^M} = \frac{\Delta E}{E} \tag{7.33}$$

Thus k approaches unity as E^H approaches zero.

Figure 7.15 shows the variation of the ΔE affect in amorphous FePC ribbons as a function of annealing condition: Z is zero-field-annealed; L and T are field-annealed, longitudinal, and transverse to the ribbon direction. The measurements were done by measuring vibration frequency of the ribbon near 400 Hz (Berry and Pritchet 1975). Annealing in the transverse direction gives an initial magnetization state that produces the most magnetostrictive strain under longitudinal flexure. At these low strain levels the modulus of the amorphous ribbon is reduced by 80% in a field of about 6 Oe. From Eqs. (7.32) and (7.33), the ΔE effect is maximum when $\mathbf{d}^2$ is maximum; $\mathbf{d}$ reaches a peak at the anisotropy field H_a, which for this sample is about 6 Oe.

7.5 SYMMETRY-INVARIANT NOTATION

So far, the forms for the ME effects have been considered only in isotropic and in cubic samples. It is of interest to know the form the magnetostriction must take in hexagonal and in other low-symmetry magnetic structures. Materials with hexagonal structures include cobalt and many cobalt-rich alloys, rare-earth metals, and some rare-earth intermetallics, as well as barium hexaferrite. A general formalism for magnetic anisotropy and magnetostriction introduced by Callen and Callen and reviewed by Lacheisserie (1994) is now described. In this formalism, the free energy functions for all crystal symmetries are expanded in a set of orthogonal harmonic functions, introduced in Chapter 6, Eq. (6.3). There, the expansion coefficients are chosen so that the anisotropy energy is invariant under all point operations of the crystal symmetry. In this formalism, one set of magnetostriction coefficients is defined and they apply

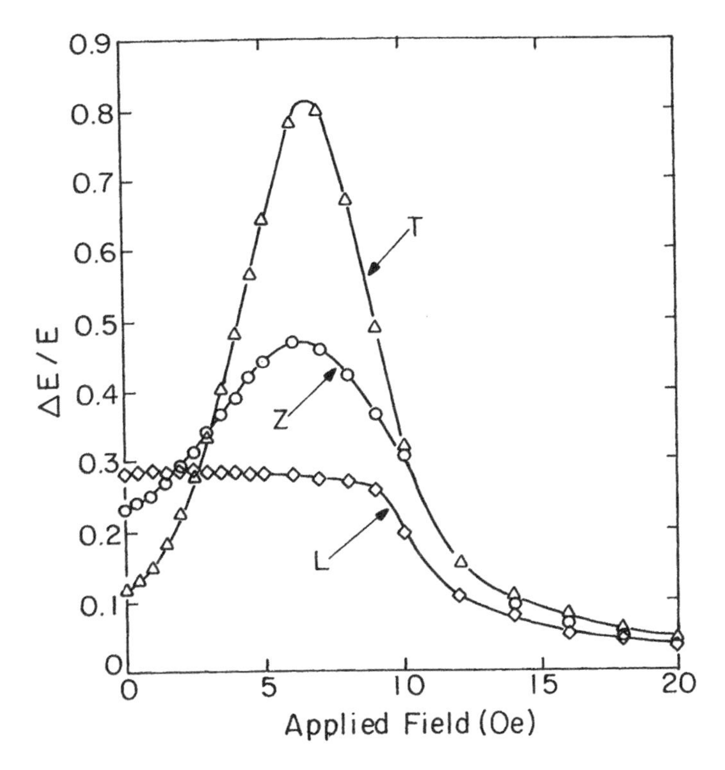

Figure 7.15 The ΔE effect measured by the vibrating reed technique on an amorphous FePC ribbon in three annealing conditions: Z, no field, L, longitudinal field; T, tranverse field (Berry and Prichett 1975).

across all symmetry classes. The correct method of forming the free energy expression is to write separately the strains and direction cosines that are the basis functions of the fully symmetric representation of the point group of interest.

Figure 7.16 illustrates the irreducible representations of the direction cosines and strains in the cubic point group. The former are illustrated with familiar atomic wavefunctions defined by the respective combinations of α_i values. The strains are isomorphous with (they show the same form as) the wavefunctions, namely, positive strain where the wavefunction has positive amplitude and vice versa. Once the correct symmetry-invariant expressions for the α_is and the e_{ij} values are formed, their direct products are taken, ensuring that the energy is invariant under all symmetry operations of the α_i and the e_{ij} values.

For brevity, the form of the magnetoelastic energy and the formula for the magnetostriction are expressed in cubic and hexagonal symmetry to lowest order in the direction cosines in each case. Higher-order terms and forms for other symmetries can be found in the thorough exposition by Callen and Callen (1963, 1965) and by Lacheisserie (1994).

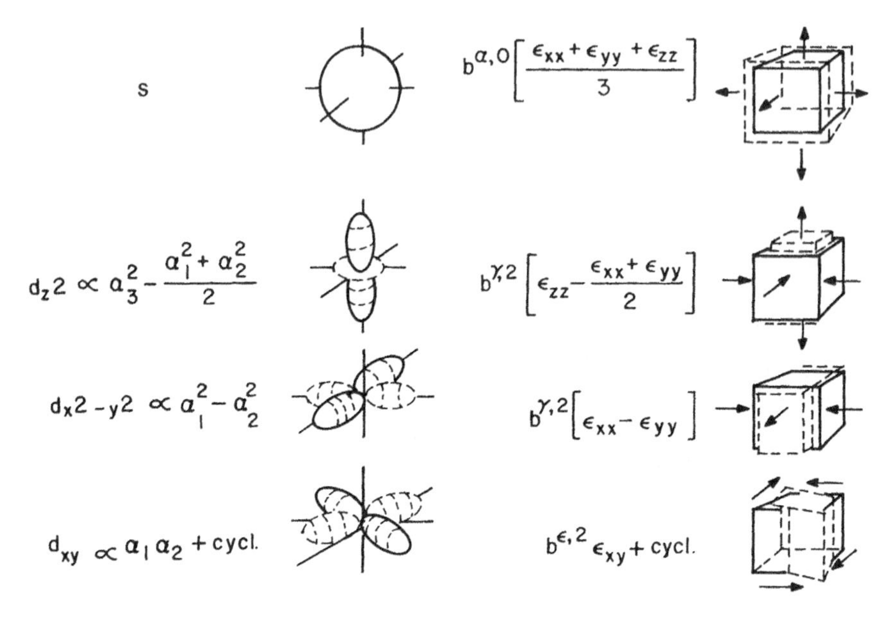

Figure 7.16 The first line shows the form of the isotopic s wave function and the isotropic, volume magnetostriction symmetry. In the next three rows, the symmetry and amplitude (solid line, $\psi_d > 0$ dotted lines, $\psi_d < 0$) of the five atomic d-functions are seen to be isomorphic with the form of the five cubic Joule magnetostriction strains [Eq. (7.33)]. This is because both are examples of irreducible representations of the cubic group.

For *cubic* symmetry the ME free energy is

$$f_{\text{me}}^{\text{cubic}} = \frac{\varepsilon_{xx}+\varepsilon_{yy}+\varepsilon_{zz}}{3}[b^{\alpha,0}+\cdots]+\frac{3}{2}\left(\varepsilon_{zz}-\frac{\varepsilon_{xx}+\varepsilon_{yy}}{2}\right)$$
$$\times\left[b^{\gamma,2}\left(\alpha_3^2-\frac{\alpha_1^2+\alpha_2^2}{2}\right)+\cdots\right]+\frac{1}{2}(\varepsilon_{xx}-\varepsilon_{yy})[b^{\gamma,2}(\alpha_1^2-\alpha_2^2)+\cdots]$$
$$+2\varepsilon_{xy}\alpha_1\alpha_2\{b^{\varepsilon,2}+\cdots\}+\text{cycl.} \tag{7.34}$$

The b terms are the new, symmetry-invariant ME coupling coefficient replacing the old B_1 and B_2. It is helpful to note that the lowest-order cubic ME energy expression is a sum of products of strain components and direction cosines, both of which have the same form, namely, that of the five irreducible representations of the cubic point group.

For *cubic* symmetry, the magnetostriction to fourth order in α_i is

$$\lambda^{\text{cubic}} = \frac{1}{3}\lambda^{\alpha,4}\left(\alpha_1^4+\alpha_2^4+\alpha_3^4-\frac{3}{5}\right)+\sum_{i=1,2,3}\left(\beta_i^2-\frac{1}{3}\right)$$

$$\times \alpha_i^2 \left[\lambda^{\gamma,2} + \lambda^{\gamma,4} \left(\alpha_i^2 - \frac{6}{7} \right) + \cdots \right]$$

$$+ \sum_{i<j=1,2,3} 2\beta_i \beta_j \alpha_i \alpha_j \left[\lambda^{\varepsilon,2} + \lambda^{\varepsilon,4} \left(\alpha_k^2 - \frac{1}{7} \right) + \cdots \right] \tag{7.35}$$

The superscript α on a magnetostriction coefficient denotes a strain that preserves crystal symmetry; the γ terms describe volume-conserving uniaxial (tetragonal) strains; ε-labeled terms distort the symmetry in the $\{100\}$ planes (see Fig. 7.16). The index after the Greek superscript indicates the order in direction cosines.

For *hexagonal* symmetry, the free energy is

$$f_{\text{me}}^{\text{uniaxial}} = b_1^{\alpha,0} \frac{\varepsilon_{xx} + \varepsilon_{yy} + \varepsilon_{zz}}{3} + b_2^{\alpha,0} \frac{\sqrt{2}}{3} \left(\varepsilon_{zz} - \frac{\varepsilon_{xx} + \varepsilon_{yy}}{2} \right)$$

$$+ \left[b_1^{\alpha,2} \frac{\varepsilon_{xx} + \varepsilon_{yy} + \varepsilon_{zz}}{\sqrt{2}} + b_2^{\alpha,2} \left(\varepsilon_{zz} - \frac{\varepsilon_{xx} + \varepsilon_{yy}}{2} \right) \right] \left(a_3^2 - \frac{1}{3} \right)$$

$$+ b^{\varepsilon,2} \left[\frac{1}{2} (\varepsilon_{xx} - \varepsilon_{yy})(\alpha_1^2 - \alpha_2^2) + 2\varepsilon_{xy} \alpha_1 \alpha_2 \right]$$

$$+ 2b^{\zeta,2} (\varepsilon_{yz} \alpha_2 \alpha_3 + \varepsilon_{xz} \alpha_1 \alpha_3) + \cdots \tag{7.36}$$

For the *hexagonal* magnetostriction (H_{I} Laue group), the result is

$$\lambda^{\text{hex}} = \frac{1}{3} \lambda_1^{\alpha,0} + \lambda_2^{\alpha,0} \left(\beta_3^2 - \frac{1}{3} \right) + \left[\frac{1}{3} \lambda_1^{\alpha,2} + \lambda_2^{\alpha,2} \left(\beta_3^2 - \frac{\beta_1^2 + \beta_2^2}{3} \right) \right] \left(\alpha_3^2 - \frac{1}{3} \right)$$

$$+ \lambda^{\varepsilon,2} \left[\frac{1}{2} (\beta_2^1 - \beta_2^2)(\alpha_1^2 - \alpha_2^2) + 2\beta_1 \beta_2 \alpha_1 \alpha_2 \right]$$

$$+ 2\lambda^{\zeta,2} (\beta_2 \beta_3 \alpha_2 \alpha_3 + \beta_1 \beta_3 \alpha_1 \alpha_3) + \cdots \tag{7.37}$$

The α- and ε-labeled magnetostrictions describe the same distortions in a hexagonal system as they do in cubic symmetry. Note that even a change in c/a ratio ($\lambda_2^{\alpha,0}$ term) preserves hexagonal crystal symmetry even though the strain is not isotropic. In hexagonal symmetry ζ-terms which shear over the c axis are also found.

The symmetry-based cubic magnetostriction coefficients can be expressed in terms of the older coefficients using the same method of assigning appropriate values to the α and β values as was done before Eq. (7.13):

$$\lambda_{100} = \tfrac{2}{15} \lambda^{\alpha,4} + \tfrac{2}{3} (\lambda^{\gamma,2} + \tfrac{1}{7} \lambda^{\gamma,4} + \cdots) \approx \tfrac{2}{3} \lambda^{\gamma,2} \tag{7.38}$$

$$\lambda_{111} = -\tfrac{4}{45} \lambda^{\alpha,4} + \tfrac{2}{3} (\lambda^{\varepsilon,2} + \tfrac{4}{21} \lambda^{\varepsilon,4} + \cdots) \approx \tfrac{2}{3} \lambda^{\varepsilon,2} \tag{7.39}$$

7.6 TEMPERATURE DEPENDENCE

The description of the symmetry-invariant notation in Section 7.5 was necessary before coming to the last major topic of this chapter, namely, the temperature dependence of magnetostriction. This necessity comes for two reasons (Clark et al. 1965, del Moral and Brooks 1974): (1) much of the modern data, especially that for rare-earth metals and alloys, is presented in the literature in terms of $\lambda^{\mu,l}$ rather than λ_{100} and λ_{111}; and (2) the theoretical explanation of the temperature dependence must be done in symmetry-invariant formalism.

The temperature dependence of magnetostriction in a number of magnetic materials is described to motivate the exposition of the theory of the temperature dependence of these symmetry-based strains. After outlining the theory, the data are interpreted.

Figure 7.17 shows the temperature dependence of the two principal magnetostriction constants in Ni. The shaded area shows the trend and scatter in data from numerous sources. The solid lines are based on calculations by Lacheisserie (1972) that will be described later. Note the sharp decrease in the magnitude of these coefficients, especially $\lambda^{\varepsilon,2}$, with increasing temperature.

Figures 7.18 and 7.19 show $\lambda^{\mu}(T)$ data for Tb and Gd, respectively. Note for Tb the steeper decrease with increasing temperature for λ^{A} compared to λ^{C} and for Gd the sharp temperature dependence including a change in sign for $\lambda^{\gamma,2}$ below $T_C = 290$ K.

Figure 7.20 shows the temperature dependence of magnetostriction in the oxide, yttrium iron garnet (YIG). Here, the form of $h_1(T)$ is reminiscent of the magnetization curve of $NiFe_{2-x}V_xO_4$ ($x = 0.86$) in Figure 4.9. The nonmonotonic behavior of h_2 with temperature is well described by the Callen–Callen model.

The temperature dependence of the isotropic magnetostriction coefficients in a number of amorphous magnetic alloys is illustrated in Figure 7.21 (O'Handley 1978).

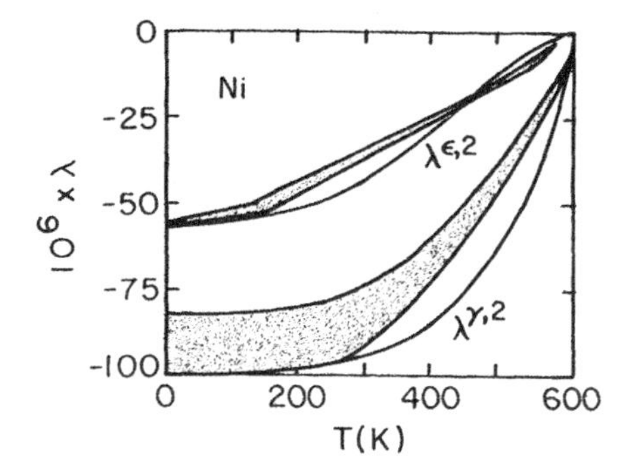

Figure 7.17 Shaded areas show range of experimental magnetostriction of Ni single crystals (Franse 1970, Bower 1971, Lee 1971), and solid lines show calculated temperature dependence (Lacheisserie 1972).

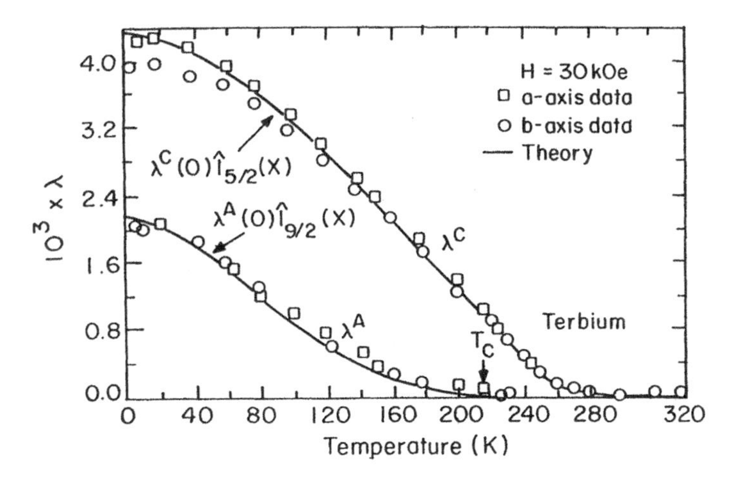

Figure 7.18 Temperature dependence of λ^A and $\lambda^C = \lambda^{\varepsilon,2}/2$ for Tb. (Rhyne and Legvold 1965).

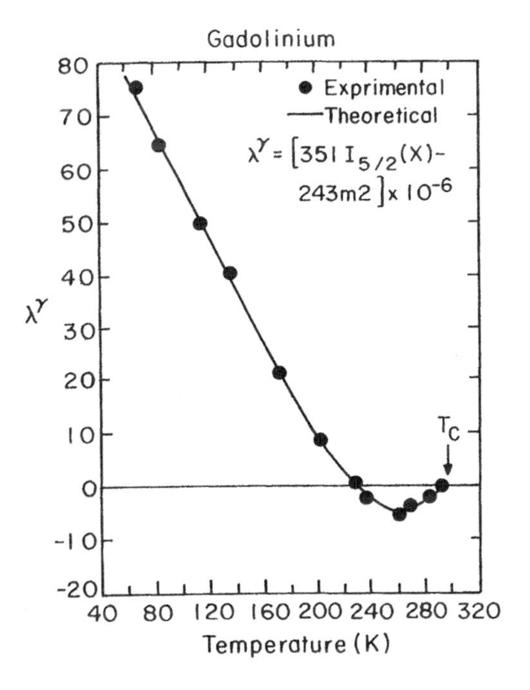

Figure 7.19 Temperature dependence of $\lambda\gamma$ for Gd. Experimental data of Coleman (1964) fit by Callen and Callen (1965) to their theory.

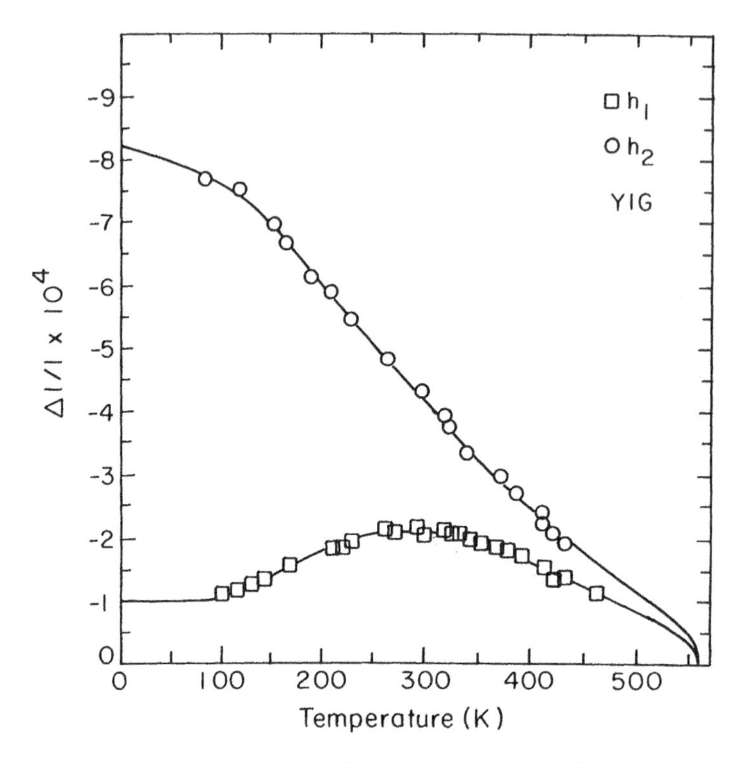

Figure 7.20 Temperature dependence of magnetostriction coefficients of yttrium iron garnet (YIG). [After Callen et al. (1963).]

In all of these examples, just as for magnetocrystalline anisotropy, $\lambda(T)$ drops much more sharply with increasing temperature than does $M(T)$. It will be shown that the theory developed by Zener (1954) and standardized in its modern form by Callen and Callen (1963, 1965) for the temperature dependence of magnetic anisotropy also applies to these strain-dependent components of magnetic anisotropy, namely, the magnetostriction.

To describe the strain dependence of the anisotropy energy, the anisotropic Hamiltonian containing two-ion and single-ion terms (see Chapter 6),

$$H = \boldsymbol{S}_i \cdot \boldsymbol{J}_{ij} \cdot \boldsymbol{S}_j + \boldsymbol{S}_i \cdot \boldsymbol{D} \cdot \boldsymbol{S}_i$$

can be expanded in powers of strain to first order, much as was done in the Néel model in Section 7.2:

$$H_{\text{me}} = e_\mu \mathbf{S}_i \cdot \frac{\partial \mathbf{J}_{ij}}{\partial e_\mu} \cdot \mathbf{S}_j + e_\mu \mathbf{S}_i \cdot \frac{\partial \mathbf{D}}{\partial e_\mu} \cdot \mathbf{S}_i + \cdots \quad \text{and} \quad H_{\text{elastic}} = \frac{1}{2} c_\mu e_\mu^2 \qquad (7.40)$$

The energy is the expectation value of this strain-dependent Hamiltonian,

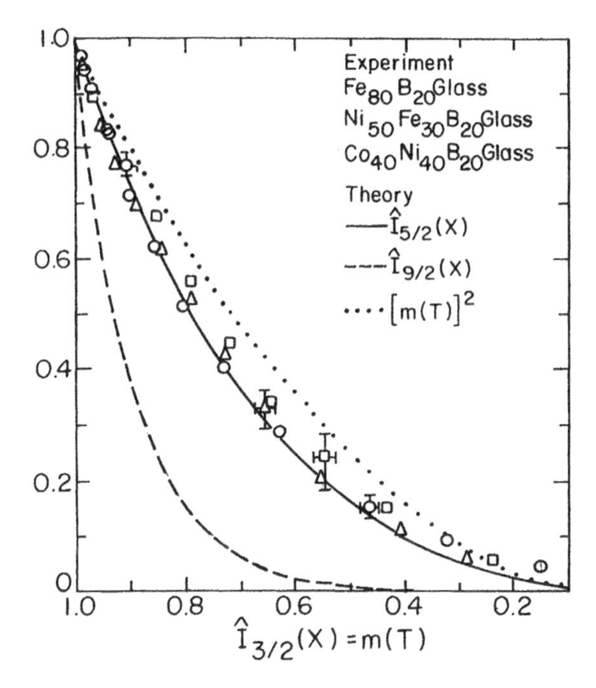

Figure 7.21 Temperature dependence of magnetostriction in three amorphous alloys after O'Handley (1978).

$E = \langle H \rangle$, and the equilibrium strain is obtained by solving

$$\frac{\partial E}{\partial e_\mu} = 0$$

for the strain components, e_μ. The results of this minimization are expressed below as sums over nearest-neighbor spins i, j.

$$\begin{array}{llll}
 & \underline{\text{Isotropic}} & \underline{\text{Anisotropic}} & \\
e_\mu = & -\dfrac{1}{c_\mu}\displaystyle\sum_{i,j}\frac{\partial \mathbf{J}_{ij}}{\partial e_\mu}\langle \mathbf{S}_i \cdot \mathbf{S}_j \rangle & -\dfrac{1}{c_\mu}\displaystyle\sum_{i,j}\frac{\partial \mathbf{J}_{ij}}{\partial e_\mu}\langle \mathbf{S}_i^z \mathbf{S}_j^z \rangle & \text{(exchange)} \\
 & -\dfrac{1}{c_\mu}\displaystyle\sum_{i}\frac{\partial \mathbf{D}}{\partial e_\mu}\langle \mathbf{S}_i^2 \rangle & -\dfrac{1}{c_\mu}\displaystyle\sum_{i,j}\frac{\partial \mathbf{D}}{\partial e_\mu}\langle (\mathbf{S}_i^z)^2 \rangle & \text{(crystal field)}
\end{array} \quad (7.41)$$

The isotropic terms here contribute to the volume magnetostriction via the two-ion exchange interaction and the Madelung part of the single-ion, crystal field term. The contributions to the Joule magnetostriction involve (1) the anisotropic exchange couplings between the z components of spins at two different sites and (2) crystal field anisotropy seen by the spin at each site.

The expectation values of the spin operators are called *spin correlation functions*. The important spin correlation functions are the same as those involved in single-ion and two-ion anisotropy which were described in Chapter 6. Suffice it to say that the exact results of the quantum statistical mechanical model of temperature dependence of single-ion ME effects can be summarized as follows:

$$\frac{\lambda^{\mu,l}(T)}{\lambda^{\mu,l}(0)} \quad \text{or} \quad \frac{K^{\mu,l}(T)}{K^{\mu,l}(0)} = \kappa^{\mu,l}(T) = \frac{\langle [\mathbf{S}_i^z(T)]^2 \rangle}{\langle [\mathbf{S}_i^z(0)]^2 \rangle} \Rightarrow \frac{I_{l+1/2}(X)}{I_{1/2}(X)} \equiv \hat{I}_{l+1/2}(X) \qquad (7.42)$$

The Bessel function, modified Bessel function, and their argument, X, were identified in Chapter 6. Everything that was said there about $\kappa_l(T)$ applies here to $\lambda_l(T)$.

Lacheisserie (1972) has shown that the nickel data available in 1971 can be fit up to 300 K with a combination of single-ion and two-ion terms (solid lines, Fig. 7.17):

$$B^{\gamma,2} = 110\{0.13\hat{I}_{5/2}(X) + 0.87\text{m}^2\} \times 10^6\,(\text{d/cm}^2)$$
$$B^{\varepsilon,2} = 149\{2.54\hat{I}_{5/2}(X) - 1.54\text{m}^2\} \times 10^6\,(\text{d/cm}^2)$$

Thus, the magnetostriction of Ni contains, in addition to a single-ion term, a significant two-ion anisotropic contribution that varies as m^2. The latter plays a larger role as T/T_c approaches unity because the single-ion terms are so small there. The two-ion term causes the sharper decrease in $\lambda^{\varepsilon,2}$ and more gradual decrease in $\lambda^{\gamma,2}$ below T_C. It is interesting that even a 3*d* metal can be described with some success by this model, which is based on localized magnetic moments. The reason for any success at all with a local moment model in a metallic ferromagnet may be due to the fact that while the 3*d* wavefunctions are periodic and spatially extended over the lattice, the charge and magnetic moment densities are fairly well concentrated at each site. The nickel magnetostriction values at room temperature and 4.2 K agree with the values in Table 7.1 using the relations in Eqs. (7.38) and (7.39). The temperature dependence of magnetostriction in iron is more complex than that of Ni (Lacheisserie and Mendia Monterosso 1983). A model taking into account the nonlocalized 3*d* character of iron has recently been applied with success to explain the nonmonotonic temperature dependence of magnetostriction in the intermetallic compounds Y_2Fe_{17} and $Y_2Fe_{14}B_1$ (Kulakowski and del Moral, 1994).

The temperature dependence of λ^A and λ^C in Tb (Fig. 7.18) follow the forms appropriate for cubic ($l = 4$) and uniaxial ($l = 2$) symmetry, respectively. The temperature dependence of $\lambda^{\gamma,2}$ in Gd follows that of a sum of a uniaxial, single-ion and two-ion terms (Callen and Callen 1965) as indicated in the figure. Here the negative two-ion term is strong enough to account for the change in sign of λ above 220 K.

The magnetostrictions of many metallic glasses appear to follow the theory for uniaxial, single-ion anisotropy. The magnetostriction of some cobalt-rich amorphous alloys shows a sign change that can be accounted for by a sum of uniaxial, single-ion, and two-ion terms that combine in ratios that are consistent over several compositions (O'Handley 1978). It is possible to integrate itinerant character into these local-moment models by inclusion of electron hopping, and thereby explain magnetostriction in metallic systems (Kulakowski and del Moral 1994, 1995).

7.7 MEASUREMENT TECHNIQUES

Strain Gauges Metal foil strain gauges are often used to measure magnetostriction in bulk samples (see Fig. 7.22). When applied to a sample using a suitable high-temperature cement (e.g. polyimide adhesive, PLD 700, BLH Electronics), the gauges can be used over a temperature range from liquid helium temperature to 670 K.

The metal foil forms a serpentine pattern with elongated legs in the direction in which strain is to be measured. A metal with a temperature-insensitive

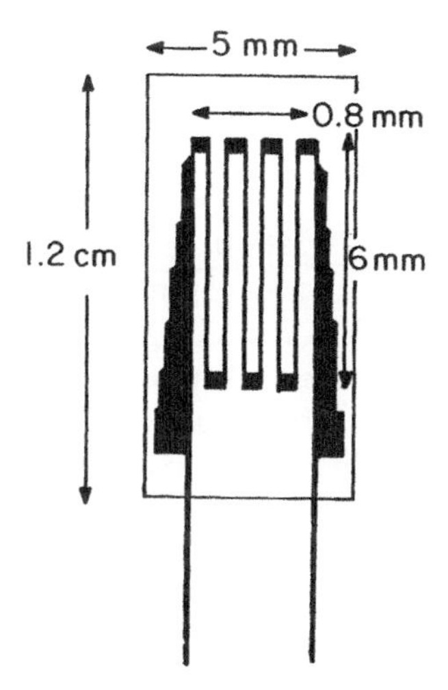

Figure 7.22 Metal foil strain gauge of the type often used in strain measurements. The gauge sketched here is patterned after type FSM-23-35-S6 (BLH Electronics). A serpentine pattern of NiCr alloy is supported by a polyimide backing. This gauge has a resistivity of 350 Ω and a useful temperature range of 0–670 K.

coefficient of resistance, $\partial \ln R/\partial T$, such as certain NiCr alloys, is often used as the strain-sensing element. The measure of the gauge efficiency, the gauge factor G, is the ratio of the fractional resistance change to the fractional length change: $\Delta R/R = G\varepsilon$. Metal foil strain gauges typically have $G \approx 2.0$. The strain causes an elongation $\Delta l/l = \varepsilon$ of the resistance foil and a $(1-2\upsilon\varepsilon)$ decrease in its cross-sectional area. These dimensional changes account for most of the resistance change in a metal foil gauge: $R = \rho l/A = \rho l_0(1+\varepsilon)/A_0(1-2\upsilon\varepsilon) \approx R^0[1+\varepsilon(1+2\upsilon)]$. Semiconductor strain gauges can have gauge factors of a few hundred, but their performance is strongly dependent on temperature. The cross-sectional dimensions of the gauge foil are greater to reduce their resistance.

The guage is cemented to the specimen of interest. The specimen must be thick enough that the presence of the gauge does not inhibit its tendency to strain. The resistance change of the gauge is measured with a sensitive bridge circuit. The voltage change across the bridge is related to the strain and gauge factor by $\varepsilon = 4\Delta V/(GV_0)$, where V_0 is the voltage applied to the bridge. When an AC voltage is applied, the change in voltage can be read with a lock-in amplifier. Often a dummy gauge in one leg of the bridge adjacent to the active gauge, can be placed near the sample (not bonded to it) to compensate for temperature variations and magnetoresistance (which increases as H^2; see Chapter 15). In most cases the largest source of error is in the bonding between the gauge and the sample.

Capacitance bridges (in which the straining material displaces a capacitor plate) are also used to measure λ. The capacitor forms part of a high-Q resonant circuit. The resonance frequency $\omega = (LC)^{-1/2}$ is a sensitive measure of the capacitance C which gives the plate spacing $d = \varepsilon A/C$, where A is the plate area. The relative change in resonant frequency thus gives the strain, $\varepsilon = 2\Delta\omega/\omega_0$. The magnetostriction may be measured at any frequency well below the capacitor resonance.

Small-Angle Magnetization Rotation The small-angle magnetization rotation method (Narita et al. 1980) has proven to be very sensitive when applied to ribbon-shaped samples such as metallic glass strips (Yamasaki et al. 1990).

The sample is magnetized along its length with a near-saturating field. A small transverse AC magnetic field causes the magnetization to oscillate through an angle $\pm\theta_0$ about the ribbon axis (Fig. 7.23).

This oscillation can sometimes be accomplished by passing a current through the ribbon; the surface magnetization is canted by the transverse H field of the current (Hernando et al. 1983). A pickup coil of N turns and cross-sectional area A can be arranged to allow measurement of the change in $M_s \cos\theta_0$, the magnetization component parallel to the ribbon axis, $V_{ac} = 10^8 NA4\pi M_s \sin\theta_0\, d\theta/dt$ (cgs units). A tensile stress applied to the ribbon either increases or decreases θ_0 depending on the sign of the magnetostriction coefficient. The effect of stress can be expressed by solving the zero-torque

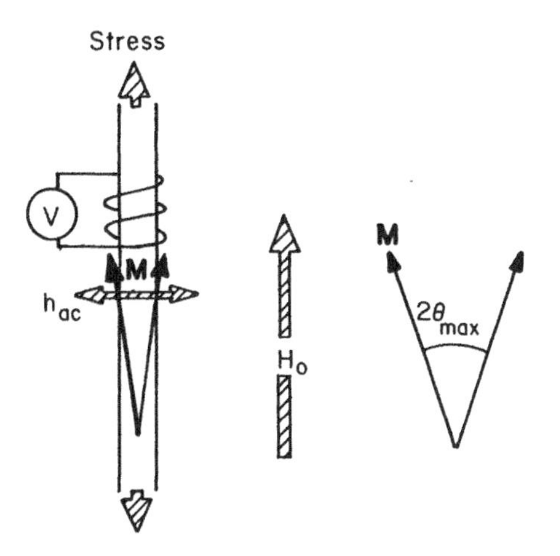

Figure 7.23 Schematic of small-angle magnetization rotation experiment showing longitudinal field H_0, transverse AC field, h_{ac}, pickup coil for measuring the small change in longitudinal magnetization and applied stress direction.

condition on the energy:

$$f = -M_s H_0 \cos\theta - M_s h_{ac} \sin\theta + \tfrac{3}{4}\lambda_s \sigma \sin^2\theta + 2\pi M_s^2 (N_\perp \sin^2\theta - N_\parallel \cos^2\theta)$$

For small θ the voltage is

$$V_{ac} = 2\pi M_s N A \sin^2\theta_{\max} \sin 2\omega t$$

with

$$\sin\theta_{\max} = \frac{h_{ac}}{H_0 + M_s(N_\perp - N_\parallel) + 3\lambda_s\sigma/M_s} \tag{7.43}$$

The strength of the second harmonic signal in the pickup coil depends on the stress applied to the sample and on the magnetostriction coefficient as expressed in Eq. (7.43). Analysis to extract λ_s is simplified if the effects of an applied stress are balanced exactly by a change in the strength of the longitudinal field, H_0, so that $V_{ac}(\sigma = 0)$ is identical to $V_{ac}(\sigma \neq 0)$. In that case

$$\Delta H_0 = \frac{3\lambda_s \sigma}{M_s} \quad \text{and} \quad \lambda_s = \frac{M_s \Delta H_0}{3\sigma}$$

where ΔH_0 is the change in H_0 needed to restore V_{ac} to its zero-stress value.

Measurements of magnetostriction in amorphous ribbons and wires by small angle magnetization are found in Yamasaki et al. (1990) and Hernando et al. (1983).

Strain-Modulated FMR Ferromagnetic resonance (FMR, Chapter 9) occurs at a microwave frequency that depends on the magnitude of the saturation magnetization and the strength of the internal field seen by the magnetization. If the magnetization is measured independently, the resonance position is then a function only of the effective internal field, $H^{\text{eff}} = H_{\text{demag.}} + H_a + H_{me}$, where, from Eq. (7.19),

$$H_{me} \propto \frac{2B_1 e}{M_s} \approx \frac{3\lambda_s \sigma}{M_s}$$

By modulating the state of strain or stress of the sample, certain magnetoelastic coupling coefficients can be determined (Zuberek et al. 1988). Care must be exercised in determining whether the sample is clamped ($H_{me} \propto \lambda_s \sigma$) or free ($H_{me} \propto Be$).

Thin-Film Techniques The magnetoelastsic coupling coefficients of thin films can be determined by varying the direction of a field to induce a magnetoelastic stress causing the substrate to bend (Klockholm 1996). The substrate deflection can be detected using either capacitance (Ciria et al. 1995) or optical techniques (Tam and Schroeder 1989, Koch et al. 1996, Weber et al. 1994, Bellesis et al. 1993). It must be recognized that a magnetic film is generally constrained by the substrates so it cannot strain freely to its saturation magnetostriction value. The magnetoelastic stress, B_i, or $b^{\mu,l}$ exerts a bending moment on the film–substrate sandwich to induce the bending. The displacement, Y, of the end of a film of length $2L$, fixed at its center (Appendix 16A), depends on the net in-plane stress σ_f exerted by the film as described by Stoney's equation

$$Y = \frac{3(1 - v_s)t_f L^2 \sigma_f}{t_s^2 E_s} \tag{7.44}$$

where t_i, E_i, and v_i are the thickness, modulus, and Poisson ratio of the film, $i = f$, and substrate, $i = s$, respectively. Klockholm *et al.* (1976) have made extensive use of this method, which is described in more detail by Tam and Schroeder (1989).

Magnetoelastic coefficients can also be determined by the *inverse method* in which an imposed strain changes the magnetic anisotropy (Sun and O'Handley 1991; O'Handley et al. 1993). Strain-modulated FMR has been used by Zuberek *et al.* (1988) to measure the magnetostriction constants of a series of [Ni/Ag] multilayers as a function of Ni thickness. The strain was applied to the film by a transducer. Their results suggest a surface contribution to the magnetostriction, $\lambda^{\text{eff}} = \lambda^{\text{bulk}} + \lambda^s/t_{\text{Ni}}$. Here λ^{bulk} and λ^s are the bulk and surface contributions to the effective magnetostriction and t_{Ni} is the thickness of the magnetic layer, in this case Ni (see Chapter 16).

The inverse method of determining magnetoelastic coupling has been used to identify the thickness dependence of the magnetoelastic coupling coefficients in Ni and permalloy films (Song et al. 1994). These results indicated not only an inverse thickness dependent contribution, but also the effect of a magnetoelastic dead layer (see Fig. 16.18):

$$B^{\text{eff}} = B^{\text{bulk}} + \frac{B^{\text{surf}}}{t - t_0}$$

It is important to recognize that the inverse methods give the magnetoelastic coupling coefficient unless the elastic constant of the film is also known. The magnitude of the strain in a film of thickness t_f on a substrate of length $2L$ and thickness $t_s > t_f$ can be related to the vertical displacement, y, in a four-point bending geometry. In a four-point fixture the bending is circular inside the inner two pressure points (see Fig. 7.24). It can be shown that in the limit of circular curvature of radius R, $e = t_s/(2R)$. Using $\cos\theta = R/A = R/(R - y)$ and $\theta = L/R$, R and θ can be eliminated to give

$$y = \frac{t_s}{2e}\left[1 - \cos\left(\frac{eL}{t_s}\right)\right] \tag{7.45}$$

Thus, knowing the vertical displacement at a given distance from the center of a film on a substrate gives the strain in the film.

In such a bending geometry the film is essentially constrained (by the substrate) across its width, $e_{yy} = 0$, and is free to strain normal to its surface, $e_{zz} \neq 0$. Using the method outlined in Appendix 7B, it can be shown that $e_{zz} = -\upsilon e_{xx}/(1 - \upsilon) \approx e_{xx}/2$. Thus the magnetoelastic energy of the film strained by e_{xx} is, from Eq. (7.8):

$$f_{me} = B_1 e_{xx}\left[\alpha_x^2 - \frac{\upsilon}{1 - \upsilon}\alpha_z^2\right] \approx B_1 e_{xx}\left(\alpha_x^2 - \frac{1}{2}\alpha_z^2\right)$$

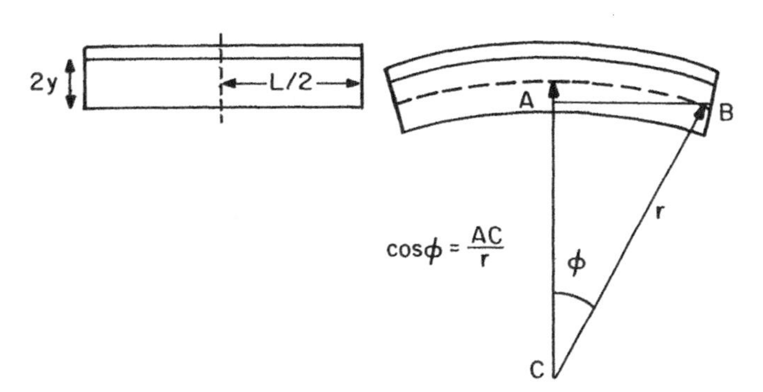

Figure 7.24 Geometry for determining strain in a thin film on a substrate subjected to a circular bend.

A measurement of $M-H$ characteristics in different states of strain generally reveals the strain-dependent part of the anisotropy provided $B_i e$ is a measurable fraction of K (see Fig. 7.12). Details for various geometries are given in O'Handley et al. (1993).

All the above methods can be used to determine specific magnetostriction coefficients or ME coupling coefficients if three conditions are met: (1) the crystallographic orientation of the sample is known, (2) the strain geometry is properly analyzed, and (3) the direction of magnetization is known throughout the experiment. Otherwise an average over some combination of ME coefficients is measured. The indirect methods in which a strain is imposed by bending the substrate are only useful if $B_1 e$ is a significant fraction of K. The direct methods in which the substrate deflection is measured as M is rotated typically require a cantilever displacement stability and sensitivity of order 1–10 nm.

7.8 SUMMARY

Magnetoelastic effects can be described by a number of different parameters: the magnetoelastic coupling coefficients B_{ij} (stresses of magnetic origin), the magnetostriction constants λ_i (strains of magnetic origin), the magnetostrictivity, **d** (the change in magnetization per unit stress or the change in strain per unit field), or the (dimensionless) magnetomechanical coupling factor k (efficiency of energy coupling from magnetic or elastic to coupled ME modes). The Joule magnetostriction λ_s is the anisotropic strain in a material associated with the direction of magnetization. In an unconstrained sample, λ is proportional to a component of the magnetic stress B_i divided by the appropriate stiffness modulus. The inverse Joule effect brings about a change in the magnetization process or in H_a^{eff}, on straining a sample.

Under small stresses a nonsaturated magnetostrictive material can appear softer than it is in its saturated, constant $\mathbf{M}$, state. This is the ΔE effect which is useful in many magnetoacoustic resonator devices.

Rare-earth metals, intermetallic compounds, and alloys with $L_z \neq 0$ can show much larger magnetoelastic effects than do most $3d$ transition metals and their alloys, but it is difficult to measure their saturation values when the anisotropy is also large.

Magnetoelastic energy has its origin in the same crystal field/spin–orbit interactions, and anisotropic exchange interactions that cause magnetocrystalline anisotropy; the ME energy is the strain-dependent part of the anisotropy. As such, the temperature dependence of ME coefficients in local moment systems can be described by the same symmetry-based model that describes the strain-independent anisotropy, $k_l(T)$.

In this and the preceding chapters, several fundamental energies that make magnetic materials different from nonmagnetic materials have been described:

- Magnetostatic
- Exchange
- Magnetocrystalline and field-induced anisotropy
- Magnetoelastic anisotropy
- Zeeman energy.

In the next chapter, these energies are combined to understand the formation of magnetic domain walls and domains, and the interaction between domain walls and defects. These effects form a bridge from the fundamental energies of magnetism to the rich variety of magnetic materials, soft and hard, that make possible numerous applications that are at the core of many sectors of our economy including energy, transportation, information technology, and communications.

APPENDIX 7A FIELD-INDUCED STRAIN IN FERROMAGNETIC SHAPE MEMORY ALLOYS

A new form of magnetic-field-induced strain has been observed in certain alloys that exhibit a martensitic transformation within the ferromagnetic phase. The field-induced strain appears to arise from a mechanism different from that responsible for magnetostriction (the rotation of the magnetization direction in a material having appreciable spin–orbit coupling). This new effect is associated with the motion of twin boundaries between regions in which the magnetization direction differs and is constrained by a large magnetocrystalline anisotropy. Because these same alloys exhibit what is called the *thermoelastic shape memory* effect, those exhibiting field-induced strain are often called *ferromagnetic shape memory alloys* (FSMAs). The shape-memory effect will be briefly described as an introduction to the magnetic-shape-memory effect.

Many materials exhibit a large crystallographic distortion upon first-order transformation from a high-symmetry, high-temperature phase (austenite) to a lower symmetry, low-temperature phase (martensite). When this distortion is diffusionless and involves an atomic shear displacement, it is referred to as a *martensitic transformation.* Examples of materials that exhibit martensitic transformations include FeC and NiTi. On cooling such a material below the martensitic transformation temperature, regions of martensite nucleate and grow in the austenitic matrix. The strain associated with the transformation causes a large elastic energy at the martensite–austenite phase boundaries. When the transformation strain is accommodated by twinning, as opposed to slip, the low-temperature phase consists of an assembly of twin variants that are arranged so as to minimize the interfacial elastic energy (Fig. 7A.1).

Gross plastic deformation (including macroscopic bending or twisting) of a mostly martensitic, twinned sample can then be accommodated by twin-boundary motion in the martensite (Fig. 7A.2*a*). On heating such a deformed sample back to the high-temperature phase, the twin variants revert to the

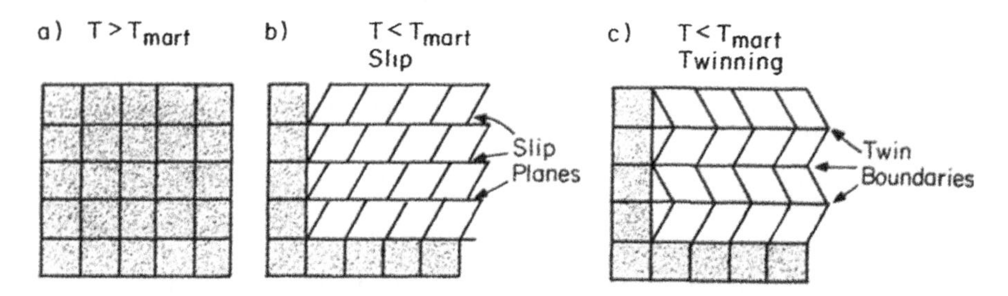

Figure 7A1 Schematic representation of an austenitic material (*a*) On cooling to $T < T_{\mathrm{mart}}$, the transformation strain associated with the appearance of the low-temperature phase can be accommodated by either slip (*b*) or twinning (*c*).

higher-symmetry phase and the deformation is erased (Wayman 1992). This is called the *thermoelastic shape memory effect.* Strains in excess of 10% and stresses measuring tens of GPa can be realized in some shape memory materials. The activation of this effect, however, is slow and inefficient because it depends on heating of the sample. In some material systems, the shape memory effect is reversible.

It is possible, in principle, to bring about this phase change by application of a magnetic field. However, the strength of the field required is prohibitively large for practical applications; the sample would have to be held within about a degree of the transformation temperature for a field of order 1 T to effect the transformation.

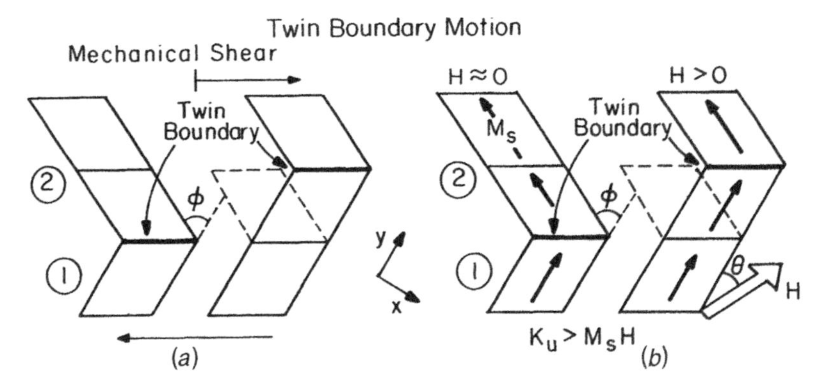

Figure 7A.2 A twinned, martensitic material can respond to a shear stress by twin boundary motion, leading to what appear to be large plastic deformations (*a*). If the martensitic phase is magnetic with a strong magnetocrystalline anisotropy that changes direction across the twin boundary, application of a magnetic field generates a Zeeman pressure on the twin boundary (*b*). This pressure tends to grow variant 1 at the expense of variant 2.

On the other hand, a magnetic field can be very effective in changing the twin structure, and hence the sample shape, if the uniaxial easy direction of magnetization changes across the twin boundary and the anisotropy energy is large.

Application of a magnetic field in a certain direction with respect to a martensitic twin boundary can create a Zeeman pressure of order $2M_sH$ on the twin boundary. If the magnetocrystalline anisotropy is sufficiently strong that the moments in unfavorably oriented twin variants cannot rotate into the field direction, the Zeeman energy may be reduced by twin boundary motion (Fig. 7A.2*b*). Unlike the thermoelastic shape memory effect, this magnetic shape memory effect occurs fully within the low-temperature (martensitic) phase. Thus it has the potential to be faster and more efficient than thermoelastic shape memory.

Large, reversible, magnetic-field-induced strains of order 0.2% were first reported in single crystals of Ni_2MnGa by Ullakko et al. (1996) (Fig. 7A.3). The measurements were done at −8°C (15°C below the martensitic transformation temperature). They also indicated that application of a field of 10 kOe caused at most a one or two degree shift in the transformation temperature. Thus, the field-induced strains do not arise from a shift in the

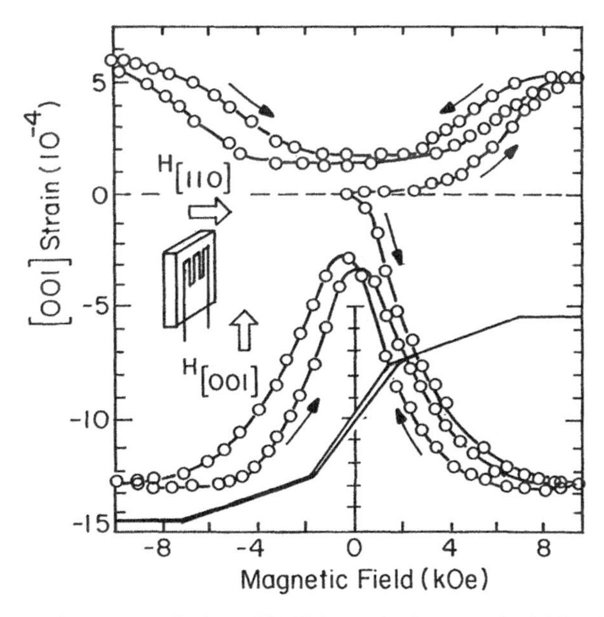

Figure 7A.3 Strain measured along [001] in a single crystal of Ni_2MnGa in fields directed parallel and perpendicular to the strain gauge. $T = -8$°C. Note that the sample contracts in a field parallel to the gauge. The same experiment done above the transformation temperature produced strains with similar field dependence but one-tenth the magnitude. [After Ullakko et al. (1996)].

martensitic transformation temperature, but occur fully within the martensitic phase. More recently, James and Wuttig (1998) have reported reversible field-induced strains of 0.5% in single crystals of $Fe_{70}Pd_{30}$ at $-17°C$. Further, their simultaneous observation of the twined microstructure showed the growth of one variant at the expense of another during field-induced deformation. It is thus concluded that twin-boundary motion is the main source of these very large, field-induced strains. More recently, a 5% shear strain was reported at room temperature in a field of 4 kOe in a NiMnGa crystal (Murray et al. 1999).

The magnetic actuation of twin boundary motion can be understood in simple terms by considering a single twin variant model as depicted in Figure 7A.2b. We define f_1 and $f_2 = 1 - f_1$ as the volume fractions of the two twin variants (assume that initially $f_1 = f_2 = \frac{1}{2}$). The Zeeman energy density of the two twin variants is given by $-M_s H[f_1 \cos\theta + f_2 \cos(\theta + \phi)]$. The motion of the twin boundary in an applied field comes at a cost in elastic energy stored in the unresponsive part of the material. The magnetoelastic free energy density after *partial* twin boundary motion may be written as (O'Handley 1998):

$$g_{\text{mag-elast}} = -M_s H[f_1 \cos\theta + f_2 \cos(\theta + \phi)] + \tfrac{1}{2} C e_0^2 [f_1^2 + f_2^2 + 2 f_1 f_2 \cos\varphi] \tag{7A.1}$$

Here the strain components have been expressed in terms of the volume frictions:

$$e_x = f_2 e_0 \sin\varphi \quad \text{and} \quad e_y = e_0(f_1 + f_2 \cos\varphi) \tag{7A.2}$$

where e_0 is the strain associated with the transformation; e_0 would be captured by an applied field if complete twin boundary motion ($f_1 = 1$, $f_2 = 0$) were achieved. The elastic stiffness constant C is an *effective* stiffness against which the twin boundary motion is occurring. It includes the resistance to twin motion from defects, work against retained austenite (Ullakko et al., 1997). Equation (7A.2) can be minimized to determine the equilibrium fractional twin boundary displacement, $\delta f = f_1 - \frac{1}{2}$, as a function of the field:

$$\delta f = \frac{p_-}{q_-} \frac{M_s H}{2Ce_0^2} \equiv \frac{p_-}{q_-} \frac{h_e}{2} \tag{7A.3}$$

where $p_\pm = \cos\theta \pm \cos(\theta + \phi)$ and $q_\pm = 1 \pm \cos\phi$ describe the twin geometry and field orientation in Figure 7A.2. The reduced field, $h_e = H/H_e$, where $H_e = Ce_0^2/M_s$, scales the response of the system by analogy with magnetization against a uniaxial anisotropy. The projection of the magnetization along the applied field direction and the diagonal components of the strain tensor can be shown to have the following form:

$$m = \frac{M}{M_s} = \frac{1}{2}\left[p_+ + \frac{p_-^2}{q_-} h_e\right], \qquad \varepsilon_x = \frac{e_x}{e_0} = \frac{\sin\varphi}{2q_-}[q_- - p_- h_e],$$

$$\varepsilon_y = \frac{e_y}{e_0} = \frac{1}{2}[q_+ + p_- h_e] \tag{7A.4}$$

These equations show that there can be a remanent magnetization, $M_s p_+/2$, or a remanent strain in the direction of a prior field depending on twin geometry. Application of a field (in this strong anisotropy regime) causes the net magnetization to increase linearly in the ratio of the magnetic pressure on the twin boundary to the elastic energy associated with complete twin boundary motion. The model has been extended to the intermediate anisotropy case in which some magnetization rotation may occur along with the twin boundary motion. This leads to nonlinearities in the m–h and e–h solutions. The nonlinear equations provide a good fit to the data of Figure 7A.3. Further, the model indicates that the most efficient direction in which to apply the field is parallel to the twin boundary.

Figure 7A.4 shows that the predicted field-induced strain e_y [from Eq. (7A.4) and $h_e = M_s H/Ce_0^2$] varies inversely with the transformation strain e_0. This can be understood from the fact that the Zeeman energy is working against an effective elastic energy during field-induced twin boundary motion. Thus, there can be greater fractional twin boundary motion when the transformation strain is smaller.

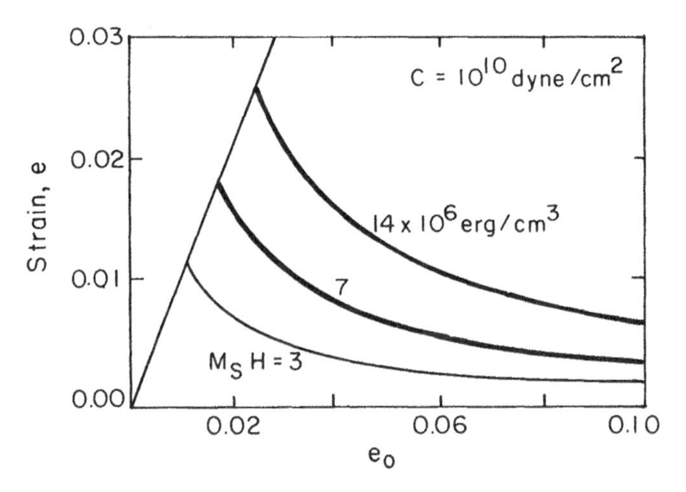

Figure 7A.4 Variation of the strain measured in the y direction as a function of the transformation strain for various values of Zeeman energy assuming $C = 10^{10}\,\mathrm{d/cm^2}$. The line extending from the origin indicates $e = e_0$.

This new class of magnetostrictive materials may become important as understanding of the phenomenon advances and improved material formulations are developed.

APPENDIX 7B REVIEW OF ELASTICITY

The elastic problem in a cubic crystal is briefly reviewed [Kittel's *Introduction to Solid State Physics* (1976, Chapter 4) is a good introduction]. The elastic energy density may be written as

$$f_{\text{el}} = \tfrac{1}{2}\sum\sum c_{ij}e_i e_j$$

where the summations each go over the six indices $1 = xx$, $2 = yy$, $3 = zz$, $4 = yz$, $5 = zx$, $6 = xy$. For cubic symmetry this reduces to

$$\begin{aligned} f_{\text{el}} &= \tfrac{1}{2}c_{11}(e_{xx}^2 + e_{yy}^2 + e_{zz}^2) + c_{12}(e_{yy}e_{zz} + e_{xx}e_{zz} + e_{xx}e_{yy}) \\ &\quad + \tfrac{1}{2}c_{44}(e_{xy}^2 + e_{yz}^2 + e_{zx}^2) \end{aligned} \tag{7B.1}$$

where the strains e_{ij} are related to the x, y, and z displacements u_i by

$$e_{ij} = \frac{1}{2}\left(\frac{\partial u_i}{\partial x_j} + \frac{\partial u_j}{\partial x_i}\right)$$

The factor of $\frac{1}{2}$ appears only in engineering texts. The c_{ij} terms are elastic stiffness constants (with units N/m^2). The inverse problem involves the components of the inverse of c, namely, s_{ij}, the elastic compliance tensor. The term c_{11} is the stiffness resisting a strain in one direction by a stress in the same direction; c_{12} is the stiffness resisting a strain in the direction that is orthogonal to the stress direction; c_{44} is the shear stiffness resisting, for example, an x-directed strain which varies in the z direction, due to an x-directed stress couple (Fig. 7B.1).

The bulk modulus B, Young's modulus E, shear modulus G, and Poisson's ratio υ, are related to the c_{ij} terms by

$$c_{11} = \frac{E(1-\upsilon)}{(1+\upsilon)(1-2\upsilon)} \qquad B = \frac{(c_{11} + 2c_{12}}{3}$$

$$c_{12} = \frac{E\upsilon}{(1+\upsilon)(1-2\upsilon)} \qquad \text{and} \qquad E = \frac{(c_{11} - c_{12})(c_{11} + 2c_{12})}{(c_{11} + c_{12})}$$

$$c_{44} = G \qquad G = c_{44}$$

$$\upsilon = \frac{c_{12}}{(c_{11} + c_{12})} \approx \frac{1}{3}$$

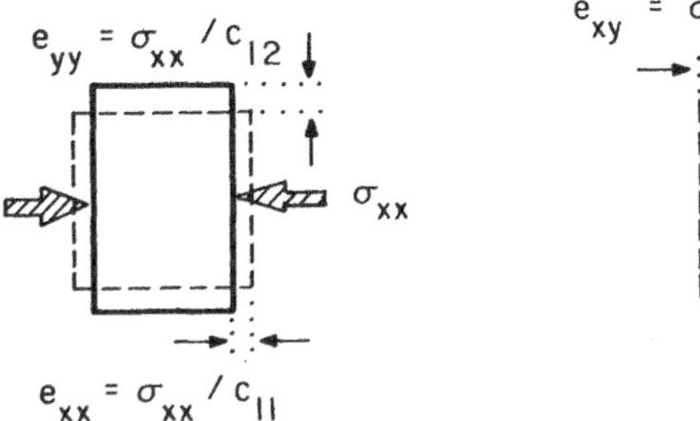

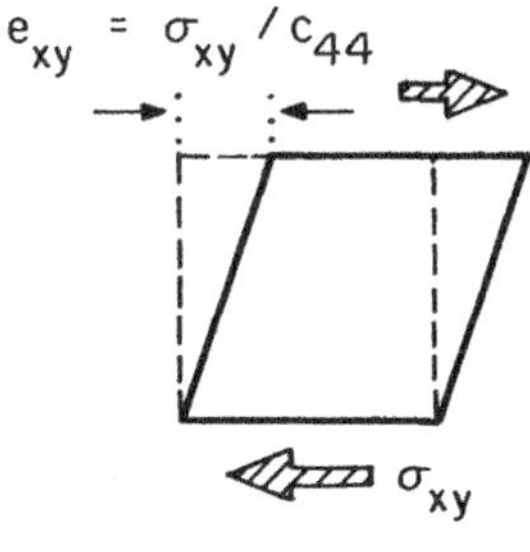

Figure 7B.1 Definition of stiffness constants in a cubic system: c_{11} resists strain in the same direction as the stress, c_{12} resists strain orthogonal to the stress, and c_{44} is a pure shear stiffness resisting strain in responses to noncollinear opposing stresses.

The compressibility is $k = 1/B$. In isotropic materials, there are only two independent elastic constants because $c_{11} = c_{12} + 2c_{44}$ giving, among other simplifications,

$$E = \frac{2c_{44}(c_{11} + 2c_{12})}{(c_{11} + c_{12})}, \qquad G = \frac{E}{2(1 + \upsilon)}$$

The elastic constants of iron and nickel are given in Table 7B.1.

It is useful to express the stress strain equations in cubic symmetry and for an elastically isotropic solid. The three coupled stress–strain equations in each case are

$$\begin{array}{cc} \underline{\text{Cubic}} & \underline{\text{Isotropic}} \\ \sigma_i = c_{11}e_{ii} + c_{12}(e_{jj} + e_{kk}) & \sigma_i = Ee_{ii} + \upsilon(\sigma_j + \sigma_k) \end{array} \tag{7B.2}$$

In a given situation, values for imposed stresses or strains are given and the Eqs. (7B.2) are solved for the remaining variables. For example, in the case of

TABLE 7B.1 Elastic Constants at Room Temperature for Fe and Ni in Units of 10^{11} N/m^2

Elastic Constant	Fe	Ni
c_{11}	2.41	2.50
c_{12}	1.46	1.60
c_{44}	1.12	1.185

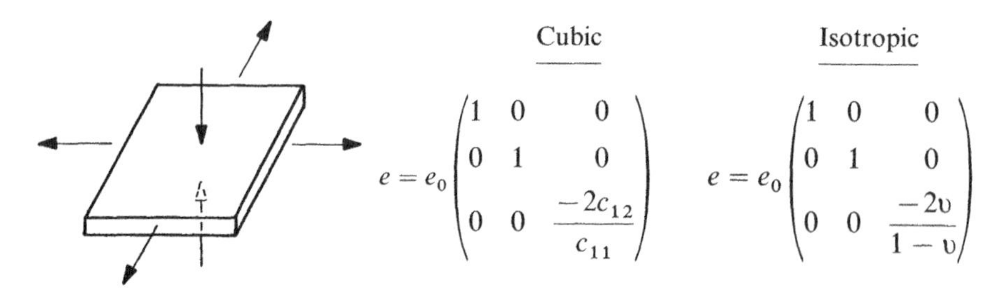

Figure 7B.2 Depiction of biaxial strain with elastically free boundary conditions in the third direction ($\sigma_z = 0$) and appropriate tensors in elastically cubic and isotropic materials.

a film subjected to a biaxial misfit strain of $\eta = e_0$, the strain and stress components are $e_{xx} = e_{yy} = e_0$, $e_{zz} \neq 0$, $\sigma_x = \sigma_y$, and $\sigma_z = 0$. Solving the three equations gives the strain tensor shown in Figure 7B.2.

Given the value of $\upsilon \approx 0.3$ for many materials, the approximation $2\upsilon/(1 - \upsilon) \approx 1$ can sometimes be used. Similarly, for an imposed uniaxial strain, $e_{xx} = e_0$, $e_{yy} = e_{zz}$, $\sigma_x \neq 0$, $\sigma_y = \sigma_z = 0$, the strain tensor is as depicted in Figure 7B.3

Note that here the lateral boundary conditions have been assumed to be free of *stress* so the material experiences a lateral Poisson contraction. When straining a thin film on a substrate, for example, by four-point bending of the substrate, the lateral in-plane *strain* is essentially zero. In this case, $\sigma_x \neq 0$, $e_{xx} = e_0$, $e_{yy} = 0$, $e_{zz} \neq 0$, $\sigma_y \neq 0$, $\sigma_z = 0$. The result is $e_{xx} = (1 - \upsilon^2)\sigma_x/E$, and $e_{zz} = \upsilon(1 + \upsilon)\sigma_x/E$, as shown in Figure 7B.4.

Finally, the implications of Eqs. (7.10) and (7.11) are illustrated. For an arbitrary magnetization direction, the strain in the x direction is *not* given by e_{xx} alone just as the strain in a solid such as represented in Figure 7B.1 is not

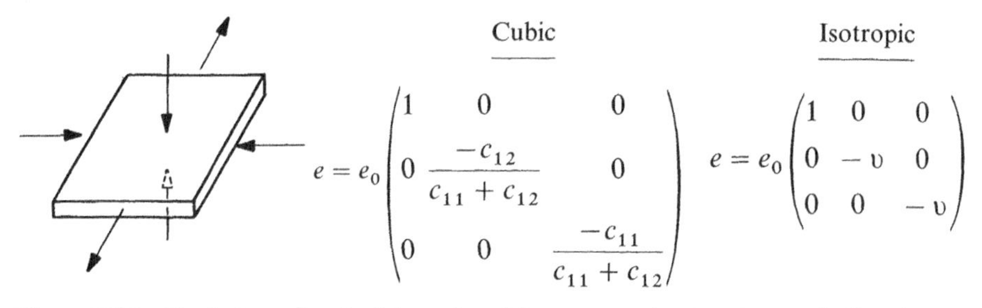

Figure 7B.3 Depiction of uniaxial strain with unconstrained orthogonal directions ($\sigma_y = \sigma_z = 0$) and appropriate strain tensors.

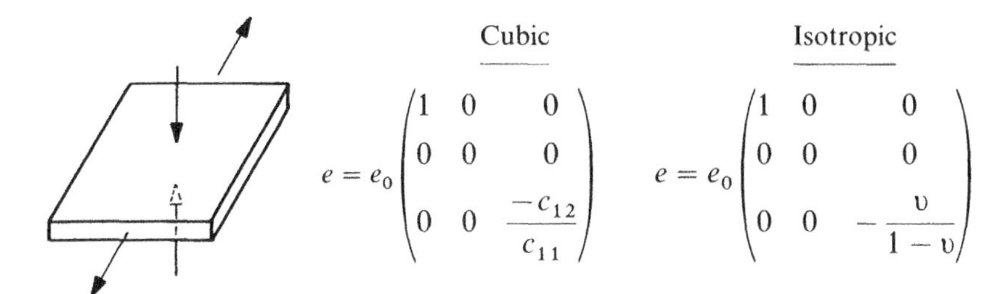

Figure 7B.4 Depiction of uniaxial strain with one constrained orthogonal direction and appropriate tensors. This case applies to the strain in a film on a substrate bent around one axis (the y axis, here).

given by e_{xx} alone. It also depends on e_{xy} and e_{xz}. Mathematically this is expressed as the vector relation

$$\boldsymbol{x} = \boldsymbol{x}_0 + \mathbf{e}\boldsymbol{x}_0 = (\mathbf{1} + \mathbf{e})\boldsymbol{x}_0$$

where $\mathbf{1}$ is the unit matrix and $\mathbf{e}$ is the deformation tensor [Eq. (7.10)]. Thus

$$\begin{pmatrix} x \\ y \\ z \end{pmatrix} = \left\{ \begin{pmatrix} 1 & 0 & 0 \\ 0 & 1 & 0 \\ 0 & 0 & 1 \end{pmatrix} + \begin{pmatrix} e_{11} & e_{12} & e_{13} \\ e_{21} & e_{22} & e_{21} \\ e_{31} & e_{32} & e_{33} \end{pmatrix} \right\} \begin{pmatrix} x_0 \\ y_0 \\ z_0 \end{pmatrix}$$
$$= \begin{pmatrix} (1 + e_{11})x_0 + e_{12}y_0 + e_{13}z_0 \\ e_{21}x_0 + (1 + e_{22})y_0 + e_{23}z_0 \\ e_{31}x_0 + e_{32}y_0 + (1 + e_{33})z_0 \end{pmatrix}$$

Thus the new x position is $x = (1 + e_{11})x_0 + e_{12}y_0 + e_{13}z_0$. The strain in an arbitrary direction from the origin to point P is illustrated in Figure 7B.5. We introduce the direction cosines of the point P by the relations $\beta_1 = x_0/r_0$, $\beta_2 = y_0/r_0$, $\beta_3 = z_0/r_0$. It can be determined that after a strain e_{ij} the x component of P' along OP is $x = x_0 + ex_0 = r_0(\beta_1 + e_{1j}\beta_j)$ with similar expressions for y and z. Hence $r^2 = r_0^2\ (1 + 2e_{ij}\beta_i\beta_j)$ neglecting products of the e_{ij}.

Thus, for small strain, $r - r_0 = r_0 e_{ij}\beta_i\beta_j$ so that the strain in any direction specified by $(\beta_1, \beta_2, \beta_3)$ is related to the components of the strain tensor by the expression

$$\frac{r - r_0}{r_0} = \frac{\partial l}{l} = \sum e_{ii}\beta_i^2 + \sum_{i<j} e_{ij}\beta_i\beta_j \tag{7B.3}$$

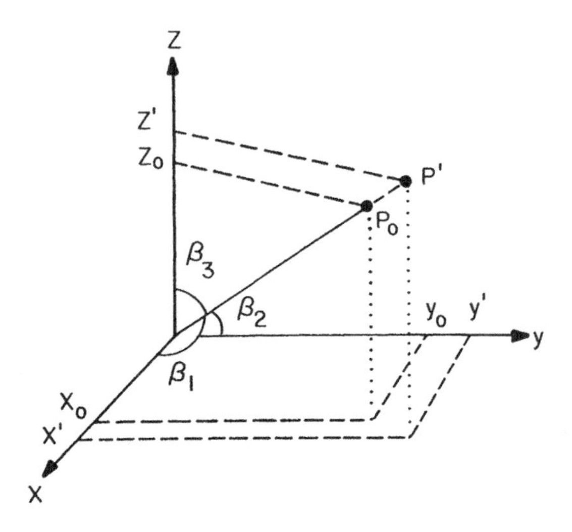

Figure 7B.5 Coordinate system for transforming strain components in principal coordinates to an arbitrary direction OP.

PROBLEMS

7.1 Show that the magnetostriction constants of Fe and Ni are not of the sign you would expect for dipole forces between atomic dipole magnets.

7.2 Consider a bar of iron with [110] along its length and the top surface is (001). It is under tensile stress along its length: $\lambda_{100} = 20.5 \times 10^{-6}$, $\lambda_{111} = -21.5 \times 10^{-6}$, $c_{11} = 2.4 \times 10^{11}$, $c_{12} = 1.4 \times 10^{11} c_{44} = 1.2 \times 10^{11}\,\mathrm{J/m^3}$.

(a) Write the strain tensor in the bar coordinates and in the crystallographic coordinates.

(b) Which terms in Eq. (7.5) are relevant to the effects of stress on the orientation of $\boldsymbol{M}$?

(c) If the stress is such as to produce a strain e of 0.1%, compare the magnitude of the magnetoelastic and magnetocrystalline anisotropy.

(d) Evaluate the appropriate energy terms to determine the direction of $\boldsymbol{M}$ at $e = 0.01$.

(e) Derive the equation of motion of the magnetization as a function of field and strain.

Plot m–H for $e = 0$ and $e = 1\%$. Describe your results qualitatively.

7.3 Derive Eqs. (7.10), the equilibrium strains, from Eqs. (7.9).

7.4 Show that setting $\lambda_{111} = \lambda_{100} = \lambda_s$ in Eq. (7.12) gives Eq. (7.1).

7.5 Consider a thin epitaxial Ni film grown coherently on Cu(100) ($a_{\text{Ni}} = 3.524$ Å, $a_{\text{Cu}} = 3.615$ Å). Assume for Ni $\lambda_{100} = -46 \times 10^{-6}$, $c_{11} = 2.5 \times 10^{11}$, $c_{12} = 1.6 \times 10^{11}$, and $c_{44} = 1.18 \times 10^{11}$ J/m^3.

(a) Compare the relative strength of crystal anisotropy and magnetoelastic anisotropy as far as determining the direction of $\boldsymbol{M}$.

(b) Assume $e_{12} \approx 0$, and find the direction of $\boldsymbol{M}$.

(c) As the film grows, describe what happens to e_{ii}.

(d) Given $e_{12} = 2e_{11}$, find the direction of $\boldsymbol{M}$.

(e) Discuss.

7.6 Show that the two alternate ways of defining the magnetoelastic coefficients are equivalent. How are the B and b terms related?

7.7 Critique the following statement by considering the energy densities in each case. "Typically, magnetic transition metals produce magnetostrictive strains e [at saturation] of order 10^{-5} to 10^{-4} which vary about the direction of magnetization roughly as $e = \frac{3}{2}\lambda_s(\cos^2\theta - \frac{1}{3})$ where λ_s is the saturation magnetostriction constant. Conversely, stresses σ imposed on the material that result in strains of order 10^{-5} to 10^{-4} (i.e., $e = \lambda_s$) contribute significantly to the total magnetic anisotropy: the uniaxial stress-induced anisotropy energy density is $\frac{3}{2}\lambda_s\sigma$."

7.8 Use Eq. (7.10) to describe the field and strain directions you would use to measure the two magnetostriction constants λ_{100} and λ_{111} on a Ni sample cut as shown in Figure 6.3. Do not assume that the sample has randomly distributed magnetization directions in zero field.

7.9 Assume a multidomain polycrystalline sample with the domain magnetizations randomly oriented as represented in the following diagrams.

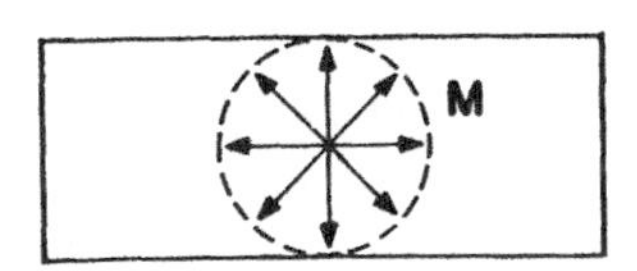

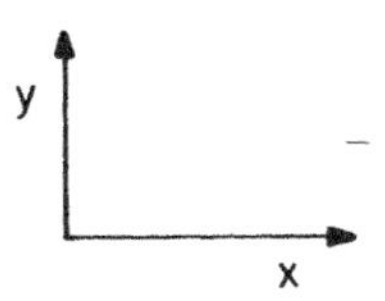

Describe how you would expect the magnetization distribution to change if the sample has positive magnetostriction λ_s and a tensile stress σ_{xx} is applied for $H = 0$. Contrast this with the change caused by application of $H = (H_0, 0, 0)$ for $\sigma_{ij} = 0$. Plot M_x against H_x for $\sigma_{xx} = 0$ and $\sigma_{xx} \neq 0$. Plot M_x against σ_{xx} for applied fields in the range $H_x \leqslant H_K$.

7.10 Consider the anisotropy constants K_1 and the polycrystalline magnetostriction constants λ_s shown in the diagram below for FCC NiFe alloys. This problem concerns the behavior of alloys having compositions A, B, and C.

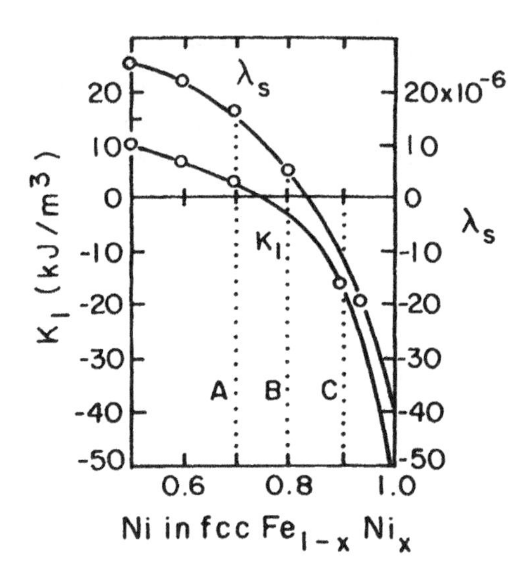

(a) Sketch and label the lowest-order crystal anisotropy surfaces in the x–y plane for thin films of compositions A and C (i.e., assume $\alpha_z = 0$).

(b) For unstrained bulk samples of compositions A, B, and C, in which crystallographic directions are the samples easily magnetized?

(c) For unstrained single-crystal samples A, B, and C, a field is applied in the [100] direction. Sketch the shape of the M–H curve in each case assuming zero coercivity. Use approximately the same field scale in each case so that the M–H curves can be compared.

(d) If the samples are subject to strains, which composition, A, B, or C, would make the best soft magnetic material? Why?

(e) For single-crystal samples A, B, and C apply a field in the [100] direction as in part **c** above but now also put the sample under tensile stress parallel to the field direction. Sketch with a dashed line over your result in **c**, the shape of the strained M–H curve in each case assuming zero coercivity.

(f) What is the approximate value of the magnetoelastic coupling coefficient B_1 for sample B? If a strain is applied along [100] to cancel the crystal anisotropy in this sample and have the magnetization lie along the x axis in zero field, what would be the nature of that strain (tensile or compressive along [100]), and what is its approximate magnitude?

BIBLIOGRAPHY

Brown, W. F., Jr., *Magnetoelastic interactions*, Springer-Verlag, Berlin, 1966.

Callen, E. R., and H. B. Callen, *J. Phys. Chem. Solids* **16**, 310 (1960).

Callen, E., *J. Appl. Phys.* **39**, 519 (1968).

Kanamori, J., in *Magnetism*, Vol. 1, Chapter 4, G. T. Rado and H. Suhl, eds., Academic Press, New York, 1963.

Tam, A. C., and H. Schroeder, *IEEE Trans.* **MAG-25**, 2629 (1989).

REFERENCES

Bellesis, G. H., P. S. Harlee III, A. Renema, and D. L. Lambeth, *IEEE Trans.* **MAG 29** (1993).

Berry, B., and W. Pritchet, *Phys. Rev. Lett.* **34**, 1022 (1975).

Bozorth, R. M., *Ferromagnetic Materials*, IEEE Press, New York, 1993.

Bozorth, R. M., and T. Wakiyama, *J. Phys. Soc. Jpn.* **17**, 1669 (1962).

Callen, E. R., and H. B. Callen, *Phys. Rev.* **129**, 578 (1963).

Callen, E. R., A. E. Clark, B. DeSavage, W. Coleman, and H. B. Callen, *Phys. Rev.* **130**, 1735 (1963).

Callen, E. R., and H. B. Callen, *Phys. Rev.* **139A**, 455 (1965).

Callen, H. B., and N. Goldberg, *J. Appl. Phys.* **36**, 976 (1965).

Ciria, M., J. I. Arnaudas, A. del Moral, G. J. Tomka, C. de la Fuente, and P. A. J. de Groot, *Phys. Rev. Lett.* **75**, 1634 (1995).

Coleman, W. E., and A. S. Pavlovic, *Phys. Rev.* **135**, A426 (1964).

Clark, A. E., B. DeSavage, and R. M. Bozorth, *Phys. Rev.* **138**, A216 (1965).

Clark, A. E., in *Ferromagnetic Materials*, Vol. 1, E. P. Wohlfarth, ed., North-Holland, Amsterdam, 1980.

Cullen, J. R., A. E. Clark, and K. B. Hathaway, Chapter 16 in *Handbook of Materials Science*, K. H. J. Buschow, ed., VCH, 1997).

del Moral, A., and M. S. S. Brooks, *J. Phys. C, Solid State Phys.* **7**, 2540 (1974).

del Moral, A., in *Magnetoelastic Effects and Applications*, L. Lanotte, ed., Elsevier Science, Amsterdam, 1993, p. 1.

Lacheisserie, E. du Tremolet de, *Phys. stat. sol.* **b54**, K135 (1972).

Lacheisserie, E. du Tremolet de, *J. Magn. Mag. Mater.* **31–34**, 837 (1983).

Lacheisserie. E. du Tremolet de, and R. Mendia Monterroso, *J. Magn. Magn. Mater.* **31–34**, 837 (1983).

Lacheisserie, E. du Tremolet de, *Magnetostriction: Theory and Applications of Magnetoelasticity*, CRC Press, Boca Raton, FL, 1994.

Franse, J. J. M., and M. Stolp, *Phys. Lett.* **32A**, 316 (1970).

Hall, R. C., *J. Appl. Phys.* **31S**, 5157 (1960).

Hernando, A., M. Vasquez, V. Madurga, and H. Kronmuller, *J. Magn. Magn. Mater.* **37**, 161 (1983).

James, R. D., and D. Kinderlehrer, *Phil. Mag.* **68**, 237 (1993); *J. Appl. Phys.* **76**, 7012 (1994).

James, R. D., and M. Wuttig, *Phil. Mag.* **77**, 1273 (1998).

Kanamori, J., in *Magnetism*, Vol. 1, Chapter 4, G. T. Rado and H. Suhl, eds., Academic Press, New York, 1963.

Kittel, C., *Introduction to Solid State Physics*, 3rd ed., Wiley, New York, 1967

Klockholm, E., *IEEE Trans.* **MAG 12**, 819 (1996).

Koch, R., N. Weber, K. Thürmer, and K. H. Rieder, *J. Magn. Magn. Mater.* **159**, L11 (1996).

Kollie, T. G., *Phys. Rev. B* **16**, 4872 (1977).

Kulakowski, K., and A. del Moral, *Phys. Rev.* **B50**, 234 (1994); *Phys. Rev.* **B52**, 15943 (1995).

Lampert, R. E., J. M. Gorres, and M. M. Hanson, *IEEE Trans.* **MAG 4**, 525 (1968).

Lee, E. W., *Rept. Progress Phys.* **18**, 5184 (1955).

Lee, E. W., and M. A. Asgar, *Proc. Roy. Soc.* **A326**, 73 (1971).

Livingston, J. D. *Phys. stat. solidi* (*a*) **70**, 591 (1982).

Miyazaki, T., T. Oomori, F. Sato, and S. Ishio, *J. Magn. Magn. Mater.* **129**, L135 (1994).

Morin, P., and D. Schmitt, in *Ferromagnetic Materials*, Vol. 5, K. H. J. Buschow, ed., North-Holland, Amsterdam, 1990, p. 1.

Murray, S. J., M. Marioni, S. M. Allen, and R. C. O'Handley, in press.

Narita, K., J. Yamasaki, and H. Fukunaga, *IEEE Trans.* **MAG-16**, 435 (1980).

Néel, L., *J. Physique* **5**, 376 (1954).

O'Handley, R. C., *Phys. Rev.* **B18**, 930 (1978).

O'Handley, R. C., OhSung Song, and C. A. Ballentine, *J. Appl. Phys.* **74**, 6302 (1993).

O'Handley, R. C., *J. Appl. Phys.* **83**, 3263 (1998).

Rhyne, J. J., and S. Legvold, *Phys. Rev.* **138A**, 507 (1965).

Sirota, N. N., A. V. Mazovko, and S. V. Sukvalo, *Dokl. Akad. Nauk Belarus.* **00**, 115 (1969).

Sirota, N. N., and A. V. Mazovko, *Dokl. Akad. Nauk Belarus.* **16**, 596 (1972).

Song, O. S., C. A. Ballentine, and R. C. O'Handley, *Appl. Phys. Lett.* **64**, 2593 (1994).

Sun, S. W., and R. C. O'Handley, *Phys. Rev. Lett.* **66**, 2798 (1991).

Tam, A. C., and H. Schroeder, *IEEE Trans.* **MAG-25**, 2629 (1989).

Tolman, C. H., *J. Appl. Phys.* **38**, 3409 (1967).

Ullakko, K., J. K. Huang, C. Kantner, V. V. Kokorin, and R. C. O'Handley, *Appl. Phys. Lett.* **69**, 1966 (1996).

Ullakko, K., J. K. Huang, and R. C. O'Handley, *Scripta Mater.* **36**, 1133 (1997).

Wayman, C. M., *MRS Bulletin*, April 1993, p. 49; *MRS Conf. Proc.*, Vol. 246, C. T. Liu, H. Kunsmann, K. Otsuka, and M. Wuttig, eds., MRS, Pittsburgh, PA, 1992.

Weber, M., R. Koch, and K. H. Rieder, *Phys. Rev. Lett.* **73**, 1166 (1994).

Yamasaki, J., Y. Ohkubo, and F. B. Humphrey, *J. Appl. Phys.* **67**, 5472 (1990).

Zener, C. *Phys. Rev.* **96**, 1335 (1954).

Zuberek, R. Z., H. Szymczak, R. Krishnan, and M. Tessier, *J. Physique* C**49**, 1761 (1998).

CHAPTER 8

MAGNETIC DOMAIN WALLS AND DOMAINS

8.1 INTRODUCTION

It is a common experience that pieces of ferromagnetic material do not always exhibit a north pole and a south pole. They often appear to be demagnetized or even nonmagnetic in the presence of another soft magnetic body. Only when placed in an external field or near a permanent magnet do soft magnetic materials begin to respond and reveal their magnetism. The demagnetization of soft magnetic materials was attributed by Weiss to the formation of magnetic domains that are regions inside the material that are magnetized in different directions so that the net magnetization is nearly zero (see Fig. 1.13). Before questions about the arrangement and size of magnetic domains can be addressed, it is important to understand the spin structure and energy density of the surfaces, called *domain walls*, that separate one domain from another. Then, the energy of different domain wall arrangements in a material will be balanced against the magnetostatic energy cost of having a single domain (no walls). In Chapter 9, the motion of domain walls will be considered.

8.1.1 Relevant Energy Densities

The previous chapters have described five types of magnetic energy density. Each of these is summarized here with succinct mathematical expressions.

Exchange energy tends to keep adjacent magnetic moments parallel to each other. Here, it is expressed in a discrete, microscopic form as well as in a continuous, macroscopic form:

$$f_{ex} = -\frac{2\mathcal{J}S^2}{a^3}\cos\theta_{ij} = A\left(\frac{\partial\theta}{\partial x}\right)^2 \xrightarrow{3D} A\sum_{i=1}^{3}\left(\frac{\nabla M_i}{M_S}\right)^2$$

Exchange energy expresses the energy cost of a change in the direction of magnetization.

Magnetostatic energy arises mainly from having a discontinuity in the normal component of magnetization across an interface. It is a form of anisotropy due to sample shape and is often uniaxial in symmetry:

$$f_{ms} = -\mu_0 M_S \cdot H_i = \frac{\mu_0}{2} M_S^2 \cos^2\theta$$

Magnetocrystalline anisotropy describes the preference for the magnetization to be oriented along certain crystallographic directions. The forms for uniaxial and cubic materials are

$$f_a = K_2 \sin^2\theta + K_4 \sin^4\theta + \cdots \text{(uniaxial)}$$

$$f_a = K_1(\alpha_1^2\alpha_2^2 + \alpha_2^2\alpha_3^2 + \alpha_3^2\alpha_1^2) + K_2\alpha_1^2\alpha_2^2\alpha_3^2 + \cdots \text{(cubic)}$$

Magnetoelastic energy is that part of the magnetocrystalline anisotropy that is proportional to strain:

$$f_{me}^c = B_1[e_{11}(\alpha_1^2 - \tfrac{1}{3}) + e_{22}(\alpha_2^2 - \tfrac{1}{3}) + e_{33}(\alpha_3^2 - \tfrac{1}{3})]$$
$$+ B_2(e_{12}\alpha_1\alpha_2 + e_{23}\alpha_2\alpha_3 + e_{31}\alpha_3\alpha_1) + \cdots$$

for a cubic material. For an isotropic material, this reduces to

$$f_{me}^{iso} \approx B_1 e_{33} \sin^2\theta = \lambda_S^2 E \cos^2\theta = \tfrac{3}{2}\lambda_S \sigma \cos^2\theta$$

Finally, the *Zeeman* energy, $F = -\mu_m \cdot B$, is the potential energy of a magnetic moment in a field or, for a large number of moments, the potential energy per unit volume is:

$$f_{Zeeman} = -\mu_0 M \cdot H$$

Until now, when magnetization behavior has been modeled in this text, the fact that M–H loops generally exhibit coercivity has been ignored. Also, energy minimization has been used to derive simple expressions for $M(H)$ assuming that the magnetization is *uniform* throughout the material, specifically $\boldsymbol{M} = M(\theta, \phi)$. In order to describe the phenomena of coercivity and irreversibility, it is necessary to consider the presence of domain walls or allow for spatial variation of magnetization, namely, $\boldsymbol{M} = M[\theta(z), \phi(z)]$. So the energy densities described above and, in particular $f_{ex}(\theta(z))$, are now used to understand magnetic domain walls and their motion. (Domain walls have their own energy density, which also must be considered.)

It is important to know how the magnetization direction varies with position from one magnetic domain to another. The answer to this question is beautifully shown in Figure 8.1 (Oepen and Kirschner 1989).

Panel (*a*) in Figure 8.1 is a domain image from the (100) surface of an iron crystal near the intersection of several domains. Outlined in white is a region along which the vertical and horizontal components of magnetization were measured as a function of position from upper left to lower right using a technique called *scanning electron microscopy with spin polarization analysis* (SEMPA). (In SEMPA the secondary electrons collected from the small area on the surface of a sample illuminated by a fine, high-intensity scanning electron beam are analyzed to determine the direction of magnetization at the surface from which they were emitted, Chapter 16.) Panel (*b*) shows the results of the spin polarization scan. The polarization in the y direction makes a transition from a negative value to a positive one on crossing the domain wall; the x polarization has a measurable value only near the center of the transition. The transition region from one domain to the next spans a range of order 100 nm. Let us look at this transition, or domain wall, in more detail.

8.1.2 180° Domain Wall: Simple Model

Magnetic domains form and domain walls are created in order to reduce the magnetostatic energy of a finite, uniformly magnetized sample. For the case of uniaxial anisotropy, $K_u \sin^2\theta(z)$, the magnetization vectors in adjacent domains are antiparallel to each other, that is, a 180° domain wall exists (see Fig. 8.2). Note that here, the plane of the wall has been chosen to be parallel to the easy axis. This choice satisfies the boundary condition (Chapter 2) requiring continuity of the normal component of $\boldsymbol{M}$ across the surface if there are to be no $\boldsymbol{H}$ fields (see Section 8.4). The choice of having the wall plane stretch between the two parallel sample surfaces that are closest to each other minimizes its area and hence its surface energy. Such a domain wall is called a Bloch wall in recognition of the seminal work of Felix Bloch in describing its structure.

If the magnetization orientation were to change abruptly from 0° at the last atomic site in one domain, to π at the first atomic site in the adjacent domain (Fig. 8.2, left), there is no cost in anisotropy energy by the creation of the domain wall. However, there is a significant cost in exchange energy from site i to site j across the domain wall. For one pair of spins straddling the wall, the exchange energy is

$$|F_{ij}| = |-2S^2 J \cos(\theta_{ij})| \approx 2S^2 J \tag{8.1}$$

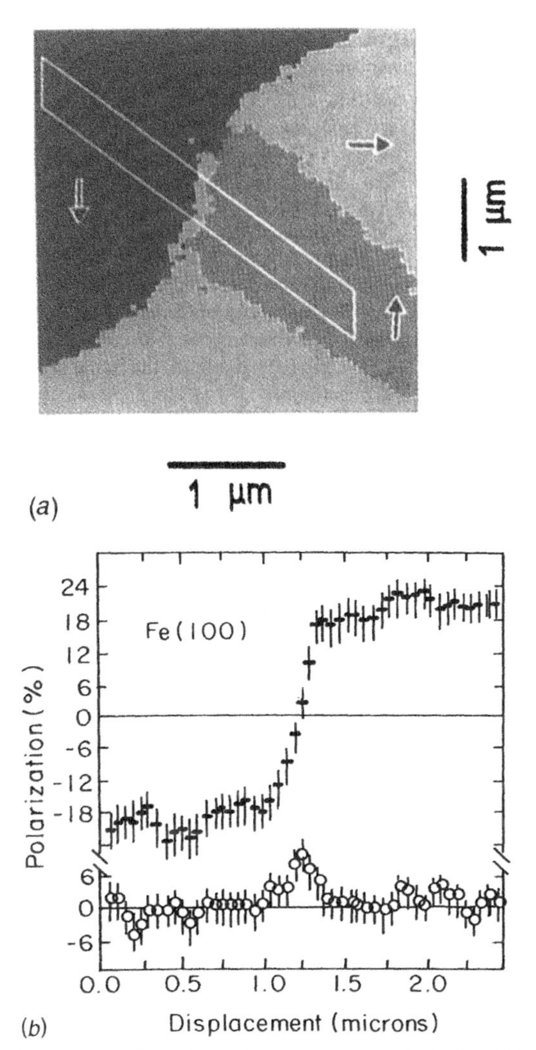

Figure 8.1 (*a*) Magnetic surface domain structure on Fe(100). The field of view is $3.5 \times 3.5\,\mu$m. The arrows indicate the measured polarization orientation in the domains. The frame shows the area over which the polarization distribution of (*b*) is averaged. (*b*) Polarization distributions across a 180° domain wall, taken from (*a*). The vertical polarization component is indicated by the crosses. The circles show the horizontal polarization distribution. [After Oepen and Kirschner (1989).]

From the chapter on exchange interactions, the value of the exchange integral may be approximated as $J \approx 0.3k_BT_C \approx 4 \times 10^{-21}$ J for a Curie temperature of 1000 K. Thus $F_{ij} \approx 10^{-20}$ J and the hypothetical domain wall in a cubic material with lattice constant a has a surface energy density σ_{ex} of approximately $F_{ij}/a^2 \approx 0.25\ \text{J/m}^2$. This surface energy is about three orders of magni-

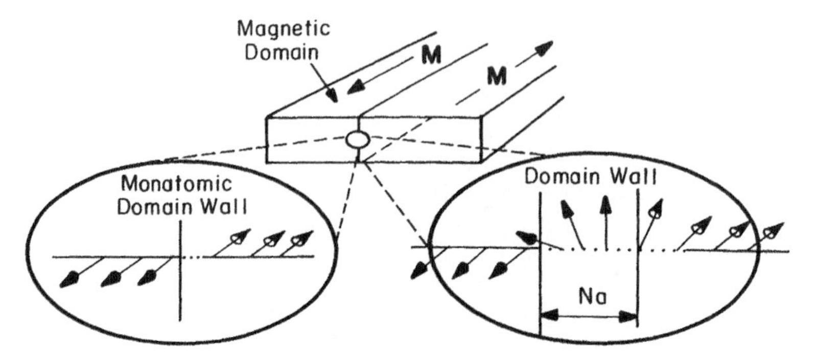

Figure 8.2 Schematic of ferromagnetic material containing a 180° domain wall (center). Left, hypothetical wall structure if spins reverse direction over one atomic distance. Right, wall structure if spins reverse direction over N atomic distances, a. In real materials, N is found to range from about 40 to nearly 10^4.

tude greater than chemical surface energies. It is also much larger than the anisotropy energy cost of having some of the spins inside the wall point in a hard direction. Thus, the material will find another, less costly way to make the transition in magnetization from one domain to another.

Clearly the exchange energy could be reduced by distributing the 180° rotation over several lattice spacings (Fig. 8.2, right). If the domain wall thickness were to span N interatomic spacings, then adjacent spins would differ by an angle approximately $\theta_{ij} \approx \pi/N$. If N is sufficiently large, $\cos(\theta_{ij})$ in Eq. (8.1) can be expanded. The lowest-order result is

$$F_{ij} \approx JS^2\theta_{ij}^2 \approx JS^2\left(\frac{\pi}{N}\right)^2 \tag{8.2}$$

to within an additive constant. From Eq. (8.2), the exchange contribution to the domain wall energy density, σ (surface energy density), can be approximated by assuming that there are N spin pairs (each with the same relative angular deviation) through the wall thickness and each line of spins occupies an area a^2 on the wall surface:

$$\sigma_{\text{exch}} \approx \frac{NF_{ij}}{a^2} = JS^2\frac{\pi^2}{Na^2} \tag{8.3}$$

As N increases, more spins are oriented in directions of higher anisotropy energy. The anisotropy energy per unit area σ_a increases with N approximately as

$$\sigma_{\text{anis}} \approx K_u Na \tag{8.4}$$

The equilibrium wall thickness will be that which minimizes the sum

$$\sigma_{\text{anis}} + \sigma_{\text{exch}} \approx JS^2 \frac{\pi^2}{Na^2} + K_u Na \tag{8.5}$$

with respect to N. Figure 8.3 shows the dependence on N of the form of energy in Eq. (8.5), $b/N + cN$. When an energy expression has this form, it is minimized for $N_0 = (b/c)^{1/2} = (JS^2\pi^2/K_u a^3)^{1/2}$. Thus the wall thickness is of order $N_0 a \approx \pi(A/K_u)^{1/2}$, where A is the exchange stiffness constant described earlier, $A = JS^2/a \approx 10^{-11}$ J/m (10^{-6} erg/cm). Thus, the wall thickness will be of order 0.2 μm in systems with small anisotropy such as many soft magnetic materials; it may be as small as 10 nm in high-anisotropy systems such as permanent magnets. The wall energy density σ_{dw} is obtained by substituting $N_0 a \approx \pi(A/K_u)^{1/2}$ in Eq. (8.5) to give $\sigma_{dw} \approx 2\pi(AK_u)^{1/2}$. Typical values for the domain wall energy density are of order 0.1 mJ/m^2 (0.1 erg/cm^2).

Thus, in most cases a 180° domain wall will have an internal structure resembling that shown in Figure 8.2, right panel. The atomic magnetic moments will make a gradual transition in orientation from one domain to the next. Figure 8.4*a* shows an expanded version of this wall. At negative infinity $\theta(z)$ approaches 0°, at the origin (center of the wall) $\theta(z) = \pi/2$, and at positive infinity $\theta(z)$ approaches π. Figure 8.4*b* represents a possible form for the spin orientation $\theta(z)$ versus position z through the wall for the 180° wall represented in Figure 8.4*a*. The exact functional form of $\theta(z)$ must be derived for arbitrary exchange stiffness and magnetic anisotropy, and more exact expressions must be sought for the domain wall thickness and energy density. The continuum or macroscopic form of exchange energy density derived earlier, $A(\partial\theta/\partial z)^2$, will be used rather than the microscopic one used in Eqs. (8.2) and (8.3). The treatment given below is a modified version of the variational method first applied to the domain wall problem by Bloch in 1932 and by Landau and Lifshitz in 1935.

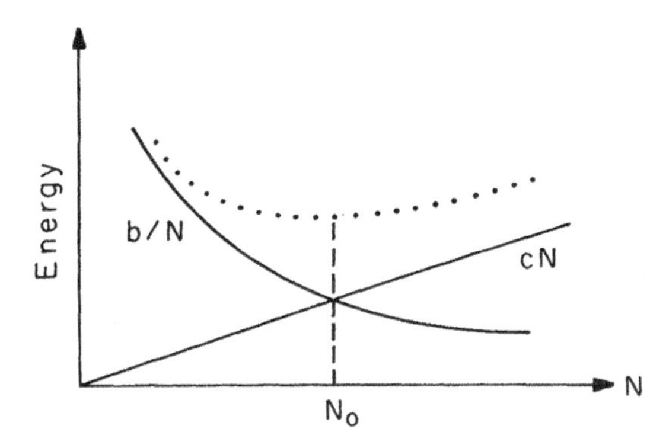

Figure 8.3 Minimization of the sum of exchange b/N and anisotropy cN energy densities occurs for $b/N = cN$, $N = \sqrt{(b/c)}$.

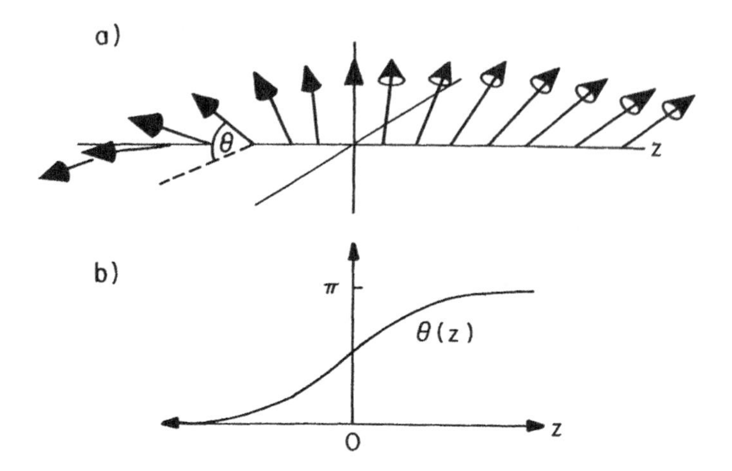

Figure 8.4 (*a*) Magnified sketch of the spin orientations within a 180° Bloch wall in a uniaxial material; (*b*) an approximation of the variation of θ with distance z through the wall.

8.2 MICROMAGNETICS OF DOMAIN WALLS

When energy minimization was used in earlier chapters to calculate M–H curves, the magnetization was assumed to be uniform throughout the sample. In order to determine the domain wall parameters δ_{dw} and σ_{dw}, θ must be allowed to be a function of position. This requires that the energy be minimized on a local or microscopic scale. This is an example of calculations of the micromagnetic type, a method first used by Landau and later generalized by W. F. Brown.

The anisotropy energy density, $f_a(\theta)$, may include magnetostatic and magnetoelastic contributions. In a one-dimensional approach to the problem the local *volume* energy density at any position z along the normal to the Bloch wall is given by a sum of anisotropy and exchange terms:

$$f = \frac{F}{V} = f_a(\theta) + A\left(\frac{\partial\theta}{\partial z}\right)^2 \tag{8.6}$$

This *volume* energy density must be integrated over the thickness of the transition region from one domain to the other to represent the total *surface* energy density of the domain wall:

$$\sigma = \int_{-\infty}^{\infty} \left[f_a(\theta) + A\left(\frac{\partial\theta}{\partial z}\right)^2 \right] dz \tag{8.7}$$

In order to calculate the stable wall profile function $\theta(z)$, σ is minimized with respect to variations of the wall profile $\delta\theta(z)$ using $\delta f = (\partial f/\partial\theta)\delta\theta$ and $\delta(\partial\theta/\partial z) = \partial(\delta\theta/\partial z)$:

$$\delta\sigma = \int_{-\infty}^{\infty} \left[\frac{\partial f_a(\theta)}{\partial\theta} \delta\theta + 2A \frac{\partial\theta}{\partial z} \frac{\partial\delta\theta}{\partial z} \right] dz = 0 \tag{8.8}$$

Integrating the second term by parts gives

$$\delta\sigma = \int_{-\infty}^{\infty} \left[\frac{\partial f_a(\theta)}{\partial\theta} - 2A \frac{\partial^2\theta}{\partial z^2} \right] \delta\theta dz + \left| 2A \frac{\partial\theta}{\partial z} \delta\theta \right|_{-\infty}^{\infty} = 0$$

The last term is zero because $\partial\theta/\partial z$ vanishes far from the wall where the magnetic moments are fixed in orientation by the anisotropy easy axis inside the domains. As for the remaining integral, because $\delta\theta(z)$ is an *arbitrary* function of z, the integral vanishes only if its argument vanishes *at every point z along the wall*:

$$\frac{\partial f_a(\theta)}{\partial\theta} - 2A \frac{\partial^2\theta}{\partial z^2} = 0 \tag{8.9a}$$

This is the *Euler equation* for a domain wall. The first term is the local torque on a spin due to the gradient in anisotropy at each point. The second term is the local torque on a spin due to the gradient in exchange energy at the spin in question. Equation (8.9) says that the net torque is zero at any point along the wall. Note that for the simple wall profile in Figure 8.4, where the torque, $f_a'(\theta)$, is given by $K_u \sin(2\theta)$, there is no torque from the anisotropy at the center of the wall ($\theta = \pi/2$). Equation (8.9) then indicates that θ'' vanishes there so θ' is constant. It should be clear that the condition of constant θ' in the neighborhood of a spin results in no net exchange torque on that spin. To the left (right) of the center of the wall ($0 < \theta < \pi/2$) both f_a' and θ'' are positive (negative), so Eq. (8.9) can be satisfied for a particular curvature θ'' (exchange torque) that exactly balances the torque due to the anisotropy.

For a uniaxial material, Eq. (8.9a) becomes

$$K_u \sin 2\theta - 2A\theta'' = 0 \tag{8.9b}$$

which transforms to the pendulum equation for $2\theta = \phi$ and $mg = K_u/A$:

$$m \cdot g \cdot \sin\phi - \phi'' = 0$$

Equation (8.9a) can be integrated to solve the wall profile problem by multiplying both sides by $d\theta/dz$:

$$\int \frac{\partial\theta}{\partial z}\left[\frac{\partial f_a(\theta)}{\partial\theta} - 2A\frac{\partial}{\partial z}\frac{\partial\theta}{\partial z}\right]dz \rightarrow \int \left[\partial f_a(\theta) - 2A\frac{\partial\theta}{\partial z}\partial\frac{\partial\theta}{\partial z}\right] = 0$$

Hence

$$f_a(\theta) - A\left(\frac{\partial\theta}{\partial z}\right)^2 = C \tag{8.10}$$

The integration constant, C is allowed to be zero by noticing that Eq. (8.10) is an energy density expression and a constant energy density cannot exert a torque on the magnetization. More rigorously, far from the wall, $\theta' = \partial\theta/\partial z = 0$ and an arbitrary constant can be added to the anisotropy energy function such that $f_a(\theta = -\infty) = 0$, hence $C = 0$ (see Problem 8.5). The first integral of Eq. (8.9a) is then

$$\frac{\partial\theta}{\partial z} = \pm\left(\frac{f_a}{A}\right)^{1/2} \tag{8.11}$$

which can be integrated to give

$$z - z_0 = \int_0^{\theta_0} \sqrt{\frac{A}{f_a(\theta)}}\, d\theta \tag{8.12}$$

For the uniaxial case, $f_a = K_u \sin^2\theta$, the integral in Eq. (8.12) has the exact solution

$$z = \sqrt{\frac{A}{K_u}} \ln\left[\tan\left(\frac{\theta}{2}\right)\right] \tag{8.13}$$

This equation, which relates the position in a domain wall to the orientation of the magnetization at that point, can be inverted to give $\theta(z) = 2\arctan[\exp(\pi z/\delta)]$, where $\delta = \pi(A/K_u)^{1/2}$. It is left as an exercise to show that this is equivalent to

$$\theta(z) = -\text{arc}\cot\left[\sinh\left(\frac{\pi z}{\delta}\right)\right] + \pi = \arctan\left[\sinh\left(\frac{\pi z}{\delta}\right)\right] + \frac{\pi}{2} \tag{8.14}$$

This is an analytic form for the profile of a 180° domain wall sketched in Figure 8.4*b*. It is useful to plot it exactly, based on the forms in Eqs. (8.13) or (8.14). The solution for the uniaxial 180° domain wall is shown in Figure 8.5 for the

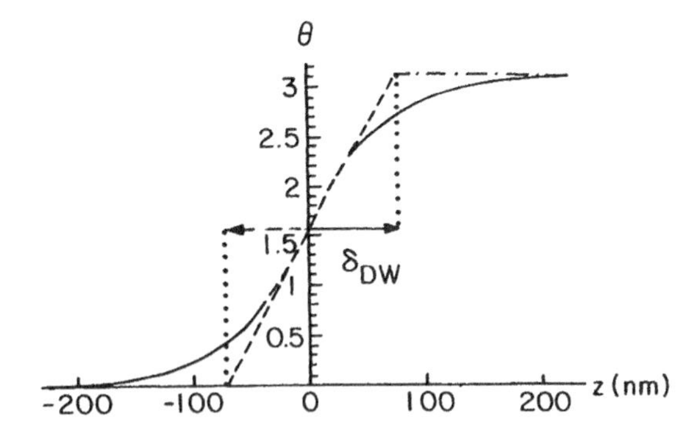

Figure 8.5 Graph of Eq. (8.14) showing θ versus z in nanometers generated using Mathematica software for a uniaxial material with $A = 1 \times 10^{-20}$ J/nm and $K_u = 5 \times 10^{-24}$ J/nm^3. The domain wall thickness, δ_{dw}, is defined on the graph.

following values for K and A (rounded values close to those of α-Fe):

$$\text{MKS:}\quad K = 5 \times 10^{3}\,\text{J/m}^3 = 5 \times 10^{-24}\,\text{J/nm}^3$$

$$A = 1 \times 10^{-11}\,\text{J/m} = 1 \times 10^{-20}\,\text{J/nm}$$

$$\text{cgs:}\quad K = 5 \times 10^{4}\,\text{erg/cm}^3 = 5 \times 10^{-17}\,\text{erg/nm}^3$$

$$A = 1 \times 10^{-6}\,\text{erg/cm} = 1 \times 10^{-13}\,\text{erg/nm}$$

It is useful to define the wall thickness in terms of the slope at its center. From Eq. (8.11) and for the uniaxial case $[dz/d\theta]_{\pi/2} = (A/K_u)^{1/2}$. This gives $dz = \pi(A/K_u)^{1/2}$ for $d\theta = \pi$ (see dashed line, Figure 8.4*a*). Hence the 180° wall thickness is given by

$$\delta_{dw} = \pi\left(\frac{A}{K_u}\right)^{1/2} \tag{8.15}$$

where δ_{dw} is the thickness over which the spins rotate to within about 27° of the domain magnetization direction.

It is useful to express the wall energy density [Eq. (8.7)] solely in terms of $f_a(\theta)$ and A. This can be done by using Eq. (8.11) to replace $(\partial\theta/\partial z)^2$ and to change the variable of integration in Eq. (8.7), giving

$$\sigma = 2\int_{-\infty}^{\infty} f_a(\theta)dz = 2\sqrt{A}\int_0^{\pi}\sqrt{f_a(\theta)d\theta}$$

For the simple uniaxial case, $f_a(\theta) = K_u \sin^2(\theta)$, the result is

$$\sigma_{dw} = 4(AK_u)^{1/2} \tag{8.16}$$

For iron and nickel, the calculated values of wall thickness δ_{dw} are 30 and 72 nm, respectively, and the values of domain wall energy density σ_{dw} are 3 and $0.7\ \mathrm{mJ/m^2}$, respectively ($1\ \mathrm{mJ/m^2} = 1\ \mathrm{erg/cm^2}$). Note that these wall energies are more than three orders of magnitude smaller than the surface energy of the hypothetical domain wall, having a thickness of one lattice constant, as calculated in Section 8.1. Some permanent magnets, which generally have very large anisotropy energies, have domain wall thicknesses under 10 nm and wall energy densities of order $30\ \mathrm{mJ/m^2}$. The softest crystalline magnetic materials, certain NiFe alloys, have wall thicknesses of order $1\ \mu\mathrm{m}$ and wall energy densities as low as $0.1\ \mathrm{mJ/m^2}$.

Equation (8.9) can be solved numerically using appropriate mathematical software.

Numerical Calculation of 180° Domain Wall Profile With access to Mathematica* software, it is possible to solve the Bloch wall probem in any of three ways (one analytic and two numeric) on a computer in a matter of minutes: (1) numerically solve the second-order Euler differential equation (8.9) using the two boundary conditions $\theta(0) = \pi/2$, and $\theta'(0) = \pi/\delta$ [Eq. (8.11) at $z = 0$]; (2) numerically solve the first-order differential equation [Eq. (8.11)] with the boundary condition $\theta(0) = \pi/2$; and (3) the integral equation [Eq. (8.12)] can be solved analytically by Mathematica. The three Mathematica programs that give Figure 8.5 are listed below. Units are Joules per nanometer.

```
        In:=
        k=5.10^-24//N
        a=1. 10^-20//N
        d=Sqrt[a/k]//N
        Out:=
        5. 10^-24
        1. 10-20
        44.7214
First:  In:=
        NDSolve[{2a t''[z]==k Sin[2 t[z]], t[0]==pi/2, t'[0]==1/d}, t, {z, -2d, 2d}]
        Out:=
        {{t->InterpolatingFunction[{-280.993, 280.993}, <>]}}
                Plot[Evaluate[t[z]/.%], (z, -2d, 2d}, AxesLabel->{"z(nm)","theta"},
                PlotLabel->"Bloch Wall"]
Second: In:=
        NDSolve[{t'[z]==(1/d) Sin[t[z]], t[0]==pi/2}, t, {z, -2d, 2d}]
        Out:=
        {{t->InterpolatingFunction[{-280.993,280.993}, <>]}}
                Plot[Evaluate[t[z]/.%], (z, -2d, 2d}, AxesLabel->{"z(nm)", "theta"},
                PlotLabel->"Bloch Wall"]
Third:  In:=
        Integrate[d/(Sin[t]),t]
        Out:=
```

*Mathematica is a technical computing program created by and registered trademark of Wolfram Research.

```
-44.7214 Log[Cos[t/2]] + 44.7214 Log[Sin[t/2]]
In:=
Solve[y==d Log[Tan[t/2]], t]
Out:=
Solve::fun:
                            0.0223607 y
{{t→2. ArcTan[1. E                    ]}}
In:=
Plot[2 ArcTan[E^(y/d)], {y, -5 d, 5 d},
        AxesLabel→{"z(nm)", "theta"},
              PlotLabel→"Bloch Wall"]
```

For all three programs the output is the same:

```
Out:=
-Graphics- (Fig. 8.5 above.)
```

In addition it is interesting to plot:

```
In:=
ParametricPlot3D[{y, -(u Cos[2 ArcTan[E^(y/d)]]),
        u Sin[2 ArcTan[E^(y/d)]]},
              {y, -200, 200}, {u, 0, 100}]
```

8.3 MAGNETOSTATIC EFFECTS ON DOMAIN WALLS

So far, exchange and anisotropy energies have been balanced to get a model of the 180° domain wall. Magnetostatic energy densities must be considered if one is interested in the wall structure near a surface or if the wall is not a 180° wall.

In Chapter 2, it was seen that Maxwell's equations demand that the perpendicular component of $\boldsymbol{B}$ be continuous across an interface [Eq. (2.3)], and in the present case across a domain wall. If no external field is applied (as for domains in the demagnetized or remanent states) and if the magnetization is uniform within the domains, then the only $\boldsymbol{H}$ fields present are those originating at surfaces containing a net magnetic pole density: "charged" surfaces. The magnetostatic fields from charged surfaces always increase the energy of the system because the field has an appreciable component opposite to the magnetization that sets up the surface pole density. Hence, charged walls occur only if some overriding energy demands them. In the absence of external fields and magnetostatic fields from surface poles, the boundary condition in Eq. (2.3) is equivalent to continuity of the normal component of magnetization across the interface: $(\boldsymbol{M}_1 - \boldsymbol{M}_2) \cdot \boldsymbol{n} = 0$. Continuity of the normal component of magnetization across a domain wall demands that the wall bisects the two directions of magnetization in the adjacent domains. When the normal components of $\boldsymbol{M}$ do not match across a domain wall, a net pole density exists on the wall and an associated magnetostatic field results.

Four situations are considered where magnetostatic fields can change the simple domain wall structure already calculated for uniaxial materials.

8.3.1 90° Walls

In materials of cubic anisotropy, when the ⟨100⟩ directions are the easy axes (i.e., $K_1 > 0$), 90° walls are possible. They can be observed on (100) faces of Fe crystals that are thin in the [100] direction (Fig. 8.1*a*, upper right). When the ⟨111⟩ directions are the easy magnetization axes as is the case in Ni, then 180°, 109°, and 71° walls are possible; the walls in the latter two cases appear to be similar to 90° walls.

There are two issues to be considered in defining the 90° domain wall. First, such a wall should have an orientation that maintains continuity of the normal component of magnetization *across* the wall. Second, the magnetization should rotate *within* the wall in such a way as to minimize the exchange and anisotropy energies. Continuity of the normal component of magnetization across the wall demands that the domain wall bisect the directions of magnetization in the two adjacent domains. This prevents the wall from being magnetically charged and hence, generating a magnetostatic field. A 90° domain wall in an arbitrary plane satisfying this pole-free criterion is shown in Figure 8.6*a*. Its normal $\boldsymbol{n}_2$ must be in the x–z plane and it makes an angle ψ with z, the normal to the plane containing the two domain magnetization. The relevant vectors may be represented in the Cartesian coordinates of Figure 8.6 as follows:

$$\boldsymbol{M}_1 = \left(\frac{M_s}{\sqrt{2}}\right)(-1, -1, 0), \qquad \boldsymbol{M}_2 = \left(\frac{M_s}{\sqrt{2}}\right)(-1, 1, 0), \qquad \boldsymbol{n}_2 = (-\sin\psi, 0, \cos\psi)$$

It is clear that $(\boldsymbol{M}_1 - \boldsymbol{M}_2)\cdot\boldsymbol{n}_2 = 0$ for any ψ, so it is possible to form a family of uncharged 90° domain walls. The domain wall will choose a ψ that minimizes the wall area, that is, the 90° wall will be perpendicular to the sample faces having the largest area. The next issue involves the orientation of $\boldsymbol{M}$ *within* the domain wall so as to minimize the domain wall energy. In order to minimize the sum of the exchange and anisotropy energies, it is enough to examine the cubic energy surface for $K_1 > 0$ (Fig. 6.6*a*). The path of the minimum energy for the magnetization orientation to change from one face to another (between adjacent ⟨100⟩ directions) is clearly rotation in the plane containing $\boldsymbol{M}_1$ and $\boldsymbol{M}_2$. This path is along the dashed line on the energy surface depicted in Figure 8.6*b*.

The cubic form of anisotropy, Eq. (6.6), can be substituted into the integral solution of the domain wall given in Eq. (8.12). For magnetization rotation in the (001) plane ($\theta = \pi/2$), this energy function reduces to $f_a(\phi) = K_1 \sin^2(\phi)\cos^2(\phi)$. Here ϕ is the angle through which the magnetization rotates in the plane normal to z (Fig. 8.6). Equation (8.12) then gives

$$z = \sqrt{\frac{A}{K_1}} \int_0^{\phi} \frac{1}{\sin\phi\cos\phi}\, d\phi = 2\sqrt{\frac{A}{K_1}} \int_0^{\phi} \frac{1}{\sin 2\phi}\, d\phi$$

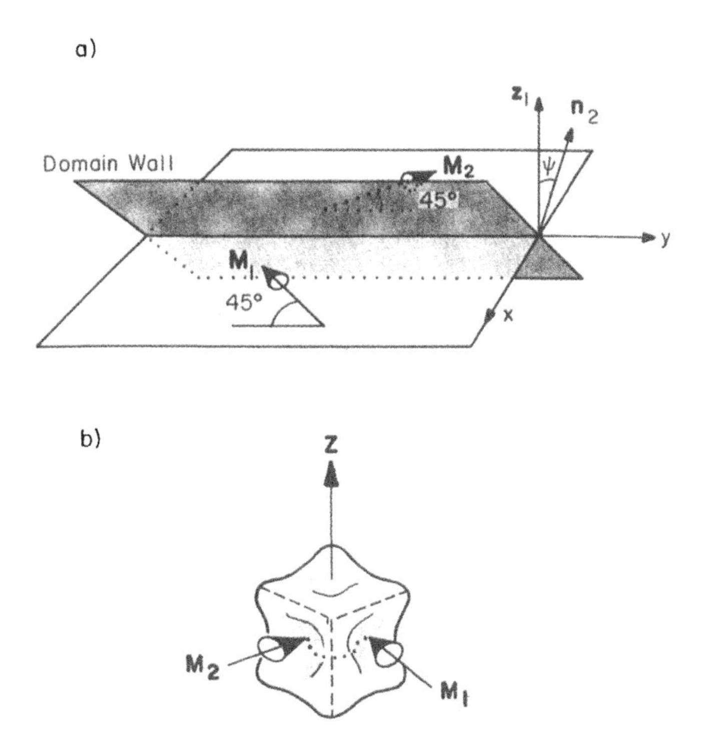

Figure 8.6 (*a*) The plane normal to z contains the domain magnetization vectors that are orthogonal to each other in cubic anisotropy [panel (*b*)]. The domain wall must lie in a plane that bisects the two directions of $\boldsymbol{M}$. A general plane satisfying this condition is shown shaded in (*a*); its normal direction, $\boldsymbol{n}_2$, must be chosen to minimize the wall energy. Panel (*b*) shows the cubic origin of the magnetization directions and the dotted line on the energy surface shows the minimal energy path between the two orientations.

which is equivalent to Eq. (8.13) for the uniaxial case. Thus, Figure 8.4 would describe $\theta(z)$ for a 90° wall if the vertical axis were to have the range 0 to $\pi/2$ instead of 0 to π. From Eq. (8.5), modified for such a 90° wall, the wall thickness becomes $\delta_{90} = \delta_{dw}/2$ and the energy density becomes $\sigma_{90} = \sigma_{dw}/2$.

But 180° walls also occur in cubic materials (i.e., ϕ) may rotate from 0 to π radians. As the magnetization rotates past the easy direction at $\pi/2$, the anisotropy energy decreases to what it was at 0° so $\theta(z)$ could flatten out here. How does a 180° wall differ from a sequence of two 90° walls? Figure 8.7 shows what would be expected on the basis of considerations so far.

If magnetostatic energy is included, the energies of the various domain walls are altered. The two domains separated by the 180° wall are equivalently strained in the magnetization direction by $\Delta l/l = \lambda_s$. There is no long-range strain incompatibility between the two domains. It is clear that two domains magnetized at 90° to each other have incompatible magnetostrictive strains. Thus, there is a strain energy associated with any region between two 90° walls.

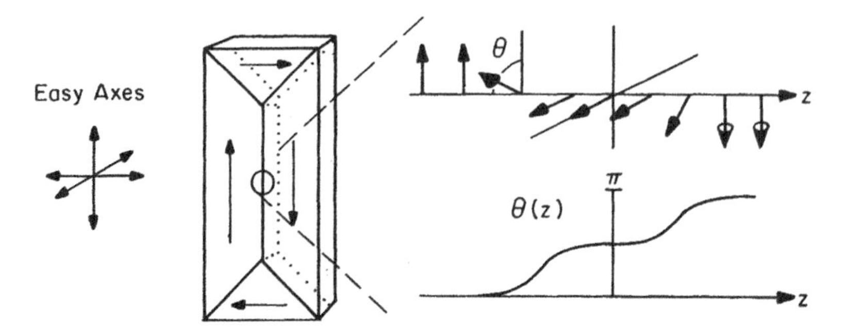

Figure 8.7 A (100) cut Fe single crystal has the possible domain pattern shown at left. The 90° walls are described in the text. The 180° wall may be thought of as a sequence of two 90° walls as shown at right. However, the inclusion of magnetoelastic energy stabilizes a 180° wall relative to the two 90° walls.

This mismatch adds a magnetoelastic anisotropy that eliminates the lingering of the magnetization near the easy direction at 90°. When magnetostriction is large, 90° domain walls are short and/or few in number; when magnetostriction is small, 90° domain walls can be longer and/or more numerous because they introduce less strain energy. However, there is an important class of materials that contain copious domain walls across which the spins rotate almost 90° even though the strain in the adjacent domains can be of the order of a few percent. In fact, $\theta \approx 2\arctan(1 \pm 3\lambda/2)$ in these materials. These materials are the magnetic shape memory materials mentioned in Chapter 7. Their domain walls coincide with crystallographic twin boundaries that allow the lattice to accommodate the large strain energy associated with the giant crystallographic strain.

8.3.2 Néel Walls

A second example of the effects of magnetostatic energy on domain walls is taken from thin films. It should be clear from Figure 8.2 that as sample thickness decreases, the magnetostatic energy of the wall that extends through the thickness of the sample increases as a result of the free poles at the top and bottom of the wall (Fig. 8.8, left). To reduce this magnetostatic energy, the spins inside the wall may execute their 180° rotation in such a way as to minimize their magnetostatic energy. If the spins were to rotate in the plane of the surface, a smaller magnetostatic energy at the internal face of the wall is accepted as the price for removing the larger magnetostatic energy at the top surface (Fig. 8.8, right). Such a wall is called a *Néel wall.* Note that on moving across a Néel wall from one domain to another, the magnetization rotates by 180° but stays in the surface plane.

Several authors after Néel [LaBonte (1969), Hubert (1969, 1970) and, more recently, Scheinfein et al. (1989, 1991)] have included magnetostatic energy

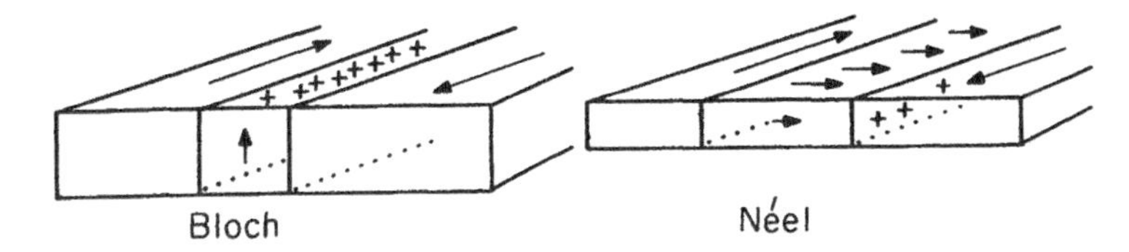

Figure 8.8 Comparison of Bloch wall, left, with charged surfaces on the external surfaces of the sample and Néel wall, right, with charged surfaces internal to the sample.

terms in modeling Bloch walls in thin films (of thickness $t \approx \delta_{dw}$). Figure 8.9*a* shows the calculated film thickness dependence of the wall energy density for Bloch and Néel walls. The Bloch wall energy density increases with decreasing film thickness because of the increased magnetostatic energy due to the appearance of charged surfaces above and below the wall. The Néel wall energy decreases with decreasing film thickness because it is proportional to the area

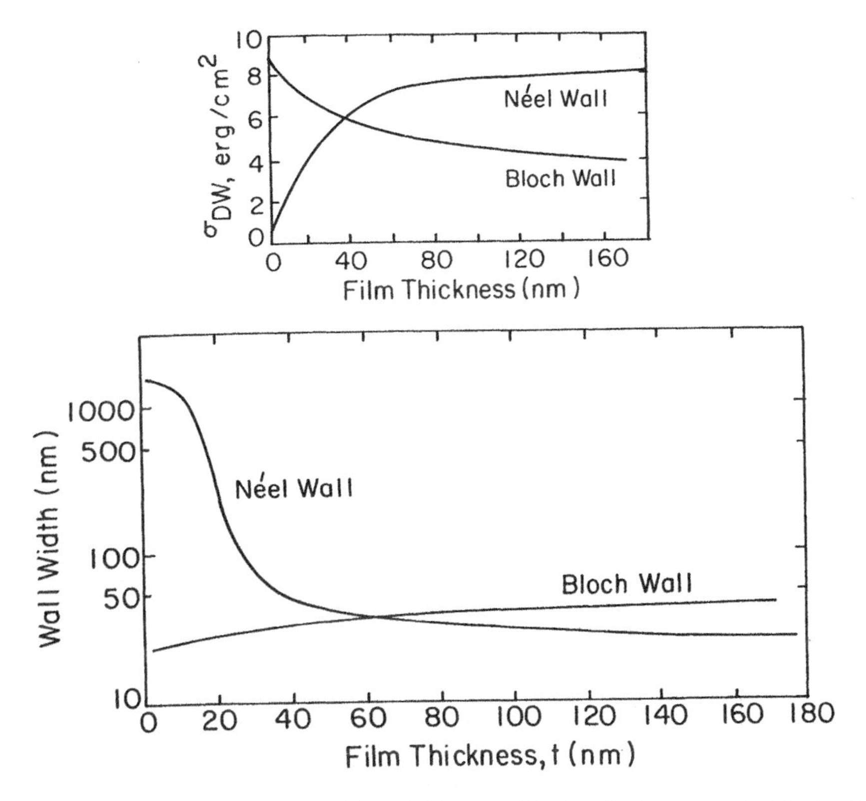

Figure 8.9 Energy per unit area (*a*) and thickness (*b*) of a Bloch wall and a Néel wall as functions of the film thickness. Parameters used are $A = 10^{-11}$ J/m, $B_s = 1$ T, and $K = 100$ J/m^3 [McGuire (unpublished)].

of the charged surfaces inside the film. Néel walls are observed to be stable in many types of magnetic films for thicknesses up to 50 or 60 nm. Figure 8.9*b* shows the calculated dependence of Bloch and Néel wall widths on film thickness. The Bloch wall thickness decreases with decreasing thickness because that reduces the magnetostatic energy associated with the charged caps of the wall. The Néel wall thickness increases with decreasing film thickness in order to minimize the magnetostatic energy associated with the charged faces of the wall. At sufficiently small film thickness, the magnetostatic energy is no longer significant and the Néel wall thickness no longer increases.

Without going into the micromagnetic calculations, it is still possible to arrive at reasonable forms for the film thickness dependence of the Néel wall thickness. The free energy density can be approximated as

$$\sigma_N = A\frac{\pi^2}{\delta_N} + K_u\frac{\delta_N}{2} + \left(\frac{2\mu_0 M^2}{\pi}\right)\delta_N \arctan\left(\frac{t}{\delta_N}\right)$$

using the expression for the demagnetization field in Eq. (2.3). Minimization of this energy density with respect to δ_N gives

$$0 = \delta_N^2\left[\frac{K_u}{2} + \frac{2\mu_0 M^2}{\pi}\arctan\left(\frac{t}{\delta_N}\right)\right] - \frac{2\mu_0 M^2}{\pi}\frac{t\delta_N}{1+(t/\delta_N)^2} - A\pi^2 \quad (8.17)$$

For $t/\delta_N \ll 1$, the limiting forms of the energy density σ_N and wall thickness δ_N follow from Eq. (8.17):

$$\sigma_N \approx \pi t M_s^2 \quad \text{and} \quad \delta_N \approx \pi\left(\frac{2A}{K}\right)^{1/2} \quad (t \ll \delta_N) \quad (8.18)$$

where the subscript N indicates parameters (energy density and thickness) for a Néel wall. These expressions should be compared with the large-film-thickness forms for Bloch walls, Eqs. (8.15) and (8.16). While the energy of a Bloch wall is determined by the product of the anisotropy and exchange stiffness, that of a Néel wall at $t \ll \delta_N$ is determined by magnetostatic energy and film thickness. Note that although it is magnetostatic energy that has forced the formation of the Néel wall, magnetostatic energy does not show up in the expression for the thin-film limit of Néel wall thickness (see Problem 8.5).

Recent experimental measurements and micromagnetic calculations of the surface magnetization distribution in materials with thicknesses greater than the Néel limit reveal that a Bloch wall can transform into a Néel wall near a surface. Figure 8.10 shows the calculated magnetization distribution near the surface of a material much thicker than one that would support a pure Néel wall (Scheinfein et al. 1989). The spins near the surface depart from their Bloch

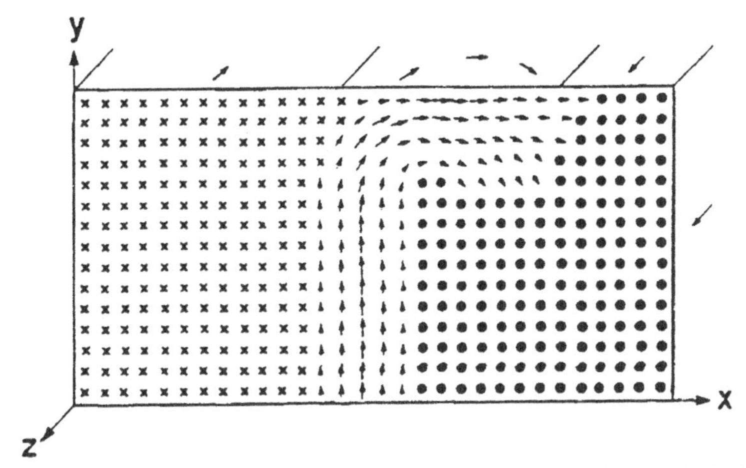

Figure 8.10 Calculated spin distribution in a thin sample containing a 180° domain wall. Note that the wall is a Bloch wall in the interior; specifically, M rotates 180° in the plane of the wall, but it is a Néel wall near the surface (to minimize magnetostatic energy), M rotates 180° passing through the wall normal (Scheinfein et al. 1989).

formation by gradually folding over to lie in the plane of the surface to reduce the magnetostatic energy. As a result, the thickness of the Néel cap where a Bloch wall intersects the surface is greater than that of a pure Bloch wall. The domain wall depicted in Figure 8.1 is also of this Néel-cap type because of the measured horizontal component of magnetization at the wall center.

Figure 8.11 shows the measured and calculated surface wall magnetization distributions for two relatively soft magnetic materials: an iron whisker and a permalloy film. (An iron whisker is a small elongated single crystal of iron, typically grown by a hydrothermal process.) The measurements were made with a SEMPA microscope similar to that used to generate Figures 1.13*a*, left and 8.1. The results of Scheinfein et al. (1991), and of Oepen and Kirschner (1989) (Fig. 8.1), show that the spin rotation near the surface is not symmetric about the center of the surface wall and is accurately described by model calculations. Apparently, even in the center of a thick iron film, the domain wall departs from the simple structure of a 180° Bloch wall in an infinite medium (Aharoni and Jakubovics 1991).

In Chapter 16, other interesting modes of magnetization in the vicinity of a surface or interface will be explored.

8.3.3 Cross-Tie Walls

Cross-tie walls provide a third example of how magnetostatic energy can influence the nature of a domain wall.

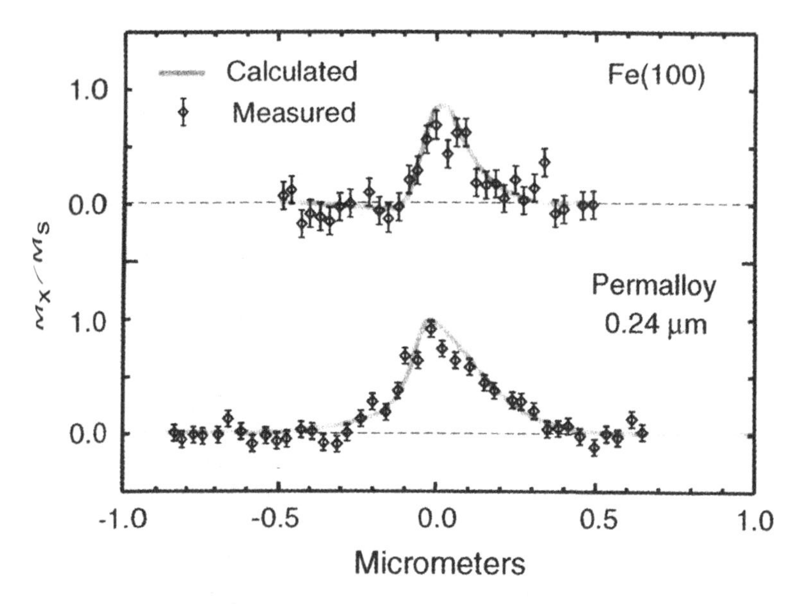

Figure 8.11 Measured (using SEMPA) and calculated surface wall profiles in a 20-μm-thick Fe whisker (top, $\delta_N \approx 0.2\,\mu$m) and a 0.24-$\mu$m-thick permalloy film (bottom, $\delta_N \approx 0.3\,\mu$m) (Scheinfein et al., 1991).

Generally 180° Bloch walls lie parallel to the easy axes along which the domain magnetization lies in order to prevent wall "charging" (Fig. 8.12, left). Néel walls have an inherent charge because of their spin structure (Fig. 8.12, right). The large magnetostatic energy associated with this magnetic charge can be reduced if the sense of polarization of the wall alternates. When this situation occurs, the wall is called a cross-tie wall. The detailed structure of the cross-tie wall, studied extensively by Craik and Trebble (1965), is an elegant example of how magnetic materials contrive to minimize their magnetostatic energy by forming domains.

Figure 8.13 shows an image of cross ties along a wall in a NiFe film taken by scanning electron microscopy with spin polarization analysis (SEMPA) (see Chapter 16). This technique reveals contrast as a result of selected components of magnetization; thus, the cross-ties may show up as curving across the wall they decorate.

Because of their magnetic charge, Néel walls can interact with each other; adjacent wall segments will attract or repel each other depending on the sense of their charge. As film thickness decreases even below the limit at which Néel walls form, the magnetostatic fields of the Néel wall decrease. In ultrathin films ($t \approx 1-10$ monolayers, which is much less than δ_N) magnetostatic considerations cease to be a factor. In such films, the wall profile function, $\theta(z)$, has been observed by Oepen to have a much slower approach to its asymptotic value than is calculated for bulk domain walls.

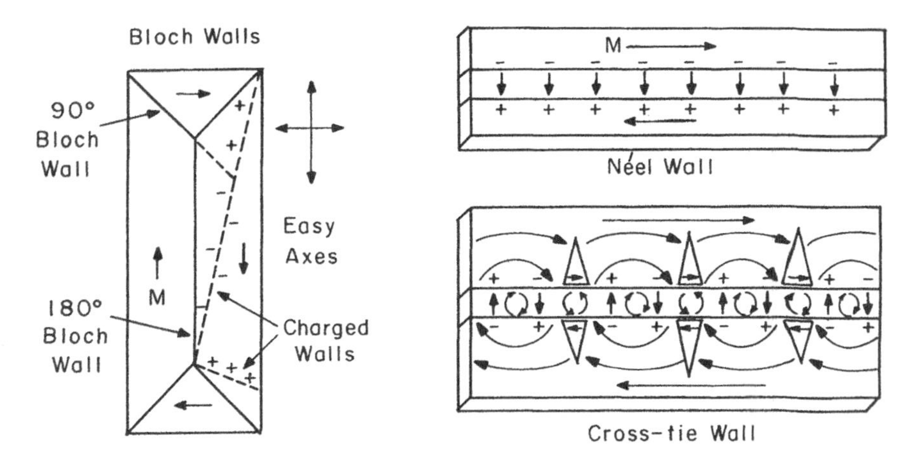

Figure 8.12 Left, Bloch wall showing how walls that do not follow the adjacent domain magnetization acquire a magnetic "charge"; Right, the charge on a Néel wall can destabilize it and cause it to degenerate into a more complex cross-tie wall.

8.3.4 Domain Walls in Ultrathin Films

As a final example of magnetostatic effects on domain walls, we consider films that are so thin that the magnetostatic energy favoring in-plane magnetization is smaller than those energies that might favor out-of-plane magnetization, such as magnetoelastic energy (Chapter 7) or surface anisotropy (see Chapter 16). In these cases, the magnetization in the domains is *perpendicular* to the

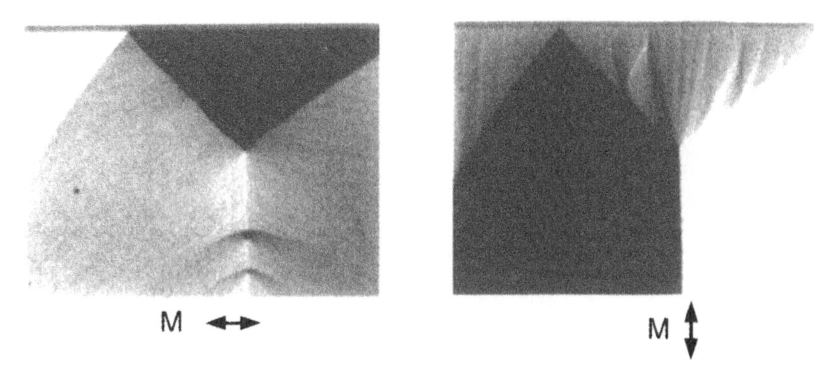

Figure 8.13 Scanning electron microscopy with spin polarization analysis (SEMPA) images of NiFe film. Panel (*a*) shows horizontal polarization contrast (white is magnetization to the right, dark to the left, and gray vertical). Panel (*b*) shows vertical polarization contrast near a triple-wall junction revealing cross ties on the domain walls. [Courtesy of Celotta et al. (1991).]

plane and that inside the 180° domain walls separating the domains lies the film plane. Figure 8.14 shows the domain pattern in a 2-nm-thick film nickel grown epitaxially on Cu(001) (Bochi et al. 1995) taken by magnetic force microscopy (see Chapter 15). The in-plane magnetization in the Bloch wall does not follow the ⟨110⟩ directions, which are the easy in-plane axes in Ni. If it did, the domain walls would appear as nearly straight line segments lying at ±45° to the borders of Figure 8.14. It happens that in Ni, the magnetocrystalline anisotropy is smaller than the energy of the walls themselves. Thus, the domain walls tend to follow curved paths so as to minimize the wall energy at the expense of magnetocrystalline anisotropy energy.

A complete understanding of the factors governing domain formation and domain wall structure in ultrathin films is not yet available.

8.4 DOMAIN WALLS NEAR INTERFACES: THE EXCHANGE LENGTH

It is appropriate here to examine the magnetization orientation transition, or pinned domain wall, that exists near an interface at which the magnetization is pinned in a direction different from the easy axis in the interior of the material (Fig. 8.15). The thickness of this transition, called the *exchange length*

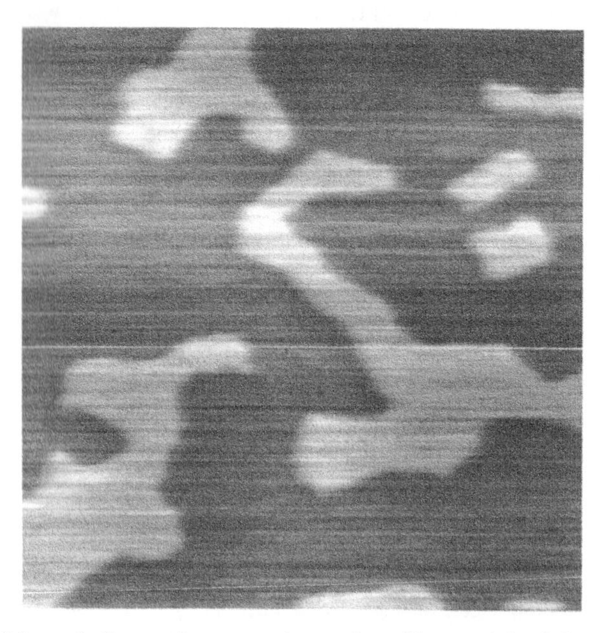

Figure 8.14 Magnetic force microscopy image (see Chapter 16) of the domains in a 20-Å-thick epitaxial Ni film in which the magnetization is perpendicular to the film plane. Field of view is 12.5 μm^2. [After Bochi et al. (1995).]

l_{ex}, is governed by the square root of the ratio of the exchange stiffness A to the energy cost of having spins near the interface oriented at a nonzero angle to the interior easy axis. The first case to be considered is that for which no perpendicular magnetization component exists near the interface. On moving away from the interface, $\mathbf{M}(x)$ rotates under the influence of the interior anisotropy energy. In this case, the mathematical form of the exchange length is given simply by

$$\mathbf{M} \text{ parallel to interface:} \qquad l_{ex}^{\parallel} = \left(\frac{A}{K_u}\right)^{1/2} = \frac{\delta_{dw}}{\pi} \qquad (8.19a)$$

This parallel case (Fig. 8.15, left) is most frequently encountered. A strong interior anisotropy energy shortens the length of the transition.

When the spin orientation at the interface has a perpendicular component (Fig. 8.15, right), the interior anisotropy energy is augmented by the magnetostatic energy associated with the charged interface; the magnetization in the interior of the material is driven toward the easy axis by $K_u + 2\pi M_{\perp}^2$:

$$\mathbf{M} \text{ perpendicular to interface:} \qquad l_{ex}^{\perp} = \left(\frac{A}{K_u + 2\pi\Delta M_{\perp}^2}\right)^{1/2} \qquad (8.19b)$$

This last equation is derived in Chapter 16 for a surface where a perpendicular magnetization component exists, and the interior anisotropy is weak. Note that when the perpendicular situation is dominated by the magnetostatic energy,

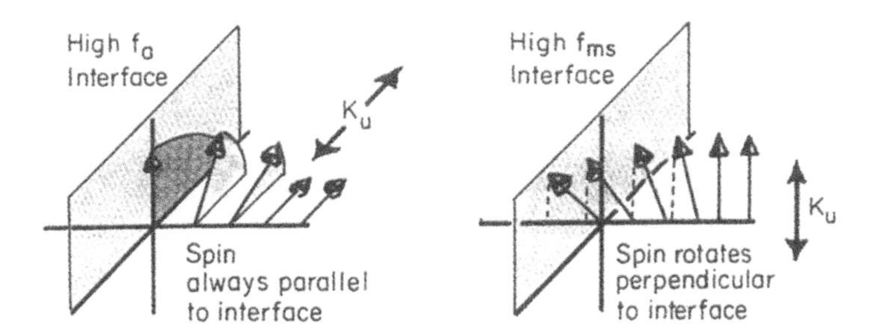

Figure 8.15 Illustration of the two cases important for determining the range of the twist in magnetization on moving from an interface at which the spins are pinned in a direction different from the interior of a ferromagnetic material. At left, the surface pinning holds the magnetization in the plane of the interface so magnetostatic energy is not an issue. At right, the surface spin pinning is such that a perpendicular component of magnetization exists near the interface. The magnetic charge at the interface gives rise to a local magnetostatic field that tends to shorten the exchange length.

TABLE 8.1 Magnetostatic and Anisotropy Energy Densities for Fe, Co, and Ni and Exchange Lengths Calculated therefrom [Eqs. (8.19a), (8.19b) ($A = 10^{-11}$ J/m Is Used in All Cases)

	$\mu_0 M_s^2/2$ (10^5 J/m^3)	K_u (10^3 J/m^3)	$l_{ex}^{\parallel}$ (nm) Parallel	$l_{ex}^{\perp}$ (nm) Perpendicular
Fe	19	48	14	2.3
Co	12	410	5	2.9
Ni	1.5	4.5	47	8.2

Eq. (8.19b) is essentially $l_{ex}^{\perp} \approx (A/2\pi\Delta M_{\perp}^2)^{1/2}$. Table 8.1 lists the magnitude of the magnetostatic energy, anisotropy energy, and the calculated exchange lengths for parallel and perpendicular ($K_u = 0$) interfaces for Fe, Co, and Ni. One important consequence of these results is that the exchange interaction, through the parameter l_{ex}, has the effect of communicating the magnetization direction in one region over distances of several nanometers to another region. This exchange coupling can effectively change the magnetic size or range of influence of certain defects.

These results will be used in Chapter 13, where ferromagnetic–antiferromagnetic and ferromagnetic–ferromagnetic interfacial exchange will be considered.

8.5 MAGNETIC DOMAINS

One of the first questions asked by many students of magnetism (including Felix Bloch in 1932) is "How big are magnetic domains?" Landau and Lifshitz pointed out in 1935 that if it were not for the dipole fields set up at the surfaces of a ferromagnetic material, there would be no domains. Domains form solely to minimize the magnetostatic energy that results when $\boldsymbol{M} \cdot \boldsymbol{n} \neq 0$ at an interface. An unbounded ferromagnet would be magnetized uniformly to saturation. So Bloch's question then becomes "What does the domain structure look like in a sample of a given shape and size, and how does it vary from the interior to the surface (where the dipole fields are strongest)?" Landau and Lifshitz solved many aspects of this problem, including the size and shape of closure domains near a surface. Kittel expanded on many of their results in his reviews of magnetic domains (1949, 1956).

Magnetic domains are regions in a ferromagnetic material within which the direction of magnetization is largely uniform. Once domains form, the orientation of $\boldsymbol{M}$ in each domain and the domain size are determined by magnetostatic, crystal anisotropy, magnetoelastic, and domain wall energy. All domain structure calculations involve minimization of the appropriately selected energies. Figure 8.16 shows the progressive subdivision of a saturated sample (*a*), with its high magnetostatic energy, into one composed of increasingly more

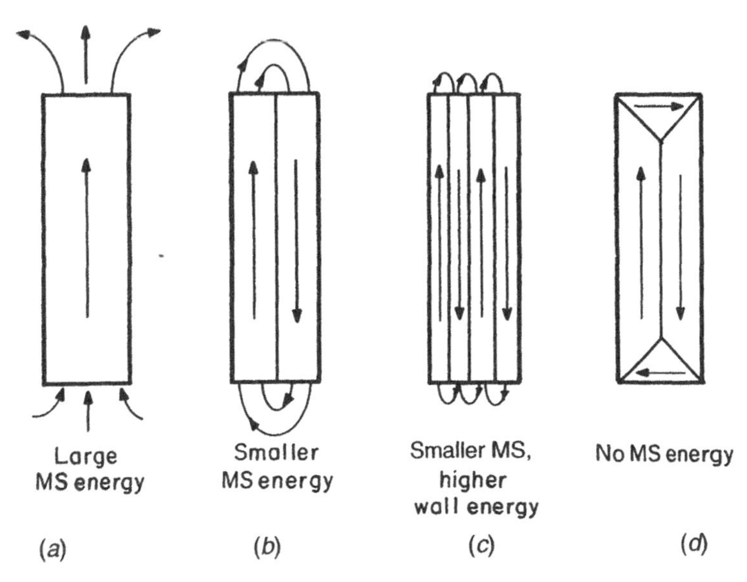

Figure 8.16 Domain formation in a saturated magnetic material is driven by the magnetostatic (MS) energy of the single domain state (left). Introduction of 180° domain walls reduces the MS energy but raises the wall energy; 90° closure domains eliminate MS energy but increase anisotropy energy in uniaxial materials and cause elastic energy due to the strain incompatibility of the adjacent 90° domains (right).

domains. The energy changes in each case are noted in the figure. First, the large magnetostatic energy due to the end poles in the case at left is reduced by allowing the formation of two domains of opposite magnetization. The integral of H^2 outside the sample decreases in the second panel because the field lines now close at the ends of the adjacent domains; the flux lines follow a path of low reluctance R_m and hence low energy $\phi^2 R_m$, where ϕ is the magnetic flux (analogous to low i^2R). This process continues from panel (*b*) to panel (*c*) with the multiplication of domains proceeding until the energy cost of adding another 180° wall is greater than the magnetostatic energy saved. There is always some magnetostatic energy cost for 180° wall formation in such a finite sample. If the material has cubic anisotropy, or if the uniaxial anisotropy is not too strong, the sample may form closure domains at its end surfaces. These closure domains allow the flux to be fully contained within the sample, thus completely eliminating the magnetostatic energy contribution. However, there is a cost to formation of these closure domains in terms of magnetic anisotropy energy (for a uniaxial material) and in terms of elastic energy if the magnetostriction is not zero. In this chapter, analytic *estimates* of these energies will be sought for samples of various geometries. Bear in mind that the demagnetizing field will be approximated in many cases, so the estimates of the energies are useful only to the extent that the sample geometry meets the assumptions of the approximate demagnetizing factors. In nonellipoidal samples the demagnetizing field approximations fail near charged surfaces.

Several magnetic domain images have already appeared in this text (Figs. 1.13, 8.1a, 8.13, 8.14). Most of the images shown here were made by magneto-optic Kerr effect microscopy, scanning electron microscopy with spin polarization analysis (SEMPA), or magnetic force microscopy (MFM). MFM and SEMPA are discussed in Chapter 16 where more domain images on thin films (Figs. 16.22, 16.23, 16.30) are presented. Historically, most magnetic domain images were generated using the Bitter solution technique (Bitter 1932). Transmission electron microscopy in the Lorentz mode (Chapman 1984, Jakubovics 1997) as well as scanning electron microscopy (Newbury et al. 1986) are also widely used in electron-transparent and opaque samples, respectively. More recently, electron holography has proven useful in many situations where the field outside as well as inside the sample are of interest (Tonomura 1991). A modern review of various method of imaging magnetic domains is given in Hubert and Schäfer (1998).

8.5.1 Uniaxial Wall Spacing

It is possible to get an estimate of the equilibrium wall spacing (domain size d) in a uniaxial sheet of magnetic material. Assume a sample of thickness t, length L (parallel to K_u), thickness W, and wall spacing d, as in Figure 8.17.

The number of domains is W/d and the number of walls is $(W/d) - 1$. The area of a single wall is tL. The total wall energy is the wall energy density multiplied by the area of a single wall, multiplied by the number of walls:

$$F_w = \frac{\sigma_{dw} tL(W - d)}{d}$$

The wall energy per unit volume is therefore

$$f_{dw} = \frac{F_w}{\text{vol}} = \frac{\sigma_{dw}(W - d)}{Wd} \approx \frac{\sigma_{dw}}{d} \tag{8.20}$$

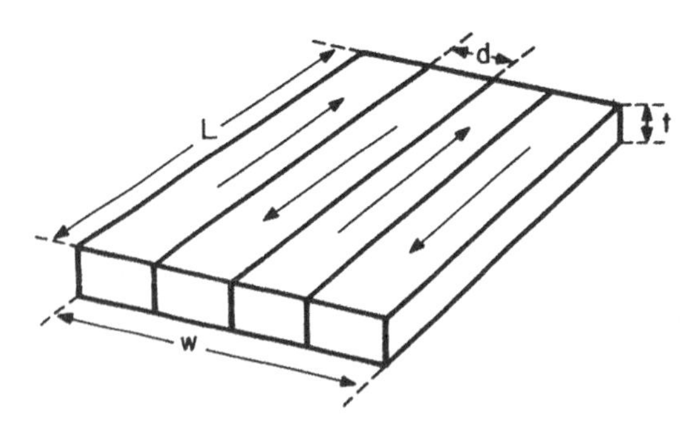

Figure 8.17 Geometry for estimation of equilibrium domain size in a thin slab of ferromagnetic material.

The demagnetization factor of each domain is not calculated as if each were isolated; their mutual magnetostatic interactions reduce the energy relative to that of independent domains. Kittel (1949) calculated the demagnetization factor for this periodic array of domains for large t to be $1.7d/L$. For small sample thickness t, Kittel's result would be reduced by a factor t/L, giving an expression for the magnetostatic energy density:

$$f_{ms} = 1.7\left(\frac{td}{L^2}\right)\mu_0 M_s^2 \tag{8.21}$$

It is important to retain the thickness dependence in the magnetostatic energy expressions. The two energies of importance, the wall energy and the magnetostatic energy, $f_{\text{wall}} + f_{\text{ms}} \approx \sigma_{dw}/d + 1.7(td/L^2)\mu_0 M_s^2$ are sketched as function of domain wall spacing in Figure 8.18.

The energy is minimized when $\partial f/\partial d = 0$. For cases where there are two terms in the free energy f that vary as d and d^{-1}, respectively, the energy minimum occurs where the two energies are equal as was seen in the derivation of domain wall energies. The equilibrium wall spacing may then be written

$$d_0 \approx L\left(\frac{\sigma_{dw}}{\mu_0 M_s^2 t}\right)^{1/2} \tag{8.22}$$

For a macroscopic magnetic ribbon described by $L = 0.01$ m, $\sigma_{dw} = 1\ \text{mJ/m}^2$, $\mu_0 M_S = 1$ T, and $t = 10\ \mu$m, the wall spacing is a little over 0.1 mm. With Eq. (8.22), the total energy density reduces to

$$f_{\text{total}} = f_{dw} + f_{ms} \approx \frac{2}{L}(1.7\sigma_{dw} t \mu_0 M_s^2)^{1/2} \tag{8.23}$$

According to Eq. (8.22), for thinner samples, the equilibrium wall spacing d_0

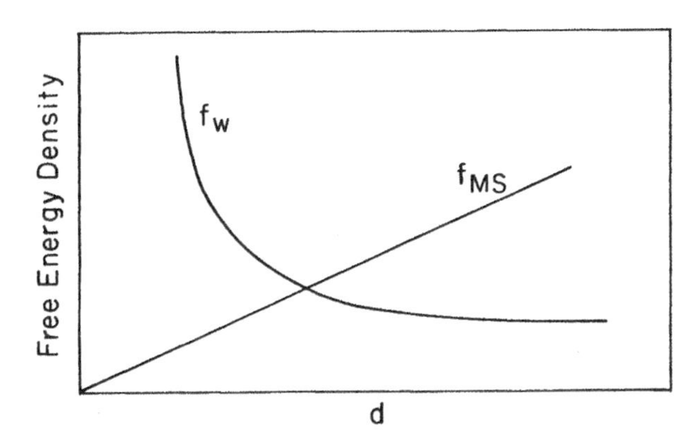

Figure 8.18 Variation of MS energy density and domain wall energy density with wall spacing d.

increases and there are fewer domains. The physical reason for this effect is that the magnetostatic energy driving domain formation has been reduced in the thin sample. Below some critical thickness, the total energy of the demagnetized state (walls present) given by Eq. (8.23) may exceed the magnetostatic energy of the single domain state [Eq. (2.22)]:

$$f_{ms}^{sd} \approx \frac{Wt}{L^2}\left[\ln\left(\frac{4L}{W+t}\right) - 1\right]\mu_0 M_s^2 \tag{8.24}$$

The forms of these two energy densities, Eqs. (8.23) and (8.24), are plotted in Figure 8.19*a*. The crossover of the two energy densities implies that there exists a critical film thickness below which the single-domain state is more stable than the multidomain state:

$$t_c \approx 4\left(\frac{L}{W}\right)^2 \cdot \frac{1.7\sigma_{dw}}{\mu_0 M_s^2} \frac{1}{\left[\ln\left(\frac{4L}{W+t}\right) - 1\right]^2} \tag{8.25}$$

Figure 8.19*b* shows the variation of this critical thickness with the length-to-width ratio of the sample for a wall energy density of 0.1 mJ/m^2 and two values of the magnetization. The singularity near $L/W = 0.68$ is an artifact of the logarithmic function that appears in the approximate demagnetizing factor.

For a film characterized by the parameters $L/W = 5$, $\sigma_{dw} \approx 0.1$ mJ/m^2, $\mu_0 M_s = 0.625$ T, typical of thin-film magnetoresistive (MR) read heads (see Chapters 15 and 17), the result is $t_c \approx 13.7$ nm. Thus, domain walls would not

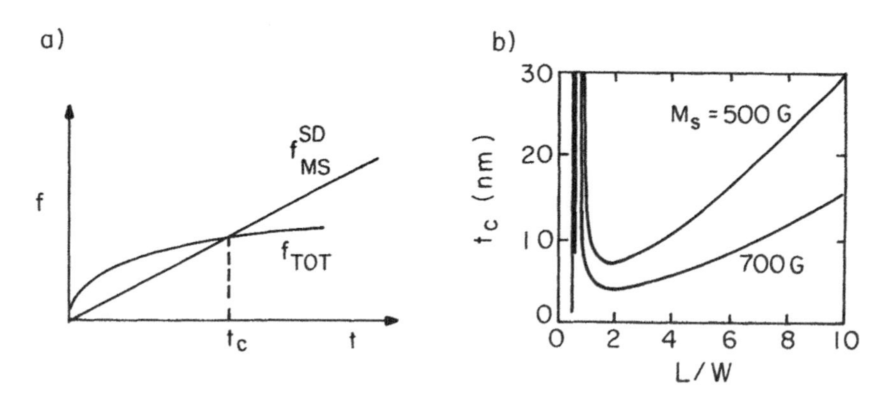

Figure 8.19 (*a*) Comparison of the thickness dependence of the free energy density for the demagnetized state f_{tot} [Eq. (8.23)] with the free energy density for the single domain state, f_{ms}^{sd} [Eq. (8.24)]. Note the crossover below which the energy is lower for the single-domain state. (*b*) Variation of the criticdal thickness with the ratio L/W for two different values of magnetization and $\sigma_{dw} = 0.1$ mJ/m^2.

be expected in such a film if it were less than about 12 nm thick. As the film gets shorter in the direction of magnetization, the multidomain state is stable to smaller thicknesses. Decreasing M_s, increasing the L/W ratio, and /or increasing wall energy density (higher K) would raise t_c.

It must be borne in mind that this model is based on the assumption that the demagnetizing field in the sample is uniform. In fact, the demagnetizing field is a function of position inside the sample and therefore domains are likely to form near charged surfaces even for film thicknesses below the t_c calculated here.

8.5.2 Closure Domains

The formation of closure domains at the ends of structures like that depicted in Figure 8.17 is now considered. Uniaxial and cubic anisotropy cases are considered separately (Kittel 1949).

The 180° domain wall energy per unit volume is, from Eq. (8.20), $f_{dw} \approx \sigma_{dw}/d$. The formation of the closure domains (Fig. 8.20) reduces the 180° wall length by a fraction d/L and replaces that length by a 90° wall of length, $2\sqrt{2}d/L$. Considering $\sigma_{90} = \sigma_{dw}/2$, the wall energy f_{dw} therefore increases by the factor $(1 + 0.41d/L)$: $\Delta f_{dw} \approx 0.41\sigma_{dw}/L$ The magnetostatic energy essentially reduces to zero, so from Eq. (8.21) $\Delta f_{ms} \approx -1.7(td/L^2)\mu_0 M_s^2$. A contribution to the anisotropy energy in a *uniaxial material* results from formation of the $2W/d$ closure domains, each of volume $d^2t/2$: $\Delta f_K \approx K_u d/L$. Hence the energy change that determines whether the closure domains will form is

$$\Delta f_{\text{tot}} \approx 0.41\frac{\sigma_{dw}}{L} + \frac{K_u d}{L} - 1.7\frac{td}{L^2}\mu_0 M_s^2 \tag{8.26}$$

If this energy change is negative, the formation of closure domains is possible (see Fig. 8.21).

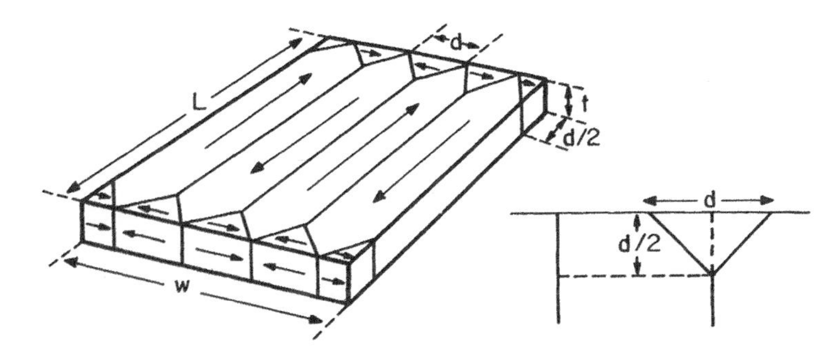

Figure 8.20 Geometry for estimation of equilibrium closure domain size in a thin slab of ferromagnetic material (cf. Fig. 8.18).

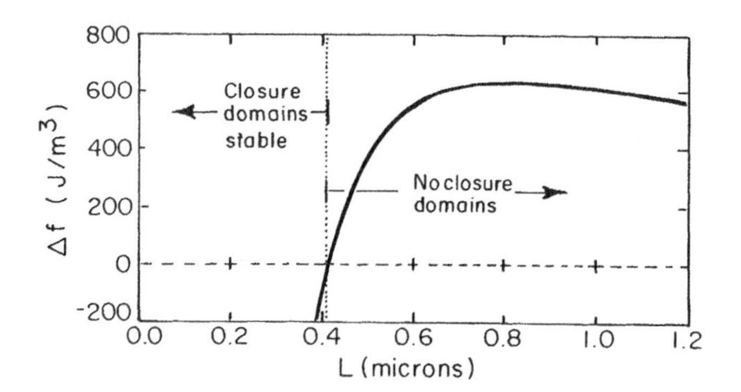

Figure 8.21 Energy density of Eq. (8.26) versus sample length L for $\mu_0 M = 0.625\,\mathrm{T}$, $\sigma = 0.1\,\mathrm{mJ/m^2}$, $K_u d = 1\,\mathrm{mJ/m^2}$, and $td = 10^{-14}\,\mathrm{m^2}$.

For the same parameters as used in Figure 8.21 except with the product td increased by four orders of magnitude, closure domains are stable below about 4 mm. Closure domain formation is favored by large magnetization, large sample thickness, small anisotropy, and small wall energy. In particular, their stability depends strongly on the ratio of td/L^2.

For L/d or $L/t \gg 1$, the spacing of the closure domains is dictated by the spacing of the larger domains. For smaller aspect ratios the energy of the closure domains becomes comparable to that of the interior domain structure and the two must be considered together in estimating the equilibrium domain spacing. If the energy balance favors closure domain formation, then the magnetostatic energy vanishes and the remaining energy terms are

$$f_{\mathrm{tot}} \approx \left(1 + 0.41\frac{d}{L}\right)\frac{\sigma_{dw}}{d} + \frac{K_u d}{L} \tag{8.27}$$

When it is energetically favorable for closure domains to form, energy minimization gives for the equilibrium wall spacing:

$$d_0^{\mathrm{clos}} \approx \left(\frac{\sigma_{dw} L}{K_u}\right)^{1/2} \tag{8.28}$$

Closure domains form when the anisotropy energy is small relative to the magnetostatic energy [cf. Eq. (8.26)]. Comparison of Eq. (8.28) with Eq. (8.22) shows that when closure domains form, the equilibrium wall spacing is allowed to be larger than it would be without closure domains: $d_0^{\mathrm{clos}}/d_0 = (\mu_0 M_s^2 t/LK_a)^{1/2}$.

For *cubic crystal anisotropy*, the formation of closure domains does not cost anisotropy energy, but it may still cost additional magnetostatic energy as it does in the uniaxial case. The closure domains strain along an axis orthogonal to the strain in the major stripe domains so there is a buildup of elastic energy, $\frac{1}{2}c_{11}e_{11}^2 = \frac{1}{2}c_{11}\lambda_{100}^2$ at the 90° domain wall (see Chapter 7). Thus, the term $(d/L)\frac{1}{2}c_{11}\lambda_{100}^2$ replaces $K_u d/L$ in the cubic case (or adds to it in the uniaxial case) in Eqs. (8.26) and (8.27). This magnetostatic term is of order 10^2 (d/L) J/m^3 for iron and 5 times larger for Ni; the magnetostatic term is of order 10^3 (d/L) J/m^3. This energy is very small on the scale of Figure 8.21. Hence, closure domain formation is still favored because of the dominance of the magnetostatic energy. Figure 1.13 is a good illustration of closure domains inside an Fe single crystal (Celotta et al., 1991).

8.6 DOMAINS IN FINE PARTICLES

Having calculated the critical thickness for single domain films [Eq. (8.25)], we are now interested in performing a similar calculation for a spherical particle. Fine particles are used in magnetic bearings, permanent magnets, flexible magnetic shields, and in magnetic recording media. In applications for which high permeability is desired, such as a flexible magnetic shield, it could be argued that domain walls should be present in the magnetic particles. On the other hand, it will be seen that particles used in permanent magnets should not have domain walls that can allow them to be easily demagnetized. Also, the presence of domain walls in particulate recording media lowers the coercivity and is a source of noise. Thus, it is important to know the size below which a particle is comprised of a single domain.

To a first approximation, it might be assumed that the critical particle diameter would be comparable to the domain wall width; that is, in a particle of diameter $d < \delta_{dw} = \pi(A/K)^{1/2}$ there can be no domain wall present. But such an approach does not take account of the magnetostatic energy that drives domain formation. Taking a more analytic approach for small particles, an energy balance is considered. For the single domain state to be stable, the energy needed to create a domain wall spanning a spherical particle of radius r, namely, $\sigma_{dw}\pi r^2 = 4\pi r^2(AK)^{1/2}$, must exceed the magnetostatic energy saved by reducing the single domain state to a multidomain state, $\Delta E_{MS} \approx \frac{1}{3}\mu_0 M_s^2 V = \frac{4}{9}\mu_0 M_s^2 \pi r^3$. (Here it has been assumed that the magnetostatic energy of the spherical particle composed of two equal domains is negligible relative to that of the single-domain state.) The critical radius of the sphere would be that which makes these two energies equal:

$$r_c \approx 9\frac{(AK_u)^{1/2}}{\mu_0 M_s^2} \quad \text{(large } K_u\text{)} \tag{8.29}$$

For iron, this method gives $r_c \approx 3$ nm; for γ-Fe_2O_3, $r_c \approx 30$ nm. These values are considerably smaller than the respective domain wall widths. This model assumes that the domain wall in a particle has the same structure as that in an infinite material (Fig. 8.22, left). This is an acceptable approximation if the anisotropy is strong enough to maintain the orientation of M_s along the easy axis despite surface poles, namely, $K_u \geqslant \mu_0 M_s^2/6$. If, on the other hand, the anisotropy is not that strong, the magnetization orientation will tend to follow the particle surface (Fig. 8.22, right). Equation (8.29) ignores the significant cost in exchange energy needed to confine the domain wall in this way, that is, in three dimensions rather than one. The three-dimensional confinement of the magnetization twist increases the exchange energy contribution considerably and Eq. (8.29) is not applicable.

The exchange energy cost should be compared with the magnetostatic energy that is saved by domain wall formation. The exchange energy density at the radius r in such a particle can be written by noting that the spins rotate by 2π radians over that radius:

$$f_{ex}(r) = A\left(\frac{\partial\theta}{\partial x}\right)^2 = A\left(\frac{2\pi}{2\pi r}\right)^2 = \frac{A}{r^2} \tag{8.30}$$

The exchange energy density can be determined over the volume of a sphere by breaking the sphere into cylinders of radius r, each of which has spins with the same projection on the axis of symmetry (Fig. 8.23).

Integrating the exchange energy density over the spherical volume in cylindrical coordinates gives

$$\langle f_{ex}\rangle = \int f_{ex}(r)r\, dr h\, d\phi = 4\pi\int_0^R \frac{\sqrt{R^2 - r^2}}{r}\, dr$$

Here, $h = 2(R^2 - r^2)^{1/2}$ has been used. Carrying out the integral with the exclusion of the singularity of radius a at the axis of symmetry, gives

$$\langle f_{ex}\rangle = \frac{3A}{R^2}\left[\ln\left(\frac{2R}{a}\right) - 1\right] \tag{8.31}$$

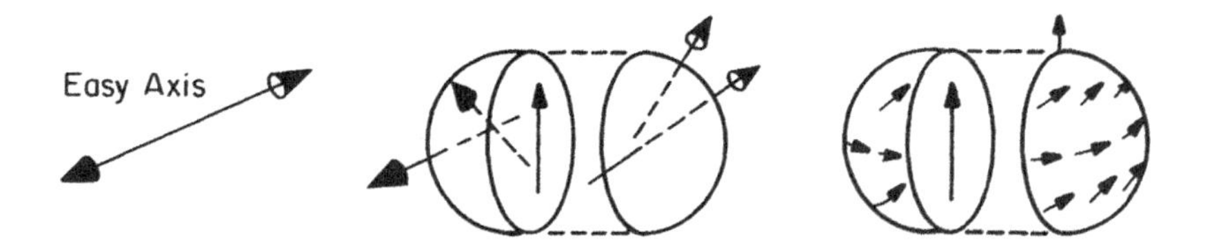

Figure 8.22 Left, model of a small ferromagnetic particle in which a domain wall similar to that in bulk material intersects the middle of the particle. Right, additional exchange energy is involved if the magnetization conforms to the surface of the particle.

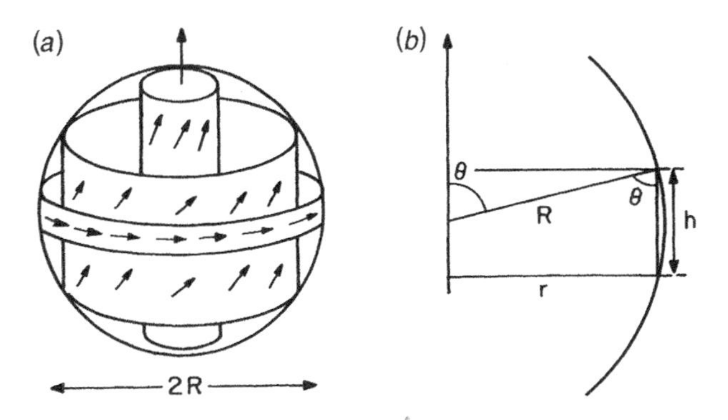

Figure 8.23 Construction for calculating the exchange energy of a particle demagnetized by curling as in Figure 8.22, right.

(This form of solution may be familiar as the continuum solution for the energy of an edge dislocation where the core singularity has been removed.) The radius of the core singularity is chosen to be the lattice constant a. The magnetostatic energy of the core magnetized along the $+z$ direction is negligible.

If this exchange energy density cost is equated to the magnetostatic energy density for a uniformly magnetized sphere, $\frac{1}{3}\mu_0 M_s^2$ (which is assumed to be saved by allowing the magnetization to curl, remaining parallel to the surface, as in Figure 8.22, right), the critical radius for single-domain spherical particles results:

$$r_c = \sqrt{\frac{9A}{\mu_0 M_s^2}\left[\ln\left(\frac{2r_c}{a}\right) - 1\right]} \quad \text{(small } K_u\text{)} \tag{8.32}$$

Equation (8.32) has the form of an exchange length but with an additional cylindrical factor that is present because of the singularity along the axis. Because Eq. (8.32) contains the critical radius on both sides of the equation, it can be graphed by solving for $\mu_0 M_s^2$ and plotting $\mu_0 M_s$ as a function of r_c. Figure 8.24 shows the values of r_c for various values of saturation magnetization according to the strong anisotropy case, Eq. (8.29) (assuming $K_u = 10^6\ \text{J/m}^3$), and the weak anisotropy case, Eq. (8.32). The critical radius for single-domain Fe particles (small anisotropy) is found to be much larger than that based on Eq. (8.29) (25 nm instead of 3 nm).

For an *acicular* iron particle of the *same volume* as a critical sphere, there is less energy to be gained by introducing a domain wall because a larger aspect ratio reduces the magnetostatic energy. Thus the critical volume for acicular particles is greater than that for a spherical particle of the same volume and material. For platelike particles with in-plane anisotropy (e.g., grains in a thin

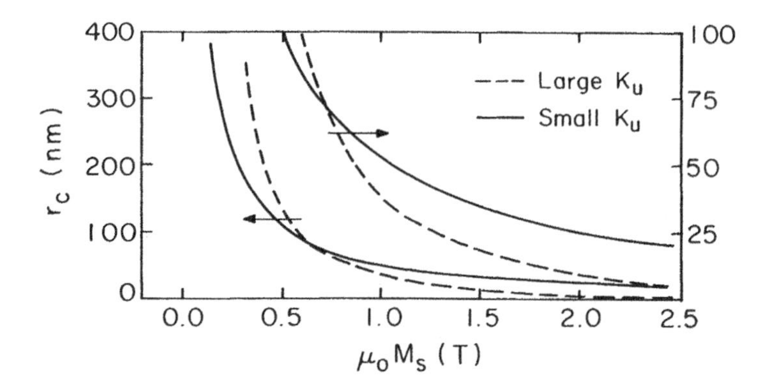

Figure 8.24 Critical radius for single-domain behavior versus saturation magnetization for spherical particles based on Eqs. (8.29) (large K_u, 10^6 J/m^3) and 8.31 (small K_u). Values of r_c for saturation magnetization greater than 0.5 Tesla are also shown magnified by a factor of 4 (right scale).

film) the critical grain diameter is also larger than that for a sphere of the same volume. However, for a platelike particle with perpendicular anisotropy, the magnetostatic energy is quite large and walls form more easily.

For more realistic calculations of single-domain particle radii, micromagnetics must be employed. The formulas shown here reveal the important factors governing this behavior and provide reasonable estimates of the magnitude of the critical radii.

In Chapter 9, the magnetization reversal process will be considered in a number of magnetic cases including these single domain particles.

Superparamagnetism One cannot reduce the size of magnetic particles indefinitely and still retain useful magnetic properties. Below a certain size, the remanent magnetization is no longer fixed in the direction dictated by particle shape or crystal anisotropy; ambient thermal energy may be large enough to cause the moment to jump between two different stable orientations of magnetization. This is a magnetic analog of Brownian motion; thermal molecular motions are random and cancel in a large system or for one particle over time: velocity $\langle v \rangle_{x \text{ or } t} = 0$ and magnetization $\langle \mathbf{M} \rangle_{x \text{ or } t} = 0$. However, on a local scale and for a short observation time, it is possible to see the effects of molecular motion, $\mathbf{v} \neq 0$, and on the scale of a few magnetic particles in short times, $\mathbf{M} \neq 0$. Typically, magnetic particles become superparamagnetic below a radius of order 20 nm. The interesting range of particle sizes for magnetic properties is above the superparamagnetic limit and below the single-domain critical radius. Let us analyze the behavior of such particles.

When a system is in metastable equilibrium, such as a magnetic system at remanence, the probability P per unit time that it will switch out of the metastable state and into the more stable demagnetized state, is given by

$$P = \nu_0 \exp\left(-\frac{\Delta f V}{k_B T}\right) \tag{8.33}$$

where ν_0 is an attempt frequency factor equal to approximately $10^9\,\mathrm{s}^{-1}$, V is the sample volume, and $\Delta f V$ is the free energy barrier that the particle moment must surmount to leave the metastable state. In a particle with strong shape or crystalline anisotropy, Δf is equal to $\Delta N \mu_0 M_s^2$ or K_u, respectively. For sufficiently small volume particles, the magnetization may leave its metastable remanent state in laboratory timescales; the remanence can be observed to decrease measurably. For still smaller particles, demagnetization occurs as quickly as the field is turned off. These particles are ferromagnetic below their Curie temperature in the sense that they have a spontaneous magnetization (given by $N'\mu_m/V$, where N' is the number of magnetic atoms in the particle). The magnetization is essentially uniform over the particle volume at any instant. However, the time-averaged magnetization appears to be zero. Application of an external magnetic field to an ensemble of such thermally demagnetized particles results in a much larger magnetic response than would be the case for a paramagnet [Eq. (3.37)] because now each local moment has magnitude $N'\mu_m$ instead of μ_m. Thus the susceptibility of a superparamagnet is increased $(N')^2$-fold. The M–H curves of superparamagnets can resemble those of ferromagnets but with two distinguishing features: (1) the approach to saturation follows a Langevin behavior and (2) there is no coercivity. Superparamagnetic demagnetization occurs without coercivity because it is not the result of the action of an applied field but rather of thermal energy.

It is important to know the minimum particle or gain size, which is magnetically stable against ambient thermal demagnetization. From the inverse of Eq. (8.33), and for a spherical particle with $K_u = 10^5\,\mathrm{J/m^3}$, the superparamagnetic radii for stability over 1 year and 1 second, respectively, are

$$r_0^{1\mathrm{yr}} \approx \left(\frac{10 k_B T}{K_u}\right)^{1/3} \approx 7.3\,\mathrm{nm}, \qquad r_0^{1\mathrm{s}} \approx \left(\frac{6 k_B T}{K_u}\right)^{1/3} \approx 6\,\mathrm{nm} \tag{8.34}$$

Stability has been defined as a flipping probability P of less than 10% over the time interval specified. Note that these values are smaller than the critical radius for single-domain behavior for small-K_u, spherical particles of magnetization less than 2.5 T (Fig. 8.24). The superparamagnetic radius is decreased in particles with stronger anisotropy. It is also important to know for a given particle size what anisotropy is required for less than 10% of the particles to switch in a given time. For one year and one second stability, a particle or radius 10 nm needs an anisotropy of 4×10^4 or $2.3 \times 10^4\,\mathrm{J/m^2}$, respectively. Below the superparamagnetic limit a particle has no memory of its remanent state after the time specified and no coercivity.

8.7 DOMAINS IN POLYCRYSTALLINE MATERIALS

Everything stated in Sections 8.4 and 8.5 applies to single crystals and single-crystal particles. The only interfaces present are the sample boundaries. In polycrystalline materials, internal surfaces (grain boundaries) exist and different grains can have easy axes with different orientations.

The factors affecting the magnetization distribution and domain structure include

1. Magnetostatics—the system avoids charged surfaces and maintains $(\boldsymbol{B}_i - \boldsymbol{B}_0) \cdot \boldsymbol{n} = 0$.
2. The magnetization $\boldsymbol{M}$ should therefore follow the crystal and shape anisotropy easy axes subject to internal fields due to charged surfaces.
3. Exchange coupling will tend to maintain the direction of $\boldsymbol{M}$ across narrow, clean grain boundaries but not across wide, contaminated grain boundaries.

Figure 8.25*a* (Shilling and Houze 1974) shows the domain structure near a grain boundary that arcs from lower left toward upper right of the figure. The magnetization in each domain is parallel to the long domain boundaries. Maxwell's equations lead to boundary conditions [Eq. (2.6)] that demand the normal component of $\boldsymbol{B} = \mu_0(\boldsymbol{H} + \boldsymbol{M})$ must be continuous across the grain boundary. Where the normal component of $\boldsymbol{M}$ is not continuous across the grain boundary, the boundary condition demands there be a field, $\boldsymbol{H}$, to maintain $\boldsymbol{B} \cdot \boldsymbol{n}$ continuous. This $\boldsymbol{H}$ field causes formation of spike domains, reducing the magnetostatic energy near the grain boundary. The effect is most pronounced in the domain whose magnetization axis has the greatest component along the field of the charged wall.

8.8 SUMMARY

At least two energies, *exchange* and *anisotropy*, are involved in determining the domain wall thickness and the *domain wall energy* density. The widths of 180° Bloch walls range from about 1 μm in low-anisotropy materials to under 10 nm in high-anisotropy materials. Where a domain wall approaches a surface, *magnetostatic* energy must be considered. Magnetostatic energy is responsible for formation of Néel and cross-tie walls.

In cubic materials where 90° walls are possible, *magnetoelastic* energy stabilizes 180° walls relative to 90° walls. Also, magnetostatic energy destabilizes 90° closure domains.

A variety of length and energy scales related to domain walls and domains have been discussed. Some of them are summarized here. Such equations are not to be memorized. Their forms, and the patterns among them, should be

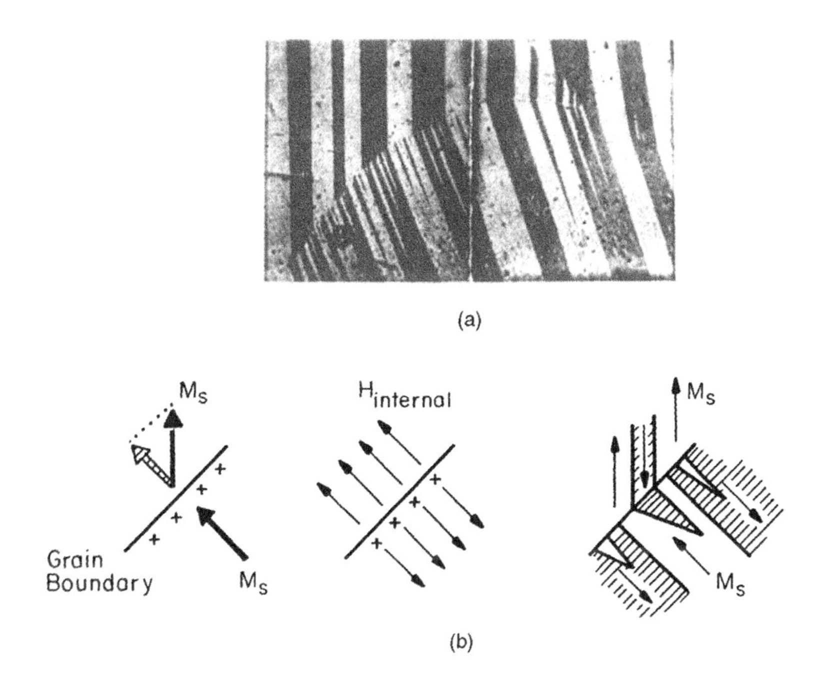

Figure 8.25 (*a*) Domains near a grain boundary in polycrystalline Fe–3%Si [after Shilling and Houze (1974)]; (*b*) process by which an imbalance in the normal component of magnetization across a boundary generates an internal field, which, in turn, favors the formation of reversal domains at the boundary.

understood. For example, note that the length scales always vary as an inverse power of either the magnetostatic or magnetic anisotropy energy density. These two energies are responsible for the magnetization departing from a state of uniform orientation throughout the sample.

For a much more extensive examination of the rich variety of domain structures found in magnetic materials, the reader is referred to the recent monograph by Hubert and Schäfer (1998).

	Length Scale	Energy Density
180° Bloch wall	$\delta_{\text{dw}} = \pi\left(\dfrac{A}{K_u}\right)^{1/2}$	$\sigma_{\text{dw}} = 4(AK_u)^{1/2}$
Néel wall	$\delta_N = \pi\left(\dfrac{2A}{\mu_0 M_s^2}\right)^{1/2}$	$\sigma_N = 2\pi(AK_u)^{1/2}$
Uniaxial domain wall spacing	$d_0 \approx L\left(\dfrac{\sigma_{\text{dw}}}{\mu_0 M_s^2 t}\right)^{1/2}$	$f_{\text{total}} \approx \dfrac{2}{L}(1.7\sigma_{\text{dw}} t \mu_0 M_s^2)^{1/2}$

	Length Scale	Energy Density
Critical single-domain film thickness	$t_c \approx 4\left(\frac{L}{W}\right)^2 \frac{1.7\sigma_{\text{dw}}}{\mu_0 M_s^2} \frac{1}{[\ln(4L/W+t)-1]^2}$	
Closure domain spacing	$d_0^{\text{clos}} \approx \left(\frac{\sigma_{\text{dw}} L}{K_u}\right)^{1/2}$	
Single-domain size for sphere		
(Large K_u)	$r_c \approx 9\frac{(AK_u)^{1/2}}{\mu_0 M_s^2}$	
(Small K_u)	$r_c = \sqrt{\frac{9A}{\mu_0 M_s^2}[\ln\left(\frac{2r_c}{a}\right) - 1}$	
Superparamagnetic radius for sphere (1 year)	$r_0^{1\text{yr}} \approx \left(\frac{9k_B T}{K_u}\right)^{1/3}$	

PROBLEMS

8.1 The figures below depict two orientations of domain walls in a uniaxial material. Explain why each is unlikely.

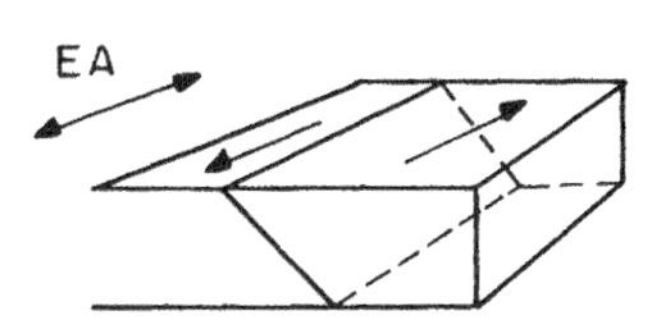

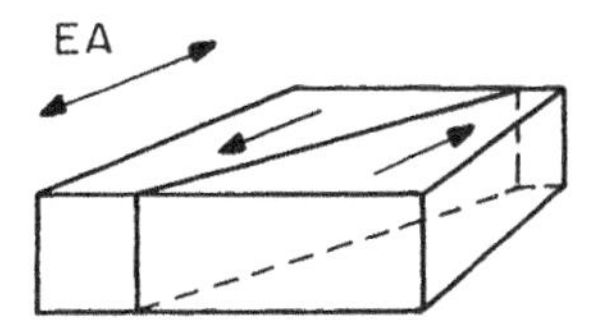

Figure 8.1P

8.2 Calculate the domain wall thickness and energy density for Co, $Nd_2Fe_{14}B$ and $Ni_{80}Fe_{20}$ from data in Chapter 6 assuming in each case $A = 2 \times 10^{-11}$ J/m.

8.3 Solve the uniaxial domain wall problem with magnetoelastic energy included. Work out the solution for three cases: (**a**) uniaxial strain along the easy axis, (**b**) perpendicular to the EA and parallel to the magnetization at the center of the wall, and (**c**) perpendicular to the EA and to the magnetization at the center of the wall.

8.4 Evaluate the constant in Eq. (8.10) for uniaxial anisotropy for the two wall profiles: $\theta(-\infty) = 0$ and $-\pi/2$ (in both cases $\theta' = 0$). Can you find a solution to the integral equation for the second set of boundary conditions?

8.5 Write the exchange, anisotropy, and magnetostatic energy expressions for a Néel wall in a film of thickness t. Compare the relative magnitude of these terms. Why does the magnetostatic energy not appear in the minimization of the wall energy per unit area to determine the wall width?

8.6 Relate the energy of the flux distribution about a sample in a state of remanent magnetization ($H_{ext} = 0$) to that of the configuration of least magnetostatic energy inside the sample. Use the i^2R power density analogy. How is this energy minimized?

8.7 **(a)** Calculate the radius r_c of a single-domain sphere for which the magnetostatic energy is exacly equal to the energy needed to create a 180° domain wall through the middle of the sphere.

(b) Calculate this radius for iron.

(c) Compare this radius with the domain wall thickness δ_{dw} for iron.

(d) Express r_c and δ_{dw} in terms of the domain wall energy density σ_{dw} and discuss the difference between these two expressions.

8.8 Calculate the superparamagnetic dimensions for an acicular iron particle with a 10:1 aspect ratio and anisotropy due only to its shape.

BIBLIOGRAPHY

Kittel, C., *Phys. Rev.* **70**, 965 (1946).

Lifshitz, E., *J. Phys. USSR* **8**, 337 (1944).

Miltat, J., in *Applied Magnetism*, R. Gerber, C. D. Wright, and G. Asti, eds., Kluwer, Boston, 1994, p. 221.

Richter, H. J., *IEEE Trans.* **MAG-29**, 2185 (1993).

Stoner, E. C., and E. P. Wohlfarth, *Trans. Roy. Soc.* **A240**, 599 (1948).

REFERENCES

Aharoni, A., and J. P. Jakubovics, *Phys. Rev. B* **43**, 1290 (1991).

Bitter, F., *Phys. Rev.* **41**, 507 (1932).

Bloch, F., *Z. Phys.* **74**, 295 (1932).

Bochi, G., H.-J. Hug, B. Steifel, A. Moser, I. Paroshikov, H.-J. Güntherodt, and R. C. O'Handley, *Phys. Rev. Lett.* **75**, 1839 (1995).

Brown, W. F., *Micromagnetics*, Kreiger, Huntington, NY, 1978.

Celotta, R. J., J. Unguris, D. T. Pierce, and M. R. Scheinfien, personal communication (1991).

Chapman, J. N., *J. Phys. D., Appl. Phy.* **17**, 623 (1984).

Craik, D. J., and R. S. Tebble, *Ferromagnetism and Ferromagnetic Domains* (North Holland, Amsterdam, 1965).

Hubert, A., *Phys. stat. solidi* **32**, 519 (1969); **38**, 699 (1970).

Hubert, A., and R. Schäfer, *Magnetic Domains*, Springer-Verlag, Berlin, 1998, p. 297.

Jakubovics, J. P., in *Handbook of Microscopy*, Vol. 1, S. Amelinckx, ed., VCH Press, Weinheim, 1997, p. 505.

Kittel, C., *Rev. Mod. Phys.* **21**, 541 (1949); Kittel, C., and J. K. Galt, *Solid State Phys.* **3**, 437 (1956).

LaBonte, A. E., *J. Appl. Phys.* **40**, 2450 (1969).

Landau, L. D., and E. Lifshitz, *Phys. Z. Soweit.* **8**, 153 (1935).

McGuire, Y., unpublished calculations of Néel wall parameters.

Newbury, D. E., D. C. Joy, P. Echlin, C. E. Fiori, and J. I. Goldstein, in *Advanced Scanning Electron Microscopy and X-Ray Microanalysis*, Plenum Press, New York, 1986, p. 147.

Oepen, H. P., and J. Kirschner, *Phys. Rev. Lett.* **62**, 819 (1989).

Scheinfein, M. R., J. Unguris, R. J. Celotta, and D. T. Pierce, *Phys. Rev. Lett.* **63**, 668 (1989); *Phys. Rev. B* **43**, 3395 (1991).

Shilling, J. W., and G. L. Houze, Jr., *IEEE Trans.* **MAG-10**, 195 (1974).

Tonomura, A., in *Physics and Engineering Applications of Magnetism*, Y. Ishikawa and N. Miura, eds., Springer-Verlag, Berlin, 1991, p. 282.

CHAPTER 9

MAGNETIZATION PROCESS

This chapter discusses the response of materials to applied fields, namely, the magnetization process. What does an M–H curve look like, and why? What does an M–H loop tell us about a material? The energies outlined in Chapters 2, 6, 7, and 8 are used to model and understand the process by which a material is magnetized. The simplest model—uniform magnetization rotation in a uniaxial material—serves as a starting point. This provides a clear picture of the difference between M–H curves for magnetization rotation and those for domain wall motion. More complicated examples of the magnetization process and two mechanisms of coercivity are then considered. Finally, AC processes are analyzed in order to get an appreciation of eddy current losses.

9.1 UNIAXIAL MAGNETIZATION

The M–H curves are reviewed for the canonical case of uniaxial anisotropy with field applied along either the hard or easy axis. Two limiting situations are considered: (1) single-domain particles (or equivalently, completely pinned domain walls) and (2) samples in which domain walls are present and move with complete freedom in the weakest field.

Consider a magnetic material with uniaxial anisotropy of any origin (magnetostatic, magnetocrystalline, magnetoelastic, or field-induced). If any domain walls are present in the sample, they are assumed to be parallel to the easy axis (EA). A field is applied first transverse to the easy direction of

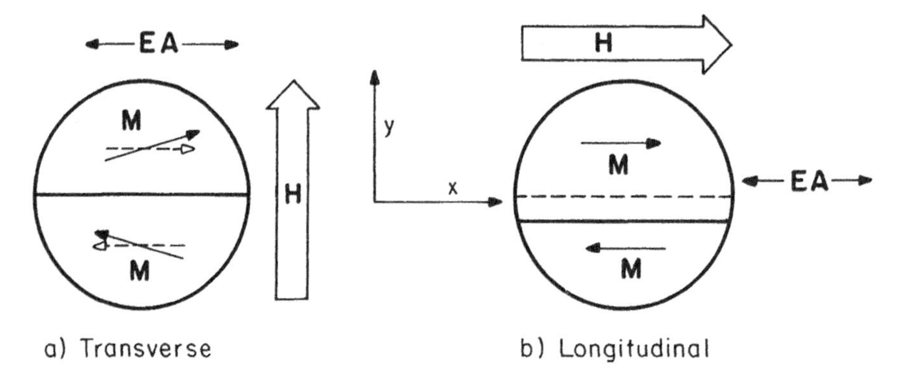

Figure 9.1. Schematic representations of a magnetic material having purely uniaxial anisotropy in the direction of the easy axis (EA). Dashed lines indicate magnetization configurations for $H = 0$. Application of a field H transverse to the EA results in rotation of the domain magnetizations but no wall motion. Application of a field parallel to the EA results in wall motion but no rotation of the domain magnetization.

magnetization (Fig. 9.1*a*), then parallel (Fig. 9.1*b*) to it. In each case, the M–H loops are derived.

9.1.1 Hard-Axis Magnetization

The energy density describing the situation for which H is perpendicular to the easy axis may be written as

$$f = K_u \cos^2\theta - M_s H \cos\theta, \qquad K_u > 0 \tag{9.1}$$

where θ measures the angle between $\boldsymbol{M}$ and the field in Figure 9.1*a*. Figure 6.6*c* shows what the anisotropy energy surface looks like with an arbitrary constant added: f_a has a minimum along an easy axis. (Draw the energy surface for $K_u < 0$ to see the easy plane in that case.)

It is helpful to sketch a cut through the total energy surface (including the Zeeman term) before seeking a mathematical solution to a magnetization problem. The shape of the energy surface gives a good qualitative picture of the solutions and serves as a guide to the appropriate quantitative expressions. The magnetization will always point in a direction of minimum energy (at least locally). The first angular derivative of the energy is the negative of torque $-\partial f/\partial\theta = T_\theta$ on $\boldsymbol{M}$ and it is zero at energy maxima and minima.

In Figure 2.12, the energy density of Eq. (9.1) is sketched for $H = 0$ and for a few values of $H > 0$. (In that figure, the uniaxial energy was of magnetostatic origin. The *angular dependence* of the uniaxial anisotropy energy density here is the same as that of the magnetostatic energy density there.) Note the zero-slope points and the stable points. It can be seen from Figure 2.12 that at

$H = 0$, the stable solution is $\theta = \pi/2$. As H increases, this stable point moves to smaller θ, indicating that the magnetization vector is aligning with the applied field. This result is derived quantitatively.

The energy density from Eq. (9.1) is minimized with respect to θ when its first derivative vanishes

$$(-2K_u \cos\theta + M_s H)\sin\theta = 0 \tag{9.2}$$

and the second derivative is positive:

$$-2K_u \cos 2\theta + M_s H \cos\theta > 0 \tag{9.3}$$

Equation (9.2) is the condition for zero torque on $\boldsymbol{M}$, and Eq. (9.3) is the stability condition. The zero-torque condition has two solutions. One of them is given by $\sin\theta = 0$:

$$\theta = 0, \pi \ldots$$

which indicates that the saturated states, $\boldsymbol{M}$ parallel or antiparallel to $\boldsymbol{H}$, are extrema. The stability condition demands that $\theta = 0$ is stable only for $H > 2K_u/M_s$ $(K_u > 0)$; $\theta = \pi$ is stable only for $H < -2K_u/M_s$ $(K_u > 0)$. This is most easily seen from Figure 2.12. The other solution to Eq. (9.2a) is given by

$$2K_u \cos\theta = M_s H \tag{9.4}$$

This is the equation of motion for the magnetization in fields below saturation $2K_u/M_s < H < 2K_u/M_s$. Note that for $H = 0$, the solution must be $\theta = \pi/2$.

The field dependence can be made clearer with a few simple substitutions. The magnetization measured in the field direction is $M = M_s \cos\theta$. When the applied field is great enough to saturate the magnetization, $\cos\theta = 1$, a definition of the anisotropy field, $H_a = H_{\text{sat}}$ results:

$$K_u = \frac{M_s H_a}{2} \tag{9.5}$$

Thus, Eq. (9.4) may be written

$$H_a M_s \cos\theta = M_s H \tag{9.6}$$

Using $\cos\theta = m = M/M_s$ for the component of magnetization in the field direction, and defining the reduced field $h = H/H_a$, Eq. (9.6) gives

$$m = h \tag{9.7}$$

for $|h| \leqslant 1$. This is the general equation for the magnetization process with the

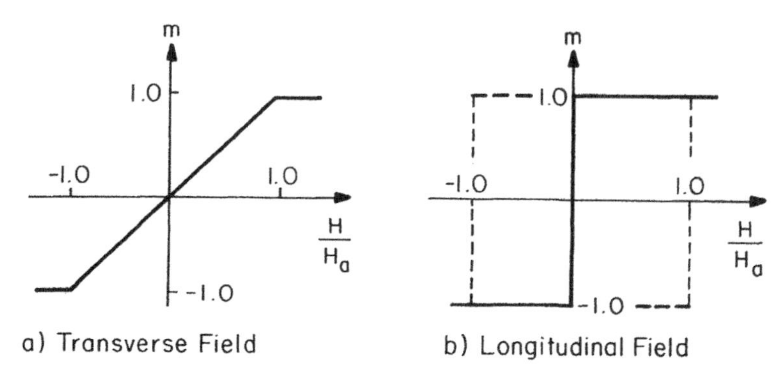

Figure 9.2. M–H loops for the two idealized cases shown in Figure 9.1: (*a*) hard-axis and (*b*) easy-axis magnetization processes.

field applied in a hard direction for a uniaxial material. The reduced magnetization m increases linearly with field up to the anisotropy field H_a, at which point $m = 1$, i.e. the material is saturated (Figure 9.2*a*).

Even if unpinned domain walls parallel to the EA are present in this hard-axis case, they do not move because there is no energy difference across the domain wall.

9.1.2 Easy-Axis Magnetization

In the uniaxial, parallel-field case (Fig. 9.1*b*), both domains have their magnetization along the easy axis so their anisotropy energies are the same. An easy-axis field exerts no torque on the domain magnetization in this case. The energy density of each domain is due solely to the applied field, M_sH and $-M_sH$, unless the direction of magnetization changes. For now, the domain wall is assumed to be pinned, or a single-domain sample is considered. To express the free energy in domain *i*, its possible angular dependence is included:

$$f_i = -M_sH\cos\theta_i + K_u\sin^2\theta_i \tag{9.8}$$

This energy density is plotted in Figure 9.3 for various values of the reduced field, $h = M_sH/2K_u$. The solutions to this problem are clear by examining the field dependence of $f(\theta)$. At zero field, stable solutions exist at both $\theta = 0$ and π; this accounts for the stability of the two types of domain shown in Figure 9.1*b*. With increasing field, these two solutions remain *locally* stable while $\theta = 0$ becomes more favored. The angular positions of the energy minima are independent of field for $h < 1$. For $h = 1$, the local stability of the solution at $\theta = \pi$ vanishes and the magnetization in those domains switches abruptly to $\theta = 0$. This m–h behavior is sketched in Figure 9.2*b*. The field at which the m–h curve crosses $m = 0$ here is called the *switching field*. It is due to rotational

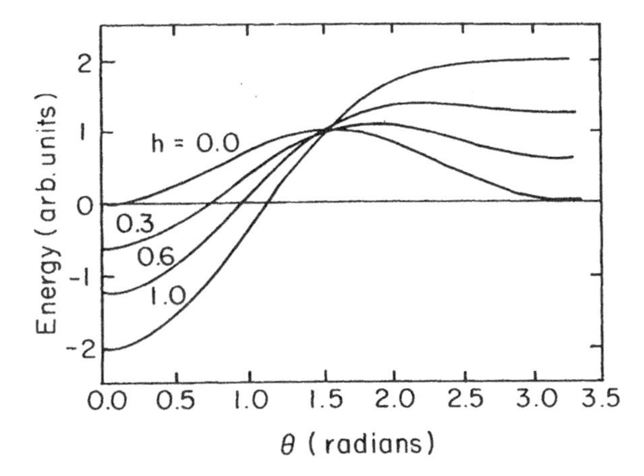

Figure 9.3. Free energy for the easy-axis magnetization process as a function of angle and applied field strength, $h = M_sH/2K_u$.

hysteresis, not domain wall motion hysteresis (the cause of coercivity), which will be examined later.

It is important to see how these solutions follow rigorously from analysis of the energy density. The zero-torque and stability conditions for the free energy in Eq. (9.8) are given, respectively, by

$$\sin\theta(M_sH + K_u \sin 2\theta) = 0 \tag{9.9}$$

$$M_sH\cos\theta + 2K_u\cos^2\theta > 0 \tag{9.10}$$

The first factor in Eq. (9.9) indicates that solutions may be found at $\theta = 0$ and π; namely, $m = 1$ and -1, respectively. Equation (9.10) indicates that $\theta = 0$ is stable only for $H > -2K_u/M_s$ while $\theta = \pi$ is stable only for $H < 2K_u/M_s$. From Figure 9.3, these solutions are only *locally* stable in the field range between $\pm 2K_u/M_s$. The stable and locally stable solutions are shown in Figure 9.2b as the solid and dashed lines, respectively. The second factor in Eq. (9.9) gives the same information as the first.

The situation is now considered in which the domain wall in Figure 9.1*b* moves easily. The field exerts no torque on the domain magnetization, but it does exert a torque on the spins making up the wall. The spins in the wall may rotate to align with H. In macroscopic terms, the difference in Zeeman energy of the two domains represents a field-induced potential-energy difference across the domain wall of $2M_sH$. The wall can lower its energy by moving so as to reduce the volume of the unfavorably oriented domain. Equivalently, the force on the domain wall, given by the potential gradient $F = -\partial U/\partial x$, is such as to move the wall down in Figure 9.1*b*. Assuming for now that wall motion is smooth and easy, the loop resembles that shown as the solid line with zero coercivity in Figure 9.2*b*. The defect processes that can impede wall motion and lead to irreversibility in the m–h curve associated with wall motion hysteresis (as opposed to the rotational hysteresis; dotted line in Fig. 9.2*b*) will be treated later.

In summary, a purely hard-axis, uniaxial magnetization process involves rotation of the domain magnetization into the field direction. This results in a linear m–h characteristic. An easy-axis magnetization process results in a square m–h loop. It is characterized in the single-domain or pinned wall limit by rotational hysteresis, $H_c = 2K_u/M_s$, or, in the free-domain-wall limit, by $H_c = 0$. The wall moves in a direction that grows the favorably oriented domain.

The important features distinguishing the M–H loops in Figure 9.2 are not only what happens when the field is increased from zero (saturation is not achieved until $H = H_a$ in one case; it is usually much easier in the other), but also what happens to M as H is decreased from above saturation. In the transverse case the system *tends to demagnetize itself*; anisotropy, from whatever source, takes over as H approaches zero and rotates $\boldsymbol{M}$ toward the nearest easy direction. If the magnetization rotation process is carried out quasistatically, the magnetization is essentially reversible; that is, the M–H loop in Figure 9.2*a* shows no coercivity. This is not so for the parallel case. If allowance had been made for any hysteresis (i.e., viscosity or irreversibility) in the wall motion, or if no walls are present, the *sample would tend to remain magnetized* as H changes sign. A square loop with a high remanence generally results from application of a field along the easy direction of magnetization; the m–h process is one of *wall motion.* A slanted loop with no coercivity, with zero remanence, and with saturation at $H = H_a$ generally results from application of a field transverse to an easy axis; the m–h process is due to *magnetization rotation.*

It is instructive to consider the permeability for a purely rotational magnetization process. From Eq. (9.4), $M = (M_s^2/2K)H$, so the rotational permeability is

$$\mu_{\text{rot}} = \mu_0 \frac{(H+M)}{H} = \mu_0\left(1 + \frac{M_s^2}{2K_u}\right) \approx \mu_0 \frac{M_s^2}{2K_u}$$
$$\mu_{\text{rot}} \approx 2\pi \frac{M_s^2}{K_u} \quad \text{(cgs)} \tag{9.11}$$

This describes the contribution to the permeability from domains or grains whose easy axes are orthogonal to the field direction. Clearly, if their anisotropy is comparable to $2\pi M_s^2$, they contribute very little. Conversely, if their anisotropy energy density is small, the M–H loop is steep and μ_{rot} can be large. In the easy-axis magnetization process, the permeability depends on the coercivity and hence on domain wall motion (Section 9.6).

9.2 FIELD AT ARBITRARY ORIENTATION TO UNIAXIAL EASY AXIS

The magnetization process in a field parallel or perpendicular to the uniaxial easy axis has been treated under conditions of pinned domain walls (or no walls) and free walls. Domain walls were allowed only when they are parallel to the easy axis.

9.2.1 Stoner–Wohlfarth Problem

The M–H loops for a field applied at an arbitrary angle θ_0 with respect to a uniaxial EA are now considered (see Fig. 9.4). Only single-domain particles are considered (at least at first) because the magnetization in two domains would respond differently to the field for $0 < \theta_0 < \pi/2$. Further, it is assumed that the particles are ellipsoidal in shape so that the particle magnetization is spatially uniform throughout the magnetization process. This is sometimes referred to as the *Stoner–Wohlfarth* (SW) problem because they were the first to report a solution (1948). The free energy of the prolate spheroid may be written

$$f = -K_u \cos^2(\theta - \theta_0) - HM_s \cos\theta \tag{9.12}$$

The particle anisotropy may include shape: $K_u = [H_a + (N_2 - N_1)M_s]M_s/2$, where N_1 and N_2 are the demagnetization factors parallel and perpendicular to the easy axis of the particle. The form of the energy is illustrated schematically in Figure 9.4, right. Note that on increasing the strength of a negative field from H_0 to H_1 to H_2, the energy minimum shifts to smaller values of θ and a discontinuity in θ can occur. The energy in Eq. (9.12) is minimized with respect to θ for

$$2K_u \sin(\theta - \theta_0)\cos(\theta - \theta_0) + HM_s \sin\theta = 0$$

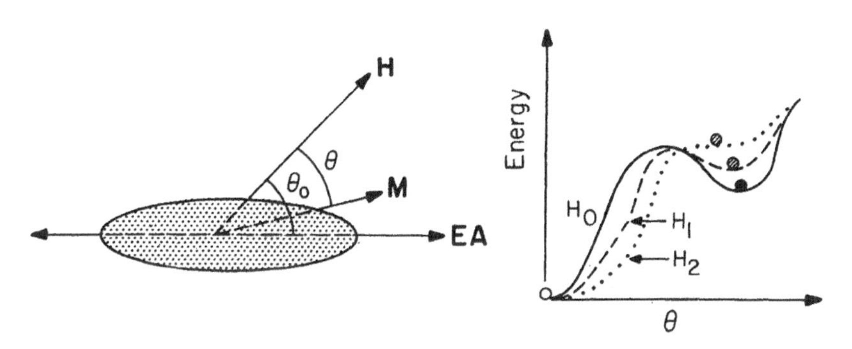

Figure 9.4. Left, coordinate system for magnetization reversal process in single-domain particle in which the shape and crystallographic easy axes coincide. Application of a field at an angle θ_0 relative to the EA causes a net magnetization to lie at some angle θ relative to the field. Right, illustration of approach to a discontinuous magnetization change at a negative field of magnitude H_2.

or, using $K_u = H_a M_s/2$ and $h = H/H_a$:

$$\sin(2[\theta - \theta_0]) + 2h \sin\theta = 0$$

Alternatively, with $m = M/M_s = \cos\theta$, the solution can be written

$$2m(1 - m^2)^{1/2} \cos 2\theta_0 + \sin 2\theta_0(1 - 2m^2) \pm 2h(1 - m^2)^{1/2} = 0 \quad (9.13)$$

This equation is readily solved for h as a function of m. The results, shown in Figure 9.5, need some explanation. Note that $\theta_0 = \pi/2$ leads to Eq. (9.4) with the linear M–H characteristic obtained in Figure 9.2*a*. As θ_0 approaches zero, the remanence increases toward $m = 1$. In the limit $\theta_0 = 0$, a square m–h loop results as in Figure 9.2*b*. For intermediate values of θ_0, it is more difficult to fully saturate the magnetization than it is for $\theta_0 = \pi/2$. The reason for this is that the applied field is working against a torque $-\partial f_a/\partial\theta$. This torque vanishes as θ approaches $\pi/2$ in the transverse case, but in the oblique case the torque does not vanish as $\boldsymbol{M}$ aligns with the field. Recall that the area between $m(H)$ and $m = 1$ is the energy needed to saturate the material. Does this energy

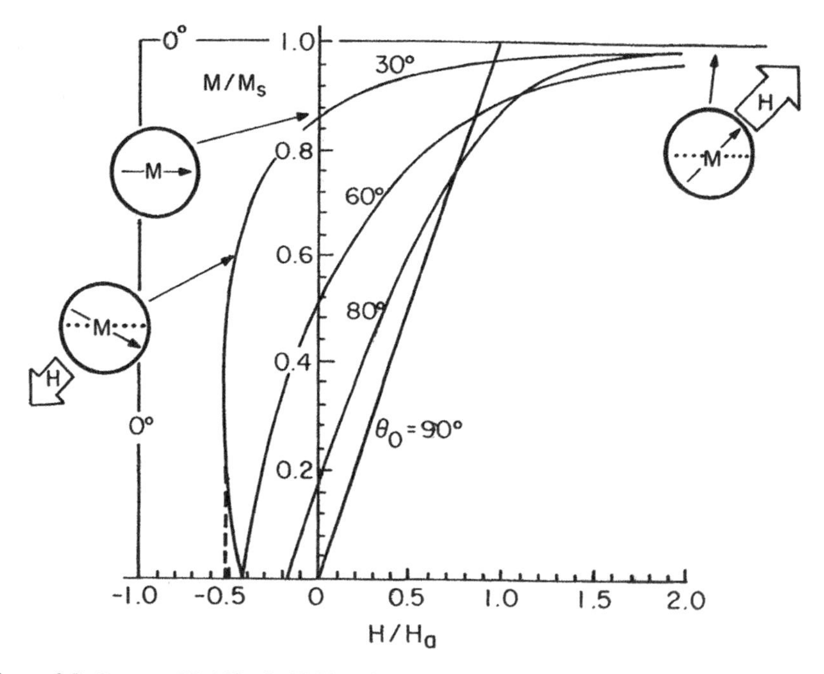

Figure 9.5. Stoner–Wohlfarth (SW) solutions: reduced magnetization versus reduced field applied at an angle θ_0 to the easy axis. The linear m–h curve represents $\theta_0 = 90°$, and the other curves of increasing remanence represent $\theta_0 = 80°$, 60°, and 30°. The magnetization process is irreversible so m–h continues for $h < 0$. Possible magnetization distributions are shown as inserts for nucleation-inhibited, single-domain particles.

diverge for $0 < \theta_0 < \pi/2$, or does it have a finite value; that is, do the $m(H)$ curves for $0 < \theta_0 < \pi/2$ ever reach $m = 1$? Problem 9.2 will provide the answer.

The single-domain, oblique magnetization process is also interesting in negative fields. For a small range of θ_0 above 0°, the magnetization reversal process leaves the magnetization in an unstable orientation. Here, a flux jump occurs at a critical field, $h_s = H_s/H_a$, called the *switching field*; h_s is defined where the m–h curve satisfies $\partial h/\partial m = 0$. This is shown in Figure 9.5 for $\theta_0 = 30°$ near $m \approx 0.4$. At that point, the magnetization switches abruptly and irreversibly to the third quadrant (along the dotted line). It does not follow the solid curve below $m \approx 0.4$ which is unstable. The switching field cannot be greater than the anisotropy field: $H_s \leqslant H_a$. The switching occurs at the field for which the free energy minimum, $df/d\theta = 0$, becomes flat, $d^2f/d\theta^2 = 0$ (Figure 9.4, right):

$$h_s \cos\theta - \cos 2(\theta - \theta_0) = 0 \tag{9.14}$$

The angle θ can be removed from Eqs. (9.14) and (9.12) most easily by eliminating $\sin 2(\theta - \theta_0)$ or $\cos 2(\theta - \theta_0)$. It is then possible to express $\sin 2\theta_0$ in terms of h_s:

$$\sin 2\theta_0 = \left(\frac{2}{h_s^2}\right)\left[\frac{1 - h_s^2}{3}\right]^{3/2}$$

Solving for the switching field gives

$$h_s = (\cos^{2/3}\theta_0 + \sin^{2/3}\theta_0)^{-3/2} \tag{9.15}$$

This result, which is really the upper limit to the switching field, is plotted as the solid line in Figure 9.6. For $45° < \theta_0 < 90°$, the switching field occurs after the magnetization has changed sign; in other words, the coercivity is less than h_s. See Figure 9.5, $\theta_0 = 80°$. The coercivity is defined by Eq. (9.12) or (9.13) at $m = 0$, namely, $h_c = \sin\theta_0 \cos\theta_0$, as shown by the dashed line in Figure 9.6. The process described by the dotted lines will be described next.

9.2.2 Magnetization Change by Curling

But single-domain particles do not always change their magnetization by coherent rotation of all moments in unison (Fig. 9.7*a*) as just described. Other modes of reversal are possible, including curling (Fig. 9.7*b*) and buckling (Fig. 9.7*c*). These switching modes for particles that are still too small to contain a domain wall, allow for easier magnetization rotation and hence lower switching fields than the Stoner–Wohlfarth limit. Of these two nonuniform modes (b, c), curling is the more important. Magnetization reversal can also be modeled by a chain of spheres interacting by dipole forces as in Figure 9.7*d*

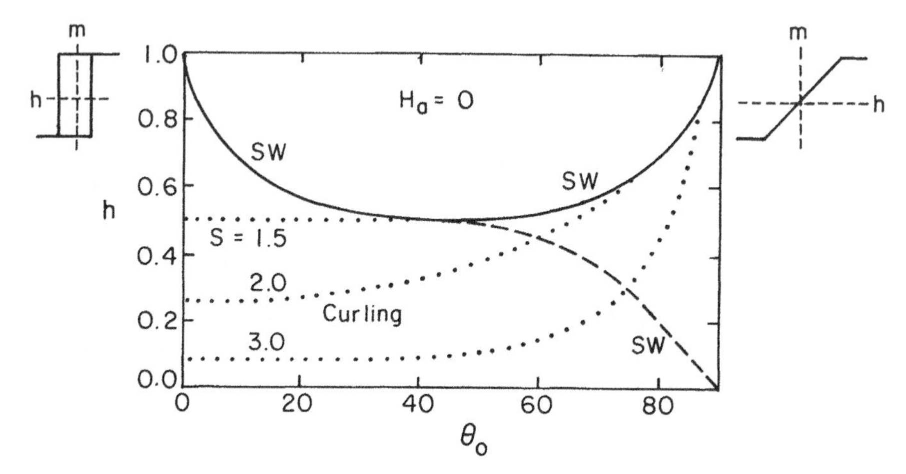

Figure 9.6. Solid line: variation of the switching or coercive field, $h_i = H_i/H_a^{\text{eff}}$, with angle between easy axis and applied field. The shapes of the m-h loops at the two extreme values of θ_0, are shown for reference. The solid line describes the switching field for the uniform rotation process (SW). For $\theta_0 > 45°$, the magnetization passes through zero (defining h_c) for fields less negative than the switching field. Dotted lines indicate switching fields for curling, Eq. (9.16), for various values of the reduced minor-axis radius, $S = b/b_0$. Magnetocrystalline anisotropy is neglected. The particle aspect ratio for the curling-mode results is 8.

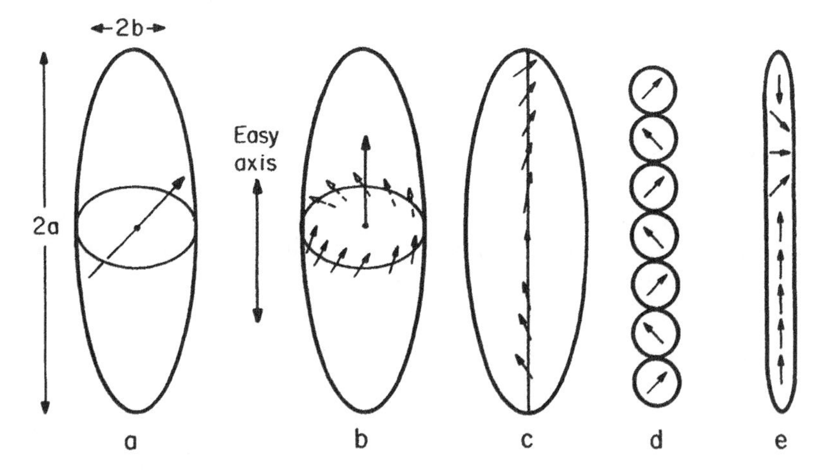

Figure 9.7. Modes of magnetization reversal in acicular fine particles. Left to right: (a) coherent rotation; (b) curling; (c) buckling; (d) fanning (in chain of spheres); and (e) domino effect. The first and third processes occur with the magnetization throughout the particle confined to a plane.

(Jacobs and Bean 1955). Knowles (1984) has found that for highly acicular particles, a likely mode of reversal is a domino-like flipping of the magnetization from one end of the particle to the other, along its long axis (Figure 9.7*e*).

In order to describe inhomogeneous switching quantitatively, one must make use of micromagnetics, the analysis of magnetization processes on a scale small enough that the exchange energy is important, and on a scale large enough that a continuum description is still appropriate. (Such techniques were used in Section 8.2 to get the Euler–Lagrange integral solution for the 180° Bloch wall.) The micromagnetic calculations for single-domain switching will not be derived here. Some results from the literature will be described (Frei et al. 1957; Aharoni 1966, 1986; Richter 1989).

Magnetization switching by curling costs exchange energy but saves magnetostatic energy by having fewer spins pointing away from the easy axis at any given stage of the reversal process. Thus, the switching field for magnetization curling in an elongated, single-domain particle of semiminor axis b is reduced from the uniform rotation value, $H_c = (N_b - N_a)M_s$, by replacing the hard-axis magnetostatic energy with the curling energy (Brown 1957):

$$H_s = \left[\frac{a}{2}\left(\frac{b_0}{b}\right)^2 - N_a\right]M_s \propto \frac{C_1}{b^2} - C_2 \tag{9.16}$$

where $b_0 = r_0/N_b^{1/2}$, is the single-domain radius [Eq. (8.32)] for an ellipsoid of revolution. The coefficient a is a function of the aspect ratio of the particle and varies between 1.08 (infinite cylinder) and 1.42 (sphere). The dotted lines in Figure 9.6 describe the θ_0 dependence of the switching field due to curling for various values of $S = b/b_0$ assuming no magnetocrystalline anisotropy. If the first term in Eq. (9.16) is too large, the curling-mode switching field may be of larger magnitude (more negative) than that for uniform rotation. In this case curling will not occur. Increasing the anisotropy of the particles pushes the curling-mode switching fields to larger values.

The mode of magnetization reversal in a particle will always be the lowest field process. Figure 9.8 illustrates the onset of curling as a viable mode of magnetization reversal in particles having larger reduced minor radii. For particle size below the crossover point, uniform rotation is the preferred mode of reversal.

When the particles become small enough for superparamagnetism to suppress their coercivity, Jacobs and Bean (1963) find

$$H_c = \frac{2K_u}{M_s}\left[1 - 5\left(\frac{kT}{K_u V}\right)^{1/2}\right]_s \propto C_1 - \frac{C_2}{b^{3/2}} \tag{9.17}$$

Here the parameters (other than C_1 and C_2) have the same meaning as in Eq. (9.16). Note that as the particle volume decreases in the superparamagnetic regime, the coercivity drops with increasing steepness for a given temperature.

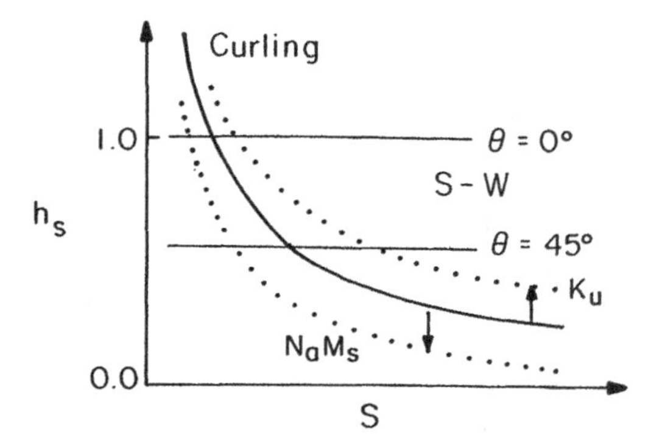

Figure 9.8. Schematic comparison of the switching fields for curling compared with the SW results shown for two angles of the field relative to the particle axis. Dotted lines for curling indicate that increased major-axis demagnetizing field $N_a M_s$ favors curling; increased particle anisotropy inhibits curling.

Equations (9.16) and (9.17) are reflected in the data of Kneller and Luborsky (1963) for the coercivities of FeCo particles (Fig. 9.9). However, the dependence on particle size in the SD regime varies more like $1/r$ than $1/r^2$.

It is very difficult to account for magnetostatic interactions between particle. White (1985) shows that a simple dipole interaction between particles causes the coercivity to *increase* with increasing packing fraction p counter to what is

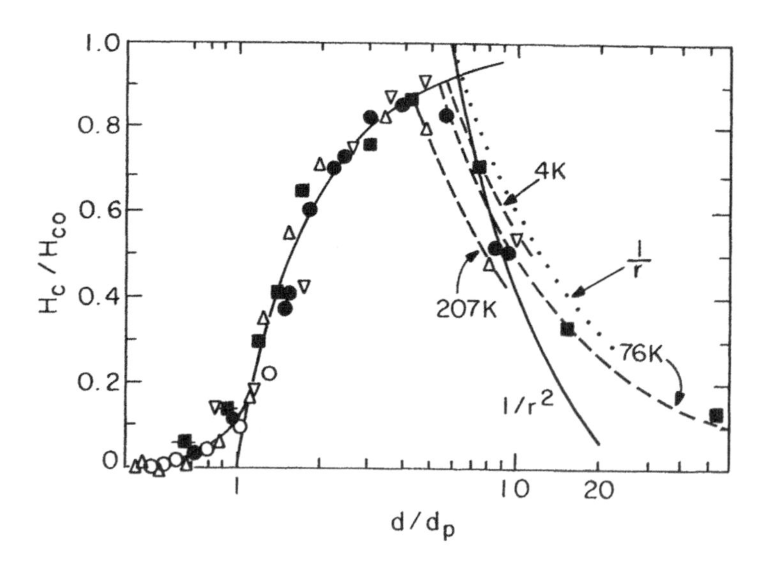

Figure 9.9. Particle-size variation of coercivity in FeCo particles (Kneller and Luborsky 1963). The solid negative-slope curve $1/r^2$ follows Eq. (9.16). The solid positive-slope curve, starting at $d/d_p = 1$, is calculated from Eq. (9.17).

generally observed (Wohlfarth 1959). Néel (1947) derived the form $H_c(p) = H_c(0)(1 - p)$, which is often followed experimentally. Apparently, as packing fraction increases, exchange coupling between the particles increases and domain walls may begin to move from particle to particle.

9.2.3 Free Domain Walls

Now let us consider particles large enough, or packed densely enough, such that domain walls can be present. Further, assume initially that the walls move with complete freedom: $H_c = 0$. The M–H characteristic displayed for $H > 0$ is the SW rotation process. On decreasing the field from positive saturation, walls are nucleated and move easily once $H < 0$. Thus, when $0 < \theta_0 < 90°$, both magnetization rotation and domain wall motion may contribute to the M–H process. There is no magnetization reversal by abrupt rotational switching because the easy wall motion has shorted out that process. The particles show zero coercivity (Fig. 9.10). Note that in the first and third quadrants of the m–h loops, however, the shape of the curves is the same as that for single-domain particles.

It is important to note that for a polycrystalline material, there is a distribution of easy axis orientations relative to the applied field direction. Each grain then responds to the field with the appropriate loop from Figure

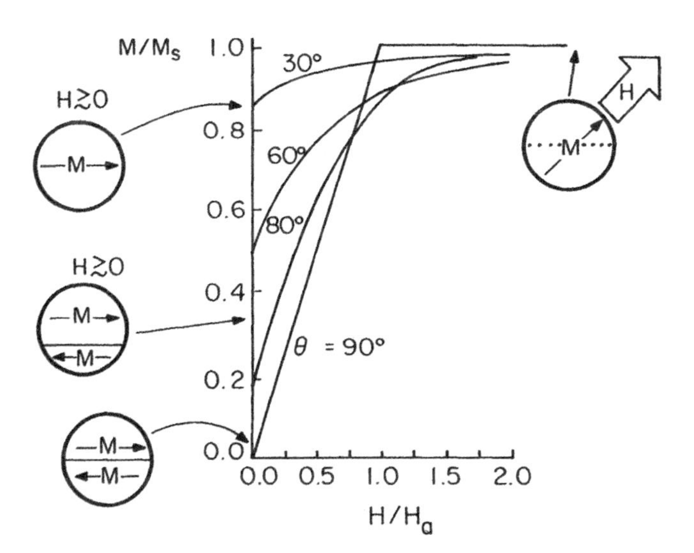

Figure 9.10. Reduced field versus reduced magnetization for a field applied at an angle θ_0 to the easy axis of particles significantly larger than the single-domain limit. The magnetization process is reversible both for magnetization rotation and for wall motion. Inserts depict possible magnetization configurations at various stages in the magnetization process (EA is horizontal in these inserts).

9.5 or 9.10 (and a coercivity to be determined later). It is easy to see then how a typical polycrystalline M–H loop may assume a somewhat rounded shape by summing over a number of grains having different values of θ_0. Still more important, caution should be exercised when trying to extract anything other than average material parameters from the shape of polycrystalline hysteresis loops.

All of these magnetization curves involving domain wall motion may be sheared over by magnetostatic effects. Stoner and Wohlfarth (1948) consider the case of the oblate spheroids, where magnetostatic energy effectively shears the loops of Figure 9.5; for a given θ_0, the oblate spheroid is slower to magnetize.

It has been seen that coercivity can be quite large in single-domain particles and that it decreases as particle size increases enough to allow curling. Above the single-domain limit, the magnetization process is determined by domain wall motion, which has not yet been covered. Before moving on to the mechanisms of domain wall-defect interactions, the reader may want to consider one more rotational magnetization process: cubic anisotropy. (See Problem 9.3.)

9.2.4 Approach to Saturation

The magnetization process is rarely as simple as suggested by the idealized forms considered here so far (Figs. 9.2, 9.5, and 9.10). In particular, a well-defined saturation of $M(H)$ is not always evident in experimental data, so it is often difficult to determine the value of the saturation magnetization by inspection. It is helpful in determining M_s to know the mathematical form of the approach to saturation. This would make it possible to fit $M(H)$ to a function that gives a quantitative indication of M_s.

The approach to saturation well below T_c can be expressed quite generally (Brown 1941) as

$$M(H) = M_s\left(1 - \frac{a}{H}\right) + \chi_{\mathrm{hf}} H \tag{9.18}$$

Here χ_{hf} is the high-field susceptibility and the term $-aM_s/H$ accounts for rotation of $\mathbf{M}$ away from the applied field as H decreases. The first of the two terms on the right-hand-side (RHS) of Eq. (9.18) follow from the Langevin equations, Eqs. (3.11) and (3.12), near saturation, namely, $s \geqslant 4$. In that case, $a = k_B T/\mu_0 \mu_m$. Note that in Eq. (9.18), $M(H)$ never reaches a state of zero slope; M_s is defined as the value of $M(H) - \chi_{\mathrm{hf}} H$ at infinite field. But χ_{hf} is generally not known *a priori*. However, for small χ_{hf} (i.e., at low temperature), a plot of M versus H^{-1} should be linear with a negative slope and an $M(H)$ intercept at M_s. See Figure 9.11 for data on $Eu_{0.99}Gd_{0.01}S$ (McGuire and Flanders 1969).

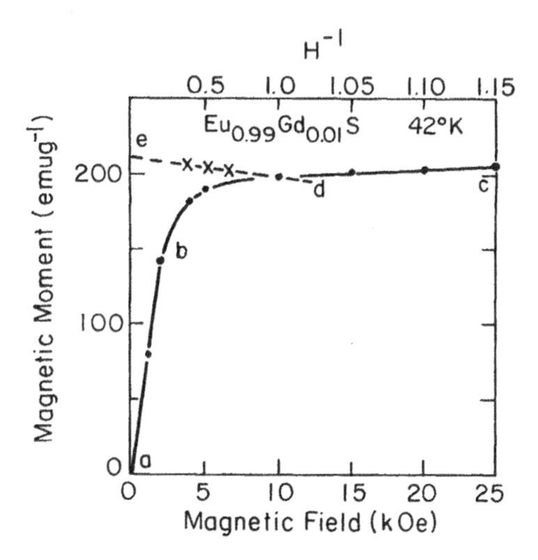

Figure 9.11. Method of determining saturation magnetization: M versus H and versus H^{-1} data for $Eu_{0.99}Gd_{0.01}S$ at 4.2 K (McGuire and Flanders 1969).

For single-domain particles, the Stoner–Wohlfarth model predicts an approach to saturation that can be derived from Eq. (9.13) by examining the limit $h \gg 1$, $m \approx 1$ (or $2m^2 - 1 \approx 1$):

$$\pm 2h\sqrt{1-m^2} \approx (2m^2 - 1)\sin 2\theta_0$$

leading to

$$m \approx \sqrt{1 - \frac{\sin^2 2\theta_0}{4h^2}}$$

and

$$M(H) = M_s\left(1 - \frac{H_a^2 \sin^2 2\theta_0}{8H^2}\right) \tag{9.19}$$

Thus, for noninteracting, single-domain particles, a plot of $M(H)$ versus H^{-2} should be a straight line extrapolating back to M_s.

Kronmüller (1980) considered the approach to saturation in a ferromagnet with isolated defects characterized by weak (magnetoelastic, in his case) anisotropy. (This situation is the inverse of the single-domain problem in which the medium is nonmagnetic and the magnetic particles are isolated and noninteracting.) He finds that here, too, the approach to saturation goes as $1 - H^{-2}$. Experimental data for an amorphous magnetic alloy at low-field ($\mu_0 H \leqslant 0.03$ T) follow Eq. (9.18) $(1 - H^{-1})$, while at higher fields the approach to saturation more closely follows the form of Eq. (9.19) $(1 - H^{-2})$.

9.3 DOMAIN WALL PINNING AND COERCIVITY

In real materials, domain walls do not move reversibly as has been assumed until now. Grain boundaries, precipitates, inclusions, surface roughness, and other defects can lower the wall energy at a particular position in the material, effectively pinning its motion, or they can place a barrier in front of the wall, inhibiting further wall motion through the defect.

For example, a *nonmagnetic* inclusion or planar defect coincident with a domain wall eliminates the need for a twist in $\boldsymbol{M}$ across the defect. This reduces the total wall energy locally by σ_{dw} times the common cross-sectional area of the defect and wall. Alternatively, a *magnetic* defect having a strong anisotropy (crystalline or magnetoelastic) relative to that of the matrix could effectively pose a barrier to a domain wall. In this case the spin is pinned in the direction of the local anisotropy, preventing the wall from moving through the defect.

These two classes of defect are depicted in Figure 9.12. A nonmagnetic inclusion, left, reduces the local wall energy while a high-anisotropy defect increases the local wall energy. Clearly, a distribution of defects in a material leaves the domain wall potential $\sigma_{dw}(x)$ highly irregular with position. As an external field is applied, a pressure $2M_sH_{ext}$ (due to the difference in potential across the wall) is exerted on the wall, tending to accelerate it. The presence of the pinning defects leads to an irregular domain wall motion consisting of a series of Barkhausen jumps as the wall skips from defect to defect. The wall

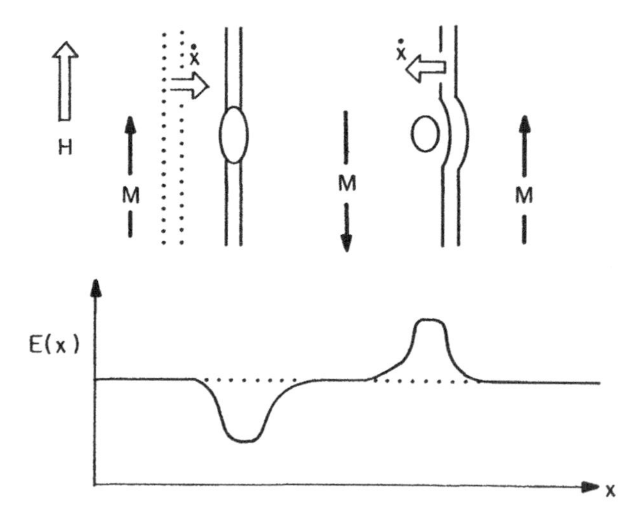

Figure 9.12. Upper panel depicts two kinds of defect and their influence on wall motion for vertical applied field: nonmagnetic inclusions locally lower the wall energy by decreasing its area; particles of different anisotropy or magnetization than the matrix present a barrier to wall motion. Below is shown the domain wall energy as a function of position in absence of an applied field.

achieves an average drift velocity (governed by the strength of the applied field as well as the density and nature of the defects) much as do charge carriers in a normal metal when an electric field is applied.

9.3.1 Large "Fuzzy" Defects

Quantitative models of domain wall motion generally distinguish two regimes of behavior based on the ratio of defect size D to wall thickness $\delta_{dw} = \pi(A/K)^{1/2}$. Several such models exist for domain wall pinning on defects whose dimensions exceed that of the domain wall, $D \gg \delta_{dw}$ (Kondorskii 1937; Kersten 1938; Dijkstra and Wert 1950). In this large defect limit, two cases must be considered. In one case, the material parameters change abruptly at the interface between the defect and the matrix. This sharply defined defect case will be treated later. In the "fuzzy defect" case considered here, the material parameters change slowly over the length scale of the wall. Thus, the domain wall can be considered to be moving in an irregular but slowly changing potential. The gradient in the wall energy density constitutes a pressure, $P = -d\sigma_{dw}/dx$, resisting wall motion. The difference in Zeeman energy $2M_sH$ across a 180° wall provides the driving pressure to overcome the gradient in wall energy. This problem can be modeled approximately to provide a useful introduction to the more rigorous treatment for abrupt interfaces.

The energy gradient $d\sigma_{dw}/dx$ may be expressed from Eq. (8.17) and the anisotropy generalized to include a magnetoelastic contribution $K_u = K_{xtl} + \frac{3}{2}\lambda_s\sigma$, where σ is the stress. Thus

$$\frac{d\sigma_{dw}}{dx} = 4\frac{d}{dx}(AK_u)^{1/2} = 2\left[\frac{\pi}{\delta_{dw}}\frac{\partial A}{\partial x} + \frac{\delta_{dw}}{\pi}\left(\frac{\partial K_{xtl}}{\partial x} + \frac{3}{2}\lambda_s\sigma\frac{\partial \sigma}{\partial x}\right)\right] \tag{9.20}$$

This says that the forces impeding domain wall motion arise from spatial variations in those properties that determine wall energy. But Eq. (8.17) for the wall energy density was derived for an infinite medium with no boundaries or interfaces. If this equation is to be useful here, it must be assumed that the interfaces between the matrix and the defect are very gradual. That assumption distinguishes this approximate treatment from the abrupt interface theory of coercivity to be discussed below. The steepest gradient in wall energy density can be taken to be the magnetic pressure responsible for the coercivity:

$$\left(\frac{d\sigma_{dw}}{dx}\right)_{max} = 2M_sH_c$$

$$H_c = \frac{(d\sigma_{dw}/dx)_{max}}{2M_s}$$

If the range of the variations is assumed to be of order D (see Fig. 9.13 for this $D > \delta_{dw}$ limit), and we can approximate the gradient as linear,

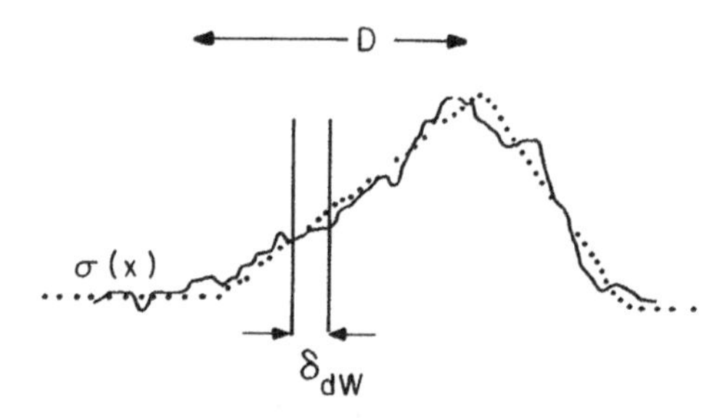

Figure 9.13. Domain wall thickness superimposed on a representation of the variation in wall energy density across a defect of full width at half maximum, D.

$d\sigma_{dw}/dx = \Delta\sigma_{dw}/D$, the coercivity can be expressed

$$H_c \approx \frac{2H_a}{\pi}\frac{\delta_{dw}}{D}\left[\frac{\Delta A}{A} + \frac{\Delta K_{xtl}}{K_{xtl}} + \frac{3}{2}\lambda_s\frac{\Delta\sigma}{K_{xtl}}\right] \tag{9.21}$$

Thus, the coercivity for $D > \delta_{dw}$ in the gradual defect interface case goes as δ_{dw}/D times a sum of fluctuation terms expressing local variations in exchange stiffness, crystal anisotropy and magnetoelastic anisotropy, respectively. The anisotropy field, $H_a = 2K/M_s$, sets an upper limit to the coercivity.

It is also possible to include in Eq. (9.21) coercive mechanisms arising from the magnetostatic energy of defects proportional to $\Delta M/M$; for example

$$\frac{\Delta(2\pi M_s^2)}{2\pi M_s^2} = \frac{2\Delta M_s}{M_s}$$

The important ingredients in a defect that make it effective in pinning a domain wall are the extent to which its magnetic properties differ from those of the matrix, namely the magnitude of

$$\frac{\Delta K}{K}, \quad \frac{\Delta A}{A}, \quad \text{and} \quad \frac{\Delta M}{M}$$

Also important are the defect dimensions relative to the domain wall thickness and the sharpness of its interface with the matrix relative to the wall thickness.

The relative importance of the three fluctuation terms in Eq. (9.21) can be appreciated by noting that while $\Delta A/A$ (or $\Delta M/M$) and $\Delta K/K$ may vary from zero to unity, the magnetoelastic term for a local 1% strain may have a magnitude of 10 in Ni or 0.7 in Fe. Clearly, the coercivity of nickel is very sensitive to strain.

The preceding limit has focused on the force on domain walls due to *gradual* changes in domain wall energy in the vicinity of *large* defects. A more rigorous model of coercivity is now treated for defects that may be large or small compared to the wall thickness, as long as the defect-matrix interface is sharp.

9.3.2 Micromagnetic Theory for Well-Defined Defects

A micromagnetic theory of coercivity based on domain wall pinning by a sharply defined planar defect was developed by Friedberg and Paul in 1975 and extended by Paul in 1982. The model is micromagnetic because it considers the boundary conditions on the magnetization at the matrix/defect interfaces and takes exchange into account. In this model the defect, of width D, can be larger or smaller than the domain wall thickness. Figure 9.14 outlines the geometry that distinguishes the anisotropy, exchange, and magnetization in the matrix, K_1, A_1, and M_1, from those parameters in the defect, K_2, A_2, and M_2. The energy density for the system in this model includes micromagnetic contributions from exchange, uniaxial anisotropy, and Zeeman energy densities. The angle θ, between the magnetization and the easy y axis, is a function of position for each term:

$$A_i\left(\frac{d\theta}{dx}\right)^2 + K_i \sin^2\theta - M_s H \cos\theta \tag{9.22}$$

This energy density is integrated over all space and minimized giving the Euler–Lagrange equation for the magnetization in each region (see Chapter 8):

$$-2A_i\left(\frac{d\theta}{dx}\right)^2 + K_i \sin^2\theta - M_s H \cos\theta = C_i\theta \tag{9.23}$$

where $i = 1, 2, 3$ corresponding to the regions in Figure 9.14. The boundary

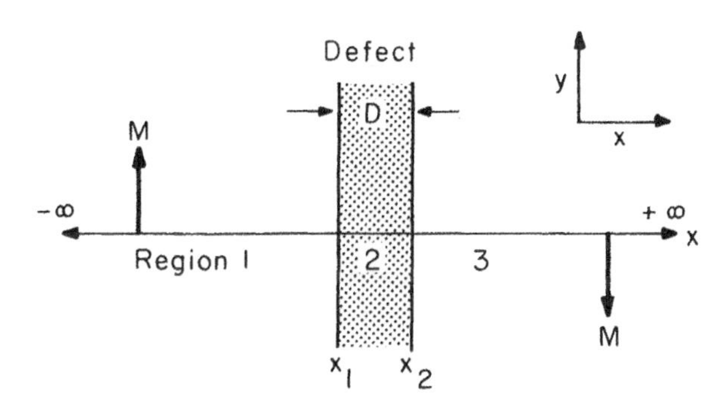

Figure 9.14. Division of a material into three regions: 1, to the left of a planar defect; 2, inside the planar defect; 3, to the right of the defect. Material properties are the same in regions 1 and 3. Defect width is $D = x_2 - x_1$, and magnetization values at $\pm$infinity indicate that a domain wall exists somewhere in between.

conditions are $d\theta/dx = 0$ at $\pm$ infinity and $\theta = 0$ and π at negative and positive infinity, respectively. Therefore $C_1 = -HM_1$ and $C_3 = +HM_1$. Continuity of θ and exchange torque, $A_i d\theta/dx$, at the interfaces, x_1 and x_2, is used (just as the wave amplitude and momentum are continuous in analogous problems in optics or quantum mechanics) to determine C_2 and to complete the solution.

The theory gives the reduced coercive field

$$h_c = \frac{H_c M_1}{K_1} \tag{9.24}$$

in terms of the following dimensionless parameters:

$F = A_2 M_2 / A_1 M_1$	defect exchange stiffness/magnetization relative to matrix
$E = A_2 K_2 / A_1 K_1$	normalized defect wall energy squared
$\delta_2 = (A_2/K_2)^{1/2}$	defect wall thickness
$w = D/\delta_2$	ratio of defect width to wall thickness

The results of this theory [see Figs. 9.15 and 9.16 (Paul 1982)] are that

1. The coercivity increases linearly in w for $w \ll 1$, varying approximately as $h_c = 0.38w\ (1 - E)/E^{1/2}$, for small defect parameter deviations from the matrix (i.e., $E, F \approx 1$).
2. h_c saturates for large values of w,
3. Nonmagnetic defects, $F \ll 1$, or low-anisotropy defects, $E \ll 1$, give larger asymptotic h_c than does $F \approx 1$ or $E \approx 1$.

The linear increase in h_c with w can be understood in terms of Figure 9.12. As the size of the defects increase in the regime $w < 1$, an average defect fills a larger fraction of the wall thickness and is, therefore, a more effective pinning site.

9.3.3 Examples

As an example of the utility of the micromagnetic theory, consider $Nd_2Fe_{14}B$ for which $4\pi M = 16$ kG, $A \approx 10^{-6}$ erg/cm, and $K_1 = 5 \times 10^7$ erg/cm^3 (see Chapter 13). The domain wall thickness in this phase is of order 40 Å. Sintered NdFeB magnets show a microstructure of 2-14-1 phases having large grains (5–10-μm) separated by a nonmagnetic $Nd_{1.1}Fe_4B_4$ and Nd-rich phases measuring 0.1 to 1 μm across (Sagawa et al. 1987). Hence, $W = D/\delta_2 > 1$ is satisfied and $E = F \approx 0$; take $F = 0.01$ giving $h_c = 1.6$ or $H_c = h_c K_1/M_1 = 40$ kOe. This coercivity exceeds the present record value by about 30%. Note that h_c is defined with an extra factor of 2 relative to the familiar anisotropy expression, $H_a = 2K_1/M$, which represents an upper limit for H_c.

For amorphous alloys the defects again appear to be larger than domain

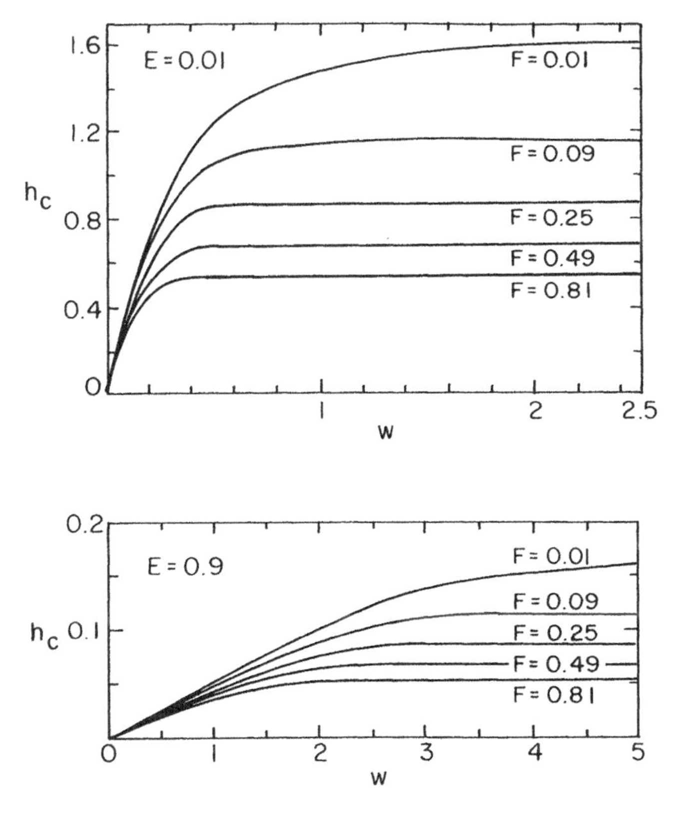

Figure 9.15. Reduced coercivity h_c versus normalized defect size w for various values of F and for small (above) and large (below) values of the square of the normalized defect wall energy density E (Paul 1982).

wall thickness and largely of magnetoelastic origin (O'Handley 1975, Egami 1976). Thus, $w \gg 1$, $F \approx 1$ and E is large. Figure 9.15 then indicates $h_c \approx 0.06$. Using $K_1 \approx 100\,\text{erg/cm}^3$ and $M_s = 1200\,\text{emu/cm}^3$, the result is $H_c = 0.006$ Oe. This value is at the lower limit of the range of coercivities observed for these alloys (see Chapter 10).

For $Sm_2(Co, Cu, Fe, Zr)_{17}$ having $B_r = 1.1$ T (11 kG), $H_c \approx 5.6 \times 10^5$ A/m (7 kOe), and $(BH)_{max} = 30$ MGOe (Ojima et al. 1977; Yonezawa et al. 1979), assume $A_1 \approx 3 \times 10^{-6}$ erg/cm and estimate $\delta_1 \approx 2.5$ nm. TEM studies (Mishra and Thomas 1979) show the important defects to have dimensions of order 4.0 nm implying $W \approx 1.8$. Figure 9.17 indicates for $E \approx 0.9$ and $F \approx 0.5$ that $h_c \approx 0.1$ and thus $H_c \approx 50$ kOe. The models of coercivity can be summarized with Figure 9.17.

For $D/\delta_{dw} < 1$ the coercivity is predicted by both models (fluctuation and micromagnetic) to increase linearly with D; the domain wall is more strongly pinned the more the defect fills the wall thickness. For large defects, two cases

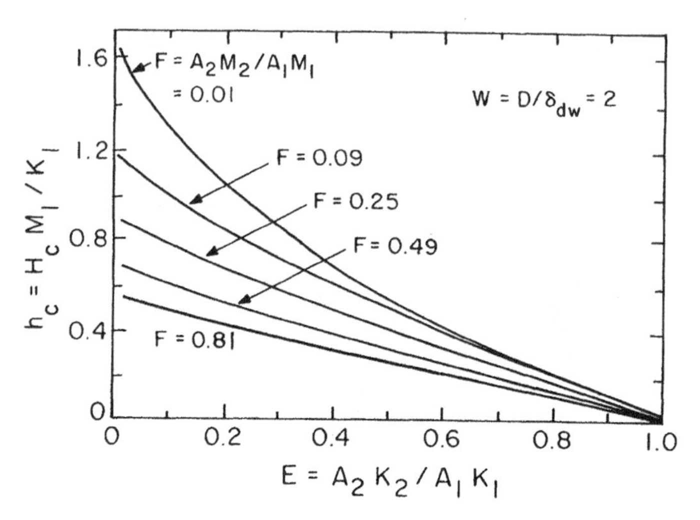

Figure 9.16. Normalized coercive force h_c as a function of the defect wall energy parameter E for various values of the defect exchange/magnetization stiffness parameter F. The normalized defect width is two for this figure. [After Paul (1982).]

have been described: the fuzzy defect case for which the coercivity decreases with increasing defect size and the sharp defect, micromagnetic theory, for which H_c is independent of defect size. Sharp defects are those defined by structure, such as grain boundaries, voids, antiphase boundaries, and dislocations. Fuzzy defects are those more often defined by strain fields or by diffusion

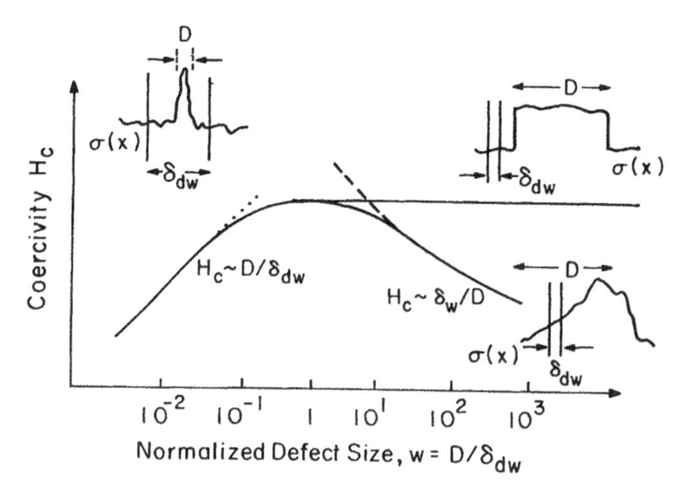

Figure 9.17. Schematic variation of coercivity with normalized defect size spanning two regions—small defects and large defects—relative to wall thickness. The predicted behavior in each case is shown.

(e.g., composition fluctuations due to spinodal decomposition). For the fuzzy defect model, the gradient in a critical parameter for wall energy gives rise to a force on the wall $F_w = -\Delta\sigma_{dw}/D$ related to the slope of $\sigma_{dw}(x)$ at the position of the wall. Theories of Kersten (1938, 1943) and Globus indicate that H_c is proportional δ_{dw}/D in this range. Clearly, then, H_c must reach a maximum at some intermediate value near $\delta_{dw}/D \approx 1$ as predicted by Dijkstra and Wert (1950). In the micromagnetic model, H_c saturates for large defects as observed by Hagedorn (1970).

There are two interesting examples that illustrate this coercivity peak. One is for hard magnetic materials (rare-earth transition metal borides) the other is for soft magnetic materials (cobalt rich amorphous alloys).

Figure 9.18 shows the maximum energy product (see Chapter 13) versus wheel speed for melt-spun $Fe_{82}Nd_{13}B_5$. (Melt spinning is a process of rapid solidification in which a jet or a thin sheet of molten metal alloy is quenched on the surface of a rapidly rotating drum.) At high wheel speed, the quench rate is high and the alloy may solidify in the amorphous state; in this limit D/δ_{dw} approaches 0. As wheel speed decreases, fine crystallites appear and H_c increases until $D/\delta_{dw} \approx 1$. For slow wheel speed, the crystallites are larger and H_c again drops. The wheel speed serves as a qualitative measure of inverse defect or grain size.

Figure 9.19 shows the coercivity of amorphous CoNbB in various stages of devitrification induced by heating the amorphous alloy for one hour at various temperatures. The defect size scale was determined from TEM studies (O'Handley et al. 1985). It is difficult to establish the D dependence of H_c below the peak because it rises so sharply. The falloff in H_c above the peak is more

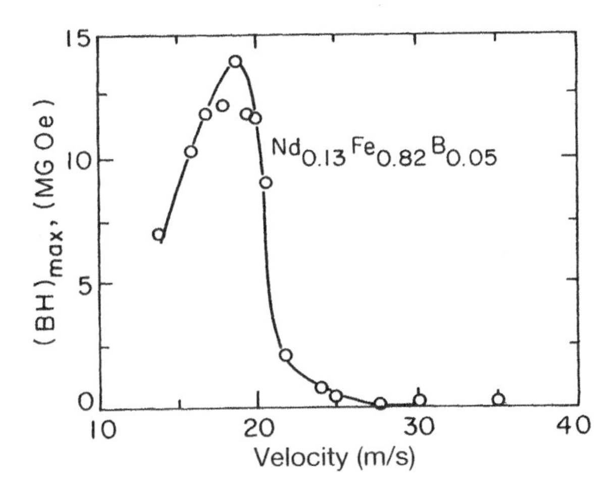

Figure 9.18. Maximum energy product of melt-spunm NdFeB alloy as a function of wheel speed showing a characteristic peak for a microstructure that optimally interferes with domain wall motion [after Croat et al. (1984)].

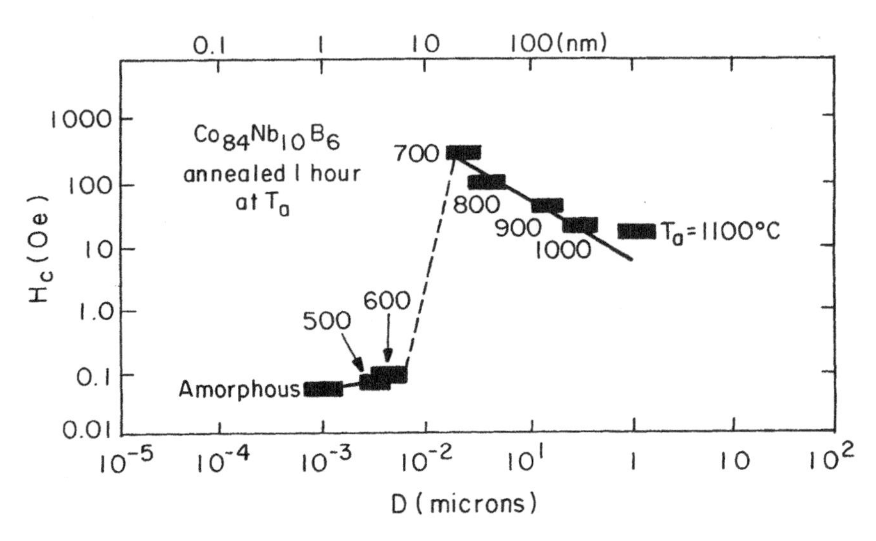

Figure 9.19. Variation of coercivity in amorphous and crystallized CoNbB as a function of mean particle size. Solid line above 10 nm goes as $1/D$. [After O'Handley et al. (1985).] Steep dashed line goes as D^6 as predicted by Herzer (1989).

gradual, consistent with a $1/D$ behavior predicted by the model for $D/\delta_{dw} > 1$.

The behavior in Figure 9.19 can be clarified by considering the microstructure of the alloy as it devitrifies. It evolves from a homogeneous, low-anisotropy ferromagnetic material by the precipitation of small single-domain particles in an amorphous matrix. The particles are characterized magnetically by their size and by their magnetocrystalline anisotropy, which is randomly oriented from particle to particle. The behavior of the single-domain particles depends also on the properties of the intergranular phase. If the intergranular material is nonmagnetic, the behavior is essentially that of isolated, single-domain particles, interacting only by their weak magnetic dipole fields. If the intergranular phase is magnetic, the behavior can be much more complicated (see Chapter 12). Either hard or soft magnetism can result, depending on the size and anisotropy of the particles as well as the exchange coupling between them. If the exchange coupling between the particles averages out the short-range fluctuations in the anisotropy from particle to particle, the direction of magnetization can be continuous from grain to grain (Fig. 9.20). In this case the behavior will be magnetically soft.

On the other hand, if the strength of the nanoparticle anisotropy exceeds the strength of the exchange coupling between them, hard magnetism can result. An example is FeNdB alloys heat-treated to form nanocrystallites of the high-anisotropy $Fe_{14}Nd_2B$ phase separated by a nonmagnetic B- and Nd-enriched phase.

The quantitative origin of this strong dependence of coercivity on nanoparticle size is outlined. More detail is given in Chapter 12 and the literature (Imry

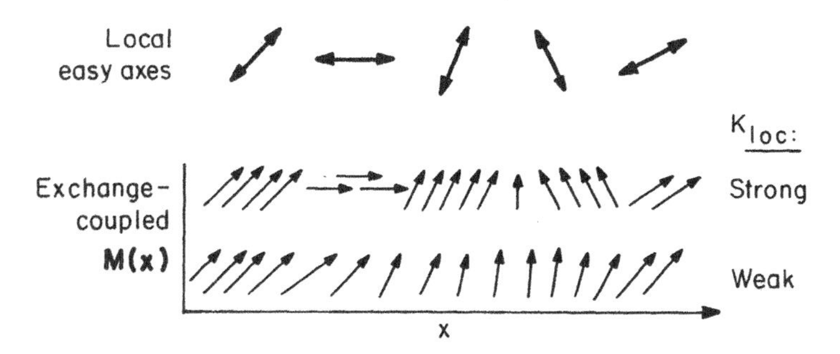

Figure 9.20. Schematic representation of the variation of local anisotropy easy axis with position (top row) and the variation of magnetization direction in response to the local anisotropy and exchange coupling (lower two rows). $M(x)$ follows closely a strong local anisotropy and is a smoother function of x for weak local anisotropy.

and Ma 1975; Alben et al. 1978; Herzer 1989, 1990, 1993). Consider the case for which the length scale of the microstructure D is smaller than the exchange length l_{ex}. The local moments experience an effective anisotropy $\langle K \rangle$ reduced by averaging the local anisotropy K_{loc} over several nanocrystalline grains:

$$\langle K \rangle \approx K_{loc} \left(\frac{D}{l_{ex}} \right)^{3/2} \tag{9.25}$$

Herzer (1989, 1993) recognized that for nanoscale inhomogeneity, it is $\langle K \rangle$ and not the local anisotropy that should appear in the equation for the exchange length, $l_{ex} = (A/\langle K \rangle)^{1/2}$. By eliminating the exchange length between these two expressions, he found that the exchange-averaged anisotropy, which controls the technical properties, scales as

$$\langle K \rangle \approx \left(\frac{K^4}{A^3} \right) D^6 \tag{9.26}$$

The sixth-power dependence predicted here for the anisotropy also applies to the coercivity, $H_c \leqslant 2\langle K \rangle / M$, and accounts well for the steep rise in coercivity in Figure 9.19 near $D = 0.01\ \mu\text{m}$. The dashed line in Figure 9.19 has a slope corresponding to D^6.

9.3.4 Additional Anisotropy

It was shown in Section 9.3.2 that for large single-domain particles, nonuniform modes of reversal such as curling or buckling become possible. When these reversal modes occur, the coercivity is *reduced* relative to its uniform reversal limit, $H_c < 2K_u/M_s$. Further, the micromagnetic model for well-defined defects (Section 9.3.2) also sets an upper limit on H_c of $2K/M_s$. However, it has been observed that in some large single-domain particle systems (e.g., Fe in a SiO_2 matrix) the coercivity can exceed the single-domain limit (Chien 1991). This behavior seems to be associated with an additional anisotropy induced by the particle interaction with its nonmagnetic matrix. This anisotropy may be due to stress or interfacial spin pinning.

This enhanced single-domain particle coercivity has been modeled in one dimension by Paul and Creswell (1993, 1995) by assuming a volume anisotropy in addition to the shape anisotropy of the particle. The volume anisotropy axis is assumed to be at an angle θ_a relative to the shape axis (z) along which the field is applied. The particle size, L, is comparable to an exchange length, l_{ex} and the spins assume an orientation $\theta(x)$ in the y–z plane. The energy density functional has the form

$$f = 2\int_0^{L/2} \left[A\left(\frac{d\theta}{dx}\right)^2 + 4\pi N M_s^2 \sin^2\theta + K_u \sin^2(\theta - \theta_a) - M_s H \cos\theta \right] dx$$

The $\theta(x)$ behavior predicted by the model is a canting of the spins in the interior of the particle while the surface spins respond less to the field, as if they were pinned. This is unusual because the anisotropy introduced is not localized at the surface. The surface spins cant less than do the interior spins because the interior spins "feel" the field also through their exchange coupling to neighbors on both sides. The surface spins are exchange-coupled on only one side. Hence the volume anisotropy is more effective in restraining the surface spins from aligning with the field. The extent of the spin canting in the interior of the particle is measured by a maximum angle θ_m at the center of the particle, $x = L/2$. θ_m depends on the particle size, shape, anisotropy, direction, and strength, as well as on the strength of the applied field. The model predicts two regimes of reversal behavior: an hysteretic, coercive regime and a zero-coercivity, switching regime. The former occurs for larger reduced particle size, L/l_{ex} and smaller anisotropy angles, θ_a. In this regime, the coercivity increases with decreasing particle size and can exceed the single-domain limit for anisotropy strength comparable to the magnetostatic energy.

9.4 AC PROCESSES

The area inside a B–H loop is exactly the energy per unit volume lost in one cycle of the hysteretic process. This can be simply demonstrated by considering the energy W dissipated in a toroidal core over one cycle, which is given by the integral of the power loss over a period:

$$W = \int_{t=0}^{t=T} i(t)V(t)dt \qquad (9.27)$$

Ampère's law can be used to relate the current to a magnetic field in a solenoid and Faraday's law relates $V(t)$ to the time rate of change of flux:

$$W = lA \int_0^T H \frac{dB}{dt} dt = lA \oint H(t) dB \tag{9.28}$$

Thus the energy lost per unit volume lA of core material is given by the area inside the B–H loop.

The area inside a quasistatic B–H loop, namely, the DC loss, is referred to as the *hysteresis loss*. If the loop is traced at increasing frequency, it is observed that H_c increases and the loop becomes more rounded. The increase in loop area is a result of eddy currents induced in the sample by the increasingly rapid change in flux density. Faraday's law [Eq. (1.6)] tells us that when the flux density changes, a voltage is induced in a circuit about the direction in which the flux changes. Further, the induced voltage is such that the resulting current opposes the initial flux change (Lenz' law). We will first assume the magnetization changes uniformly in the cross section of the material (Fig. 9.21). This is the "classical" theory of eddy current loss. Later, the loss associated with domain wall motion will be considered. In this case the flux change is concentrated in the wall and the eddy currents are concentrated more about the wall. The power loss is calculated from the Joule heating of the sample, i^2R, where i is the eddy current induced by the flux change, $J = \sigma E \propto \sigma dB/dt$.

9.4.1 Classical Eddy-Current Loss

Our treatment for classical eddy current loss is similar to that of Williams, Schockley, and Kittel (WSK) (1950). The magnetization in a bar of square cross section is initially in the z direction and reverses uniformly under application of an external field B_z (Fig. 9.21).

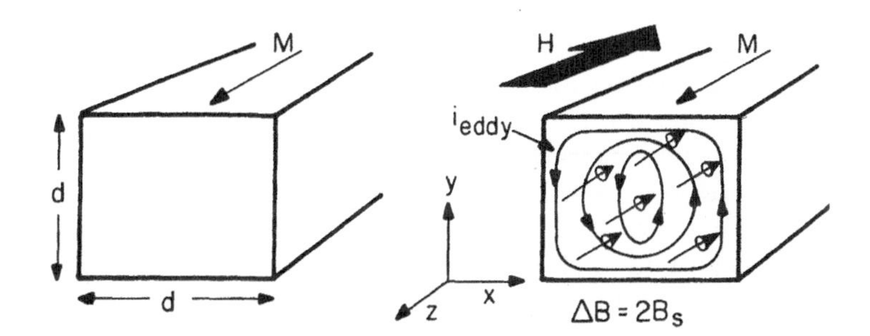

Figure 9.21. Cross section of a sheet of magnetic material, showing (left) the initial state of magnetization before application of a magnetic field H and (right) the direction of change in flux density that induces eddy currents about the flux change. Flux change direction is indicated by small arrows inside the sample in direction of applied field.

The field is applied at a low enough frequency that the classical skin depth is much greater than the smallest sample dimension, $\delta_{cl} \gg a$. [The classical skin depth, $\delta_{cl} = (\mu\sigma\omega)^{-1/2}$, measures the exponential decay length of an AC electric field in a material due to the induced eddy currents: $\delta_{cl} = 36\,\mu\text{m}$ for a permeability of $100\,\mu_0$, a conductivity of $10^7\ (\Omega\text{m})^{-1}$, and a frequency of 100 kHz. If there are no induced currents, the electric field does not decay as it penetrates the material.] By Faraday's law [Eq. (1.6)], the electric field induced around the direction of flux change is given by

$$(\nabla \times E)_z = -\frac{\partial B}{\partial t} = i\omega B_m$$

(ignoring the time dependence) and the corresponding current density is

$$(\nabla \times J)_z = \frac{i\omega B_m}{\rho} \tag{9.29}$$

Consideration of the symmetry of the problem (Fig. 9.21) suggests the solutions

$$J_x = -Ay, \quad J_y = Ax, \quad \text{with} \quad A = \frac{i\omega B_m}{2\rho}$$

The time-averaged classical power loss per unit volume is given by averaging the square of the current density over the cross section:

$$\frac{P_{\text{class}}}{\text{vol}} = \frac{1}{d^2}\iint J^2 \rho\, dx\, dy = \frac{4A^2\rho}{d^2}\int_0^{d/2}\int_0^{d/2}(x^2 - y^2)dx\, dy$$
$$\frac{P_{\text{class}}}{\text{vol}} = \frac{\omega^2 B_m^2 d^2}{48\rho} \tag{9.30}$$

This is the power loss per unit volume at low frequency for a *uniform* magnetization process. At high frequency the classical skin depth may decrease to the sample dimension and the interior of the sample will not be fully magnetized. The increased eddy currents responsible for this steeper drop in electric field increase the classical core loss at high frequencies compared to its low-frequency value, Eq. (9.30). At the same time, without an increase in drive field, the sample will no longer be magnetized to the same extent.

The most important result of these calculations is the strong dependence of the loss per unit volume on B_m, d, and ω. Magnetic losses are often expressed *per cycle*, in which case they have units J/m^3 or J/kg so that loss per cycle times frequency gives power loss density. The classical loss per cycle [Eq. (9.30) divided by frequency] is *linear* in the frequency (Fig. 9.22) and is added to the DC hysteresis loss per cycle given by the area inside the B–H loop over one quasistatic cycle.

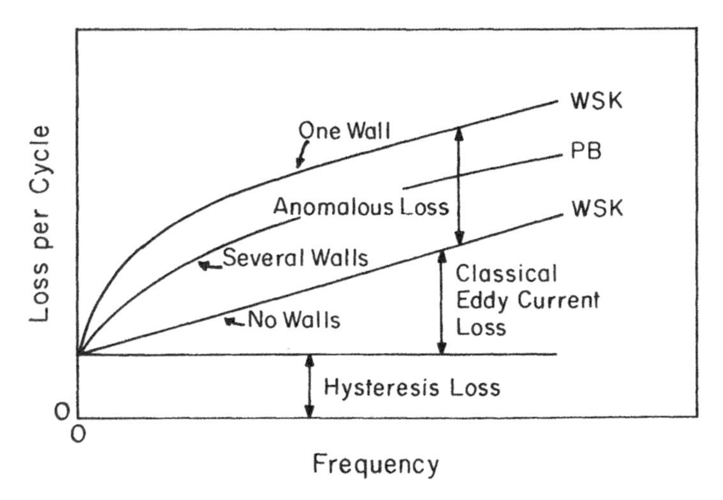

Figure 9.22. Partition of losses in a ferromagnetic metal. To the DC hysteresis loss is added the classical eddy-current loss per cycle, for uniform magnetization (Williams, Schockley, and Kittel) and an anomalous loss term due to domain walls. A single wall (WSK) generates more loss than do an array of domain walls (Pry and Bean 1958).

The magnitude of the actual measured AC loss is always greater than this classical limit. The excess of the measured loss over the classical loss is called the *anomalous eddy-current loss* (Fig. 9.22). The reason for this large loss is that the magnetization change generally is not uniform but is concentrated near domain walls. Essentially the magnetic response of the material away from the domain wall is zero because there $\mu = \mu_0$; all the flux change is concentrated in the narrow region of inhomogeneous magnetization that defines the wall.

9.4.2 Eddy-Current Loss about a Single Domain Wall

A single-domain wall is assumed in a bar of rectangular cross section with the magnetization and field directions as indicated in Figure 9.23, left. The field drives the wall from left to right; as the wall passes a fixed point, the flux density changes at that point by $2B_s$ and is directed into the paper. *The flux change is zero outside the wall.* The voltage induced by dB/dt gives rise to a current whose path is related to the choice of dA in Eq. (9.30). Though the eddy currents are now induced by a highly localized flux change, they are distributed over a region of the sample larger than the domain wall (Fig. 9.23, right).

Following WSK, note that outside the wall the induced electric field must satisfy

$$\nabla^2 \boldsymbol{E} = 0, \quad \nabla \times \boldsymbol{E} = 0, \quad \text{and} \quad \nabla \cdot \boldsymbol{E} = 0 \tag{9.31}$$

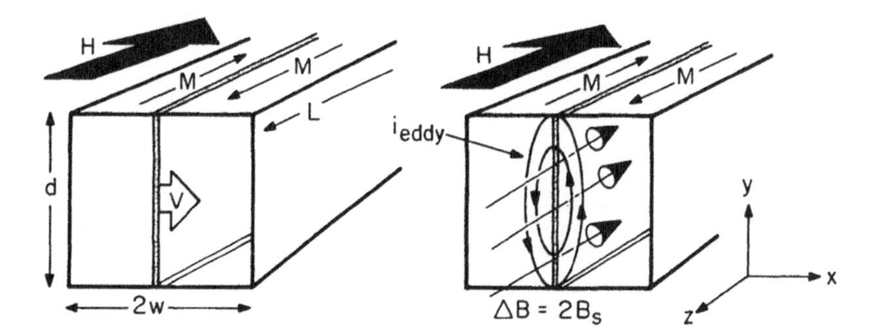

Figure 9.23. Cross section of a strip of magnetic material with one 180° domain wall (left) and illustration (right) of flux density change $2B_s$ directed into ribbon along the wall that induces eddy currents about the moving wall and about dB/dt.

as well as giving rise to zero current normal to the surface of the bar:

$$i_x(\pm W, y) = 0 \quad \text{and} \quad i_y\left(x, \pm\frac{d}{2}\right) = 0 \tag{9.32}$$

Inside the domain wall

$$\nabla \times E = -\frac{\partial B}{\partial t} \quad \text{and} \quad \nabla \times B = \mu_0 J \tag{9.33}$$

Because the flux change is no longer uniform across the sample, $\partial B/\partial t$ is better expressed in terms of the velocity v, of the domain wall of thickness δ_{dw}, moving past a point:

$$v = \frac{\partial x}{\partial t} = \frac{\delta_{\mathrm{dw}}}{\partial t}$$

The second relation in Eq. (9.33) indicates $J_z = 0$ and $\mu_0 J_y = 2B_s/\delta_{\mathrm{dw}}$. Hence, on either side of the wall,

$$J_y = \pm\frac{B_s v}{\rho} \tag{9.34}$$

Equations (9.31)–(9.34) are solved with

$$\begin{aligned} i_x &= -\sum_{n-\mathrm{odd}} D_n \sin\left(\frac{n\pi y}{d}\right) \sinh\left[(W - x)\frac{n\pi}{d}\right] \\ i_i &= -\sum_{n-\mathrm{odd}} D_n \cos\left(\frac{n\pi y}{d}\right) \cosh\left[(W - x)\frac{n\pi}{d}\right] \end{aligned} \tag{9.35}$$

where

$$D_n = \pm\left(\frac{B_m v}{\rho}\right)\frac{1}{n\pi\cosh(Ln\pi/d)}$$

with + for $n = 1, 5, 9\ldots$ and − for $n = 3, 7\ldots$.

Calculating the time-averaged power loss per unit volume as was done in Eq. (9.30), the result is

$$\frac{P_{1-\text{wall}}}{\text{vol}} = \frac{32v^2B^2d}{\pi^3\rho W}\sum_{n-\text{odd}}\frac{\tan(n\pi W/d)}{n^3} \tag{9.36}$$

The summation converges rapidly to 0.97 for $d = 2W$ and can generally be approximated as unity. WSK noted that if the loss is equated to the rate at which power is supplied per unit length of sample, $M_s Hvd$, an expression for wall velocity results:

$$v = \frac{\pi^2\rho}{32B_m d}H \tag{9.37}$$

This linear dependence of wall velocity on field is widely observed. The wall velocity can be understood in more general terms from an equation of motion

$$m\ddot{x} + \beta\dot{x} + kx = 2H_{\text{ext}}\cdot M \tag{9.38}$$

where m is domain wall mass, a measure of its inertia (because the wall responds instantaneously well below the GHz range, it can be ignored), β is a viscosity per unit area due to eddy currents, k is a restoring force density due to the average of the local inhomogeneities over the area of the wall, and $2H_{\text{ext}}\cdot M_s$ is the pressure on the wall due to the applied field H_{ext}. Experiments show that below microwave frequencies, the wall responds instantaneously to an applied field with a constant, terminal velocity. Thus, the inertial term can be neglected and Eq. (9.39) is solved for the wall velocity

$$\dot{x} = \frac{1}{\beta}\left(2M_s H_{\text{ext}} - \left\langle\frac{\partial\sigma(x)}{\partial x}\right\rangle\right) = \frac{1}{\beta}2M_s H^{\text{eff}} \tag{9.39}$$

where kx has been written as $-\langle\partial\sigma_{\text{dw}}(x)/\partial x\rangle$, the average of the position-dependent domain wall energy density gradient. The effective field is given by H_{ext} reduced by the viscosity effects

$$H^{\text{eff}} = H_{\text{ext}} - \frac{1}{2M_s}\left\langle\frac{\partial\sigma(x)}{\partial x}\right\rangle \tag{9.40}$$

Equation (9.39) says that the wall velocity depends on a balance between the applied field and a viscous drag field, H_0:

$$v \propto \frac{1}{\beta}(H_{\text{ext}} - H_0) = \frac{1}{\beta} 2M_s H^{\text{eff}} \tag{9.41}$$

This linear relation between wall velocity and H_{ext} (Fig. 9.24) has been verified in a variety of materials (Sixtus and Tonks 1931, DeBlois 1958, Becker 1963, Gyorgy 1963, O'Handley 1975). The viscous drag field H_0 is not the coercive field. These measurements are made by determining the wall velocity in a constant applied field. H^{eff} becomes zero when the average gradient in the pinning potential exactly balances the effect of the applied field. The slope of the velocity–field curve gives the inverse of the viscous damping term, which can be related either to $\partial\sigma_{\text{dw}}(x)/\partial x$ at low frequencies, to eddy-current damping at moderate frequencies, or to magnetic relaxation at very high frequencies.

The energy dissipated per unit wall area by the action of this viscous force $F_v = -\beta v = -2\pi M_s \cdot H^{\text{eff}}$ [Eq. (9.39)] on the wall moving from point A to point B is given by the integral of $\boldsymbol{F}_v \cdot \boldsymbol{dx}$, or

$$\frac{E_\beta}{\text{area}} = \int_A^B \beta \dot{x} dx = \int_{t_1}^{t_2} \beta \dot{x}^2 dt$$

The power loss per unit volume around a given wall is, therefore

$$P = \frac{1}{W_{\text{AB}} \Delta t} \int_{t_A}^{t_B} \beta \dot{x}^2 dt \tag{9.42}$$

where W_{AB} is the distance a wall travels during a half-cycle; if the material is driven to saturation each half-cycle, W_{AB} is the average distance between domain walls, W.

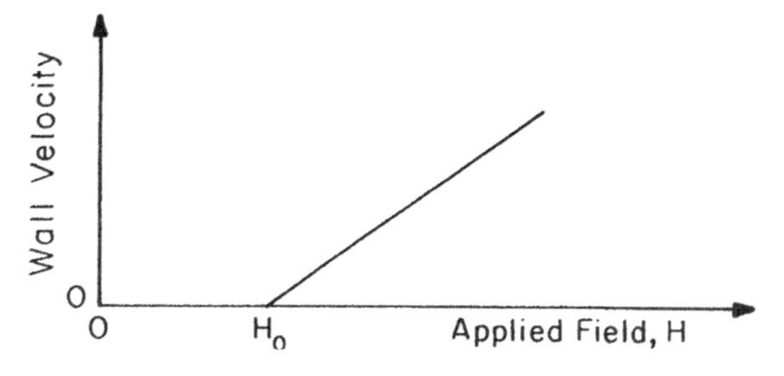

Figure 9.24. Variation of domain wall velocity with applied field. H_0 is the threshhold field for wall motion and is generally smaller than the coercive field. References in text indicate a variety of materials in which such behavior is observed.

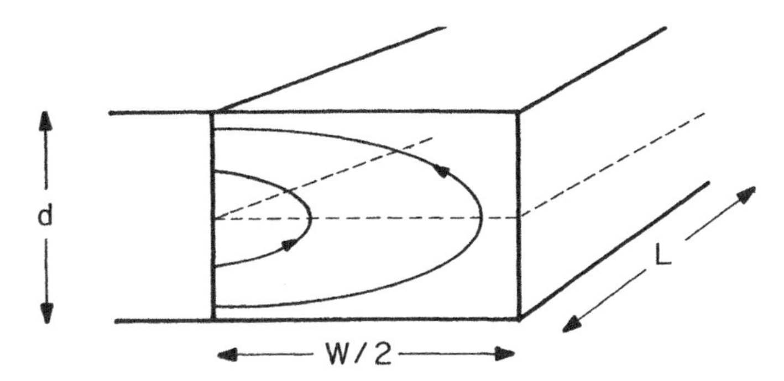

Figure 9.25. Interior of material showing dimensions and sense of eddy currents ahead of the moving domain wall of Figure 9.23.

It is important to see that the power loss density for wall motion goes as the square of the wall velocity while for the uniform magnetization process, $P \propto \omega^2$ [Eq. (9.30)]. For sinusoidal flux conditions the velocity is $dx/dt = (B_m/B_s)\omega W_{AB} \cos(\omega t)$. Using this expression in Eq. (9.42), the power loss density is seen to go as $\beta W_{AB}\omega^2 B_m^2$; thus, the smaller the wall spacing, the lower the loss. An estimate of the integral in Eq. (9.42) can be had by assuming the flux change to occur over the wall cross-sectional area $\delta_{dw}d$. The time required for the flux at a point to change due to the passage of a wall of thickness δ_{dw} and velocity v is $\Delta t = \delta_{dw}/v$, where $v = W_{AB}\omega/\pi$. Thus the induced voltage about the wall, $V = A\Delta B/\Delta t$, is given by $V = 2B_m\omega W_{AB}d/\pi$. The voltage induced about a moving domain wall is comparable to that for the classical case, $4B_m\omega W_{AB}d$. Here, sinusoidal flux variation has been assumed, which is not always the case. (Sometimes the H field is sinusoidal $e^{i\omega t}$, so $B(t) = \mu(H)H$ contains odd harmonics of the drive frequency.) The resistance of the current path, $R \approx \rho l_i/A_i$, is given by $R \approx 2\pi(\rho/W)$, where the path length for the current is $l_i \approx \pi d$ and the cross-sectional area through which it flows is $A_i \approx Wd/2$ (see Fig. 9.25). This resistance is very much reduced from the classical value, $4\rho/d$, where the current path follows the sample cross-sectional perimeter.

Thus the micro-eddy-current power loss per unit volume for a single domain wall, $P = V^2/RdW_{AB}L$, is given by

$$\frac{P_{\text{micro}}}{\text{Vol}} = \frac{4\omega^2 B_m^2 W_{AB} d}{2\rho\pi^3} = \frac{P_{\text{class}}}{\text{Vol}} \frac{W_{AB}}{2\pi^3 d} \tag{9.43}$$

The functional dependence is the similar to that of the classical case [Eq. (9.30)], but the magnitude here is roughly $W_{AB}/2\pi^3 d$ larger than the classical result. The difference is the anomalous loss shown in Figure 9.22.

9.4.3 Multiple Domain Walls

Note in Figure 9.26 that the introduction of more walls ($1 \rightarrow N$) decreases the wall spacing to $W_{AB} = W/N$, and hence is expected to lower the single-wall loss toward the classical value.

In Eq. (9.43), W_{AB} can be replaced with W/N to approximate the effects of multiple walls. However, it must be remembered that if the sample is not driven to saturation, W is less than the domain wall spacing. At large wall spacing most of the current flows in a volume considerably smaller than the volume between walls. At small W/N the two volumes approach each other. This effect tends to make eddy-current loss smaller than estimated at large wall spacing. What actually happens as the number of walls increases is that the eddy currents from two adjacent walls interfere with each other and reduce the micro-eddy-current loss.

Pry and Bean (PB) (1958) extended the WSK treatment of a single domain wall to a sample with several walls present across its width; the equilibrium wall spacing was $2W$ and the sample thickness, d.

In the limit of many domain walls (i.e., small W/d), the PB loss reduces to the uniform case [classical Eq. (9.30)]. In the limit of few domain walls (i.e., large W/d), the PB result approaches that for a single domain wall (WSK).

Pry and Bean solved the same set of equations (9.31)–(9.33), as did WSK, but with two added conditions: (1) the normal component of current density J_x is continuous across a wall, and (2) the current adjacent to a given wall is determined by contributions from the motion of *all* walls. The PB result for large wall spacing W_{AB} is proportional to the classical loss [Eq. (9.32)]:

$$\frac{P_{\text{PB}}}{P_{\text{class}}} = 1.628\left(\frac{2W}{d}\right)$$

Figure 9.27 summarizes the PB results normalized to the classical (uniform) loss case and the single-wall expression of WSK.

While all of these AC magnetization models predict the loss to vary as $\omega^2 B_m^2$, experiments usually show the loss to increase with a smaller power dependence of the frequency, often closer to $\omega^{1.6}$. The dependence on maxi-

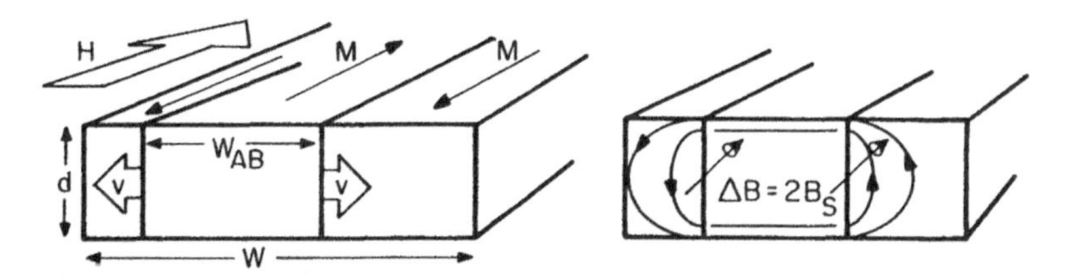

Figure 9.26. Similar to Figure 9.23 but now showing interference of eddy currents from two nearby walls moving in opposite directions.

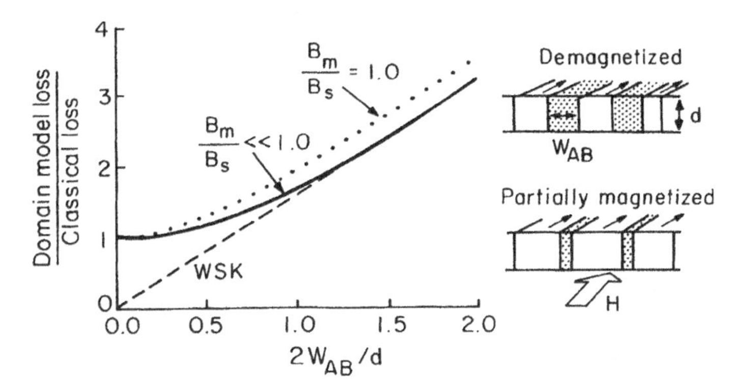

Figure 9.27. Ratio of loss in the domain wall (PB) model to that in the classical (WSK, no walls) model for complete, $B_m/B_s = 1$, and partial, $B_m/B_s \ll 1$, driving.

mum flux density B_m, to which the magnetic material is driven during the loss measurement, is more variable. The losses will increase rapidly if one attempts to drive the sample to saturation because increased drive field at fixed-frequency causes the magnetic material to switch during a shorter fraction of the cycle, effectively increasing the drive frequency.

Figure 9.28 compares the 60 Hz losses of three types of high-induction magnetic material as a function of B_m. Excessive losses generate heat that must be dissipated; as temperature increases, the saturation induction of the material decreases. Also shown in the figure is the excitation power loss. That refers to the i^2R losses in the copper windings about the magnetic material. If the anisotropy of the material is too large, very large currents will be required to drive the magnetic core to the designed flux density. Note the sharper increase in excitation loss as saturation is approached. Low-induction uses of magnetic materials include chokes, inductors, and signal transformers. High-induction applications are in power-line transformers, motors, and generators.

9.5 MICROWAVE MAGNETIZATION DYNAMICS AND FERROMAGNETIC RESONANCE

How does the magnetization of a ferromagnetic material respond when the drive-field frequency approaches the natural precession frequency of the moment in a magnetic field? In Chapter 3, it was found that the Larmour frequency, $\omega_L = \gamma \boldsymbol{B}$, for an orbital magnetic moment ($\mu_m = \gamma L$) is of order 10^{10} Hz for fields of about 0.2 T. What this means is that a magnetic moment wil *not* be able to follow an AC drive field of frequency greater than the resonance frequency of the magnetic moment.

The torque on an orbital magnetic moment is given by Eq. (3.2), $\boldsymbol{T} = -\mu_m \times \boldsymbol{B}$. But the torque is defined as the time rate of change of the

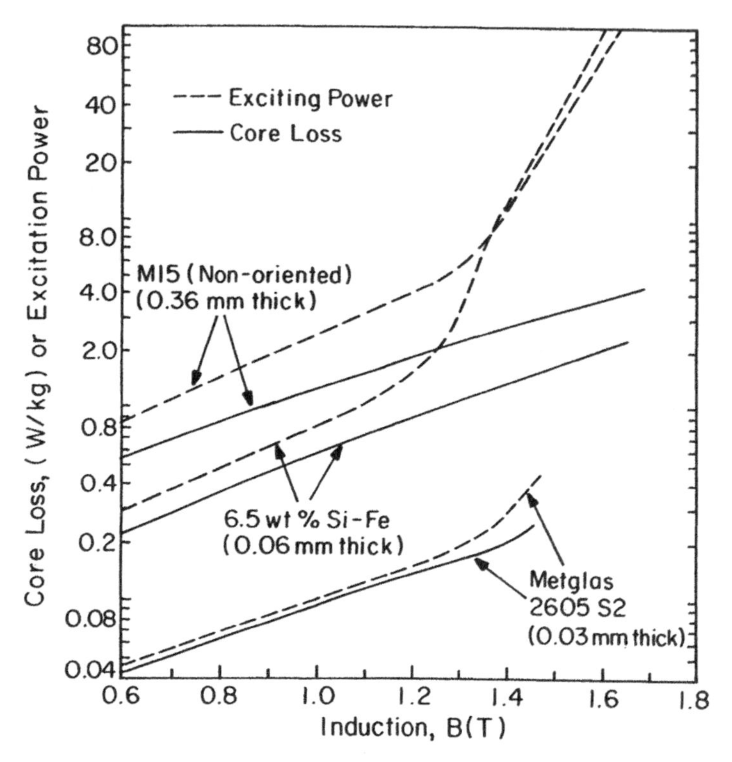

Figure 9.28. Core loss per unit mass at 60 Hz in M15-grade, nonoriented SiFe, in rapidly quenched 6.5% Si-Fe, and in the amorphous alloy, Metglas alloy 2605-S2. [After Fish (1990).]

angular momentum, $\boldsymbol{T} = d\boldsymbol{L}/dt$ (angular form of $\boldsymbol{F} = d\boldsymbol{p}/dt$). Thus the time rate of change of the angular momentum is given by

$$\frac{d\boldsymbol{L}}{dt} = -\mu_0 \boldsymbol{\mu}_m \times \boldsymbol{H}_i$$

where $\boldsymbol{H}_i$ is the internal field. For a ferromagnet with $\boldsymbol{M} = (N/V)\langle \mu_m \rangle$, we obtain

$$\frac{d\boldsymbol{M}}{dt} = -\gamma \mu_0 \boldsymbol{M} \times \boldsymbol{H}_i \tag{9.44}$$

(Recall that for an electron, $\gamma < 0$.) This last equation indicates that the magnetization precesses around the field like a top; its potential energy with respect to the field does not change.

Equation (9.44) can be solved by assuming the magnetization and the field to be made up of static parts and small time-dependent components:

$$\boldsymbol{H} = \boldsymbol{H}_i + \boldsymbol{h}e^{i\omega t} \quad \text{and} \quad \boldsymbol{M} = \boldsymbol{M}_s + \boldsymbol{m}e^{i\omega t}$$

Substitution of these fields into Eq. (9.44) leads, after some manipulation, to the solution

$$\boldsymbol{m} = \frac{\gamma^2\mu_0^2[(\boldsymbol{M}_s\cdot\boldsymbol{H}_i)\boldsymbol{h} - (\boldsymbol{H}_i\cdot\boldsymbol{h})\boldsymbol{M}_s + (i\omega/\gamma)(\boldsymbol{M}_s\times\boldsymbol{h})]}{\omega_0^2-\omega^2} \tag{9.45}$$

where $\omega_0 = \gamma H_i$. This solution can be expressed in terms of the tensor permeability, $\boldsymbol{B} = \mu\boldsymbol{H}$:

$$\mu = \begin{pmatrix} \mu & -i\kappa & 0 \\ i\kappa & \mu & 0 \\ 0 & 0 & 1 \end{pmatrix}$$

where $\mu = 1 + \omega_0\omega_M/(\omega_0 - \omega^2)$ and $\kappa = \omega_0\omega_M/(\omega_0 - \omega^2)$ and $\omega_M = 4\pi\gamma M_s$. The transverse permeability shows a dispersive form in its real component and a sharp loss peak in its imaginary component.

Equation (9.44) does not describe the fact that the moment eventually aligns with the field on the timescale of, say, a magnetometer measurement. What is missing is a relaxation process that allows the moment to lower its energy with respect to the field. If it is assumed that the rate of relaxation is proportional to the amount by which the moment is out of equilibrium, $[-\boldsymbol{M}\cdot\boldsymbol{B} - (-MB)]/B = M_s - M_z$, then a loss term must be added to Eq. (9.44):

$$\frac{\partial M_z}{\partial t} = -\gamma(\boldsymbol{M}\times\boldsymbol{B})_z - \frac{M_s - M_z}{\tau_1} \tag{9.46}$$

The longitudinal relaxation time τ_1 is a measure of the rate at which the magnetization aligns with the field. Similarly, the transverse components of the magnetization relax toward zero, but with a different relaxation time:

$$\frac{\partial M_{x,y}}{\partial t} = -\gamma(\boldsymbol{M}\times\boldsymbol{B})_{x,y} - \frac{M_{x,y}}{\tau_2} \tag{9.47}$$

Assuming that resonance occurs in a field of 0.2 T, τ is expected to have a magnitude of order $(\gamma B)^{-1} \approx 5\times10^{-11}$ s.

Equations (9.46) and (9.47) are the Bloch–Blombergen forms of the equation of motion for the magnetization. Two different relaxation times are needed because there are two different ways in which the magnetization may give up its original projection on the field. This is summarized in Figure 9.29.

The RF field creates a uniform precession of the magnetic moments about the applied field $\boldsymbol{B}$. The fact that the transverse components of the magnetization are in phase throughout the material means that this state can be

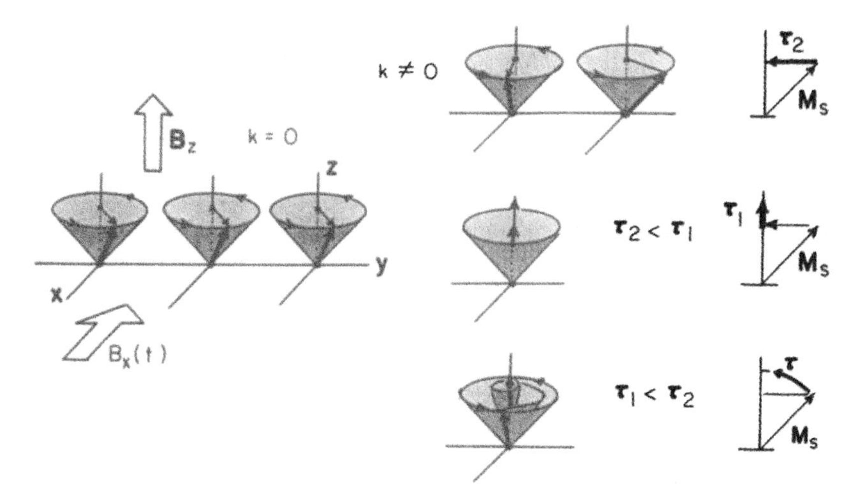

Figure 9.29. Left, illustration of uniform precession of magnetization in presence of static field B_z and an orthogonal microwave field B_x. Right, three modes of decay of the uniform precession mode: top, decay of $k = 0$ mode to $k \neq 0$ mode with transverse relaxation time τ_2; center, reduction in number of $k \neq 0$ modes with characteristic longitudinal relaxation time τ_1. This process increases magnetization component along the field. When the longitudinal process is faster than the transverse process, the Landau–Liftshitz relaxation shown at lower right occurs.

described microscopically as a $k = 0$ spin wave superimposed on the z component of the magnetization. [A spin wave, or magnon, is a quantized excitation of the spin system having a well-defined momentum, $|k| = 2\pi/\lambda$, and energy, $h\omega = 4JS(1 - \cos ka)$ where J is the exchange integral, S is the magnitude of the local spin, and a is the lattice constant. See Chapter 3, Section 3.7.1, or Kittel (1986).]

This uniform precession state can relax by two quite different processes. The first process, shown at upper right in Figure 9.29, is the decay of the $k = 0$ magnon into a $k \neq 0$ magnon. This process increases the entropy of the system (and so is more likely at elevated temperatures) without decreasing the magnetic potential energy. It is described by the *transverse relaxation time* τ_2. There is evidence to suggest that this process occurs at defects where magnetostatic fields can dephase the $k = 0$ precession.

Another relaxation process destroys two $k \neq 0$ magnons (created in the transverse relaxation process shown in Fig. 9.29) while creating only one new magnon and a phonon. This process, described by the *longitudinal relaxation time* τ_1, represents a transfer of energy from the spin system to the lattice. It therefore results in a decrease of the magnetic potential energy; the average magnetization aligns with the field. If $\tau_2 \ll \tau_1$, these two processes occur essentially in sequence as illustrated in Figure 9.29, right, middle. If $\tau_1 \leqslant \tau_2$, the system relaxes with constant $|M|$ along the arclike path illustrated in Figure

9.30, right, lower panel. This constant $|M|$ situation is described by the Landau–Liftshitz (1935) (see References list in Chapter 8) form of the equation of motion:

$$\frac{\partial \boldsymbol{M}}{\partial t} = -\gamma(\boldsymbol{M} \times \boldsymbol{B}) - \frac{\boldsymbol{M} \times (\boldsymbol{M} \times \boldsymbol{B})}{M^2\tau} \tag{9.48}$$

Note that here the relaxation always carries the tip of the magnetization vector along a path that is perpendicular to its length. As soon as $k \neq 0$ magnons are created, they combine and transfer energy to the lattice.

This classical phenomenology of magnetic resonance can be expressed quantum mechanically in terms of transitions between two energy levels split by the application of a magnetic field (cf. Fig. 3.15). A magnetization with an arbitrary orientation will have an energy $-\boldsymbol{M} \cdot \boldsymbol{B}$ with respect to the applied field. It can relax to a lower energy state, defined by the selection rule $\Delta m_s = \pm 1$, via the processes described above. The microwave field will tend to excite the magnetization to higher energy states of precession. Resonance occurs when $h\omega = \Delta E$.

A magnetic resonance experiment generally involves sweeping the magnetic field while the sample is exposed to a microwave field of fixed frequency. For

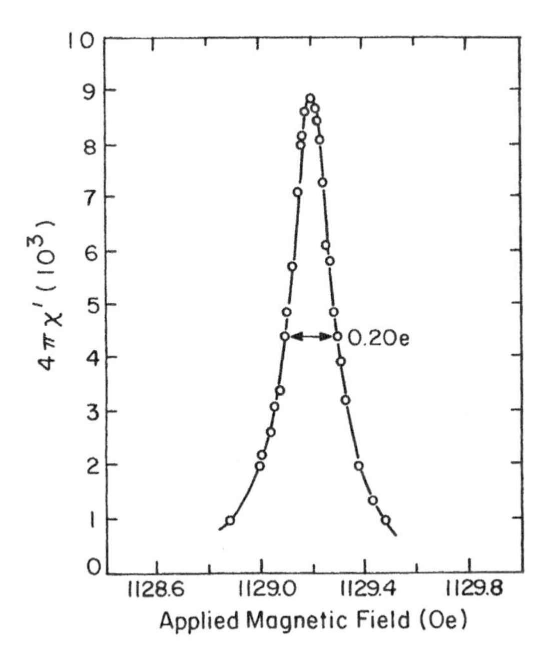

Figure 9.30. Ferromagnetic resonance line for a polished YIG sphere at room temperature in a microwave field of 3.33 GHz [after LeCraw and Spencer (1962).]

weak applied field strengths, the magnetization is not saturated, $M < M_s$, and the level splitting is smaller than $h\omega$. In this case, the natural resonance frequency of the magnetization is lower than the drive field and there the magnetic response is weak and 180° out of phase with the microwave field. In a strong DC field that splits the energy levels more than $h\omega$, the magnetization follows the field with a small phase lag. Right at resonance, the magnetization lags the microwave field by 90°. The half-width of the resonance line at half power maximum gives the relaxation time: $(\Delta\omega)_{1/2} = 1/\tau_2$.

Because of the large magnetization in a ferromagnet, the resonance condition is altered by magnetostatic fields associated with the sample shape. The internal field components become $B_j^{\text{in}} = B_j^{\text{out}} - N_j M_j$, $j = x, y, z$. In the limit of a weak microwave field, $M_z \approx M_s$, the Bloch–Blombergen equations become

$$\frac{\partial M_s}{\partial t} = -\gamma[M_y(B_z - N_z M_z) + M_z N_y M_y] = -\gamma M_y[B + (N_y - N_z)M_s]$$

$$\frac{\partial M_y}{\partial t} = -\gamma M_x[B + (N_x - N_z)M_s]$$

Assuming the transverse magnetization components to vary as $e^{i\omega t}$, these equations are solved for

$$\omega^2 = \gamma^2[B + (N_x - N_z)M_s][B + (N_y - N_z)M_s] \tag{9.49}$$

Knowing the sample shape, it is possible by a resonance experiment to determine a relaxation time from the linewidth and the saturation magnetization or gyromagnetic ratio. If there is appreciable anisotropy, that can be determined if M_s is known (a uniaxial anisotropy behaves like a demagnetizing field).

Figure 9.30 (LeCraw and Spencer 1962) shows the magnetic resonance absorption line of a polished yttrium iron garnet (YIG) sphere. Note the very narrow linewidth of 0.2 Oe at a resonance field of 1129.2 Oe. This implies a very small imaginary component to the transverse, microwave permeability.

For ferromagnetic metals, while the relaxation mechanisms described by τ_1 and τ_2 are still operative, the linewidth is generally dominated by eddy-current damping. The largest eddy currents come from the uniform precession. Eddy-current heating takes energy from the spin system directly to the lattice. The skin depth of the microwave field plays a major role in determining the eddy-current loss of ferromagnetic resonance. A resonance line for a metallic Gd film is shown in Figure 9.31.

In metals for which the skin depth is less than the sample dimensions, the gradient in the microwave field near the surface generates copious spin waves that enhance the damping.

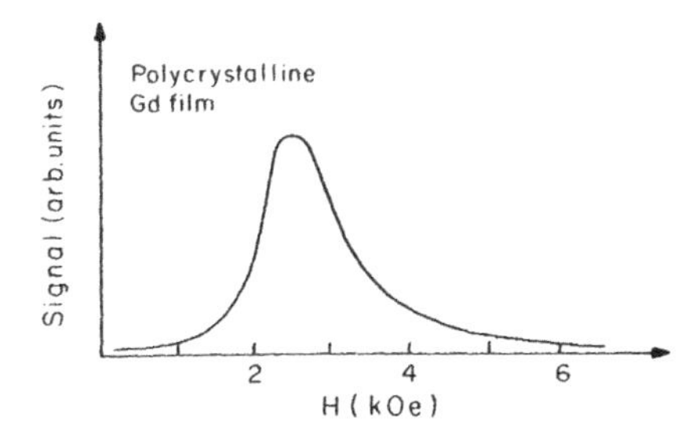

Figure 9.31. Ferromagnetic resonance line for a Gd thin film at room temperature in a microwave field of 8.8 GHz [O'Handley and Juretschke (unpublished).]

9.6 SUMMARY

The M–H loop has been shown to be the result of a complex magnetization process involving domain wall motion (when the field is parallel to the easy axis of a sample or grain) and domain magnetization rotation (when the field is perpendicular to the easy axis). For intermediate angles of the field relative to the easy axis, the M–H loop is more rounded and slower to saturate. Cubic systems can show discontinuous magnetization jumps for certain conditions. Demagnetizing fields can shear over a loop that is intrinsically vertical and can lower the slope of a hard-axis loop that intrinsically would saturate at $H = H_a$.

Various defects and their interaction with domain walls were considered. The strength of the pinning effect of a defect depends on two factors: (1) the difference between the magnetic properties of the defect and the matrix and (2) the ratio of the defect size to the domain wall size. For small defects, the coercivity increases roughly linearly with defect size; for defects larger than the wall thickness, the coercivity is either constant (sharp defect) or varies as the inverse of the defect size (fuzzy defect). These models are qualitatively supported by coercivity data in a variety of magnetic systems.

As magnetic materials are driven to higher frequencies, the magnetization process is more rapid and Faraday's law demands that a voltage be induced in the sample, in a closed path normal to the flux change. This voltage induces eddy-currents if the material is not an insulator. When domain walls are present, the eddy-current loss is anomalously large compared to the classical loss for uniform magnetization. The worst case is for a single domain wall, where power loss goes as wall velocity squared. Introduction of more domain walls leads to eddy-current cancellation and the loss drops toward that for uniform magnetization. While the models predict magnetic core loss to vary as $\omega^2 B_m^2$, the frequency dependence of the loss is often closer to 1.6 and the induction power dependence increases from about 1.6 as the material is driven closer to saturation.

At microwave frequencies, eddy-current losses dominate in metals and other microscopic relaxation processes play a role in insulators.

PROBLEMS

9.1 Solve for the magnetization versus field for **(a)** a thin film of amorphous iron boron silicon (assume $\mu_0 M_s = 1.6\,\mathrm{T}$ and $K = 0$) with the field applied normal to the film surface and **(b)** a single-crystal sphere of Ni with the field applied along the [111] direction.

9.2 Does the energy required for complete saturation in the Stoner–Wohlfarth model as described by Eq. (9.13) diverge for $0 < \theta_0 < \pi/2$ or take on a finite value?

9.3 Consider the magnetization process in a single-domain particle having cubic anisotropy using a field applied along an easy axis orthogonal to the initial magnetization. (Because of the uniqueness of this initial condition, i.e., the special way the sample is prepared before application of the measuring field, the result you derive will not be typical for cyclic magnetization.) Write and plot the energy density against θ, then find the shape of the m–h curve. Locate the critical parameters by combining the equilibrium condition with the condition that the solution to $f'(\theta) = 0$ also be an inflection point.

9.4 Consider the two dimensional magnetization of a thin film with fourfold in-plane anisotropy $f_a = K_1 \cos^2 2\theta$ with an external field and stress σ_{xx} applied collinearly as shown below:

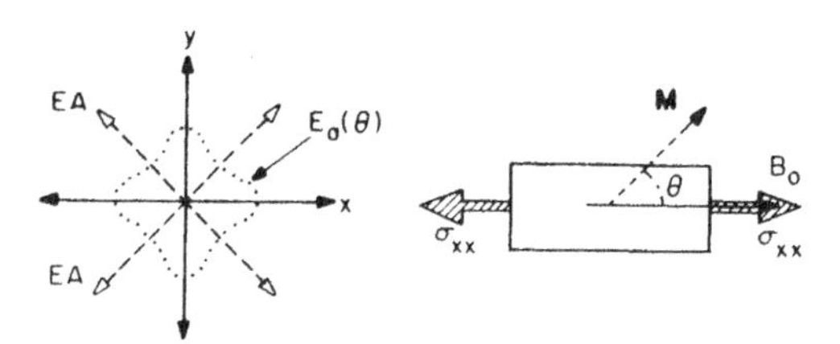

Solve for the equation of magnetization and sketch the results for $B_1 e_{xx} > 0$ and <0. Compare with the results in Problem 9.3.

9.5 Consider a thin film with in-plane cubic anisotropy K_1 and a superimposed uniaxial anisotropy with easy axis along one of the fourfold EAs. A field is applied in-plane, perpendicular to the uniaxial easy axis.

(a) Write the expression for the free energy.

(b) Sketch the energy surfaces for the various terms.

(c) Write the equation of motion and sketch m versus H.

(d) How does m–H differ for $K_u >$ or $< 5K_1$?

9.6 Work out the steps to derive the field dependence of magnetization $M(H)$ for a material with uniaxial magnetic anisotropy and H applied perpendicular to the easy direction of magnetization.

9.7 Explain why the coercivity of amorphous alloys goes through a minimum as the magnetostriction constant λ_s goes through zero. Give formula(s) to support your explanation.

9.8 Derive an expression for rotational permeability in cubic anisotropy, and compare it with Eq. (9.8) for uniaxial anisotropy.

9.9 Determine the g factor of the YIG sphere in Figure 9.30 using Eq. (9.15).

BIBLIOGRAPHY

Köster, E., and T. C. Arnoldussen, in *Magnetic Recording Handbook*, C. D. Mee and E. D. Daniel, eds., McGraw-Hill, New York, 1989, p. 101.

Richter, H. J., *IEEE Trans.* **MAG-29**, 2185, (1993).

Schilling, J. W., and G. L. Houze, Jr., "Magnetic properties and domain structure in grain-oriented 3% Si-Fe," *IEEE Trans.* **MAG-10**, 195 (1974).

Shtrikman, S., and D. Treves, in *Magnetism*, Vol. III, G. T. Rado and H. Suhl, Academic Press, New York, 1963, p. 395.

Sparks, M., *Ferromagnetic Relaxation Theory*, McGraw Hill, New York, 1964).

Wohlfarth, E. P., in *Magnetism*, Vol. III, G. T. Rado and H. Suhl, Academic Press, New York, 1963, p. 351.

REFERENCES

Aharoni, A., "Perfect and imperfect particles," *IEEE Trans.* **MAG-22**, 478 (1986).

Aharoni, A., "Magnetization curling," *Phys. stat. solidi* **16**, 3 (1996).

Alben, R., J. Budnick, and G. S. Cargill III, in *Metallic Glasses*, J. J. Gilman and J. J. Leamy, ed., American Society for Metals Park, OH, 1978, p. 304.

Becker, J. J., *J. Appl. Phys.* **34**, 1327 (1963).

Brown, W. F., *Phys. Rev.* **60**, 139 (1941).

Brown, W. F., *Phys. Rev.* **105**, 1479 (1957).

Chien, C.-L., *J. Appl. Phys.* **60**, 5267 (1991).

Croat, J. J., et al., *J. Appl. Phys.* **55**, 2078 (1984).

DeBlois, R. W., *J. Appl. Phys.* **29**, 459 (1958).

Dijkstra, L. J., and C. Wert, *J. Appl. Phys.* **79**, 979, 1950.

Egami, T., and P. J. Flanders, AIP Conference Proceedings, No. 29, American Institute of Physics, New York, 1976, p. 220.

Fish, G., *Proc. IEEE* **78**, 947 (1990).

Frei, E. H., S. Shtrikman, and D. Treves, *Phys. Rev.* **106**, 446 (1957).

Friedberg, S., and D. I. Paul, *Phys. Rev. Lett.* **34**, 1234 (1975).

Gyorgy, E. M., in *Magnetism*, Vol. III, G. T. Rado and H. Suhl, eds., Academic Press, New York, 1963, p. 525.

Hagedorn, F. B., *J. Appl. Phys.* **41**, 2491 (1970).

Herzer, G. *IEEE Trans.* **MAG-25**, 3327 (1989).

Herzer, G., "Grain size dependence of coercivity and permeability in nanocrystalline ferromagnets," *IEEE Trans.* **MAG-26**, 1397 (1990).

Herzer, G., "On the theoretical understanding of nanocrystalline soft magnetic materials," *J. Mater. Eng. Perform.* **2**, 193 (1993).

Imry, Y., and S. K. Ma, *Phys. Rev. Lett.* **35**, 1399 (1975).

Jacobs, I. S., and C. P. Bean, *Phys. Rev.* **100**, 1060 (1955).

Jacobs, I. S., and C. P. Bean, in *Magnetism*, Vol. III, G. T. Rado and H. Shul, eds., Academic Press, 1963, p. 271.

Kersten, M., in *Problem of the Technical Magnetization Curve*, ed., Springer-Verlag, Berlin, 1938, p. 42.

Kersten, M., *Grundlagen einer Theorie der Ferromagnetichen Hysterese und der Koertzitworkraft*, S. Hirzel, Leipzig, 1943; reprinted by J. W. Edwards, Ann Arbor, MI, 1943).

Kittel, C., *Introduction to Solid State Physics*, 6th ed., Wiley, New York, 1986, p. 430.

Kneller, E., and Luborsky, F. E., *J. Appl. Phys.* **34**, 656 (1963).

Knowles, J. E., *IEEE Trans.* **MAG-20**, 2588 (1984).

Kondorsky, E., *Phys. Z. Sowjetunion* **11**, 597 (1937).

Köster, E., and T. C. Arnoldussen, in *Magnetic Recording Handbook*, C. D. Mee and E. D. Daniel, eds., McGraw-Hill, New York, 1989, p. 101.

Kronmüller, H., *J. Phys. Coll.* **C8**, 618 (1980).

LeCraw, R. C., and E. G. Spencer, *J. Phys. Soc. Jpn.* **17**, Suppl. B1, 401 (1962).

McGuire, T. R., and P. J. Flanders, in *Magnetism and Metallurgy*, Vol. 1, A. E. Berkowitz and E. Kneller, eds., Academic Press, New York, 1969, p. 123.

Mishra, R., and G. Thomas, *Proc. 4th Internatl. Workshop on Rare-Earth Permanent Magnets*, Hakone, Japan, 1979, p. 301.

Néel, L., *Compt. Rend.* **224**, 1550 (1947).

O'Handley, R. C., and H. J. Juretschke (unpublished).

O'Handley, R. C., *J. Appl. Phys.* **46**, 4996 (1975).

O'Handley, R. C., J. Megusar, S. Sun, Y. Hara, and N. J. Grant, *J. Appl. Phys.* **57**, 3563 (1985).

Ojima, T., S. Tomegawa, and T. Yonezama, *IEEE Trans.* **MAG-13**, 1317 (1977).

Paul, D. I., "General theory of the cohesive force due to domain wall pinning," *J. Appl. Phys.* **53**, 1649 (1982).

Paul, D. I., and A. Creswell, *Phys. Rev.* **B48**, 3803 (1993).

Pry, R. H., and C. P. Bean, *J. Appl. Phys.* **29**, 532 (1958).

Richter, H. J. *J. Appl. Phys.* **65**, 3597 (1989).

Sagawa, M., S. Hirosawa, H. Yamamoto, S. Fujimura, and Y. Matsuura, *Jpn, J. Appl. Phys.* **26**, 785 (1987).

Sixtus, K. J., and L. Tonks, *Phys. Rev.* **37**, 930 (1931).

Stoner, E. C., and E. P. Wohlfarth, *Trans. Roy. Soc.* **A240**, 599 (1948).

White, R. M., *Introduction to Magnetic Recording*, IEEE Press, New York, 1985, p. 13.

Williams, H. J., W. Shockley, and C. Kittel, *Phys. Rev.* **80**, 1090 (1950).

Wohlfarth, E. P., *Adv. Phys.* **8**, 87 (1959).

Yonezawa, T., A. Fukuno, and T. Ojima, *Proc. 4th Internatl. Workshop on Rare-Earth Permanent Magnets*, Hakone, Japan, (1979), p. 407.

CHAPTER 10

SOFT MAGNETIC MATERIALS

This chapter discusses the properties of a variety of soft magnetic materials, including both metals and ceramics. Frequent reference will be made to earlier chapters where the domain structure and fundamental property requirements for various materials are outlined and explained. This chapter includes some material on amorphous soft magnetic materials. In later chapters, the scientific basis of amorphous magnetic alloys (Chapter 11), nanocrystalline materials (Chapter 12), and magnetic thin films (Chapter 16) will be considered.

Soft magnetic materials are generally used in applications where high permeability ($\mu = B/H$) is required. Referring to the magnetization curves for single-crystal iron (Fig. 6.1)—the prototypical soft magnetic material—it can be seen that the most magnetization is achieved for the least applied field when the field is applied along one of the $\langle 100 \rangle$ directions.

In soft magnetic materials, the flux density is dominated by the contribution from M. Hence, the permeability is expected to be maximized when anisotropic ($K_i \neq 0$) polycrystalline materials are textured so that crystallographic easy directions of magnetization lie predominantly along the direction of the applied field.

Figure 10.1 reviews the magnetization process for a single crystal of iron. Assuming a simple domain structure in the demagnetized state (a), the initial magnetization process involves domain wall motion. This relatively easy process accounts for the initial permeability and the steep rise in magnetization for $H \approx H_c$ (states a–b). Beyond point b the magnetization in the closure domains must be rotated from the easy [010] directions to the easy [100]

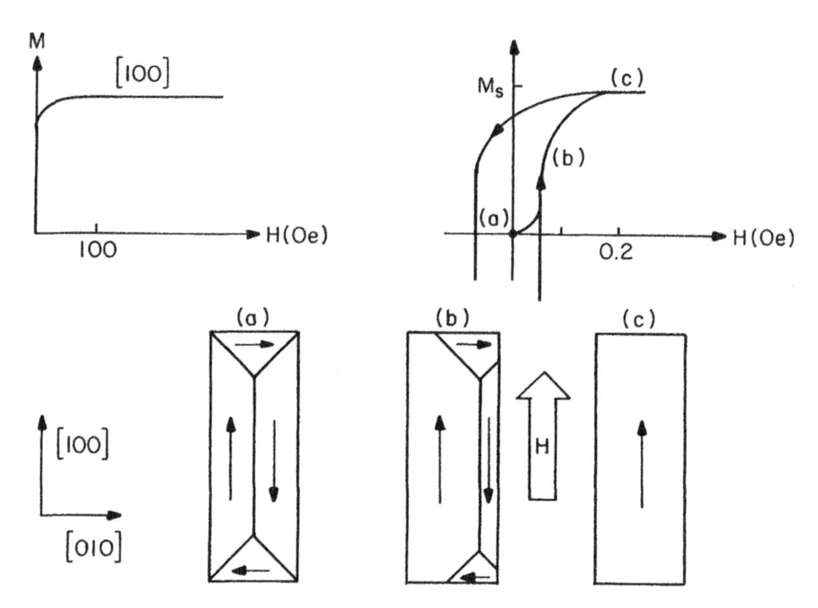

Figure 10.1 Above, typical magnetization curves for iron along [100] displayed on high-field and low-field scales; below, schematic of domain structure changes with applied field.

direction. This magnetization rotation process generally requires more energy than wall motion [Eq. (9.8) Problem 9.8].

10.1 IRON AND MAGNETIC STEELS

The most common application of magnetic steels is as cores in power and distribution transformers. The energy loss in these transformers generally consists of magnetic core loss and pure Joule heating loss in the copper coils. The core loss arises from (1) eddy currents induced in the core by the uniform changes in B, (2) microscopic eddy currents localized near moving domain walls, and (3) acoustic losses due to magnetostrictive deformation of the core under changing flux in the so-called supplementary domain structure (90° walls, closure domains). These three types of loss can be minimized by decreasing lamination thickness, increasing resistivity, decreasing domain size, and decreasing magnetostriction. Coil losses are i^2R losses in the electrical windings that provide the field needed to drive the flux changes in the core and in the secondary winding that carry the output power. The more easily the core can be driven to the desired flux density, the lower the coil loss. Coil loss is therefore reduced in materials with higher remanence, lower magnetic anisotropy, or better crystallographic alignment. (Obviously, it is desirable to keep the coil resistance low also.)

10.1.1 Iron and Silicon Steels

Pure iron is the prototypical soft magnetic material. It has a very high saturation flux density, $B_s = 2.2\,\text{T}$, and its cubic anisotropy leaves it with a relatively small magnetocrystalline anisotropy, $K_1 = +4.8 \times 10^4\,\text{J/m}^3$, and small magnetostriction constants, $\lambda_{100} = +21 \times 10^{-6}$, $\lambda_{111} = -20 \times 10^{-6}$. Domain images from a (100) iron single crystal taken by SEMPA are shown in Figure 1.13. Note how the magnetization follows the easy $\langle 100 \rangle$ directions leading to 90° and 180° walls.

It should be clear from Figure 10.1 that high anisotropy energy would suppress permeability in a nonoriented polycrystalline material. The higher the field needed to achieve a given level of flux density, the more coil loss or i^2R loss there will be in the copper windings of an inductive device. Also, high anisotropy increases the domain wall energy density, making walls harder to move. When the domain walls are harder to move, the coercivity increases and energy is lost in driving the material around a hysteresis loop.

Further, if the material has appreciable magnetostriction, localized strain fields due to internal stresses become localized magnetoelastic anisotropy fields that can hold the magnetization in a particular direction, thus pinning domain wall motion.

Iron is notoriously susceptible to degradation of its magnetic properties by impurities. Impurities may give rise to magnetic effects through their strain fields if magnetoelastic coupling is nonzero. Alternatively, if the impurities condense or form nonmagnetic precipitates, they eliminate the domain wall energy over the length of the precipitate and thus may pin a domain wall. Table 10.1 shows the major impurities in standard Armco iron (mill sheet from ingot).

Of those impurities listed, carbon, nitrogen, and oxygen are interstitial impurities and have a relatively high mobility in the BCC lattice of iron. They are the major impurities responsible for magnetic aftereffects, or disaccommo-

TABLE 10.1 Impurity Content of Armco Iron

Element	wt%
C	0.015
N	0.007
O	0.15
Si	0.003
P	0.005
S	0.025
Mn	0.03
Al	0.003

Source: Littmann (1971).

dation, which degrades permeability with time at certain temperatures and frequencies (see Chapter 14). Removal of impurities from common grades of iron by hydrogen annealing (typically at 750–800°C) can lead to permeabilities as high as $10^5\mu_0$. Hydrogen annealing of high-purity iron can give permeabilities as high as $10^6\mu_0$. Processing of iron in a fluorine atmosphere can lead to the presence of some FeF_2 and tends to scavenge impurities. Such processing, although impractical, gives the best soft magnetic properties achievable in iron (Nesbitt, unpublished).

From the materials point of view, a good high-frequency material also has high electrical resistivity, small dimensions normal to the direction of magnetization, and many domain walls.

It is useful to begin by modifying iron with small amounts of selected elements to increase resistivity or make other favorable property changes. Figure 10.2 shows that for iron, silicon, or aluminum additions produce the most dramatic increase in resistivity. Details of the iron-rich side of the Fe-Si phase diagram shown in Figure 10.3 indicate silicon is soluble in α-Fe up to about 4 wt%. Beyond that limit, the brittle intermetallic Fe_3Si (B2 or DO_3) phase may also be present. The addition of Si to iron causes changes in a number of other important magnetic properties (Fig. 10.4). Values of some properties of 3% Si-Fe are compared with those of α-iron in Table 10.2. Note that silicon not only increases the electrical resistivity but also significantly reduces the magnetic anisotropy and has little effect on magnetostriction up to 4%. Unfortunately the saturation flux density is decreased by about 10% for 3% Si additions.

The introduction of Si into iron also produces dramatic effects in lowering

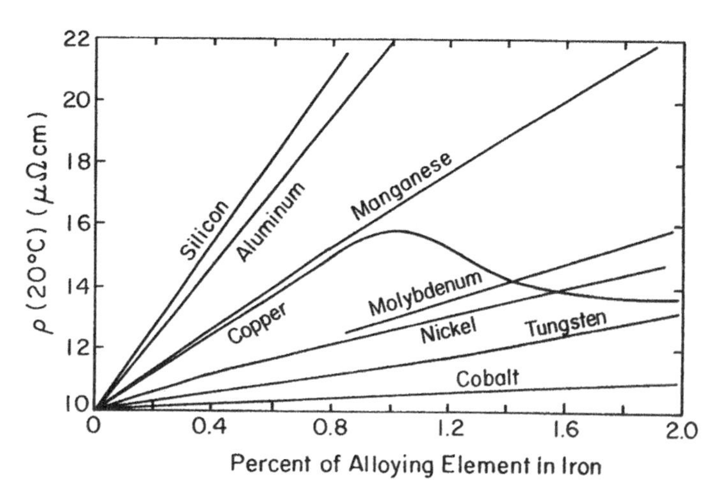

Figure 10.2 Variation in electrical resistivity of iron with addition of selected impurities. [After Bozorth (1994).]

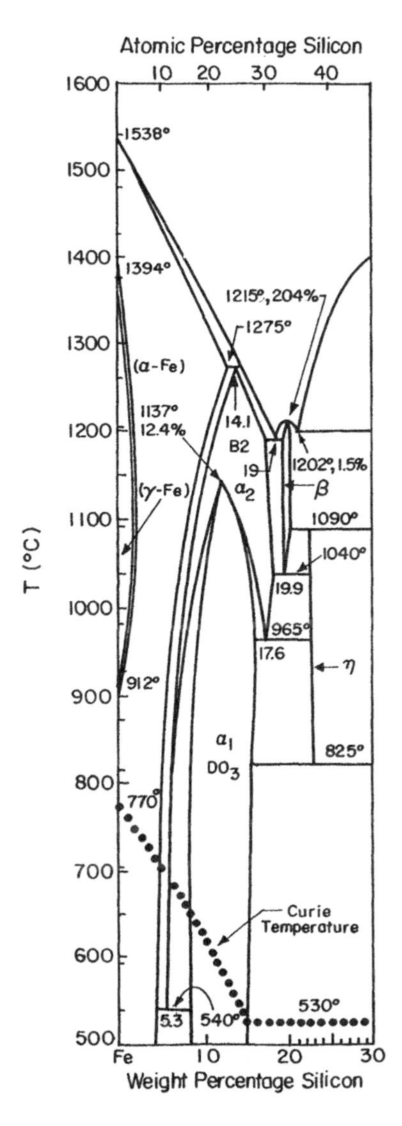

Figure 10.3 Fe-Si phase diagram (*Metals Handbook*, 1973) showing solid solubility of up to 4 at% Si beyond which the B2 phase or the more brittle, ordered DO_3 phase may be present.

the core loss. Figure 10.5 shows a monotonic decrease in core loss of low-carbon steel with increasing Si content (increasing resistivity) and lower core loss for thinner-gauge sheet. These changes are consistent with the simple model core loss presented in Chapter 9. Even though silicon reduces the saturation flux density, Figure 10.6 shows that the higher-silicon-content steel

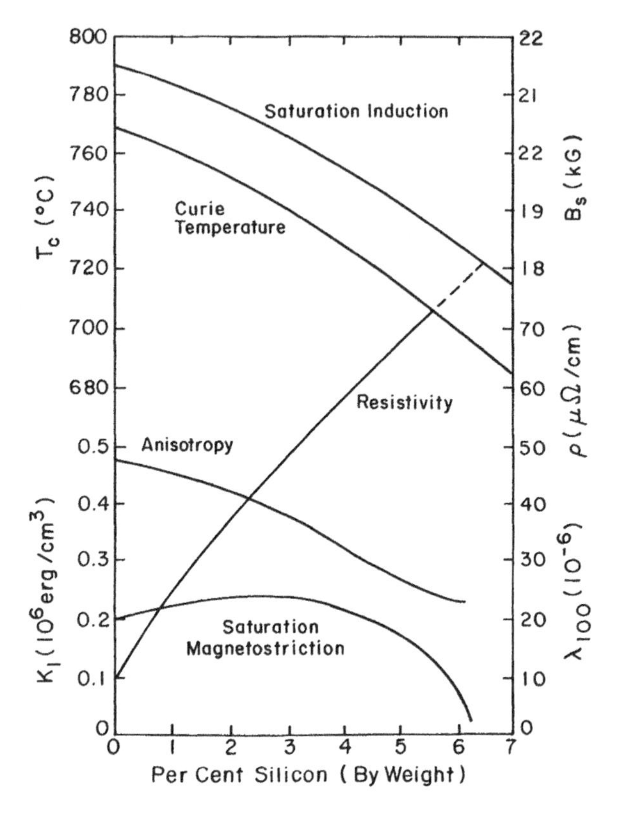

Figure 10.4 Variation of physical properties of iron with Si content [After Littman (1971).]

TABLE 10.2 Comparison of Physical Properties of Pure Iron with Those of 3% SiFe

Property	BCC Fe	3% SiFe
$4\pi M_s$ (kG)	22	20
ρ_m (g/cm^3)	7.87	7.65
ρ_e ($\mu\Omega$ cm)	10.1	48
a_0 (Å)	2.866	2.857
T_c (°C)	771	745
K_1 (10^5 erg/cm^3)	5	3.6
λ_{100} (10^{-6})	25	23
λ_{111} (10^{-6})	−20	−4

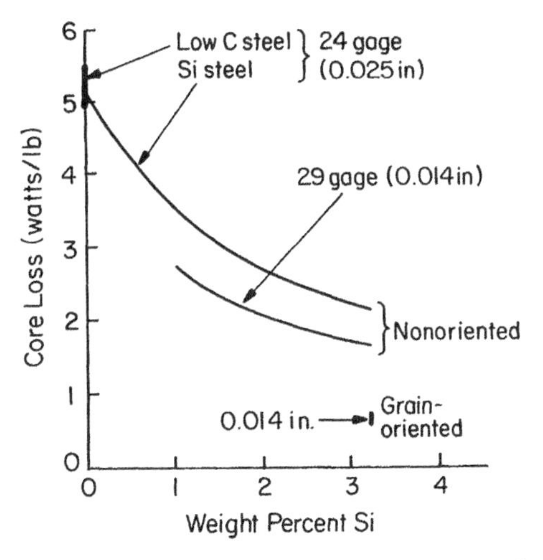

Figure 10.5 Core loss at 60 Hz and driving to 15 kG for low-carbon and silicon steels.

can be driven to considerably higher flux densities at a given level of loss. These results are for nonoriented steels that are favored in applications where the field direction changes with respect to the material, such as motors or generators.

In other applications the orientation of the magnetic laminations is fixed relative to the field axis. Further reductions in loss can be realized by texturing the material so that more of the grains have a [100] direction closer to the direction of the applied field. See "Grain-oriented" silicon steel in Figure 10.5. The permeability at a small applied field such as 10 Oe is a reliable measure of

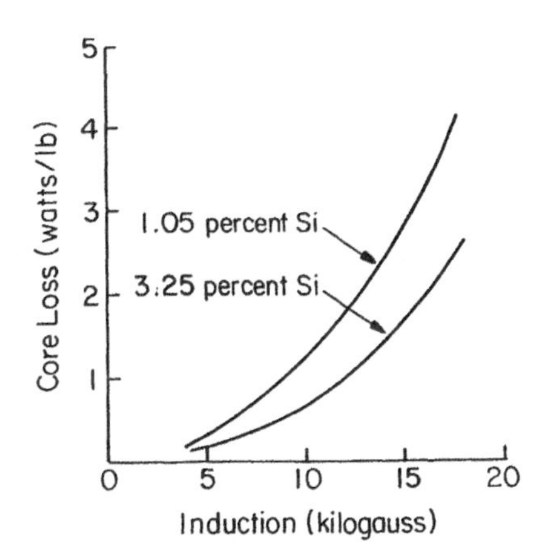

Figure 10.6 Variation of 60-Hz core loss with maximum induction for nonoriented silicon steel, 0.014 in thick.

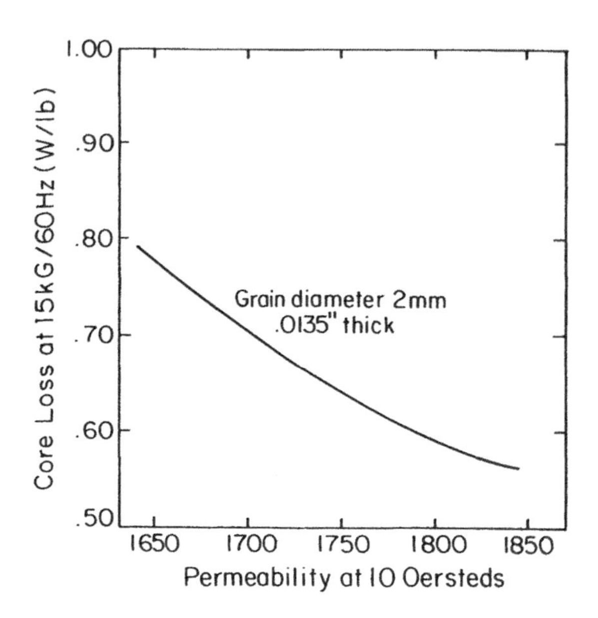

Figure 10.7 Relationship between 60-Hz, 15-kG core loss and permeability at 10 Oe in (110) [001] grain oriented 3% SiFe (Shilling and Houze 1974).

texture. The effects on core loss of increasing texture and permeability are illustrated by Figure 10.7 (Shilling and Houze 1974). Higher initial permeability correlates with lower core loss. One way of achieving this grain orientation was discovered by Goss in 1933. He found that rolling steel causes slip along {110} planes. The surface of the rolled sheet then contains more than a random number of {110} planes and the [001] direction is found to be predominantly along the roll direction after annealing at 800°C. This so-called Goss or "cube-on-edge" texture favors low field magnetization because the easy ⟨100⟩ directions have a large component along the field (roll) direction.

Control of grain size is also important for optimizing core loss. Figure 10.8 shows how core loss is minimized at a particular grain size. If the grain size is too large, there are fewer domain walls and micro-eddy-current loss is high. If the grain size is too small, the internal stresses and abundant grain boundary pinning sites increase the loss. In 3% Si steel, optimal grain size is near 0.7 mm.

Domain structure in textured Si-Fe can be refined without grain size reduction by achieving a small out-of-plane orientation θ of the [001] easy axis (Fig. 10.9*a*–*b*). The domain structure is refined by formation of closure domains visible on the top surface of the sheet. Further, as [001] tilts out of the sheet plane more, the [100] and [010] cube direction have a greater projection in the plane and a "herringbone" pattern of obliquely magnetized domains appears (*c*, *d*). For an extensive discussion of grain oriented Si-Fe (see Shilling and Houze (1974).

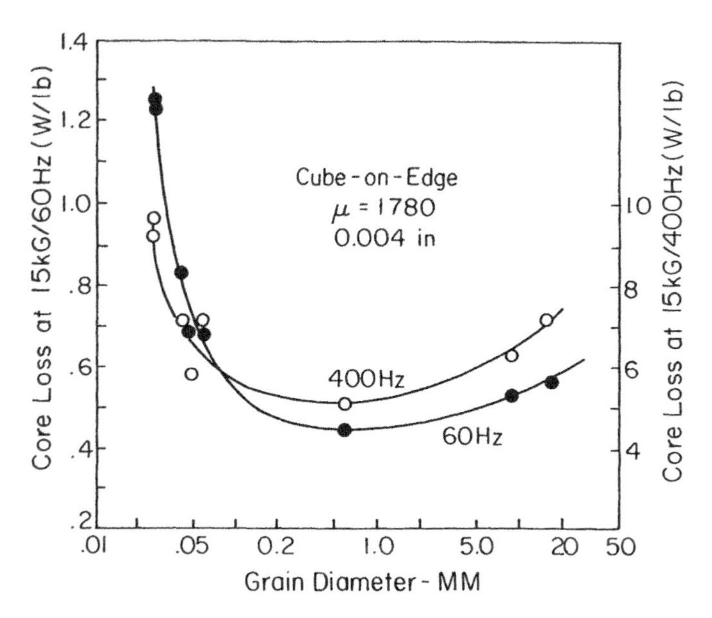

Figure 10.8 Core loss versus grain size for two different frequencies in 3.15% SiFe sheets of comparable texture and purity (Littman 1967).

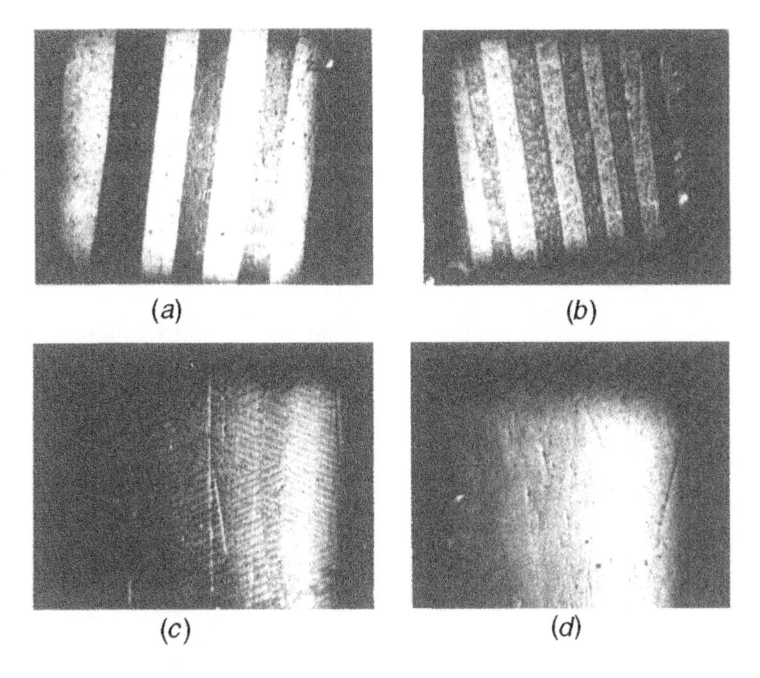

Figure 10.9 Domain structure in demagnetized 3% SiFe single crystal at 5× magnification: (*a*) [110] in plane; (*b*) [001] out-of-plane angle $\theta = 2°$; (*c*) $\theta = 4°$; (*d*) $\theta = 8°$ (Shilling and Houze 1974).

More recent developments in low-loss magnetic steels involve increasing the number of domain walls by laser scribing the surface of steel. A commercial example is Orientcore Hi B (Suzuki et al. 1972). Laser scribing leads to local strain fields that refine the domain structure.

10.1.2 Sendust

Addition of small amounts of silicon improves the soft magnetic properties of materials, but it does not follow that more is better. More than 3% silicon certainly increases resistivity and further reduces K and λ. However, magnetic materials must be able to be formed and handled. As silicon content increases, ductility decreases because of the appearance of the intermetallic DO_3 phase Fe_3Si. At 4% Si the elongation at yield drops to less than half of that for low C steel.

But there are some applications that make use of a brittle material if the magnetic properties are sufficiently improved. Those applications make use of powders to form "potted cores" and tape heads. The performance advantage comes not only with increased Si content but with Al additions as well. Figure 10.10 shows the paths of the zero anisotropy and magnetostriction lines in $Fe_{1-x-y}Si_xAl_y$. A famous and appropriately named magnetic material

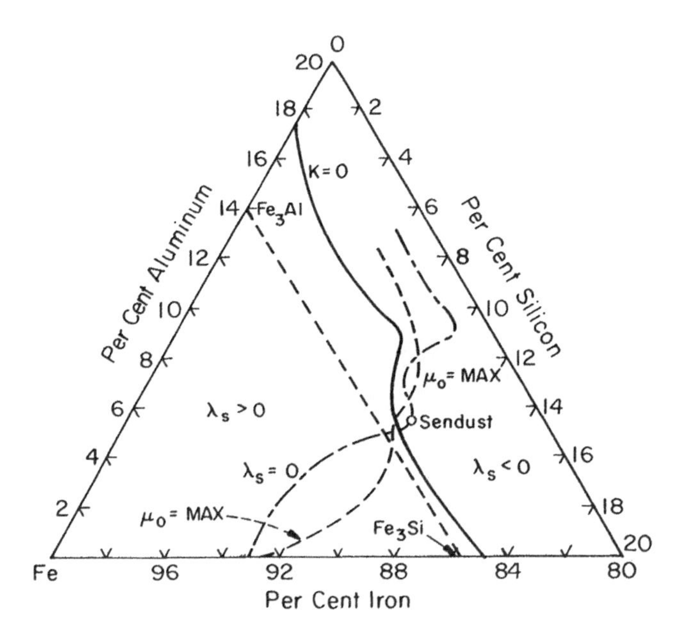

Figure 10.10 Iron-rich corner of ternary Fe-Si-Al diagram (wt%) showing fields of positive and negative magnetostriction, the courses of the zero-anisotropy line and maximum-permeability line. The Sendust composition is defined by the intersection of the $\lambda_s = 0$ and $K = 0$ lines (Bozorth, IEEE Press, copyright 1994).

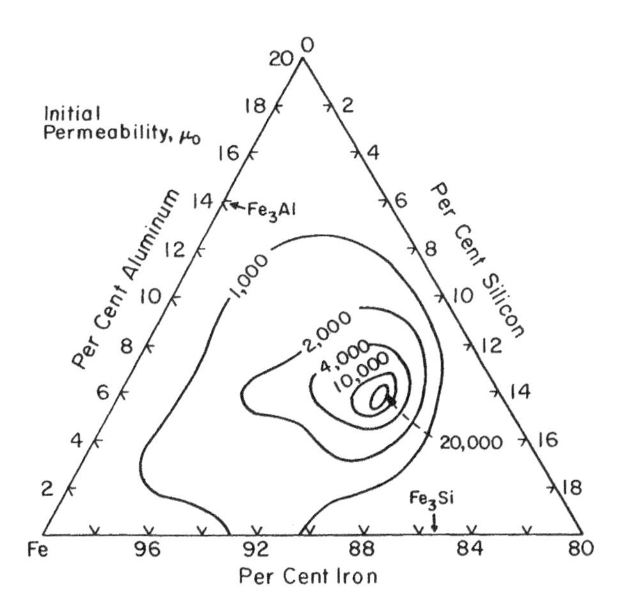

Figure 10.11 Contours of initial permeability on same Fe-Si-Al ternary diagram as shown in Figure 10.10 (Bozorth, IEEE Press, copyright 1994).

"Sendust" occurs near the intersection of these two lines around the composition 10wt% Si, 5wt% Al-Fe. (Sendust was named by its discoverers at Tohoku University in Sendai, Japan, to reflect the fact that its very brittle nature usually meant that the material had to be made and used in powder or dust form.) Figure 10.11 shows how the permeability peaks near this zero-anisotropy and zero-magnetostriction composition. The high Si and Al content reduces the saturation induction to about 1.2 T, so Sendust is not used in power transformer applications. Its mechanical hardness and magnetic softness makes it well suited to use in some magnetic recording heads.

10.2 IRON–NICKEL ALLOYS (PERMALLOYS)

Magnetic iron-nickel alloys are generally called permalloys. Originally *Permalloy* was the registered trademark for certain nickel–iron alloys, but it has now become a generic term. There are three major Fe-Ni compositions of technical interest:

1. 78% nickel permalloys (e.g., Supermalloy, Mumetal, Hi-mu 80). What makes the 78% nickel permalloys so important is the fact that magnetostriction and magnetocrystalline anisotropy both pass through zero near this composition (see Figs. 6.9 and 7.5). These alloys are used where the highest initial permeability is required.

2. 65% nickel permalloys (e.g., A Alloy, 1040 Alloy). The 65% nickel permalloys show a strong response to field annealing while maintaining $K_1 \approx 0$.
3. 50% nickel permalloy (e.g., Deltamax). What makes the 50% nickel permalloys important is their higher flux density ($B_s = 1.6\,\mathrm{T}$) as well as their responsiveness to field annealing, which gives a very square loop.

All the FCC iron-nickel alloys with Curie temperatures in excess of 400°C respond very well to magnetic field heat treatments so that B–H loops with a variety of shapes can be achieved. (It will be seen in Chapter 14 that for field annealing to be effective, there must be atomic mobility over a range of a few lattice constants while the material is in the magnetic state.) Figure 10.12 (Pfeifer and Radeloff 1980) summarizes the compositional variation of the important parameters over the FCC phase range from 35 to 100% Ni. The ordering reaction near the Ni_3Fe compound is important because it makes possible strong induced uniaxial anisotropy K_u to balance the crystal anisotropy (see Chapter 14).

The permalloys are sufficiently ductile to be rolled to thicknesses of 2.5 μm (0.1 mil) or drawn to wires having diameters as small as 10 μm.

10.2.1 78% Ni Permalloys

Models of permeability invariably predict μ to be proportional to M_s^2/K_{eff}, where K_{eff} includes a variety of sources of anisotropy. Following English and Chin (1967), four sources of anisotropy can be identified:

1. Crystal structure (magnetocrystalline anisotropy, Chapter 6)
2. Mechanical stress or strain (magnetoelastic anisotropy, Chapter 7)
3. Heat treatment with or without an applied field (thermomagnetic anisotropy, Chapter 14)
4. Cold work or plastic deformation (slip-induced anisotropy, Chapter 14)

Thermomagnetic anisotropy rarely exceeds $10^2\,\mathrm{J/m^3}$. It is not a factor if T_C is below the temperature at which appreciable diffusion can occur (e.g., in FeNi alloys away from the peak T_C near 70% Ni). Slip-induced anisotropy can be as great at $10^4\,\mathrm{J/m^3}$ in 60% Ni permalloy. It can be relieved by annealing. Thus, these two sources of anisotropy can be neutralized by annealing low-T_C compositions above T_C and cooling with sufficient speed that there is insufficient time to develop thermomagnetic anisotropy below T_C. The remaining sources of anisotropy may be eliminated by seeking K_1 and $\lambda \approx 0$.

Because K_1 can be altered by annealing more strongly than can λ_{100}, the usual practice is to select a composition for which $\lambda_s = 0$ and then anneal to achieve as near as possible $K_1 = 0$. If the selected composition has $\lambda_{100} = 0$, then K_1 slightly positive is better than K_1 slightly negative because then the

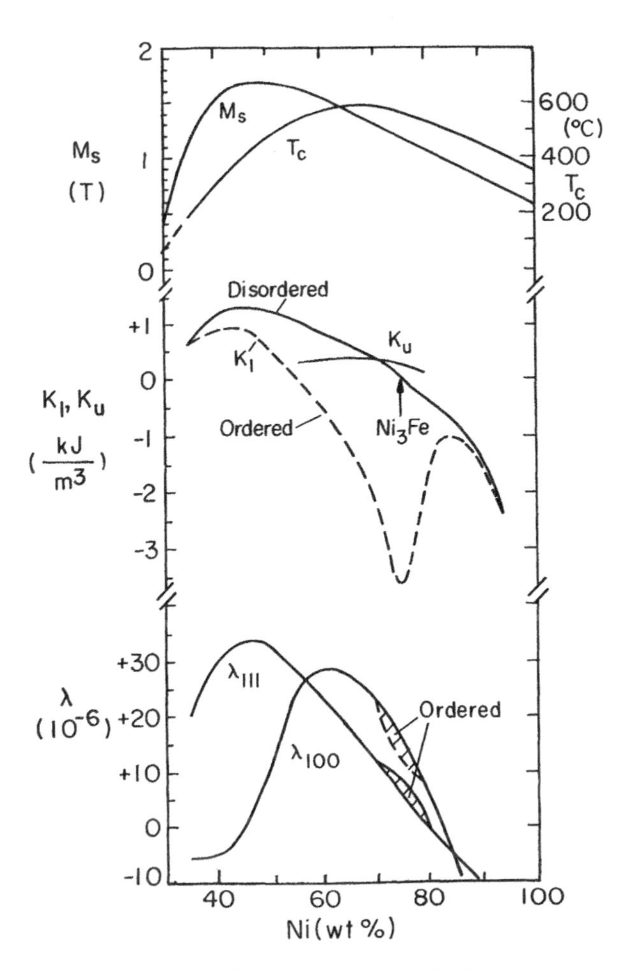

Figure 10.12 Variation of saturation magnetization, Curie temperature, magnetocrystalline anisotropy, and magnetostriction constants with Ni content in the FCC FeNi alloys. Bold marks on the composition axes indicate the technically important alloy ranges. Values of K_u were determined after tempering treatment at 450°C (After Pfeifer and Radeloff 1980).

easy anisotropy direction will be one of zero magnetostriction. Conversely, if $\lambda_{111} = 0$, then K_1 slightly negative is preferred to K_1 slightly positive.

From Figure 10.12 it is clear that λ_s and K_1 are not exactly zero simultaneously in Fe-Ni alloys. If compositions having $K_1 = 0$ cannot be found for $\lambda_s = 0$, then a heat treatment is designed to minimize K_1. Pfeifer (1996) and English and Chin (1967) mapped out the composition and heat treatment effects in the Fe-Ni-Mo range of interest. Their results are summarized in Figure 10.13.

Here λ_s is the polycrystalline average magnetostriction [Eq. (7.16)]. In the $\lambda_s = 0$ composition range $K_1 < 0$ for the binary Fe-Ni disordered alloy.

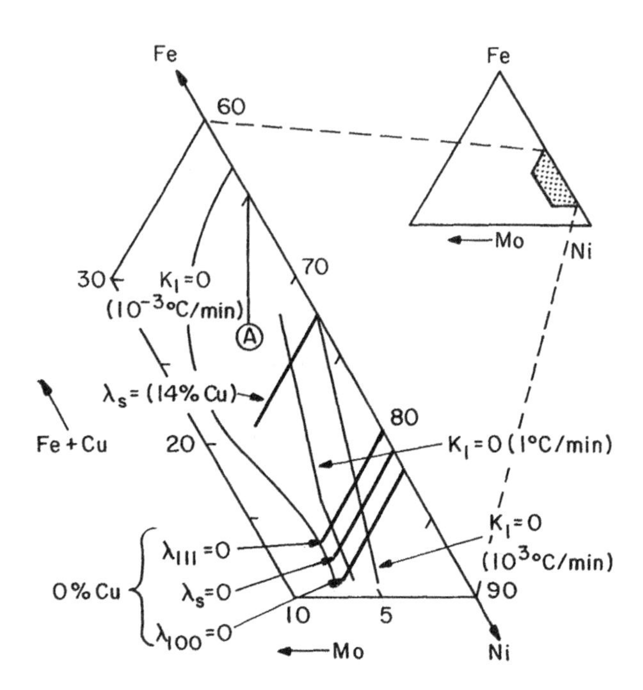

Figure 10.13 Portion of pseudoternary (Fe + Cu)–Ni-Mo phase diagram showing lines of zero magnetostriction (bold, solid) and lines of zero anisotropy (shaded). Values of λ and K are negative to the right side of the various lines. The $\lambda_s = 0$ lines depend mainly on Ni content and not on the makeup of the remainder of the (Fe + Cu)-Ni-Mo alloy. Line A defines the compositions of 65% Ni permalloys, which are strongly sensitive to field annealing. The intersection of the K and λ lines below and to the right of A is the composition of Alloy 1040. The cooling rates for the different zero-anisotropy lines distinguish ordered (slow-cooled) from disordered (quenched) structures. [Adapted from English and Chin (1967).]

Ordering (below 500°C) expands the $K_1 < 0$ field; stronger ordering is achieved by slower cooling (10^{-3} K/s). Cu and/or Mo additions tend to suppress ordering. Ordering also can increase M_s by up to 6%. With proper use of alloying additions, especially Mo, Cu, and/or Cr, and careful heat treatment, the holy grail of zero λ_s and zero K_{eff} can be achieved in a single composition. Mo additions also increase electrical resistivity of iron–nickel alloys.

Replacing iron with copper up to 14% shifts the $\lambda_s = 0$ composition from near 81% nickel to 72% nickel. Cu and Mo additions tend to suppress M_s and T_C because the number of empty d states is decreased by Cu and Mo and the width of the d band is increased [density of states (DOS) is decreased] by Mo.

Intersection of the $\lambda_s = 0$ and $K_1 = 0$ lines can result in permeabilities in the range 10^5–$10^6\,\mu_0$.

Figure 10.14 shows the frequency dependence of permeability in 4-79 Mo permalloy of different thickness. Eddy-current losses, which reduce permeabil-

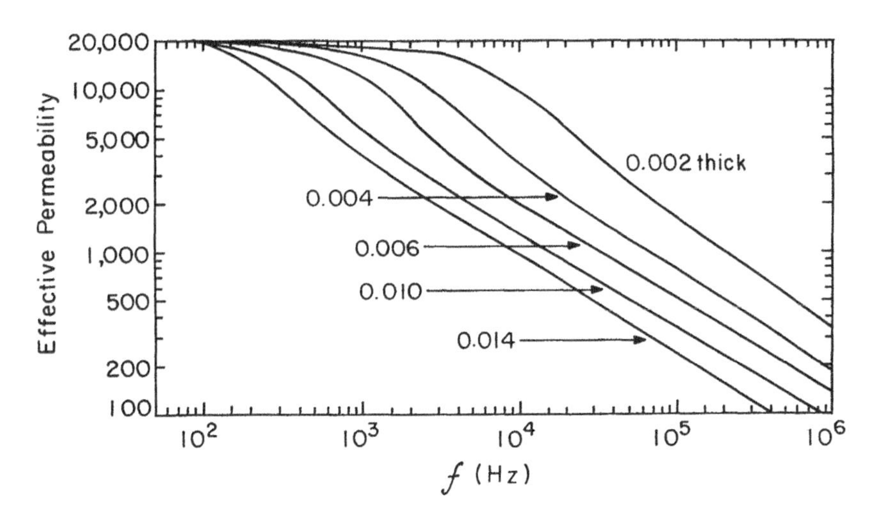

Figure 10.14 Frequency dependence of permeability in 4-79 Mo permalloy sheets of different thicknesses labeled in inches.

ity, are suppressed in thinner samples (Chapter 9). It is not possible to reduce the thickness of permalloy indefinitely without introducing unacceptably large fractional thickness variations. [Amorphous magnetic ribbons (see Chapter 11) can be made in thicknesses down to approximately 12 μm]. Experimental samples of good quality as thin as 3 μm have been melt-spun in vacuum. Thin films having thickness in the range 0.2–1 μm of either permalloy or amorphous metals can be used to frequencies of up to 10 MHz.

10.2.2 Other Permalloys

In some cases it is desirable to have a very square loop with high remanence ratio B_r/B_s and high maximum permeability μ_{max}. First, these characteristics can be achieved only when the sample presents a closed flux path; second, a square loop requires a dominant easy anisotropy axis in the field direction. This anisotropy can be achieved either crystallographically (by texture) if $K_1 \neq 0$, by stress-induced anisotropy, or by thermomagnetically induced anisotropy.

Two classes of alloy that respond particularly well to heat treatment (i.e., $T_C > 500°C$ and K_1 or λ is small) are 65% Ni permalloys (alloy A in Fig. 10.13) and 50% Ni permalloys (Deltamax).

In some applications that require a constant permeability (e.g., loading coils or pulse transformers) a sheared B–H loop is desirable. This can be achieved by several means, including inducing an easy axis transverse to the operating field direction by crystallographic texture or transverse field-induced anisotropy, as well as by using the material in powder form where the magnetostatic energy shears the loop.

When a B–H loop is limited by nucleation and the nucleation field exceeds the wall motion field, a very square loop results. This phenomenon is observed in crystalline or amorphous wires and is suitable for magnetic switches, sensors, or harmonic generation.

10.2.3 Zero-Magnetostriction Compositions

The occurrence of $\lambda = 0$ is primarily a function of composition, unlike $K_1 = 0$, which also depends to a significant extent on chemical ordering via heat treatment. It is desirable, and should be possible, to describe the course of the $\lambda = 0$ lines with a model based on composition parameters. Early attempts at doing so (Rossmann and Hoffman 1968) were largely empirical. Basically they joined Ni-Fe and Ni-M (M = nonmagnetic metal) binary compositions of zero anisotropy or magnetostriction on a ternary phase diagram. There was no physical basis for the choice other than observation.

On the other hand, Ashworth et al. (1969) and Berger (1976, 1977), in a series of articles, show that $\lambda = 0$ when the effective orbital angular momentum vanishes at E_F in the alloy d band. This singular point, $\langle L_z \rangle = 0$, can be identified in a split-band model of a strong magnetic material (no majority-spin holes), Figure 5.13, as the d-band occupation for which E_F lies between two minority-spin subbands. When E_F lies in such a "gap," $\langle L_z \rangle = 0$ and a moment still exists because there are empty minority-spin d states above E_F. The split-band model can be applied to alloys of any two or more transition metals for which $\Delta Z \geqslant 2$.

In Chapter 5, it was shown that for $Fe_{1-x}Ni_x$ alloys, the condition for $\langle L_z \rangle = 0$ is the same as

$$n_h = n_{\text{states}} \tag{10.1}$$

where n_h is the number of holes in the alloy and n_{states} is the number of minority states in the low-Z (higher-energy) manifold. Clearly, n_h is given by $10 - n_{3d}$ where n_{3d} is the number of $3d$ electrons. The number of minority states in the low-Z manifold is 5 times the low-Z atomic concentration. Both n_h and n_{states} are functions of composition only. For $Fe_{1-x}Ni_x$ alloys, this equation is satisfied for $x = 78.6\%$, the permalloy composition. This model is now applied to the ternary Fe-Ni-Mo system for which $\lambda_s = 0$ is plotted in Figure 10.13.

In ternary $Fe_{1-x-y}Ni_xMo_y$ alloys (band model in Figure 10.15), each Mo atom ($4s^1 3d^5$) contributes 5 holes to n_h for the alloy:

$$n_h = 2.8(1 - x - y) + 0.6x + 5y$$

There may be some confusion in counting states in the manifolds above E_F because it may not be immediately clear whether the Mo majority states lie above or below E_F.

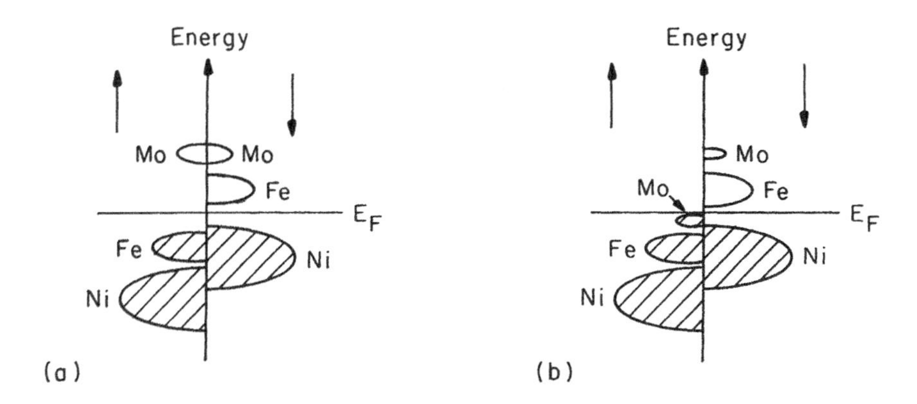

Figure 10.15 Split-band models for FeNi alloy with small Mo concentration. The energy of each group of *d* states is determined by the strength of the nuclear potential or equivalently by the atomic number *Z*. Spinup and spindown states are shown separately. States above the Fermi energy are unoccupied. Two possible models are shown: one for paramagnetic Mo *d* band and one for exchange-split Mo *d* band.

In Figure 10.15 case (*a*), the number of empty states in the three manifolds shown, $Mo^{+,-}$ and Fe^- is given by $n_{states} = 10y + 5\ (1 - x - y)$ because all 10 Mo states lie above E_F. In case (*b*) the number of empty states in the Mo^- and Fe^- manifolds is $n_{states} = 5y + 5(1 - x - y)$. The results of solving Eq. (10.1) are quite different for the two pictures:

$$x = \frac{2.2}{2.8} + y \qquad \text{(Fig 10.15}a\text{)} \tag{10.2a}$$

$$x = \frac{2.2(1 - y)}{2.8} \qquad \text{(Fig. 10.15}b\text{)} \tag{10.2b}$$

These lines are plotted on the ternary diagram shown in Figure 10.16. Reference to the data in Figure 10.13 shows that the correct assumptions are those of Figure 10.15*a*. Thus the Mo *d*-bands are not split by an exchange interaction to give a magnetic moment and both spin components of the Mo *d*-band are found at an energy above E_F. The composition for $\lambda = 0$ in $Fe_{0.95-x}Ni_xMo_{0.05}$ is calculated from Eq. (10.2a) to be 83.6% Ni. This is close to that found in Figure 10.13.

It is possible also to calculate the moment in each model:

$$n_B = n_h^- - n_h^+$$

$$n_B = 5(1 - x - y) + 5y - 5y = 5(1 - x - y) \qquad \text{(Fig. 10.15}a\text{)} \tag{10.3a}$$

$$n_B = 5(1 - x - y) + 5y = 5(1 - x) \qquad \text{(Fig. 10.15}b\text{)} \tag{10.3b}$$

The model of Figure 10.15*b* predicts that the moment of an iron–nickel alloy

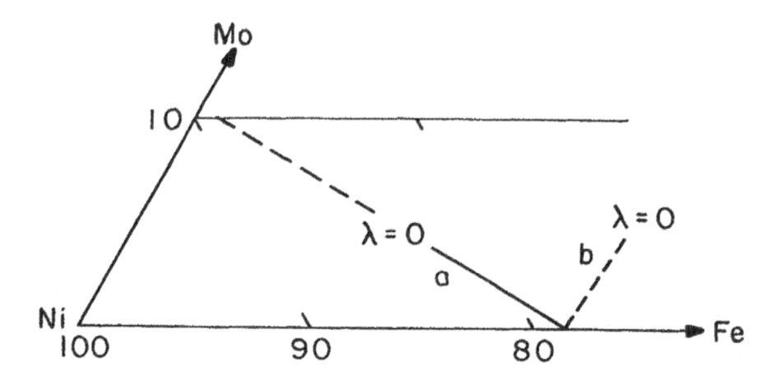

Figure 10.16 Predicted lines of zero magnetostriction for Mo additions to permalloy. The two lines labeled (*a*) and (*b*) correspond to the two different assumed band structures in Figure 10.15. Comparison with the data in Figure 10.13 shows that assumption (*a*) is more realistic.

is independent of molybdenum content. This is clearly counter to observation. All of the Mo *d* states must lie above E_F. This is explained in more detail by the Friedel VBS model (Chapter 5). Scenario (a) is clearly the appropriate one based on the composition dependence of the $\lambda_s = 0$ line and on the observed strong dependence of alloy moment on Mo content.

In Figure 10.13 it is noted that the addition of Cu shifts the $\lambda_s = 0$ composition from 81% to 72% Ni. In terms of the split-band model, Cu shifts the minority *d*-band weight to lower energies. Thus, *FE* coincides with the singular point $\langle L_s \rangle = 0$ for less Ni, and more Fe, content.

10.3 IRON–COBALT ALLOYS (PERMENDUR)

The equiatomic BCC FeCo alloys (Permendurs; Fig. 10.17*a*) have very high saturation induction ($B_s \approx 24$ kG) and relatively low magnetic anisotropy, K_1 (disordered) $\approx -1 \times 10^5$ J/m^3 and K_1 (ordered) ≈ 0 (Chin and Wernick 1980). The magnetostriction, however, is substantial: $\lambda_{111} = 25 \times 10^{-6}$ and $\lambda_{100} = 150 \times 10^{-6}$, giving a polycrystalline average of $\lambda_s \approx 60 \times 10^{-6}$. While the anisotropy (including stress-induced) sets the upper limit for the permeability and the lower limit for the coercivity of these alloys, grain size is the primary factor determining the technical magnetic properties actually attained. Hiperco 50 (an FeCo alloy made by Carpenter Technology) has a yield strength σ_y of about 460 MPa, which increases for grain sizes below 30 μm.

Because equiatomic BCC FeCo shows an order–disorder transformation to the CsCl structure below 730°C, the anisotropy, magnetostriction, and mechanical properties depend strongly on annealing and on cooling rates. The ordering can be suppressed by addition of vanadium (vanadium–Permendur or Supermendur: $Fe_{49}Co_{49}V_2$), which renders the alloys ductile enough to be rolled into thin sheets. Unfortunately, rolling can give rise to a duplex texture: (112) [110] and (001) [110], neither of which contains a $\langle 111 \rangle$ easy axis

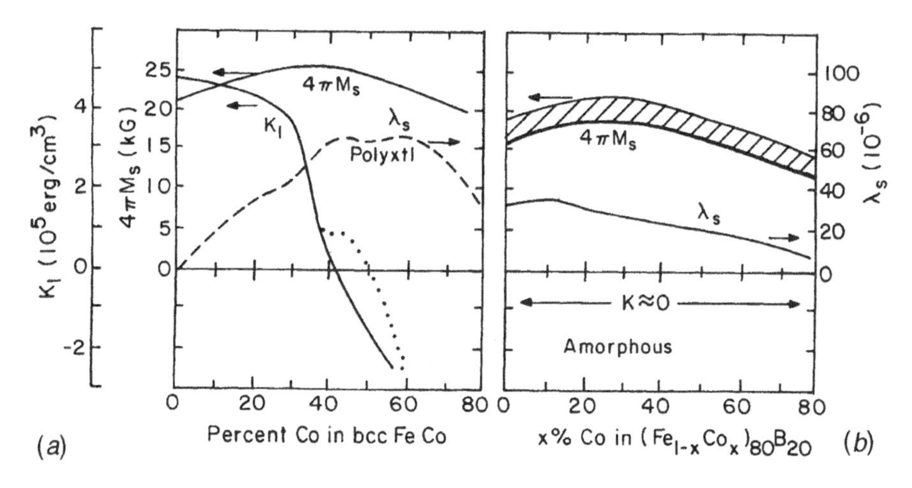

Figure 10.17 (*a*) Magnetic properties of BCC Fe-Co alloys. Anisotropy and magnetostriction after Hall (1960); dotted line is the anisotropy for ordered CsCl-ordered phase. (*b*) Magnetic properties of amorphous FeCoB alloys for which the anisotropy is essentially zero over the entire composition range (O'Handley et al. 1979).

($K_1 < 0$) in the plane of the sheet and the latter containing the [100] high-magnetostriction axis. Table 10.3 compares the properties of some FeCo-based alloys. Permendur shows a resistivity of under 10 $\mu\Omega \cdot$cm, while 2% vanadium–Permendur and Supermendur show $\rho \approx 25\,\mu\Omega \cdot$cm, which is more favorable for higher-frequency applications. Recent studies of high-temperature properties of Hiperco-50 indicated a degradation of magnetic properties over time at 450°C (Li, 1996).

These soft FeCo alloys are typically hot-rolled in the FCC phase ($T > 912$°C) and then quenched for cold rolling in the BCC phase. Optimal magnetic properties are then achieved by heat treatment near 850°C to develop the disordered BCC phase, although treatment at temperatures near 695°C has been shown to improve mechanical strength and ductility (Thornburg 1969).

TABLE 10.3 Room Temperature Properties of FeCo-Based, High-Induction Alloys

	B_s (kG)	μ_{max} (10^3)	H_c (Oe)	ρ ($\mu\Omega \cdot$cm)	σ_y (GPa)
Permendur (50% FeCo)	24.5	5	2	7	—
Hiperco-50 2% V-FeCo (ordered)	24	17	5	40	0.36–0.63
Amorphous Fe-Co-B	19	50	0.5	130	2.4

The cooling rate after this anneal is important; the best permeability and coercivity are achieved for intermediate cooling rates. Ni additions up to 4.5% improve the mechanical ductility of vanadium–Permendur without significantly degrading magnetic performance for heat treatments over a broad temperature range.

Iron–cobalt alloys of the Permendur family are used in applications where the highest flux density (hence lowest weight) is more important than AC loss or cost. Thus these alloys are used in transformers and generators on aircraft power systems. Such systems generally operate at 400 Hz instead of 60 Hz in order to increase the power per unit mass of the devices.

10.4 AMORPHOUS ALLOYS

The attraction of Sendust and the permalloys is that two of the major sources of anisotropy (crystallography and stress) can be effectively nulled by careful selection of composition and heat treatment.

Amorphous metallic alloys are materials that are rapidly quenched from the melt, so their atomic structures lacks the long-range order of a crystalline solid. Without *long-range* order, amorphous alloys have no magnetocrystalline anisotropy. They retain a fair degree of short-range order, comparable to the order that characterizes the liquid state. Thus amorphous metallic alloys based on transition metals can show a very easy magnetization process. Magnetoelastic, thermomagnetic, and slip-induced anisotropy remain as sources of anisotropy in amorphous alloys. Also, the high electrical resistivity of amorphous alloys (120–150 $\mu\Omega \cdot$cm) compared to Si-Fe and iron–nickel alloys (30–50 $\mu\Omega \cdot$cm) makes them attractive for high-frequency operation. Chapter 11 describes some of the more fundamental aspects of these fascinating materials. Here, attention is given to the soft magnetic properties of amorphous alloys.

Reasonably strong magnetization can be realized in a variety of amorphous alloys based on iron, cobalt, and/or nickel. However, as will be seen in Chapter 11, the saturation magnetization of an amorphous transition metal–metalloid (T-M) alloy is generally smaller than that of a crystalline alloy of the transition metals alone. This moment reduction is due to the presence of the metalloid atoms, B, P, Si, and so on, which are needed to stabilize the glassy state. It is not a result of the absence of long-range order.

10.4.1 High-Induction Amorphous Alloys

Not long after the discovery of high permeability in ferromagnetic amorphous alloys based on Fe-P-C in 1964 by Duwez and Lin, work began in several labs to develop higher-induction metallic glasses suitable for distribution transformers. High-induction amorphous alloys meet many of the criteria outlined in Section 10.1 for high-power transformers.

Figure 10.17*b* shows the variation of $4\pi M_s$ and magnetostriction for amorphous $(FeCo)_{80}B_{20}$ alloys for comparison with the properties of crystalline Fe-Co alloys at left. The lower values of magnetostriction in amorphous alloys compared to crystalline Fe-Co alloys makes them less sensitive to stress-induced anisotropy. Amorphous Fe-Co-B alloys containing less than 20% glass former and exhibiting B_s up to 19.6 kG (still only 80% of the flux density of the Permendurs) have been made (shaded area, Figure 10.17*b*). Other advantages that metallic glasses bring to high-induction applications are their inherently high values of electrical resistivity and yield stress (2.4 GPa for $Fe_{80}B_{20}$ compared to less than 0.7 GPa for Hiperco-50 and V-Permendur; see Table 10.3).

High-induction amorphous Fe-B alloys ($B_s \approx 16$ kG) are used commercially in 60 Hz, 25-kVA distribution transformers because of their very low core and excitation losses ($L < 0.1$ W/kg at 15 kG, 60 Hz). In these applications, continuous operation at 100°C with transients to 200°C has not been found to degrade performance. Crystallization typically occurs near $T_x \approx 400$–450°C and is a function of composition. In some amorphous Fe-B-Si alloys, T_x exceeds 500°C (O'Handley et al. 1979). Amorphous Fe-Co-B alloys containing less than 20% glass former and exhibiting B_s up to 19.6 kG have been made (shaded area, Fig. 10.17*b*). It should be noted that reducing boron content in amorphous Fe-B alloys lowers the Curie temperature. (This is believed to be associated with the close-packed nature of the amorphous structure, unlike the more open BCC structure of α-Fe.) This decrease in T_C can be offset significantly by replacement of Fe with Co (O'Handley et al. 1979).

Figure 10.18 summarizes the core loss under sine-flux conditions* in several early amorphous alloys (Hatta et al., 1978; Luborsky et al., 1978; O'Handley et al., 1979). These results are compared to Armco M-4-grade 3% Si-Fe, Orientcore HiB (Suzuki et al. 1972), and Nippon Steel's laser scribed 3% Si-Fe. Laser irradiation of 3% Si-Fe was exploited by Nippon Steel to dramatically refine the domain structure and thereby reduce core loss.

Note that the core loss increases rapidly with maximum flux level and diverges as B_{max} approaches B_s of the particular alloy. Core loss is found to increase approximately linearly with frequency.

*When a magnetic core is operated under AC conditions, it makes a difference whether it is driven by a sinusoidal H field (sine H) or driven so as to produce a sinusoidal voltage in the windings (sine B). The latter implies a single frequency for dB/dt in the material. But sinusoidal H drive results in components of $dB/dt = (\partial B/\partial H)dH/dt$ at the drive frequency, ω, as well as at 3ω, 5ω.... This is because $B = \mu H$ and μ is nonlinear ($\mu = \mu_0 + H^2\mu_2 + H^4\mu_4 \ldots$), especially at high flux density. Sine-H drive results in much more eddy current loss than sine-B because a significant component of the flux is being driven at multiples of the fundamental frequency. Most transformers have such a large inductance that they draw very little current from the source and sine-B conditions are maintained. Many laboratory test coils are not well designed to be optimally filled by the sample. Also the power sources used in the laboratory may not be sufficiently "stiff" to maintain constant drive voltage (sine-B) throughout the drive cycle. The back EMFs are small, and sine-H (sine current) conditions obtain unless a feedback system is employed to maintain sinusoidal voltage.

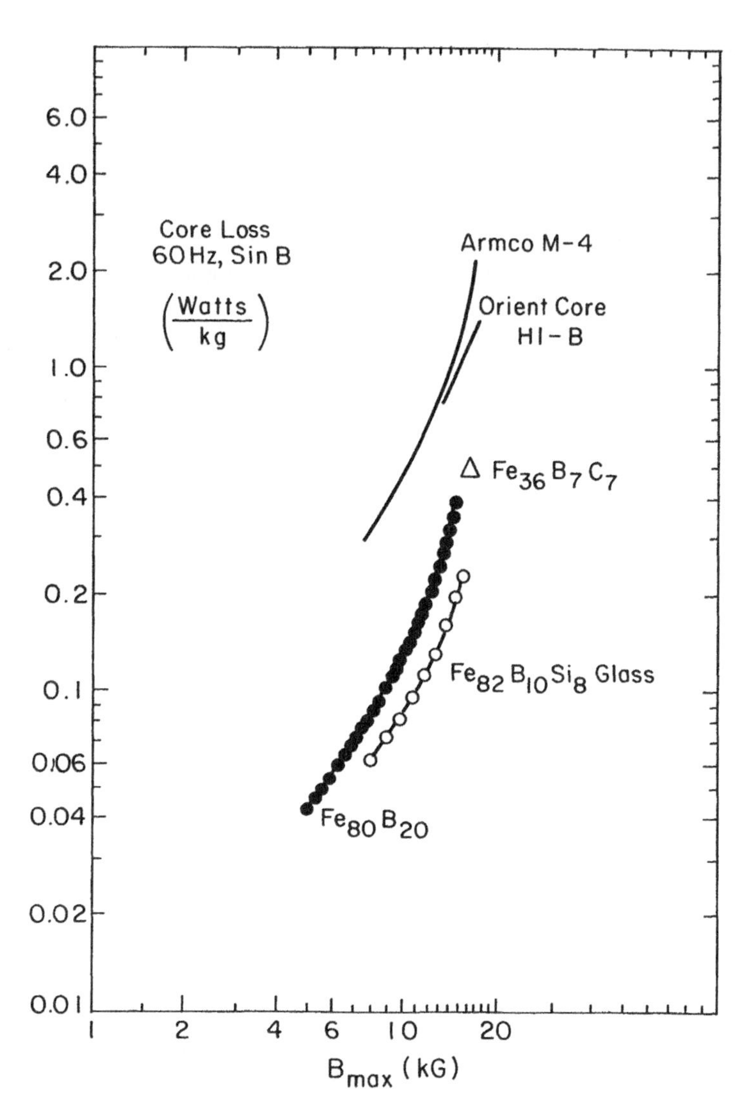

Figure 10.18 Variation of 60-Hz core loss with maximum flux density (under sine flux conditions) for selected amorphous and crystalline alloys.

The reduced loss of amorphous alloys has been estimated to be due mostly to their high electrical resistivity (120–150 $\mu\Omega \cdot$cm) compared to that of 3% Si-Fe (40 $\mu\Omega \cdot$cm). One problem with amorphous alloys paradoxically has been their very low-coercivity DC B–H loops, which rapidly degrade at higher operating frequencies. This degradation is due to the large domain size, which may be traced to their lack of internal surfaces (grain boundaries) at which new domain walls may be nucleated. In this regard, Gyorgy once described metallic glasses as behaving like "lousy single crystals." With very few domain walls, wall velocity increases sharply with increasing frequency, and with it, micro-eddy currents increase (Chapter 9).

One solution to this problem was successfully demonstrated in the early 1980s by several groups. Figure 10.19 shows the results that Datta et al. (1982) achieved by annealing $Fe_{79}B_{16}Si_5$ to different stages of crystallization in order to minimize 50 kHz loss. Annealing for 20 min at 450°C leaves the DC B–H loop with an attractive square character (curve a). In this case, the 50 kHz core and coil losses are moderate, and the domain structure (Fig. 10.19, upper right) is marked by relatively few walls. Annealing for 40 min at 450°C leads to a rounded B–H loop (curve b). Both core and excitation losses at 50 kHz decrease dramatically even though the DC B–H loop is somewhat rounded. This heat treatment leads to incipient crystallization of a small volume fraction of α-Fe phase crystallites (Fig. 10.19, lower right). The α-Fe crystallites are seen by Lorentz microscopy to have nucleated additional domain walls and increased the angular dispersion of domain magnetization vectors. Annealing

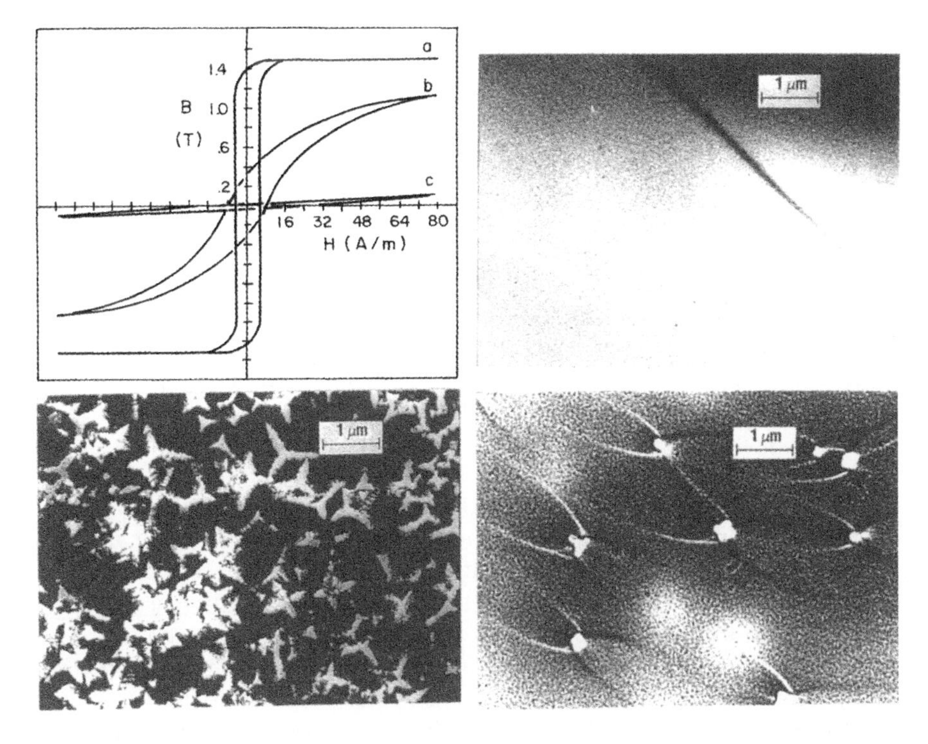

Figure 10.19 Upper left, B–H loops representing three annealing conditions leading to (a) square, (b) round, and (c) sheared loops (Datta et al. 1982). Clockwise, from upper right: transmission electron micrographs illustrating the microstructures of the heat-treated amorphous alloys having square, round, and sheared B–H loops, respectively. (Micrographs courtesy of DeCristofaro).

for one hour at 450°C resulted in a highly sheared loop (curve *c*) and a dramatic increase in excitation loss. Core loss remains low on this case because the magnetization process is characterized less by domain wall motion, and more by magnetization rotation. These characteristics are the result of copious crystallization (Fig. 10.19, lower left), which leaves the material highly stressed and inhomogeneous. These techniques for lowering 50 kHz losses are analogous to the Goss texturing and laser scribing used to lower 60 Hz loss in crystalline Si steels: they promote a finer grain structure by microscopic magnetostatic effects.

Some special high-induction applications place more importance on operating induction level than on cost. In such applications iron–cobalt alloys offer some advantages (Section 10.3). It is useful here to recall what can be achieved for such applications by amorphous alloys. Figure 10.17*b* shows that while cobalt enhances the flux density of amorphous Fe-B(–Si) alloys, the presence of the metalloids prevents the flux density from reaching the levels achieved in Permendur (Fig. 10.17*a*). The Fe-Co-based amorphous alloys do offer several advantages over crystalline Fe-Co alloys. First, the amorphous alloys have no magnetocrystaline anisotropy, whereas K_1 is a strong function of composition in the crystalline counterparts. Amorphous alloys also show advantages in terms of electrical resistivity, mechanical strength, and readily obtained thin gauge suitable for high-frequency rotating machinery.

10.4.2 Other Amorphous Alloys

Because the magnetocrystalline anisotropy of amorphous alloys is zero, most soft magnetic properties are controlled by stress and by the magnetostriction coefficient. Figure 10.20 shows that for $Co_xFe_{1-x}B_{20}$ amorphous alloys, the coercivity of the as-prepared amorphous ribbon passes through a minimum close to the composition for which $\lambda_s \approx 0$. As the Fe:Co ratio is varied, λ_s goes from positive (Fe-rich) to negative (Co-rich). Where $\lambda \approx 0$, H_c reaches a minimum. Such zero-magnetostriction alloys are important for some of the same reasons permalloy is important.

The ternary diagram in Figure 10.21 shows the compositional variation of magnetostriction over a field of Fe-Co-Ni amorphous alloys containing 20% boron. Zero-magnetostriction compositions are found along the solid line.

What is seen in Figure 10.21 is that the magnetostriction is of order 30×10^{-6} for iron rich glasses and drops to zero with cobalt additions near Fe:Co $\approx$ 5:75. This ratio is close to the Fe/Co ratio for zero magnetostriction in crystalline Fe-Co alloys (Fig. 7.7). A line of zero magnetostriction runs near the Co-Ni side of the phase diagram, but alloys rich in nickel have low magnetization and are difficult to make by melt spinning when boron is the major glass former. The compositions of zero magnetostriction can be predicted with a simple split-band model (like that in Fig. 10.15) in certain cases (O'Handley and Berger, 1978). Changes in glass former from all boron to mixtures including Si, C, or P have only small effects on λ_s. Changes in

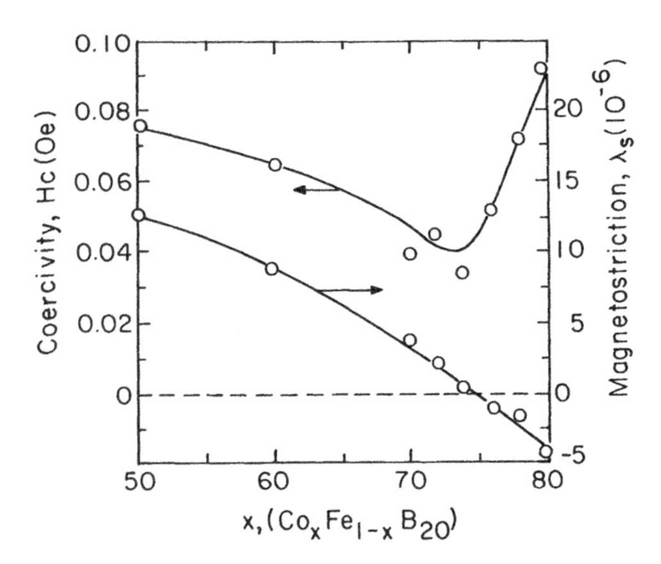

Figure 10.20 Variation of coercivity (left scale) and magnetostriction (right scale) with Fe:Co ratio in amorphous $(CoFe)_{80}B_{20}$ alloys (O'Handley et al. 1976).

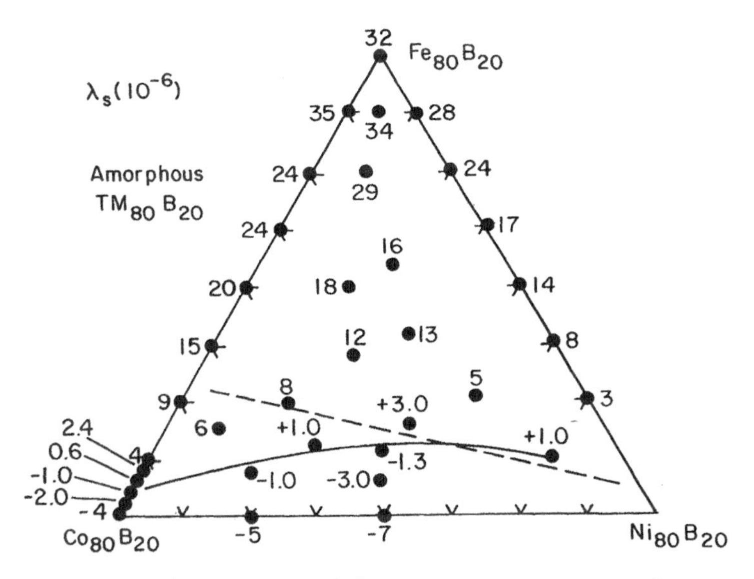

Figure 10.21 Saturation magnetostriction at room temperature for amorphous $(FeCoNi)_{80}B_{20}$ alloys. Solid line shows course of zero-magnetostriction compositions, and dashed line shows predictions based on split-band model (O'Handley 1978).

metal:metalloid ratio can have significant effects on λ_s in the iron-rich glasses (O'Handley et al. 1977).

There are also important $\lambda_s = 0$ compositions in the $(FeCoNi)_{100-x}TE_x$ system. Here, TE represents an early transition metal element such as Zr, Ta, Nb, or Hf; metalloids need not be present in such metallic glasses (see Chapter 11). Figure 10.22 is a pseudoternary magnetostriction phase diagram for the amorphous $(FeCoNi)_{90}Zr_{10}$ system (Ohnuma et al. 1980). Contours of equal magnetostriction are shown. Along each axis of the ternary diagram the variation of magnetostriction in the pseudobinary system is shown. Note the small magnetostriction of the iron-rich zirconium-stabilized glasses and the zero magnetostriction near $Co_{80}Ni_{10}Zr_{10}$. The low magnetostriction in the Fe-rich Zr-stabilized glasses is due mainly to the low Curie temperature there. These amorphous $TL_{90}TE_{10}$ (late transition metal–early transition metal) materials are characterized by good mechanical hardness, corrosion resistance, and—in the Fe-Co rich members—large values of $4\pi M_s$. Amorphous alloys based on this system are sometimes used in thin-film recording heads and as shields in magnetoresistance heads (Chapter 17).

Because of the technological importance of low magnetostriction, considerable efforts have been made to reduce the magnetostriction of amorphous

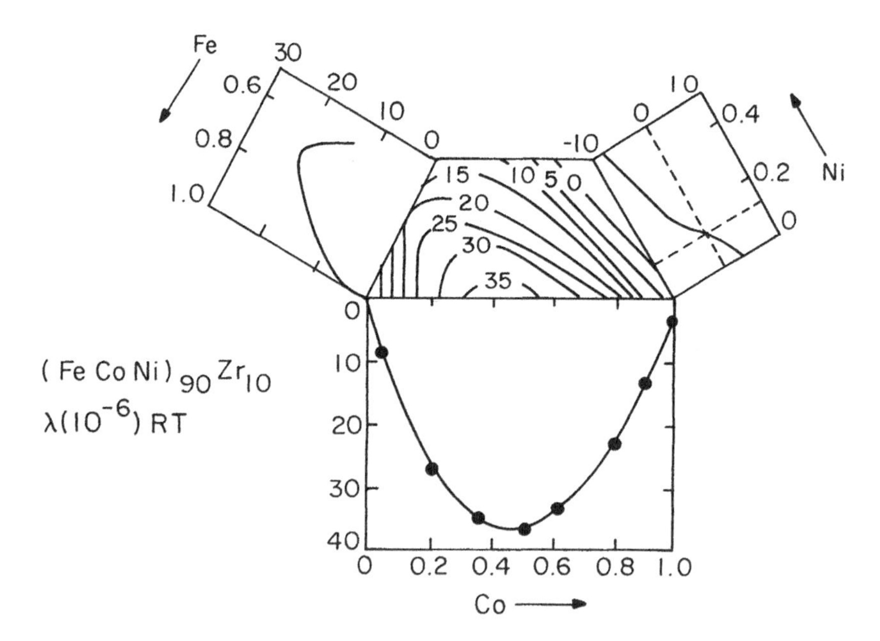

Figure 10.22 Quasiternary diagram for amorphous $(FeCoNi)_{90}Zr_{10}$ alloys showing contours of equal linear magnetostriction λ_s. Off each ide of the ternary diagram, the magnitude of λ_s is plotted for a range of the binary alloys along that side (Ohnuma et al. 1980).

alloys based on iron. Iron has a higher saturation magnetization and is relatively abundant in nature compared to cobalt; the magnetostriction of $Fe_{80}B_{20}$ is approximately 32×10^{-6} at room temperature. Room temperature magnetostriction of $(FeNi)_{80}B_{20}$ glasses was shown to scale with M^2 (O'Handley 1977) so that $\lambda_s = 0$ could be approached only with a loss of magnetization. Nevertheless, it has been found that with substitutions of Mo, Cr, or Nb for Fe, it is possible to achieve $\lambda_s < 5 \times 10^{-6}$ while retaining reasonable values of magnetization (Hasegawa and Ray 1978, Inomata et al. 1983, Corb et al. 1983, Yoshino et al. 1984). In these Fe-TE-metalloid alloys the decreased room temperature magnetostriction is due partially to the suppressed Curie temperature.

When the highest permeability is required, the amorphous cobalt-rich alloys sometimes represent an attractive alternative to crystalline 78% Ni permalloy. One advantage that amorphous alloys have over permalloy is their much higher hardness and greater yield stress. The high yield strength of amorphous metallic alloys generally makes them much more resistant to plastic deformation and slip-induced anisotropy than crystalline alloys. This is important in many applications.

Because amorphous magnetic alloys are generally made by rapid solidification, they are presently limited to thin ribbon and sheet form, typically 20–50 μm in thickness. This limits their use in some applications while it is an advantage in others.

10.5 SOFT FERRITES

The most widely used soft ferrites are manganese–zinc ferrite and nickel–zinc ferrite. These materials are based on the spinel structure of Fe_3O_4 discussed in Chapter 4:

$$(Fe^{3+})_A(T^{2+}Fe^{3+})_BO_4$$

where T = Ni or Mn. Recall that the addition of zinc drives the trivalent iron from the A to the B sublattice because Zn^{2+} has a greater affinity for tetrahedral coordination than does Fe^{3+}. With more Fe^{3+} ($\mu_m = 5\,\mu_B$ per Fe^{3+}) on the same (B) sublattice, the magnetic moment per formula unit increases. This increased magnetization is important, but the major shortcoming of ferrites remains their low magnetization compared to that of 78% Ni permalloys. Their major advantages are high electrical resistivity and low cost, due to both the abundance of the raw materials and the relative ease of processing to final form.

Figure 10.23 compares the temperature dependence of magnetization in NiZn ferrites and MnZn ferrites. While MnZn ferrites have more magnetization at room temperature than do NiZn ferrites, their Curie temperatures are generally lower.

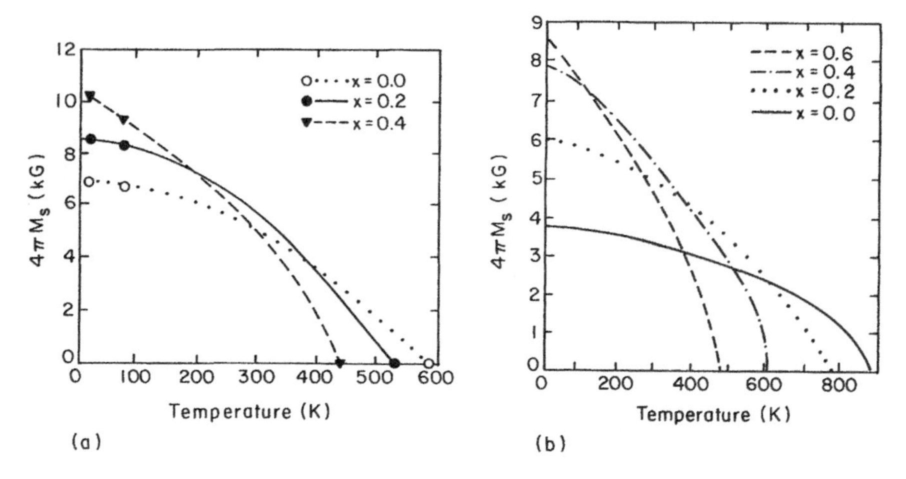

Figure 10.23 Temperature dependence of magnetization in (*a*) manganese–zinc ferrites, $Mn_{1-x}Zn_xFe_2O_3$ [after Guillaud and Roux (1949)] and (*b*) NiZn ferrites, $Ni_{1-x}Zn_xFe_2O_3$ [after Pauthenet (1952)].

The addition of zinc to manganese ferrite results in greater magnetization. It also causes the $\lambda_{111} \approx 0$ line to move from a composition field where $K_1 < 0$ (where [111] is the easy axis), through a region of $K_1 > 0$ compositions and back to compositions of $K_1 < 0$ (see Fig. 10.24). These sort of data are very helpful in designing compositions for specific applications. Polycrystalline

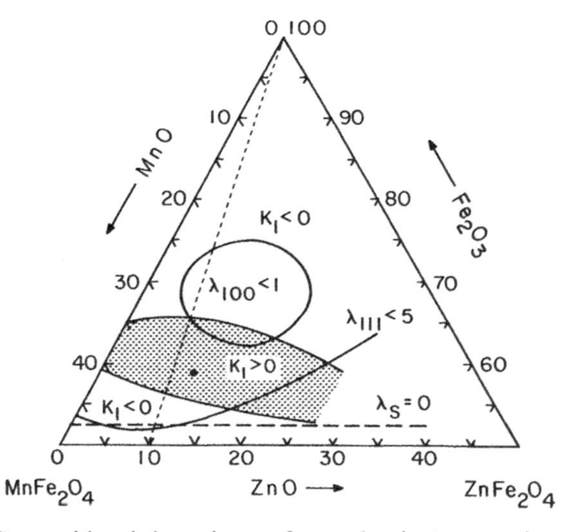

Figure 10.24 Compositional dependence of crystal anisotropy and magnetostriction constants in the mixed oxide system (MnZnFe)–Fe_2O_4 at 20°C (Ohta 1963).

samples in a $K_1 > 0$ field (easy axis [100]) show highest permeability when $\lambda_{100} = 0$. These compositions exist along the intersection of the shaded area and the boundary of the $\lambda_{100} < 1$ region. Similar considerations show where to find compositions with $K_1 < 0$ and λ_{111} small.

These effects lead to high permeability in $(MnO)_{0.26}(ZnO)_{1.22}(Fe_2O_3)_{0.52}$, close to the composition where the $\lambda_s = 0$ line intersects the $K_1 = 0$ line (Figure 10.25).

Figure 10.26 shows the variation of magnetostriction and permeability with Fe_2O_3 content in $[(MnO)_{0.7}(ZnO)_{0.3}]_{1-x}(Fe_2O_3)_x$. These compositions are found along the nearly vertical dashed line in Figure 10.24. In these compositions, K_1 is small so the permeability peaks near the composition at which $\lambda_s = 0$.

Figure 10.27 shows the variation of permeability and anisotropy with temperature in $(MnO)_{0.31}(ZnO)_{0.11}(Fe_2O_3)_{0.58}$ (the dot in Figure 10.24). Note that at temperatures where K_1 passes through zero, the permeability peaks. When $K_1 = 0$, the permeability is still limited by magnetoelastic anisotropy $(3/2)\lambda_s\sigma$.

A peak in permeability occurs also in nickel zinc ferrites near a maximum in λ_s (Fig. 10.28). In this case the permeability is dominated by a large K_1, not λ.

The attraction of the ferrites is their high electrical resistivity. They can be used at frequencies well above 1 MHz, whereas magnetic metals, depending on their thickness, can rarely be used above 100 kHz except in thin film form.

10.6 SUMMARY

Figure 10.29 shows that while the ferrites are suitable for high-frequency applications, they do not provide the flux density that the permalloys or

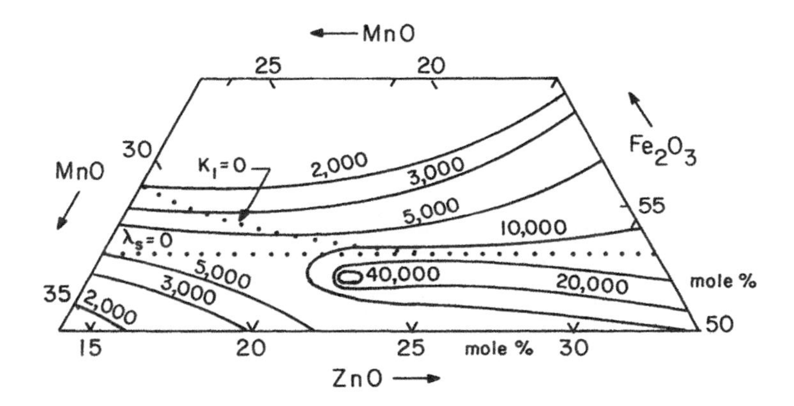

Figure 10.25 Constant permeability contours and lines of zero crystal anisotropy and magnetostriction for the (MnZnFe)–Fe_2O_4 system (Roess 1971).

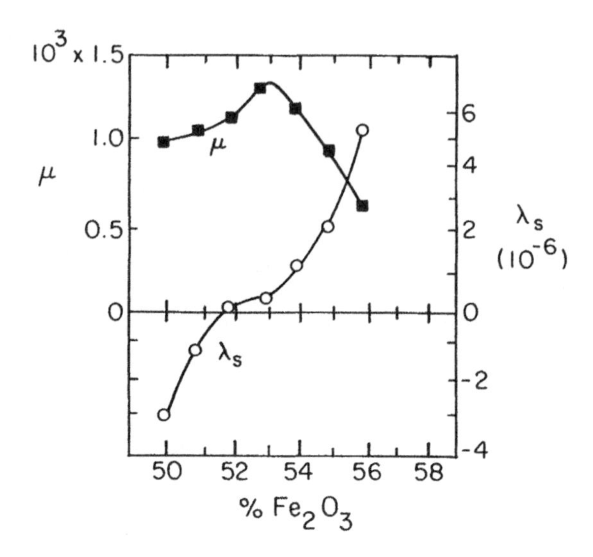

Figure 10.26 Variation of permeability and magnetostriction with iron oxide content in $[(MnO)_{0.7}(ZnO)_{0.3}]_{1-x}[Fe_2O_3]_x$ ferrites. [After Guillaud (1957).]

amorphous alloys do. This figure serves as a useful summary comparing the properties of three of the major classes of soft magnetic materials described above.

Iron-rich crystalline alloys and selected amorphous alloys provide sufficiently large saturation flux density for use in high-power-density applications such

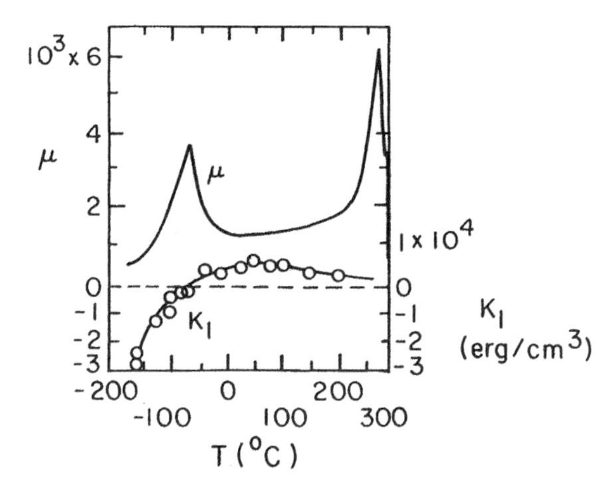

Figure 10.27 Variation of permeability and anisotropy constant with temperature in the MnZn ferrite $(MnO)_{0.31}(ZnO)_{0.11}(Fe_2O_3)_{0.58}$. [after Slick (1980).]

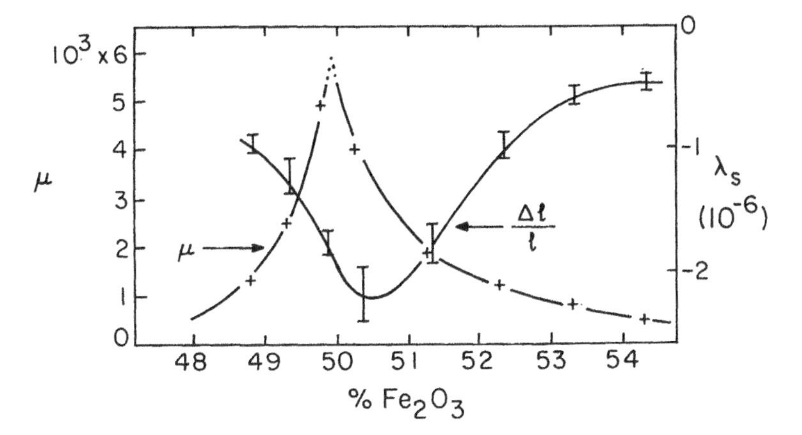

Figure 10.28 Dependence of permeability and magnetostriction on Fe_2O_3 content for NiZn ferrites with NiO:Zn:O ratio of 15:55 (Guillaud et al. 1957).

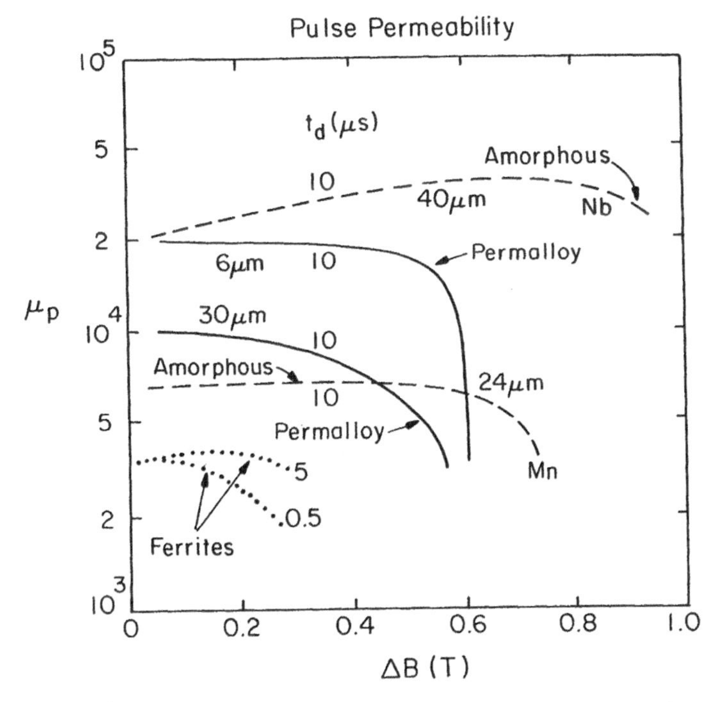

Figure 10.29 Pulse permeability versus maximum flux swing for three classes of soft magnetic materials: amorphous metallic alloys (CoFeNbBSi and CoMnFeMoBSi), crystalline Mo permalloy (NiFeCuMo), and ceramic MnZn ferrites; data for two thicknesses are given in each case (Boll and Hiltzinger 1983).

as 60-Hz distribution transformers. While core loss and coil loss can be low in the amorphous alloys, they cannot be driven to the same flux levels as can Si steels.

PROBLEMS

10.1 Assume that a 180° domain wall exists in a demagnetized, uniaxial magnetic material.

(a) Sketch what happens to the domain magnetization and domain wall in the two cases described below for $H > 0$ but less than saturation: (i) applied field parallel to the easy axis and (ii) applied field perpendicular to the easy axis.

(b) Sketch the M–H loops in each case.

(c) Describe how a defect might pin or impede domain wall motion.

BIBLIOGRAPHY

Bozorth, R. M., *Ferromagnetism*, Van Nostrand, New York, 1953; IEEE Press, New York, 1993.

Chin, G. Y., and J. H. Wernick, in *Ferromagnetic Materials* Vol. 2.

Fish, G. E., "Soft magnetic materials," *Proc. IEEE* **78**, 947 (1990).

Guillaud, C., C. Villers, A. Maraid, and M. Paulus, *Solid State Physics*, Vol. 3, M. Desirant and J. L. Michiels, eds., Academic Press, New York, p. 71.

Hasegawa, R., ed., *Metallic Glasses: Magnetic, Chemical and Structural Properties*, CRC, Boca Raton, FL, 1983.

Luborsky, F. E., in *Amorphous Metallic Alloys*, F. E. Luborsky, ed., Butterworths, London, 1983.

Luborsky, F. E., J. D. Livingston, and G. Y. Chin, in *Physical Metallurgy* vol. 2, R. W. Cahn and P. Haasen, eds. Elsevier, Amsterdam, 1983, p. 673.

O'Handley, R. C., in Amorphous Metallic Alloys, F. E. Luborsky, ed., Butterworths, London, 1983, p. 257.

REFERENCES

Ashworth, H., D. Sengupta, G. Schnakenberg, L. Shapiro, and L. Berger, *Phys. Rev.* **185**, 792 (1969).

Berger, L., in *American Institute of Physics Conf. Proc. No. 34*, T. Mizoguchi, ed., 1976, p. 355.

Berger, L., *Physica* **B91**, 31 (1977).

Boll, R., and H. R. Hiltzinger, *IEEE Trans.* **MAG-19**, 1946 (1983).

Bozorth, R. M., *Ferromagnetism*, IEEE Press, Piscataway, NJ, 1994.

Chin, G. Y., and J. H. Wernick, in *Ferromagnetic Materials*, Vol. 2, E. P. Wohlfarth, North Holland, Amsterdam, 1980, p. 55.

Corb, B. W., R. C. O'Handley, and N. J. Grant, *Phys. Rev.* **B27**, 636 (1983).

Datta, A., N. J. De Cristofaro, and L. A. Davis, *Proc. Rapidly Quenched Metals IV*, T. Masumoto and K. Suzuki, eds., Japan Institute of Metals, Sendai, 1982, Vol. II, p. 107.

Duwez, P., and S. Lin, *J. Appl. Phys.* **38**, 4096 (1967).

English, A. T., and G. Y. Chin, *J. Appl. Phys.* **38**, 1183 (1967).

Graham, C. D., in *Magnetism and Metallurgy*, A. E. Berkowitz and E. Kneller, eds., Academic Press, New York, 1969, p. 723.

Guillaud, C., and M. Roux, *Compt. Rend.* **242**, 2525 (1949).

Guillaud, C., *Proc. IEEE* **104B**, 165 (1957).

Hall, R. C., *Trans. Met. Soc. AIME* **218**, 268 (1960).

Hasegawa, R., and R. Ray, *J. Appl. Phys.* **49**, 4174, (1978).

Hatta, S., T. Egami, and C. D. Graham, Jr., *IEEE Trans.* **MAG-14**, 1013 (1978).

Inomata, K., T. Kobayashi, M. Hasegawa, and T. Sawa, *J. Magn. Magn. Mater.* **31–34** (1983).

Lin, L., *J. Appl. Phys.* **79**, 4578 (1996).

Littmann, M., *J. Appl. Phys.* **38**, 1104 (1967).

Littmann, M., *IEEE Trans.* **MAG-7**, 48 (1971).

Luborsky, F. E., J. J. Becker, P. G. Frischmann, and L. Johnson, *J. Appl. Phys.* **49**, 1769 (1978); *IEEE Trans.* **MAG-14**, 1008 (1978).

Metals Handbook, Vol. 8, American Society of Metals, Metals Park, OH, p. 306.

O'Handley, R. C., L. I. Mendelsohn, R. Hasegawa, R. Ray, and S. Kavesh, *J. Appl. Phys.* **47**, 4660 (1976).

O'Handley, R. C., E. A. Nesbitt, and L. I. Mendelsohn, *IEEE Trans.* **MAG-12**, 942 (1976).

O'Handley, R. C., M. C. Narasimhan, and M. O. Sullivan, *J. Appl. Phys.* **50**, 1633 (1977).

O'Handley, R. C., *Phys. Rev.* **B18**, 930 (1978).

O'Handley, R. C., and L. Berger, in *Physics of Transition Metals*, Institute of Physics Conf. Series, No. 39, p. 477 (1978).

O'Handley, R. C., C.-P. Chou, and N. DeCristofaro, *J. Appl. Phys.* **50**, 3603 (1979).

Ohnuma, S. et al., *IEEE Trans.* **MAG-16**, 1129 (1980).

Ohta, K., *J. Phys. Soc. Jpn.* **18**, 685 (1963).

Pauthenet, R., *Ann. Phys.* **7**, 710 (1952).

Pfeifer, F. and C. Radeloff, *J. Magn. Mag. Mtls.* **19**, 190 (1980).

Pfeifer, F. *Z. Metall.* **57**, 295 (1966).

Roess, E., *Proc. Int. Conf. Ferrites, Kyoto*, Y. Hoshino et al., eds., Univ. Tokyo Press, 1970, p. 187.

Rassmann, G., and H. Hoffman, *J. Appl. Phys.* **39**, 603 (1968).

Shilling, J. W., and G. L. Houze, Jr., *IEEE Trans.* **MAG 10**, 195 (1974).

Slick, P. I., in *Ferromagnetic Materials*, E. P. Wohlfarth, ed., North-Holland, Amsterdam, 1980, p. 189.

Suzuki, T., H. Nakayama, T. Yamamoto, and S. Taguchi, *IEEE Trans.* **MAG-8**, 321 (1972); S. Taguchi, T. Yamamoto, and A. Sakakura, *IEEE Trans.* **MAG-10**, 123 (1974).

Thornburg, D. R., *J. Appl. Phys.* **40**, 1579 (1969).

Yoshino, H., K. Inomata, M. Hasegawa, M. Kobayashi, and T. Sawa, *J. Appl. Phys.* **55**, 1751 (1984).

CHAPTER 11

AMORPHOUS MATERIALS: MAGNETISM AND DISORDER

In Chapter 10, the principles of soft magnetic behavior were considered and illustrated with examples. These examples included some data on metallic glasses, or amorphous magnetic alloys. In the present chapter, amorphous magnetic alloys are examined more from the point of view of the influence of atomic and chemical disorder on magnetism. These new materials now represent a major class of magnetic materials finding applications in distribution transformers, power supply transformers, inductors, and security labels.

11.1 INTRODUCTION

Amorphous alloys are materials having a noncrystalline structure that is produced by some form of nonequilibrium processing, often rapid solidification from the melt. Cooling rates of order 10^5 C/s are generally required. This high rate of heat removal often dictates that the sample have at least one small dimension to facilitate thermal transfer. Rapid solidification precludes the development of long-range topological and chemical order in the alloy. In order to stabilize the glassy state, it is generally necessary to alloy the metallic elements with glass formers such as boron, carbon, silicon, or phosphorus. Other strongly interacting metallic species suffice in some cases (Section 11.5.1). While the structure of amorphous alloys has been referred to as *random dense-packed*, the atoms of a metallic glass are not arranged randomly as in a gas. The chemical interactions among the constituents enforce a degree of short-range order that is similar to that in the liquid phase of the material. The absence of long-range order leaves amorphous materials devoid of microstructure such as grain boundaries, precipitates, or phase segregation. (The terms *amorphous*, *glassy*, and *noncrystalline* are used interchangeably here, as in the literature.)

The history of amorphous magnetic alloys is relatively short compared to that of oxide glasses. Various researchers appear to have stumbled across noncrystalline metallic alloys in their quest for new materials. One example is the early observations on NiP electrodeposits (Brenner and Riddell 1946). However, it is generally accepted that until Pol Duwez began his extensive research on metastable and amorphous alloys at the California Institute of Technology in the late 1950s, the intrinsic scientific interest and the technological potential of such materials were not widely appreciated.

A ferromagnetic amorphous CoP alloy was first reported in 1965 (Mader and Nowick 1965) and splat-quenched, glassy ferromagnets with attractive soft ferromagnetic properties were reported by Duwez' group in 1966 (Tsuei and Dewez, 1966; Duwez and Lin 1967). The subsequent growth of interest in amorphous magnetic alloys was rapid.

Under equilibrium processing conditions, local atomic arrangements are essentially determined by the thermodynamics of interatomic (chemical) interactions. However, under nonequilibrium conditions (e.g., rapid heat removal, rapid solidification front velocity, or solid-state treatment in a time–temperature regime where the time needed for diffusion to the equilibrium crystalline state exceeds that required to reach metastable states), the kinetics of the fabrication process itself can override the tendency for the atoms to order locally the way they do in the equilibrium ground state.

Essentially all of the successful methods for fabrication of amorphous metals remove heat from the molten alloy at a rate that is fast enough to preclude crystallization of the melt. Thus, certain metallic alloys can be rendered amorphous by thin-film deposition techniques such as sputtering, or by melt spinning, splat quenching, or gas atomization of fine particles. It is also possible to make glasses from the solid crystalline state by careful exploitation of fast diffusion and chemical thermodynamics (Schultz, 1990). The critical cooling rate required for glass formation can be reduced if the kinetics of atom transport in the melt are slowed down. This is the case for processing at low temperatures. It also occurs at eutectic compositions where the two or more species constituting the alloy have strong chemical interactions favoring two competing short-range orderings.

Amorphous alloys of magnetic interest are based either on 3*d* transition metals (T) or on rare-earth metals (R). In the first case, the alloy can be stabilized in the amorphous state with the use of glass-forming elements or metalloids (M), such as boron, phosphorus and silicon: $T_{1-x}M_x$, with approximately $15 < x < 30$ at%. Examples include $Fe_{80}B_{20}$, $Fe_{40}Ni_{40}P_{14}B_6$, and $Co_{74}Fe_5B_{18}Si_3$.

The late transition metals (TL = Fe, Co, Ni) can be stabilized in the amorphous state by alloying with early transition metals (TE) of 4*d* or 5*d* type

(Zr, Nb, Hf): $TE_{1-x}TL_x$, with x approximately in the range 5–15 at%. Examples include $Co_{90}Zr_{10}$, $Fe_{84}Nb_{12}B_4$, and $Co_{82}Nb_{14}B_4$.

Rare-earth metals can be stabilized by alloying with transition metals and metalloids: $R_{1-x-y}T_xM_y$ with x in the approximate range 10–25 at% and y from 0 to 10 at%. Examples include $Co_{80}Gd_{20}$ and $Fe_{75}Tb_{25}(B)$.

The $3d$ transition-metal-based amorphous alloys are generally soft magnetic materials, while the rare-earth-based amorphous alloys can be tailored to span a range from hard (permanent) magnets to semihard materials suitable for use as magnetic recording media.

Amorphous metallic alloys of the technologically useful variety discussed here may be represented as $TL_{1-x}(TE, R, M)_x$. They are typically made up of $1 - x = 60$–90 at% late transition metal (TL; e.g., Fe, Co, Ni), in which the balance x is some combination of early transition elements (TE; e.g., Cr, Mo, Nb), rare earths (R; e.g., Gd, Tb, Sm), and/or metalloids (M; e.g., B, Si, C). These are the approximate compositional limitations defining room temperature ferromagnetism in these materials.

Amorphous magnetic alloys lack long-range atomic order and consequently exhibit: (1) high metallic resistivity (100–200 $\mu\Omega \cdot$cm) due to electron scattering from atomic disorder, (2) no macroscopic magnetocrystalline anisotropy (residual anisotropies, due mostly to internal stress, typically averaging 10–100 J/m^3 for $3d$-based alloys but they can approach 10^7 eJ/m^3 for certain rare-earth containing alloys), and (3) no microstructural discontinuities (grain boundaries or precipitates) on which magnetic domain walls can be pinned. As a result, ferromagnetic metallic glasses based on $3d$ transition metals are generally good "soft" magnetic materials with both low DC hysteresis loss and low eddy-current dissipation. In addition, they are characterized by high elastic limits (i.e., they resist plastic deformation) and, for certain compositions, they show good corrosion resistance. Amorphous magnetic alloys containing appreciable fractions of R metals show magnetic anisotropy and magnetostriction that can be varied up to very large values by changing the R composition. These characteristics, combined with the expectation that metallic glasses can be economically mass-fabricated in thin gauges, have led to broad commercial interest.

11.2 STRUCTURE AND FUNDAMENTAL MAGNETIC PROPERTIES

This section describes observations of some of the fundamental (M_s, T_C, K, and λ_s) and technical (H_c, domains) magnetic properties that characterize amorphous magnetic alloys.

11.2.1 Atomic Structure

It was recognized quite early that the local atomic arrangement in amorphous alloys is not completely random. Dense packing of hard spheres with no chemical interaction (Polk 1972) still imposes a degree of local structural order typified by the Bernal polyhedra (Fig. 11.1a). In amorphous alloys, the short-range order is likely similar to that of Bernal polyhedra (*a*) and (*b*) constructed of transition metal atoms and having metalloid atoms at their centers. In polyhedron (*c*), the trigonal prism of T atoms may have M atoms off each face and possibly at the prism center. The role of chemistry in determining the local order was demonstrated by EXAFS measurements on $Pd_{80}Ge_{20}$ glasses (Hayes et al. 1978). In that eutectic glass, the germanium environment was found to be very similar to that in crystalline PdGe and, in particular, contained essentially no Ge–Ge nearest neighbors.

The short-range order of metallic glasses generally results in X-ray diffraction patterns similar to that in Figure 11.1 right, center. Fourier transformation of such data leads to radial distribution functions or to pair correlation functions indicating T-T pairs and T-M pairs but no M-M nearest neighbors. Coordination numbers for metallic glasses are usually close to 12, as in close packed crystals. Hence the structure of amorphous alloys is often referred to as dense random packing.

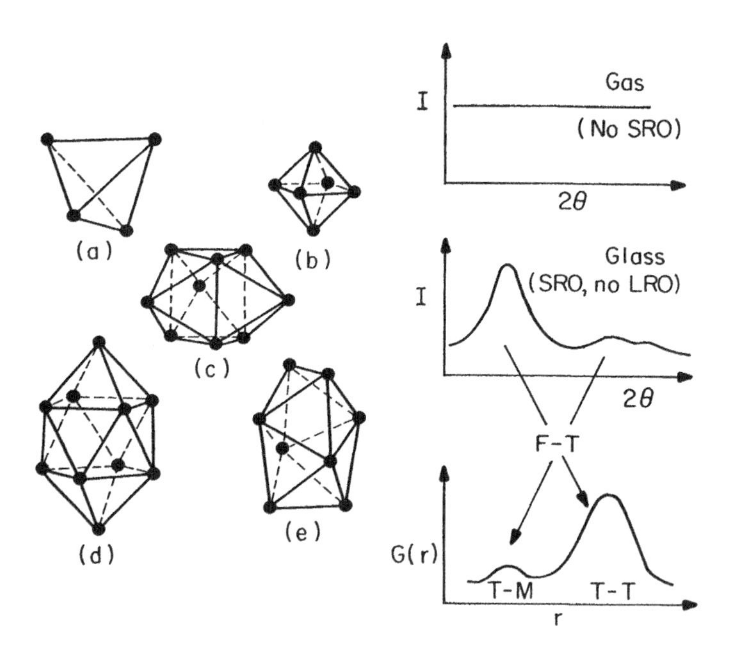

Figure 11.1 Left, structure of the Bernal polyhedra that are believed to represent the short-range ordering of atoms in many metallic glasses. Right, schematic of scattered x-ray intensity versus scattering angle 2θ for a completely random arrangement of atoms (gas), and for an amorphous alloy with short-range order (glass). The pair correlation function $G(r)$, derived from $I(\theta)$ by Fourier transform, indicates the statistical distribution of nearest-neighbor distances.

11.2.2 Magnetic Moments and Curie Temperatures

Figure 11.2 shows the variation of saturation moment per transition metal atom (4.2 K) as a function of T content for amorphous alloys based on boron, $T_{80}B_{20}$, and on phosphorus, $T_{80}P_{20}$. The variation of magnetic moment in crystalline alloys (cf. Fig. 5.1) is shown as a dotted line for reference. Amorphous $T_{80}B_{20}$ alloys show magnetic moments that are shifted relative to the Slater–Pauling curve in a way that is consistent with data for crystalline TB and T_2B compounds and alloys (Cadeville and Daniel, 1966, Mizoguchi et al. 1973, O'Handley 1983).

Reasonably large magnetic moments can be realized in a variety of amorphous alloys based on iron, cobalt, and/or nickel. The generally reduced moments of amorphous alloys compared to crystalline alloys reflect the presence of the nonmagnetic M atoms such as B, P, and Si, which are needed to stabilize the glassy state. The effect on the magnitude of the local magnetic moment due to the absence of long-range order is negligible.

The effects of early transition metal species (Cr, V, Nb...) on the saturation moment of amorphous magnetic alloys based on Fe, Co, and Ni can be understood in terms of the virtual-bound-state model (Chapter 5, O'Handley, 1981).

Figure 11.3 shows the variation of T_C in amorphous $(FeNi)_{80}B_{20}$, $(FeCo)_{80}B_{20}$, and other assorted alloys. The Curie temperatures of crystalline FeNi alloys are shown by the dotted lines. The compositional dependence of the Curie temperature is not readily described by fundamental theories.

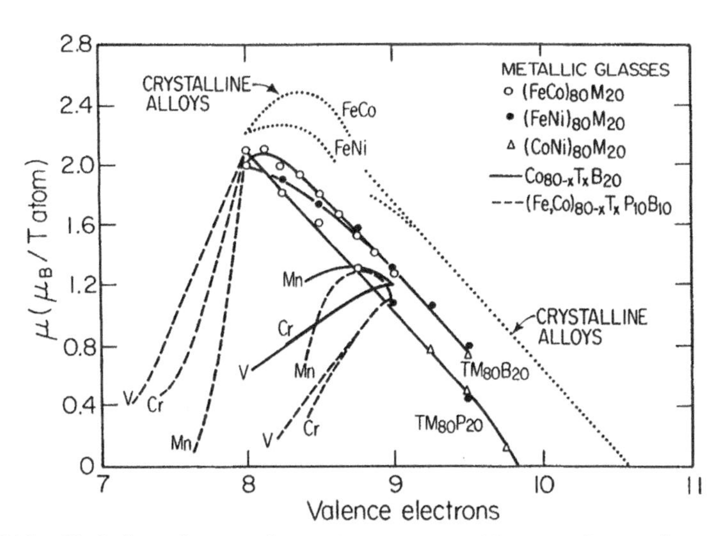

Figure 11.2 Variation of magnetic moment per transition metal atom in crystalline and amorphous alloys as a function of number of valence electrons n_v. The values $n_v = 8, 9$, and 10 correspond to Fe, Co (or $Fe_{0.5}Ni_{0.5}$), and Ni, respectively. The data for crystalline materials are based on Figure 5.1.

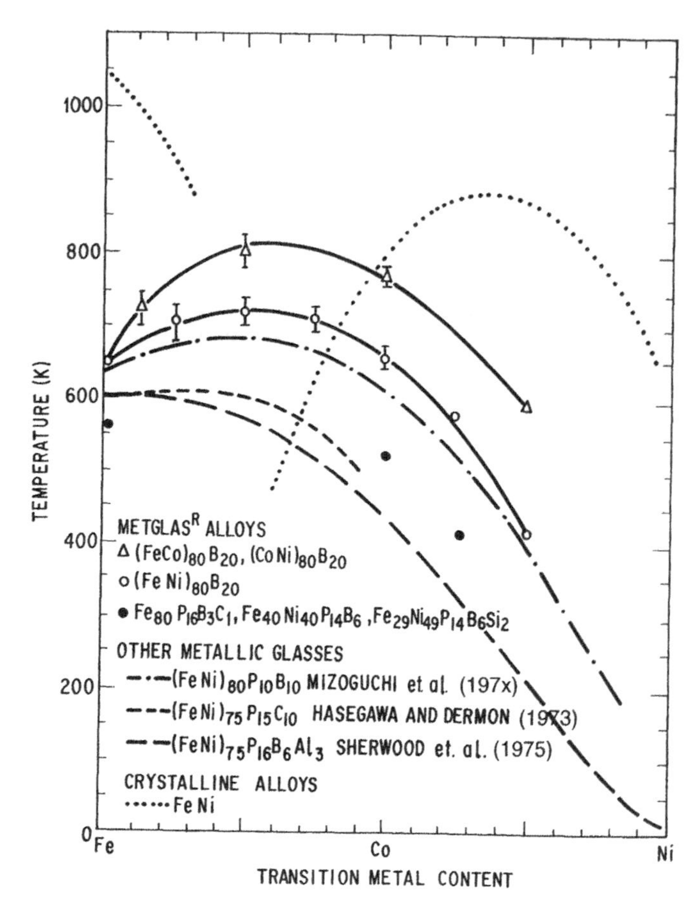

Figure 11.3 Curie temperatures of TM-B alloys versus transition metal content. Solid line over (FeNi)B data is a mean field fit to the data. Dotted lines show trends for crystalline FeNi alloys.

Molecular field theory can be used to get phenomenological parameters that describe the compositional dependence of T_C in certain alloy systems. The solid line on the (FeNi)B data in Figure 11.3 is fit to the data by adjusting the exchange parameters: $\mathcal{J}_{\text{FeFe}} = 1$, $\mathcal{J}_{\text{FeNi}} = 1.5$ and $\mathcal{J}_{\text{NiNi}} = 0.2$ (O'Handley and Boudreaux 1978). On the other hand, the compositional dependence of the magnetic moment in alloys is amenable to analysis by either electronic structure calculations or simple band models (see Section 11.5.2).

The saturation moments and Curie temperatures vary as the T/M ratio deviates from 80/20. Figures 11.4 and 11.5 show how the saturation magnetic moment per TM atom and the Curie temperature vary with metal:metalloid ratio in amorphous Fe-based and Co-based systems, respectively. In both cases the moments increase with decreasing metalloid content, extrapolating to

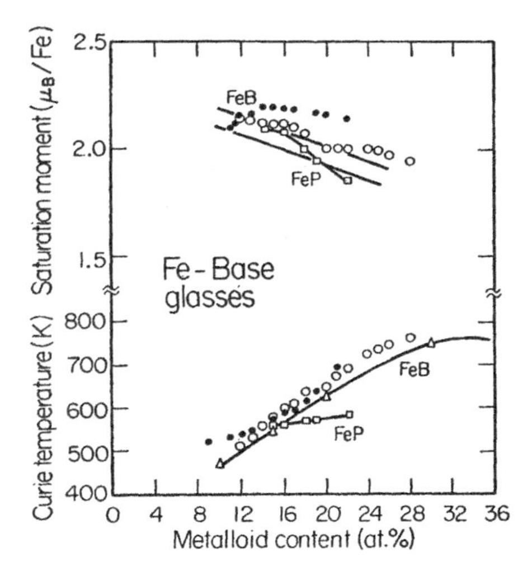

Figure 11.4 Variation of magnetic moment and Curie temperature with metal–metalloid ratio in binary FeB and FeP metallic glasses (O'Handley 1983).

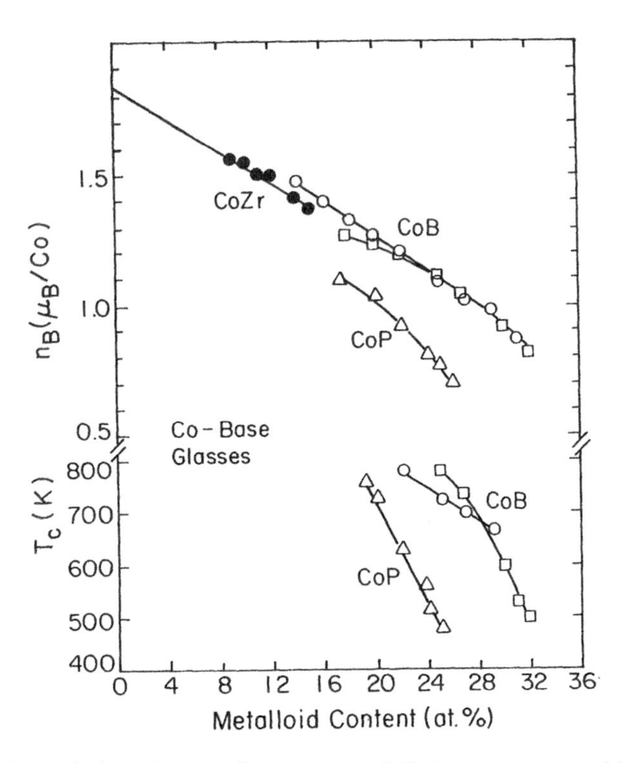

Figure 11.5 Variation of magnetic moment and Curie temperature with metal–metalloid ratio in binary Co-B, Co-P, and Co-Zr metallic glasses (O'Handley 1983).

values close to those of BCC Fe (2.2 μ_B/atom) and HCP Co (1.7 μ_B/atom), respectively. Magnetic moment data for CoZr-based glasses are included in Figure 11.5 (Ohnuma et al. 1980). These alloys exhibit low magnetostriction (Fig. 10.22) combined with high magnetization and hardness as well as good corrosion resistance.

The Curie temperatures of most cobalt-rich metallic glasses exceed their crystallization temperatures, so their trend with metalloid content is shown over a limited metalloid range (Fig. 11.5). T_C for cobalt-rich glasses *increases* for *decreasing* metalloid content much as does the magnetic moment. The behavior of the Curie temperature for the Fe-based glasses is quite different (Fig. 11.4): T_C *decreases* with *decreasing* metalloid content for both Fe-B and Fe-P glasses. This demands an explanation.

Figure 11.6 shows the variation of T_C in amorphous $Fe_{1-x}B_x$ alloys over a wider range of boron content. The Curie temperatures are seen to vary nonmonotonically with x following the pattern established by T_C values of the three crystalline compounds FeB, Fe_2B, and Fe_3B (Chien and Unruh 1981). This suggests that T_C is a function of the local coordination of magnetic atoms, not just the overall content of magnetic atoms. Further, it suggests a correlation between the local coordination in $Fe_{1-x}B_x$ glasses and that in Fe_nB

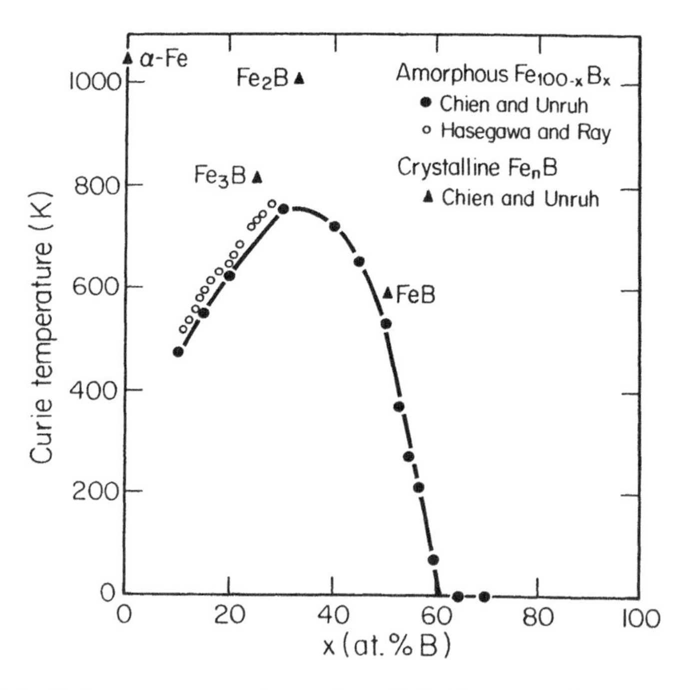

Figure 11.6 Curie temperatures of amorphous FeB alloys versus boron concentration and for relevant crystalline FeB phases from Chien and Unruh (1981) and Hasegawa and Ray (1978).

compounds. The extrapolations of the Curie temperature data for amorphous Fe-B and Fe-P alloys toward zero metalloid content in Figures 11.4–11.6, leads to T_C values well below that of BCC iron ($T_C = 1044$ K). This is presumably because the local coordination of an amorphous alloy (approximately 12 nearest-neighbor transition metal atoms) is not like that of a BCC structure (8 nearest neighbors). Instead, the amorphous structure is more like that of an FCC crystal. FCC iron is antiferromagnetic, a fact often explained in terms of the Bethe–Slater curve (Fig. 5.3) and the smaller interatomic spacing in this close-packed structure.

Changes in magnetization and Curie temperature with M-atom type (Fig. 11.7) are generally weaker than changes with TM/M ratio or with TM makeup.

Phosphorous is seen to have a stronger effect than boron in suppressing both the magnetic moment and T_C in Fe-based and Co-based amorphous alloys. For Fe-based glasses, C, Si, and Ge are less effective, in that order, in suppressing magnetism. These trends may be related to the size of the metalloid; larger metaloids expand the amorphous structure, moving Fe pairs away from their antiferromagnetic exchange range and generally reducing Fe–Fe coordination from that of a random dense-packed structure to one that is less densely packed. The effects of metalloid type on magnetic moment and Curie temperature in cobalt-based glasses are different. Only carbon enhances

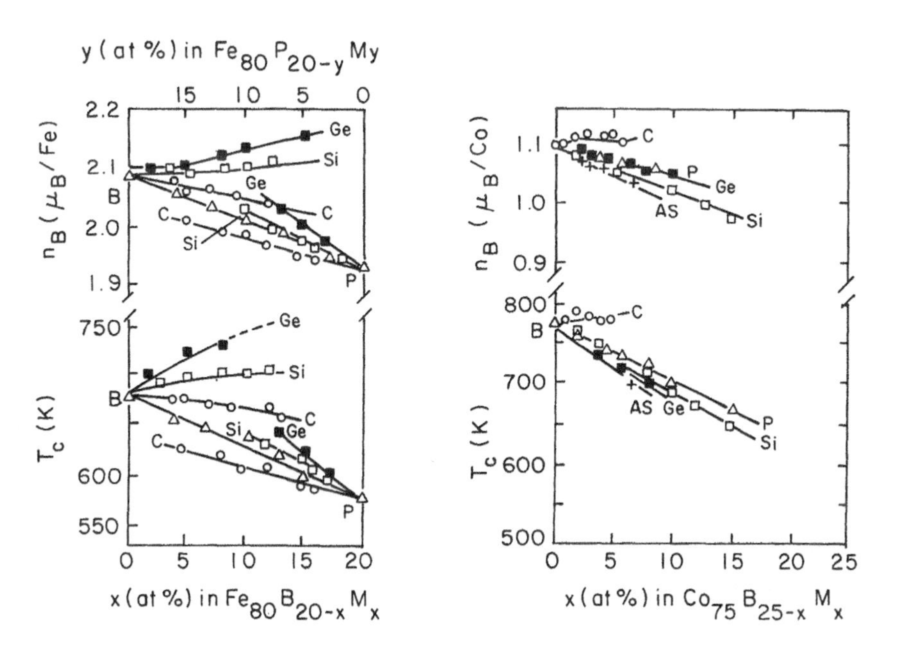

Figure 11.7 Variation of saturation moment and Curie temperature with metalloid content for amorphous alloys based on Fe [left, Mitera et al. (1978) and Kazama et al. (1978*a*, 1978*b*)] and Co [(right, Kazama and Fujimori 1982).]

n_B or T_c. These trends in average magnetic moment with metalloid type will be modeled in Section 11.5.2.

11.2.3 Magnetic Anisotropy and Magnetostriction

Clearly, magnetic anisotropy due to long-range crystalline order is not a factor in amorphous alloys. Thus, there is no fundamental anisotropy constant (other than perhaps K_0) that can be attributed to an amorphous alloy of a particular composition as was done for crystalline magnetic materials in Chapter 6. However, the *local* atomic order of amorphous magnetic alloys, which varies randomly in direction, gives rise to a random anisotropy that plays an important role in their magnetic properties. This will be covered in Section 11.4. The ease or difficulty of reaching saturation in a given direction in an amorphous magnet is also affected by sample shape, by strain-induced anisotropy (Egami et al. 1975, O'Handley 1975), or by field-induced anisotropy (Chapter 14). These factors still operate in noncrystalline materials.

Because strain-induced anisotropy depends on the magnitude of λ_s, the coercivity of amorphous alloys is expected to scale with $|\lambda_s|$. The magnetostriction of amorphous alloys was described in Chapters 7 and 10 and has been reviewed by Lacheisserie (1982), O'Handley (1983), Lachowicz and Szymczak (1984), and, for random-anisotropy, amorphous rare-earth materials, del Moral (1993).

11.3 DOMAINS AND TECHNICAL PROPERTIES OF AMORPHOUS ALLOYS

11.3.1 Domains

Figure 11.8 compares the domain structure of polycrystalline permalloy (Tanner, unpublished) with that of an amorphous iron-rich alloy (Celotta et al. 1997). The SEM image of the polycrystaline $Ni_{81}Fe_{19}$ sample shows several domains with rectilinear domain walls in one grain. (The magnetization directions in the other grains are not at the optimum angle to show magnetic contrast in this titled specimen, type II contrast, image.) The domains in the amorphous sample were imaged using scanning electron microscopy with spin polarization analysis (SEMPA) (Unguris et al. 1991). Note the absence of rectilinear domain walls in the noncrystalline material. This reflects the absence of long-range crystalline order in the metallic glass and consequently the absence of magnetocrystalline anisotropy. The magnetization tends to follow the local easy axis, which, in this case, is largely dictated by internal stress.

11.3.2 Coercivity

We saw in Chapter 9 that for a defect to strongly impede wall motion, it should have magnetic properties very different from those of the matrix and it should

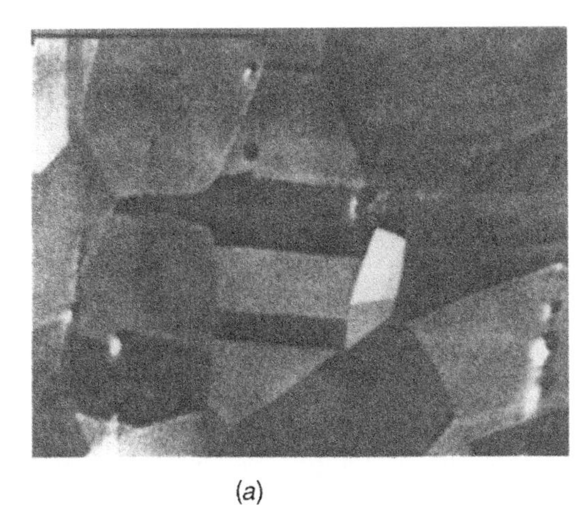

(a)

(b)

Figure 11.8 Magnetic domain pattern in polycrystalline permalloy foil (*a*), contrasted with that in an amorphous FeBSi alloy (*b*). Permalloy image taken by type II contrast (45° tilt of sample stage) in SEM; the grain size is 30 μm (courtesy of Tanner). The amorphous alloys was imaged by scanning electron microscopy with spin polarization analysis (SEMPA) after Celotta et al. (1997); the field of view in (*b*) is 1 mm.

have dimensions comparable to the domain wall width (approximately 0.2–2 μm). For the most part, amorphous alloys are homogeneous; specifically, there are no grains, no grain boundaries and no precipitates of any appreciable size. Because these alloys are rapidly quenched from the melt, most impurities tend to remain in solution rather than precipitating out. Thus chemical or structural inhomogeneities (except for surface roughness, pinholes, and strain fields) have a scale less than 2 or 3 nm. Because the domain walls are wide in amorphous alloys and the defects are narrow, there is little pinning of domain walls on defects in amorphous materials and H_c can be very small, or equivalently, the permeability can be very large. Transition-metal-based amorphous alloys make excellent soft magnetic materials, as was seen in Chapter 10.

Decades ago it was appreciated that single-crystal, soft magnetic metals typically show a low density of highly mobile domain walls. Soft magnetic crystals often show good magnetic characteristics under quasistatic conditions; however, these properties degrade rapidly with increasing frequency. The increase in H_c with frequency was understood to be a consequence of the concentration of eddy currents around the few active walls involved in the magnetization process (see Chapter 9, Section 9.4). Increasing the number of walls (by grain refinement in crystalline materials) lowered these anomalous losses and improved high-frequency magnetization performance (see Figure 10.19). Well-quenched magnetic glasses behave much like single crystals in this respect. They show a low density of domain walls because of the shortage of nucleation sites. Thus, their magnetic response deteriorates strongly with increasing frequency (O'Handley 1975). Figure 10.19*a* shows the magnetic properties and core loss of an amorphous iron-base alloy. While the DC loop is attractive, the core loss at 50 kHz is high. Techniques for optimizing AC magnetic response in magnetic glasses involve careful precipitation of finely dispersed phases whose properties are such that their benefits in terms of domain wall nucleation outweigh their pinning and induced anisotropy effects (Fig. 10.19*b*) (Datta et al. 1982, Hasegawa et al. 1982). It was shown that fine, low-anisotropy precipitates that strain the amorphous matrix only weakly are useful in this regard and their volume density should be small (1–2 vol%). More copious crystallization (Fig. 10.19*c*) leads to stresses and domain wall pinning, degrading both DC and AC properties.

The atomic disorder of amorphous alloys increases the electrical resistivity to values of order 120 $\mu\Omega \cdot$cm because conduction electrons probe the material on a scale of about one nanometer. This high electrical resistivity suppresses eddy currents, which are induced when rapid flux changes occur. This makes amorphous alloys attractive for higher-frequency operation, provided the number of active domain walls can be kept high.

11.4 MAGNETISM AND SHORT-RANGE ORDER

11.4.1 Ingredients of Short-Range Order

We first address the issue of the differences and similarities between the fundamental magnetic properties of amorphous and crystalline alloys as depicted primarily in the saturation moments (Fig. 11.2) and Curie temperatures (Fig. 11.3). Early interpretations of the Slater–Pauling-like curves of magnetic moment variation with composition (or electron concentration) invoked charge transfer from the valence band of the glass former to the 3d band of the T species to explain what appears to be a shift of the data for the glasses relative to that for pure crystalline alloys. The assumption was that the addition to the alloy of metalloid p electrons would tend to fill the 3d band, thus reducing the local magnetic moment. However, a net charge transfer between atomic sites cannot be supported in a metal because the conduction electrons will move to screen out the electric fields created by charge transfer. Further, the metalloid p states are found at energies below E_F and thus will not give up their electrons to fill empty d states. The naive charge transfer models soon gave way to a more realistic understanding of variations in magnetic moment with composition in terms of $(sp)-d$ bonding. This $(sp)-d$ bonding or hybridization is evident in theoretical treatments and experimental data on electronic structure and bonding of metallic glasses.

Hund's rules (Chapter 4) describe the formation of magnetic moments on *isolated atoms.* Orbitals fill so as to maximize the sum of their spin quantum numbers. The basis for this rule is found in the Coulomb interactions that lower the energy of electrons whose motions are correlated by occupying orbitals of different angular momentum. Essentially Hund's rule can be described as an intraatomic interaction that favors parallel spins. But what happens to this intraatomic interaction as atoms come together to form solids?

We saw in Chapter 5 that the fundamental characteristics of magnetic materials, namely the local atomic moment and the Curie temperature, are mostly functions of the short-range order (described in local environment models by the number, type, distance, and symmetry of the nearest neighbors). Thus, the fact that the absence of long-range order in amorphous magnetic alloys does not destroy magnetism should be no surprise. The decreased values of μ_T and T_C of 3d-based metallic glasses has more to do with the presence of nonmagnetic glass-forming species than with the absence of long-range order. Are there any effects on magnetism due to the absence of long-range order?

11.4.2 Exchange Fluctuations

It is well known (Moorjani and Coey 1984, Mizoguchi et al. 1977) that many antiferromagnetic (AF) compounds become ferromagnetic (F) in the amorphous state. Two factors may be contributing to this. The first is the frustration of perfect AF coupling in a disordered structure. For atoms with two of their nearest neighbors also being nearest neighbors to each other (e.g., odd-numbered rings of nearest neighbors), frustration of AF coupling sets in. This leads to fluctuations in the strength of nearest-neighbor interactions and allows for local departures from AF exchange. The second factor is the positive slope of $\mathcal{J}(r)$ near the equilibrium atomic spacing for iron, assuming that the exchange interaction $\mathcal{J}(r)$, varies according to the Bethe–Slater curve. The increase in mean atom spacing characteristic of the amorphous state then drives the system from AF to F coupling. Fluctuations in short-range order also cause the strength of the exchange interaction to vary spatially. This leads to local variations in the saturation moment.

The effect of local disorder on the exchange interaction and hence on the reduced magnetization curve $m(T/T_C) = M(T/T_C)/M(0)$ was considered theoretically by Handrich (1969). He explained the exchange about its mean value $\langle \mathcal{J}_{ij} \rangle$ (where $\langle \cdots \rangle$ is an average over the random bonds):

$$\mathcal{J}_{ij} = \langle \mathcal{J}_{ij} \rangle + \Delta \mathcal{J}_{ij} \tag{11.1}$$

in terms of the exchange fluctuation $\Delta \mathcal{J}_{ij}$. The Brillouin function [Eq. (3.34)] is then suppressed:

$$\frac{M(T)}{M(0)} \equiv m(T) = \frac{1}{2}\{B_s[x(1+\delta)] + B_s[x(1-\delta)]\} \tag{11.2}$$

Here δ is the root-mean square (RMS) exchange fluctuation

$$\delta^2 = \frac{\langle (\Sigma_j \Delta \mathcal{J}_{ij})^2 \rangle}{(\Sigma_j \langle \mathcal{J}_{ij} \rangle)^2} \tag{11.3}$$

and x is the effective field variable $x = \mu_0 \mu_m H / k_B T$. This effect of random exchange on $m(T/T_C)$ is shown in Figure 11.9a. Many observers have interpreted reduced magnetization data $m(T/T_C)$ for amorphous alloys in terms of this model (Tsuei and Lillienthal 1976). Figures 11.9*b* and 11.9*c* show the suppression of $m(T/T_C)$ by random exchange in two different alloy systems

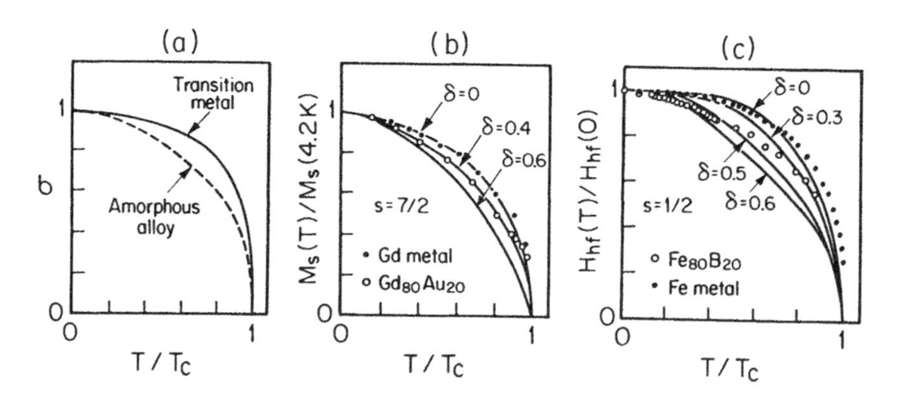

Figure 11.9 Reduced magnetization versus temperature $m(T/T_C)$ for crystalline and amorphous alloys: (*a*) Brillouin function and fluctuation-modified reduced magnetization [Eq. (11.2)] for amorphous alloy. Panels (*b*) and (*c*) show data for crystalline and amorphous Gd-based and Fe-based metals compared with theoretical curves [After Kaneyoshi (1985).]

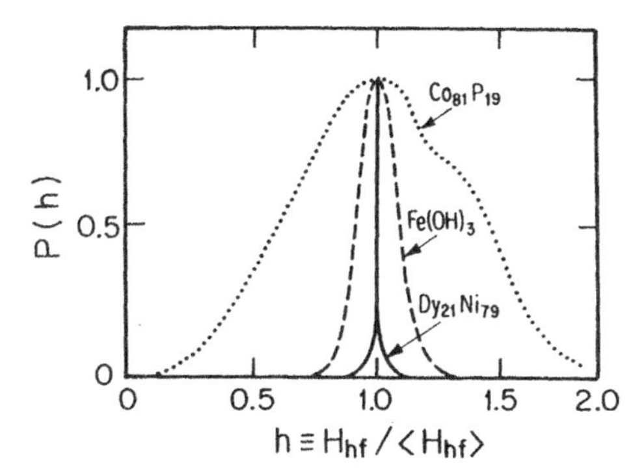

Figure 11.10 Hyperfine field (magnetic moment) distribution for Dy in amorphous DyNi compared with that for Co in amorphous CoP alloys. The Fe distribution in the insulating compound $Fe(OH)_3 0{\cdot}9H_2O$ is shown for comparison (Coey, 1978).

(Kaneyoshi 1985). However, Alben et al. (1978) have cautioned against such interpretations, noting that chemical disorder alone can suppress $m(T/T_C)$.

It is observed that rare-earth moment distributions, as measured by nuclear magnetic resonance or Mössbauer spectroscopy, remain quite narrow in amorphous alloys whereas those of $3d$ species are strongly broadened (Figure 11.10). In the case of $4f$ amorphous alloys, the more highly localized nature of the R magnetic moment leaves its magnitude relatively insensitive to its environment (Kaneyoshi and Tamura 1984). The more delocalized $3d$ orbitals that are responsible for magnetism in transition metals are more strongly perturbed by the irregularity of their local environment.

11.4.3 Random Anisotropy

Magnetic anisotropy arising from long-range crystallinity is clearly absent in amorphous alloys. However, the same "crystal field" or more accurately "local-field" that gives rise to magneto*crystalline* anisotropy is effective in noncrystalline materials on a scale of a few nanometers (Harris et al. 1973; Imry and Ma 1975). The orientation and strength of this local anisotropy in amorphous alloys varies with position, hence the term *random anisotropy*. It is important to determine the degree to which this random *local* anisotropy field affects *macroscopic* magnetic behavior or, on the other hand, is averaged out because of fluctuations in orientation of the local "easy axes" (see Fig. 11.11).

The local crystal field energy can be represented by a uniaxial (dipolar) term of strength K_{loc}. The orientation of the easy direction of this uniaxial anisotropy in an amorphous alloy fluctuates with a correlation length l determined by the local structure: l is one or two nanometers for amorphous alloys. Further, the local magnetic moments are assumed to be coupled to each other by an exchange interaction of stiffness A expressed by the form $A[\nabla \boldsymbol{m}(r)]^2$, where

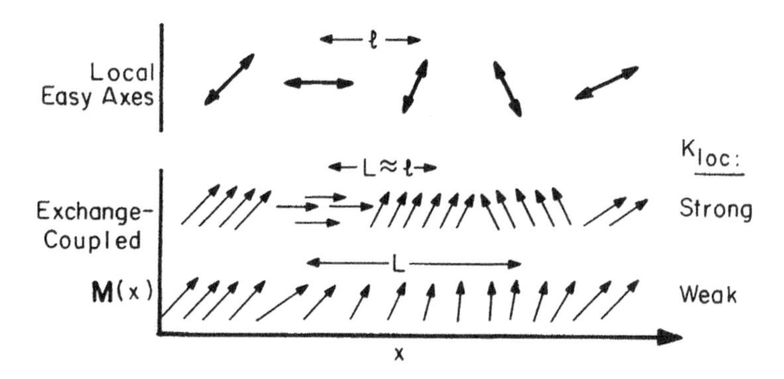

Figure 11.11 Schematic representation of the variation of local anisotropy easy axis with position and the variation of magnetization direction in response to the local anisotropy and exchange coupling. $\boldsymbol{M}(x)$ closely follows a strong local anisotropy and is a smoother function of x for weak local anisotropy.

$\boldsymbol{m}(r) = \boldsymbol{M}(r)/M_s$ is the local reduced magnetization (Imry and Ma 1975, Alben et al. 1978).

Assuming a strength for the local anisotropy K_{loc}, it is important to know the orientational correlation length L of the local magnetic moments. In other words, how closely can the exchange-coupled magnetic moments follow the short-range changes (over distance, l) in local easy-axis orientation given by the random unit vector $\boldsymbol{n}(\boldsymbol{r})$? Mathematically the problem reduces by minimizing two competing terms in the free energy F, namely, the exchange, promoting long-range correlation in the magnetization direction, and the random crystal field, favoring short-range fluctuations in the magnetization direction. The exchange and random uniaxial anisotropy contributions to the free energy are given by

$$f = \frac{F}{V} = A[\nabla \boldsymbol{m}(\boldsymbol{r})]^2 - K_{\text{loc}}\{[\boldsymbol{m}(\boldsymbol{r})]^2 - \tfrac{1}{3}\} \tag{11.4}$$

Clearly the first term scales as A/L^2 provided $L \gg l$. The strength of the random local uniaxial anisotropy expressed by the second term can be evaluated using random-walk considerations (Alben et al. 1978). The net anisotropy from a random collection of $N = (L/l)^3$ local anisotropies of strength K_{loc} goes as $N^{-1/2}$. The *macroscopic* anisotropy can then be expressed as $K_{\text{loc}}(l/L)^{3/2}$, a scaling down of the *local* anisotropy by the ratio $(l/L)^{3/2}$. Energy minimization of F with respect to L then gives

$$L = \frac{16A^2}{9K_{\text{loc}}^2 l^3} \approx \frac{10^4}{K_{\text{loc}}^2} \tag{11.5}$$

where $A = 10^{-11}$ J/m and $l \approx 2$ nm have been assumed for amorphous alloys. For a 3d-based amorphous alloy, K_{loc} is of order 5×10^4 J/m^3 while for 4f-rich alloys it is of order 10^6 J/m^3. Hence for these two cases, the correlation length L, over which the magnetization orientation is roughly constant despite fluctuating local anisotropy, is approximately

$$L = 20\,\mu\text{m} \quad 3d\text{-based}$$

$$L = 20\,\text{nm} \quad 4f\text{-based}$$

These results indicate that in $3d$-based amorphous alloys, exchange stiffness can maintain local moment orientational correlation up to 20 μm despite changes in the local anisotropy direction. The weak local anisotropy hardly perturbs the stiff spin system. However, in $4f$-based amorphous alloys the local magnetic moment may fluctuate over a much shorter range, possibly approaching a few nanometers (following the local anisotropy field). Pictures of ferromagnetic domains in $3d$-based amorphous alloys (e.g., Fig. 11.8) support the first estimate (Chen et al. 1975; Hasegawa et al. 1976, Livingston and Morris 1985) and the dispersion of rare-earth moment directions observed in many R-T amorphous alloys supports the second (Coey, 1978).

The technical implications of these estimates of L lie in the wide range of macroscopic magnetic anisotropies and coercivities that can be realized in amorphous and nanocrystalline alloys (Chi and Alben 1977; Alben et al. 1978). When $L \gg l$, the effective macroscopic anisotropy is small and soft magnetism is observed. When L/l decreases toward unity, the magnetization vector is more strongly constrained by the orientation of the local anisotropy and saturation of the magnetization is harder to achieve.

Sellmyer and Nafis (1985) have gathered the effects of exchange fluctuations and local field disorder into a simple model Hamiltonian:

$$H = -\sum_{ij} (\mathcal{J}_0 + \Delta\mathcal{J}_j)\boldsymbol{J}_i \cdot \boldsymbol{J}_j - \sum_i D(\boldsymbol{n}_i \cdot \mathbf{J}_i)^2 \tag{11.6}$$

In the first term, the isotropic exchange interaction between magnetic moments at two sites (expressed in terms of total angular momenta J_i and J_j) is expanded to include the effects of exchange fluctuations $\Delta\mathcal{J}_{ij}$. The second term allows for disorder in the distribution of the axes of the local crystal field of strength D by introduction of the local unit vector $\boldsymbol{n}_i$ [cf. Eq. (11.4)]. By defining the dimensionless variables $t = k_B T/\mathcal{J}_0$, $d = D/\mathcal{J}_0$, and $\delta = (\Delta\mathcal{J})/\mathcal{J}_0$, a ternary phase space can be constructed which illustrates the rich variety of magnetic orderings that can be realized in disordered magnets through the introduction of exchange fluctuations and random magnetic anisotropy (Sellmyer and Nafis 1985) (Fig. 11.12). The existence of the multicritical point (MCP) and the "reentrant" aspect of the phase diagram along the δ axis [i.e., the replacement of the ferromagnetic (F) phase by a metamagnetic (M) or spin glass (SG) phase on cooling] have been observed in many systems (Rao et al. 1982; Geohgean and Bhagat 1981). These are features that are unique to disordered magnets.

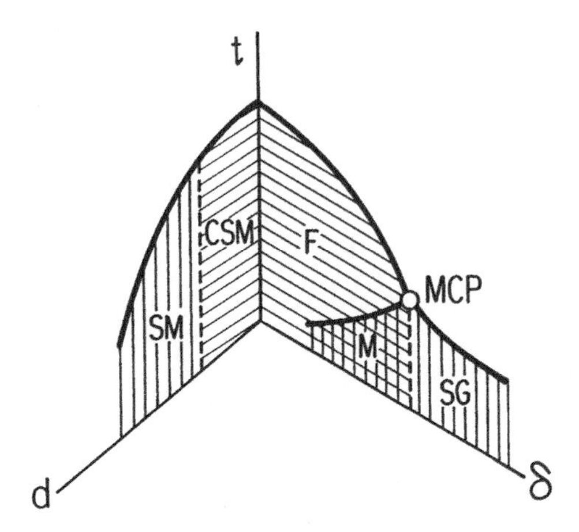

Figure 11.12 Phase diagram in t, d, δ space showing the rich variety of magnetic structures [spero- and sperimagnets (SM), canted speri(o)magnets (CSM), ferromagnets (F), metamagnets (M), and spin glasses (SG)] created by interplay of random anisotropy and exchange (Sellmyer and Nafis 1985).

When the reduced crystal field energy d approaches zero as in 3d-based amorphous alloys (see Fig. 11.12), local anisotropy is basically not felt by the strongly coupled spins. Andreev (1978) noted that then one can have either collinear ferromagnetism ($S_i \cdot S_j \approx |S_i||S_j|$ (Fig. 11.13a) or, in a two magnetic sublattice alloy, collinear ferrimagnetism ($0 \leqslant S_i \cdot S_j < |S_i||S_j|$) (Fig. 11.13$b$). There is no strongly preferred direction for the net magnetization. If, at the same time that $d \to 0$, strong exchange fluctuations exist (δ large), then the spin glass (SG) state results and $\Sigma S_i = 0$ in the absence of a field.

11.4.4 Dispersed-Moment Structures

Coey (1978) has added to these categories by considering first the additional effects of a strong local anisotropy D_i (as in 4f-based amorphous alloys) competing with exchange and then allowing for more than one magnetic species. Random anisotropy can promote spatial dispersion of the local moments, and the additional presence of more than one magnetic species can lead to a rich variety of magnetic structures with dispersed parallel (speromagnetic) or dispersed antiparallel [sperimagnetic (SM)] magnetic sublattices. For smaller values of d, canted speromagnetism (CSM) can be observed.

When the spin on one sublattice (in Fig. 11.13, 1 = R species) couple more strongly to the local crystal field than with each other ($d \gg 1$) or than with the other species, $D_1/\mathscr{J}_{11} = D_1/\mathscr{J}_{12} \gg 1$, speromagnetism $\mathscr{J}_{12} > 0$ (early R-T amorphous alloys) (Fig. 11.13c) or sperimagnetism $\mathscr{J}_{12} < 0$ (late R-T amorphous alloys) (Fig. 11.12d) can result. The spins on the second sublattice as well as the first may also be dispersed in direction either by strong coupling to

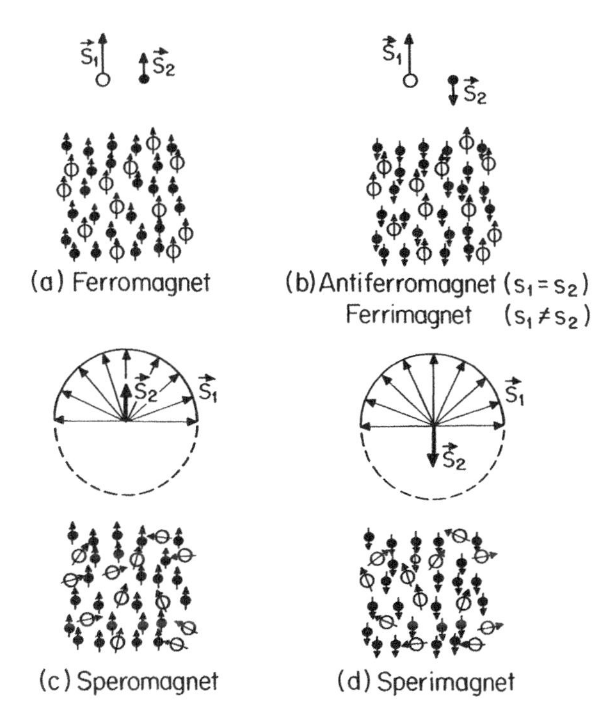

Figure 11.13 Schematic representation of some of the magnetic structures identified in Figure 11.15. (*a*) $\mathscr{J}_{12} > 0$, ferromagnet ($\mathscr{J}_{12}/D_k \gg 1$, $k = 1$ or 2); (*b*) $\mathscr{J}_{12} < 0$, antiferromagnet if $|S_1| = |S_2|$ or ferrimagnet if $|S_1| \neq |S_2|$; (*c*) $\mathscr{J}_{12} > 0$, speromagnet ($\mathscr{J}_{11}/D_1 \ll 1$, $\mathscr{J}_{22}/D_2 \gg 1$); (*d*) $J_{12} < 0$ sperimagnet ($\mathscr{J}_{11}/D_1 \ll 1$, $\mathscr{J}_{22}/D_2 \gg 1$).

their own local crystal field $\mathscr{J}_{12}/D_2 \ll 1$ and $\mathscr{J}_{22}/D_2 \ll 1$ or by strong coupling to the spins of a dispersed sublattice 1, $\mathscr{J}_{12}/D_2 \gg 1$ and $\mathscr{J}_{12}/\mathscr{J}_{22} \gg 1$. Intermediate cases also exist.

Examples of such behavior abound in the literature. Most 3*d*-based alloys exhibit the magnetic structure shown in Fig. 11.13*a* with full saturation achieved in relatively weak applied fields. The behavior shown in Fig. 11.13*b* is exhibited by Gd-Co or Gd-Fe (Orehotsky and Schroeder 1973, Taylor et al. 1978).

Speromagnetism is observed in Fe-Nd and Co-Nd (Taylor et al. 1978). Difficulty in saturating these materials (closing down the cone angle with an applied field) can be considered to have the same origin as the magnetic hardness of Fe-Nd-B permanent magnets (see Chapter 13). Sperimagnetism is typified by Fe-Tb (Rhyne et al. 1974) or Co-Dy (Coey et al. 1976, Jouve et al. 1976). Weaker coupling of the RE sublattice to the random local field occurs for smaller values of atomic orbital angular momentum.

The special case of $Co_{80}Gd_{20}$ amorphous alloys is worth mention because of its technical importance first as a bubble material then as a prototype for magnetooptic recording (Chaudhari et al. 1973). While Gd is an S-state ion (zero angular momentum), it has sufficient spin–orbit-induced angular momentum so that $\mathscr{J}_{11}/D_1$ is close to unity. Also, although Gd can be thought of as separating the light from the heavy R elements, its moment couples antiferromagnetically with that of cobalt, $\mathscr{J}_{12} < 0$. At low temperatures, $N_{Gd}\mu_{Gd} > N_{Co}\mu_{Co}$, but the magnetization of the Gd sublattice drops more steeply with increasing temperature than does that of the Co sublattice. Thus a magnetic compensation temperature exists for which $N_{Co}\mu_{Co} + N_{Gd}\mu_{Gd} = 0$, where N_i is the number of atoms with moment μ_i. More recent interest in higher coercivity perpendicular media, exclusively for magnetooptic memory, has focused on the higher anisotropy TbFe sperimagnetic system (Connell and Bloomberg 1985).

11.4.5 Induced Anisotropy

It is possible to alter the technical properties of a metallic glass by field annealing. Dramatic micrographs of the magnetic domains in amorphous ribbons annealed in longitudinal and transverse fields illustrate the magnetic consequences of field annealing (Fig. 11.14). Notice how the domain walls in field-annealed amorphous ribbons follow the direction of field-induced anisotropy in order to minimize magnetostatic energy. [In asquenched amorphous ribbons, the domain patterns are more often wavy, as shown in Fig. 11.8; the

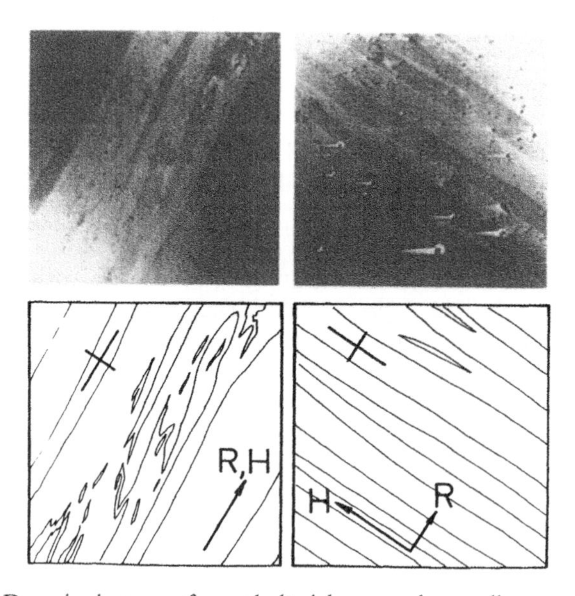

Figure 11.14 Domain images of a cobalt-rich amorphous alloy as cast, left, and annealed in a transverse field, right (Chen et al., 1975). The ribbon length direction and applied field direction during annealing are indicated by R and H, respectively.

magnetization there tries to follow the average anisotropy (Eq. 11.5) and any stress-induced anisotropy.] The atomic mechanism behind field-induced anisotropy is essentially that the local atomic order is biased by field annealing to have a small degree of directional order. This effect will be discussed in Chapter 14.

Briefly, when a magnetic material is heated below its Curie and crystallization temperatures but at a temperature high enough to allow substantial short-range atomic mobility, the thermal motion of the atoms may result in a slight biasing of the local structure toward an atomic configuration that is more stable with respect to the local direction of magnetization (Fig. 11.15). If a field is applied during the annealing process, then the local atomic rearrangements may result in a long-range correlation of a small fraction of the bond directions with the magnetization direction. On cooling, the magnetization of the entire sample will have a tendency to orient in the direction it had during the annealing process. The metalloids play a role in this process because of their high mobility and strong chemical interaction with the T metals (Allia and Vinai, 1978; Becker, 1978). If they assume a nonrandom orientational distribution around the T sites, they may favor magnetization in a particular direction.

The role of pair ordering in field annealing was illuminated by pioneering studies of induced anisotropy in the Fe-Ni-B series of glasses (Luborsky, 1977; Fujimori, 1977; Takahashi and Miyazaki, 1978).

Annealing in the absence of a field also induces a local magnetic anisotropy that changes in direction as the direction of magnetization changes from domain to domain. Further, where a domain wall exists during annealing, it will be stabilized or pinned in that position by the same mechanism that stabilizes the axis of magnetization during field annealing. Such wall pinning is illustrated by the $M-H$ loops of amorphous cobalt-rich alloy ribbons, shown in Figure 11.24.

Uniaxial magnetic anisotropy can be induced also by stress annealing. In contrast to field-induced anisotropy, stress-induced magnetic anisotropy can be

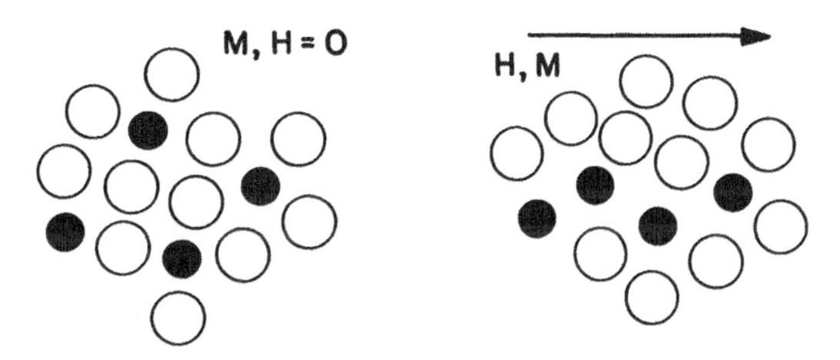

Figure 11.15 Schematic representation of short range order in a binary amorphous alloy. Left, no applied field and no net magnetization. The bond directions of strongly interacting magnetic pairs have a random distribution. Right, magnetization ordered under influence of field, results in bond directional order.

effected even above the Curie temperature. This characteristic indicates that the source of stress-induced anisotropy lies in viscoelastic effects rather than in magnetoelastic effects. The magnitude of stress-induced anisotropy can be much greater than that of magnetic-field-induced anisotropy (see also Chapter 14).

Essentially, a magnetic anisotropy due to field annealing or stress annealing refects a perturbation of the otherwise random local anisotropy in a way that provides a weak long-range order.

11.5 ELECTRONIC STRUCTURE

Modern theory of electronic structure indicates that the nature of electronic states is more a function of local atomic arrangements than of long-range atomic order.

Evidence has been presented indicating that amorphous materials are not completely random in their atomic structure. A degree of short-range order exists, albeit an order that fluctuates randomly in its orientation and strength. Because the compositions of amorphous alloys are often close to eutectic compositions, their short-range order is characterized by a competititon between the two bracketing stable phases that are responsible for the existence of the eutectic. (It is this competition that inhibits crystallization to one of the two adjacent stable crystalline phases. Further, the local order of the amorphous solid may or may not resemble that of nearby crystalline phases.) Hence the electronic structure of an amorphous alloy on a local scale (Fig. 11.16, left, center) need not resemble that of any related crystalline composition. Further, because of the long-range orientational disorder in the local structural units, the features of the electronic structure that may exist on a local scale are angle-averaged on a macroscopic scale (Fermi surface, Fig. 11.16, right).

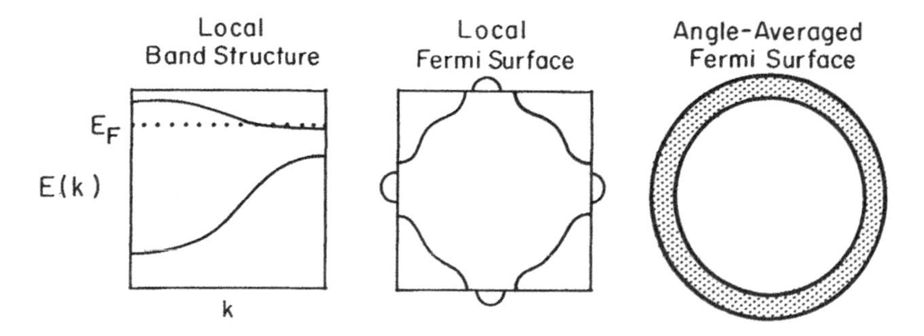

Figure 11.16 Left, schematic representation of electronic structure in a small region (1 nm^3) of an amorphous alloy and its corresponding Fermi surface, center. At right is displayed the effect of averaging local Fermi surfaces having random orientations at different locations in an amorphous alloy.

The electronic structure of amorphous alloys shows subtle effects due to the noncrystallinity of the material. However, the chemical effects due to the presence of glass-forming species are often more dramatic. It is important, therefore, when discussing features of electronic structure that characterize the glassy state, to distinguish, where possible, chemical effects due to the necessary presence of glass-forming species from the effects due solely to the structural and/or chemical disorder.

11.5.1 Split *d* Bands and *p–d* Bonds

***d–d* Splitting** The valence bands of an alloy $A_{1-x}B_x$ are often split into two resolvable components having different energies when their atomic number difference is greater than or equal to two (e.g., polar *d–d* bonding). The lower-energy states are identified primarily with electrons having greater probability density at the site of the more attractive species (i.e., the anion, which for two metals of the same row is the one with higher atomic number).

Figure 5.14 (top panel) shows UPS (ultra-violet photoelectron spectroscopy) data for amorphous PdZr (Oelhafen et al. 1979). The lower energy (greater binding energy) feature reflects the chemical stabilization due to the more attractive core potential at the Pd site compared with that at the Zr site. The calculated state densities shown in the lower panel (Moruzzi et al. 1978), clearly identify the two UPS peaks as due separately to Pd and Zr.

The connection between split bands and chemical bonding can be seen in UPS spectra on a variety of amorphous TE-TL alloys. The splitting of *d*-band features in an alloy correlates with the valence difference between the two transition metal species [Fig. 11.17; Oelhafen et al. (1980)]. (The deviation of Cu-based glasses from linear dependence would be remedied by using electronegativity difference as the abscissa instead of valence difference.) The general increase in binding energy with T valence difference correlates with the compound heat of formation as well as with the stability of the glassy phase. This implies that an increasing polar character stabilizes bonding in amorphous alloys as it does in crystalline alloys.

***p–d* Splitting** An important aspect of the behavior of transition metal *d*-bands that differs between amorphous TE-TL alloys and T-M alloys is revealed by photoemission studies. In the former case the individual (TE and TL) components of the *d* bands become narrowed slightly on alloy formation, whereas in T-M alloys the *d* bands are broadened compared to those of the pure metals. In the latter case, XPS data (Amamou and Krill 1980) show the 3*d* bandwidth of amorphous Fe-B and Fe-P alloys (and crystalline compounds) to be significantly greater than that of pure Fe (Fig. 11.18).

This difference between amorphous TE-TL and T-M electronic state densities arises from the nature of the bonding responsible for alloy formation in the two cases: *d–d* bonding in TE-TL alloys and *sp–d* as well as *d–d* bonding

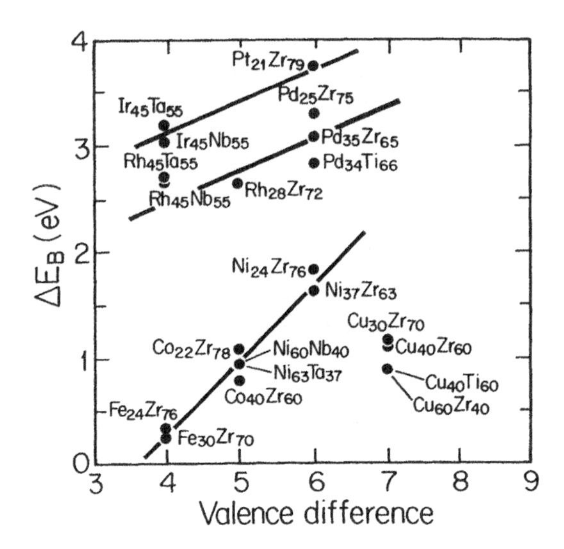

Figure 11.17 Binding energy shifts ΔE_B of the late T-species d-band maximum on alloying as a function of the late–early T metal valence difference. The lines suggest that $3d$, $4d$, and $5d$ late T species define three separate alloy groups (Oelhafen et al. 1980).

in T-M alloys. In TE-TL alloy formation, the d–d bonding is such that electronic wavefunctions associated with the lower-energy DOS peak (dominated by the strongly attractive TL core potential) have small amplitude at the TE sites. Conversely, the higher-energy wavefunction has greater amplitude at the TE sites. Once the stabilization of the states has occurred as a result of alloying [the one-center integral Γ_{ii} of Johnson and Tenhover (1983)], the system behaves to an extent like two interpenetrating but only weakly interacting sublattices. The electrons occupying the two energy ranges of high state density respond mainly to the core potentials of their own sublattice. Thus, the like-atom overlap (TE-TE or TL-TL covalency) is reduced because of the large average spacing of similar atoms and, therefore, the two components of the DOS are narrower than they would be in a pure metal; that is, the two center integrals Γ_{ij} are smaller than in the pure metal (Johnson and Tenhover 1983).

In amorphous T-M alloys, $(sp)-d$ bonding predominates. The hybridization between metalloid sp states and metal d states reduces the degree of localization of the d electrons and tends to broaden the d band into a manifold of states that is more free-electron like, less like those of a tight-binding approximation. This $(sp)-d$ hybridization broadens the d-electron contribution to the DOS in T-M alloys to differing degrees depending on the extent of hybridization. It also gives rise to the discrete $(sp)-d$ bonding states seen in Fe-based alloys at -9.5 eV and lower energies but not seen in α-Fe [Fig. 11.18; Amamou and Krill (1980)].

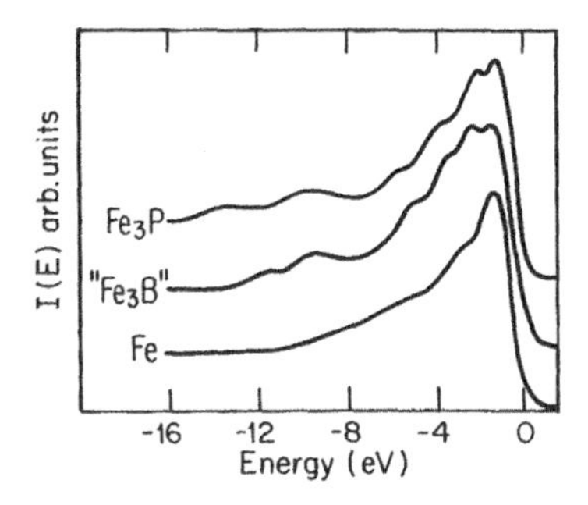

Figure 11.18 XPS valence band spectra for the TM glassy alloys $Fe_3B_{0.9}P_{0.1}$ ("Fe_3B") and Fe_3P compared with that for crystalline Fe (Amamou and Krill 1980).

Calculations on finite clusters of atoms allow for the direct study of the effects of local order on electronic structure (Messmer 1981). The self-consistent-field, $X - \alpha$, scattered wave, molecular orbital method was used to model the electronic structure of tetrahedral clusters of Fe, Ni, and metalloid atoms. At the left in Fig. 11.19, a Ni_2Fe_2 cluster is shown and below are displayed the

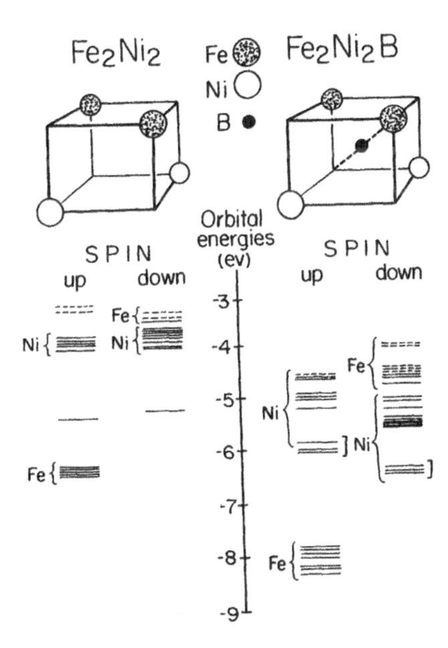

Figure 11.19 Spin-split electronic states calculated for Fe_2Ni_2 tetrahedral clusters without (left) and with (right) a central boron atom present. Dashed lines indicate unoccupied states. States labeled with curly brackets are localized near the species indicated. Unlabeled states near -5.2 eV (left) mark the locations of FeNi hybrid states. Square brackets to right of eigenstates mark NiB bonding states (Messmer 1981). Fe and Ni atomic d states lie at -8.2 eV, while boron atomic $2p$ levels lie at -3.9 eV.

calculated spin-split eigenstates, identified by their site localization. Adding central B or P atoms to these clusters reveals the chemical bonding effects important in T-M metallic glasses. The formation of metal–metalloid bonding states is found to decrease the amount of *d*–band character and hence decreases the magnetic moment. The occupied (*sp*)–*d* hybrid states (square brackets at −6, −6.5 eV below the vacuum level) appear at lower total energy than do the corresponding states in the pure metal clusters, accounting for the stabilization on alloying. Fe-P and Ni-B bonds were found to be stronger than were Fe-B and Ni-P bonds. The *p*–*d* bonding states are the small-cluster analogs of the *p*–*d* hybrid peak observed by XPS at −9.5 eV (Fig. 11.18).

Electronic states have been calculated for clusters of 1500 atoms arranged as determined by dense random packing (Fujiwara 1984). The location of spectral features arising mainly from boron *s* states, from iron-*spd* bonded with boron *p* states, and from iron-*d* states, are consistent with those observed by photoemission in Figure 11.18.

Table 11.1 compares the binding energies of the major bonding features in the Fe-B system determined by different methods. Calculational results are listed for small clusters, large clusters, and band structures. Experimental results are also listed. All three calculational approaches, namely, band structures based on simple close-packed structures, cluster calculations on large random clusters as well as on small, high-symmetry clusters, agree quite well on the basic chemical physics of bonding and electronic structure. Small clusters always show narrower bands and weaker binding energies for bonding states because of the smaller number of interactions. Also, all the calculations are consistent with the experimental XPS data on "Fe_3B."

These calculations and observations on transition metal–metalloid alloys suggest the mechanisms behind the general observation that chemical bonding weakens magnetism. When magnetic *d* orbitals are involved in bonding, some states are shifted from the *d* band to low-lying bonding orbitals. These bonding orbitals are fully occupied and therefore do not contribute to the magnetic moment. The remaining states in the *d* band become somewhat delocalized as

TABLE 11.1 Binding Energy (in eV below E_F) for Boron *s* and Fe-*d*–B-(*sp*) Hybrid Bonding States as Determined by Four Different Calculations and by XPS Studies

Method	Boron *s* States	Fe-*d*–B-(*sp*) Hybrid	Reference
Fe_4B cluster	10.4	7	Collins et al. (1988)
Fe_2Ni_2B cluster	—	6.5	Messmer (1981)
1500 atom $Fe_{80}B_{20}$	12.5	6.8	Fujiwara (1984)
ASW Fe_3B band structure	—	8–10	Moruzzi (unpublished)
Experimental Fe_3B XPS	11.5	9.5	Amamou and Krill (1980)

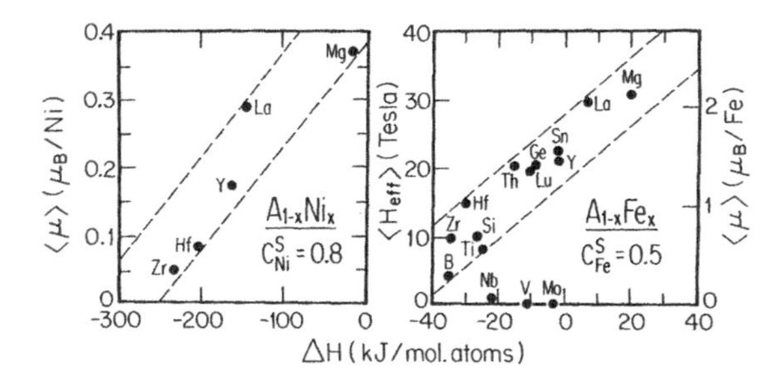

Figure 11.20 Left, average Ni moment as a function of heat of alloying in various Ni alloys of compositions such that the coordination sphere about an average Ni atom is about 80% covered, $CS_{Ni} = 0.8$; right, average effective hyperfine field or average Fe moment as a function of heat alloying for amorphous Fe-based alloys such that $CS_{Fe} = 0.5$ (Buschow 1984).

if they corresponded to *d* states of a lighter T species (Fig. 5.18). The increased kinetic energy of the electrons in a broadened *d* band makes it more difficult for intraatomic exchange to operate and the magnetic moment decreases. Figure 11.20 illustrates the competition between bonding and magnetism in systems based on transition metals and nonmagnetic metals. The stronger the bonding (the more negative the heat of alloy formation), the smaller are the magnetic moments or hyperfine fields (Buschow 1984).

11.5.2 Modeling Magnetic Moment Variations

It is not always possible to determine from first-principle electronic structure calculations what the magnetic moment per transition metal atom is in an amorphous alloy. There are some simple empirical models that relate the average transition metal magnetic moment $\langle\mu_T\rangle$ to the metalloid type and concentration. Some of these models are reviewed by O'Handley (1987).

It is useful here to mention one of these models because of the physical insight it provides. Corb et al. (1983) recognized that it is the metalloid-*sp*/metal-*d* bonding that is largely responsible for moment reduction in 3*d*-based metallic glasses. He formulated a model based on the assumption that the extent of *sp*–*d* bonding is proportional to the number of T atoms, Z^T_M, that can fit around a metalloid atom of a given size. Hence the moment suppression of the average T atom is proportional to Z^T_M:

$$\mu_T = n_B\left(\frac{1 - Z^T_M N_M}{5N_T}\right) \tag{11.7}$$

The coefficient of Z^T_M is a measure of how much bonding each T atom experiences with the M atoms. In Eq. (11.7), n_B is the magneton number of the pure metallic T species and $N_M/N_T = x/(1-x)$ for $T_{1-x}M_x$ alloys. The factor of 5 reflects the assumption that each T atom forms one bond with M and hence loses one of its five 3*d* orbitals to nonmagnetic bonding states. The implication is that smaller metalloids (e.g., B), which can be surrounded by fewer T atoms because of their size, are less effective in moment suppression. Larger glass formers (e.g., P or Si) can be surrounded by more *T* species and thus a given number of them, N_M, is more effective in moment reduction.

Figure 11.21 shows the success of this simple concept in describing moment suppression in various cobalt-based crystalline amorphous alloys. The solid lines express Eq. (11.7). The coordination number Z_M^T is determined from that of a related crystalline phase as follows.

Amorphous CoB: local order resembles that of Co_3B (the structure of Fe_3C, cementite), thus $Z_B^{Co} = 6$.

Amorphous CoP: local order resembles that of Co_3P(BCT), thus $Z_P^{Co} = 9$

HCP Co(P): local order resembles that of HCP Co, thus $Z_P^{Co} = 12$ at low x

Crystalline CoP: local order resembles that of BCT Co_3P, thus $Z_P^{Co} = 9$ at large x

In summary, these calculations and data flesh out our physical insight into the effects on magnetism of changes in local environment (*number*, *type*,

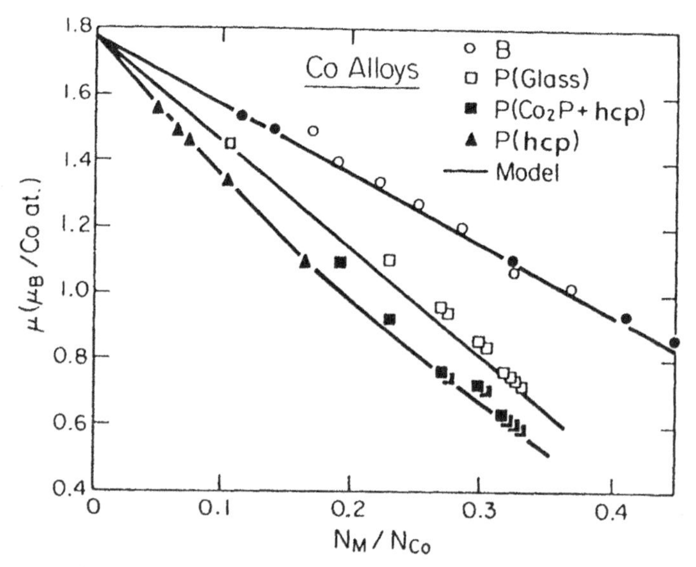

Figure 11.21 Average magnetic moment per magnetic atom as a function of concentration of nonmagnetic metalloid species M to transition metal species, N_M/N_T. Solid lines are calculated from Eq. (11.7) for appropriate M coordination (Corb et al. 1983).

distance, and *symmetry* of nearest neighbors; see Chapter 5). They prescribe the principal ingredient favoring magnetism in alloys, whether amorphous or crystalline: (1) strongly positive heat of formation to favor like-atom clustering rather than ordering and (2) minimum p–d hybridization (smaller TM coordination as is usually found for metalloids of small radius; this tends to reduce p–d bonding).

11.5.3 Electron Transport

Amorphous magnetic alloys have electrical resistivities falling usually in a range from 100 to 150 $\mu\Omega \cdot$cm (3 times to 5 times those of most crystalline magnetic alloys). The loss of structural and chemical order beyond a length of about one nanometer contributes to electron scattering and limits the electronic mean free path to a distance of order 1 nm. Metallic glasses are characterized by small positive or negative temperature coefficients of resistivity (Mooij, 1973; Tsuei, 1986). In this sense, metallic glasses bear a close resemblance to liquid metallic alloys of similar composition. Figure 11.22 shows the resistivity of Pd-Cu in the crystalline, amorphous and liquid phases (Güntherodt and Künzi 1978). This similarity led to the application of Ziman liquid metal theory to the understanding of electrical transport in metallic glasses. That theory is based on the scattering of electrons by the random structure. The randomly oriented scattering vector of the structure has a magnitude given by the $k = 4\pi \sin\theta/\lambda$ value corresponding to the location of the peak in the structure factor, $S(k)$. One consequence of liquid metal theory is a simple explanation of the systematics of occurrence of positive or negative temperature coefficients of resistance, depending on the values of the Fermi

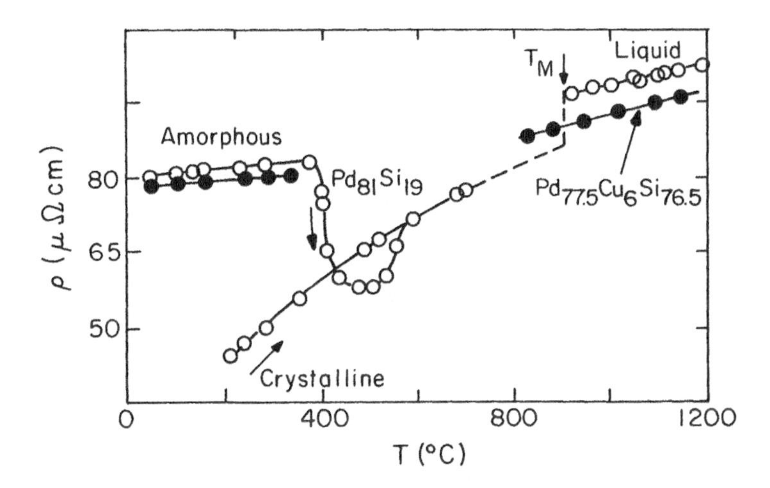

Figure 11.22 Electrical resistivity of $Pd_{81}Si_{19}$ and $Pd_{77.5}Cu_6Si_{16.5}$ in the amorphous, crystalline, and liquid states (Güntherodt and Künzi 1978).

wavevector relative to the peak of the structure factor [Fig. 11.23; Güntherodt and Künzi (1978)]. Alloy systems having their Fermi energy near the peak of the structure factor exhibit lower electrical resistivity at elevated temperatures [less scattering due to reduced peak in $S(k)$], while those with E_F on the wings of $S(k)$ show increased ρ with increasing temperature. Other aspects of electron transport in amorphous solids are competently reviewed by Harris and Strom-Olsen (1983), Coté and Meisel (1981), and Rao (1983).

In a nonmagnetic material the application of a magnetic field perpendicular to the primary current can alter the longitudinal resistance $\Delta\rho/\rho$ and induce a Hall voltage in a direction orthogonal to both the current and applied field. In nonmagnetic materials these anisotropic effects can be understood as consequences of the Lorentz force on the primary current carriers. In a ferromagnetic material these ordinary anisotropic transport effects are generally overshadowed by phenomena with similar geometrical dependences but arising from the much stronger spin–orbit interaction between the current carrier (orbit) and the local magnetization (spin). Amorphous magnetic alloys show weak anisotropic magnetoresistance $(\rho_{\parallel} - \rho_{\perp})/\rho_{\perp} = \Delta\rho/\rho$ compared to crystalline transition metal alloys. (Here $\parallel$ or $\perp$ refer to the orientation of the magnetization relative to the direction of the current.) This weak magnetoresistance is due in part to the large value of electrical resistivity that characterizes metallic glasses. Metallic glasses do show a strong spontaneous Hall effect, $V_H = R_s(\boldsymbol{J} \times \boldsymbol{M})$, where $\boldsymbol{J}$ is the current density vector, $\boldsymbol{M}$ is the magnetization vector, and R_s is the spontaneous Hall coefficient (O'Handley 1978).

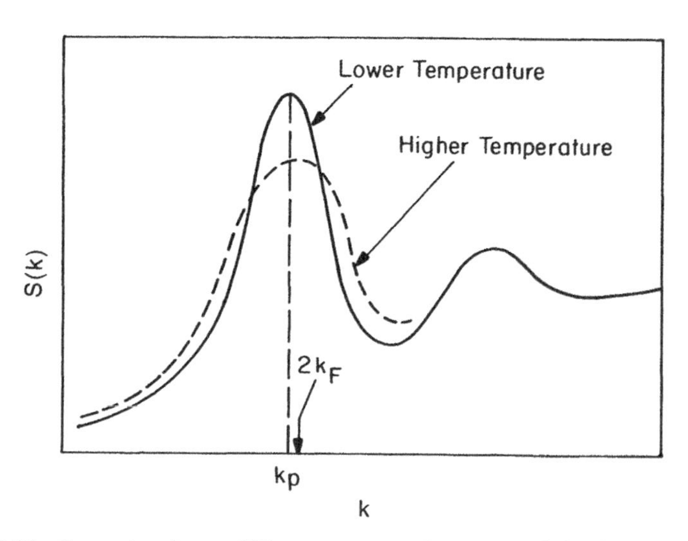

Figure 11.23 Scattering factor $S(k)$ versus scattering vector $\boldsymbol{k}$ for lower and higher temperatures in an amorphous alloy (Güntherodt and Künzi 1978).

11.6 APPLICATIONS

Amorphous metallic alloys lack long-range atomic order and consequently exhibit many characteristics important for a variety of applications. Some of their attractive technical characteristics are listed below.

1. Amorphous magnetic alloys exhibit high electrical resistivities ($\rho \approx 100$–$200\,\mu\Omega\cdot\text{cm}$) due to electron scattering from atomic disorder. High resistivity is important in suppressing eddy currents during high-frequency magnetization reversal.
2. They show no macroscopic magnetocrystalline anisotropy. Residual anisotropies typically amount to $10\,\text{J/m}^3$ for $3d$-based alloys but can approach $10^6\,\text{J/m}^3$ for certain rare-earth-containing alloys. Hence magnetization rotation is relatively easy in the former class of amorphous alloys; anisotropy fields H_a of a few oersteds are readily achieved.
3. Amorphous alloys have no microstructural discontinuities (grain boundaries or precipitates) on which magnetic domain walls can be pinned. Hence magnetization by wall motion is relatively easy. Coercive fields H_c of a few millioersteds are readily achieved.
4. They exhibit high elastic limit due to the absence of an ordered atomic lattice through which dislocation motion would be relatively easy. Highly elastic behavior (i.e., little plastic deformation) renders metallic glasses less prone to magnetic degradation during handling. However, many metallic glasses are brittle after annealing and subject to considerable flow and to anelastic relaxation at elevated ($150°\text{C} < T < 400°\text{C}$) temperatures (Chen 1982, Berry 1978, Greer 1984).
5. Many amorphous metals can show attractive corrosion resistance that is now understood to result from a variety of factors, including the absence of grain boundaries along which contaminants can enter. The fine scale of oxide/corrosion layers renders the materials less susceptible to pitting (Hashimoto, 1983, Hashimoto and Masumoto, 1983) and the general presence of strong oxide formers in solution as glass formers (B, Si, P) affords a ready source of material for surface passivation (Cotell and Yurek 1986).

As a result, ferromagnetic metallic glasses based on $3d$ transition (T) metals are generally good "soft" magnetic materials with both low DC hysteresis loss and low eddy-current dissipation. In addition, they are characterized by high elastic limits (i.e., they resist plastic deformation), and, for certain compositions, they show good corrosion resistance. Amorphous magnetic alloys containing appreciable fractions of rare-earth (R) metals show magnetic anisotropy and magnetostriction that can be varied almost continuously with composition up to very large values. These characteristics, combined with the fact that metallic glasses can be economically mass-fabricated in thin gauges, has led to broad commercial interest.

11.6.1 Distribution Transformers

The technological development of amorphous magnetic alloys was motivated largely by their application as cores in distribution transformers (O'Handley 1975, Luborsky 1983, Hasegawa 1983, Fish 1990). Power distribution transformers rated at about 25 kVA contain approximately 100 kg of magnetic material. Close to one million distribution transformers are installed annually in the United States alone. Amorphous alloy transformer cores have lower losses and thus save power and reduce the cooling requirements of the transformer construction (O'Handley et al. 1979). However, the metastable nature of the amorphous state was a cause for some initial concern for a device that is expected to operate for a minimum of 20 years: What would happen to a transformer if the core crystallized and the losses increased dramatically? The stability of many commercial amorphous magnetic alloys is now known to be sufficient for hundreds of years of operation at 200°C (O'Handley et al. 1979). Installations of amorphous alloy core transformers began in the early, 1980s and field performance of the amorphous cores has improved with aging.

Figure 10.19 compares the 60-Hz core loss versus maximum induction level during operation for amorphous Fe-based alloys with that for various commercial crystalline alloys. While the amorphous alloys show dramatically lower loss, they cannot be driven to as high a flux level because of their reduced saturation magnetization compared to crystalline iron alloys.

When the highest permeability is required, as in smaller transformers and other inductive elements, the amorphous cobalt-rich alloys represent an attractive alternative to 78% Ni permalloy. One advantage they have over permalloy is their much higher hardness. The high yield strength of amorphous metallic alloys generally makes them much more resistant to plastic deformation and slip-induced anisotropy than crystalline alloys.

Iron-rich amorphous alloys can be used in high-frequency applications if they are heat treated to develop a small volume fraction of domain-nucleating, α-Fe precipitates of nanocrystalline dimensions [Fig. 10.19; Datta et al. (1982)].

11.6.2 Electronic Article Surveillance Sensors

A growing application of magnetic metallic glasses is in the field of electronic article surveillance (EAS), the process of placing remotely detectable tags on items to locate them, to control inventory, or to deter theft. EAS requires an interrogation zone (usually defined by a magnetic antenna pair) near the exit from an area to be secured. When the magnetic field in the interrogation zone is perturbed by an *active* tag, the system is alerted. The tags of interest here most often consist of a small strip of amorphous magnetic alloy. Magnetic tags change the characteristics of the field in the interrogation zone in *frequency* or in *time* (O'Handley 1993). We describe each technique briefly.

When a material in the presence of an external AC field shows a response at a frequency different from that of the drive field, it is referred to as *harmonic* (or *subharmonic*) generation. Harmonic tags multiply the drive frequency due to the nonlinear permeability of the tag. Materials for such tags must have very low coercivities and large, nonlinear permeabilities. Cobalt-rich, zero-magnetostriction alloys are most often used for harmonic tags. The transmit antennas can operate in a continuous-wave mode at one frequency while listening continuously for the tag response at a different frequency. Thus the transmit signal is not a source of noise for the receive antenna.

A novel type of harmonic tag is based on the concept of domain wall pinning. By annealing a strip of soft amorphous alloy in zero field, the domain walls are stabilized in their demagnetized locations (Schafer et al. 1991). After annealing, a small field is required to free or depin the walls. When this field threshold is exceeded, the walls snap to a new position with a resultant sharp change in magnetization. This pinned-wall behavior (Fig. 11.24) is the same effect in a small sample comprised of a few magnetic domains, as the well-known *perminvar effect* in larger multidomain samples.

Some magnetic materials respond to a pulsed excitation field in a way that persists for a time after the excitation field is off. Resonant magnetoelastic tags operate in this way. When a magnetoelastic tag is excited by a primary magnetic field signal for a period of time, it stores magnetic energy in a coupled magnetoelastic mode. Once the excitation field is turned off, the magnetoelastic tag "rings down" in a characteristic way that allows its signal to be separated in time from the drive signal as well as being distinguished from the signals of most other possible magnetic objects passing through the interrogation zone. The requirement here is for a magnetically soft material having nonzero magnetostriction. Amorphous alloys are very suitable for these magnetoelastic resonant devices because of their very low acoustic losses, their high electrical resistivity (low eddy current losses), and the ability to tailor their magnetostriction to optimize magnetomechanical coupling (Chapter 7).

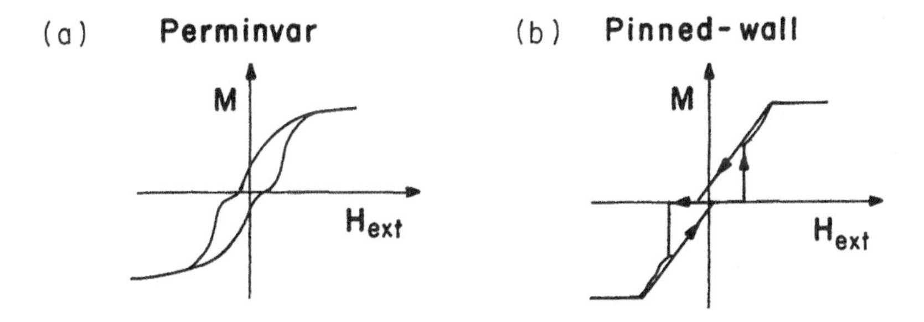

Figure 11.24 (*a*) Typical Perminvar loop; (*b*) pinned wall loop of a small amorphous ribbon (O'Handley 1993).

11.7 SUMMARY

Far from being random in their atomic structure, amorphous metallic alloys are characterized by well-defined short-range order. The unusual magnetic behavior of Fe metalloid glasses was reviewed. First noted was the disparity between the high Curie temperature of pure α-Fe and the decreasing Curie temperature toward the low-metalloid-content limit of the glass forming range in amorphous $Fe_{100-x}B_x$. This is now understood to be a consequence of the dense packing in the glassy phase leading to increased antiferromagnetic exchange interaction with increasing Fe content.

Rapidly solidified alloys exhibit a range of properties that challenge our ability to process and understand materials. Further, they offer a variety of new technical opportunities in diverse magnetic applications. Their fundamental magnetic properties can be understood as consequences of their electronic structure, which, in turn, reflects their short-range chemical and topological order. Short-range order is a consequence of the chemical interactions between the constituents of a material.

Many of the technical advantages of metallic glasses stem from the fact that these materials are quenched from the melt; most impurities remain in metastable solid solution, there are no grain boundaries or precipitates, and little segregation, and a broad range of compositions can be fabricated by the same process, yielding a continuous spectrum of property values. The electrical resistivity is high for metallic alloys, suppressing eddy currents in AC applications. Further, the exchange interaction tends to average out any local magnetic anisotropy, ensuring soft magnetic characteristics in many alloy compositions. Magnetostriction remains one of the major factors inhibiting easy magnetization in some metallic glasses.

Many good reviews of amorphous alloys are available, covering a variety of physical properties. Moorjani and Coey (1984) present a wealth of information on metallic as well as on insulating disordered magnetic systems. Alben et al. (1978) treat several fundamental issues in amorphous ferromagnetism, notably the effects of random anisotropy on coercivity, spin wave demagnetization, and chemical versus structural disorder effects on the temperature dependence of magnetization. Petrakovskii (1981) gives a lucid survey of some fundamental issues related to the effects of amorphous structure on the magnetic state of a system. O'Handley [in Luborsky (1983)] covers a number of topics, such as charge transfer, bonding, and crystal field anisotropy, that emerge from the body of the data available on transition-metal base glasses. Wohlfarth [in Luborsky (1983)] examines many aspects of magnetism in light of itinerant electron theory. Durand (1983) focuses on the effects of structural disorder on transition metal amorphous alloys [in Beck and Güntherodt (1983)] and elsewhere (Hasegawa 1983) gives a thorough treatment of the local structure of amorphous alloys as seen by various local magnetic probes. Kaneyoshi [in Hasegawa (1983); see also Kanayoshi (1985)] reviews theoretical issues related to exchange and moment fluctuations as well as spin wave states. The electronic structure of metallic glasses, as determined by various electron spectroscopies and compared with results of augmented spherical wave (ASW) calculations (Moruzzi et al. 1978) on model structures, is thoroughly reviewed by Oelhafen (Beck and Güntherodt 1983). Electronic structure, particularly for some superconducting alloy compositions, is reviewed by Johnson and Tenhover (Hasegawa 1983). O'Handley (1987) reviews some issues of the physics of ferromagnetic amorphous alloys.

Topics of a more applied nature are covered in several good reviews. Luborsky (1983) treats anisotropy, moment variations, coercivity, and losses. Hasegawa (1983) treats a number of basic properties that affect magnetic applications at low frequency. Soohoo reviews several issues important to the use of amorphous rare-earth-containing thin films in magnetic recording [in Hasegawa 1983)]. Fujimori (1983) treats magnetic anisotropy and specific applications are reviewed by Boll, Hiltzinger, and Warlimont [in Hasegawa (1983)].

APPENDIX: MAGNETISM IN QUASICRYSTALS

Quasicrystals (QCs) are a class of noncrystalline materials first observed in the AlMn system (Shechtman et al., 1984). They show sharp x-ray diffraction peaks as do crystals, and the electron microdiffraction patterns can exhibit fivefold or other symmetry not allowed by classical crystallography. The diffraction peaks do not index to any known crystalline phase; instead they occur at intervals that are multiples of the "golden ratio" $r = (1 + \sqrt{5})/2$ (Bancel et al. 1985). The scattering can be explained by structures of randomly assembled icosahedra (Shechtman and Blech 1985, Stephens and Goldman 1986) or by Penrose tilings composed of two rhombohedral bricks arranged so as to preserve long-range five-fold orientational order with translational quasiperiodicity rather than periodicity (Levine and Steinhardt, 1984, 1986). Another structural model for quasicrystals proposed by Guyot and Audier (1985) describes the Al-Mn QC structure as a quasiperiodic distortion of the α-Al-Mn-Si crystal structure. In this structure there are no central atoms present in the icosahedral units. All three of these models give diffraction patterns in good agreement with scattering studies. The actual decoration of the Penrose lattice or of the icosahedral glass structure with various species of the QC alloy is an important issue not adequately resolved yet. The QC composition $(Ti_{1-x}V_x)_2Ni$ (Zhang et al. 1985) appears to have a Ni atom in the icosahedral site (Zhang et al. 1986).

Quasicrystals appear to be metastable phases clearly distinct from the amorphous state. Some metallic glasses devitrify directly to the crystalline state, while others can pass through the quasicrystalline (icosahedral) state before crystallizing (Lillienfeld et al. 1985; Poon et al. 1985). Much of the research on quasicrystals began as an effort to model the structure of amorphous materials (Nelson, 1983; Sadoc, 1980; Sethna, 1983; Levine and Steinhardt, 1984, 1986).

Quasicrystals (QCs) appear to form at compositions where the preference for local icosahedral order is strong. This is the case for many Frank-Kasper (1958) phases. A classic example is the α-Al-Mn-Si structure for which the MacKay icosahedron is the important building block (Guyot and Audier 1985). One means of achieving the QC structure rather than long-range crystalline order is by some form of nonequilibrium processing, such as melt spinning (Shechtman et al. 1984), ion implantation (Budai and Aziz 1986), ion mixing by energy beam irradiation (Knapp et al. 1985), or solid state diffusion (Knapp et al. 1986). Thus, the QC state appears to be an intermediate phase between the glassy and crystalline states, in terms of both the degree of atomic order and the extent of nonequilibrium processing needed to achieve that order.

Symmetry requires (McHenry et al. 1986, O'Handley et al. 1991) that an atom in an icosahedral environment have fully degenerate d orbitals (no crystal field splitting, unlike the splitting in cubic symmetry sites; see Fig. 6.14) and any orbitals of f symmetry would be split into a fourfold degenerate state and a threefold degenerate state, unlike the f splitting in cubic symmetry (2–2–3). Consequently the DOS should be highly peaked and unusual magnetic and other physical properties could result, depending on the position of these peaks relative to E_F as well as on the bonding or antibonding nature of the states at E_F.

The high degree of symmetry in the icosahedral group implies small values of the magnetic anisotropy. This is evident from an expansion of anisotropy energy in harmonic functions such as in Eq. (6.4) where the first three lowest symmetry terms, $l = 0, 2, 4$, vanish for icosahedral symmetry. Furthermore, the multiplicity of symmetry axes in an icosahedral sample can be shown to give rise to a variety of new types of domain wall and to lead to easy magnetization by wall motion or rotation processes (McHenry et al. 1987a).

Most reported quasicrystals (e.g., Al-Mn–(Si), Al–(Fe, Cr, U, Mo, Ru), Pd-U-Si, Al-Li-Mg-Cu, $(Ti_{1-x}V_x)_2Ni(x < 0.1)$, or $(Al, Zn)_{49}Mg_{32}$) are not strongly magnetic. However, the possibility of studying magnetic systems in which a magnetic ion may reside in the high-symmetry icosahedral site has driven a search for strongly magnetic icosahedral phases (McHenry et al. 1986). This search is supported by the observation of enhanced magnetic susceptibility in icosahedral (I) Al-Mn alloys relative to the orthorhombic Al_6Mn crystalline phase (Hauser et al. 1986, Youngquist et al. 1986). While crystalline Al_6Mn is nonmagnetic and has no Mn atom at the center of its McKay icosahedron, I Al-Mn alloys show the presence of a local moment that increases with the square of the Mn concentration (Hauser et al. 1986). However, Mössbauer studies (Swartzendruber et al. 1985) show an appreciable electric field gradient, that is, no icosahedral symmetry, for Fe atoms at the Mn sites in I Al-Mn. Also, NMR studies (Warren et al. 1985) have found no tendency toward stronger magnetism in QC Al-Mn relative to the crystalline state. Some results of a program to produce strongly magnetic icosahedral quasicrystals have appeared (O'Handley et al. 1981b, McHenry et al. 1987).

Quasicrystals continue to be a source of scientific curiosity but have yet to show properties that would recommend them for any significant applications.

BIBLIOGRAPHY

Beck, H., and H. J. Güntherodt, eds., *Glass Metals*, Vol. II, Springer-Verlag, Berlin, 1983.

Gilman, J. J., and H. J. Leamy, eds., *Metallic Glasses*, ASM, Metals Park, OH, 1978.

Güntherodt, H. J., and H. Beck, eds., *Glassy Metals*, Vol. I, Springer-Verlag, Berlin, 1981.

Hansen, P., in *Ferromagnetic Materials*, Vol. 4, K. H. J. Buschow, ed., Elsevier, Amsterdam, 1991, p. 289.

Hasegawa, R., ed., *Glassy Metals: Magnetic, Chemical and Structural Properties*, CRC Press, Boca Raton, FL, 1983.

Liebermann, H. H., ed., *Rapidly Solidified Alloys*, Marcel Dekker, New York, 1993.

Luborsky, F. E., ed., *Amorphous Metallic Alloys*, Butterworths, London, 1983.

Moorjani, K., and J. M. D. Coey, *Magnetic Glasses*, Elsevier, Amsterdam, 1984.

O'Handley, R. C., "Physics of ferromagnetic amorphous alloys," *J. Appl. Phys.* **62**, R15 (1987).

REFERENCES

Alben, R., J. I. Budnick, and G. S. Cargill III, *Metallic Glasses*, J. J. Gilman and H. J. Leamy, eds., ASM, Metals Park, OH, 1978, p. 304.

Allia, V., and F. Vinai, *IEEE Trans. Magn.* **MAG-14**, 1050 (1978).

Amamou, A., and G. Krill, *Solid State Commun.* **33**, 1087 (1980).

Andreev, A. F., *Soviet Phys. JETP* **47**, 411 (1978).

Bancel, P. A., P. A. Heiney, P. W. Stephens, A. I. Goldman, and P. M. Horn, *Phys. Rev. Lett.* **54**, 2422 (1985).

Beck, H., and H. J. Güntherodt, eds., Glassy Metals, Vol. II, Springer-Verlag, Berlin, 1983.

Becker, J. J., *IEEE Trans. Magn.* **MAG-14**, 938 (1978).

Berry, B., in *Metallic Glasses*, J. J. Gilman and H. J. Leamy, eds., ASM, Metals Park, OH, 1978.

Brenner, A., and G. Riddell, *J. Res. Natl. Bur. Stand.* **37**, 31 (1946).

Budai, J. D., and M. J. Aziz, *Phys. Rev. B* **33**, 2876 (1986).

Buschow, K. H. J., *J. Phys.* **F13**, 563 (1984).

Cadeville, M. C., and E. Daniel, *J. Phys.* (Paris) **27**, 29 (1966).

Celotta, R. J., J. Unguris, M. J. Scheinfein, and D. T. Pierce, personal communication (1997).

Chaudhari, P., J. J. Cuomo, and R. J. Gambino, *Appl. Phys. Lett.* **22**, 337 (1973).

Chen, H. S., S. D. Ferris, H. J. Leamy, R. C. Sherwood, and E. M. Gyorgy, *Appl. Phys. Lett.* **26**, 405 (1975).

Chen, H. S., *Rapidly Quenched Metals IV*, T. Masumoto and K. Suzuki, eds., Jpn. Inst. Metals, Sendai, 1982, p. 495.

Chi, M. C., and R. Alben, *AIP Conf. Proc.* **29**, 147 (1977).

Chien, C. L., and K. M. Unruh, *Phys. Rev. Lett.* **B24**, 1556 (1981).

Coey, J. M. D., J. Chappert, J. P. Rebuillat, and T. S. Wang, *Phys. Rev. Lett.* **36**, 1061 (1976).

Coey, J. M. D., *J. Appl. Phys.* **46**, 1646 (1978).

Collins, A., R. C. O'Handley, and K. H. Johnson, *Phys. Rev. B* **38**, 3665 (1988).

Corb, B. W., R. C. O'Handley, and N. J. Grant, *Phys. Rev. B* **27**, 636 (1983).

Coté, P. J., and L. V. Meisel, in *Glassy Metals*, Vol. II, H. Beck and H. J. Güntherodt, eds., Springer-Verlag, Berlin, 1983, p. 141.

Cotell, C., and G. Yurek, *Oxid. Met.* **26**, 363 (1986).

Datta, A., N. J. De Cristofaro, and L. A. Davis, in *Rapidly Quenched Metals IV*, T. Masumoto and K. Suzuki, eds., Jpn. Inst. Metals, Sendai, 1982, p. 107.

del Moral, A., in *Magnetoelastic Effects and Applications*, L. Lanotte, ed., Elsevier Science, Amsterdam, 1993, p. 1.

Durand, J., in *Metallic Glasses: Magnetic, Chemical and Structural Properties*, R. Hasegawa, ed., CRC Press, Boca Raton, FL, 1983.

Duwez, P., and S. Lin, *J. Appl. Phys.* **38**, 4096 (1967).

Egami, T., P. J. Flanders, and C. D. Graham, AIP Conf. Proc. **24**, 697 (1975).

Fish, G., *Proc. IEEE* **78**(6), 947 (1990).

Frank, F. C., and J. F. Kaspar, *Acta Crystallogr.* **11**, 184 (1988).

Fujimori, H., in *Amorphous Metallic Alloys*, F. E. Luborsky, ed., Butterworths, London, 1983.

Fujiwara, T., *J. Non-Cryst. Solids* **61–62**, 1039 (1984).

Geohagen, J. A., and S. M. Bhagat, *J. Magn. Magn. Mater.* **25**, 17 (1981).

Greer, A. L., *J. Non-Cryst. Solids* **61–62**, 737 (1984).

Güntherodt, H. J., and Künzi, H. V., in *Metallic Glasses*, J. J. Gilman and H. J. Leamy, eds., ASM, Metals Park, OH, 1978, p. 247.

Guyot, P., and M. Audier, *Philos. Mag. B* **52**, L15 (1985).

Handrich, K., *Phys. stat. solidi* **B32**, K55 (1969).

Harris, R., M. Plischke, and M. J. Zuckermann, *Phys. Rev. Lett.* **31**, 160 (1973).

Harris, R., and J. O. Strom-Olsen, in *Glassy Metals*, Vol. II, H. Beck and H. J. Güntherodt, eds., Springer Verlag, Berlin, 1983, p. 325.

Hasegawa, R., L. E. O'Handley, R. R. Tanner, and S. Kavesh, *Appl. Phys. Lett.* **29**, 219 (1976).

Hasegawa, R., and R. Ray, *J. Appl. Phys.* **49**, 4174 (1978).

Hasegawa, R., G. Fish, and V. R. V. Ramanan, in *Rapidly Quenched Metals IV*, T. Masumoto and K. Suzuki, eds., *Jpn. Inst. Metals*, Sendai, 1982, p. 929.

Hasagawa, R., ed., *Glassy Metals: Magnetic, Chemical, and Structural Properties*, CRC Press, Boca Raton, FL, 1983.

Hashimoto, K., in *Amorphous Metallic Alloys*, F. E. Luborsky, ed., Butterworths, London, 1983, p. 471.

Hashimoto, K., and T. Masumoto, *Phys. Rev. B* **33**, 3577 (1983).

Hauser, J. J., H. S. Chen, and J. V. Waszczak, *Phys. Rev. B* **33**, 3577 (1986).

Hayes, T. M., J. W. Allen, J. Tauc, B. C. Giessen, and J. J. Hauser, *Phys. Rev. Lett.* **40**, 1282 (1978).

Imry, Y., and S.-K. Ma, *Phys. Rev. Lett.* **35**, 1399 (1975).

Johnson, W. L., M. Atzmon, M. Van Rossum, B. P. Dolgin, and X. L. Yeh, in *Rapidly Quenched Metals*, S. Steeb and H. Warlimont, eds., Elsevier, North-Holland, New York, 1985, p. 1515.

Jouve, H., J. P. Rebuillat, and R. Meyer, *AIP Conf. Proc.* **29**, 97 (1976).

Kaneyoshi, T., and I. Tamura, *Phys. status solidi B* **123**, 525 (1984).

Kaneyoshi, T., unpublished notes prepared for International Symposium on Magnetism of Amorphous Materials, Balatlonzeplak, Hungary, Sept. 1985.

Kazama, N. S., M. Mitera, and T. Masumoto, *Proc. 3rd Internatl. Conf. Rapidly Quenched Metals*, B. Cantor, ed., Metals Society, London, 1978a, Vol. 2, 164.

Kazama, N. S., N. Heiman, and R. L. White, *J. Appl. Phys.* **49**, 1706 (1978b).

Kazama, N. S., and H. Fujimori, in *Rapidly Quenched Metals IV*, T. Masumoto and K. Suzuki, eds., Jpn. Inst. Metals, Sendai, 1982, p. 709.

Knapp, W. A., and D. M. Follstadt, in *Rapidly Solidified Alloys and Their Mechanical and Magnetic Properties*, B. C. Giessen, H. Taub, and D. E. Polk, (eds., MRS, Pittsburgh, 1986, p. 233.

Lacheisserie, E., du Tremolet de, *J. Magn. Magn. Mater.* **25**, 251 (1982).

Lachowicz, H. K., and H. Szymczak, *J. Magn. Magn. Mater.* **41**, 327 (1984).

Levine, D., and P. Steinhadt, *Phys. Rev. Lett.* **53**, 2477 (1984).

Levine, D., and P. Steinhardt, *Phys. Rev. B* **34**, 596, 617 (1986).

Lillienfeld, D. A., M. Hastasi, H. H. Johnson, D. G. Ast, and J. W. Mayer, *Phys. Rev. Lett.* **55**, 1587 (1985).

Livingston, J. D., and W. D. Morris, *J. Appl. Phys.* **57**, 3555 (1985).

Luborsky, F. E., ed., *Amorphous Metallic Alloys*, Butterworths, London, 1983.

Luborsky, F. E., in *Amorphous Magnetism II*, Hasegawa and Levy, Plenum Press, New York, 1977, p. 345.

Madar, S., and A. Nowick, *Appl. Phys. Lett.* **17**, 57 (1965).

McHenry, M. E., M. E. Eberhart, R. C. O'Handley, and K. H. Johnson, *Phys. Rev. Lett.* **56**, 81 (1986).

McHenry, M. E., R. C. O'Handley, W. Dmowski, and T. J. Egami, *Appl. Phys.* **61**, 3225 (1987).

McHenry, M. E., and R. C. O'Handley, *J. Mater. Sci. Eng.* **99**, 377 (1988).

Messmer, R. P., *Phys. Rev. B* **23**, 1616 (1981).

Mitera, M., M. Naka, T. Masumoto, and N. Watanabe, *Phys. status solidi A* **49**, K163 (1978).

Mizoguchi, T., T. Yamauchi, and H. Miyajima, in *Amorphous Magnetism*, H. O. Hooper and A. M. de Graaf, eds., Plenum Press, New York, 1973, p. 325.

Mizoguchi, T., T. R. McGuire, R. Gambino, and S. Kirkpatrick, *Physica B* **86–88**, 783 (1977).

Mooij, J. H., *Phys. Stat. solidi A* **17**, 521 (1973).

Moorjani, K., and J. M. D. Coey, *Magnetic Glasses*, Elsevier, Amsterdam, 1984.

Moruzzi, V. L., J. F. Janak, and A. R. Williams, *Calculated Electronic Properties of Metals*, Pergamon Press, New York, 1978, p. 52.

Nelson, D. R., *Phys. Rev. Lett.* **50**, 982 (1983).

Oelhafen, P., E. Hauser, and H.-J. Güntherodt, *Solid State Commun.* **35**, 1017 (1980).

Oelhafen, P., in *Glassy Metals*, Vol. II, H. Beck and H. J. Güntherodt, eds., Springer Verlag, Berlin, 1983, p. 283.

O'Handley, R. C., *J. Appl. Phys.* **46**, 4996 (1975).

O'Handley, R. C., *Phys. Rev. B* **18**, 2577 (1978).

O'Handley, R. C., *Solid State Commun.* **21**, 1119 (1981).

O'Handley, R. C., in *Amorphous Metallic Alloys*, F. E. Luborsky, ed., Butterworths, London, 1983, p. 257.

O'Handley, R. C., "Physics of ferromagnetic amorphous alloys," *J. Appl. Phys.*, **62**, R15 (1987).

O'Handley, R. C., *J. Mater. Eng. Perform.* **2**, 211 (1993).

O'Handley, R. C., and D. S. Boudreaux, *Phys. stat. solidi A* **45**, 607 (1978).

O'Handley, R. C., C.-P. Chou, and N. De Cristofaro, *J. Appl. Phys.* **50**, 3603 (1979).

O'Handley, R. C., and M. O'Sullivan, *J. Appl. Phys.* **52**, 1841 (1981).

O'Handley, R. C., R. A. Dunlap, and M. E. McHenry, in *Ferromagnetic Materials*, Vol. 4, K. H. J. Buschow, ed., Elsevier, Amsterdam, 1991, p. 453.

Ohnuma, S. et al., *IEEE Trans.* **MAG-16**, 1129 (1980).

Orehotzky, J., and K. Schroeder, *J. Appl. Phys.* **43**, 2413 (1973).

Petrakovskii, G. A., *Sov. Phys. Usp.* **24**, 511 (1981).

Polk, D., *Acta Met.* **20**, 485 (1972).

Poon, S. J., A. J. Drehman, and K. R. Lawless, *Phys. Rev. Lett.* **55**, 2324 (1985).

Rao, K. V., J. Gerber, J. W. Halley, and H. S. Chen, *J. Appl. Phys.* **53**, 7731 (1982).

Rao, K. V., in *Amorphous Metallic Alloys*, F. E. Luborsky, ed., Butterworths, London, 1983, p. 401.

Rhyne, J. J., J. H. Shelling, and N. C. Koon, *Phys. Rev. B***10**, 4672 (1974).

Sadoc, J. F., *J. Phys.* (Paris) *Colloq.* **41**, C8-236 (1980).

Schafer, R., W. Ho, J. Yamasaki, A. Hubert, and F. B. Humphrey, *IEEE Trans.* **MAG-27**, 3678 (1991).

Schultz, L. *Philos. Mag.* **61**, 453 (1990).

Sellmyer, D. J., and S. Nafis, *J. Appl. Phys.* **57**, 3584 (1985).

Sethna, J. P., *Phys. Rev. Lett.* **50**, 2198 (1983).

Shechtman, D., I. Blecht, D. Gratias, and J. W. Cahn, *Phys. Rev. Lett.* **53**, 1951 (1984).

Shechtman, D., and I. A. Blech, *Met. Trans.* **16A**, 1005 (1985).

Stephens, P. W., and Goldman, A. I., *Phys. Rev. Lett.* **56**, 1168 (1986).

Swartzendruber, L. J., D. Shechtman, L. Bendersky, and J. W. Cahn, *Phys. Rev. B* **32**, 1382–1383 (1985).

Takahashi, M., and T. Miyazaki, *Jpn. J. Appl. Phys.* **17**, 361 (1978).

Taylor, R. C., T. R. McGuire, J. M. D. Coey, and A. Gangulee, *J. Appl. Phys.* **49**, 2885 (1978).

Tsuei, C. C., and P. Duwez, *J. Appl. Phys.* **37**, 435 (1966).

Tsuei, C. C., and H. Lilienthal, *Phys. Rev. B* **13**, 4899 (1976).

Tsuei, C. C., *Phys. Rev. Lett.* **57**, 1943 (1986).

Unguris, H., R. J. Celotta, and D. T. Pierce, *Phys. Rev. Lett.* **67**, 140 (1991).

Warren, W., H. S. Chen, and J. J. Hauser, *Phys. Rev. B* **32**, 7614 (1985).

Yoshino, H., K. Inomata, M. Hasegawa, M. Kobayashi, and T. Sawa, *J. Appl. Phys.* **55**, 1751 (1984).

Youngquist, S. E., P. F. Miceli, D. G. Wiesler, H. Zabel, and H. Frazer, *Phys. Rev. B* **34**, 2960 (1986).

Zhang, Z., H. Q. Ye, and K. H. Kuo, *Philos. Mag.* **A52**, 149 (1985).

Zhang, Z., and K. H. Kuo, *Philos. Mag.* **B54**, L83 (1986).

CHAPTER 12

MAGNETISM IN SMALL STRUCTURES: EXCHANGE COUPLING AND NANOCRYSTALS

Composite materials in which one of the component microstructures has one, two, or three nanoscale dimensions allow new properties and functions to be realized that are not achievable in simpler materials. Nanostructured soft magnetic materials often are composed of single-domain crystalline particles in an amorphous matrix. The prototype of this class of materials is the nanocrystalline magnet FeSiBNbCu (α-Fe_3Si particles in a matrix of residual amorphous phase). In these materials, the properties can vary widely, depending on the size of the nanocrystalline particles as well as the dimensions and magnetic properties of the intervening amorphous matrix. A glimpse of this behavior was seen in Figure 9.19. There the coercivity was shown to increase by five orders of magnitude as amorphous Co-Nb-B was heat treated to grow crystalline Co_nB particles of increasing size in a nonmagnetic Nb-B-rich matrix. A geometric classification (based on the number of nanometer-scale dimensions of one of the phases) and examples of nanostructured materials are given in Figure 12.1. Panels (*a*)–(*c*) depict systems having three small dimensions: (*a*) granular solids in which one or both phases are magnetic, (*b*) quasigranular films made by heat treating multilayers of immiscible solids, and (*c*) nanograined layers grown on columnar films. Panels (*d*) and (*e*) illustrate systems having two small dimensions: (*d*) columnar thin films and engineered nanowires; (*e*) acicular particulate recording media. Panel (*f*) represents four different types of thin-film structures whose unique properties arise from one nanoscale dimension: $[Fe/Cr]_N$ multilayers, spin valves, spin switches, and spin-tunnel junction. Some aspects of these materials and, in some cases, the

Dimensions of nanometer scale	Material examples

3 small dimensions

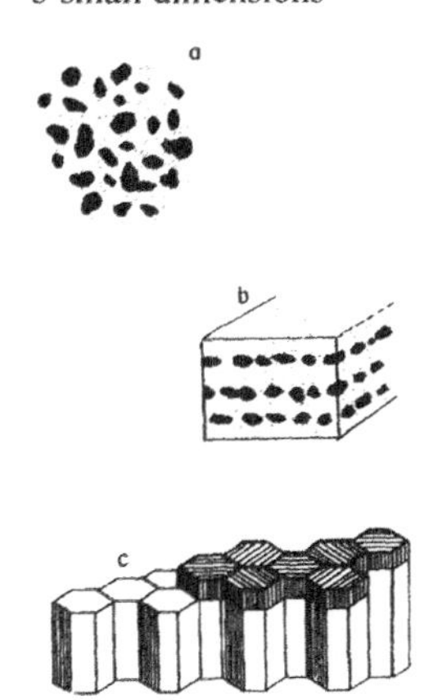

a. Nano-Fe-B-Si-Nb-Cu, $(Fe\text{-}Co)_{80}M_9C_{10}$

a. Composite nanostructured permanent magnets α-Fe/Fe-Nd-B

a. Granular GMR materials Fe/Cu, Fe-Ag, Co-Cu

b. Quasi-granular GMR films $[Fe\text{-}Ni/Ag]_N$

c. Thin film magnetic recording media: Co-Cr-Ta/Co-Cr

2 small dimensions

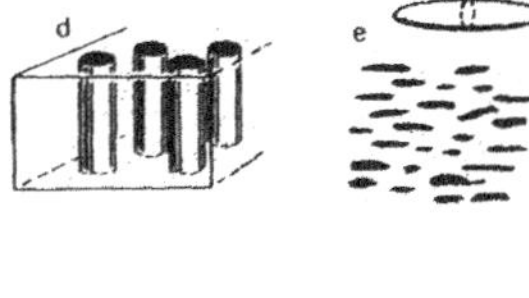

a. Columnar structures

a. Nanowires

b. Particulate media

1 small dimension

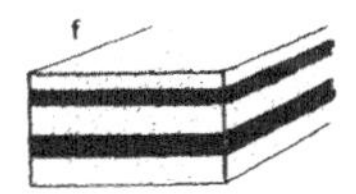

Magnetic multilayers: $[Fe/Cr]_N$

Spin valves: Co/Cu/Ni-Fe

Spin switches: Ni-Fe/Cu/Ni-Fe

Figure 12.1 Classification of nanostructured materials by the number of their small dimensions.

devices based on them, are discussed in the remaining chapters. This chapter focuses mainly on magnetic size effects and their manifestations in nanocrystalline magnetic materials.

What are the interactions that make magnetism so sensitive to nanoscale structure in these materials? Some of the effects are simply *magnetostatic*. Others also involve as well the limitations that size puts on the ability of the magnetization to change its direction across the thin dimension of the structure (*exchange energy* and *single-domain behavior*). Also relevant is the effect of thermal energy in demagnetizing particles that are so small that the total energy defining their quiescent magnetization state (K_uV) is less than some multiple of k_BT (*superparamagnetism*). The concept of *random anisotropy*, introduced in Chapter 11, is also relevant here in treating the variation in direction of magnetization from particle to particle. Another effect observed only in samples with at least one small dimension is the coupling between spins on different sides of a physical interface (*exchange coupling*). Because this phenomenon is critical to understanding the magnetic behavior of nanocrystalline materials and the behavior of many thin-film magnetic structures and devices (see Chapters 15–17), a thorough introduction to this topic will be given here. There are also transport effects in which the direction of spin in one magnetic layer is communicated across an interface by a current carrier to produce a spin-dependent scattering in another magnetic layer. This subject will be covered in Chapter 15 on transport properties in magnetic materials.

These various competing interactions should ideally be treated by a micromagnetic approach where all interactions are considered locally, simultaneously and self-consistently. Here, the physical principles behind the ingredients of such calculations are introduced and the consequences of these interactions are illustrated with examples from nanostructured materials.

For several reasons it is appropriate to cover the remarkable properties of nanostructural magnetic materials after Chapter 10 (soft magnetic materials) and Chapter 11 (amorphous magnetic materials), and prior to the chapter on permanent magnets (Chapter 13):

1. Nanocrystalline magnetic materials were first made by devitrification of amorphous alloys and hence are a logical sequel to the chapter on amorphous magnetism.
2. Depending on the size and composition of the nanocrystallites and the intergranular material, either soft or hard magnetic properties can be engineered. These properties can be very different from what would be expected based on the constituent magnetic elements present. In other words, the nanoscale structure can be as important in determining the properties of these materials as is the overall chemistry.
3. Nanocrystalline materials form a conceptual and practical bridge between soft, amorphous magnetic alloys and permanent magnets. This will be appreciated when the concept of random anisotropy is presented in the context of finely structured magnetic materials.
4. Nanocrystalline materials provide an excellent context in which to review the relations between microstructure and properties in magnetic materials prior to the following chapters on permanent magnets, transport properties, surface magnetism, and magnetic recording.

12.1 REVIEW

It is appropriate to begin with a review of some concepts that were discussed in earlier chapters and are key to understanding nanocrystalline behavior.

For homogeneous magnetic particles, free of defects, the rotational coercivity is governed by magnetic anisotropies (including shape). However, H_c can be appreciably smaller if mobile domain walls are present. Because there are generally more domain walls in samples having a larger thinnest dimension, t, [Eq. (8.22)], it may be expected that $H_c \propto 1/t^n$ (see Fig. 12.2 for spherical particles with $r > r_c$). (This decrease in H_c with increasing d assumes a constant pinning defect density.)

12.1.1 Single-Domain Particles

It was shown in Chapter 8 that below a certain thickness, a thin film should be composed of a single domain. Also, magnetic particles are comprised of a single domain if it costs more energy to create a domain wall than to support the magnetostatic energy of the single-domain state. Expressions were derived for the single-domain particle radius, r_c, in the strong and weak anisotropy limits. The magnetization processes in such particles were reviewed in Chapter 9. The coercivity of an assembly of identical, *noninteracting*, single-domain particles has been shown to go as $H_c \approx a/r^2 - b$ [Eq. (9.16)] as the superparamagnetic limit is approached from larger radius. Later in this chapter, a derivation will be given for the coercivity of randomly oriented, single-domain particles in an *exchange-coupling* matrix. With exchange coupling, it will be shown that $H_c \approx r^6$! These forms are illustrated in Figure 12.2 in the range $r_0 < r < r_c$. The presence of exchange coupling between particles reduces the coercivity by extending the range of the magnetization fluctuations in particles that switch more easily. Single-domain particles will be seen to play a central role in the history and present engineering of hard magnetic materials, including thin film magnetic recording media.

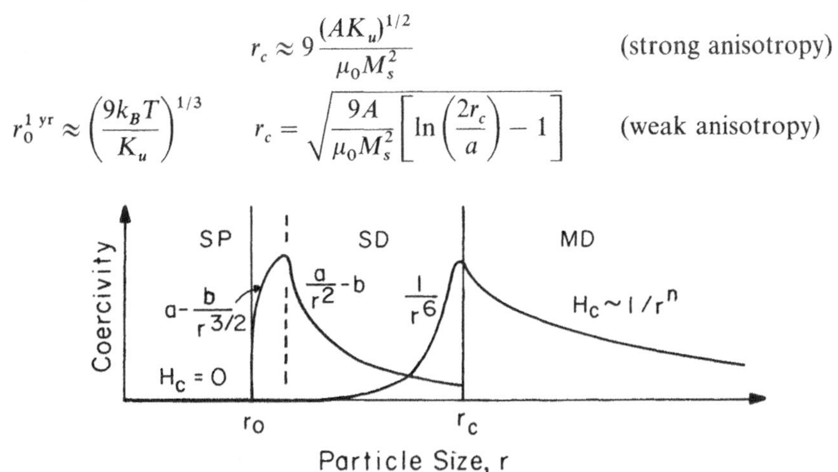

Figure 12.2 Overview of the size dependence of coercivity exhibited by magnetic particles: $H_c = 0$ below the superparamagnetic (SP) particle size limit r_0; single-domain behavior (SD) between r_0 and the single-domain limit, r_c; and multidomain behavior (MD) for $r > r_c$. In the SD regime, H_c is given by Eq. (9.16) or for noninteracting particles, and it goes as r^{-6} for exchange-coupled SD particles.

Single-domain particles can show a broad range of coercivities from zero to $2K_u/M_s$. The lower limit applies when the particles become so small that thermal energy is sufficient to flip the magnetization direction over the energy barrier established by the uniaxial anisotropy of the particle. The upper limit is approached for particles close to the upper limit of the single-domain particle size.

12.1.2 Superparamagnetism

In Chapter 8, the critical radius r_0 for superparamagnetic behavior was derived and in Chapter 9, the magnetization process of such particles was described. Their magnetization curve is like that of a paramagnet with a giant local moment. Thus $M(H)$ saturates in achievable fields below the blocking temperature and exhibits zero coercivity. When thermal relaxation is added to single-domain behavior for particles near the superparamagnetic limit, it was shown in Chapter 9 that $H_c \approx a - b/r^{3/2}$.

These three ranges of particle behavior—multidomain (MD), single domain (SD), and superparamagnetic (SP)—are summarized in Figure 12.2. This figure should not be confused with a similar figure in Chapter 9, where the effects of defects (nonmagnetic and otherwise) on coercivity of a ferromagnetic matrix were described. It was found that H_c increases linearly with defect size D in the range $D < \delta_{dw}$, the width of a domain wall. The coercivity reaches a maximum when $D/\delta_{dw} \approx 1$. Beyond that limit, H_c may decrease as $1/D$ or remain constant, depending on the nature of the interface between the defect and the matrix. That dependence of coercivity on defect size for a magnetic material is to be distinguished from the case considered here, namely, H_c for magnetic particles in a nonmagnetic matrix. The operative mechanisms in the two cases are entirely different.

12.1.3 Random Anisotropy

In Chapter 11 it was shown that when spins in a continuous magnetic material experience a randomly oriented local magnetic anisotropy and are exchange-coupled to each other, two important effects occur. First, as long as the length scale of the microstructure is smaller than the exchange length, the magnetization experiences an anisotropy reduced from its local value K_{loc} by exchange-averaging over the random local anisotropy:

$$\langle K \rangle \approx K_{loc}\left(\frac{l}{L}\right)^{3/2} \tag{12.1}$$

Second, the magnetization shows an orientational coherence over a length $L = l_{ex}$ that can range from the length scale of the random anisotropy l to much larger values (Figure 9.20) depending on the ratio of the exchange stiffness A to the strength of the local anisotropyK_{loc}:

$$L \approx l_{ex} = \frac{16A^2}{9K_{loc}^2 l^3} \tag{12.2}$$

In the present context, the random local anisotropy will be that of each single-domain nanoparticle. When nanocrystalline particles are part of a composite, their behavior depends not only on particle size and anisotropy but also on the properties of the intergranular phase. If the intergranular material is nonmagnetic, the nanocrystals behave essentially like single-domain particles, interacting only by their weak magnetic dipole fields; the results of Section 9.2 apply. If the intergranular phase is magnetic, the behavior will be governed by Eqs. (12.1) and (12.2). Either hard or soft magnetism can result, depending on the strength of the magnetic anisotropy and the physical size of the particles as well as on the strength of the exchange coupling between them. Before the magnetic properties of nanocrystalline materials can be analyzed, it is necessary to understand exchange coupling in more detail.

12.2 EXCHANGE COUPLING

Exchange coupling refers to a preference for specific relative orientations of the moments of two different magnetic materials when they are in intimate contact with each other or are separated by a layer thin enough ($\leqslant 60$ Å) to allow spin information to be communicated between the two materials. One material is generally magnetically softer and the other harder (or is antiferromagnetic). Exchange coupling is manifest as a displacement of the loop of the soft material along its field axis. The term *exchange coupling* does not apply to dipole coupling, namely, the longer-range mutual influence of the magnetic moments of two nearby materials due to their dipole fields. Such magnetostatic coupling can arise from interfacial charges due to roughness of the boundary between two materials.

12.2.1 Ferromagnetic–Antiferromagnetic Exchange Coupling

The ferromagnetic–antiferromagnetic (F/A) form of exchange coupling is considered first. In 1956 Meiklejohn and Bean observed that the M–H loops of fine cobalt particles ($2r \approx 20$ nm) could be displaced on the field axis by more than 1 kOe (Figure 12.3) if the particles were cooled in a magnetic field.

This effect was traced to the presence of a thin layer of CoO on the surface of the particles. CoO is antiferromagnetic (A) with a Néel temperature of $T_N = -3°C$. Above this temperature CoO is paramagnetic, and below T_N the cobalt moments are ferromagnetically coupled in $\{111\}$ planes with the moments in adjacent $\{111\}$ planes antiparallel. The spins align along one of the

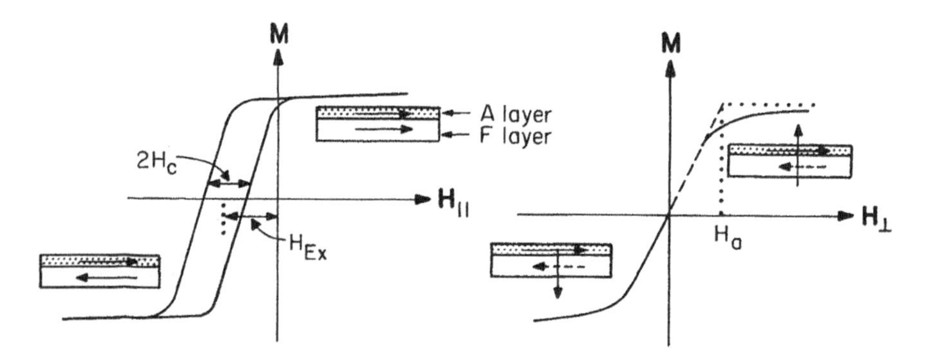

Figure 12.3 Schematic representation of effect of exchange coupling on $M-H$ loop for a material with an antiferromagnetic (A) surface layer and a soft ferromagnetic (F) interior. The arrow in the A layer is *not* the axis along which its two sublattices are oppositely magnetized; it is the direction of the exchange field exerted by the A layer on the soft, F layer. Coercivity and exchange fields are defined on an easy-axis loop, left. The anisotropy field is defined on a hard-axis loop, right.

⟨100⟩ directions in the rocksalt structure of this oxide. The Curie temperature of cobalt exceeds 1000°C. When the oxidized cobalt particles are cooled through T_N in the presence of an external field, the CoO antiferromagnetism gets "turned on" at T_N in the presence of the magnetized Co. As a result, *the magnetic moments in the CoO chose an axis of magnetization that minimizes their energy of interaction with the Co moment across the interface.* For $T < T_N$, the cobalt magnetization, which for generality is identified here as M_F, is now observed to be biased to the direction it had when the A layer was cooled through its Néel temperature. This bias field (not the A magnetization axis) is represented by the arrow in the A layers in Figure 12.3. As long as the memory of this direction is not erased by heating above T_N, the particle retains a preference for magnetization in that direction. A stronger negative field is required to demagnetize the sample than if it had been cooled in zero field. Meiklejohn and Bean referred to this field-displaced loop as exhibiting *exchange anisotropy*, and it is said to result from an *exchange coupling* between the moments of the Co and CoO. The vertical scale of the $M-H$ loop of the sample refects only the cobalt magnetization because the antiferromagnetic CoO is essentially unmagnetized by weak fields.

Figure 12.3 shows the definitions of the coercivity and exchange field in systems exhibiting exchange anisotropy. When the field is applied in the same direction used during cooling ($H_{||}$) the hysteresis loop is shifted toward the negative field direction by H_{ex}. The coercivity is half the width of the $M-H_{||}$ loop at $M = 0$. No exchange shift in the loop is observed if the external field is applied orthogonal to the direction of the field present during cooling. The hard axis loop serves to define the anisotropy field H_a of the system. The coercivity vanishes in $H_\perp$.

Meikeljohn and Bean also observed that the exchange field vanished if the thickness of the CoO layer on the Co particles was below a certain threshold. Further, these thin oxide systems exhibited a rotational hysteresis, which would not be expected from a simple Stoner-Wohlfarth rotational magnetization model. How can this be understood?

Initially, models of exchange coupled materials considered the spins to be either parallel or antiparallel to their neighbors across the interface. The actual microscopic moment arrangement for exchange-coupled F (soft, ferromagnetic) and A (antiferromagnetic) layers is now believed to resemble that illustrated in Figure 12.4. *The exchange coupling across the interface is such that the moments in the A material lie on an axis that is orthogonal to the F moment at the time of cooling through* θ_N (Jungblut et al. 1994, 1995). This is the lower energy configuration because it is easier to cant the A moments into the direction of an orthogonal field than one that is parallel to the preferred A axis. That is, for an antiferromagnet, $\chi_\perp > \chi_\parallel$. This magnetic moment configuration gives rise to an exchange field parallel to the direction of the F moment during field cooling.

Once the exchange coupling is established by field-cooling through T_N, a preferred *direction* of magnetization (not just a preferred axis) exists at the interface. This can be appreciated as follows. On application of a weak or moderate field to an exchange-coupled system, the soft material will tend to follow the field subject to the coupling at the interface. It is important first to establish that *for semiinfinite media* on each side of the interface, most of the twist in magnetization caused by the field would occur in the soft material (Fig. 12.4, right). (This twist occurs over a length comparable to a domain wall thickness.) There are two reasons why the twist occurs mostly in the soft layer.

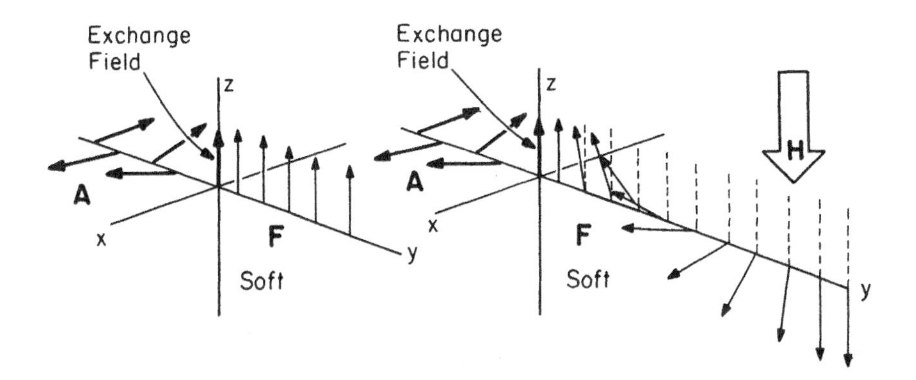

Figure 12.4 Schematic representation of soft ferromagnetic (F) layer ($y > 0$) exchange coupled to an antiferromagnetic (AF) layer ($y < 0$). Left, direction of the spins near the interface between the AF and F layers after cooling through Néel temperature of AF material. Right, effect on spins in soft material of a field applied opposite to the exchange field.

First, the magnetic anisotropy in the F layer is small. Second, the external field exerts no net torque on the A moments because they sum to zero and are held to a particular crystallographic direction by a strong magnetocrystalline anisotropy. When the applied field is reduced, the energy stored in the twist (domain wall) causes M_F to return to alignment with the hard material sooner than it would if there were no exchange coupling at the interface. The additional applied field energy needed to create an interfacial magnetization twist in the soft material shows up as a shifted $M-H$ loop (toward negative H in the case shown here).

If the thickness of the F material is less than the thickness of a domain wall or an exchange length, then it cannot support a magnetization twist. In this case the soft magnetization may respond to the field by storing energy in the exchange coupling at the interface or by twisting the A moments away from their preferred axis. Here, the model in Figure 12.5 applies.

In this thin film regime, if the A material were not sufficiently thick, the exchange torque exerted on it at the interface with the soft magnetic component could cause the magnetic moments in the entire A layer to flip to an orientation with lower total energy. The unidirectional symmetry of the exchange coupling would then be broken and uniaxial symmetry would remain. This can be seen in Figure 12.6 (Mauri et al. 1987) for permalloy films exchange coupled to the antiferromagnet, FeMn. Note that H_{ex} vanishes for $t_{FeMn} < 50$ Å.

To analyze this loss of exchange coupling, K_F and K_A are defined as the respective uniaxial anisotropies in the two layers of thickness t_F and t_A, and J is the exchange energy per unit area of interface coupling the two media.

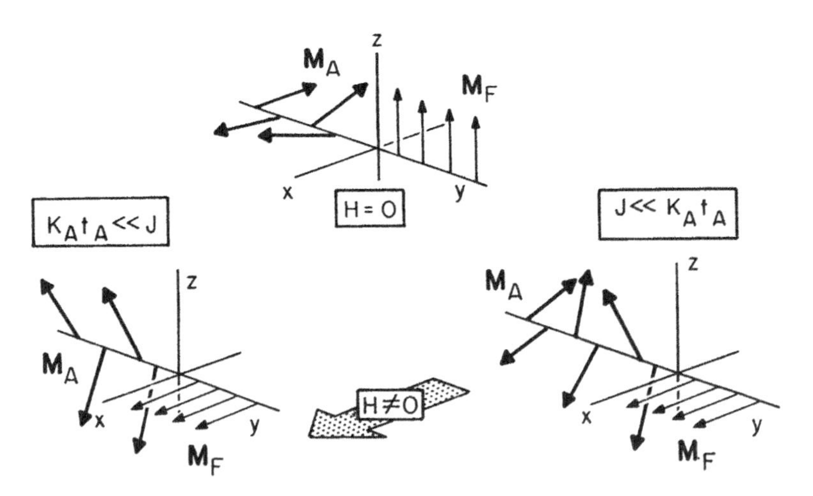

Figure 12.5 Above, the interfacial moment configuration in zero field. Below, left, the weak-antiferromagnet limit, moments of both films respond in unison to field. Below, right, in the strong-antiferromagnet limit, the A moments far from the interface maintain their orientation.

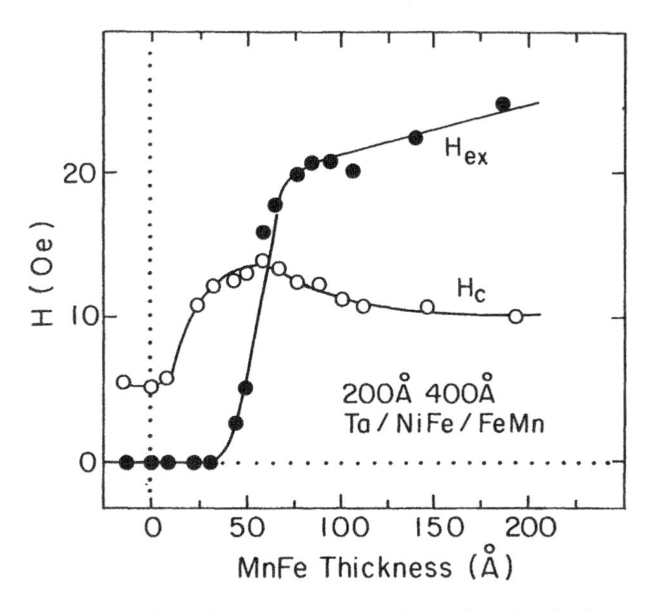

Figure 12.6 Exchange field and coercivity as functions of FeMn thickness, after Mauri et al. (1987).

In the weak-antiferromagnet limit, $K_A t_A \ll J$, the F and A layers respond in unison to the field. This condition defines the A thickness range for which no exchange anisotropy is observed:

$$t_A \leqslant \frac{J}{K_A} = t_A^c \tag{12.3}$$

For the FeMn system shown in Figure 12.6, $t_A^c \approx 50\,\text{Å}$. This suggests that $J \approx 0.1\ \text{mJ/m}^2$ if $K_A \approx 2 \times 10^4\ \text{mJ/m}^3$. In this case, the moments in the A layer are pulled away from their preferred axis by their strong coupling to M_F (Fig. 12.6, lower left) and unidirectional anisotropy does not exist. The M–H loop has a coercivity dictated by the harder (both processes must occur) of the two processes: coherent rotation of M_F against its anisotropy K_F

$$H_c^{\text{weak}-\text{A}} \leqslant 2\frac{K_F}{M_F} \tag{12.4a}$$

or rotation of the M_A against K_A by the interfacial torque (Fig. 12.5, lower, left):

$$H_c^{\text{weak}-\text{A}} \leqslant 2\frac{K_A t_A}{M_F t_F} \tag{12.4b}$$

Equation (12.4a) suggests $H_c^{\text{weak}-A} \approx 1$ Oe, whereas Eq. (12.4b) indicates a coercivity of order 100 Oe. In SD particles, Eq. (12.4b) would apply. The observed coercivity in the exchange-coupled FeMn/FeNi films is about 5 Oe. If the F film has a tendency to break up into domains, it can do so with little cost in energy because of its exchange coupling to the A layer, because A is weakly pinned. Magnetization reversal by wall motion will always be an easier process than either of these coherent rotation processes. Thus wall motion could be responsible for lowering the coercivity from its Eq. (12.4b) value in FeMn/NiFe films.

On the other hand, in the strong-antiferromagnet limit, $K_A t_A \gg J$, when the soft film responds to a field, it causes an exchange twist in the A layer near the interface (Fig. 12.5, lower right). (Remember that $t_F < \delta_{\text{dw}}$, so a twist cannot be supported in the soft layer. If, instead of generating a twist in the A layer, the exchange coupling were broken, then the unidirectional exchange coupling would be lost. Observation shows that it is not.) If the exchange coupling is established by cooling through T_N with the F layer saturated, then the A layer is locked in a single domain state and the F layer will not demagnetize in zero applied field. The slightly larger coercivity observed in this exchange-coupled case (Fig. 12.5) is difficult to explain quantitatively because of its small magnitude. However, Eq. (12.4b) for the weak antiferromagnet indicates that the full thickness of the A layer is responsible for the coercivity in that case. In the strong-antiferromagnet case of interest here, Figure 12.5, lower right, shows that only a thin layer of A near its interface with F is rotated to a high-energy configuration. Thus, the coercivity can be modeled in the case $K_A t_A > J$ as

$$H_c^{\text{strong}-A} = 2\frac{K_A \delta_A}{M_F t_F} \approx \frac{\sigma_{\text{dw}}^A}{2M_F t_F} \tag{12.5}$$

where $\sigma_{\text{dw}}^A = (AK_A)^{1/2}$ is the energy of a magnetization twist or domain wall in the antiferromagnet. It will be shown below that δ_A is of order 1 nm. Choosing $K_A \approx 10^4\ \text{J/m}^3$ and $M_F t_F = 0.02$ A (2×10^{-3} emu/cm^2), Eq. (12.5) gives the coercivity of 10 Oe observed in Figure 12.5.

Mauri et al. (1987) derived an expression for the M–H loop of the soft film in the exchange-coupled regime ($t_A > t_A^c$). The free energy of the system in Figure 12.5, with an external field applied along the $+z$ direction, is given by

$$\frac{\text{F}}{\text{Area}} = -M_F H t_F \cos\theta + K_F t_F \sin^2\theta - J\cos\theta \tag{12.6}$$

Here, the orientation of the magnetization in the soft film is defined by its angle with respect to the $+z$ axis, θ. The soft magnetization M_F is subject to an external field, H (a negative field is shown in Fig. 12.3), a uniaxial anisotropy with easy axis parallel to z, and the exchange coupling favoring $\theta = 0$. (The exchange term accounts for the effects of all the energies in the A layers.) Minimization of this energy leads to the expression for the easy-axis magnetization:

$$\sin\theta\left[\cos\theta + \frac{M_F H}{2K_F} + \frac{J}{2K_F t_F}\right] = 0 \qquad (12.7)$$

There are stable solutions at $\theta = 0$ and π corresponding to $\pm M_F$. The magnetization makes a transition between these two states when $m = \cos\theta = 0$ for

$$H_{ex} = -\frac{J}{M_s t_F} \qquad (12.8)$$

In other words, for ferromagnetic exchange coupling, the easy-axis loop is displaced to the left by H_{ex} [cf. Eq. (12.3)]. This expression predicts that the exchange field should decrease for thicker F layers. Figure 12.7 (Mauri et al., 1987) shows that this t_F^{-1} dependence is observed for permalloy coupled to FeMn. These H_{ex} data are fit with Eq. (12.8) for $M_F = 5 \times 10^9$ A/m (500 emu/cm^3 and $J = 0.075$ mJ/m^2. The H_c data in Figure 12.6 are described by Eq. (12.5) for $\sigma_{dw}^A = 0.12$ mJ/m^2.

Cain and Kryder (1988) have shown that H_a also varies almost as t_F^{-1} (Fig. 12.7). This situation also can be modeled by writing the energy for the

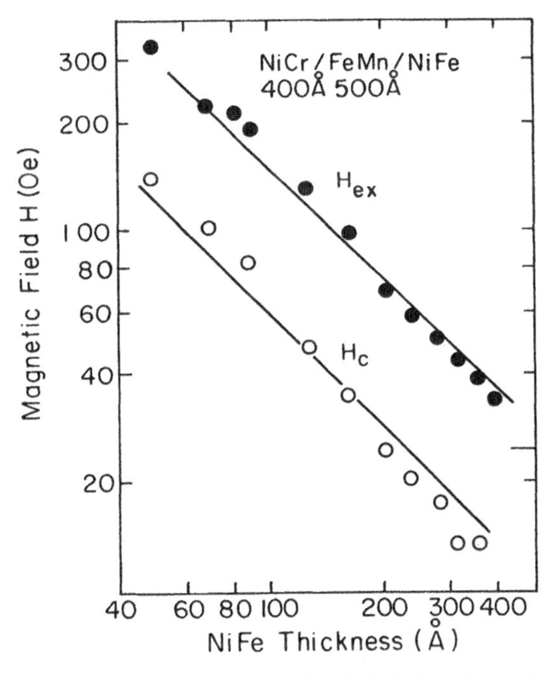

Figure 12.7 Log–log plot of exchange field (closed circles) and coercivity (open circles) versus NiFe thickness. Solid lines show inverse t_F dependence [After Mauri et al., 1987).]

hard-axis magnetization process. With the field along the x direction in Figure 12.5, this gives

$$\frac{F}{\text{Area}} = -M_F H t_F \cos\theta + K_F t_F \cos^2\theta - J\sin\theta \tag{12.9}$$

with θ defining the angle between M_F and H. The zero-torque condition gives for the hard-axis magnetization process:

$$\cos\theta = m_x = \frac{M_F t_F}{2K_F t_F + J} H \equiv \frac{H}{H_a} \tag{12.10}$$

where

$$H_a = \frac{2K_F t_F + J}{M_F t_F} = H_{a0} + \frac{J}{M_F t_F} \tag{12.11}$$

The exchange energy had been estimated to be $J \approx 0.1\ \text{mJ/m}^2$. Thus $J > 2K_F t_F$ for $t_F < 10\,\text{nm}$ if $K_F = 5 \times 10^3\ \text{J/m}^3$. Only in this weak $K_F t_F$ limit does H_a come close to varying inversely with t_F (see Fig. 12.8). For much thicker F layers, or generally $2K_F t_F > J$, H_a is predicted to become independent of t_F.

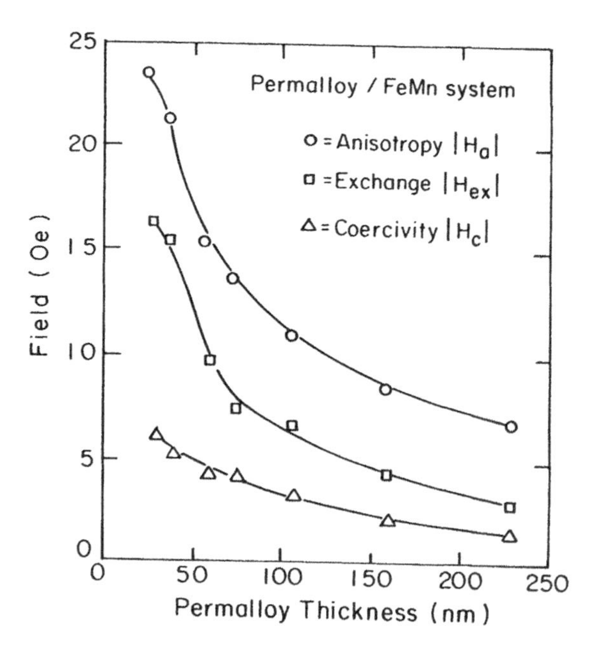

Figure 12.8 Anisotropy, exchange, and coercive fields as functions of permalloy thickness. The dotted line goes as $1/t_f$. [After Cain and Kryder (1988).]

Thus, several parameters are important in exchange coupling: the *strength* of the interface exchange, J, the *anisotropy–thickness* products of the two materials, and the *magnetization–thickness* product of the soft layer.

Koon (1997) reported micromagnetic calculations on body-centered tetragonal, A/F exchange-coupled films sharing a (110) interface. His results show in more quantitative detail the implications of A/F exchange coupling. The results described here assume antiferromagnetic interactions across the interface as well as antiferromagnetic interactions in the A layer and ferromagnetic interactions in the soft layer. The calculations show first that far from the interface the equilibrium orientation of $\boldsymbol{M}_F$ is orthogonal to the easy axis in the A layer: $\phi = 90°$ (Fig. 12.9). Close to the interface the spin orientations are more complex. The F spins try to remain parallel to each other ($J_F > 0$) and orthogonal overall to the A axis while being antiparallel to their A nearest neighbors (half way across a body diagonal ($J_{A/F} < 0$). Similarly, the A spins try to maintain their nearest-neighbor A coupling ($J_A < 0$) and remain orthogonal to the overall F axis while being antiparallel to the F spins to which they couple. The incompatibility of these three requirements on each interface spin is called "frustration." It leads to a canting of the interface spins ($\phi > 90°$) as they try to balance the competing interactions. The canting of the interface spin decays within 5 or 6 monolayers (MLs) of the interface. Koon obtained similar results for F exchange across the interface, but in that case the canting from $\phi = 90°$ is such as to favor $\phi < 90°$ near the interface.

Figure 12.10 shows one of the results of Koon's calculations illustrating the switching behavior of antiferromagnetically coupled A and F layers as functions of the A layer thickness, t_A. These more thorough micromagnetic results confirm the simple model just described and add further details. For small t_A, switching of the F layer is reversible and requires only the energy needed to rotate the A moments over the energy barrier between their uniaxial [001]

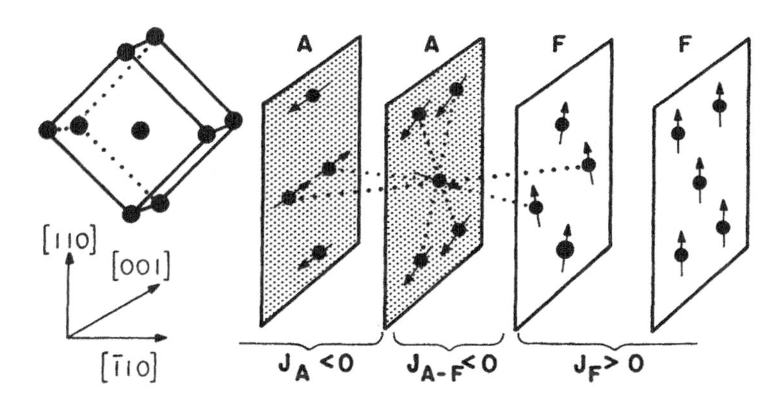

Figure 12.9 Local-moment orientations in an antiferromagnet (A) and ferromagnet (F) that share a [110] plane across which antiferromagnetic exchange coupling exists. [Adapted from Koon (1997)].

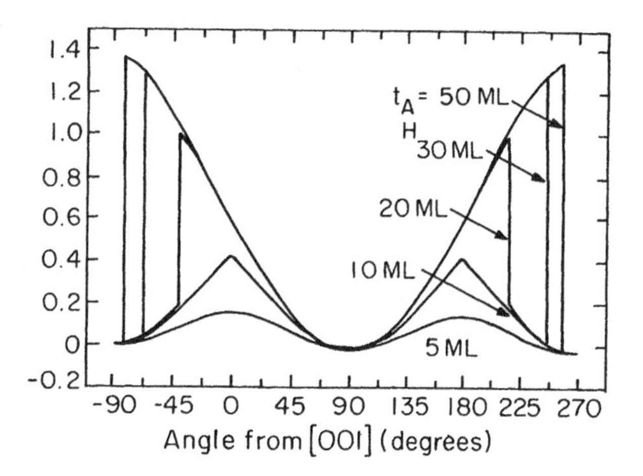

Figure 12.10 Energy per unit area of the exchange-coupled system in Figure 12.9 as a function of the angle ϕ between the F moments and the [001] direction (the easy axis in the A layer) for different A layer thicknesses t_A. [After Koon (1997).]

anisotropy energy minima [see Eq. (12.4b)]. This behavior was observed by Mauri et al. below $t_A = 50$ Å (Fig. 12.6). For $t_A \gg \delta_A$, the A domain wall thickness, application of a field rotates the F moments and stores energy in an A magnetization twist (antiferromagnetic domain wall) near the interface (Fig. 12.5, lower right). The A moments in the layer farthest from the interface are least affected by torque from the F moments. After removal of the applied field, the F moments spring back to their initial orientation 90° from [001] due to the energy stored in the A wall. This unidirectional behavior is not complete for intermediate values of $t_A K_A$; for t_A not too large, the A magnetization can still be switched discontinuously as a result of the torque from the F moments at the interface. This switching behavior provides a reasonable description of the high-field loss that Meiklejohn and Bean noted as a characteristic of exchange coupling in many Co–CoO samples.

The calculated interfacial configuration shown in Figure 12.9 and Koon's results in Figure 12.10 explain many of the features of exchange coupling that had eluded earlier models (Malozemoff 1987) and provides a more solid foundation for the insights of the model due to Mauri et al. (1987). These include the following:

1. Exchange coupling does not vanish but is strongest when the A layer closest to the interface is compensated as in Figure 12.9 (zero net spin).
2. The effect of surface roughness is always to weaken exchange coupling.
3. For small A layer thickness, $t_A < \delta_A \approx 9$ ML, the exchange coupling is reversible and uniaxial rather than unidirectional.
4. For thicker A layers, the rotation of the F moment involves a discontinu-

ous jump, hysteresis. This additional loss process for intermediate t_A may explain the tendency for H_c to peak near the transition from uncoupled ($t_A < t_A^c$) to coupled behavior (Fig. 12.6).

5. For $t_A \gg t_A^c$, the magnetization process of the F layer is unidirectional and reversible.
6. The effective exchange coupling is much smaller than the magnitude of the microscopic exchange integral (used in primitive estimates of H_{ex}) would indicate, because a domain wall is formed parallel to the interface allowing the F layer to reverse more easily.

Despite the clear, intuitive picture this model gives of the structural phenomenology of F/A exchange coupling, uncertainties still exist concerning the strength of the exchange coupling (items 1, 2, and 6, listed above). A complementary microscopic model that addresses the strength of J from a statistical perspective has been described and supported with data by Takano et al. (1997). This model starts from an assumed Heisenberg exchange across an atomically smooth F/A interface (cf. Eq. 12.8):

$$H_{ex} = \frac{\mathscr{J}_{ex} \boldsymbol{S}_F \cdot \boldsymbol{S}_A}{a^2 M_s t_F} \tag{12.12}$$

The parameters $\boldsymbol{S}_F$ and $\boldsymbol{S}_A$ are the spins in the F and A layers, respectively, $\mathscr{J}_{ex}$ is their Heisenberg exchange (not to be confused with the phenomenological interfacial exchange energy density J, in Eq. 12.8) and a is their spacing. The challenge is to explain why the observed exchange fields are typically of order one percent of this Heisenberg exchange field, or equivalently, why J_{ex} (energy/area) $\ll \mathscr{J} \boldsymbol{S}_A \cdot \boldsymbol{S}_F / a^2$. Takano et al. studied the thermoremanent moment (TRM) of CoO/MgO multilayers, which is due to uncompensated Co moments at the CoO/MgO interfaces. (These uncompensated spins select a direction close to their easy axis that minimizes their energy with respect to the direction of $\boldsymbol{M}_F$ established by the applied field on cooling through the Neél temperature. See Fig. 12.9). The magnitude of the uncompensated interfacial spin moment shows the same temperature dependence as does the strength of the exchange coupling in permalloy/CoO bilayers. The CoO grows coherently on MgO (both have the FCC rocksalt structure with lattice constants differing by about 1.5%: $a = 0.427$ and 0.421 nm, respectively).

The magnitude of the TRM of the CoO/MgO multilayers was found to scale with the number of CoO interfaces but remained independent of the CoO layer thickness, indicating that it was due to uncompensated interfacial spins rather than those at internal defects such as grain boundaries. The thermoremanent cobalt moment appeared on cooling below 295 K and had a constant magnitude from about 250 K down to 50 K (Fig. 12.11, upper panel). The TRM moment value at the plateau corresponded to approximately 1% of the total interfacial moment, that is the sum of the Co^{2+} moments in one monolayer.

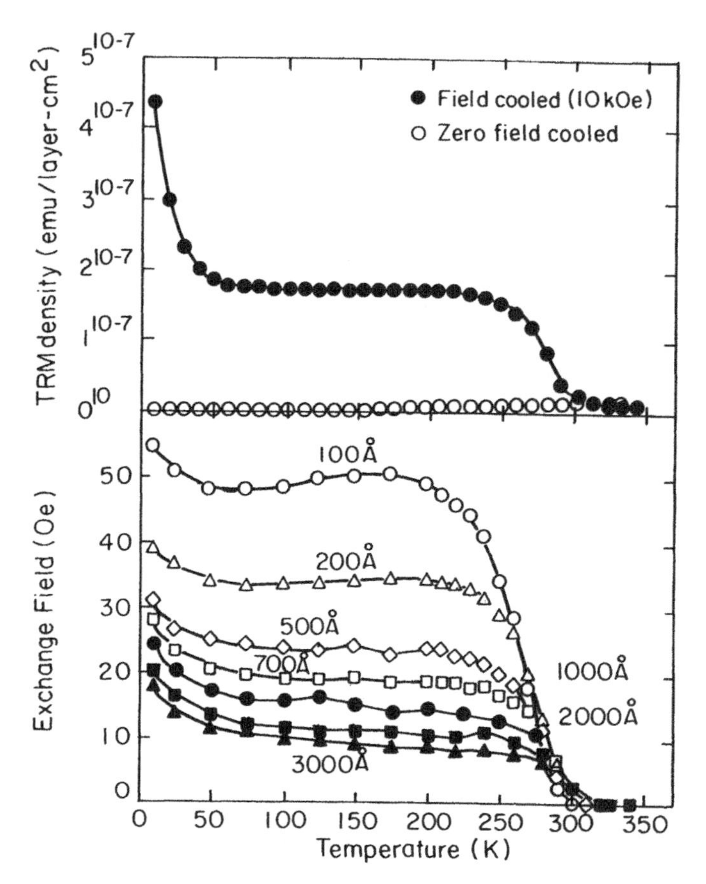

Figure 12.11 Upper panel: Thermoremanent magnetization (TRM) as a function of temperature for [CoO/MgO] multilayers in field cooled (FC) and zero-field cooled (ZFC) states. Lower panel: exchange field versus temperature for NiFe/CoO bilayers having different CoO layer thicknesses. After Takano et al. 1977.

When the exchange coupling of the NiFe/CoO bilayer films was measured, it also showed the same temperature dependence: onset at 295 K and constant magnitude from about 250 K to 50 K (Fig. 12.11, lower panel). Further, the exchange coupling data suggest that the density of uncompensated spins is proportional to the radius of curvature of the CoO grains which in turn can be controlled by the CoO layer thickness (Fig. 12.11, lower panel). Thus, $H_{ex} \propto (\text{grain diameter})^{-1}$ was observed.

Takano et al. made a statistical analysis of the number of uncompensated spins $\langle N \rangle$ on a surface inclined at an angle θ to the ferromagnetically coupled (111) planes (Fig. 12.12). For an ideal stepped surface, $\langle N \rangle = 0$ for any θ. However, for rough surfaces typical of a small grain size, $L < 10$ nm, the model predicts $N \approx 0.6\, L/d_{110}$, so $H_{ex} \propto N/L^2 \propto L^{-1}$ as observed.

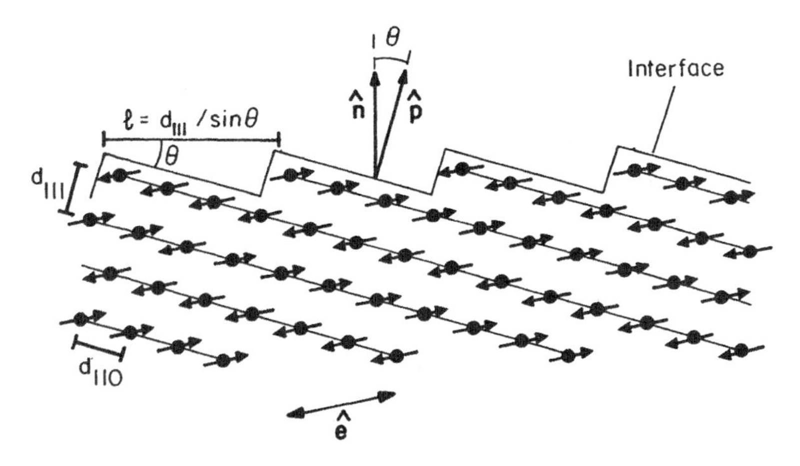

Figure 12.12 Schematic of CoO surface with normal *n* having (111) parallel-spin planes inclined at an angle θ to the physical surface. The easy axis of the CoO is given by *e*. After Takano et al. 1997.

Exchange fields were numerically calculated to range from 31 Oe to 107 Oe for grain size ranging from 48 nm to 1.2 nm. The experimentally observed grain diameter in the CoO films, $L = 12$ nm, indicates $H_{ex} = 50$ Oe, consistent with the observed value of 48 Oe. The strength of the exchange coupling varies inversely with the grain size. The focus of this model on *uncompensated* interfacial spins appears counter to the mechanism pictured in Figure 12.9.

It has recently been shown (Gökenmeijer et al. 1997) that the exchange coupling between an antiferromagnet and a ferromagnet need not result from a nearest-neighbor exchange mechanism such as that pictured in Figure 12.9. They show that the exchange coupling between permalloy and CoO drops off exponentially with the thickness of an intervening layer of Cu, Au, or Ag (Fig. 12.13): $H_e = H_{e0} \exp(-t/L)$ where L is a measure of the range of the exchange coupling. The effects of exchange coupling are still seen to be measurable above 5 nm of Ag. Further, they showed that the exchange coupling is *unidirectional* while the coercivity is *uniaxial* in the orientation of an applied field relative to the setting-field direction. However, neither H_{ex} nor H_c is a simple harmonic function; they both contain higher order terms, perhaps related to the rotational hysteresis that characterizes exchange coupling. Clearly, all the details of the mechanisms of exchange coupling are not yet understood.

12.2.2 Ferromagnetic–Ferromagnetic Coupling

Examples of F-A exchange coupling across a nonmagnetic layer were seen in Figure 12.13. A more complex dependence on spacer thickness is observed for F-F coupling (see Section 12.2.3). Of interest here are the consequences of

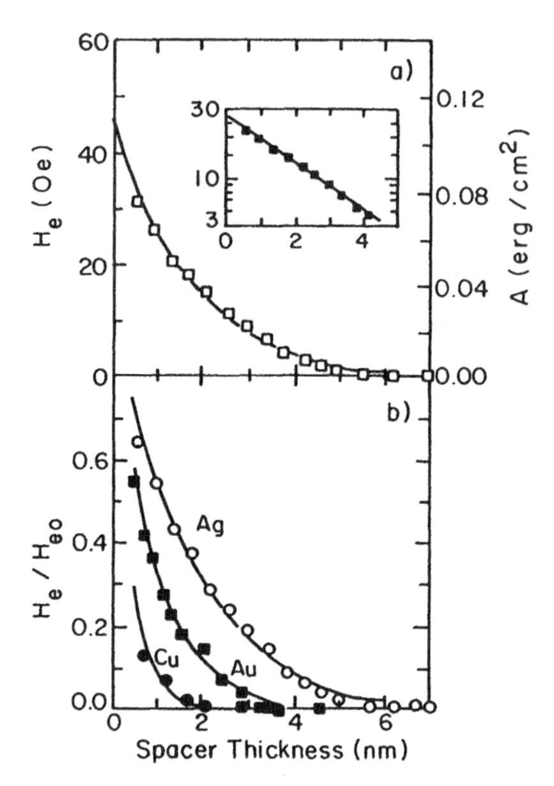

Figure 12.13 (a) Variation of the NiFe–CoO exchange field and coupling strength with Ag thickness in nanometers. The inset shows a semilogarithmic plot of this data. (b) Normalized exchange field versus spacer thickness for three different metals. The solid lines are fits to $\exp(-t/L)$. [After Gökenmeijer et al. (1997)].

exchange coupling on the spins near the interface between two ferromagnetic media. Exchange coupling across a ferromagnetic-ferromagnetic interface is assumed to arise from a direct, such as Heisenberg-like, $\pm \mathcal{J}_{ij} S_i S_j$, exchange interaction, or from an indirect, RKKY-like (Section 5.6), exchange interaction.

In Chapter 8, Section 8.4, the thickness of the twist, or pinned domain wall, near the interface of a *single* ferromagnetic material was considered. Two cases were identified. In one case, the spins rotate in the plane of the interface and, hence, the exchange length in each medium was given by the square root of the ratio of the exchange stiffness to the anisotropy in the ferromagnetic medium [Eq. (8.19a)]. In the other case, the anisotropy at the interface is assumed to cause a component of the magnetization to be perpendicular to the interface. Under such circumstances, the magnetostatic energy associated with the discontinuity of the magnetization across the interface combines additively with the anisotropy in the ferromagnetic medium to determine the exchange length [Eq. (8.19b)]. The results are shown schematically in Figure 8.15.

Here, the interest is in the form and thickness of the twists or pinned domain walls that exist on either side of the interface between *two exchange-coupled ferromagnetic materials.* In each medium, the exchange length l_{ex} is governed by the ratio of the exchange stiffness A to either the anisotropy K_u or the anisotropy plus magnetostatic energy, $K_u + 2\pi\Delta M_s^2$, depending on whether the magnetization twist is in the plane of the interface or perpendicular to it (Fig. 12.14). A mathematical derivation is not given here. Instead, it is suggested that the form of magnetization on either side of the interface is similar to the partial domain wall solutions given in Eqs. (8.19). The actual solution should satisfy continuity of the magnetization orientation and its first derivative (torque) across the interface. [It should be pointed out that the solutions leading to Eqs. (8.19) do not simultaneously satisfy continuity of both $\theta(z)$ and $\partial\theta(z)/\partial z$ across the interface.]

When the magnetization at the interface has no perpendicular component, it is the anisotropy energy of the spins in each medium that governs the exchange length there:

$$\boldsymbol{M} \text{ parallel to interface:} \quad l_{ex}^{\|(i)} = \left(\frac{A^{(i)}}{K_u^{(i)}}\right)^{1/2} \qquad i = \text{medium } 1,2 \quad (12.13a)$$

The spins in the stronger (weaker) anisotropy medium vary more abruptly (more gradually) from their orientation at the interface to the value they assume in the interior. The spin orientation at the interface will be biased toward that of the stronger anisotropy material.

When there is a perpendicular component of magnetization at the interface, there will be an asymmetric magnetic charge distribution near the interface if the gradients of the magnetizations in the two media differ from each other [$K_u^{(1)} \neq K_u^{(2)}$]. This charge distribution produces a magnetostatic field distribution near the interface which will tend to equalize the forms of the magnetiz-

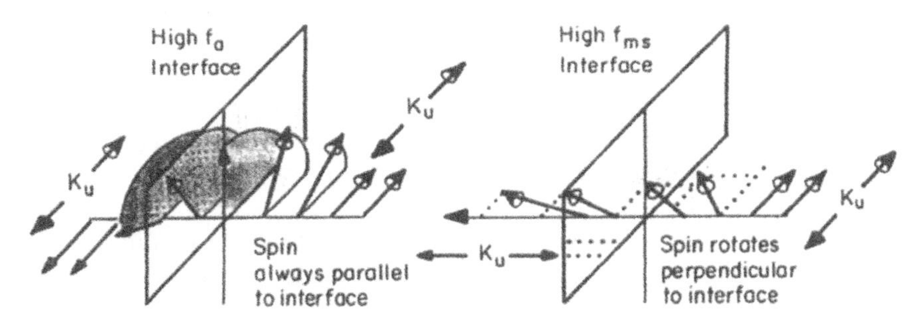

Figure 12.14 Illustration of the two limiting cases important for determining the range of the effect of exchange coupling. Left, both media have easy magnetization axes parallel to the interface between them. Right, one of the media has a preferred direction of magnetization perpendicular to the interface. The range of the exchange twist in $\boldsymbol{M}$ is greater in the weak anisotropy medium.

ation distributions on either side of the interface. The mathematical forms of these two lengths may be approximated as

$$\boldsymbol{M} \text{ perpendicular to interface: } l_{\text{ex}}^{\perp(i)} = \left(\frac{A^{(i)}}{K_u^{(i)} \pm 2\pi \, \Delta M_{\perp}^2} \right)^{1/2} \qquad (12.13\text{b})$$

The appropriate sign is that which tends to equalize the denominators in the exchange lengths of the two media.

The parallel case (Fig. 12.14, left) is most frequently encountered. The perpendicular case (Fig. 12.12, right) would apply to exchange coupling between a perpendicularly magnetized film and a longitudinally magnetized one.

Table 8.1 lists the magnitude of the magnetostatic energy, anisotropy energy, and the calculated exchange lengths for Fe, Co, and Ni interfaces. Note that for high-anisotropy materials, the parallel exchange length can be very small. This result was assumed in the previous section on F/A interfacial exchange; in the strong antiferromagnet case, a domain wall measuring about 5 or 6 monolayers in thickness was found near the interface in the A layer in Koon's micromagnetic calculations.

An important consequence of this result is that the exchange interaction, through the parameter l_{ex}, has the effect of communicating the magnetization direction in one region over distances of several nanometers into another, exchange-coupled region.

12.2.3 Oscillatory Exchange Coupling

Multilayers of rare-earth metals and yttrium were shown to exhibit an exchange coupling of the RE layers through the Y layers that oscillates in sign with the thickness of the Y layer (Erwin et al. 1986). Baibich et al. (1988) observed a very large field dependence to the resistance in $[\text{Fe/Cr}]_N$ multilayers, where N is the number of Fe/Cr bilayers (see Chapter 15). The iron layers are ferromagnetic and Cr layers, antiferromagnetic. They observed that adjacent iron layers are antiferromagnetically coupled to each other through the Cr layers with a strength that decreases with increasing Cr thickness over the range they investigated, 0.9–1.8 nm. This observation initiated extensive research into transition metal–noble metal multilayers. Parkin et al. (1990) observed that the Fe/Fe coupling oscillates smoothly between ferromagnetic and antiferromagnetic as the thickness of the spacer layer varies. Figure 12.15 shows the Cr thickness dependence of the field needed to saturate $[\text{Fe/Cr}]_N$ multilayers in a hard $\langle 110 \rangle$ in-plane direction. The large saturation field for 10, 24, and 45 Å of Cr implies strong antiferromagnetic coupling between the Fe (or Co) layers; a small saturation field implies ferromagnetic coupling between the Fe layers.

Parkin et al. (1990) showed that the multilayer system [Co/Cu] also is characterized by an exchange coupling of the Co layers that oscillates from

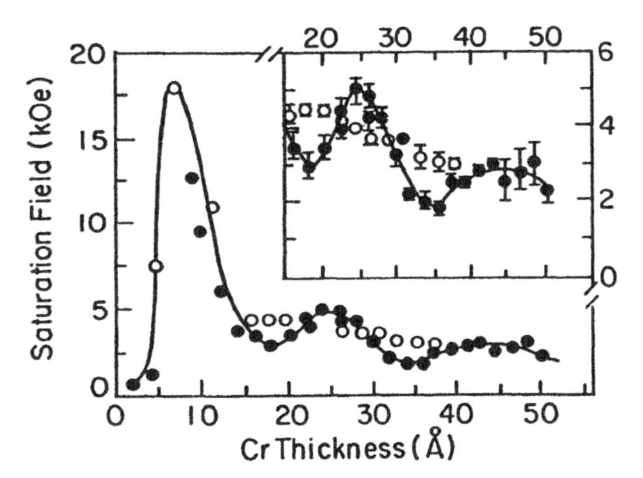

Figure 12.15 Field needed to saturate the magnetization at 4.2 K versus Cr thickness for Si(111)/100 Å Cr/[20 Å Fe/t_{Cr} Cr]$_N$/50 Å Cr, deposited at temperatures of 40°C, $N = 30$ (triangle and square) and at 125°C, $N = 20$ (circle). The insert shows on an expanded field scale the oscillations in the thick Cr region [After Parkin (1990)].

antiferromagnetic to ferromagnetic as the thickness of the nonmagnetic Cu layer increases. The oscillation period is typically of order 1 nm and oscillations have been observed out to spacer thicknesses of 5 or 6 nm.

This oscillating exchange coupling is beautifully illustrated by a series of experiments done by Unguris et al. (1991) using a wedge-shaped Cr layer on a single-crystal Fe whisker, capped with another Fe layer (Fig. 12.16, below). When the magnetic domain pattern in the top Fe film is imaged [this group used scanning electron microscopy with spin polarization analysis (SEMPA)], the pattern shown in Figure 12.17 is observed. The demagnetized Fe whisker substrate consists of two domains magnetized along the whisker length and separated by a 180° domain wall. The iron film above each domain is exchange coupled through the Cr layer to the magnetization in the whisker below. For zero Cr thickness the Fe film is magnetized parallel to the underlying domain. As t_{Cr} increases, the exchange coupling is directly observed to oscillate to antiferromagnetic and back to ferromagnetic with a period corresponding to 11 Cr atomic layers.

The physical mechanism responsible for this long-range oscillation in exchange coupling in two-dimensional magnetic nanostructures is believed to be related to the spin polarization of conduction electrons as described by the RKKY model (Chapter 5). It can also be explained in terms of the symmetry of the electronic wavefunction trapped in the quantum well defined by the layered structure. (Wavefunction spatial symmetry is necessarily connected to its spin symmetry by the Pauli exclusion principle, which demands that the total wavefunction be antisymmetric; Chapter 4).

No oscillations have yet been observed in F/A exchange coupling (Section 12.2.1).

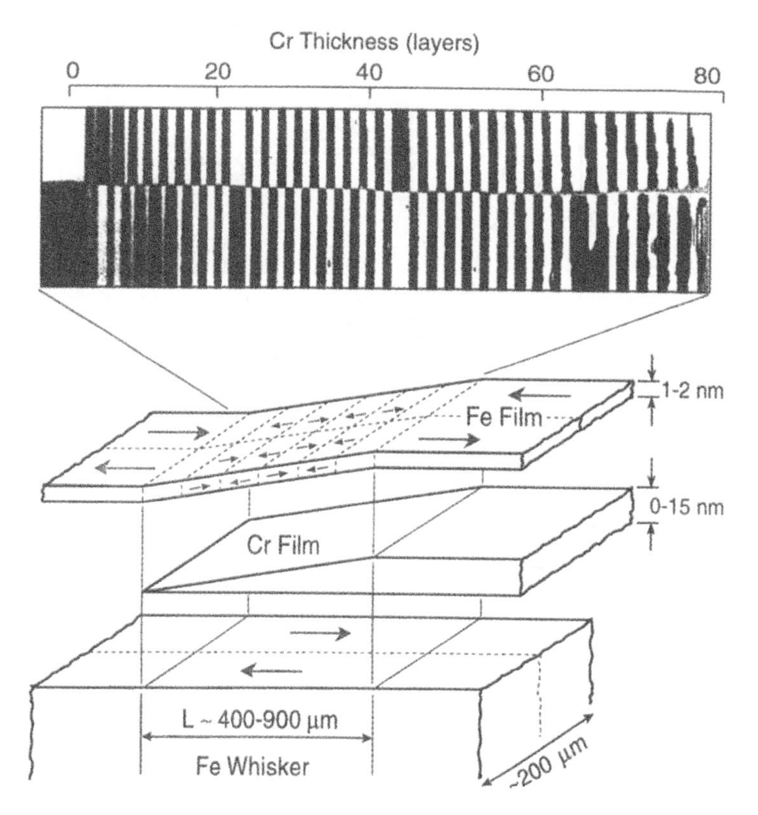

Figure 12.16 Below, structure of Fe film/Cr wedge/Fe whisker illustrating the Cr thickness dependence of FeFe exchange—whisker has simple domain pattern illustrated, and magnetic domain pattern in Fe overlay is also depicted; above, SEMPA image of domain pattern generated from top Fe film (Unguris et al. 1991).

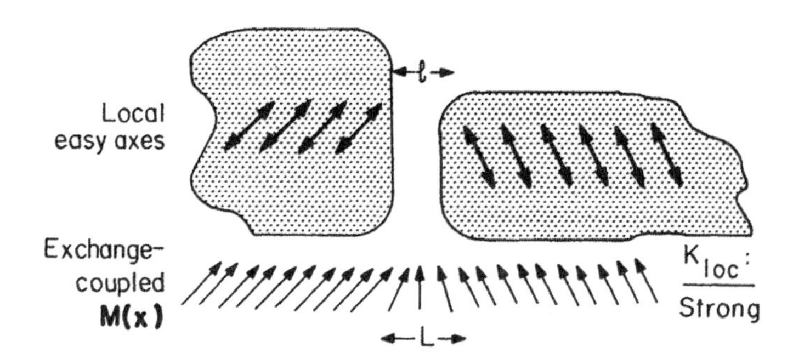

Figure 12.17 Application of the random anisotropy model to a nanocrystalline material where the nanocrystal anisotropy is greater than the anisotropy of the amorphous intergranular layer. The weak nanocrystalline anisotropy is described by Fig. 11.12.

12.2.4 Random Anisotropy in Nanostructured Materials

The preceding sections have described how exchange coupling, either F-A or F-F, can exist between two distinct materials and in both cases can operate over distances of several nanometers of intervening nonmagnetic material. The implications of such interactions for nanostructured materials are very important. The spins in magnetic particles or layers of different materials can be coupled with different strengths or even different signs depending on the nature of their exchange interaction and the thickness and properties of the material between them.

At the beginning of this chapter, the concept of random anisotropy in *continuous* magnetic systems such as amorphous alloys was reviewed. It is now illustrated how interparticle exchange and random anisotropy combine to affect the magnetic properties of *discrete* magnetic systems such as nanocrystalline alloys. In the random anisotropy situation, the effective anisotropy is reduced by averaging K_{loc} over the exchange length of the magnetic moments, Eq. (12.1). In nanocrystalline materials, K_{loc} applies over the entire nanocrystal volume, and hence the appropriate length scale for the random anisotropy variation l is given by the thickness of the interparticle layer. The variation in magnetization direction occurs over the intergranular material between the particle surfaces if K_{loc} is strong compared to the anisotropy of the intergranular, amorphous matrix K_{am} (Fig. 12.17), or over a larger length if K_{loc} is comparable to K_{am} (Fig. 11.11).

Herzer (1993) suggested that it is the average anisotropy of a nanocrystalline material, $\langle K \rangle$, and not the local anisotropy K_{loc}, that should appear in the expression for the exchange length, Eqs. (12.13a) or (12.13b). A reason for this is that the exchange length is defined largely in the amorphous matrix *between* the nanocrystalline particles, where the direction of $\boldsymbol{M}$ changes most, not in the interior of the particles where K_{loc} is uniform (Fig. 12.17). By eliminating the exchange length between Eqs. (12.13a) and Eq. (12.2), Herzer (1989) found that the exchange-averaged anisotropy of a nanostructured material [(Eq. 11.4)], which controls the technical properties and in particular the coercivity, scales as the sixth power of l, the thickness of the interparticle layer:

$$\langle K \rangle \approx 0.32 \left(\frac{K_u^4}{A^3} \right) l^6 \tag{12.14}$$

Because coercivity is limited by the relation, $H_c \leqslant 2\langle K \rangle / M_s$, nanostructural materials may show a range of coercivities controlled by the sixth-power dependence on microstructure length scale. The sixth-power dependence fits very well the steep rise in coercivity first reported in a nanostructured system with variable length scale (Fig. 9.19). There the steep line goes as D^6; assuming $l \propto D$, the data are consistent with l^6. Thus *nano*structured materials may show coercivities far lower than a *micro*crystalline sample of the same phase depending on the scale l of the nanostructure.

12.3 NANOSTRUCTURED MAGNETIC MATERIALS

Nanocrystalline alloys are comprised primarily of crystalline grains having at least one dimension on the order of a few nanometers. Each magnetic nanocrystallite is a single-domain particle that may or may not be exchange coupled to other nanocrystallites. Thus such materials provide an ideal context in which to demonstrate the principles of magnetism in small structures.

12.3.1 Processing of Magnetic Nanostructures

Nanocrystalline alloys can be fabricated by quenching certain alloy compositions from the melt at a rate *insufficient* to achieve a homogeneous noncrystalline structure. Fine-grain structures may result from the large undercooling and kinetically frustrated grain growth. Alternatively, nanocrystalline materials may be made by heat-treating an amorphous alloy precursor. Amorphous alloys inherently have a uniform distribution of constituent elements, a condition suitable for the formation of nanocrystalline alloys. Elements are added to the glass-forming alloy to promote massive nucleation (insoluble species such as Cu) and other elements are added to inhibit grain growth (stable, refractory species such as Nb or TaC).

The major volume fraction of nanocrystalline alloys is a random distribution of crystallites having dimensions of order 2–40 nm. These crystallites are nucleated and grown from the amorphous phase, rejecting insoluble species during the growth process. The residual amorphous phase therefore has a chemistry different from that of the parent amorphous alloy. A widely studied nanocrystal-forming amorphous alloy has nominal chemistry $Fe_{73.5}Cu_1Nb_3Si_{13.5}B_9$. The presence of Cu, insoluble in Fe, promotes massive nucleation, and the Nb retards grain growth. The nanocrystalline phase in these alloys is α-Fe_3Si (B2 if disordered, DO_3 if ordered), which occupies some 70–80 vol%. The amorphous grain boundary phase in the fully mature nanocrystalline alloy has a thickness of about 1 nm.

Two-dimensional films and multilayers, sometimes with nanoscale patterning in the film plane, can also be considered as nanostructured magnetic systems. Thin-film microfabrication techniques, including sputter deposition, electrodeposition, metal-organic chemical vapor deposition, and electron–beam evaporation/condensation, are used.

12.3.2 Examples of Nanocrystalline Alloys

Example 12.1: Co-Nb-B The amorphous Co-Nb-B system shows near-zero magnetostriction and relatively strong magnetization, combined with good mechanical strength and corrosion resistance.

An early example of the effects of nanocrystalline structure on magnetic properties was achieved in amorphous Co-Nb-B subjected to various annealing schedules (O'Handley et al. 1985). Figure 9.19 shows the coercivity of this alloy in various stages of devitrification corresponding to various nanostructure length scales. The size scale of the nanocrystallites was determined from TEM studies. It is difficult to establish the size dependence of H_c below the peak because H_c rises so sharply. However, the data are consistent with the l^6 power law [(Eq. 12.14)] derived by Herzer (1989). The falloff in H_c above the peak is more gradual, consistent with the $1/D$ behavior predicted for "fuzzy" defect of sizes greater than the domain wall width. Herzer has verified the l^6 dependence in a variety of other nanostructures (Herzer 1990).

In addition to having a strong impact on DC coercivity as illustrated above, nanocrystalline precipitates can also be used to enhance AC performance of otherwise amorphous alloys. Datta et al. (1882) reported that the high frequency losses of Fe-rich amorphous metallic alloys could be reduced by a mild heat treatment that produced a few volume percent of primary, α-Fe crystallites measuring about 100 nm in diameter. The low volume density of these large nanocrystalline particles precludes their magnetic interactions other than with the amorphous matrix. In this material they serve to nucleate domain walls in the amorphous matrix by virtue of the magnetization discontinuity at their interface. The increase in the number of domain walls greatly greduces the AC loss (Chapter 9).

Example 12.2: Fe-B-Si-Nb-Cu Iron-rich amorphous and nanocrystalline alloys generally show larger magnetization than those based on Co and much greater than those containing Ni. However, iron-rich amorphous alloys generally have fairly large magnetostriction, limiting their permeability. Formation of a nanocrystalline iron-rich alloy can lead to a dramatic reduction in magnetostriction, thus favoring easy magnetization.

The prototype iron-rich nanocrystalline magnetic alloy is $Fe_{74}Si_{15}B_7$-Cu_1Nb_3 (Yoshizawa et al. 1988). The Cu and Nb are added to what is otherwise a common glass-forming composition in order to enhance nucleation (Kataoka et al. 1989, Hono et al. 1992) and to retard growth of those nuclei, respectively. Annealing of the parent amorphous phase (typically at 550°C for 60 min) results in nucleation of α-Fe_3Si in the vicinity of local Cu concentrations. Growth of α-Fe_3Si is sluggish, inhibited by Nb buildup at its grain boundaries. The amorphous intergranular phase is enriched in Nb and hence has a lower Curie temperature than do the Fe_3Si nanocrystals.

Yoshizawa et al. (1988) report a grain size of 10–15 nm for the nanocrystalline alloys and attractive soft magnetic properties including coercivities in the range of 0.01 Oe, $4\pi M_s \approx 10$–12 kG, and permeabilities of order 10^5. The strong saturation magnetization is a result of rejection of moment-suppressing Nb from the Fe_3Si nanocrystals. Yoshizawa also observed that the magnetostriction of the nanocrystalline alloy is sharply reduced (from $\lambda_s = 20 \times 10^{-6}$ to 3×10^{-6}) on annealing.

These properties can now be understood in terms of Eq. (12.14). The local magnetic anisotropy strength of α-Fe_3Si is of order 8×10^4 erg/cm^3, about

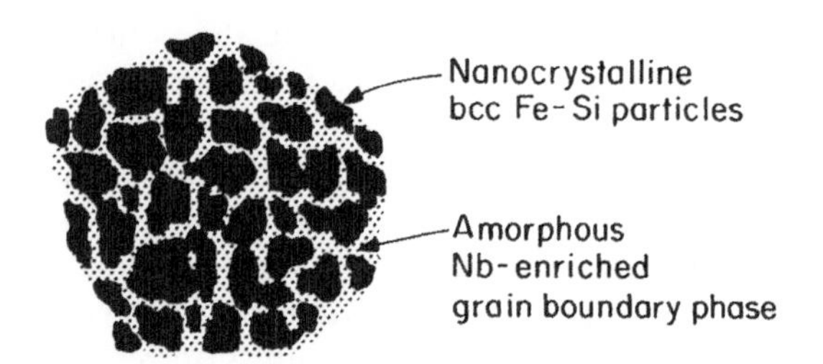

Figure 12.18 Schematic microstructure of nanocrystalline alloy, FeBSiNbCu.

20% of that of α-Fe. Hence $\langle K \rangle$ is of order 50 erg/cm^3, which implies an upper limit to the coercivity of $H_c \approx 0.1$ Oe, in fair agreement with that observed.

The fact that the intergranular amorphous phase acts as an exchange coupling medium between the single-domain Fe_3Si particles is strongly indicated by experiments of Skorvanek and O'Handley (1995). They studied the effects of variations in Nb content on magnetic properties. Nb has a strong suppressing effect on ferromagnetism and hence excessive Nb buildup in the intergranular phase can lead to a breakdown of the exchange coupling between the Fe_3Si nanocrystals (Fig. 12.18). With 6% Nb in the starting alloy, the T_C of the amorphous phase is reduced so that the magnetic properties of the SD Fe_3Si particles can be observed in the exchange coupled $T < T_C$ (amorphous) and decoupled $T > T_C$ (amorphous) states (Fig. 12.19). The peak in coercivity at the decoupling temperature may have an origin similar to that at the decoupling thickness in Figure 12.6; that is, as the exchange coupling weakens,

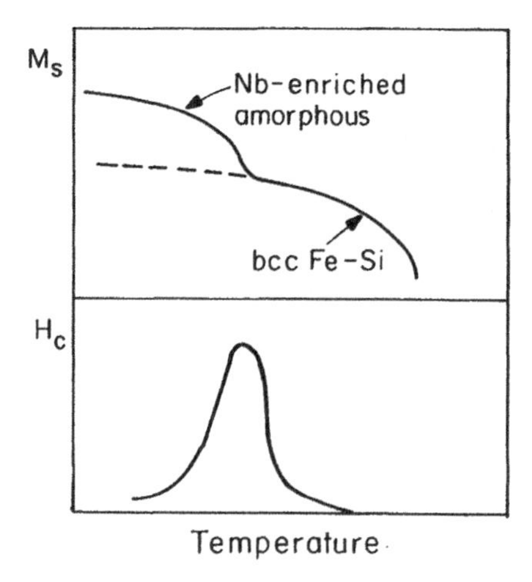

Figure 12.19 Schematic summary of the results of measurements on nanocrystalline FeBSiNbCu alloys having increased Nb contents (Skorvanek et al. 1995).

the particles no longer switch in unison for the entire magnetization cycle. Discontinuous response for some particles generates hysteresis in the system. The vanishing of H_c for T_C (amorphous) $< T < T_C$ (α-Fe_3Si) reflects the superparamagnetic behavior of these SD particles.

Example 12.3: (Fe, Co)$_{81}$M$_9$C$_{10}$(M = Ta, Hf, Zr, Nb, . . .) These thin-film alloys offer strong magnetization density and the possibility of low magnetostriction in the nanocrystalline form, as do certain other iron-rich alloys. The inclusion of refractory metals increases the strength, stability, and corrosion resistance of these alloys.

Hasegawa et al. (1991, 1993) and Hasegawa and Saito (1989) made use of spinodal decomposition in metastable amorphous transition metal carbon alloys (Taylor, 1989) to form nanocrystalline alloys of the general formula $(Fe, Co)_{81}Ta_9C_{10}$. Annealing at 550°C for 20 min results in primary crystallization of α-Fe (or α-FeCo) particles measuring 5–10 nm in diameter and dispersed transition metal carbide nanocrystals (generally at triple junctions; see Fig. 12.20). The grain size of the carbide nancrystals is 1–4 nm. In optimally prepared materials, there is no evidence of a residual amorphous phase between the primary nanocrystals as there is for the FeSiBNbCu nanocrystalline system. Hence, the primary nanocrystals share grain boundaries, making grain-to-grain exchange coupling stronger. The softest magnetic properties are obtained for the smallest nanocrystalline grain sizes.

In these alloys the stable carbide grain boundary phase inhibits grain growth just as Nb does in Fe-B-Si-Cu-Nb.

Because of the significant tantalum concentration in these alloy systems, the parent amorphous phase is only weakly magnetic. On annealing, the appearance of primary Fe or Co results in a significant increase in magnetization (Fig. 12.21). The change in magnetization has been taken as a measure of the fraction x of amorphous material transformed and was shown to follow Johnson–Mehl–Avrami kinetics very well for the Fe-Ta-C system (Fig. 12.22).

Hasegawa et al. (1993) studied the compositional dependence of various properties of the carbide-dispersed nanostructured materials and determined that the structures and magnetic performance depend on composition in a way

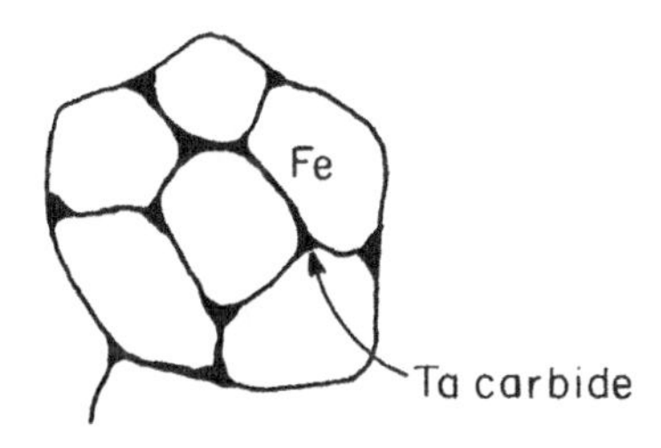

Figure 12.20 Schematic of microstructure in nanostructured magnetic materials based on tantalum carbide precipitation.

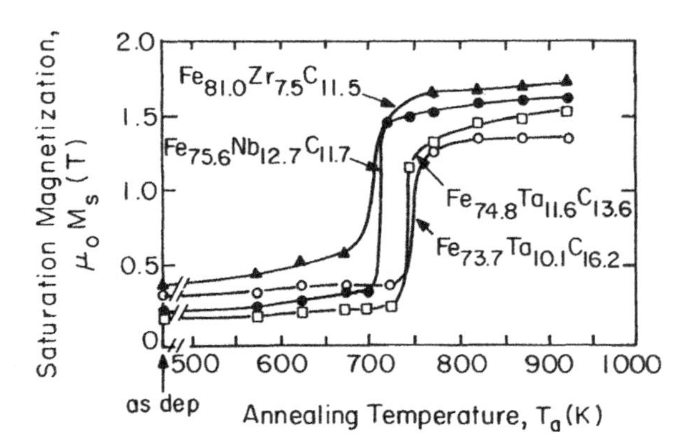

Figure 12.21 Annealing temperature dependence of saturation magnetization M_s measured at room temperature; annealing time at each temperature is 1.2×10^3 s (Hasegawa et al. 1993).

consistent with the exchange-coupled model described in Section 12.2. Too much Fe results in larger nanocrystalline grains and loss of the exchange-averaging effect described by Eq. (11.4). Excessive Ta concentration leads to the decoupling of the Fe nanocrystals because an amorphous nonmagnetic intergranular material, rich in Ta, appears. Grain decoupling brings an increase in H_c. Excessive C results in the formation of BCT Fe_3C, which has a much

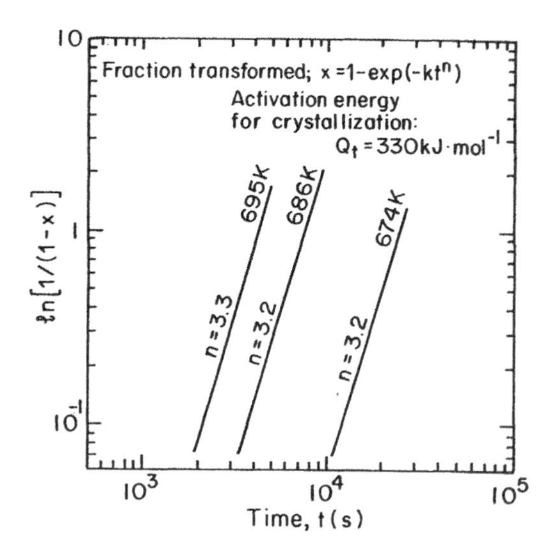

Figure 12.22 Johnson–Mehl–Avrami plots of crystallization of ferromagnetic phases in $Fe_{81.4}Ta_{8.3}C_{10.3}$ films; activation energy Q was obtained by an Arrhenius plot of the time to 50% crystallization (Hasegawa et al. 1993).

greater magnetic anisotropy than the primary BCC phase. Too much Ta and C in combination gives finer Fe nanocrystals but results in a volume fraction of carbide phase large enough to cause the Fe grains to be decoupled.

It is important to note that zero magnetostriction can be achieved in iron-rich, carbide-dispersed alloys by nanostructuring. In nanocrystalline alloys, the sum of the volume-weighted magnetostrictions of the nanocrystalline grain phase and of the amorphous grain boundary phase can result in a net zero magnetostriction even when the starting amorphous alloy shows strong magnetostriction. Figure 12.23 shows the evolution of magnetostriction from amorphous $Fe_{81.4}Ta_{8.3}C_{10.3}$ ($\lambda_s \approx +2.5 \times 10^{-6}$) through various annealing temperatures to $\lambda_s \approx 0$ for 20 min at 870°C (Hasegawa et al. 1993). The positive magnetostriction of the residual amorphous phase is balanced by the negative magnetostriction of the growing α-Fe nanocrystalline phase. In the Fe-Cu-Nb-B-Si nanocrystalline system, the α-Fe_3Si nanocrystals have $\lambda_s < 0$ and similarly serve to reduce the magnetostriction from that of the parent amorphous phase. This absence of magnetostriction makes such nanocrystalline alloys attractive in numerous soft magnetic applications, including high-frequency electronic components. Along with the approach to zero magnetostriction, the amorphous → nanocrystalline transformation also results in increased magnetic permeability, making nanocrystalline alloys even more attractive in soft magnetic applications.

In these carbide-based nanocrystalline alloys, Ta can be replaced by other species that have low solubility in iron and that are strong carbide formers, such as Hf, Nb, and Zr. Nanostructures based on Ta or Hf show superior thermal stability, and those based on Zr or on Hf show superior saturation induction. Extensive studies were done, therefore, on Hf-based systems.

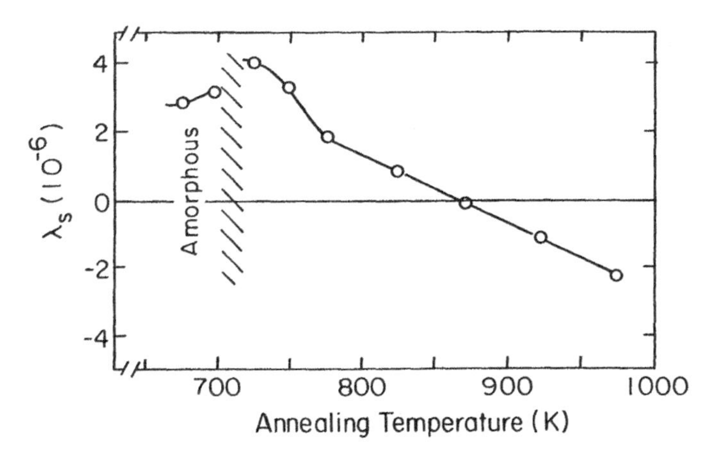

Figure 12.23 Annealing temperature dependence of magnetostriction λ_s for $Fe_{81.4}$-$Ta_{8.3}C_{10.3}$ films; annealing time at each temperature, 1.2×10^3 s (Hasegawa et al. 1993).

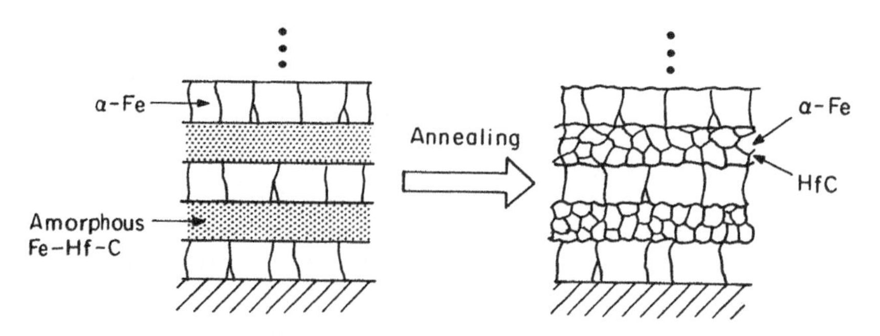

Figure 12.24 Schematic of cross section of Fe/Fe-Hf-C multilayers (Hasegawa et al. 1993).

In an effort to increase the saturation induction of these versatile alloys, Hasegawa et al. (1993) added Co to the Fe-Hf-C system and achieved a saturation induction over 18 kG while retaining good, soft magnetic properties. Higher saturation indication (20 kG) and good properties were also obtained in novel multilayered nanostructures (Hasegawa et al. 1991), such as Fe/Fe-Hf-C (Fig. 12.24).

Two other classes of magnetic nanocrystalline materials show potential for increased the saturation flux density. Alloys of the type $Fe_{88}Zr_7B_4Cu_1$ (Suzuki et al. 1991) crystallize to high-magnetization α-Fe nanocrystals rather than α-Fe_3Si nanocrystals. Willard et al. (1998) have succeeded in making bulk α-(FeCo) phase nanocrystalline alloys based on $Fe_{44}Co_{44}Zr_7B_4Cu_1$ and showing saturation flux densities in excess of 20 kG and permeabilities of order 2000–4000.

Example 12.4: CoCrTa–(Pt) Recording Media CoCr alloys form the HCP phase below 18% Cr, and exhibit strong magnetic anisotropy with high coercivity. They play a key role in magnetic recording media. Modern thin-film recording media (see Chapter 17) are designed to meet several performance criteria. They must have a large enough coercivity that they do not self demagnetize. They must remain sufficiently magnetized in the absence of an external field to produce a detectable signal. The grain size must be small enough that typically 10^3 grains constitute a recorded bit (measuring approximately 1.5 μm × 0.15 μm × 30 nm thick). This ensures low statistical noise at the transition between bits. Finally, the grains must be magnetically isolated from each other (i.e., minimal exchange coupling, unlike the soft magnetic cases described above) so that a magnetic transition (bit) is stable (high H_c) and its edges are sharply defined (low noise).

These criteria are met in CoCr alloys doped with Ta or Pt. The Co-Cr-Pt film is sputter deposited on a seed layer of nonmagnetic Co-Cr that grows in a columnar morphology (Fig. 12.25). The Co-Cr–(Pt, Ta) grows in the HCP structure with its easy c axis in the plane of the film.

The magnetic film is typically 30 nm thick with grains of about the same

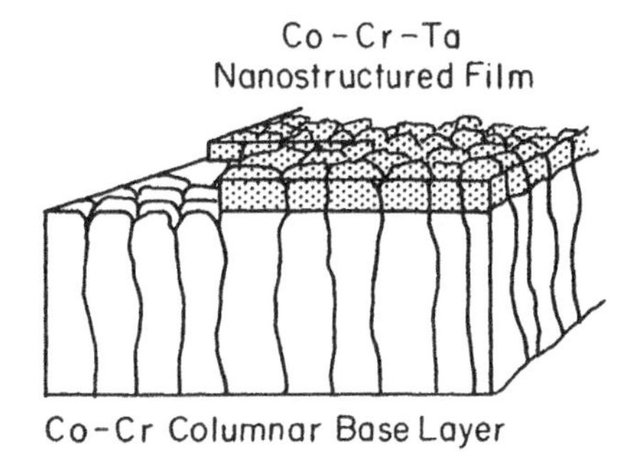

Figure 12.25 Representation of thin-film magnetic recording medium based on CoCr/Ta deposited on CoCr.

size. The Pt or Ta additions retard grain growth and promote segregation of nonmagnetic refractory metals and oxides to the grain boundaries. More will be said about these materials in Chapter 17.

Example 12.5: Nanostructured Permanent Magnets High-energy permanent magnets derive their utility from the large magnetic fields they produce in the face of opposing fields (Chapter 13). Their ability to remain magnetized in their own demagnetizing field is a result of their large coercivity. Recent advances in Nd-Fe-B permanent magnets (see Chapter 13) have led to record energy product of order 50×10^6 G·Oe. Achievable B_r values are limited by the saturation induction, 16 kG, of the $Nd_2Fe_{14}B$ phase. The large anisotropy energy of the 2–14–1 phase provides more than adequate intrinsic coercivity, ${}_iH_c$, over 20 kOe in some cases.

The next phase of improvement in these materials may come from increases in B_r by exchange-coupling the high-anisotropy $Nd_2Fe_{14}B_1$ grains to a larger-magnetization α-Fe intergranular phase. Improvements in energy product can be realized by making nanostructured composites of fine (20–40-nm) $Nd_2Fe_{14}B_1$ grains separated by a thin layer (5–10 nm) of α-Fe (Fig. 12.26). However, the particles must not be exchange coupled to each other to the extent that they switch cooperatively and reduce H_c (as shown in Figure 12.19 when the intergranular, amorphous material becomes magnetic on cooling through its Curie temperature).

While α-Fe has a saturation induction of order 22 kG, it has only a weak magnetic anisotropy and will not remain magnetized in a fixed direction in the face of fields of any significant magnitude. When α-Fe is in contact with $Nd_2Fe_{14}B$ grains, the Fe magnetization may be exchange-coupled to that of the magnetically hard nanocrystalline grain; the coupling tends to hold the Fe moments parallel to those of the nearest grain over an exchange length of order 10 nm (Table 12.1 and Figure 12.27). In this system, the intergranular Fe that is exchange coupled to the Fe-Nd-B nanostructured grains is less likely to result in soft magnetic behavior than exchange coupling does in the cases considered in Sections 3.2.1–3.2.3, because of the strength of the local (grain)

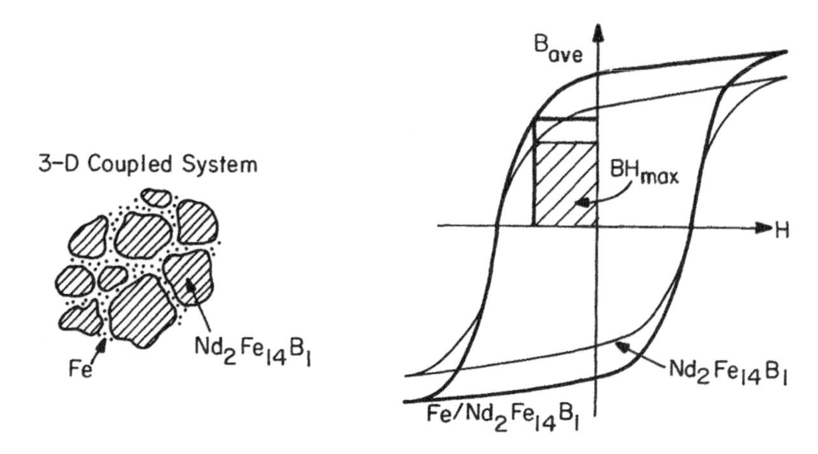

Figure 12.26 Left, schematic of 3D nanostructured composites; right, *B-H* loop of NdFeB (fine) compared with that for an exchange-coupled α-Fe/NdFeB magnet (bold).

anisotropy. In the present case the random anisotropy of the nanocrystals is approximately 10^7 erg/cm^3, so a twist in magnetization from grain to grain occurs in the Fe intergranular material if it is thicker than about 20 nm.

Figure 12.28 shows the results of one-dimensional (1D) model calculations for $Nd_2Fe_{14}B$ coated with a layer of α-Fe. The results show that *large* particles coated with 5 nm of α-Fe will experience no useful increase in average magnetic induction, $B_{\text{ave}} = (t_h B_h + t_{\text{Fe}} B_{\text{Fe}})/(t_h + t_{\text{Fe}})$ because the surface layer is a small volume fraction of the material. Here B_h and B_{Fe} are the flux densities of the hard, $Nd_2Fe_{14}B_1$, phase and α-Fe, while t_h and t_{Fe} are their thicknesses, respectively. However, if the $Nd_2Fe_{14}B$ grains can be refined to nanometer dimensions, $t_h \leqslant 200$ Å, then 50 Å of α-Fe can have a significant impact on the magnetization (Figs. 12.28). The effect of the α-Fe layer is obviously greater in a three-dimensional (3D) nanostructure because there the intergranular α-Fe is a larger volume fraction of the composite.

TABLE 12.1 Magnetostatic and Anisotropy Energy Densities for Fe, Co, and Ni and Exchange Lengths Calculated from Eqs. 12.13[a]

	$2\pi\Delta M_s^2$ (10^5 J/m^3)	K_u (10^3 J/m^3)	l_{ex} (nm) Parallel	l_{ex} (nm) Perpendicular
Fe	19	48	14	2.3
Co	12	410	5	2.9
Ni	1.5	4.5	47	8.2

[a] $A = 10^{-11}$ J/m(10^{-6} erg/cm) is used in all cases. $\Delta M_\perp$ is the difference in magnetization across the interface; the values given here assume $\Delta M_\perp = M_s$.

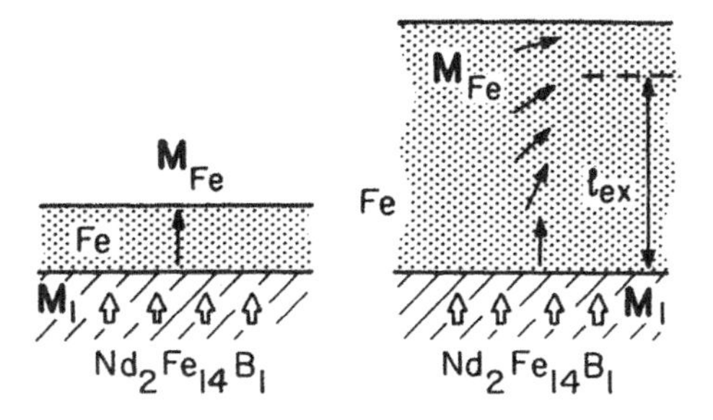

Figure 12.27 Planar model of exchange coupling between soft Fe and hard $Nd_2Fe_{14}B$. When the layer adjacent to the permanent magnet grains is thinner than the exchange length, the iron magnetization adds to that of the hard magnet (left). When the iron thickness is greater than l_{ex}, no additional moment enhancement is achieved (right).

The concept of using exchange coupling to enhance the energy product of fine-grained, two-phase permanent magnets was first described by Kneller and Hawig (1991). It has been shown to work for α-Fe/$Nd_2Fe_{14}B$ by Withanawasam et al. (1994) who achieved $Nd_2Fe_{14}B$ grain sizes of approximately 30 nm and an α-Fe intergranular phase by heat treatment of melt-spun material. Significant contributions to micromagnetic modeling of such composite magnets have been made by Schrefl and Fischer (1994) and Skomski (1994).

Example 12.6: Nanotubes and Nanowires Reference was made in Figure 12.1 to nanowires. Essentially these structures are made by creating nanoscale holes

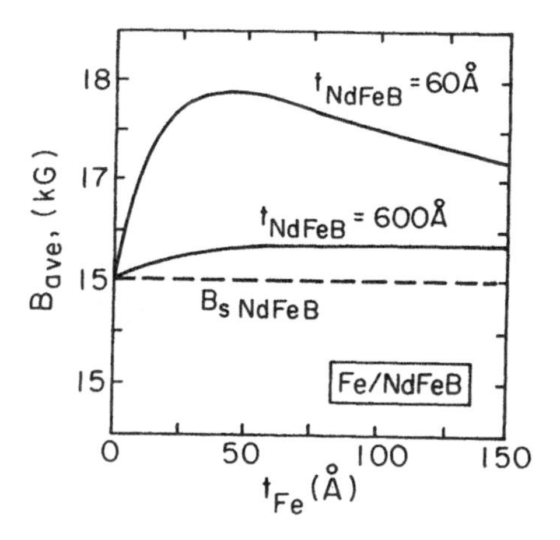

Figure 12.28 Calculated enhancement of magnetization in α-Fe/$Nd_2Fe_{14}B$ nanocomposite films when Fe thickness is comparable to the exchange length (O'Handley, unpublished).

in polymer films and filling them by electrodeposition with a magnetic material (Piraux et al. 1994; Blondel et al. 1994). In these cases, the pores were filled with alternating layers of magnetic metals and noble metals. This was done to create multilayered magnetic materials in which the current flows across all interfaces rather than parallel to the interfaces. It will be seen in Chapter 15, on transport in magnetic materials, that this "current perpendicular to the plane" geometry is important for achieving strong magnetic scattering. This new class of nanostructured magnetic materials holds potential for other novel magnetic effects associated with the presence of magnetically isolated columns.

12.4 SUMMARY

Exchange coupling between dissimilar magnetic particles or layers, in contact or separated by a suitable spacer layer of a few nanometers, can give rise to new phenomena. While magnetic exchange is an inherently atom-scale phenomenon, it has longer-range consequences because it can constrain magnetic moments to be parallel over a length scale of several nanometers. In particular, when exchange coupling operates in the presence of a random anisotropy the technical magnetic properties, limited by the exchange-averaged anisotropy $\langle K \rangle$, depend on the length scale of the random anisotropy to the sixth power.

Nonmagnetic nanostructured materials appear to have no mechanisms analogous to exchange coupling, except perhaps their long-range stress fields, that introduces a new characteristic length scale to material properties. The presence of the magnetic exchange interaction renders nanostructured magnetic materials versatile compared to less complex magnetic materials and distinguishes them from nonmagnetic nanostructured materials.

PROBLEM

12.1 Consider a magnetic thin film of dimensions L W t (length, width, thickness) that is oxidized so as to form a hard magnetic layer on its exposed surface. Assume that the thickness of this hard layer is negligible compared to t and that the magnetizations are exchange-coupled across the hard–soft interface. The hard layer is saturated by a strong field in the positive direction. At $H = 0$, the hard layer holds the magnetization of the film in a saturated state: $M_r = M_s$. As the field is increased in the negative direction, a 180° domain wall moves across the width of the film and creates in its wake a static 180° twist in the magnetization at the interface between the soft and hard materials. Write a simple equation that balances the force on the wall due to the field with that due to the creation of the interfacial twist. Graph the result and discuss.

REFERENCES

Baibich, M. N., J. M. Broto, A. Fert, F. Nguyen Van Dan, F. Petroff, P. Etienne, G. Creuzet, A. Friedrick, and J. Chazelas, *Phys. Rev. Lett.* **61**, 2472 (1987).

Blondel, A., J. P. Meier, B. Doudin, and J.-Ph. Ansermet, *Appl. Phys. Lett.* **65**, 3019 (1994).

Cain, W. C., and M. Kryder, *IEEE Trans.* **MAG-24**, 2609 (1988).

Datta, A., N. J. De Cristafaro, and L. A. Davis, in *Rapidly Quenched Metals*, vol. II, T. Masumoto and K. Suzuki, eds., *Jpn. Inst. Metals, Sendai*, 1982, p. 1007.

Erwin, R., et al., *Phys. Rev. Lett.* **56**, 259 (1986).

Gökenmeijer, N. J., T. Ambrose, and C. L. Chien, *Phys. Rev. Lett.* **79**, 4270 (1997).

Hasegawa, N., N. Kataoka, H. Fujimori, and M. Saito, *J. Appl. Phys.* **70**, 6253 (1991).

Hasegawa, N, M. Saito, N. Kataoka, and H. Fujimori, *J. Mater. Eng. Perform.* **2**, 181 (April 1993).

Herzer, G., *IEEE Trans.* **MAG-25**, 3327 (1989).

Herzer, G., *IEEE Trans.* **MAG-26**, 1397 (1990).

Herzer, G., *J. Magn. Magn. Mater.* **112**, 258 (1992).

Herzer, G., *J. Mater. Eng. Perform.* **2**, 193 (April 1993).

Hono, K., K. Hiaraga, Q. Wang, A. Inoue, and T. Sakurai, *Acta Met. Mater.* **40**, 2137 (1992).

Jungblut, R., R. Coehorn, M. Johnson, J. aan de Stegge, and A. Reinders, *J. Appl. Phys.* **75**, 6659 (1994).

R. Jungblut, R. Coehoorn, M. T. Johnson, Ch. Sauer, P. J. van der Zaag, A. R. Ball, Th. G. S. M. Rijks, J. aan de Stegge, and A. Reinders, *J. Magn. Magn. Mater.* **148**, 300 (1995).

Kataoka, N., A. Inoue, T. Masumoto, Y. Yoshizawa, and K. Yamauchi, *Jpn. J. Appl. Phys.* **28**, L1820 (1989).

Kneller, E., and R. Hawig, *IEEE Trans. Magn.* **27**, 3588, 1991.

Koon, N. C., *Phys. Rev. Lett.* **78**, 4865 (1997).

Malozemoff, A. P., *Phys. Rev. B* **35**, 3679 (1987).

Mauri, C., H. C. Siegmann, P. S. Bagus, and E. Kay, *J. Appl. Phys.* **62**, 3047 (1987).

Meiklejohn, W. H., and C. P. Bean, *Phys. Rev.* **102**, 1413 (1956).

O'Handley, R. C., J. Megusar, S.-W. Sun, Y. Hara, and N. J. Grant, *J. Appl. Phys.* **57**, 3563 (1985).

Parkin, S. S. P., *Phys. Rev. Lett.* **64**, 2304 (1990).

Piraux, L., J. M. George, J. Depres, C. Leroy, C. Ferain, R. Legras, K. Ounadjela, and A. Fert, *Appl. Phys. Lett.* **65**, 2484 (1994).

Schrefl, T., R. Fischer, J. Fidler, and H. Kronmüller, *J. Appl. Phys.* **76**, 7053 (1994).

Skomski, R., *J. Appl. Phys.* **76**, 7059, 1994.

Skorvanek, I., and R. C. O'Handley, *J. Magn. Magn. Mater.* **140–144**, 467 (1995).

Skorvanek, I., C. K. Kim, and R. C. O'Handley, in *Science and Technology of Rapid Solidification Processing*, M. Otooni, ed., Kluwer, Amsterdam, 1995, p. 309.

Suzuki, K., A. Makino, N. Kataoka, A. Inoue, and T. Masumoto, *Mater. Trans. Jpn. Inst. Metals* **32**, 93 (1991).

Takano, K., R. H. Kodama, A. E. Berkowitz, W. Cao, and G. Thomas, *Phys. Rev. Lett.* **79**, 1130 (1997).

Taylor, K. A., L. Chang, G. B. Olson, G. D. W. Smith, M. Cohen, and J. B. Vander Sande, *Met. Trans.* **A20**, 2712 (1989).

Unguris, J., R. J. Celotta, and D. T. Pierce, *Phys. Rev. Lett.* **67**, 140 (1991).

Willard, M. A., D. E. Loughlin, M. E. McHenry, D. Thoma, and K. Sikafus, *J. Appl. Phys.* **84**, 1 (1998).

Withanawasam, L., A. L. Murphy, and G. C. Hadjipanayis, *J. Appl. Phys.* **76**, 7065 (1994).

Yoshizawa, Y., S. Oguma, and K. Yamauchi, *J. Appl. Phys.* **64**, 6044 (1988).

CHAPTER 13

HARD MAGNETIC MATERIALS

13.1 INTRODUCTION

Permanent magnets are used to produce strong fields without having to apply a current to a coil. Hence they should exhibit a strong net magnetization (they may be either ferromagnetic or ferrimagnetic). It is also important that the magnetization be stable in the presence of external fields. These two conditions indicate that the B–H loop should have large values of remanent induction B_r and coercivity H_c, respectively. Permanent magnets have coercivities in the range of 10^4 to 10^6 A/m.

Permanent magnets are used extensively in motors, generators, loudspeakers, bearings, fasteners, and actuators (such as for positioning a recording read in a hard disk drive). The permanent magnet market is shared between hard ferrites (a shrinking market share), cobalt rare-earth magnets and $Nd_2Fe_{14}B_1$ (a growing share). There also exist smaller markets for a variety of speciality hard magnets such as the mechanically ductile and magnetically semihard magnets, Arnokrom or Crovac ($H_c = 20$–90 Oe) which can be rolled to thin sheets. Finally, it must not be overlooked that the magnetic recording media used for information storage, both tape and disk, are also based on permanent magnets. The materials science of recording media is covered in Chapter 17.

A mechanical analogy may be useful to put in perspective the different modes of operation of hard and soft magnetic materials. Soft magnetic materials exhibit a kind of magnetic flexibility (low stiffness); application of even a small magnetic force (H) results in a large and mostly reversible response (B). (The irreversible component of the response ideally should be minimized). This analogy breaks down because the mechanical displacement of materials eventually diverges at large stress while the response of soft magnets eventually saturates at large field. Permanent magnetic materials exhibit a very stiff and, ideally, *elastic* behavior; they can produce a force without physical contact. Hard magnets lose their "permanent" magnetization if the opposing force is comparable to their coercive field. In that case their magnetic state is changed "plastically" and they no longer produce the same field.

When soft magnetic materials were discussed, it mattered little whether an M–H loop or a B–H loop was used to describe the material properties; the shape is essentially the same because $\boldsymbol{B} = \mu_0 (H + M) \approx \mu_0 \boldsymbol{M}$. However, for magnetic materials with large coercivities the distinction between $\boldsymbol{B}$ and $\boldsymbol{M}$ is more than simply a scale factor of μ_0 (or 4π in cgs). At $H = 0$ (remanence) the B–H and M–H curves are equal, but for fields that oppose the direction of magnetization of a hard magnet, a smaller external field is required to give $B = 0$ than to give $\boldsymbol{M} = 0$ (Fig. 13.1*a*, and 13.1*b*). Thus, it is necessary to distinguish the B–H loop coercivity, ${}_BH_c$ from the intrinsic, M–H loop coercivity, ${}_iH_c$. It is this intrinsic coercivity that measures the intrinsic hardness of a permanent magnet, independent of its shape. However, it is the area inside the second quadrant of the B–H loop that describes the available energy density of a material. The value of the intrinsic coercivity is always greater than the coercivity from the corresponding B–H loop. The permanent magnet literature deals with both M–H loops and B–H loops, and their differences must be appreciated.

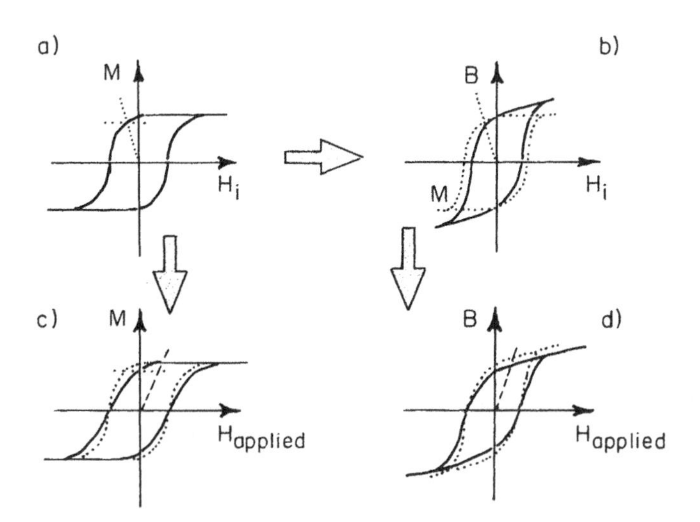

Figure 13.1 (*a*) The M–H_i loop is an intrinsic characteristic of the material (no shape effects). It appears for reference as the dotted loops in panels (*b*) and (*c*). The solid loops in (*b*) and (*c*) show the change in loop shape on going from M to B (*a*–*b*) and H_i to $H_{applied}$ (*a*) to (*c*) showing sample shape effects]. Panel (*c*) shows B–H_{appl}.

Recall from Chapter 2 that the form of the hysteresis loop is very sensitive to sample shape whether plotted as $M-H_{appl}$ or $B-H_{appl}$ but not when plotted versus internal field, $M-H_i$ or $B-H_i$. When permanent magnets are used in "open circuit" applications (which is nearly always the case in order that the fields they produce may be used) their technical properties are strong functions of the shape of the magnet. (Closed-circuit applications could involve the use of a permanent magnet to bias a soft element in a magnetic circuit.) Thus, it is important to distinguish between intrinsic or internal field behavior (Fig. 13.1*a*, *b*) (as would be measured in a closed magnetic circuit in Figs. 13.2*a*), and extrinsic or applied field behavior, Figure 13.1*c*, and 13.1*d* (as would be measured in an open circuit (Fig. 13.2*b*). The more the measurement technique or application design resembles a closed circuit (fewer free surfaces normal to $\mathbf{M}$), the more closely the material performance will reflect the intrinsic properties of the magnet; more open-circuit applications and measurements involve properties that also reflect the sample shape.

The load line construction introduced in Chapter 2 is commonly used when describing permanent magnets. It allows the use of intrinsic data (Fig. 13.1*a*, or 13.1*b*) for determination of the properties of samples of a particular shape. The load line has a slope given by $-(1\text{-}N)/N$. This line, drawn on Figure 13.1*a*, intersects the intrinsic $M-H_i$ curve at the point that is the remanent point in an $M-H_{appl}$ curve, that is, it indicates the remanence actually achieved in a given shape having average demagnetizing factor N.

The maximum energy density of a permanent magnet $(BH)_{max}$ is determined by the point on the second-quadrant branch of the $B-H_i$ loop that gives the largest area for an enclosed rectangle. It may be found at the point on the loop that extends furthest into the hyperbolic contours of constant BH product (shown by dashed lines in Figure 13.3*a*).

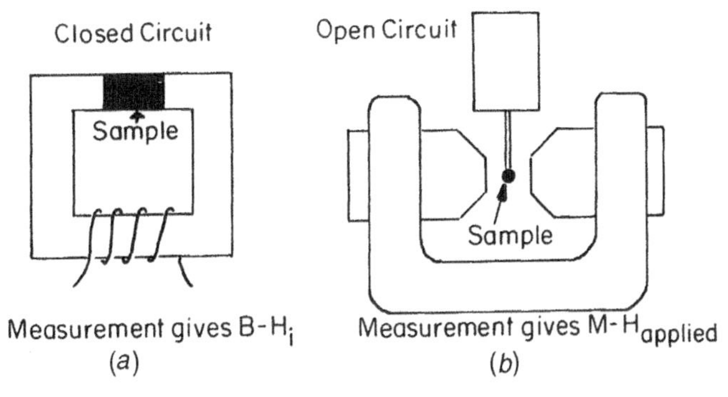

Figure 13.2 Schematic of two ways of measuring properties of permanent magnets. (*a*) Closed circuit inductive measurements give $B-H_{appl} \approx B-H_i$ with negligible demagnetization effects from sample surfaces (Fig. 13.1*b*); (*b*) open-circuit magnetometer measurements give $M-H_{appl}$ (not $=M-H_i$) loops with full sample shape effects (Fig. 13.1*c*).

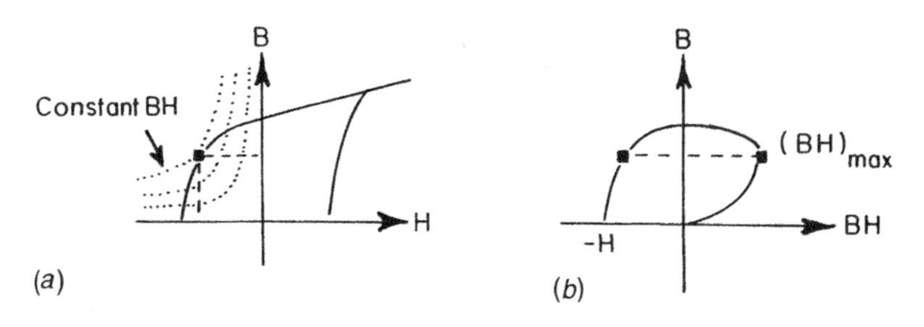

Figure 13.3 (*a*) Partial B–H loop showing contours of constant B–H in second quadrant and $(B-H)_{max}$ rectangle enclosed; (*b*) a common way of calculating and displaying $(BH)_{max}$ from second-quadrant data.

From the second quadrant of the B–H_i loop (*a*), one can calculate the BH product point by point and plot its value on a horizontal axis (Fig. 13.3*b*, first quadrant) for each value of B. The location of $(B-H)_{max}$ is the point at which the material characteristics of a permanent magnet are most efficiently used. How does one design a magnet to make optimal use of its maximum energy product? Either the shape must be chosen for a given material so that the load line intersects the intrinsic B–H loop at the optimal point or, given a shape, a material must be selected having the maximum energy product at the intersection with the load line.

The shapes of second quadrant M–H loops for some common permanent magnets are compared in Figure 13.4*a*. The extremes are Alnico, which shows a rather large remanence but a relatively weak coercivity; and cobalt-platinum, which shows a weak remanence and a fairly large coercivity. The permanent

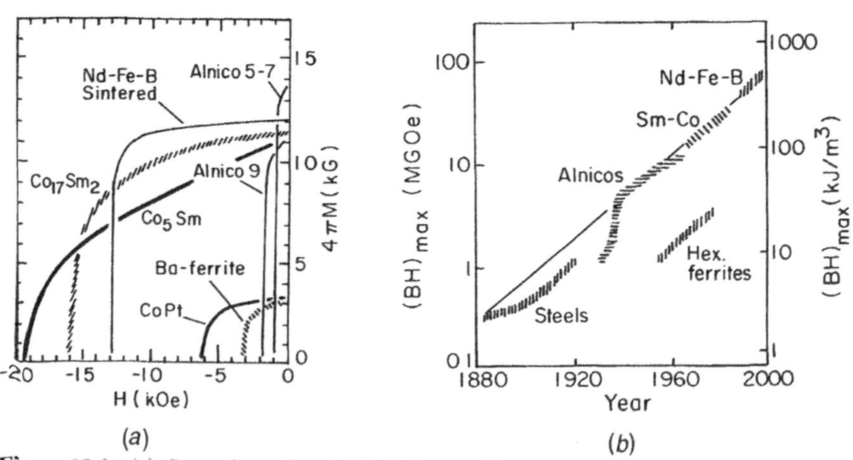

Figure 13.4 (*a*) Second quadrant M–H loops of some common permanent magnets. (*b*) Increase in $(BH)_{max}$ of permanent magnets over recent decades.

magnets with the highest energy products are based on $Fe_{14}Nd_2B_1$. Figure 13.4*b* shows the evolution of the maximum energy product in permanent magnets over recent decades. In this chapter, the different classes of permanent magnets are described and then their properties are explained in terms of single-domain-particle behavior or nucleation-limited or domain wall pinning behavior. The magnets covered include the Alnico phase-separated magnets; oxide magnets based on hexagonal ferrites; and intermetallics based on Co_5Sm, $Co_{17}Sm_2$, and $Fe_{14}Nd_2B_1$; as well as a number of specialty permanent magnets such as CoPt (order–disorder), FeCrCo, and MnAlC.

Note the scales on Figure 13.4*b*: MGOe and kJ/m^3. The left-hand scale is the result of a BH product or, inside the material, a $4\pi MH$ product. To reconcile this unit with the right-hand scale, the 4π must be removed to get MH in erg/cm^3: 10^6 GOe $\div 4\pi = 8 \times 10^4$ $erg/cm^3 = 8$ kJ/m^3. For example, if B and H at the maximum energy point are 10^3 G and 10^3 Oe, respectively, they convert to 0.1 T and 8×10^4 A/m, respectively. This product, whether considered as $\mu_0 M \times H$ or $M \times \mu_0 H$, is 8 kJ/m^3.

The physical mechanisms or ingredients of magnetic hardening begin with strong, uniaxial magnetic anisotropy. In a *random* polycrystalline sample of close-packed Co ($K_u = 4.7 \times 10^5$ J/m^3), the magnetization in zero field is randomly oriented along the dispersed easy axes. On reducing an applied field to zero after saturation, the magnetization rotates to the nearest easy-axis direction in the hemisphere centered about the field direction, leaving the remanence somewhat suppressed: $M_r = M_s\langle\cos\theta\rangle \approx 0.5M_s$. To reduce the magnetization to zero in this case requires one of three processes: (1) the magnetization in some of the grains rotates uniformly against the anisotropy in which case $H_c \leqslant 2K_u/M_s$ (Fig. 9.7), (2) the magnetization rotates incoherently (i.e., by curling or buckling), in which case H_c is less than that for the coherent case, or (3) if the grain size is sufficiently large that domain walls can be accommodated and if the domain walls are not strongly pinned, the sample may be demagnetized in a field considerably less than $2K_u/M_s$ by nucleation of reversal domains followed by domain wall motion. The mathematics describing these situations, called the *Stoner–Wohlfarth theory*, were covered in Chapter 9. From the point of view of permanent magnets, the magnetic anisotropy sets the upper limit to ${}_iH_c$. Three major factors can cause ${}_iH_c$ to fall short of its maximum value: (1) a dispersion in easy-axis (grain) orientations, (2) the presence of mobile domain walls, or (3) exchange coupling between single-domain particles. The design of different permanent-magnet materials involves, among other challenges, (1) optimizing K_u (by crystallography, chemistry, and/or particle shape), (2) maximizing B_r by introducing texture (preferred orientation) into the grain structure, (3) eliminating domain walls (by making single-domain particles) or pinning domain wall motion (by introducing certain defects), and (4) minimizing exchange coupling between single-domain particles (nonmagnetic grain boundaries).

It was shown in Chapter 6 how magnetocrystalline anisotropy depends on crystal structure and material chemistry ($\langle L_z\rangle \neq 0$). Chapters 2 and 8 illustrated how the shape and size of particles can preclude domain wall formation, and Chapter 9 showed how defect size and properties can change the strength of domain-wall defect interactions.

This chapter describes how these ingredients are combined in various material systems.

Early Permanent Magnets The first known permanent magnet, lodestone, was an impure form of iron oxide, mostly ferrimagnetic magnetite (Fe_3O_4) with fine regions more strongly oxidized to antiferromagnetic γ-Fe_2O_3. It exhibits a coercivity of several tens of Oe and a modest saturation induction of 0.3–0.4 T. Although it is the first magnetic material used in an application (by the Chinese in a compass), it is no longer used commercially as a permanent magnet.

Pure metallic iron is an ideally soft magnetic material. Iron has a large saturation magnetization and relatively small values of cubic anisotropy and magnetostriction. However, the presence of impurities destroys the soft magnetism. Carbon impurities render iron hard, both magnetically and mechanically. Figure 13.5 shows the phase diagram for the FeC system. Note the low carbon solubility in α-Fe and the eutectic near 17 at% C before reaching the stable cementite phase, Fe_3C.

For carbon concentrations above the carbon solubility limit, cementite (Fe_3C) can coexist in equilibrium with α-Fe. Quenching from a temperature at which carbon is soluble in α-Fe to a lower temperature where that concentration is insoluble can result in the formation of martensitic Fe_3C precipitates. Fine martensitic Fe_3C precipitates can pin dislocations, rendering the material *mechanically* hard; they can also pin domain walls, rendering the material *magnetically* hard with a coercivity of order 4 kA/m (50 Oe). Iron carbon magnets have been used since the mid nineteenth century.

Tungsten steel (7–8% W), first used in the 1940s, can have coercivities of about 7 kA/m (90 Oe). Co-Mo and Co-Cr steels show coercivities more than twice those of W steels and energy products of order 8 kJ/m^3 (1 MG·Oe). Equiatomic FeCo alloys show saturation induction of about 2.4 T. They can be made either magnetically hard or soft depending on heat treatment. When an FeCo alloy is cooled slowly through the order–disorder transformation temperature, 800°C, it becomes mechanically brittle and can show coercivities of order 10^4 A/m.

In the early 1960s efforts were made to fabricate permanent magnets based on the Stoner-Wohlfarth model prediction of large coercivity for elongated single-domain (ESD) particles. Luborsky (1961) used electrodeposition to create FeCo particles 10–20 nm in diameter. Coercivities realized (160 kA/m) were considerably less than the $\mu_0 M_s$ predicted for coherent rotation. Later, Luborsky and Morelock (1964) made FeCo whiskers from the vapor state and achieved more uniform ESD shapes and higher coercivities.

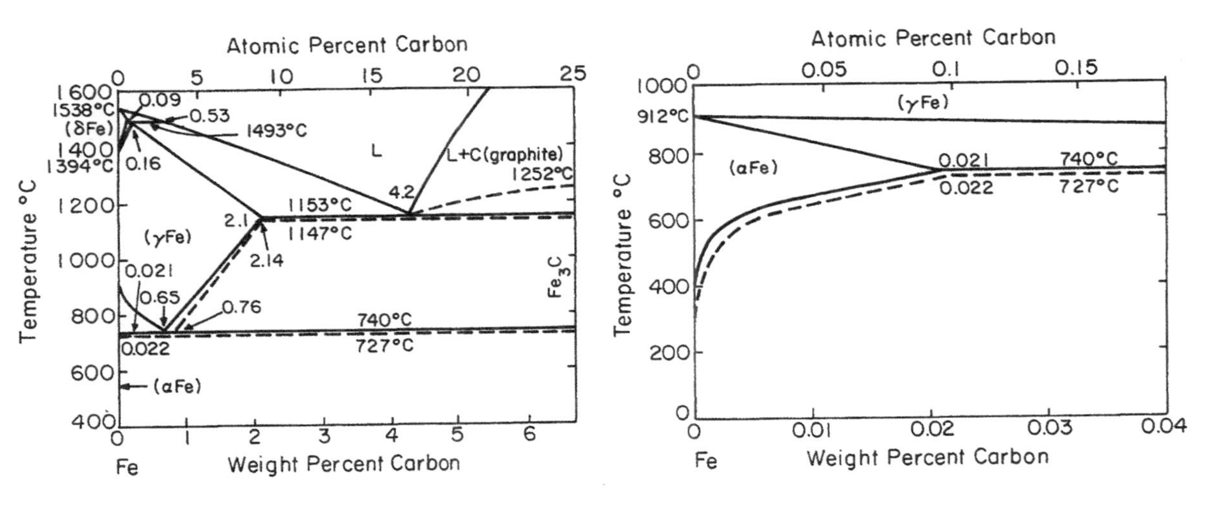

Figure 13.5 Phase diagram for FeC (*Metals Handbook*).

13.2 ALNICO AND FeCrCo MAGNETS: SPINODAL DECOMPOSITION

These two classes of permanent magnet are grouped together because they both make use of spinodal decomposition to achieve a two-phase microstructure of magnetic, α, and nonmagnetic, α', phases.

13.2.1 Alnico

In 1932, Mishima reported the development of attractive hard magnetic properties in a new class of alloy derived from the intermetallic Heusler compound Fe_2NiAl. Often called *Mishima alloys,* they offered a significant improvement in coercivity over the permanent magnets in wide use at the time, which were based on Fe-C, Fe-W, and Fe-Co. Coercivities of early Alnico magnets were typically ${}_BH_c \approx 5 \times 10^4$ A/m (600 Oe). However, because of their reduced saturation magnetization relative to those of the hard magnetic steels, they did not show better energy products: $(BH)_{\max} \approx 8\ \text{kJ/m}^3$ (1 MG·Oe). Alnicos were widely used in motors and loudspeakers after the late 1930s and are still used in many applications today, although they have been replaced by hexagonal ferrites in many low-end applications and by rare-earth intermetallics in some high-performance applications.

The processing and performance of alnico magnets is a beautiful example of magnetic materials science and also provides a useful application of many of the concepts developed in our chapters on magnetostatics (Chapter 2) and magnetization processes in small particles (Chapters 9 and 12).

Alnico magnets derive their hard magnetism from the shape of the particles formed on spinodal decomposition of the Heusler phase into an iron-rich phase (α) and a BCC NiAl-rich phase (α'). Spinodal decomposition is a form of precipitation distinct from nucleation and growth; it occurs below a miscibility gap in a phase diagram (see pseudobinary FeNiAl phase diagram, Figure 13.6) and generally results in a periodic array of α and α' phases.

A BCC solid solution is seen to exist above 1000°C for Fe_2NiAl. Below that temperature spinodal decomposition proceeds at a rate governed by temperature and composition. Any local fluctuation in iron concentration below the spinodal line *increases* the stability of the system, and the fluctuation grows in amplitude. The thermodynamics of spinodal decomposition are such that periodic composition profiles are favored with a well-defined wavelength λ, which is a function of the $\alpha - \alpha'$ interface energy.

Cahn (1963) pointed out that the elastic energy associated with the $\alpha - \alpha'$ phase boundary is minimized if the wavevectors $\boldsymbol{k}$ of the periodic fluctuations obey the following conditions:

$$\boldsymbol{k} \parallel \langle 100 \rangle \quad \text{for} \quad 2\,C_{44} > C_{11}\text{-}C_{12}$$

$$\boldsymbol{k} \parallel \langle 111 \rangle \quad \text{for} \quad 2\,C_{44} < C_{11}\text{-}C_{12}$$

The first condition applies in the alnico alloys and $\boldsymbol{k} \parallel \langle 100 \rangle$ is, in fact,

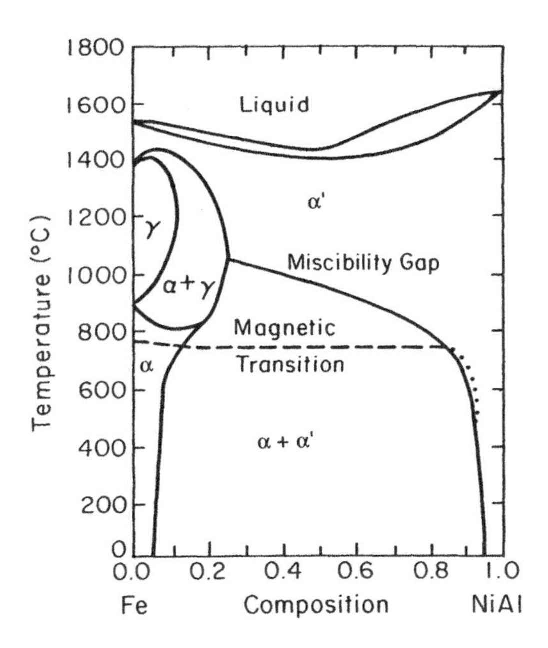

Figure 13.6 Pseudobinary phase diagram for Fe/NiAl showing miscibility gap that gives rise to spinodal decomposition, which is essential to formation of Alnico permanent magnets (McCurrie 1982).

observed. The early stage of spinodal decomposition is characterized by the appearance of spherical concentrations of α and α' centered on two interpenetrating simple cubic lattices. As phase separation continues, the amplitude of the composition fluctuation increases, preserving λ, and the α and α' regions grow along the three $\langle 100 \rangle$ directions and then join (Fig. 13.7). The result is a microstructure composed of two interpenetrating networks of elongated rods of Fe(α) and NiAl(α').

Figure 13.8 shows the microstructure of Fe_2NiAl after the development of phase elongation along the $\langle 100 \rangle$ directions a suggested in Figure 13.7b.

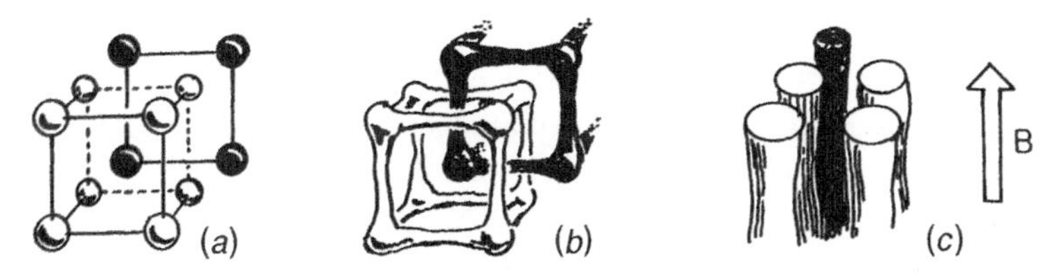

Figure 13.7 Schematic of the evolution of the $\alpha - \alpha'$ microstructure in alnico: (*a*) spheroidal precipitates of α on simple cubic lattice and α' on interpenetrating simple cubic lattice; (*b*) particles elongate along $\langle 100 \rangle$ during tempering; (*c*) one direction of elongation is favored if the decomposition occurs in a saturating magnetic field.

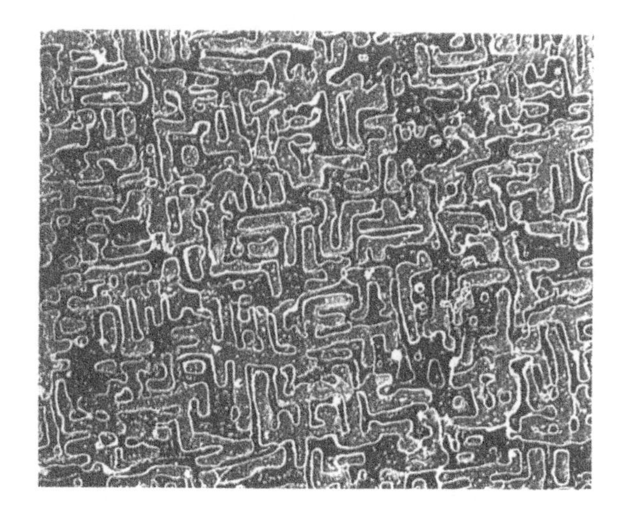

Figure 13.8 Microstructure of Fe_2NiAl after slow cooling through the spinodal followed by 2 h at 850°C, then quenching. The lighter-colored rod-shaped features are the magnetic Fe-rich phase, and their periodicity is about 80 nm [After DeVos, 1969).]

Replacement of some of the Fe with Co was found to increase the α-phase saturation magnetization, leading to higher B_r, and to increase the magnetization difference $\Delta M = M_\alpha - M_{\alpha'}$ between α and α'. Because H_c is due to magnetostatic anisotropy [cf. Eq. (2.24)]

$$H_c \propto \Delta M(N_2 - N_1) \tag{13.1}$$

the coercivity is expected to increase with increasing cobalt content (Fig. 13.9).

Cobalt also increases the Curie temperature of the α phase. It will be seen later that this improves the efficacy of magnetic field heat treatment in Alnicos 5–9. In addition, cobalt tends to stabilize the unwanted γ phase and slow the kinetics of the spinodal decomposition. Spinodal decomposition was found to be accelerated by the addition of up to 3.5% Cu and by reduction of the Ni and Al content. Finally, the addition of up to 4 or 5% Ti helps offset the deleterious effects of cobalt; while it lowers B_r, the increase it brings to ${}_BH_c$ is enough to increase $(BH)_{max}$.

The first Alnico alloys developed (Alnicos 1–4) were random polycrystalline alloys which therefore showed an equal distribution of elongated particles along the three ⟨100⟩ directions as in Figure 13.7*b*. Later Alnicos (5–9) made use of field annealing to establish a preferred direction of rod orientation (Fig. 13.7*c*) and hence show anisotropic magnetic properties (i.e., much larger values of remanence and $(BH)_{max}$ along the rod direction).

Anisotropic Alnicos A thermomagnetic treatment for the development of anisotropic alnicos (those designated as Alnicos 5–9) was discovered in 1938.

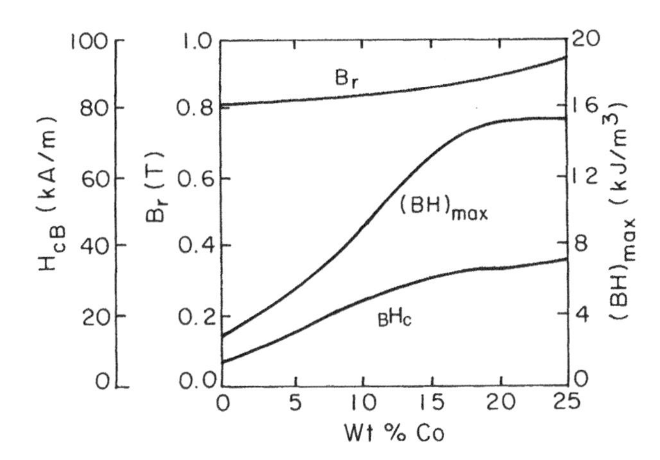

Figure 13.9 Variation of remanence, coercivity, and maximum energy product with cobalt substitution for iron in $Fe_xCo_{1-x}NiAl$ [after McCurrie (1982)].

The idea is to establish an energetic preference for particle elongation along the $\langle 100 \rangle$ direction closest to an applied field (Fig. 13.7*c*). The internal magnetic field near an α–α' interface is proportional to the surface magnetic charge density resulting from a difference in perpendicular magnetization across the interface, $H_i \propto \Delta M = M_\alpha - M_{\alpha'}$. The two phases are characterized by magnetization–thickness products $M_\alpha t_\alpha$ and $M_{\alpha'} t_{\alpha'}$. The magnetostatic contribution to the surface energy then goes as $\sigma_m \approx (\mu_0/2)\Delta M \langle M \rangle (t_\alpha + t_{\alpha'}) \cos^2\theta$, where $\langle M \rangle = (M_\alpha t_{\alpha'} t)/(t_\alpha + t_{\alpha'})$ is the thickness-averaged magnetization. Here θ is the angle between the magnetization direction and the normal to the interface, which is assumed here to be planar. For the interface energy averaged over a magnetized particle, the factor $\cos^2\theta$ is replaced by an effective demagnetization factor. For $\mu_0 \Delta M \approx \mu_0(M_1 - M_2) \approx 0.1$ T, at the initial stages of spinodal decomposition, the surface energy can be estimated to be $\sigma_m \leqslant 0.4\ \mathrm{mJ/m^2}$ ($0.4\ \mathrm{erg/cm^2}$) for $0.1\ \mu$m particles characterized by a demagnetization factor $N \approx 0.1$. This magnetic contribution is significant compared to typical chemical interface energies (which are of order $1\ \mathrm{mJ/m^2}$).

Interfaces that are parallel to the magnetization direction ($\theta = 90°$) have zero magnetostatic energy (Fig. 13.10*a*). If the α–α' interface is diffuse, the interfacial charge is spread out (Fig. 13.10*c*) and the magnetostatic energy is less than that for a sharp interface.

Whereas the nonmagnetic interface energy γA tends to favor spherical particles (A is the surface area of the particle), the magnetostatic energy $f_{ms}V$ is the driving force for particle elongation (V is the particle volume); $f_{ms}V \approx \mu_0 M_s^2 \Delta N V$ is a measure of the extensive magnetostatic energy saved by elongation. The energy reduction is proportional to the change in demagnetization factor ΔN during growth. If the particle grows elongated in its direction

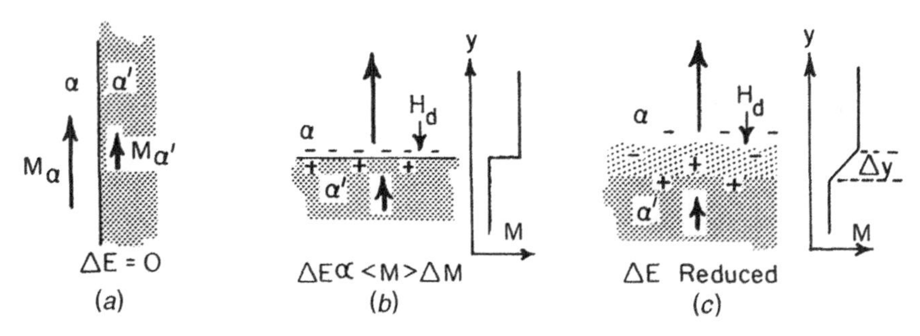

Figure 13.10 Schematic representation of the effects of magnetization on interfacial energy. (*a*) magnetization parallel to an interface produces no surface magnetic charge; (*b*) perpendicular magnetization discontinuity across an interface results in a charged surface and generates magnetostatic fields that raise the energy; (*c*) a diffuse interface distributes the magnetic charge and reduces the magnetostatic energy.

of magnetization, then its magnetostatic energy is reduced from that of a sphere to a smaller value characteristic of its elongated shape. The rate of elongation is therefore proportional to

$$R_{\text{elongation}} \propto f_{\text{ms}} V - \gamma A \tag{13.2}$$

Larger particles (smaller surface:volume ratio) are more strongly driven to large aspect ratios. To develop long particles while they are still small enough to display single-domain behavior, it is helpful to maximize f_{ms} by using as much cobalt as possible; this increases T_C and ΔM.

The anisotropic processing described here for alnico permanent magnets, as well as other magnetic processing effects, are discussed more generally in Chapter 14.

Figure 13.11 shows that the application of a field is indeed effective in bringing about a preferred orientation to the $\langle 100 \rangle$ elongated particles. The preference for orientation is greatest for higher Co content alnicos where ΔM is greater.

Optimal parameters for Alnico 5 are $(BH)_{\text{max}} = 10.8\,\text{kJ/m}^3$ (1.35 MG·Oe) perpendicular to the texture and $13.4\,\text{kJ/m}^3$ (1.8 MG·Oe) parallel to the texture. Increasing cobalt content has stronger benefits for anisotropic Alnicos; for 23% Co $(BH)_{\text{max}} = 41.6\,\text{kJ/m}^3$ (5.4 MG·Oe) can be achieved.

Even though a particular direction for α elongation is established by the field applied during decomposition and tempering, there remains in polycrystalline samples a dispersion in the orientation of $\langle 100 \rangle$ directions and hence the texture is not optimized. It was found that the properties of anisotropic Alnicos could be enhanced by increasing texture in the microstructure. This could be accomplished by chilling the end faces of the casting mold to promote directional solidification (Fig. 13.12).

Figure 13.13 shows the second quadrant B–H curves for anisotropic Alnicos 5–9 (a directed-growth variation of Alnico 8), as well as for single-crystal Alnico 5 (dotted). Table 13.1 summarizes the compositions and properties of

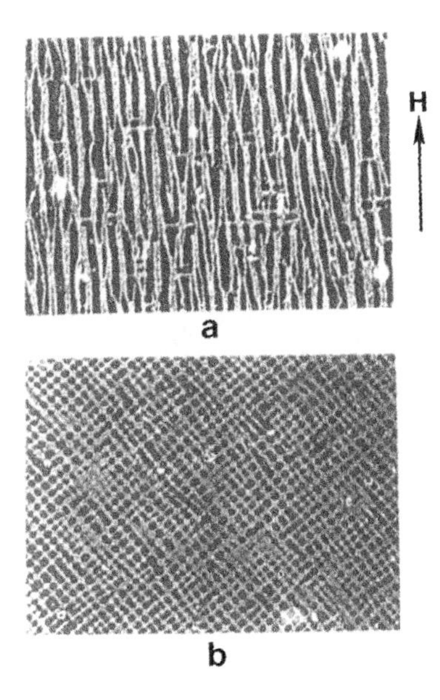

Figure 13.11 Microstructures of Alnico 8 after isothermal heat treatment in a magnetic field for 9 h at 800°C: (*a*) view perpendicular to the direction of applied field; (*b*) view along direction of applied field showing microstructure with $\lambda \approx 32$ nm (De Vos 1966).

the data in Figure 13.13. Note the dramatic improvement in magnetic properties, particularly $(BH)_{max}$, that comes with grain alignment by directional growth (DG) and development of single-crystal Alnico 5 (dotted line). Alnico 9 achieves its higher coercivity with greater cobalt content and, despite a lower remanence, has in DG form an energy product comparable to that of single-crystal Alnico 5.

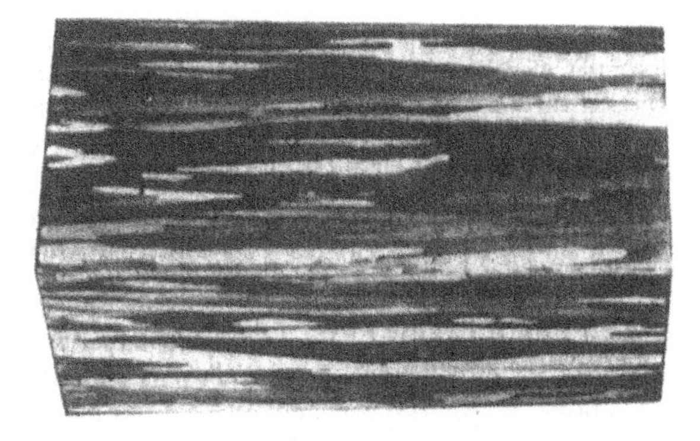

Figure 13.12 Grain structure in a directionally solidified Alnico 9 before spinodal decomposition [After Gould 1964).]

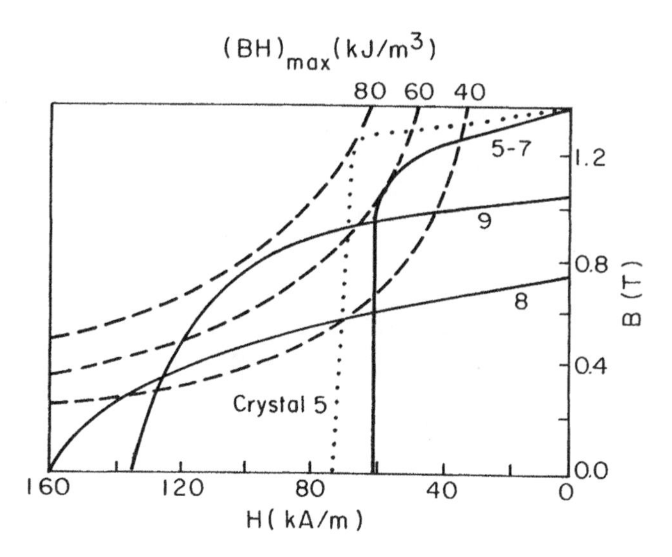

Figure 13.13 Second-quadrant *B–H* curves for selected anisotropic Alnicos; maximum *BH* products can be determined from the dashed *BH*-constant hyperbolas (McCurrie 1982).

With the development of highly textured α–α' phase separation in the Alnicos (Fig. 13.12), it is possible to attempt to describe the magnetic behavior in terms of the Stoner–Wohlfarth model for single-domain, anisotropic particles (Chapter 9).

Typical α-FeCo particles in Alnico 5 measure 1500 Å long by 400 Å in diameter. The shape anisotropy of such particles is related to Eq. (2.25):

$$f_{\text{ms}} = \frac{\mu_0 \Delta N \langle M \rangle \Delta M_s}{2} \tag{13.3}$$

The magnetostatic energy is the energy of the average magnetization $\langle M \rangle$ in the field from the magnetization discontinuity at the interface, ΔM. For simplicity, choose $M_\alpha = 3M_{\alpha'} = 2.1\ \text{T}/\mu_0$ so that $\langle M \rangle \approx \Delta M_s \approx 1.4\ \text{T}/\mu_0$. $\Delta N = N_\perp - N_\parallel$ can be calculated from Eqs. (2.18) and (2.17) for prolate spheroids with $m \approx 4$. These numbers give $K_u \approx 1.1 \times 10^5\ \text{J/m}^3$ and thus $H_a \approx 1.8 \times 10^5$ A/m (2240 Oe). Nesbitt and Williams (1955) used torque magnetometry to determine the strength of the uniaxial anisotropy of Alnico 5 to be $10^5\ \text{J/m}^3$.

Figures 13.14*a*, and 13.*b* compare the measured orientation dependence of H_c (McCurrie and Jackson 1980) with the calculations of Shtrikman and Treves (1959) using the curling form of the SW model. The rough similarity in the shape of the experimental and calculated curves for large *S* values

TABLE 13.1 Compositions and Magnetic Properties of Anisotropic, Thermomagnetically Treated Alnicos 5 and 8 as Well as Single-Crystal Alnico 5

		Composition (wt%) (balance Fe)					Magnetic Properties		
Alnico	Character	Ni	Al	Co	Cu	Other	B_r (T)	$_BH_c$ (kA/m)	$(BH)_{max}$ (kJ/m^3)
5	Random grain	12–15	7.8–8.5	23–35	2–4	0–0.5 Ti 0–1 Nb	1.2–1.3	52–46	40–44
DG[a] 5	Directed grain 5	13–15	7.8–8.5	24–25	2–4	0–1 Nb	1.3–1.4	62–56	56–64
5 xtl	Single crystal	14	8	25	3		1.4	68	80
8	Random grain	14–15	7–8	37–40	3	7–8 Ti	0.74–0.78	150–170	44–48
DG 8(9)	Directed grain 8	14–16	7–8	32–36	4	0.3S	1.0–1.1	140–110	60–75

[a]Directional growth.

Source: After McCurrie (1982).

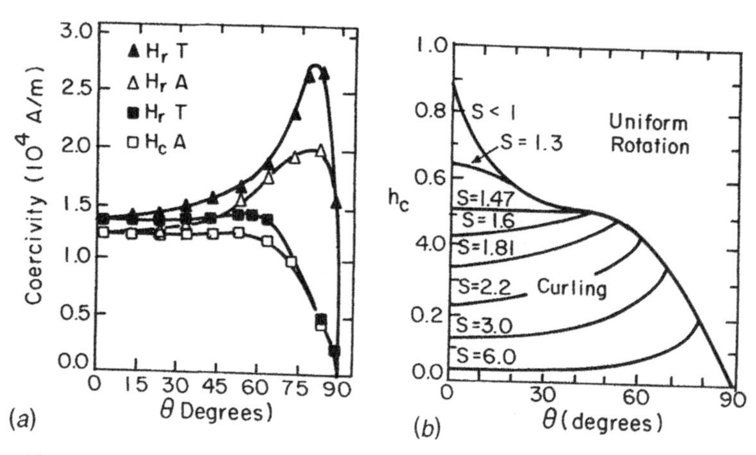

Figure 13.14 (*a*) Dependence of the coercivity on the angle θ between the preferred axis and the measurement direction measured after saturation H_c and from remanence H_r for anisotropic (columnar) Alnico 9 (A) and for single-crystal Alnico 9 (T) (McCurrie and Jackson 1980); (*b*) reduced coercivity versus θ calculated for single-domain particles of various effective radii S (Shtrikman and Treves 1959).

suggests that many particles exceed the single-domain size for this composition: $S = r/r_0 \gg 1$ with $r_0 = (4\pi A/\mu_0)^{1/2}/M_s$. (Compare with Chapter 9.)

13.2.2 Fe-Cr-Co

BCC Fe-Cr-Co alloys develop hard magnetic characteristics by spinodal decomposition as do the Alnicos (Kaneko et al. 1971). However, the product phases in Fe-Cr-Co are relatively ductile Fe-rich (α) and Cr-rich (α') phases compared to FeCo and NiAl. Magnetic properties of Fe-Cr-Co magnets can be enhanced by field-annealing to introduce an elongated morphology (small Mo additions help here) and low temperature tempering to enhance the α–α' phase separation. Coercivities of 65 kA/m and energy products approaching 80 kJ/m^3 have been achieved.

Ductility is a major reason for the interest in Fe-Cr-Co magnets; it allows a certain amount of machining on parts, and it has been exploited by using uniaxial deformation to enhance alignment and thus increase H_c and the energy product (Kaneko et al. 1976). The low cobalt content of this family of magnets is also of commercial interest.

One would expect the properties of Fe-Cr-Co to be optimized when $M_\alpha - M_{\alpha'}$ is maximized (nonmagnetic Cr-rich phase). However, Lorentz electron microscopy results suggest that some high-coercivity Fe-Cr-Co alloys are composed of two magnetic phases, so domain wall pinning may play a role in the coercivity mechanism.

Two permanent magnets based on FCC spinodal decomposition are known to exist: Cu-Ni-Fe (Cunife) and Cu-Ni-Co (Cunico). In both cases the Cu-poor phase (γ) is strongly magnetic while the Cu-rich phase (γ') is nonmagnetic. The very good ductility of these alloys allows strong texture to be developed by uniaxial deformation (Kikuchi and Ito 1972).

13.3 HEXAGONAL FERRITES AND OTHER OXIDE MAGNETS

Hard ferrites constitute the major fraction of world tonnage in permanent magnets. They are used where energy per unit weight and cost are important considerations.

The hexagonal ferrites based on $BaO{\cdot}6\,[Fe_2O_3]$ have the magnetoplumbite structure (Kojima 1982), which is given the notation BaM and includes PbM and SrM. There are several compounds between BaO and Fe_2O_3, but the 1:6 ratio is the most important for magnetic purposes. The hexagonal ferrites have a strong uniaxial anisotropy, $K_u = 3 \times 10^5\ \mathrm{J/m^3}$, with the preferred axis of magnetization along the hexagonal c axis. Because their saturation magnetization is small, $4\pi M_s \approx 4$–$5\ \mathrm{kG}$, their c-axis anisotropy is sufficient to overcome the magnetostatic energy, $2\pi M_s^2 \approx 8 \times 10^4\ \mathrm{J/m^3}$, and give perpendicular magnetization in either the flat hexagonal platelets that can be grown from solution (Fig. 13.15) or in (0001) textured thin films. Consequently, the hexagonal ferrites have received some attention also as potential perpendicular recording media. The smooth surface, nonmetallic electrical properties, superior corrosion resistance, and high volume density of magnetic material achievable in thin films give them some advantages over tape-cast media for high-density recording. The small saturation magnetization of BaM or SrM is a disadvantage in many applications.

The phase diagram of the BaO–Fe_2O_3 system in air ($p_{O_2}) = 0.2\ \mathrm{atm}$) is shown in Figure 13.16 (Goto and Takada 1960); there is still a difference of

Figure 13.15 Micrograph showing the hexagonal platelet morphology of BaM grown from a solution of 88% Fe_2O_3 and 12% BaO. The hexagonal c axis is perpendicular to the plates (courtesy of J. Adair).

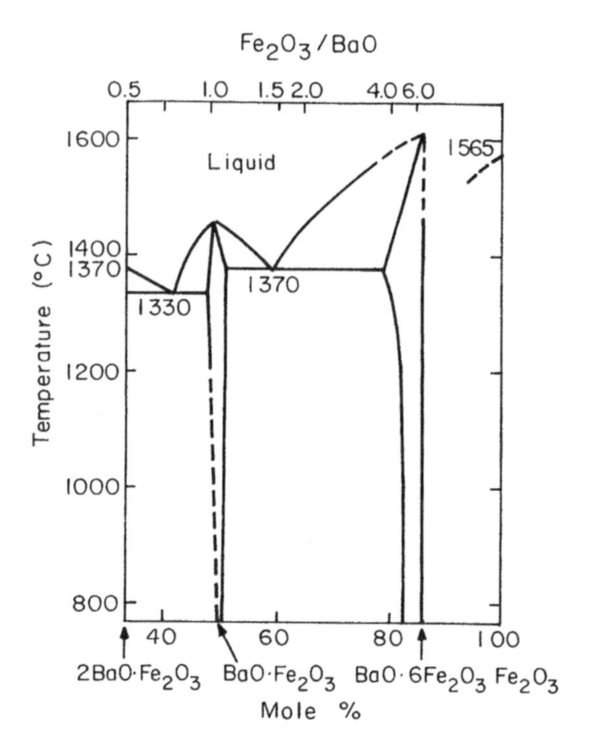

Figure 13.16 Phase diagram of the $BaO-Fe_2O_3$ system. In the solid-phase regions the oxygen partial pressure is 0.2 atm, and in the liquid phase region it is 1 atm [After Goto and Takada (1960). Reprinted with permission of The American Ceramic Society, P.O. Box 6136, Westerville, Ohio 43086-6136. Copyright (1960) by The American Ceramic Society. All rights reserved.]

opinion on solubility limits of some phases. Note the narrow range of solubility of BaM. It is often formed with an admixture of the 1:1 phase or of Fe_2O_3. The $SrO-Fe_2O_3$ and $PbO-Fe_2O_3$ phase diagrams are similar to that of $BaO-Fe_2O_3$.

The crystal structure of the magnetoplumbites is shown in perspective in Figure 13.17. There are two formula units per unit cell. The oxygen ions (open circles, A, B, and C layers) form an HCP lattice with stacking sequence ABAB... or ACAC.... One of the oxygen ions in every five layers is replaced by the divalent cation Ba^{2+}, Sr^{2+}, or Pb^{2+} (solid circle). The iron ions occupy octahedral, tetrahedral and trigonal interstitial sites as described in detail by Kojima (1982).

The magnetic moments of the iron ions (ideally Fe^{3+} has a moment of $5\,\mu_B$/ion) lie along the c axis and are coupled to each other by a superexchange interaction. As with the spinel ferrites, the Fe-O-Fe bonds that are closer to 180° are more strongly antiferromagnetic; some of the weaker antiferromagnetic couplings are forced to be parallel. The unit cell has 24 Fe^{3+} ions, 16

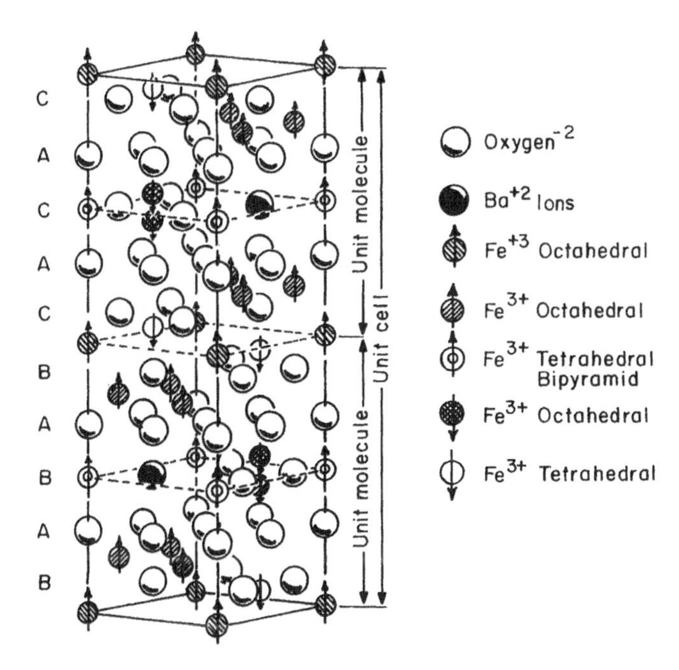

Figure 13.17 The hexagonal crystal structure of BaO-6Fe_2O_3. [Adapted from Kojima (1982).]

having one orientation and 8 having the other. Thus the magnetic moment can be estimated to be 20 μ_B/formula unit or 40 μ_B per unit cell, corresponding to a saturation induction at 0 K of $\mu_0 M_s = 0.66$ T ($4\pi M_s = 6.6$ kG). The saturation induction at room temperature is quoted as 0.475 T.

The magnetic properties of BaM, SrM, and PbM differ little from each other (Table 13.2). The magnetization–temperature curves for BaM and SrM are nearly linear as a result of the ferrimagnetic ordering (Fig. 13.18).

The anisotropy field for hard-axis magnetization, $H_a = 1.2 \times 10^6$ A/m (15 kOe), suggests an anisotropy energy density of 2.5×10^5 J/m^3. The temperature dependence of the measured magnetic anisotropy is also shown in Figure 13.18 as a dashed line. The temperature-dependent behaviors shown for

TABLE 13.2 Some Fundamental Physical Properties of Ba, Sr, and Pb Hexaferrites

	Lattice Constants (nm)		Mass Density	$\mu_0 M_s$(RT)	T_c
	a	c	(g/cm^3)	(T)	(K)
BaM	0.589	2.32	5.3	0.48	740
SrM	0.587	2.31	5.11	0.48	745
PbM	0.588	2.30	5.68	0.43	725

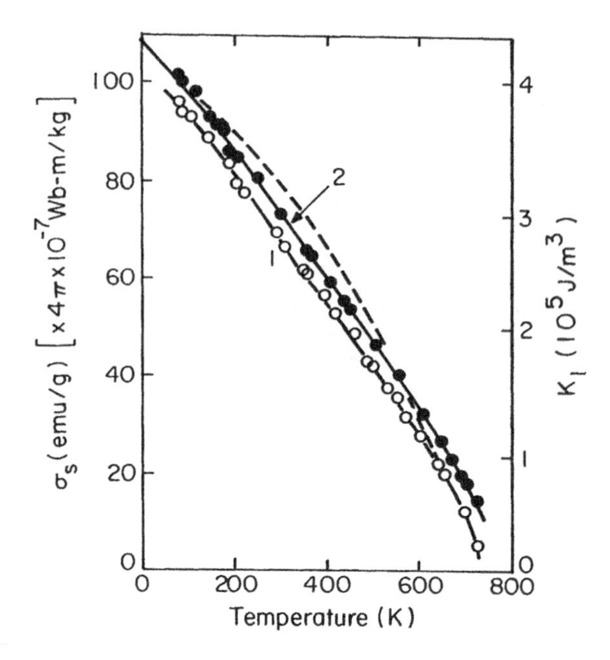

Figure 13.18 Specific magnetization versus temperature for BaO-6F$_2$O$_3$(1) and SrO-6Fe$_2$O$_3$(2); the dashed line shows the temperature dependence of K_1, which is similar for BaM and SrM (Shirk and Buessem 1969).

M_s and K_u do not correlate in the straightforward way predicted by Callen and Callen (1966) for ferromagnetic materials, $K(T)/K(0) = [M(T)/M(0)]^{l(l+1)/2}$, where $l = 2$ for a uniaxial structure. The hexagonal magnetostriction coefficients of BaM range from 11 to 15×10^{-6}.

The anisotropy of hexagonal ferrites has its origin in the crystal structure. The coercivity of the magnetoplumbites is limited by nucleation; once a domain wall exists in a particle, it moves with ease through the particle under an applied field. However, wall motion does not appear to propagate from grain to grain. Thus, the coercivity of hexagonal ferrites can be described by the mechanism of single-domain particle magnetization with the anisotropy provided not by particle shape as in the Alnicos but purely by magnetocrystalline anisotropy. In fact, the particle shape detracts from the effectiveness of the crystal anisotropy, yielding, for a random distribution of particles

$$_iH_c \approx 0.48\left(\frac{2K_u}{\mu_0 M_s} - NM_s\right) \approx 4 \times 10^5 \frac{A}{m} \tag{13.4}$$

This is essentially an angle-averaged form of the Stoner–Wohlfarth result, Eq. (9.16), with magnetocrystalline anisotropy replacing shape anisotropy. Because the intrinsic coercivities observed usually fall below this value, various modes of incoherent rotation are believed to play a role. Such effects are clearly more significant in larger particles.

Figure 13.19 shows the domain pattern of a thin plate of barium ferrite grown on a single-crystal substrate. The field of view is $90 \times 75\ \mu m$ on the basal plane. The demagnetized state (Fig. 13.19*a*) is characterized by an interpenetrating maze pattern consisting of equal areas of magnetization parallel and antiparallel to [0001]. At a field of 3.6 kOe parallel to the c axis (Fig. 13.19*b*), the pattern is dominated by magnetization in one direction. Increasing the field to 3.9 kOe (Fig. 13.19*c*) causes the magnetization to break up into an array of magnetic bubbles ($-\boldsymbol{M}$) in a matrix of positive magnetization.

Figure 13.20 shows the demagnetization curves of two anisotropic (1 and 2) and one isotropic (3) grades of commercial barium ferrite. The dashed lines show the Stoner-Wohlfarth (SW) model curves for isotropic and oriented materials using representative material parameters. The fact that the oriented ferrites show lower remanence than the oriented SW calculation reflects the

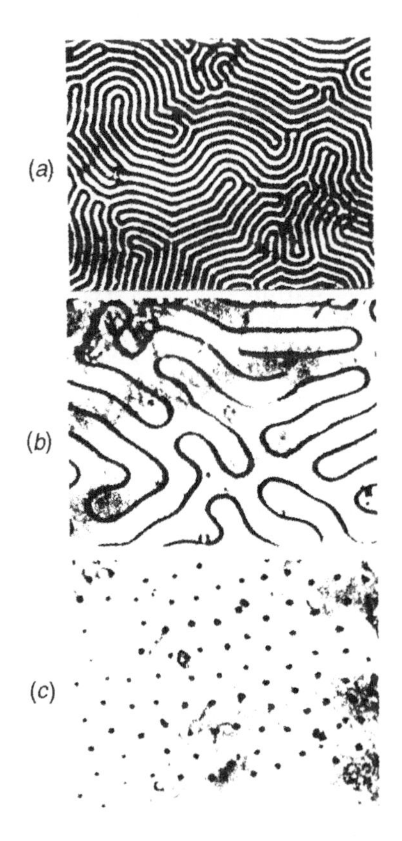

Figure 13.19 Magnetic domain pattern on the basal plane of hexagonal barium ferrite particle taken by Faraday rotation technique: (*a*) Demagnetized state; (*b*) $H = 3600$ Oe; (*c*) $H = 3900$ Oe (Kojima and Goto 1965).

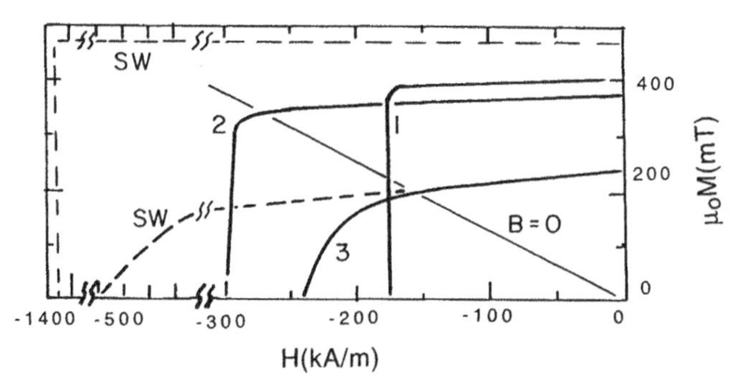

Figure 13.20 Demagnetization curves for two anisotropic (1, 2) and one isotropic commercial grade of hard ferrite. The dashed lines are the predictions of the Stoner–Wohlfarth (SW) model for isotropic and fully oriented materials using $\mu_0 M_s = 0.47\,\mathrm{T}$ and $K_1 = 3.46 \times 10^5\,\mathrm{J/m^3}$ [After Stäblein (1982).]

residual dispersion in *c*-axis orientation of the real material. The shortfall of the intrinsic coercivity in both isotropic and anisotropic cases relative to the SW model indicates the presence of mechanisms other than coherent rotation.

Table 13.3 summarizes typical values of important parameters for isotropic and anisotropic hexagonal ferrites. The two columns after ${}_iH_c$ give the fractional change in magnetization and intrinsic coercivity with temperature. These parameters are important for many applications.

The manufacture of hexagonal ferrites, detailed by Stäblein (1982), is typical of ceramic materials. The starting materials (generally oxides and carbonates such as $BaCO_3$, $SrCO_3$, PbO, and α-Fe_2O_3) are mixed and reacted in the solid state in air at temperatures between 1000 and 1300°C (this process is called *calcination*). The resulting hexaferrite is recrushed and pulverized to a particle size suitable for the next steps, which depend on the grade of magnet desired. These steps are either (1) field alignment of fine single-domain particles in an aqueous suspension as the slurry is cast and pressed prior to sintering, (2) dry compression in a magnetic field prior to sintering if optimal particle alignment is not essential, or (3) compaction of the unaligned particles for bonding or sintering. Fine particle size (optimally $< 1.0\,\mu\mathrm{m}$) ensures more single-domain

TABLE 13.3 Magnetic Properties of Isotropic and Anisotropic BaM

Grade	$(BH)_{max}$ (kJ/m^3)	B_r (T)	${}_BH_c$ (kA/m)	${}_iH_c$ (kA/m)	$\Delta M_s/M_s\Delta T$ (%/K)	$\Delta_i H_c/{}_iH_c\Delta T$ (%/K)	Mass Density (g/cm^3)
Isotropic	6.5–9	0.19–0.22	125–145	210–270	−0.2	0.2–0.5	4.6–5.0
Anisotropic	20–30	0.32–0.4	125–250	130–340	"	"	"

Source: After Stäblein (1982).

particles (and hence higher H_c; in addition, fine particle size enhances sintering which is driven by surface energy reduction). The addition of 0.5% SiO_2 aids sintering and results in greater products of B_r times ${}_iH_c$.

Hexagonal ferrites remain stable in air to temperatures well over 1400°C. At reduced oxygen pressure the stable temperature range decreases. Because the ions in hexagonal ferrites are in their highest states of oxidation, these materials have a significant advantage in terms of chemical stability over metallic magnets.

13.4 RARE EARTH-TRANSITION METAL INTERMETALLICS

Hard magnets based on $SmCo_5$ boast the highest uniaxial anisotropies of any class of magnet, $K_u \approx 10^7$ J/m^3. On the other hand, phases based on Sm_2Co_{17} exhibit higher flux density and Curie temperature. The more recently developed magnets based on $Fe_{14}Nd_2B$ exhibit the highest energy products achieved so far in permanent magnets. Before discussing these materials, four important aspects of magnetism in rare-earth metals, alloys, and intermetallic compounds are reviewed because they will be needed in Sections 13.4.1 and 13.4.2.

1. The variation across the lanthanide series of the quantum numbers, the calculated effective moments $g\mu_B\sqrt{J(J+1)}$, saturation moments, $\mu_m = g\mu_B m_j$, and of the observed moments are shown in Table 13.4. Except for Gd, the R species are characterized by $L \geqslant S$. Hund's third rule states that $\boldsymbol{L}$ and $\boldsymbol{S}$ combine subtractively in the first half of the $4f$ series. For light R species the net moment is reduced and is antiparallel to $\boldsymbol{L}$ (because the negative electronic charge makes $\mu_B < 0$). $\boldsymbol{L}$ and $\boldsymbol{S}$ combine additively in the second half of the series, so there the net moment is enhanced and is antiparallel to both $\boldsymbol{L}$ and $\boldsymbol{S}$. The effective magnetic moments of the heavy R species reach a higher maximum value than do those of the light R species. The ordering temperatures are low or negative for the light R elements; these materials show spiral or canted spin structures rather than ferromagnetism. Figure 13.21 shows the variation of the measured effective moment and the magnetic ordering temperatures across the rare-earth metal series. The variation of the rare-earth moments across the series is fairly well described by the quantum numbers in Table 13.4. The variation of the magnetic ordering temperatures is systematic in the heavy R metals, as will be described below.

2. The Curie temperatures of the rare-earth metals and of various rare-earth alloys vary in a systematic way with the de Gennes factor, $G = (g-1)^2 J(J+1)$ as shown in Figure 13.22 (Chikazumi 1997). The de Gennes factor has its basis in the molecular field expression for the Curie temperature [Eq. (4.21)], proportional to $S(S+1)$; the term $(g-1)^2$ in the de Gennes factor singles out the spin ($g = 2$) part of the moment from $J(J+1)$. Thus the Curie or Néel temperatures scale with the spin part of the magnetic moment,

TABLE 13.4 Electronic Structure as well as the Calculated and Experimental Values of the Effective and Saturation Magnetic Moment in Bohr Magnetons for Rare-Earth Ions and Metals

	Electronic State ($^{+}3$ Ion)				Effective Moment (μ_B)			Saturation Moment (μ_B)	
					Theory	Observed			
Element	n_{4f}	S	L	J	(Hund)[a]	$^{+}3$ Ion	Metal	gJ	Observed
Y	0	0	0	0	0	0	0	0	0
La	0	0	0	0	0	0	0	0	0
Ce	1	1/2	3	5/2	2.54	2.52	2.51	2.14	—
Pr	2	2/2	5	8/2	3.58	3.60	2.56	3.2	—
Nd	3	3/2	6	9/2	3.62	3.50	3.3–3.7	3.27	—
Pm	4	4/2	6	8/2	3.68	—	—	2.4	—
Sm	5	5/2	5	5/2	0.85	—	1.74	0.72	—
Eu	6	6/2	3	0	0.0	—	8.3	0.0	—
Gd	7	7/2	0	7/2	7.94	7.8	7.98	7.0	7.55
Tb	8	6/2	3	12/2	9.72	9.74	9.77	9.0	9.34
Dy	9	5/2	5	15/2	10.64	10.5	10.65	10.0	10.2
Ho	10	4/2	6	16/2	10.60	10.6	11.2	10.0	10.34
Er	11	3/2	6	15/2	9.58	9.6	9.9	9.0	8.0
Tm	12	3/2	5	12/2	7.56	7.1	7.6	7.0	3.4
Yb	13	1/2	3	7/2	4.53	4.4	0	4.0	—
Lu	14	0	0	0	0.0	0	0	0	—

[a]Hund's rule moment is $g\sqrt{J(J+1)}$.

Source: Data taken from Chikazumi (1997).

as well as the square of the total angular momentum. The variation of the Curie temperature with R species for the Co_5R, $Co_{17}R_2$, $Fe_{14}R_2B$, and R_2Fe_{17} intermetallic series are shown in Figure 13.23. The variation of T_C with R species in the two iron-based series is well described by the effective spin or de Gennes factor, $G = (g - 1)^2 J(J + 1)$, which decreases on either side of Gd in the series. Another remarkable feature of this figure is the general decrease in T_C as Fe concentration in the compounds increases. The opposite is true in the analogous CoR compounds. In simplest terms, this trend is consistent with a Bethe–Slater curve (Fig. 5.3) interpretation that suggests a stronger tendency for antiferromagnetic Fe—Fe exchange at Fe—Fe distances smaller than about 2.5 Å. This is consistent with the observation that pressure generally causes T_C to decrease in iron-based compounds.

3. For rare earth–transition metal (R–TM) intermetallics, it is important to understand the nature of the coupling between the rare-earth moment ($\mu_R = g\mu_B m_J$) and that of the transition metal species ($\mu_{TM} = g\mu_B m_s$). The magnetic moments of transition metals are observed to couple ferromagnetically with light rare-earth moments ($\boldsymbol{J} \cdot \boldsymbol{s} > 0$) and antiferromagnetically with

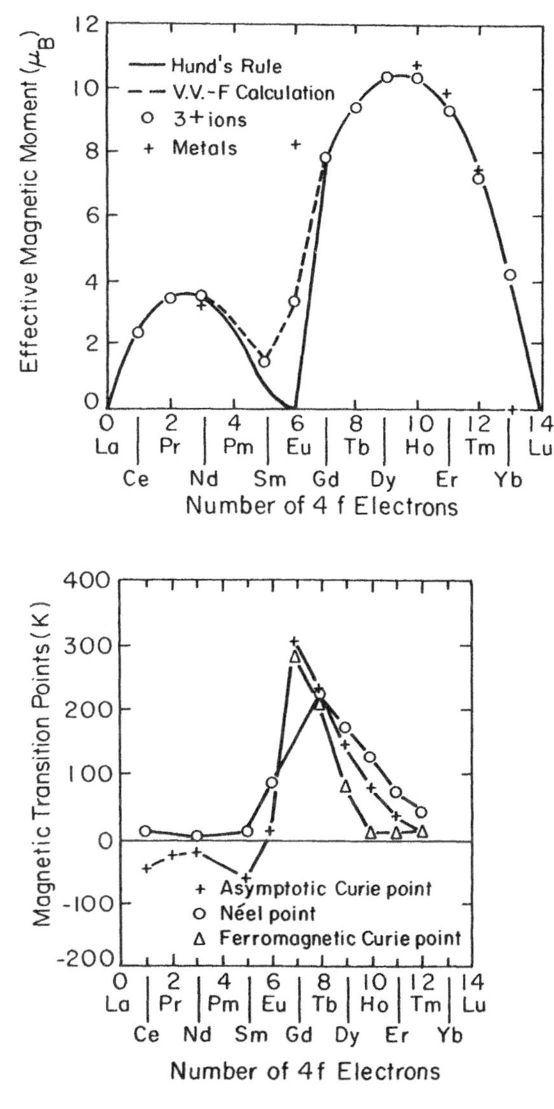

Figure 13.21 Above, variation of the Hund effective moment and the moments observed for +3 ions and metals of the rare-earth (R) series; below, variation of magnetic ordering temperatures for R metals. [After Chikazumi (1997).]

heavy rare-earth moments ($\boldsymbol{J} \cdot \boldsymbol{s} < 0$). Recall that in the first half of a shell, $J = L - S$ (total moment and spin moment are antiparallel) and in the second half, $J = L + S$. The spin–spin coupling between R and TM species is always antiferromagnetic (Fig. 13.24, left). It has been suggested that R–TM coupling is due to the antiferromagnetic conduction-electron-mediated

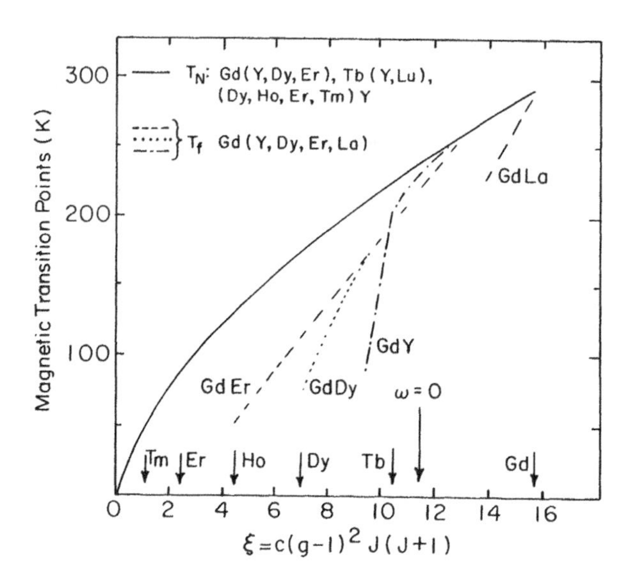

Figure 13.22 Dependence of magnetic transition temperatures of R metals and R–R alloys on the de Gennes factor, $g = (g - 1)^2 J(J + 1)$ [after Chikazumi 1997).]

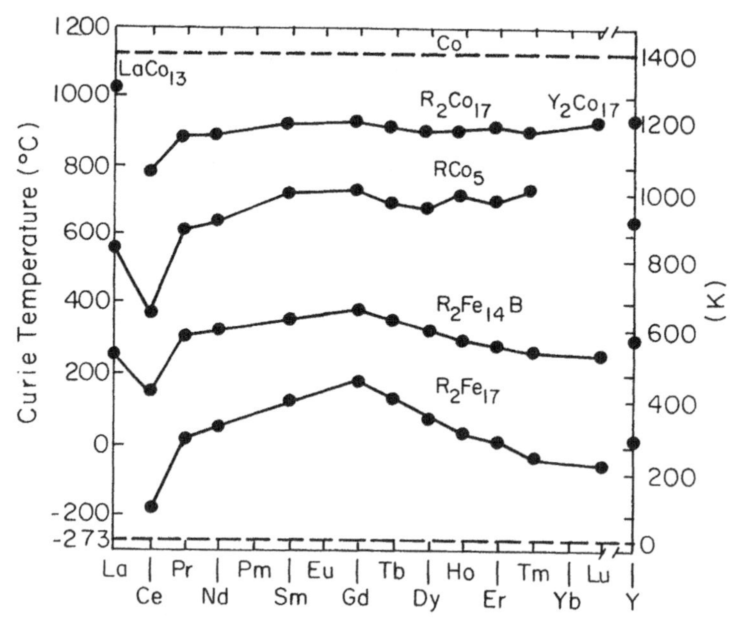

Figure 13.23 Variation of the Curie temperature with R species in four series of R-TM intermetallics of importance as permanent magnets [after Strnat (1988).]

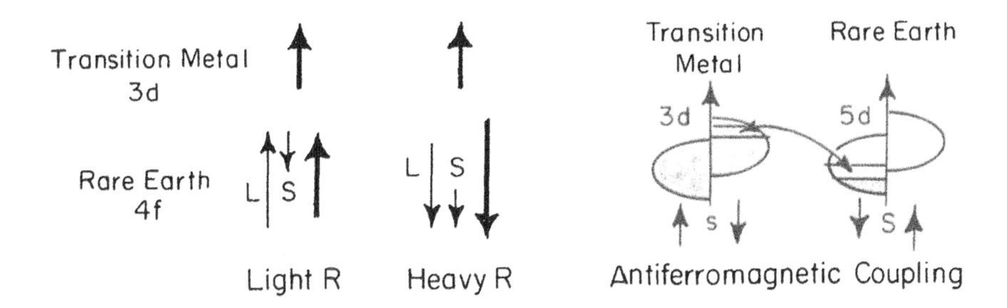

Figure 13.24 Simplified schematic representation of spin and angular momentum coupling at rare-earth site and antiferromagnetic exchange coupling between R and TM spins. Right, schematic band structure that accounts for antiferromagnetic R–TM spin coupling.

exchange as described by the RKKY model. However, the applicability of the RKKY model to this problem is questionable because variations in R–TM separation do not change the sign of the coupling. Alternatively, Campbell (1972) focuses on the $5d^2$ conduction electrons of the rare-earth (whose spin is always parallel to that of the $4f$ electrons) and their interaction with the

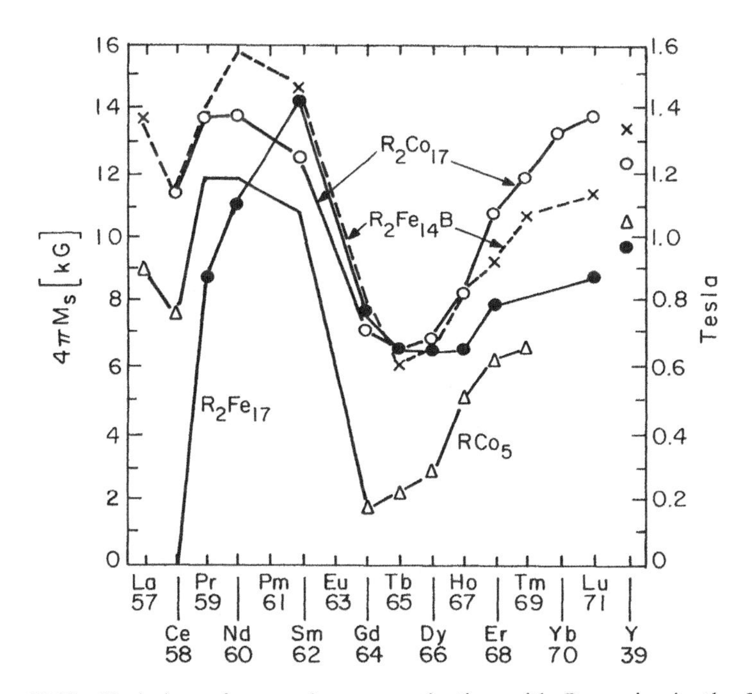

Figure 13.25 Variation of saturation magnetization with R species in the R–TM intermetallic series of Figure 13.23 (Strnat 1988).

symmetry-compatible $3d^n$ electrons of the transition metal. Exchange between these two sets of d states is invariably antiferromagnetic with respect to the d electrons involved; the mechanism is analogous to the electron hopping mechanisms in superexchange. There are only majority-spin electrons in the R $5d$ orbitals; there are only minority-spin holes in the TM $3d$ orbitals, and spin is conserved in the strongest hopping process (Fig. 13.24, right). Because the $4f$–$5d$ exchange at the R site is ferromagnetic, the antiferromagnetic $5d$–$3d$ interaction explains the ferromagnetic coupling of light R to TM moments and vice versa for heavy R species. The net result of R-TM exchange coupling is generally larger magnetic moments for intermetallics of light rare earths and transition metals. The $4\pi M_s$ data in Figure 13.25 for four R–TM intermetallics of importance as permanent magnets bear this out. In the first half of each series the relatively small R moments couple ferromagnetically with the TM moments. In the second half of the series, larger R moments couple antiferromagnetically to the TM moments. The net moment of each compound of course depends on the magnetic moment of each species, on the concentration of each species in a given alloy, and on T/T_C.

4. Finally, for permanent magnet applications, it is important that the crystal system be uniaxial rather than cubic and that the anisotropy give rise to an easy *axis* ($K_u > 0$) as opposed to an easy *plane* ($K_u < 0$) (see Chapter 6). The latter leads to easy demagnetization and low coercivity because the sample can demagnetize without M rotating to a hard direction. The rare-earth metals show complex spiral spin structures in many cases as well as easy-axis magnetization (Gd from 240 to 293 K), and easy plane magnetization (Gd from 170 to 220 K or Tb and Dy up to 222 and 83 K, respectively) (Chikazumi 1997, p. 186). Thus it is not surprising that their intermetallic alloys also show a rich variety of spin structures. This important effect limits the range of alloy substitutions available in designing new R–TM permanent magnets. For most of the RCo_5 intermetallics with light R species (R = Y, La, Ce, Pr, Nd, Sm), $K_1 > 0$ (easy axis). The situation is more complex in the R_2Co_{17} and in the $Fe_{14}R_2B$ magnets, as will be seen.

13.4.1 Cobalt/Rare-Earth Magnets

CoR magnets are based mainly on the two phases RCo_5 and R_2Co_{17}. Although the first structure is more widely known, most CoR magnets are, in fact, multiphase composites of these two structures and sometimes other phases. The Co-Sm phase diagram is shown in Figure 13.26. Co_5Sm is stable only above 805°C. Thus, its use is limited to temperatures below which kinetics preclude transformation to the more stable Sm_2Co_7, Sm_2Co_{17}, and Co phases. The hexagonal $CaCu_5$ structure of $SmCo_5$ is shown in Figure 13.27. The R species form hexagonal nets with a smaller Co hexagon in the same plane and similar hexagonal Co layers (rotated by $\pm 30°$) are stacked between the R nets.

This uniaxial structure together with the orbital angular moment of the R species is the source of the magnetocrystalline anisotropy of the RCo_5 system. $SmCo_5$ exhibits a magnetocrystalline anisotropy of $10^7\,J/m^3$. Figure 13.28 shows the easy- and hard-axis magnetization curves for the two principal Co-Sm magnets an $Fe_{14}Nd_2B$. Single-phase $SmCo_5$ can exhibit room temperature intrinsic coercivities, ${}_iH_c$, of 3.2 MA/m (40 kOe) and maximum energy products of over 200 kJ/m^3 (24 MG · Oe). This, combined with their relatively high Curie point ($T_C = 685°C$), makes them suitable for a wide range of applications. Iron substitutions for cobalt lead to a change from easy *axis* magnetization to easy *plane* magnetization in RCo_5.

Because of the large magnetic anisotropy of $SmCo_5$, a 180° Bloch wall in this material should have a width of only 3.1 nm. Further, the domain wall energy density is 40 mJ/m^2 (40 erg/cm^2), 100 times that of a soft material. It is not surprising, therefore, that the magnetization process in single-phase RCo_5 intermetallics is limited by reversal domain nucleation (Livingston 1973). Once

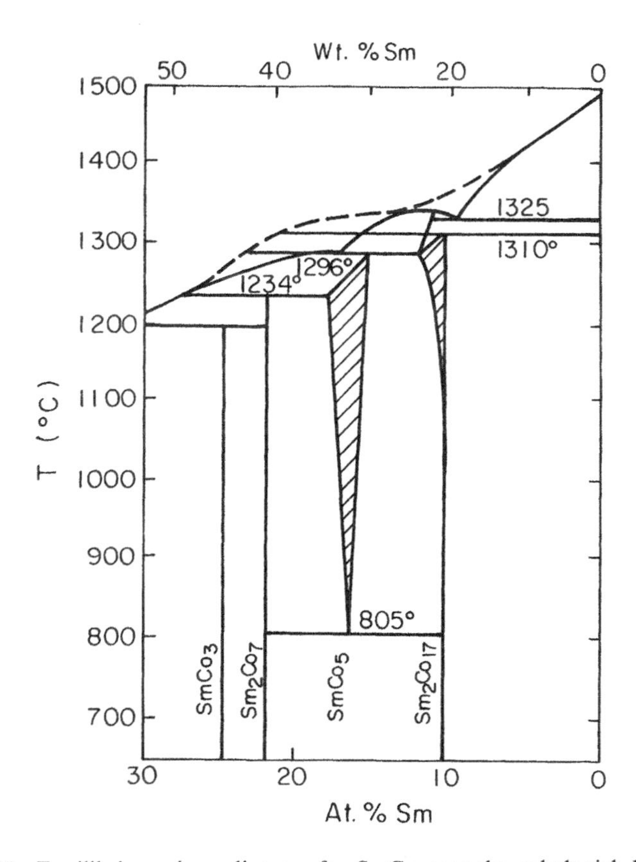

Figure 13.26 Equilibrium phase diagram for SmCo near the cobalt-rich limit [After den Broder and Buschow, 1977).]

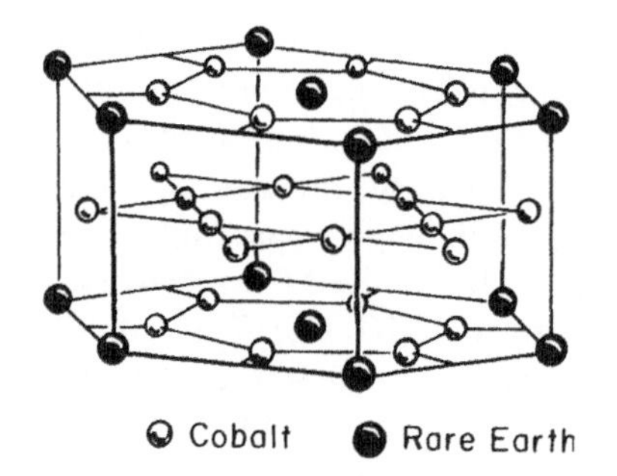

Figure 13.27 Hexagonal $CaCu_5$-type crystal structure of $SmCo_5$ (Barrett and Massalski 1980).

nucleation occurs, domain walls move relatively easily until they reach a grain boundary, as in Ba ferrites. Hence, initial efforts to produce cobalt/rare-earth magnets focused on the fabrication of single-domain $SmCo_5$ particles. It was discovered by Nesbitt et al. (1968) that small substitutions of Cu for Co would lead to the precipitation of a nonmagnetic phase which increased the coercivity. They showed that heat treatment of $R(CoCu)_5$ magnets results in precipitation of a dispersion of fine ($d \approx 10$ nm) second-phase, Cu-rich particles in a R_2Co_{17} matrix having a grain size of order 10 μm (Fig. 13.29). The coercivity mechanism becomes domain wall pinning on the small nonmagnetic $SmCu_5$ particles.

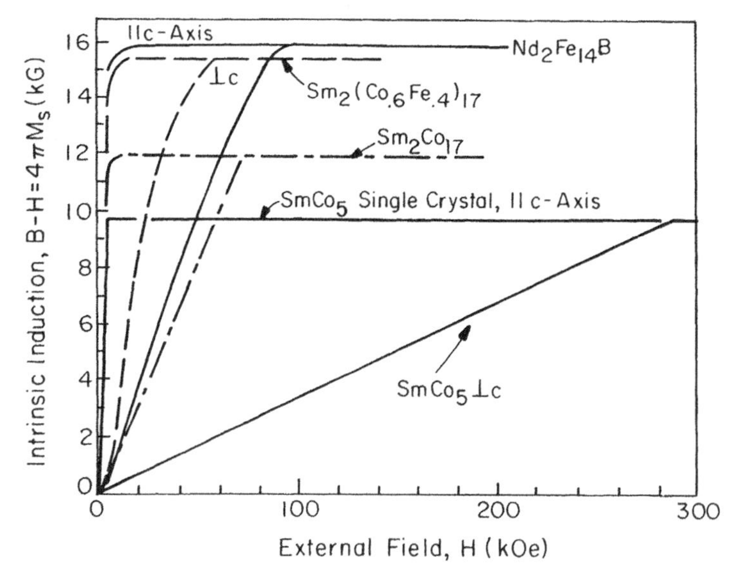

Figure 13.28 Easy-axis and hard-axis magnetization curves for RCo 1–5 and 2–17 compounds. [After Strnat 1988).]

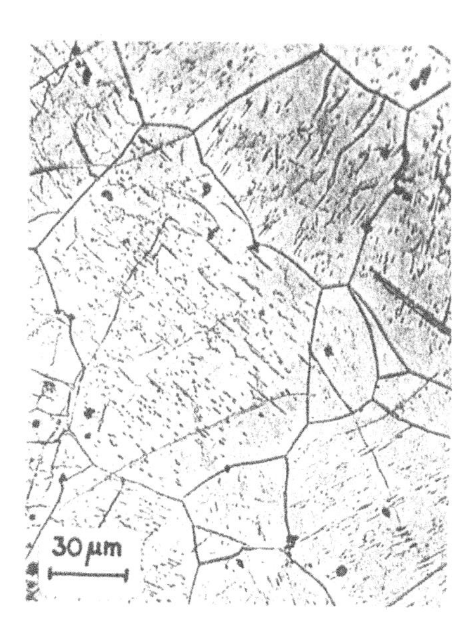

Figure 13.29 Precipitation microstructures of 1–5-type RE–Co magnets, bulk-hardened with copper: $SmCo_{3.5}Cu_{1.0}Fe_{0.5}$ homogenized at 1100°C/3 h, quenched and aged at 525°C. The fine (10-nm) dark precipitates are platelets of Cu-rich, $Sm(CuCo, Fe)_5$. [After Strnat (1988)].

Tawara and Strnat (1976) showed that magnetic hardening by Cu precipitates could be extended in $Sm(CoCu)_x$ from $x = 5$ to $x \approx 7.2$. Further extension to $x \approx 8.5$ by inclusion of other transition elements led to the first practical R_2TM_{17} magnets (Nagel 1976).

The rhombohedral Th_2Ni_{17}-type crystal structure of R_2Co_{17} is shown in Figure 13.30. A similar hexagonal Th_2Zn_{17}-type also exists. These structures have the same cobalt hexagonal nets that occur in the RCo_5 compounds but fewer R atoms in the adjacent layers. Whereas R atoms are located at the centers of alternate hexagonal Co rings of the 1–5 compound (Fig. 13.27), a TM atom occupies an axial site in the midplanes between certain of these hexagonal rings (Fig. 13.30). The magnetocrystalline anisotropies of R_2Co_{17} phases ($3–4 \times 10^6$ J/m^3) are generally less than those of the corresponding 1–5 phases ($11–20 \times 10^6$ J/m^3), but the 2–17 saturation magnetizations and Curie temperatures are generally greater (cf. Fig. 13.28). See Table 13.5.

The rhombohedral crystal structure of the 2–17 compounds allows incorporation of greater Fe content than does the 1–5 hexagonal structure. Coercivities of order 800 kA/m (10 kOe) and energy products of order 240 kJ/m^3 (30 MG·Oe) have been reported in single-phase $Sm_2(CoFe)_{17}$-based alloys. Because the rhombohedral 2-17 phase has a lower anisotropy than the 1–5 phase, heat treatment and inclusion of nonmagnetic atoms such as Cu ($\approx$7 at%) and Zr ($\approx$2 at%) to promote optimal phase segregation are

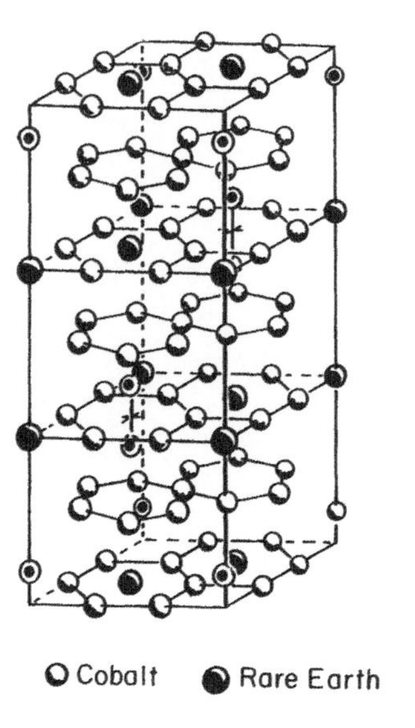

Figure 13.30 Rhombohedral, Th_2Zn_{17}-type crystal structure on which R_2Co_{17} magnets are based. [After Strnat (1988)].

generally used to achieve higher coercivities. The Curie point and saturation flux density are substantially improved by partial Fe, Zr, and Cu substitutions for Co. These substitutions also promote formation of a cellular microstructure based on a 2–17 rhombohedral phase, a 1–5 intergranular phase, and a Zr-rich platelet phase (Fig. 13.31). This cellular microstructure provides the pinning sites needed to obstruct domain wall motion in 2–17 magnets. In contrast, high coercivity in the 1–5 magnets is a result of inhibited nucleation.

The cellular microstructure of 2–17 magnets is generally sensitive to heat treatments; the maximum use temperature is limited to about 300°C. However, 2–17 permanent magnets of the modified composition $Sm_2(CoFeZrCu)_{17}$

TABLE 13.5 Comparison of Some Magnetic Properties for $SmCo_5$, $Sm_2(CoFe)_{17}$, and $Fe_{14}Nd_2B$ Permanent Magnets at 25°C

	$\mu_0 M_s$ (T)	T_C (°C)	K_u (MJ/m^3)	${}_iH_c$ (MA/m) Isotropic	${}_iH_c$ (MA/m) Aligned	$(BH)_{max}$ (MG·Oe) 2D	$(BH)_{max}$ (MG·Oe) 3D
$SmCo_5$	1.0	685–700	10	0.8–1	2.9	14–16	18–24
$Sm_2(CoFe)_{17}$	1.2–1.5	810–970	3.3	1–1.3	2.4	16–20	24–30
$Fe_{14}Nd_2B$	1.6	312	5	—	1.2–1.6	34–45	

Figure 13.31 Precipitation microstructure of bulk-hardened 2–17 magnets of the high-H_c type: $Sm(Co_{0.67}Fe_{0.23}Cu_{0.08}Zr_{0.02})_{8.35}$ fully heat treated to ${}_iH_c = 23$ kOe. The section shown contains the *c* axis as indicated; rhomboid cells of the twinned 2–17 matrix are seen. The boundary phase surrounding the cells is of 1–5 type, and thin bands of "*z* phase" are visible in the basal plane, crossing many cells and walls (Courtesy A. E. Ray.)

are attractive for high-temperature applications because of their high values of the ferromagnetic Curie point ($T_C = 810°C$), saturation magnetization ($\mu_0 M_0 = 1.5$ T), and, at 25°C, $(BH)_{max} = 24–30$ MG·Oe. A further advantage of this two-phase, 2–17 system when it incorporates a significant fraction of iron is reduced cost relative to Sm_2Co_{17}.

Single-phase $TbCu_7$-type thin film magnets of $Sm(CoFeCuZr)_7$ compositions can be sputter deposited to exhibit either high intrinsic coercivities or, as aligned films, to exhibit enhanced energy products. Coercivity in excess of 16 kOe at room temperature has been observed in fields of ± 18 kOe in single-phase $TbCu_7$-type films crystallized from amorphous deposits. Relatively thick sputtered films exhibit strong in-plane *c*-axis alignment for a wide range of film thicknesses up to 120 μm. An important feature of these in-plane *c*-axis $TbCu_7$-type films is that their magnetic properties appear to be insensitive to thermal treatments up to nearly 500°C (Cadieu 1995).

Table 13.5 compares some room temperature properties of Co_5Sm and Co_2Sm_{17} magnets with those of the $Fe_{14}Nd_2B$ class.

Figure 13.32 describes the change in preferred magnetization direction from easy axis to easy plane as various transition metals are substituted for Co in R_2Co_{17} magnets (Ray and Strnat 1972).

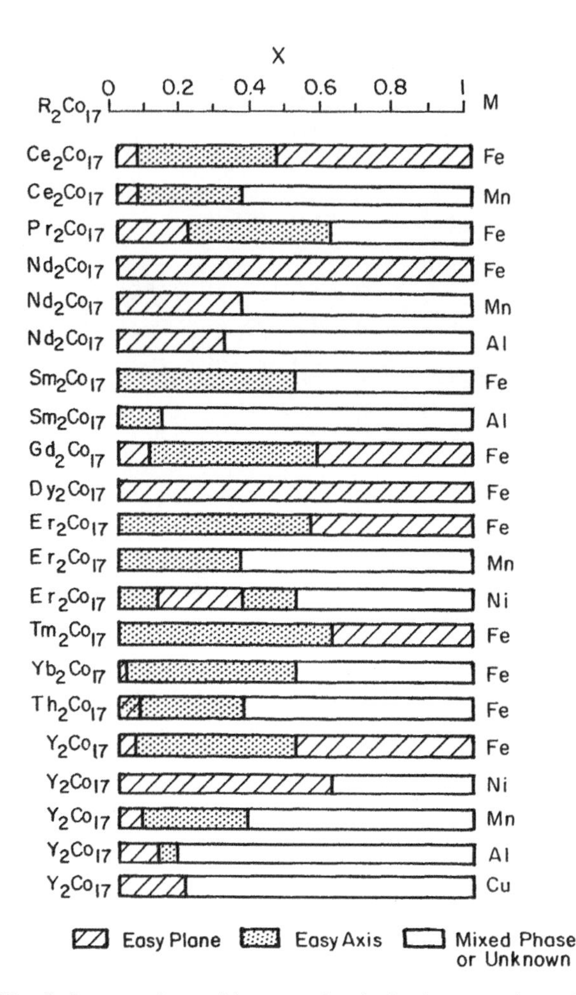

Figure 13.32 Influence of transition metal substitutions on the anisotropy (magnetic symmetry) of R_2Co_{17}-based intermetallics of composition $R_2(Co_{1-x}M_x)_{17}$. [After Strnat, (1988) and Ray and Strnat (1972).]

The 2–17 magnets are used in multiphase form to enhance coercivity by domain wall pinning. Without domain wall pinning, the coercivity would be relatively small. On the other hand, the stronger anisotropy of RCo_5 magnets makes nucleation-limited coercivity viable. Nucleation-limited coercivity would not be strong enough in a single-phase 2–17 magnet because of its larger magnetization and smaller anisotropy relative to $SmCo_5$.

Due to the large concentration of Co in even the two-phase Co-R magnets, their cost per kilogauss of saturation flux density remains very high. This provided the impetus for development of less expensive high-energy magnets based on iron.

13.5 RARE-EARTH INTERMETALLICS BASED ON $Fe_{14}Nd_2B_1$

Fe-Nd-B magnets represent an important class of rare-earth intermetallic compounds that advance the technology of permanent magnets beyond that of the Co-R magnets. Their development came as a result of the cost and limited world supply of cobalt.

The Fe-based rare-earth intermetallics were studied independently by Das and Koon (1981) while exploring rare earth additions to amorphous FeB-based alloys (FeLaB), by Croat et al. (1981) and Hadjipanayis et al. (1983). Commercial FeNdB magnets based on sintering (Sagawa et al. 1984) and melt spinning (Croat et al. 1984) were rapidly developed.

The attractive permanent-magnet properties of $Fe_{14}Nd_2B_1$ magnets arise from several factors:

1. The large uniaxial magnetic anisotropy ($K_u = +5 \times 10^6\ \mathrm{J/m^3}$) of this tetragonal phase.
2. The large magnetization ($B_s = 1.6\ \mathrm{T}$) owing to the ferromagnetic coupling between the Fe and Nd moments.
3. The stability of the 14–2–1 phase, which allows development of a composite microstructure characterized by 14–2–1 grains separated by nonmagnetic B-rich and Nd-rich phases, which tend to decouple the magnetic grains.

The tetragonal $Fe_{14}Nd_2B_1$ structure (Fig. 13.33) contains four formula units and has unit cell dimension of $a = 0.88\ \mathrm{nm}$ and $c = 1.22\ \mathrm{nm}$. There are six distinct Fe sites, two distinct Nd sites, and one B site. The unit cell consists of eight layers vertically stacked. All the Nd and B sites are found in layers at $z = 0$ and $\frac{1}{2}$. Most of the Fe atoms reside in puckered hexagonal nets between the Nd- and B-rich layers. The Fe net puckering is associated with an attractive Fe-B bonding that places the B at the center of a trigonal prism of Fe(e) and Fe(k) atoms. The tetragonal structure of $Fe_{14}R_2B_1$ phases is related to the hexagonal $CaCu_5$ structure of $SmCo_5$ (Fig. 13.27) via the hexagonal nets (Givord et al. 1984). Also, the Fe(j) sites above and below the hexagonal groups of Fe are very similar to the TM(c) sites in R_2TM_{17} compounds (Fig. 13.30). The Fe(j) sites are the most highly coordinated by transition metal atoms and, not surprisingly, show the largest magnetic moments (Herbst et al. 1985). Details of the structure and local atomic environments are found in the review by Herbst (1991).

The 14–2–1 unit cell shows the usual lanthanide contraction when Nd is replaced by other R species.

The saturation magnetization versus temperature of the $Fe_{14}R_2B$ series (Hirosawa et al. 1986) is shown in Figure 13.34. The data are separated into two groups, the light rare-earth and the heavy rare-earth species. The $M_s(T)$ behavior in the former appears to have the normal shape of a Brillouin

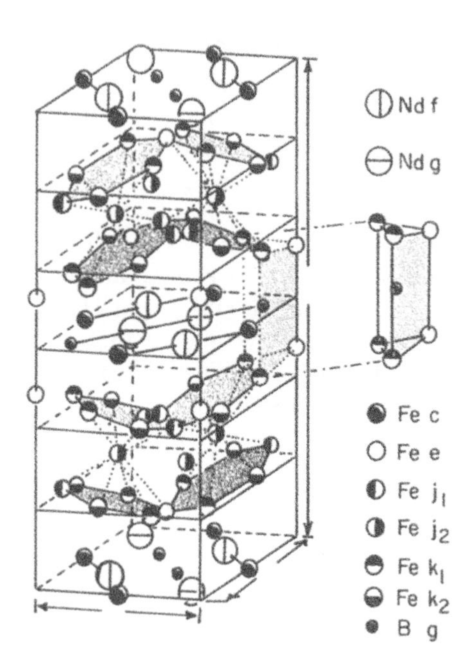

Figure 13.33 Tetragonal unit cell of $Nd_2Fe_{14}B$, the prototypical structure of the $R_2Fe_{14}B$ compounds; the *c*/*a* ratio in the figure is exaggerated to emphasize the puckering of the hexagonal iron nets (Herbst et al. 1984).

function. In contrast, the $M_s(T)$ behavior of the late R species resembles the $M(T)$ form characteristic of ferrimagnets for which the magnetization of the higher T_C sublattice (Fe in this case) is greater than that of the lower T_C sublattice (R). Thus, there are no compensation temperatures in the $Fe_{14}R_2B_1$ series (see Figs. 4.9 and 4.11). To interpret these data, recall first that the magnetic moments of the six iron sites (averaging 2.1 μ_B/Fe) are ferromagnetically coupled to each other. Figure 13.34 then suggests a ferromagnetic coupling of the net TM moment to the R moment in the first half-series and an antiferromagnetic coupling in the second half-series. The same was found to be true of CoR intermetallics and it is generally the case for other rare-earth transition metal intermetallics.

The experimentally determined moments at the R sites in $Fe_{14}R_2B$ compounds correlate very well with the theoretical saturation moments, $g\mu_B m_J$ (Herbst 1991). The Curie temperatures of $Fe_{14}R_2B$ compounds shown in Figure 13.25 vary systematically with the number of 4*f* electrons on the R species. Here the T_C data for the $Fe_{14}R_2B$ series are compared with that for several other TM-R intermetallics.

Fuerst et al. (1986) have shown that a molecular field analysis describes very well the temperature dependence of the net iron sublattice moment, 14 μ_{Fe}, and the net Nd sublattice moment, 2 μ_{Nd}, and their ferromagnetic sum, the

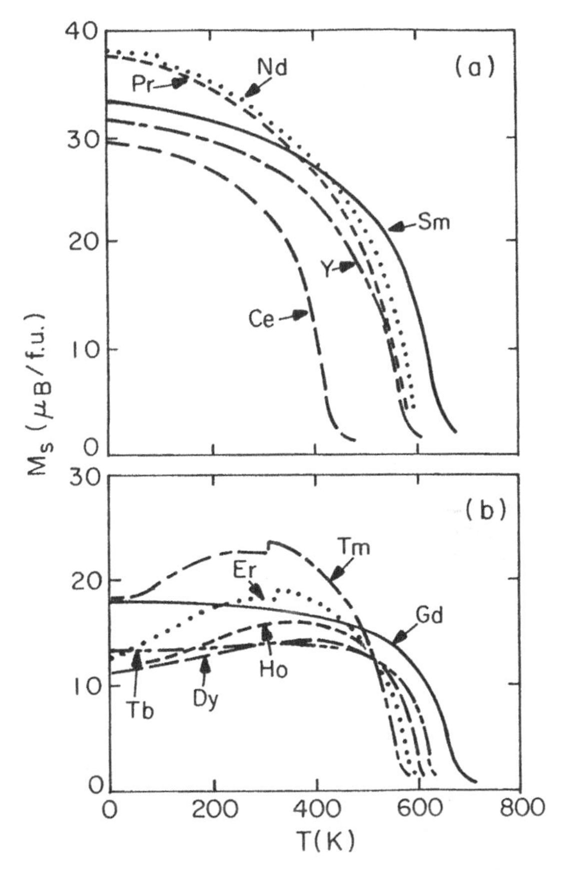

Figure 13.34 Saturation magnetization M_s versus temperature T for $R_2Fe_{14}B$ compounds: (*a*) light rare earths (R and Fe magnetic moments ferromagnetically coupled); (*b*) heavy rare earths (antiferromagnetic RFe coupling) [Adapted from Hirosawa, et al. (1986).]

measured $M_s(T)$. The molecular fields seen at the Nd(R) and Fe(F) sites can be represented as in Eq. (4.27):

$$\begin{aligned} H_R &\propto 2\lambda_{RR}\mu_R(T) + 14\lambda_{RF}\mu_F(T) \\ H_F &\propto 14\lambda_{FF}\mu_F(T) + 2\lambda_{RF}\mu_R(T) \end{aligned} \tag{13.5}$$

The proportionality constant is $N_A\mu_B\rho/A$, where N_A is Avogadro's number, μ_B is the Bohr magneton, ρ is the mass density, and A is the weight per formula unit. The temperature dependences of μ_F and μ_R are assumed to follow their respective Brillouin functions. The dimensionless molecular field coefficients, λ_{ab}, were found by Fuerst et al. (1986) to be $\lambda_{FF} = 5.9 \times 10^3$, $\lambda_{FR} = 2.2 \times 10^3$, and $\lambda_{RR} = 0.33 \times 10^3$. This hierarchy of molecular field coefficient strengths is typical of most R–TM intermetallics.

The strong magnetic anisotropy of the $Fe_{14}R_2B$ compounds is a function of the low symmetry of the structure as well as the orbital angular momentum of the R ions and their low-symmetry atomic environments. However, it is noteworthy that in those members of the series that have no magnetic moment at the R site (viz., R = La, Lu, Ce, Y, Th), the anisotropy is still very strong: K_1 is of order 10^6 J/m^3 (H_a is of order 25 kOe). This value should be compared with the anisotropy of α-Fe, $K_1 = 5 \times 10^4$ J/m^3 or H_a of order 500 Oe (Fig. 6.1). This difference points to the fact that magnetic anisotropy of the Fe sublattice depends on crystal field symmetry at the Fe sites (low in the iron hexagonal nets) as well as on the orbital angular momentum of the magnetic species (weak here). Herbst (1991) also points out that the uniaxial anisotropy in this subset of $Fe_{14}R_2B$ compounds is positive; that is, an easy c-axis direction is preferred by the $3d$ moment (perpendicular to the mean plane of the hexagonal units).

Of the other members of the series, those for which R = Pr, Gd, Tb, Dy show easy axis magnetization and R = Sm shows easy basal plane magnetization over the entire temperature range below T_C. The compounds based on R = Nd, Ho, Er, Tm, Yb transform from easy plane (or canted moments for the first two ions) to easy-axis magnetization at a spin-reorientation temperature below T_C.

This behavior can be understood in terms of the Stevens factor α_J introduced in Chapter 6 [Eq. (6.12)]. In Chapter 6, it was shown that the crystal field splitting parameter D of R ions oscillates from positive to negative values (α_J oscillates from negative to positive values) with increasing atomic number across each half of the R series (Fig. 6.17). Positive and negative values of α_J imply prolate and oblate $4f$ charge distributions, respectively. Easy-axis magnetization is generally associated with oblate, $\alpha_J < 0$ ($D > 1$) $4f$ orbitals. Thus the α_J values of the $4f$ ions suggest a mechanism by which the Fe anisotropy couples to the $4f$ orbital shapes: the $4f$ magnetization prefers to lie perpendicular to the plane of neighboring, oblate $4f$ orbitals.

The spin reorientation transition in $Fe_{14}Nd_2B$ at 135 K (indicated by a small kink in the data shown in Figure 13.34) was described in Chapter 6, Figure 6.11. Similar phenomena occur in the R = Ho, Er, Tm, Yb members of the series.

Processing of 14–2–1 sintered magnets is similar to that for Co_5Sm but cannot be done in air: cast, crush, mill to 10 μm or less, magnetically align, press, sinter at 1100°C and rapidly cool. Starting compositions are generally slightly enriched in Nd and B to provide a liquid grain boundary phase at the solidification temperature of the 14–2–1 phase. This enhances density and prevents the magnetic particles from exchange coupling to each other. The final microstructure then consists of 10–20-μm grains of the 2–14–1 phase with a Nd-rich grain boundary phase and $Fe_4Nd_{1.1}B_4$ inclusions usually at the grain boundary junctions in the sintered microstructure (Fig. 13.35). The efficacy of

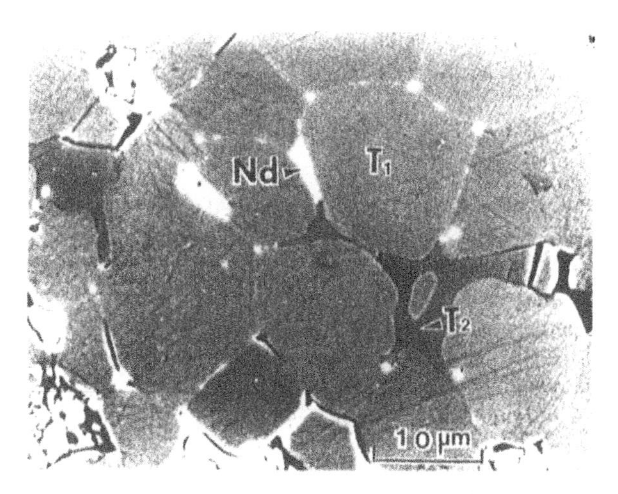

Figure 13.35 X-ray composition micrograph of a sintered $Nd_{0.15}Fe_{0.77}B_{0.08}$ magnet; T_1, T_2, and Nd denote $Nd_2Fe_{14}B$, $Nd_{1+e}Fe_4B_4$, and a Nd-rich phase, respectively (Sagawa et al. 1987).

these precipitates in pinning domain walls is suggested quantitatively by Paul's model, described in Chapter 9.

Magnetization reversal in sintered 14–2–1 magnets occurs by nucleation and growth of reversal domains. Kronmüller et al. (1988) present a micromagnetic model for misoriented grains (no exchange), that leads to a coercivity of the form

$$_iH_c = \frac{2K_1}{M_s} a_K a_\psi - N_{\text{eff}}(4\pi M_s) \tag{13.6}$$

Here a_K describes the micromagnetic effects of anisotropy, wall width, and inhomogeneity size; a_ψ describes the effects of grain misalignment. These two factors differ depending on whether wall motion is limited by pinning or nucleation. The authors find excellent agreement between measured values of $_iH_c$ and the nucleation form of their model for temperatures up to about 450 K. Note the similarity of Eq. (13.4) (mks) with Eq. (13.6) (cgs-emu) except for the added generality due to the anisotropy and grain orientation factors in the latter.

Melt spinning is also used to prepare Fe-Nd-B magnets. While amorphous Fe-Nd-B ribbons can be made, they do not show hard magnetic properties; they are essentially homogeneous and isotropic. A fine microstructure and preferred grain orientation are required for optimal permanent magnet characteristics. Melt-spun Fe-Nd-B can be quenched at a slower rate than required to produce amorphous ribbons, producing a nanocrystalline microstructure optimal for most effective wall pinning on grains (Fig. 9.18). This process produces an isotropic magnet with low remanence. Alternatively, melt-spun Fe-Nd-B can be slightly overquenched relative to the peak in Figure 9.18. This allows the magnetic hardness *and grain alignment* to be developed during subsequent heat treatment, hot pressing, or die upsetting. Grain size is of order 100 nm in melt-spun Fe-Nd-B, considerably finer than that of sintered Fe-Nd-B or Co-Sm. Crushed, melt-spun ribbons can be formed to bulk magnets by one of three methods, (1) epoxy bonding, (2) hot isostatic pressing, or (3) die upsetting (uniaxial hot compression) at about 700°C. Magnetic properties of melt-spun magnets improve from case (1) to case (3) because of increased density (1–3) and crystallographic alignment (2–3) (Fig. 13.36). The microstructure of the die upset product is characterized by platelike grains typically 60 nm thick and 300 nm in diameter with the *c*-axes (normal to the plate) strongly aligned in the compression direction. (Here the alignment of magnetic easy axes is a consequence of the crystallographic alignment induced by hot pressing. In sintered magnets, an applied magnetic field tends to line up the moments, and thus the *c* axes, of the particles before fixing the grain orientation by sintering.) Die upsetting results in a 50% increase in remanence

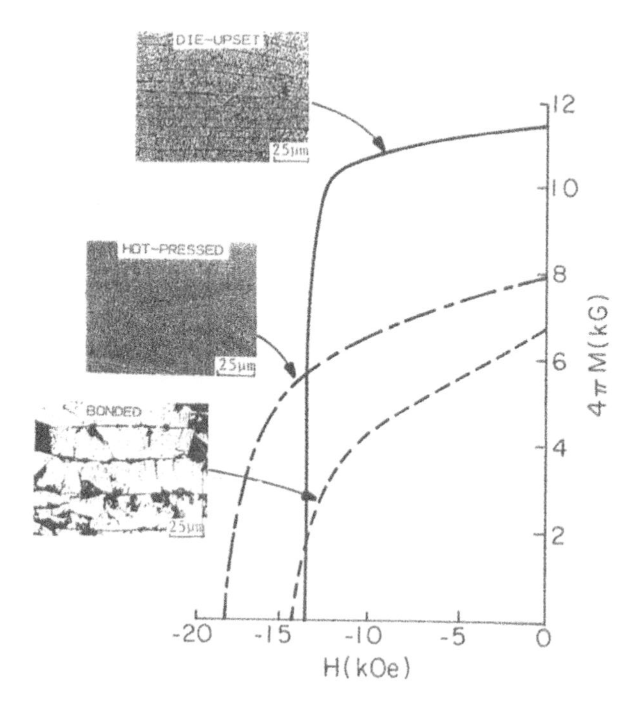

Figure 13.36 Room temperature demagnetization curves and optical micrographs of bonded, hot-pressed, and die-upset NdFeB magnets prepared from melt-spun ribbons [after Herbst (1991)].

($B_r \approx 1.2$ T) relative to hot-pressed, melt-spun material ($B_r = 0.8$ T). Compared to sintered Fe-Nd-B magnets, melt-spun Fe-Nd-B magnets generally contain less Nd-rich grain boundary phase because the rapid solidification process precludes significant phase separation. This improves their corrosion resistance. The origin of H_c in melt-spun Fe-Nd-B magnets is more complicated and somewhat less well understood than in sintered Fe-Nd-B.

High-energy ball milling has also been shown to be effective in producing permanent magnets having properties comparable to those of melt-spun ribbons (Schultz et al. 1989).

At room temperature $Fe_{14}Nd_2B$ magnets exhibit the best energy products of any practical permanent magnets. However, the Curie temperature of $Fe_{14}Nd_2B_1$ (315°C) is too low for some applications. Replacement of 50% of the Fe by Co gives a Curie temperature of 627°C but much lower anisotropy (and coercivity) at room temperature (Fig. 13.37). This reduced anisotropy is due to several factors. The rare-earth transition metal exchange coupling is weaker for Co-Nd than for Fe-Nd which leads to a smaller Nd moment at room temperature (hence smaller net magnetization). The smaller net magnetization leads to reduced anisotropy energy and coercivity. Additionally, the transition-metal anisotropy changes from positive for Fe to negative for Co (Herbst 1991). This is clear from Figure 13.38, where the magnetic ordering versus temperature is displayed for $Fe_{14}R_2B$ and $Co_{14}R_2B$ phases. Additions of up to 3% Al, Ga, and/or Mg minimize the loss of anisotropy and coercivity. Replacing some Nd with Dy (which has a greater anisotropy) leads to a significant enhancement in coercivity at a cost in saturation induction (moments of late R species such as Dy couple antiferromagnetically to the transition metal moment).

The magnetic properties of $Fe_{14}Nd_2B$ can be improved at elevated temperature by partial substitution of *both* the rare earth and transition metals. Partial replacement of Fe by Co gives a higher Curie temperature without incurring negative transition metal sublattice anisotropy or significantly lower transition metal sublattice magnetization. Partial replacement of Nd by Tb leads to a larger uniaxial anisotropy at elevated temperatures. This occurs because Tb, like Nd, favors easy-axis magnetization in 2–14–1 phases, it has a larger anisotropy energy density, and it has a stronger exchange coupling than Nd to maintain the favorable anisotropy up to temperatures as high as 300°C. An undesirable feature of Tb is the decrease in compound magnetization due to the antiferromagnetic coupling with the transition metal sublattice. However, as indicated by the data in Figure 13.34 (Hirosawa et al. 1986), the magnetization penalty is not severe at temperatures near T_C.

Development of other high-energy product magnets based on, or related to the 2–14–1 structure has been active (Buschow 1988, 1991, Herbst 1991). It has been suggested that allowing for an intergranular layer of soft iron in a fine-grained Fe-Nd-B magnet may afford an increase in magnetization, remanence, and energy product (Kneller and Hawig, 1991). Initial results suggest

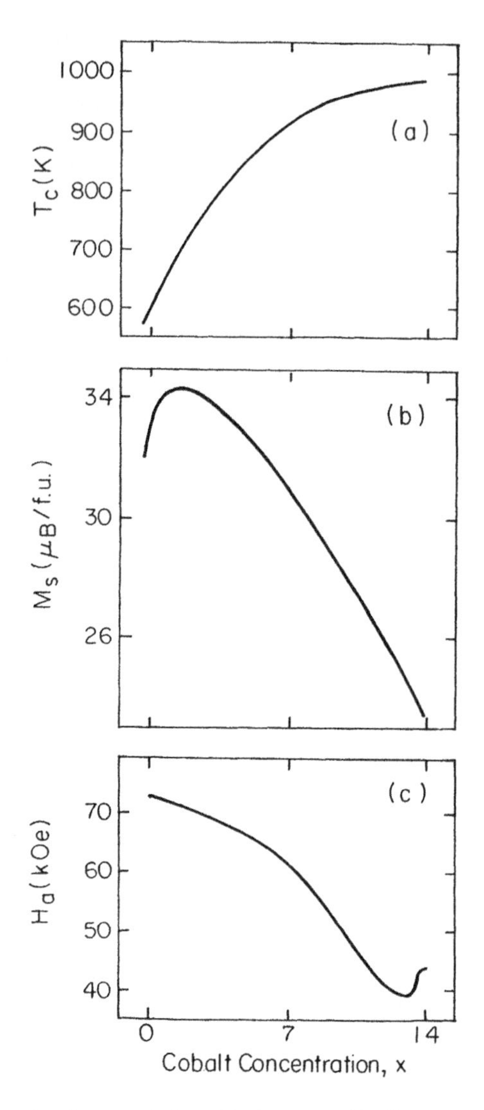

Figure 13.37 Magnetic characteristics of $Nd_2Fe_{14-x}Co_xB$ pseudoternary compounds as functions of cobalt concentration x: (*a*) Curie temperature T_C and (*b*) saturation magnetic moment at 295 K (Fuerst et al. 1986); (*c*) room temperature anisotropy field. [After Grössinger et al. (1988).

some promise in this regard (Withanawasam et al. 1994). These magnets are variously referred to as exchange-coupled magnets or spring magnets.

Recently, Coey and Sun (1990) reported the development of higher T_C and sustained energy product in $Fe_{17}Sm_2N_\varepsilon$ compounds (ε is a variable quantity). Stability may be an issue with these important materials (Li and Coey 1991).

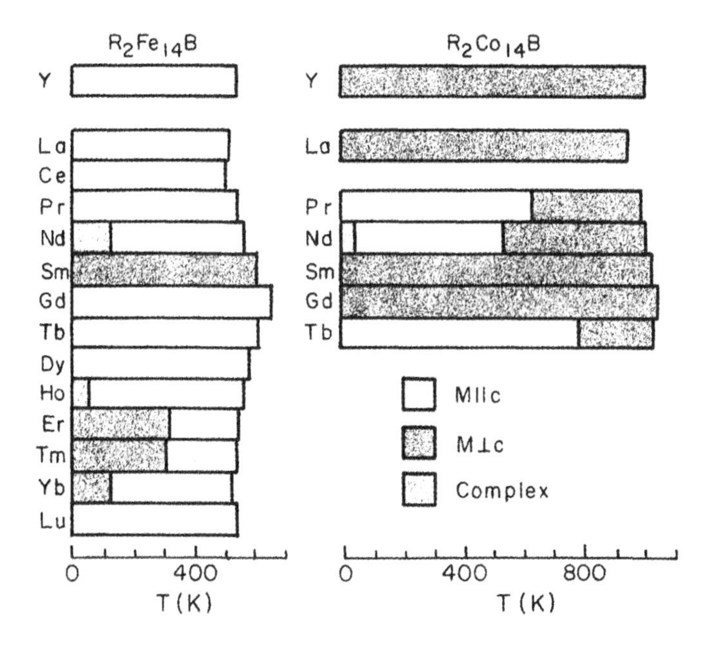

Figure 13.38 Temperature dependence of magnetic order in $R_2Fe_{14}B$ and $R_2Co_{14}B$ compounds. Results from Coey (1986) and De Mooj and Buschow (1988), respectively.

13.6 OTHER PERMANENT MAGNETS

CoPt CoPt has a saturation induction of order 0.2–0.4 T, a coercivity typically near 340 kA/m (4.3 kOe), and a Curie temperature of 550°C. It is a face-centered cubic ferromagnet above 820°C and below that temperature orders to a CuAu structure (Fig. 13.39) for compositions from 35 to 65% Co.

This atomic ordering results in a tetragonal distortion of the material. The ordered phase has a uniaxial anisotropy energy density of order 5×10^6 J/m^3, and coercivities in excess of 6 kOe have been achieved. The coercivity, however, is not greatest in the fully ordered phase; rather, H_c peaks when there is a fine-grained mixture of ordered and disordered phases. Gaunt (1966) believes that domain walls are pinned at the grain boundaries. Because grain-boundary pinning means no domain walls in the grains, this is equivalent to the existence of single-domain behavior. CoPt typically has $(BH)_{max} = 70$ to 95 kJ/m^3 (8.5–12 MG·Oe) and exhibits very good corrosion resistance relative to most other metallic hard magnetic materials. It plays a role in many of the thin-film magnetic recording media used in hard disks (Chapter 17).

FePt and FePd alloys also go through similar order–disorder transformations to the CuAu structure. Even though these materials are tetragonal in their ordered states (uniaxial along [001] axes), the high strain associated with ordering results in twinning to three *variants* that pack together to give an

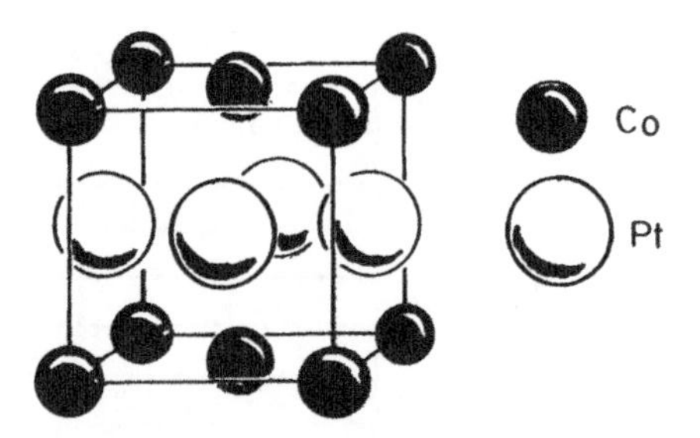

Figure 13.39 Ordered structure of CoPt (CuAu structure) below 820°C.

isotropic or unaligned permanent magnet and relatively low remanence. The remanence and energy product can be increased by field annealing or by plastically deforming the material to align the [001] directions.

MnAlC Magnets The ordered FCT, CuAu phase of MnAl(τ) is ferromagnetic and metastable, forming by a martensitic transformation from the high-temperature FCC phase (Kojima et al. 1974). It has a uniaxial magnetic anisotropy of order 10^6 J/m^3 and thus is of interest as a permanent magnet. The addition of carbon leads to greater stability of the τ phase and the ability to induce [001] directional order by extrusion up to 700°C. Coercivities of 2 to 3 kOe and energy products of order $(BH)_{max} = 64$ kJ/m^3 (8 MG·Oe) are achieved in this largely single-phase magnet.

Spinel Oxides Spinel cobalt ferrite, CoO·Fe_2O_3 (see Chapter 4) shows a trigonal distortion and hence strong magnetocrystalline anisotropy relative to the undistorted spinels. High coercivity (up to 320 kA/m) has been speculated to be associated with pinning of domain walls on stacking faults. Pinning on stacking faults (planar defects typically 1 nm in thickness) would require an extremely strong anisotropy in the defect to pin a domain wall that would be about 10 times thicker. While not used commercially, cobalt ferrite appears to play the role of a compass in certain magnetotactic bacteria.

13.7 ANALYSIS AND SUMMARY

Figure 13.4 shows how the maximum energy product has increased with time as new classes of materials have been discovered, developed, and optimized. Energy product is not the only important design criterion. Energy product per dollar and energy product per weight are also important design criteria.

Magnetization reversal takes place by various processes in different permanent magnets (Zijlstra 1980). See Table 13.6. For fine-particle magnets (particle size < single-domain particle size) coherent rotation dominates. As particle size increases, curling, buckling, or other incoherent modes of reversal become

TABLE 13.6 Summary of Factors Limiting Coercivity in Various Classes of Permanent Magnets

Magnet Microstructure	Coercivity-Limiting Process	Examples	Source of Coercivity
Single-domain particle: $r < r_{sd}$	Coherent rotation	ESDs[a], melt-spun FeNdB	Shape, magnetocrystalline anisotropy
$r \approx r_{sd}$	Incoherent rotation	Alnico	Shape, magnetocrystalline anisotropy
$r > r_{sd}$	Nucleation	Single-phase $SmCo_5$ hexagonal ferrites, and sintered $Fe_{14}Nd_2B_1$	
$r > r_{sd}$	Pinning	Two-phase Sm_2Co_{17}, melt-spun $Fe_{14}Nd_2B$, Alnicos	Magnetocrystalline anisotropy, pinning sites (precipitates, defects)

[a]Elongated, single-domain particles.

possible. In still larger particles that remain magnetically isolated, domain walls must be nucleated in each grain for magnetization reversal to occur. If the average grain volume is less than the average volume per defect (of those defects capable of nucleating reversal domains), then the coercivity remains large. If reversal domain nucleation occurs, then the only way to maintain high coercivity is to have strong wall pinning sites. The term *reversal domain nucleation* should be understood to include the release from grain boundaries of residual domains not fully removed by a strong applied field. Reversal domain nucleation generally occurs on large defects having either reduced anisotropy or a combination of shape and magnetization relative to the matrix that provides large local magnetostatic energy.

The difference between nucleation-limited and pinning-limited behavior in permanent magnets composed of grains larger than single-domain particles is determined experimentally by noting the initial magnetization curve from a thermally demagnetized state (Kronmüller, Durst, and Sagawa 1988). (Thermal demagnetization is the most reliable way to introduce domain walls into the grains of a permanent magnet). If the coercivity is controlled by pinning, the domain wall will not move significantly with application of a field until $H \approx H_c$ and the initial magnetization curve will be flat. If the coercivity is limited by nucleation but wall motion is easy, the initial magnetization curve will be steep. On this basis, sintered Fe-Nd-B magnets appear to be nucleation limited because domain walls introduced by thermal demagnetization move quite easily. This simple concept is born out by measurements on R_2Co_{17} (Fig. 13.31).

From the model of domain wall pinning described in Chapter 9, it is clear that the defects that are most effective in pinning domain walls are those whose magnetic properties differ most from those of the matrix and whose dimensions are comparable to the domain wall width, $\delta = \pi\sqrt{(A/K_u)}^{1/2}$. While domain wall thicknesses in soft materials were found to vary from about 50 nm to a few thousand nanometers, those in hard magnetic materials can range from a few nm to about 10 nm. Clearly, then, point defects and grain boundaries play a much bigger role in wall pinning in permanent magnets than in soft magnets where long-range strain fields and larger precipitates are more effective. Large defects in permanent magnets provide the opportunity for nucleation of reversal domains.

The properties of Alnico alloys are dominated by the high aspect ratio of the fine α-phase particles and hence are often described in terms of single-domain particle behavior where the anisotropy is provided by shape ($H_c \leqslant \Delta N M_s \approx \frac{1}{2} M_s$). However, quantitative comparison of properties with model prediction is complicated by particle interactions: α′ may not be nonmagnetic, α-phase particles are often multiply connected; domain walls are observed in Bitter pattern studies. (Domains are also seen in ESD magnets and here, as well as in the Alnicos, they may separate *interaction domains* in which particles align by their magnetostatic interactions.)

PROBLEMS

13.1 **(a)** Calculate for a spherical particle the critical radius below which the particle is a single-domain particle, that is, it cannot be demagnetized since it cannot support a domain wall [Hint: Balance the wall energy against the magnetostatic energy of the particle without a wall, $\frac{1}{2} M_s H_d = \frac{2}{3} \pi M_s^2$].

(b) What is the critical radius for $Nd_2Fe_{14}B$ where $A = 10^{-6}$ erg/cm, $M_s = 1274$ G, and $K = 5 \times 10^7$ erg/cm^3?

BIBLIOGRAPHY

Croat, J. J., *J. Mater. Eng.* **10**, 7 (1988).

Croat, J. J., and J. F. Herbst, *Mater. Res. Bull.* **13**, (6) 27 (1988).

Durst, K. D., and H. Kronmüller, *J. Magn. Magn. Mater.* **68**, 63 (1987).

Kneller, E., and R. Hawig, *IEEE Trans.* **MAG-27**, 3588 (1991).

Livingston, J. D., *Prog. Mater. Sci.* **243**, 9012 (1981).

Luborsky, F. E., J. D. Livingston, and G. Y. Chin, in *Physical Metallurgy*, R. W. Cahn and P. Haasen, ed., Elsevier, Amsterdam, 1996, p. 2501.

Nesbitt, E. A., and J. H. Wernick, *Rare Earth Permanent Magnets*, Academic Press, New York, 1973.

Wohlfarth, E. P., in *Magnetism*, G. T. Rado and H. Suhl, eds., Academic Press, New York, 1963, 395.

REFERENCES

Barrett, C., and T. B. Massalski, *Structure of Metals*, Pergamon Press, Oxford, New York, 1980, p. 266.

Buschow, K. H. J., in *Ferromagnetic Materials*, Vol. 4, E. P. Wohlfarth, ed., Elsevier, Amsterdam, 1988.

Buschow, K. H. J., *Rept. Prog. Phys.* **54**, 1123 (1991).

Cadieu, F., invited talk, Electrochemical Society, Chicago, 1995.

Cahn, J., *Trans. AIME* **242**, 166 (1963).

Callen, H. B., and E. Callen, *J. Phys. Chem. Sol.* **27**, 1271 (1966).

Campbell, I. A., *J. Phys.* **F2**, L47 (1972).

Chikazumi, S., *Physics of Ferromagnetism*, Clarendon Press, Oxford, UK, 1997.

Coey, J. M. D., *J. Less-Common Met.* **126**, 21 (1986).

Coey, J. M. D., and H. Sun, *J. Magn. Magn. Mater.* **87**, L251 (1990).

Croat, J. J., *Appl. Phys. Lett.* **39**, 357 (1981); *J. Magn. Magn. Mater.* **24**, 125 (1981).

Croat, J. J., J. F. Herbst, R. W. Lee, and F. E. Pinkerton, *J. Appl. Phys.* **55**, 2078 (1984).

Das, B. N., and N. C. Koon, *Met. Trans.* **14A**, 953 (1981).

De Mooij, S. B., and K. H. J. Buschow, *J. Less-Common Met.* **142**, (1988).

den Broder, F. J. A., and K. H. J. Buschow, *J. Less-Common Met.* **29**, 67 (1977).

De Vos, K. J., Ph.D. thesis, Tech. Hoch Sch., Eindhoven, 1966.

De Vos, K. J., in *Magnetism and Metallurgy*, A. E. Berkowitz and E. Kneller, eds., Academic Press, New York, 1969, p. 473.

Fuerst, C. D., J. F. Herbst, and E. A. Alson, *J. Magn. Magn. Mater.* **54–57**, 567 (1986).

Gaunt, P., *Phil. Mag.* **13**, 579 (1966).

Givord, D., H. S. Li, and J. M. Moreau, *Solid State Commun.* **50**, 497 (1984).

Goto, Y., and T. Tadaka, *J. Am. Ceram. Soc.* **43**, 150 (1960).

Gould, J. E., *Cobalt*, **23**, 82 (1964).

Grössinger, R., R. Krewenka, H. Buchner, and H. Harada, *J. Phys.* (Paris) **49**, C8-659 (1988).

Hadjipanayis, G., R. C. Hazelton, and K. R. Lawless, *Appl. Phys. Lett.* **43**, 797 (1983); *J. Appl. Phys.* **55**, 2073 (1984).

Herbst, J. F., J. J. Croat, F. E. Pinkerton, and W. B. Yelon, *Phys. Rev.* **B29**, 4176 (1984).

Herbst, J. F., J. J. Croat, and W. B. Yelon, *J. Appl. Phys.* **57**, 4086 (1985).

Herbst, J. F., *Rev. Mod. Phys.* **63**, 819 (1991).

Hirosawa, S., Y. Matsuura, H. Yamamoto, S. Fujimura, M. Sagawa, and H. Yamauchi, *J. Appl. Phys.* **59**, 873 (1986).

Kaneko, H., M. Homma, and K. Nakamura, *AIP Conf. Proc.* **5**, 1088 (1971).

Kaneko, H., M. Homma, M. Okada, S. Nakamura, and N. Ikuta, *AIP Conf. Proc.* **29**, 620 (1976).

Kikuchi, S., and S. Ito, *IEEE Trans.* **MAG-8**, 344 (1972).

Kojima, H., and K. Goto, *J. Appl. Phys.* **36**, 538 (1965).

Kojima, H., T. Ohtani, N. Kato, K. Kojima, Y. Sakamoto, I. Konno, M. Tsukahara, and T. Kubo, *AIP Conf. Proc.* **24**, 768 (1974).

Kojima, H., in *Ferromagnetic Materials*, Vol. 3, E. P. Wohlfarth, ed., North Holland, Amsterdam, 1982, p. 305.

Kronmüller, H., K. D. Durst, and M. Sagawa, *J. Magn. Magn. Mater.* **74**, 291 (1988).

Li, H.-S, and J. M. D. Coey, in *Ferromagnetic Materials*, Vol. 6, K. H. J. Buschow, ed., North Holland, Amsterdam, 1991.

Livingston, J. D., *AIP Conf. Proc.* **10**, 643 (1973).

Luborsky, F.E., *J. Appl. Phys.* **32**, 1715 (1961).

Luborsky, F. E., and C. R. Morelock, *J. Appl. Phys.* **35**, 2055 (1964).

Metals Handbook, Vol. 8, ASM, Metals Park, OH.

McCurrie, R. A., and S. Jackson, *IEEE Trans.* **MAG-16**, 1310 (1980).

McCurrie, R. A., *Ferromagnetic Materials*, Vol. 3, E. P. Wohlfarth, ed., North Holland, Amsterdam, 1982, p.107.

Nagel, H., *AIP Conf. Proc. No. 29*, J. J. Becker et al., eds., *Amer. Inst. of Phys.*, NY, 1976, p. 603.

Nesbitt, E. A., and H. J. Williams, *Phys. Rev.* **80**, 112 (1955).

Nesbitt, E. A., R. H. Williams, R.C. Sherwood, F. Buehler, and J. H. Wernick, *Appl. Phys. Lett.* **12**, 361 (1968).

Ray, A. E., and K. Strnat, *IEEE Trans.* **MAG-8**, 861 (1972).

Sagawa, M., S. Fujimura, M. Togawa, and Y. Matsuura, *J. Appl. Phys.* **55**, 2083 (1984).

Sagawa, M., S. Hirasawa, H. Yamamoto, S. Fujimuri, and Y. Matsuura, *Jpn. J. Appl. Phys.* **26**, 785 (1987).

Schnittke, K., and J. Wecker, *J. Magn. and Magn. Mater.* **80**, 115 (1989).

Shirk, B., and W. R. Buessem, *J. Appl. Phys.* **40**, 1295 (1969).

Stäblein, H., in *Ferromagnetic Materials*, Vol. 3, E. P. Wohlfarth, ed., North Holland, Amsterdam, 1982, p. 441.

Shtrikman, S., and D. Treves, *J. Phys. Rad.* **20**, 286 (1959).

Stoner, E. C., and E. P. Wohlfarth, *Phil. Trans. Roy. Soc.* (London) **A240**, 599 (1948).

Strnat, K. J., in *Ferromagnetic Materials*, Vol. 4, E. P. Wohlfarth, ed., Elsevier, Amsterdam, 1988.

Tawara, Y., and K. J. Strnat, *IEEE Trans.* **MAG-12**, 954 (1976).

Withanawasam, L., A. L. Murphy, and G. C. Hadjipanayis, *J. Appl. Phys.* **76**, 7065 (1994).

Zijlstra, H., in *Ferromagnetic Materials*, Vol. 3, E. P. Wohlfarth, ed., North-Holland, Amsterdam, 1982, p. 37.

CHAPTER 14

MAGNETIC ANNEALING AND DIRECTIONAL ORDER

14.1 INTRODUCTION

The objective in magnetic annealing, rolling, or other magnetic processing technique, is generally to tailor the shape of the B–H loop for a particular application. These processes are effective because they alter the anisotropy of the sample by some mechanism. If constant permeability is needed for a wide range of applied fields, a sheared loop is sought (Fig. 14.1, left). Such a loop can result when the dominant magnetic anisotropy is transverse to the direction of the applied field. A sheared loop can be achieved by heat treatment in a magnetic field that is oriented transverse (T) to the direction of the field to be used during operation. A sheared loop can also be produced in some cases by rolling. If a large flux change at low applied field is needed, as in many magnetic switching and power applications, a square B–H loop as shown in Figure 14.1, right, is desirable. Such a loop can be achieved by heat treating in a longitudinal (L) field or by rolling. These effects often involve a small change in the angular distribution of certain atomic-pair bonds in the material.

From the calculations of idealized M-H loops in Chapter 9, it is possible to identify the simplest magnetic domain structure responsible for the loops in Figure 14.1 as those shown in Figure 14.2. The domain pattern of a T-annealed sample shows magnetization lying largely transverse to the sample axis; each domain is separated by a 180° domain wall and the net magnetization is zero in the demagnetized state. The domain pattern of an L-annealed sample shows the magnetization lying along the sample axis, again with 180° domain walls separating the domains magnetized in opposite directions such that the net magnetization is zero in the demagnetized state. The domain structures of

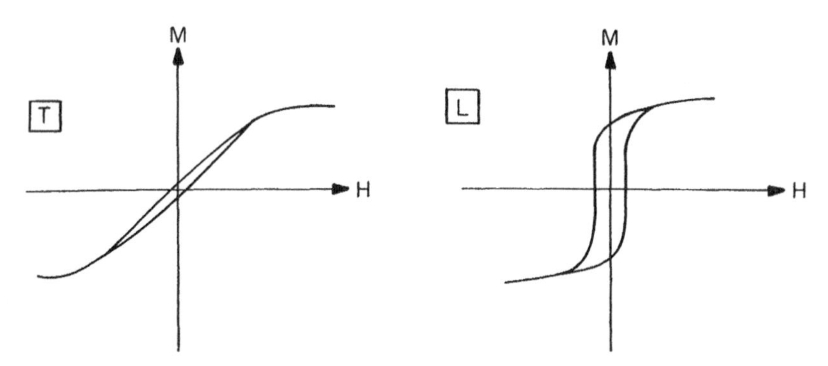

Figure 14.1 Schematic of the two extremes of field-induced anisotropy: transverse (T) and longitudinal (L) with loops taken in a longitudinal field.

field-annealed amorphous alloys, where there is no strong crystalline anisotropy, were shown in Figure 11.4.

The B–H curve for completely T-annealed material is linear in H and reaches saturation at a field H_a that is proportional to the strength of the induced anisotropy energy density K_u. The permeability is low and the remanence, coercivity, and hysteresis loss, ideally, are zero. This sheared-over loop suggests that the magnetization process occurs by rotation of the domain magnetization into the field direction. The magnetostriction and magnetoresistance (Chapter 15) of this state are close to their maximum values for the composition.

The B–H loop of completely L-annealed material is square with a remanence ratio close to one (depending on sample shape). The square B–H loop of an L-annealed sample suggests that the magnetization process is dominated by 180° domain wall motion. This loop indicates a high maximum permeability and a hysteresis loss that depends on the strength of the coercivity. (H_c is governed by the ease of wall motion which is a function of frequency and material properties; see Chapter 9). The magnetostriction and magnetoresistance are reduced well below their peak values for the composition. Field annealing has little effect on the fundamental properties of a magnetic material (M_s, T_C).

The use of magnetic fields in heat-treating ferromagnetic materials to induce magnetic anisotropy is commonplace. A variety of soft magnetic materials

Figure 14.2 Schematic of domain structures typical of transverse (T) and longitudinally (L) field annealed samples. These domain patterns, exposed to fields along the sample length, give rise to M–H loops like those in Figure 14.1 T and L, respectively.

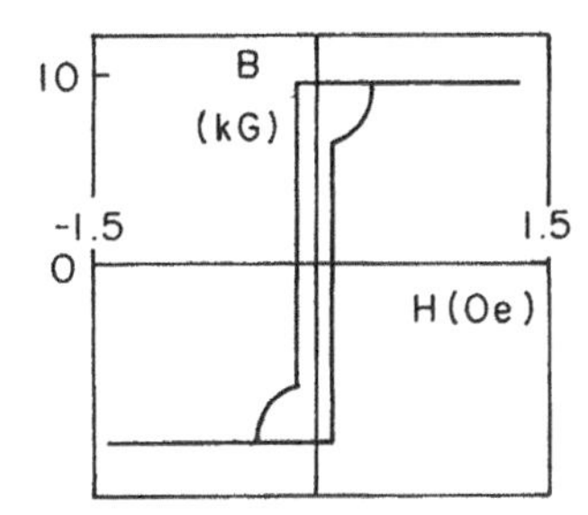

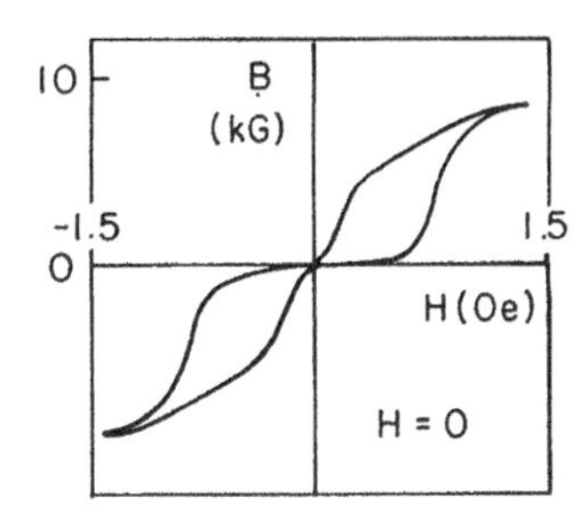

Figure 14.3 *B–H* loop of an Fe-Ni-Co perminvar alloy after annealing in a longitudinal field (left) and in zero magnetic field at 900°C (right) [After Bozorth, copyright IEEE Press, 1994].

respond to some form of magnetic field processing to meet certain performance specifications. These include magnetic oxides (Slonczewski 1963, Slick 1986), commercial magnetic alloys such as 50% Ni permalloy [Chapter 10, here; Chin and Wernick (1981)], Co-Cr or Fe-Ni films for magnetic recording media or thin-film heads, respectively (Chapter 17), and finally, amorphous magnetic alloys [Chapter 11, here; Fujimori 1983).] The development of orientation in permanent magnets is not always a matter of field annealing. Permanent magnets sintered from powders (e.g., rare earth–transition metal intermetallics) get their oriented characteristics by physically aligning the particles with a magnetic field before sintering, or by die upsetting or extrusion of compacted powders. Alnico magnets develop a uniaxial, textured microstructure by spinodal decomposition in a magnetic field (Chapter 13).

Before proceeding to a discussion of the mechanisms that cause field-induced anisotropy, three important observations related to field annealing are worth mentioning: (1) pure elements do not respond to heat treatment whereas many alloys do, at least to some degree; (2) for field annealing to be effective, the field must be applied below the Curie temperature of the material and, for some effects, must be sufficient to saturate the material; and (3) ternary Fe-Co-Ni alloys, known generally as *perminvar alloys* (based on $Ni_{45}Fe_{30}Co_{25}$), for reasons that will become clear later, respond strongly to field annealing. When perminvars are heat-treated *without* a magnetic field, a very unusual loop shape results (Fig. 14.3): the waist of the loop near $M = 0$ is constricted (this is often referred to as a *wasp-waisted loop*). This effect, due to the stabilization of domain walls, was illustrated in Figure 11.24.

The explanation of magnetic annealing to be developed should be able to describe these effects as well as those depicted in Figures 14.1–14.3.

14.2 MECHANISMS OF FIELD ANNEALING

The easy direction of magnetization in a material is governed mainly by the magnetic anisotropy, due to its shape, its crystal structure, and its state of stress. It is also a function of the crystallographic texture and microstructure of the sample, which depend on the method of fabrication or postfabrication processing. Finally, magnetic anisotropy (strength and axis orientation) can also be a function of annealing in the presence of an applied magnetic field or stress. How does magnetic field annealing induce an easy axis? First, it is the direction of magnetization in the sample during the heat treatment and not simply the direction of the applied field that is responsible for field-induced magnetic anisotropy. Magnetic annealing leaves a preferred direction of magnetization in the sample generally by rearranging atoms on a local scale in such a way as to favor magnetization in a given direction. At an annealing temperature T_a sufficiently high for atomic mobility, yet not so high that the material is no longer magnetic ($T_a < T_C$), some atom pairs orient themselves relative to the direction of magnetization so that their magnetic anisotropy energy is minimized. Once the temperature is reduced to a level at which significant diffusion can no longer occur, if the field is removed, the frozen-in atomic pair directional ordering may be sufficient to override other anisotropies and hold the magnetization in the direction it had during annealing.

It was initially postulated that the improved permeability and loop squareness associated with field annealing could be due to the relief of stress between grains in polycrystalline materials. The origin of this intergranular stress is as follows. Adjacent grains which have their easy axes (and hence their magnetization directions) canted with respect to each other are magnetostrictively strained in such a way that they no longer fit together. The stress exerted by one grain on the other induces a magnetoelastic anisotropy and hence makes it more difficult to magnetize the sample. However, it is now known that this effect is small relative to the magnetocrystalline anisotropy for most 3*d* metals and alloys, as was shown in Chapter 7. It is a significant effect in many rare-earth alloys. Further, it is possible to induce anisotropy by field-annealing zero-magnetostriction alloys and single-crystal alloys. Hence, the effects of magnetic annealing cannot be accounted for by stress relief alone. Nevertheless, field annealing does have implications for magnetostrictive behavior.

The domain patterns characterizing the L and T limits have special implications for the extent to which magnetostriction is a factor during the magnetization process. Figure 14.2(T) depicts an idealized domain structure in which the sample width is increased and its length is decreased by magnetostriction (if $\lambda_s > 0$). Figure 14.2(L) illustrates a sample elongated along the ribbon axis and contracted across it (if $\lambda_s > 0$). (These strains would be opposite if $\lambda_s < 0$). Given these idealized domain patterns, it is evident that magnetizing the sample in Figure 14.2(T) to saturation along its length rotates the direction of magnetization and changes the magnetostrictive strain of the sample. During the magnetization process in Figure 14.1(T) the sample strains by $(\frac{3}{2})\lambda_s$. Conversely, going from the L domain pattern in Figure 14.2(L) to saturation along the sample length does *not* change the magnitude of the sample strain (domains both parallel and antiparallel to the ribbon axis in Figure 14.2(T) have the same strain along the ribbon axis). The longitudinal magnetization process in Figure 14.1(L) does not alter the state of strain of the sample regardless of the value of the saturation magnetostriction constant of the material, that is, from Eq. (7.1), $\Delta l/l = 0$.

The microscopic effects accompanying magnetic field heat treatment on materials may vary widely. At temperatures for which short-range atomic diffusion is possible, the presence of a magnetic field can have some of the following consequences:

1. The direction of bonds between similar or dissimilar atomic species may take on an asymmetric distribution (directional pair ordering), even in single-phase materials. Typical examples are the field annealing of polycrystalline iron–nickel alloys or a variety of amorphous alloys.
2. If interaction 1 is strong enough and time and temperature are adequate, atoms of a mobile, minority species may coalesce into fault planes to lower the free energy. This appears to be the mechanism responsible for annealing in many Co-Ni-containing crystalline alloys such as perminvar or permalloy and has recently been suggested to apply to certain cobalt-rich amorphous alloys.
3. The application of a magnetic field to any material that is magnetic or which has components with magnetic properties affects the energy of the system. Hence magnetic fields can have orienting effects on microstructural features that exhibit anisotropy in their ferromagnetic, paramagnetic, or diamagnetic response.
4. In two-phase alloys, a textured or anisotropic microstructure may evolve during annealing. The foremost example is the growth of oriented columns of Ni_3Al in an Fe-rich matrix during phase separation in the fabrication of Alnico permanent magnets (Chapter 13).

The first effect, directional pair ordering, is the most subtle and it is the mechanism behind much of the field annealing that is practiced. Atomic pair ordering, first suggested by Néel, is the mechanism to be considered first. Mechanisms 2–4 will be described later in this chapter.

14.3 DIRECTED-PAIR ANISOTROPY

14.3.1 Pair Interaction Energies: Isotropic and Anisotropic

Different species interact chemically with each other in different ways. In a binary alloy $A_{1-x}B_x$, the affinity of species A and B may be expressed as a negative value for their interaction energy, E^{AB}. If the A–B affinity is greater than the affinity of either A for A or B for B, annealing will tend to increase A–B coordination: $E^{AB} < (E^{AA}$ or $E^{BB}) < 0$. This rearrangement of species on a crystal lattice is called *chemical ordering*. In contrast, a preference for A–A or B–B coordination (E^{AA}, $E^{BB} < E^{AB} < 0$) is called *segregation*. A nonmagnetic system will arrange its bond coordination, subject to packing constraints, so as to minimize the energy:

$$E = N^{AA}E^{AA} + N^{BB}E^{BB} + N^{AB}E^{AB} + N^{BA}E^{AB} \tag{14.1}$$

Here, N^{AB} is the coordination of species A by B; $N^{AA} + N^{AB}$ gives the total coordination of species A. [The appendix to this chapter reviews the concepts and formalism of chemical order–disorder phenomena, arising from Eq. (14.1), in parallel with magnetic order–disorder phenomena.] In a magnetic material, the strength of this chemical bond interaction can depend on the orientation of the A–A, B–B, and A–B bonds relative to the magnetization direction. To quantify this magnetization dependence, the bond energies can be expanded in polynomials of the angle between the bond direction and the magnetization direction as in Eq. (6.7):

$$E^{AB}(\varphi) = l^{AB}(\cos^2\varphi_i - \tfrac{1}{3}) + q^{AB}(\cos^4\psi - \tfrac{6}{7}\cos^2\psi + \tfrac{3}{35}) + \cdots \tag{14.2}$$

Here, l^{AB} is the strength of the uniaxial part of the anisotropic pair interaction. Only the dipole term need be retained.

It is important to consider some microscopic aspect of directional bonding. An isolated *substitutional* atom has the symmetry of the crystal lattice on which it sits. However, an *interstitial* atom or a unique *pair of atoms* on the crystal lattice can have a symmetry lower than that of the crystal because they define a particular direction in the crystal. Chikazumi (1950) first referred to this as "directional order," and Slonczewski (1963) distinguished the two cases as "monatomic directional ordering" (interstitial atom) and "diatomic directional ordering" (unlike atom pairs, A–B, one or both of which may be impurities). If these directional bond interactions depend on the direction of magnetization, as suggested above, a change in the direction of magnetization can alter the energy of the system. Hence, the direction of the magnetization during an annealing process may influence the bond orientational order by changing the number distribution of the bonds in various directions. Once this directional order is frozen in after annealing, the bond orientational order imposes a preferred axis of magnetization on the sample.

The microscopic mechanism behind the phenomenological parameters l^{AB} may be the same as that for magnetocrystalline anisotropy itself; the electrons involved in bonding, the $3d$ electrons, are also responsible for the magnetic moment. Spin–orbit coupling is the mechanism that links the magnetization direction, related to $\boldsymbol{S}$, to the bond direction, related to $\boldsymbol{L}$, as was described in Chapter 6. In the present context of directed-pair anisotropy, attention is focused on either interstitial atoms, impurities (interstitial or substitutional), or mobile atom pairs. In all cases, the ability to induce anisotropy depends on the *mobility* of these unique, local atomic arrangements. The spin–orbit interaction (or similar microscopic anisotropy mechanisms) can influence bond directional order in two ways:

1. First, just as there is a macroscopic magnetic coupling between the direction of $\boldsymbol{M}$ and the crystallography, $(K_1\alpha_i^2\alpha_j^2 \ldots)$, so there is also a microscopic coupling between the direction of $\boldsymbol{M} = M_s(\alpha_1, \alpha_2, \alpha_3)$, and the direction of the local atomic bonds. The first-order expansion of this local interaction, in the spirit of Eq. (14.2), can be written as follows:

$$E_{\text{mag}} = D \sum_i N_i(\alpha_i^2 - \tfrac{1}{3}) \tag{14.3}$$

Here, the sum is over the i nearest-neighbor directions and N_i is the number of bonds in the ith direction. The sum of the N_i values gives the total number of impurity atom bonds or ordered pairs. When the crystal field coefficient D is positive (as is usually the case), this interaction favors bonds ordered in the plane perpendicular to the local direction of magnetization (Fig. 14.4).

2. Next, there is a coupling between the lattice strain (due to magnetoelastic or other stress) and the bond orientational order:

$$E_{\text{strain}} = \varepsilon \sum_i N_i e_{jj} \tag{14.4}$$

When the local magnetoelastic parameter ε is negative (as is usually the case), this interaction favors atoms in sites that are opened up by the strain (i.e., bond order is parallel to $\boldsymbol{M}$ for $\lambda_s > 0$) (see Fig. 14.5). For magnetostrictive strains, $e_{jj} \propto \lambda_s \cos^2\theta$, where θ is the angle between $\boldsymbol{M}$ and the strain, Eq. (14.4) becomes

$$E_{\text{strain}} = \varepsilon\lambda_s \sum_{i=1}^{Z} N_i(\cos^2\theta - \tfrac{1}{3}) \tag{14.5}$$

Following Néel's analysis [see Slonczewski (1963) or Chikazumi and Graham 1969)], the internal energy of the magnetic system now depends on the number of bonds in each of the i nearest-neighbor directions:

$$E(\varphi_i, T) = \sum_{i=1}^{Z} N_i E(\varphi_i, T) = \sum_{i=1}^{Z} N_i l(\cos^2\varphi_i - \tfrac{1}{3}) + \cdots \tag{14.6}$$

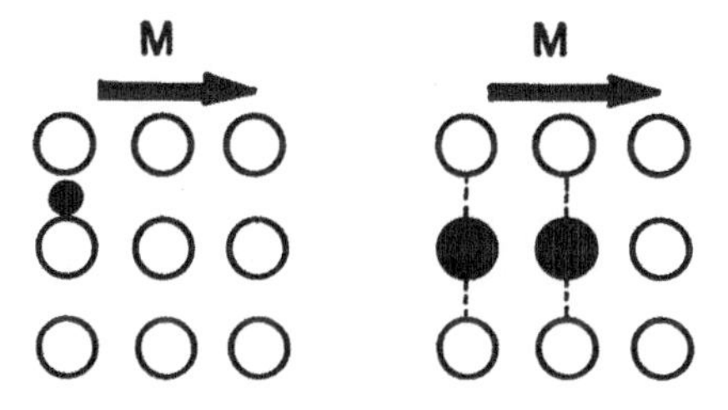

Figure 14.4 Interstitial impurity bonds or those of atom pairs, A–B, align perpendicular to the magnetization direction for $D > 0$ under the interaction in Eq. (14.3).

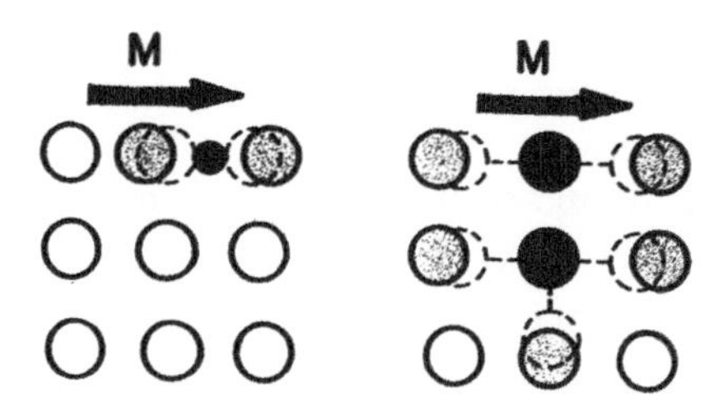

Figure 14.5 Impurity bonds or A–B atom pair bonds favor alignment with the magnetization direction for $\varepsilon < 0$ and $\lambda_s > 0$ according to Eq. (14.5).

Here, l_i (which is a function of temperature) is the strength of this uniaxial part of the interaction [(e.g., interactions (14.3) or (14.5)]. The sum is taken over the Z nearest-neighbor directions, $i = 1, 2\ldots, Z$, about the interstitial or the Z possible directions for an A–B bond, and ϕ_i is the angle between the ith bond direction and the magnetization direction. The term $\cos\phi_i$ can be written as $\alpha_1\beta_{1i} + \alpha_2\beta_{2i} + \alpha_3\beta_{3i}$ where the α terms are the direction cosines of the magnetization and the β_i terms are the direction cosines of the bonds.

14.3.2 Field-Induced Anisotropy

The energy in Eq. (14.6) can be used to calculate the equilibrium number of atoms in directed sites at the annealing temperature T_a using the Boltzmann statistical form of the concentrations (see Chapter 4):

$$N_i = N \frac{\exp[-E_i\varphi_{ai}, T_a)/k_B T_a]}{\Sigma_i^Z \exp[-E_i(\varphi_{ai}, T_a)/k_B T_a]} \tag{14.7}$$

The $\varphi_{\alpha i}$ terms indicate the direction of $\boldsymbol{M}$ relative to the bond direction at the annealing temperature T_a. Defining $l_a = l(T_a) \ll k_B T$, N_i can be expressed at T_a, to first order in the interaction energy, $E_{ai} = l_a(\cos^2\varphi_{\alpha i} - \frac{1}{3})$ as

$$N_i(T_a) \approx \frac{N}{Z}\left[1 - \left(\frac{E_i\varphi_{ai}, T_a)}{k_B T_a} - \frac{1}{Z}\sum_i^Z \frac{E_i(\varphi_{ai}, T_a)}{k_B T_a}\right) + \cdots\right] = \frac{N}{Z}\left[1 - \frac{E - \langle E \rangle}{k_B T_a}\right] \tag{14.8}$$

That is, the deviation of the number of bonds in a given direction from N/Z is proportional to the deviation of the energy in that direction from the mean energy per neighbor. Using Eq. (14.6), Eq. (14.8) can be written as follows:

$$N_i(T_a) = \frac{N}{Z}\left[1 - \frac{l_a(\cos^2\varphi_{ai} - \frac{1}{3})}{k_B T_a}\right] \tag{14.9}$$

This equation gives the equilibrium bond ordering for a given direction of magnetization during annealing. Thus, if a site has nearest neighbors distributed over six directions (an octahedral site), the number of bonds at T_a along any one direction will deviate slightly from $N/6$ by an amount that depends on the strength and sign of the magnetic interaction l_a, the relative orientation of $\boldsymbol{M}(T_a)$, and the annealing temperature T_a.

After establishing the bond distribution given by Eq. (14.9), the sample is returned to the temperature T. At $T < T_a$, the energy is given by Eq. (14.6) with φ_i being the angle between the induced anisotropy direction and $\boldsymbol{M}$ at the *observation* temperature. Substitution of Eq. (14.9) into (14.6) gives

$$E_{\text{ind anis}} = -\frac{N}{Z}\sum_{i=1}^{Z}\frac{l_a l \cos^2\varphi_{ai}\cos^2\varphi_i}{k_B T_a} + \text{constant} \tag{14.10}$$

The field-induced anisotropy energy is a function of the magnetization orientation at the annealing temperature as well as its orientation at the measurement temperature. Slonczewski (1963) shows that the form of the free energy in Eq. (14.10) can be deduced on the basis of symmetry. First, application of time reversal invariance to the magnetization at *both* T and T_a requires that angular functions show up quadratically; second, invariance of the energy to all cubic symmetry operations on $\boldsymbol{M}$ at both temperatures leads to the general form of Eq. (14.10). Chikazumi and Graham (1969) tabulate the numerical coefficient resulting from the sum in Eq. (14.10) over different structures.

Equation (14.10) says that the induced anisotropy varies inversely with annealing temperature and has an angular dependence that is uniaxial in the measuring direction about the induced easy axis. Further, its strength depends on two factors: (1) the number of interacting pairs and (2) the product of the strengths of their pair interactions at the measurement temperature and at the annealing temperature. Because each of these interactions is dipolar in form, it can be assumed that their strengths are approximately quadratic in the magnetization at the respective temperatures:

$$l \sim M^2(T), \qquad l_a \sim M^2(T_a)$$

It is expected that the part of Eq. (14.10) that is independent of φ_i and φ_{ai}, namely, the strength of the field-induced anisotropy in a binary alloy A_xB_{1-x}, should vary as

$$K_u \propto \frac{1}{k_B T_a}\left[\frac{M_s(T_a)}{M_s(0)}\right]^2\left[\frac{M_s(T)}{M_s(0)}\right]^2 x^2(1-x)^2 \tag{14.11}$$

Thus, if the anisotropy is due to an interaction between major constituents, the strength of induced anisotropy varies as $x^2(1-x)^2$. However, for dilute alloys $A_{1-x}B_x$ with $x \ll 1$, the compositional dependence is proportional to x^2.

The angular part of Eq. (14.10) can be written as $\cos^2(\varphi - \varphi_a)$, where φ is the direction of $\boldsymbol{M}$ at observation and φ_a is the direction of the field-induced easy axis. Hence Eq. (14.10) may be written simply as

$$u_a = -K_u \cos^2(\varphi - \varphi_a) \tag{14.12}$$

where K_u is the strength of the field-induced anisotropy from Eq. (14.11); φ_a may not coincide with the direction of $\boldsymbol{M}$ during annealing in a single crystal (Chikazumi 1956).

Figure 14.6*a* shows the dependence of the strength of the field-induced anisotropy on annealing temperature in crystalline $Ni_{1-x}Fe_x$ alloys (Ferguson 1958). The annealing temperature enters the field-induced anisotropy through $l_a \approx M^2(T_a)$ as well as appearing explicitly in the denominator of Eq. (14.11). The strength of the anisotropic pair ordering depends on the strength of the magnetization at the annealing temperature $[M_s(T_a)]^2$ and not the strength of the field. Clearly there can be no field-induced anisotropy in a sample quenched from an annealing temperature above T_C. The results in Figure 14.6*a* follow the predicted dependence, but the data in panel (*b*), for an amorphous alloy, fall below the theoretical prediction for temperatures below about 280°C. This weakening of the induced anisotropy for lower annealing temperatures is a kinetic effect associated with the short-range diffusion that is involved. At lower temperatures the equilibrium pair distribution is not achieved in the times allowed for annealing.

The fact that induced anisotropy is predicted to vanish for $T_a \geqslant T_C$ presents a challenge in explaining anisotropy induced by a field present during film deposition. Conventional wisdom would say that $M = 0$ during condensation of the film, so no anisotropy should be induced during that process. Further,

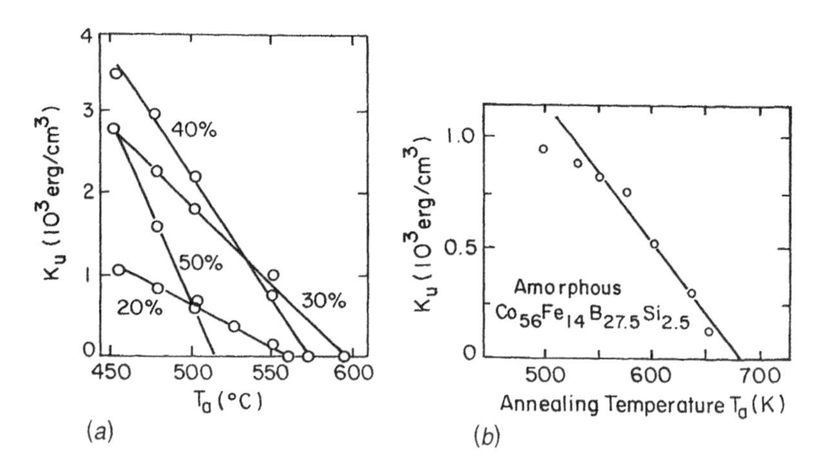

Figure 14.6 Dependence of strength of field-induced anisotropy in (*a*) Fe_xNi_{1-x} alloys (Ferguson 1958), where the curves are labeled by Fe content; and (*b*) amorphous CoFeBSi (Fujimori et al., 1978) on annealing temperature. The solid lines follow $M^2(T_a)/T_a$ as suggested by Eq. (14.11).

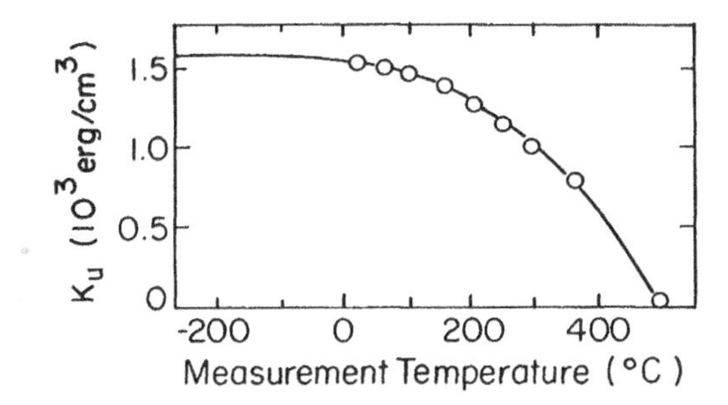

Figure 14.7 Temperature dependence of strength of field-induced anisotropy in $Ni_{69}Co_{31}$. The solid line varies as $M^2(T)$ (Yamamoto et al. 1961).

even if the Curie temperature is high enough that some diffusion can occur at that temperature, the film has so little mass that it should cool quickly through the ferromagnetic temperature range below T_C, again making it unlikely that significant pair ordering could be established. Hence, it is difficult to understand the presence of anisotropy induced during film deposition unless the substrate is held at a sufficiently high temperature.

The dependence of K_u on *measurement* temperature is shown in Figure 14.7. As predicted by Eq. (14.11), $K_u \propto M_s^2(T)$.

Evidence of the composition dependence of K_u represented in Eq. (14.11) is shown in Figure 14.8 for crystalline $Fe_{1-x}Ni_x$. The anisotropy in NiFe follows the compositional dependence $x^2(1 - x^2)$ of Eq. (14.11) reasonably well. The departure of the NiFe from the n_{Ni}^2–n_{Fe}^2 curve is due to variations in Curie temperature and magnetization with composition. To take these into account, all the composition-dependent terms in Eq. (14.11) should be grouped on the

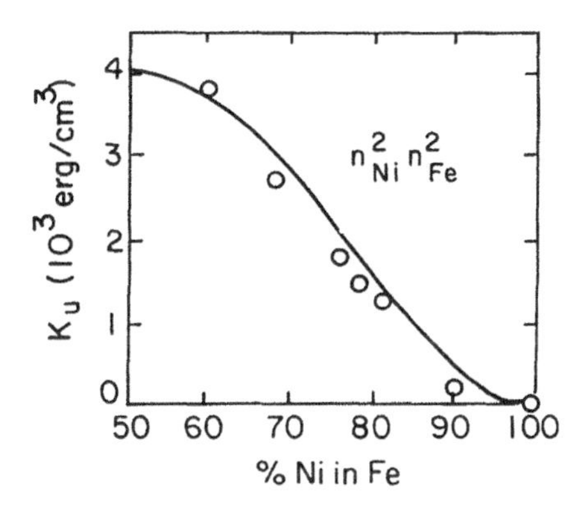

Figure 14.8 Variation of strength of field-induced anisotropy with composition; on left, K_u in $Fe_{1-x}Ni_x$ varies as the concentration squared of each component (solid line) (Chikazumi and Oomura 1955).

left side of the equation. Luborsky and Walter (1977) have done this sort of analysis for amorphous $(Fe_{1-x}Ni_x)_{80}B_{20}$ alloys (see Fig. 14.14*b*, below).

Example 14.1 The interstitial concentration parallel and perpendicular to the magnetization can be estimated for annealing at 425°C. From Eq. (14.9), $N_i/N \approx \frac{1}{3}[1 - l(\cos^2\phi_a - \frac{1}{3})/k_B T_a]$ and the departure from equilibrium concentration goes as $1 - C/T_a$. It will be shown later that $l \approx 5 \times 10^{-16}$ erg, which then gives

$$\frac{N_{\parallel}}{N} \cong \frac{1}{3}(1 - 3.0 \times 10^{-3})$$

$$\frac{N_{\perp}}{N} \cong \frac{1}{3}(1 + 1.5 \times 10^{-3})$$

Thus, for annealing at 425°C, the equilibrium deviations from an isotropic interstitial distribution are measured in tenths of a percent. More interstitial bonds are found in the plane perpendicular to $\boldsymbol{M}$ than in the direction of $\boldsymbol{M}$ for $l_a > 0$.

Slonczewski (1963) summarizes the experimental results and theoretical interpretation with regard to the strong induced anisotropy in cobalt-substituted magnetite, $Co_xFe_{3-x}O_4$. The mechanism for the induced anisotropy is spin–orbit interaction for the monatomic directional ordering of the Co^{2+} ions. The ground state of the Co^{2+} ion in an octahedral (B) site is orbitally degenerate. The trigonal arrangement of next-nearest-neighbor atoms (see Figure 6.17) lowers the symmetry and lifts the degeneracy when the angular momentum $\boldsymbol{L}$ and hence the spin $\boldsymbol{S}$ are parallel to the trigonal, $\langle 111 \rangle$, axis.

14.4 MAGNETIC AFTEREFFECTS

Although kinetic factors can limit the effectiveness of field annealing at lower temperatures, there are some impurities that are so small that they are mobile even at room temperature. For example, carbon and nitrogen interstitials are sufficiently mobile in α-Fe to be able to hop to adjacent interstitial sites at room temperature. When such impurities are present, the same pair ordering as described above can occur at room temperature. This pair ordering can give rise to time-dependent effects in the magnetic properties as the direction of magnetization is changed. These so-called magnetic aftereffects are now described.

In 1887 Lord Rayleigh observed that application of a stress to low-carbon steel has two effects on different timescales: (1) an instantaneous elastic strain and (2) an additional time-dependent strain (Fig. 14.9). The time constant of this additional strain, or aftereffect, depends strongly on temperature. This time-dependent response is a manifestation of what is called *anelasticity*.

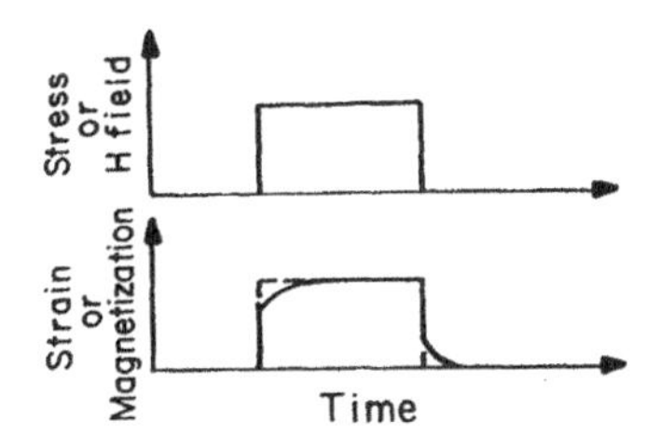

Figure 14.9 Schematic illustration of an elastic strain aftereffect (below) when a material is subjected to the stress shown above. Similarly aftereffects occur in magnetization or in magnetostriction.

Snoek (1941) explained this phenomenon using an atomic model of carbon interstitials in the BCC iron lattice. Carbon atoms can occupy X, Y, or Z octahedral interstices in the BCC lattice of iron (Fig. 14.10). For low concentrations of carbon and in the absence of strain or magnetization, the X, Y, and Z sites are equally populated by carbon to a concentration C_0. (The carbon can represent any mobile interstitial species.)

The application of a strain, along the x axis, for example, favors occupation of the X interstices because of the slightly larger volume those sites take on in the strained state. For carbon to jump from Y or Z sites to the X sites requires thermal energy represented by an activation energy Q; the jump process is a funcion of time and temperature. Figure 14.11 shows a schematic representation of the potential seen by an interstitial on the top surface of the BCC structure shown at upper right. The potential maxima correspond to the corner and body-centered atom positions. The one-dimensional potential energy diagram at lower right shows how the minima, of equal depth in the absence of magnetization or stress, can take on different values in the presence of an applied stress. The characteristic time for this jump is given by

$$\tau = \tau_0 \exp\left(-\frac{Q}{k_B T}\right) \tag{14.13}$$

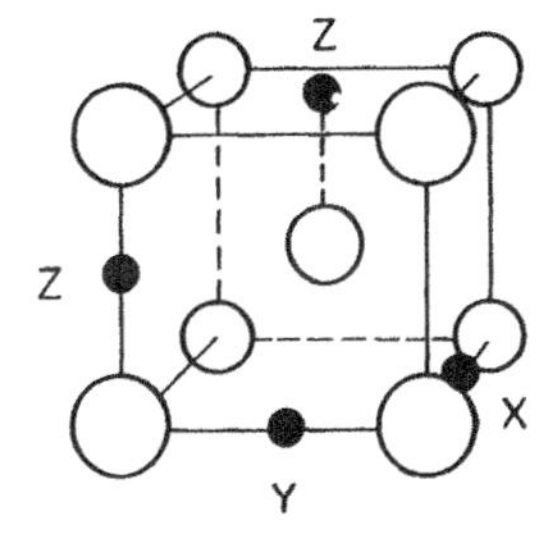

Figure 14.10 Octahedral interstices preferred by carbon, nitrogen, or boron atoms in body centered cubic structure. Hopping of interstitials between sites is responsible for the typical Snoek effect and for magnetic aftereffects.

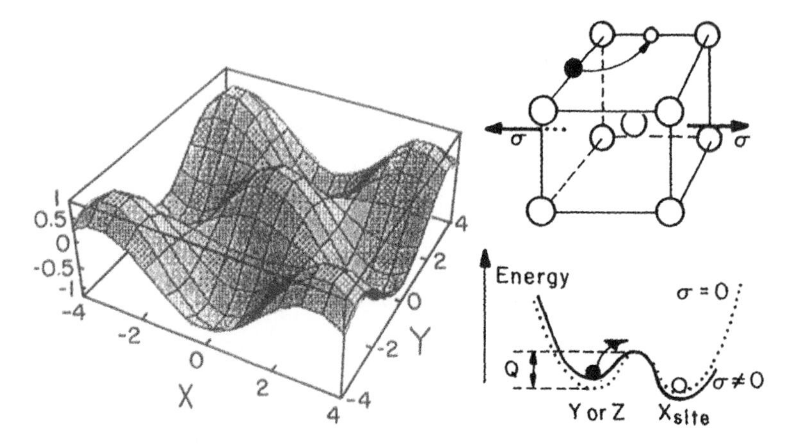

Figure 14.11 Left, representation of potential in the x–y plane (arbitrary units) seen by an interstitial on the top surface of BCC structure shown at upper right. Below, right: double-well potential involved in the thermally activated process shown above. The X site is favored by the tensile stress along the x direction.

With adequate time and temperature there will be a small excess carbon concentration C_x at the X sites. This interstitial mobility is responsible for the time-dependent strain depicted in Figure 14.9. On removal of the stress, there is an instantaneous elastic contraction along x, followed by a gradual anelastic contraction as the carbon interstitials return to an isotropic distribution. This elastic aftereffect is called the *Snoek effect* in recognition of the explanation given for it by Snoek in 1941.

The magnitude of the time-dependent, anelastic strain due to impurity redistribution can be determined by considering the free energy density with pure elastic, impurity elastic [Eq. (14.4)], and entropy terms as follows (Snoek 1941):

$$\begin{aligned} f_{\text{elastic}} &= (\tfrac{1}{2})c_{11}(e_{xx}^2 + e_{yy}^2 + e_{zz}^2) + c_{12}(e_{xx}e_{yy} + e_{yy}e_{zz} + e_{zz}e_{xx}) \\ &\quad + (\tfrac{1}{2})c_{44}(e_{xy}^2 + e_{yz}^2 + e_{zx}^2) + \varepsilon(C_X e_{xx} + C_Y e_{yy} + C_Z e_{zz}) \\ &\quad + \left[\frac{RT}{2VC_0}\right](C_X^2 + C_Y^2 + C_Z^2) \end{aligned} \tag{14.14}$$

where c_{ij} are elastic constants; ε is the energy caused by the strain on occupation of X, Y, or Z sites at concentrations C_i (i = X, Y, or Z) in excess of C_0; and $3C_0$ is the total atomic concentration of interstitials and $\Sigma C_i = 0$. The last term in Eq. (14.14) describes the effect of entropy; it is a minimum for isotropic distribution ($C_i = 0$). V is the molar volume of iron, $7.1 \times 10^{-6}\ \text{m}^{-3}$, and $R = N_A k_B = 8.31$ J/(mol·deg) is the gas constant, while N_A is Avogadro's number. The energy ε is found to be -9.9 and $-9.2 \times 10^{10}\ \mathrm{J/m^3}$ for carbon and nitrogen impurities, respectively, in BCC Fe. The elastic constants of iron are given in the second appendix to Chapter 7. Minimization of f_{elastic} with respect to the e_{ij} and C_i determines the equilibrium strains and concentration anisotropies for a given C_0 and temperature. The fact that $\varepsilon < 0$ implies that the energy is lowered if an excess X occupation ($C_X > 0$) is accompanied by a positive strain $e_{xx} > 0$. Thus, the mechanical aftereffect (time-dependent strain) is always in the direction of the stress. (If the carbon concentration were sufficient that the dipole strain field of one interstitial, say, X, biased nearby carbons to occupy X sites also, then, in that region, the iron might undergo a tetragonal distortion and locally move toward a new, body-centered tetragonal structure similar to the intermetallic compound cementite, Fe_3C.)

As implied by Figure 14.9, there are also aftereffects associated with certain magnetic processes. Consider, for example, BCC Fe with a small carbon concentration; the easy axes are the $\langle 100 \rangle$ directions. Annealing in the presence of a field sufficiently strong to align the magnetization causes a redistribution of the carbon interstitials to accommodate the field direction. If the sample is rotated quickly to another easy direction orthogonal to the first, the carbon interstitials redistribute and cause an observable time rate of change in the anisotropy in the new direction. Ratheneau and de Vries (1969) conducted this experiment and measured the anisotropy aftereffect by monitoring the oscillation period of the sample in a torque magnetometer (Chapter 6). The results for this anisotropy aftereffect are shown in Figure 14.12*a* for Fe–0.015 wt% C. The induced anisotropy axis changes from the original direction to the new direction of the field as interstitials move to accommodate the new direction of magnetization. The increase in induced magnetic anisotropy with time results in increased torque and, therefore, decreased oscillation period.

An aftereffect can also appear in the magnetostrictive strain accompanying the magnetization process; Figure 14.12*b* shows the magnetostrictive aftereffect observed in Fe–0.008 wt.% C. With initial application of a field, the normal magnetostrictive strain appears immediately. With time, this strain *decreases* as interstitials redistribute themselves to minimize their energy; that is, the magnetostrictive aftereffect is negative (opposite the effect observed for mechanically induced strain). Thus interstitials do not move into the sites opened up by the magnetostriction [which would be consistent with Snoek's mechanism given by ε in Eq. (14.14)]. Insted, they move to reduce the instantaneous strain in the direction of $\boldsymbol{M}$. Attempting to describe the observed magnetostrictive aftereffects of Figure 14.12 with Eq. (14.14), gives the wrong sign, and the calculated effect is too small. The magnetostrictive aftereffect demands a new microscopic mechanism.

It turns out that the energy of an interstitial in certain directed sites depends also on the local direction of magnetization as described in Eq. (14.3) (Néel, 1952). The converse, namely, that the direction of magnetization depends on the energy of directed pairs, is at the root of magnetic anisotropy. To describe

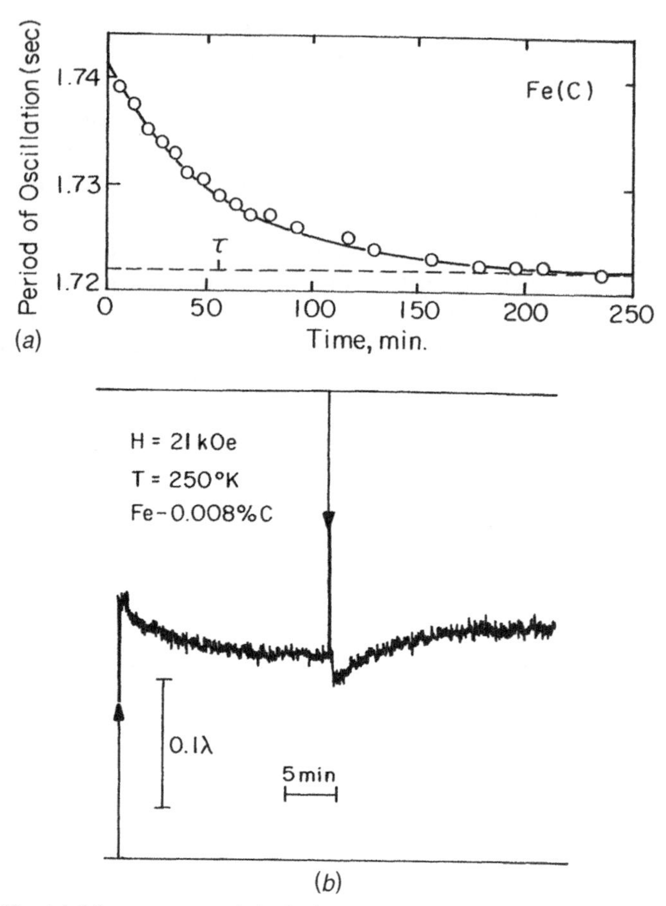

Figure 14.12 (*a*) Measurement of the induced anisotropy aftereffect for Fe–0.015% C using torque magnetometer oscillation period with a field along an easy axis at 90° to the induced anisotropy direction: (*a*) time-dependent change in hard-axis magnetization; (*b*) time-dependent change in magnetostrictive strain [After Ratheneau and de Vries (1969).]

this and other magnetic aftereffects, magnetic terms must be added to the elastic terms in Eq. (14.14) as follows:

$$
\begin{aligned}
f = f_{\text{elastic}} + f_{\text{mag}} = f_e &+ D\{(\alpha_1^2 - \tfrac{1}{3})C_X + (\alpha_2^2 - \tfrac{1}{3})C_Y + (\alpha_3^2 - \tfrac{1}{3})C_Z\} \\
&+ \varepsilon(C_X e_{xx} + C_Y e_{yy} + C_Z e_{zz}) \\
&+ K_1[\alpha_1^2\alpha_2^2 + \alpha_2^2\alpha_3^2 + \alpha_3^2\alpha_1^2] + K_2\alpha_1^2\alpha_2^2\alpha_3^2 \\
&+ B_1[(\alpha_1^2 - \tfrac{1}{3})e_{xx} + (\alpha_2^2 - \tfrac{1}{3})e_{yy} + (\alpha_3^2 - \tfrac{1}{3})e_{zz}] \\
&+ B_2(\alpha_1\alpha_2 e_{xy} + \alpha_2\alpha_3 e_{yz} + \alpha_3\alpha_1 e_{zx}) \qquad (14.15)
\end{aligned}
$$

The pure magnetic anisotropy and magnetoelastic terms (in K_i and B_i) are familiar (Chapters 6 and 7): $K_1 = 4.2 \times 10^4$, $K_2 = 1.5 \times 10^4$, $B_1 = -3.3 \times 10^6$, $B_2 = 7.3 \times 10^6$ J/m^3 for Fe. The terms proportional to D describe the magnetic anisotropy energy difference for impurity site imbalances given by the C_i. Experiments show that D is of order $+7.1 \times 10^6$ J/m^3 for carbon impurities and $+4.5 \times 10^6$ J/m^3 for nitrogen impurities in Fe.

A positive value for D means that the direction of magnetization has fewer impurity bonds (i.e., $\alpha_i \approx 1$, $\alpha_j \approx \alpha_k \approx 0$ implies $C_i < 0$, $C_{j,k} > 0$) (see Fig. 14.4). Conversely, any excess concentration, $C_i > 0$, is preferably located along a direction of small magnetization component, $\alpha_i \approx 0$. The important point is that the direction of magnetization during annealing results in different energies for different local atomic arrangements or, conversely, different local atomic arrangements can alter the preferred direction of magnetization.

It is possible to use the free energy in Eq. (14.15) to calculate the anisotropy induced by a given impurity concentration C_X after annealing at a temperature T_a. This calculation also requires the equilibrium strain due to the impurities, so the magnetoelastic problem must be solved first. Minimization of the free energy (de Vries et al. 1959) with respect to strain e_{xx} and impurity concentration C_X gives, respectively:

$$e_{xx} = -\frac{\varepsilon C_X + B_1\alpha_1^2}{c_{11} - c_{12}} \tag{14.16}$$

(which, without impurities e_{xx}^0, reduces to the magnetostrictive strain calculated in Chapter 7) and

$$C_X = \frac{-(D\alpha_1^2 + \varepsilon e_{xx})}{A} \tag{14.17}$$

where the definition $A = RT/VC_0$ has been made. Eliminating the concentration and solving for the strain

$$e_{xx} = \alpha_1^2 \frac{(\varepsilon D/A) - B_1}{(c_{11} - c_{12}) - \varepsilon^2/A} \tag{14.18}$$

gives for the magnetostrictive aftereffect

$$\frac{e_{xx} - e_{xx}^0}{e_{xx}^0} = \frac{\Delta\lambda}{\lambda} = \frac{\varepsilon}{AB_1}\left[\frac{\varepsilon B_1}{(c_{11} - c_{12})} - D\right] \tag{14.19}$$

where a small term has been omitted. If $|\varepsilon B_1/(c_{11} - c_{12})| > |D|$, then the magnetostrictive aftereffect is always positive. If $|D| > |\varepsilon B_1/(c_{11} - c_{12})|$, then the magnetostrictive aftereffect depends on the sign of $-\varepsilon D/B_1$. In α-Fe, the magnetostrictive aftereffect is negative (Fig. 14.12*b*), consistent with the independently determined parameters, $B_1 < 0$, as $\varepsilon < 0$ and $D > 0$. For

γ-Fe-Ni(C), the magnetostrictive aftereffect is positive for λ_{110} (Masters and Wuttig 1974) and can be interpreted consistently within the framework of this model.

For the impurity-induced magnetic anisotropy, the direction cosines that give the orientation of $\boldsymbol{M}$ during and after the anneal, α_i, and β_1 respectively, must be distinguished. When this is done, Eqs. (14.17) and (14.18) can be used in Eq. (14.15) to give the impurity-induced anisotropy:

$$K_u = \frac{A}{2}\left(D - \frac{\varepsilon B_1}{c_{11} - c_{12}}\right)^2 [(\alpha_1^2 - \beta_1^2)^2 + (\alpha_2^2 - \beta_2^2)^2 + (\alpha_3^2 - \beta_3^2)^2] \quad (14.20)$$

Figure 14.12*a* shows this magnetic anisotropy aftereffect at 239 K for 0.015 wt% C in α-Fe as measured by oscillation period (Ratheneau and de Vries 1969). The magnetization aftereffect vanishes for the magnetization in the easy direction. Note that the anisotropy aftereffect cannot be used to determine the signs of ε and D because their difference is squared. This is not the case for the magnetostrictive aftereffect. Only with the addition of Néel's anisotropy term D could the negative magnetostrictive aftereffect in Fe-C be explained (B_1 is negative for Fe-C).

The log of the decay time of induced anisotropy can be plotted as a function of inverse temperature (Arrhenius plot) to determine the activation energy Q of the atomic jumps involved. Figure 14.13 shows the anisotropy activation energy for boron mobility in amorphous $Co_{56}Fe_{14}B_{25}Si_2$ alloys to be of order 1 eV.

Any two dissimilar atoms, at least one of which is magnetic, can show the same form of anisotropy energy due to their orientation relative to the direction of magnetization. These interactions are particularly strong if both atoms are magnetic. The mechanisms involved in magnetic aftereffects are the same as those described in Section 14.3 for field-induced anisotropy. The principal difference between magnetic aftereffects and field-induced anisotropy is the temperature range over which atomic mobility occurs. Because of the lower mobility of Fe in Ni (or vice versa) compared to C or N in Fe, the field-induced anisotropy due to pair ordering of the former species must be induced at elevated temperatures.

While carbon and nitrogen interstitials in α-Fe show comparable impurity–elastic coefficients, $\varepsilon \approx -10^{11}\ \mathrm{J/m^3}$, their impurity anisotropy coefficients, $D = 7.1$ and $4.5 \times 10^6\ \mathrm{J/m^3}$, respectively, differ significantly. Levy (1965) has proposed a local crystal field model of the interstitial in the tetragonal field of its positively charged Fe neighbors. The formalism is similar to that used to describe magnetocrystalline anisotropy in insulators (Chapter 6). The crystal field potential in which the impurity resides is expanded in spherical harmonics. The isotropic term in the potential describes the bonding or heat of solution of the impurity in the site and is proportional to the impurity radius. The dipolar term describes the local uniaxial interaction of impurity p orbitals with the distorted crystal field. Both of these terms may be weakly perturbed

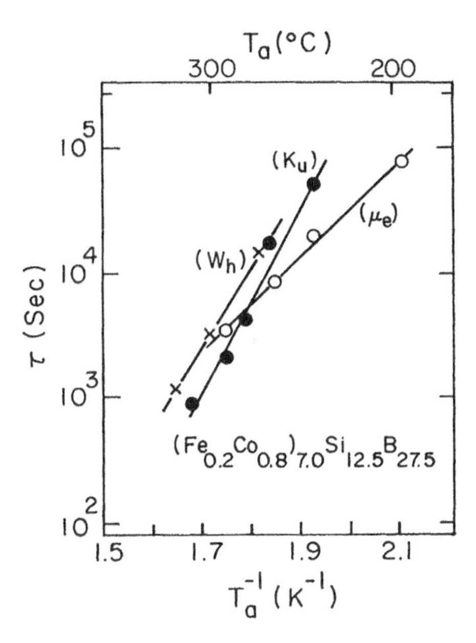

Figure 14.13 Arrhenius plots of the relaxation time of the magnetic anisotropy aftereffect in an amorphous metallic alloy measured from the anisotropy aftereffect (K_u), effective AC permeability (μ_e), and hysteresis loss (W_e). Activation energies (in eV) and frequency prefactors (in 10^9 s^{-1}) were determined as follows: from K_u, 1.4, 1.9; from μ_e, 0.75, 1.1; from W_e, 1.2, 1.3 [After Fujimori, 1983)].

by changes in the direction of magnetization on the surrounding Fe atoms. The coupling of the isotropic term to the direction of the iron moment is responsible for the impurity–elastic interaction given by ε. The coupling of the dipole potential term to the direction of magnetization of the surrounding Fe atoms gives rise to the impurity–magnetic term, D.

Levy's model leads to the following predictions:

1. Interstitials in p^0, p^3, and p^6 orbital configurations should have $D = 0$. In these cases, any magnetic aftereffects will arise from ε alone and will, therefore, be positive. This suggests that C and N interstitials may be in $2p^1$ and $2p^2$ configurations, perhaps due to their bonding with the surrounding Fe atoms.

2. The relative strength of the impurity elastic effects for N and C in Fe, $\varepsilon_N/\varepsilon_C \approx 0.93$, goes as the ratio of the ionic radii of the impurities: $r_N/r_C \approx 0.93$). The relative strength of the magnetic coupling to the dipolar potential term (giving rise to D) is 2:1 for p^1 (C) and p^2 (N) configurations. This may explain the stronger impurity magnetic effect of C relative to N.

The theory of magnetic aftereffects is reviewed by Kronmüller (1968) and specifically for amorphous magnetic alloys (Kronmüller 1983).

14.5 STRESS ANNEALING

Many materials respond to annealing under stress, and in some cases the stress is not intentionally applied but is a consequence of the environment of the material during heat treating. When a material is annealed under stress, three effects are in operation: (1) the material may experience an elastic strain due to the stress, (2) it may also experience an irreversible strain called creep during annealing, and (3) the stress-induced anisotropy $\lambda_s \sigma$ at the annealing temperature may align the magnetization and so give rise to a field-induced anisotropy. It is important to know which of these effects dominate and how the direction of the net induced anisotropy relates to the applied stress.

First, find the concentration [Eq. (14.17),] that minimizes the energy in Eq. (14.15) for a plastic flow given by $e_{zz} = e_0 > 0$. On the basis of known experimental values of $D \approx 10^6$–10^7 J/m^3 and $\varepsilon \approx -10^{11}$ J/m^3, the result for $e_0 > 10^{-3}$ is due largely to the impurity elastic term:

$$C_z \approx -\varepsilon e_0 \frac{C_0 V}{RT}$$

For a 1% elongation, a directed-pair concentration of $3C_0 \approx 20\%$, a value for $R/V = 1.2 \times 10^6$ J/m^3 K, and $T = 380°\text{C} = 553$ K, the concentrations are $C_z \approx 10^{-3}$ and thus, $C_x = C_y = -0.5 \times 10^{-3}$.

The magnitude of various anisotropic terms in Eq. (14.15) are now compared. The $\varepsilon C_i e_{ii}$ terms are of order 10^6 J/m^3 but give no magnetic anisotropy; $D\alpha_i^2 C_i$ is of order $10^4 \alpha_z^2$ J/m^3 and, $B_1 \alpha_z^2 e_{zz}^{\text{elastic}}$ is of order $10^2 \alpha_z^2$ J/m^3 (because it is defined for elastic strains). For the C_i given above, the strongest terms from Eq. (14.15) are $f \approx 1.5 \times 10^{-3} D \cos^2\theta + K_1 \alpha_i^2 \alpha_j^2 + \cdots$ constant. For Fe, $D = 7 \times 10^6$ J/m^3, and $K_1 = 4.5 \times 10^4$ J/m^3. Thus, aside from the magnetocrystalline anisotropy, which may average to a much smaller value in a polycrystalline sample, magnetization is favored at 90° to the plastic flow direction, and H_a is of order $H_a \approx 10^5/M_s \approx 100$ Oe.

Plastic flow is the most important effect in stress annealing, it enters the magnetic anisotropy through Néel's D term giving an easy axis or plane *orthogonal* to the elongation direction.

14.6 MORE ON MECHANISM OF ANISOTROPY

The directional ordering of transition metal pairs after field or stress annealing, believed to constitute a redirection of approximately one bond per thousand, cannot be seen directly by X-ray or electron diffraction or other direct structural probes. What has been found is that annealing in FCC alloys containing Co and/or Ni (e.g., permalloys, perminvars), is dependent on the presence of oxygen. Beginning with work by Heidenreich et al. (1959), it was found that field annealing of high-purity FCC alloys in reducing atmospheres induces little anisotropy. However, when oxygen is present in excess of 10 ppm, field-induced anisotropy can be significant. How does oxygen play a critical role in field annealing?

The mechanism appears to be associated with the directional interaction between cobalt (or Ni) and oxygen. Transmission electron microscopy evidence suggests that initially, oxygen atoms coalesce between $\{111\}$ planes, forming an oxygen fault. These local concentrations of one species are called Guinier–Preston (GP) zones. They show up as streaks in electron diffraction patterns and can be directly imaged as lines or bands perpendicular to the $\langle 111 \rangle$ directions. With increasing time and temperature, the oxygen concentration profile across fault planes becomes more sharply defined.

The atomic arrangement about the oxygen plane is similar to that about $\{111\}$ oxygen planes in antiferromagnetic CoO or NiO of rocksalt structure. It is also similar to that in the ferrimagnetic spinel $(Co\text{-}Ni\text{-}Fe)_3O_4$. The favored direction of magnetization is perpendicular to the fault plane. The stacking fault energy in FCC is lowest for $\{111\}$ planes. Thus, when the field is applied in a direction that favors impurity faults in $\{111\}$ planes (i.e., the field direction is perpendicular to the $\{111\}$ planes), the strongest anisotropy will be induced. Experiment bears this out: The strength of field-induced anisotropy decreases for field directions $\langle 111 \rangle$, $\langle 110 \rangle$, and $\langle 100 \rangle$ (Nesbitt et al., 1959).

The mechanism by which anisotropy is induced in BCC iron-rich alloys (e.g., 3% Si-Fe) is assumed to be directional ordering of Si-Fe pairs. However, there is no detailed structural evidence here comparable to that for FCC alloys.

14.7 FIELD-INDUCED ANISOTROPY IN AMORPHOUS ALLOYS

It has been assumed that in amorphous alloys, bond orientational order is responsible for field-induced anisotropy. The usual model can be summed up by the schematic shown in Figure 11.15. It is interesting to consider a statement made by Berry in describing the first field annealing of an amorphous alloy (Berry and Pritchett 1975):

> It is well known that the structure-sensitive ferromagnetic properties of certain crystalline solid-solution alloys can be significantly modified and made uniaxially anisotropic by annealing in a magnetic field. This behavior has been successfully explained in terms of a directionality induced in the state of short-range order of the alloy by the magnetization $\boldsymbol{M}$, a process which necessitates local atom motion below the Curie point, T_C. Although developed to explain the behavior of crystalline alloys, this mechanism is actually of wide generality and should also apply to amorphous alloys, since only short-range directional order is involved.

There is now some structural evidence that the mechanism of field-induced anisotropy in cobalt-rich amorphous alloys is similar to the faulting mechanism in crystalline FCC alloys (Kim and O'Handley 1996). TEM studies show that in cobalt-rich amorphous alloys, heat treatment allows metalloids (notably boron and silicon) to segregate to the surface, driven by their high heat of oxide formation. Thin superficial oxide layers (B_2O_3, SiO_2) or patches result for temperatures and times of order 300°C, 30 min. The underlying amorphous material is depleted of glass-stabilizing metalloids so its crystallization temperature drops. Small (typically 300-Å) FCC crystallites form by primary crystallization. These crystallites are marked by a high density of $\{111\}$ faults similar to those associated with field annealing of crystalline perminvars. The faults are expected to form on those $\{111\}$ planes most nearly normal to the direction of the magnetization during annealing. There is yet no experimental evidence of preferred orientation of these faults.

The field annealing mechanism in Fe-rich alloys remains to be understood in such detail. By default, a simple directional ordering is assumed. There is indirect evidence supporting this assumption.

Because of the dependence of induced anisotropy on the number of pairs in an A_xB_{1-x} alloy, the strength of induced anisotropy varies as $[x(1-x)]^2$. The composition dependence of induced anisotropy for amorphous $(Fe_{1-x}[Ni, Co]_x)_{80}B_{20}$ shown in Figure 14.14*a*, however, differs from that for crystalline NiFe (Fig. 14.8) (Fujimori 1983). Note that while K_u is greater near $x = 0.5$ as expected, it does not vanish quadratically near $x = 0$ or 1. It is believed that this is due to contributions to the directed-pair energy by transition metal–metalloid pairs. This weaker Fe–B or Ni–B pair energy is expected to be less stable because of the greater mobility of the small B atoms in the amorphous matrix compared to that of Ni or Fe. There is some evidence for this.

Luborsky and Walter (1977) have analyzed the induced anisotropy data for amorphous $(Fe_{1-x}Ni_x)_{80}B_{20}$ alloys (Fig. 14.15*b*). (Because of the role of B in the field-induced anisotropy, the value of K_u measured for $Fe_{80}B_{20}$ has been subtracted from the field-induced anisotropy for the rest of the compositions.) All composition-dependent terms in Eq. (14.11) have been taken to the left-hand side (LHS) of the equation. It is clear that the data presented in this way fall well above the solid curve $n_{Fe}^2 n_{Ni}^2$ (normalized to the value at $x = 0.5$), varying more like $x(1-x)$. This may be due to chemical ordering of Fe and Ni within the amorphous structure.

Iwata (1961) has modified the Néel theory to account for the effects of chemical ordering or segregation on the induced anisotropy. The dashed line in Figure 14.14*b* is Iwata's theory for the effect of the normalized Fe–Ni interaction $2V/k_BT$ on K_u. The parameter V represents the effective nonmagnetic interaction energy between *like* pairs of atoms:

$$V = V_{AA} + V_{BB} - 2V_{AB}$$

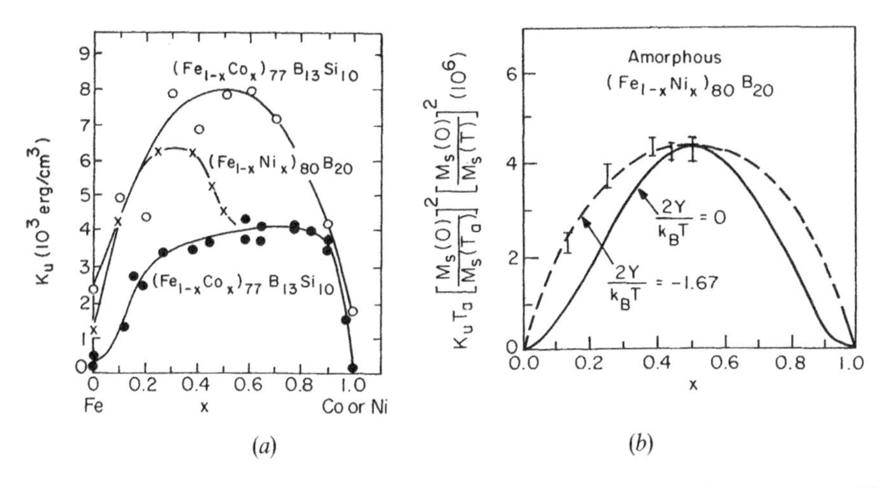

Figure 14.14 (*a*) Concentration dependence of induced anisotropy in various metallic glasses (Fujimori 1983); (*b*) in amorphous FeNi-based alloys a similar dependence is observed but account must be taken of the chemical interactions, given by V, between Fe and Ni (Luborsky and Walter 1977).

A value of $2V/k_BT = -1.67$ (implying segregation, i.e., species A and B separating) fits the data well. The solid line is the bare Néel form $x^2(1-x)^2$ for an ideal solid solution, $V = 0$. The strong tendency toward segregation evidenced in Figure 14.14*b* is interpreted to mean that Fe–Ni pairs are less available for directed pair induced anisotropy than if $V = 0$ or $V > 0$ (favoring Ni-Fe pairs).

14.8 MICROSTRUCTURAL EFFECTS

In conventional field-induced anisotropy, four steps have been identified: (1) the applied field orients the magnetization; (2) the direction of magnetization establishes a thermodynamically preferred orientation for certain *local* atomic arrangements; (3) given sufficient time and temperature, the kinetic process of establishing this preferred orientation occurs; and finally, (4) once a degree of local directional ordering is frozen in by cooling to lower operating temperatures, this ordering determines a preferred direction for the magnetization. The local atomic arrangements involved here require only small atomic displacements and constitute a small volume fraction of the material. Hence, they are not readily detected by direct structural observations.

However, in some cases field-induced anisotropy may be due to directly observable microstructural effects brought about by field annealing. One case already seen is the development of an oriented, two-phase structure in alnico magnets (Figs. 13.7, 13.11). The spinodal decomposition responsible for the periodic phase separation in Alnico, as well as in certain ductile Fe-Co-Cr magnets, has been analyzed by Cahn (1962, 1963, 1968) (Section 13.2.1). The elongated two-phase microstructure provides a shape anisotropy to the material. Similarly striking is the field-induced reorientation of $Fe_{16}N_2$ platelets in certain Fe-N alloys (Sauthoff and Pitsch 1987). In this system, the α''-$Fe_{16}N_2$ phase forms as an array of platelets with their normals distributed at random among the three $\langle 100 \rangle$ axes of the α-Fe matrix (Fig. 14.15*a*). This orientation is favored because of elastic energy as well as the fact that the high-magnetic-anisotropy platelets ($K_u = 8 \times 10^5$ J/m^3) prefer to be magnetized along their normal, *c*-axis direction and the platelet magnetization tends to coincide with the magnetization direction of the α-Fe matrix, which has $\langle 100 \rangle$ easy axes. When a magnetic field is applied, one of these $\langle 100 \rangle$ directions (the direction that has the greatest projection along the field) is energetically favored. If the field is applied at a temperature for which there exists sufficient atomic mobility for microstructural change (about 180°C in this case), then the platelets tend

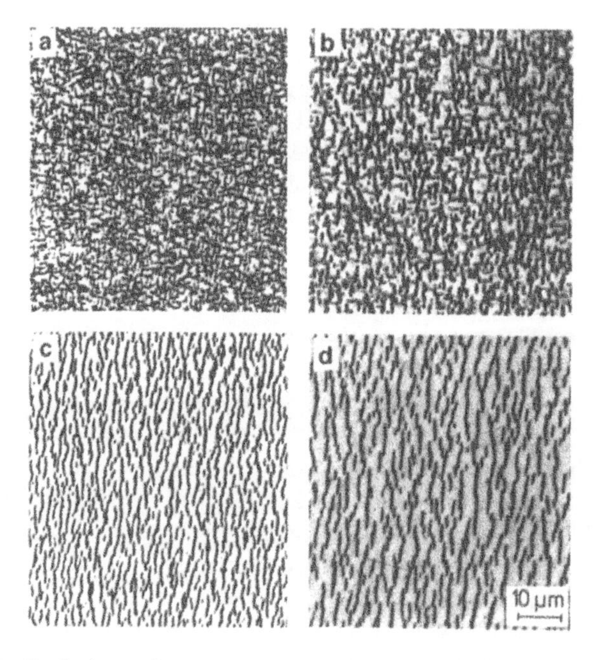

Figure 14.15 Optical metallography of the temporal evolution of the microstructure of α''-$Fe_{16}N_2$ plate-like precipitates in α-Fe matrix on field annealing at $T = 190$°C ($H = 1.71$ MA/m, parallel to scale marker in panel (*d*)). The four panels indicate the degree of texture after aging times of (*a*) 1, (*b*) 5, (*c*) 10, and (*d*) 70 hs. The plates coarsen so as to be oriented with their normals, which are the easy axes, parallel to the applied field direction [After Sauthoff and Pitsch (1987).]

to form with their normals along that proximal $\langle 100 \rangle$ direction (Fig. 14.15*b*). The Fe(N) system is technically important because of the very large magnetic moment (nearly $3\,\mu_B$/Fe) observed for the $Fe_{16}N_2$ phase (Kim and Takahashi 1972).

This section analyzes some of the *anisotropic* microstructural effects associated with processing materials in a magnetic field. Also included is an explanation of the *isotropic* stabilization of certain phases because of their magnetic energy. The latter plays no role in magnetic anisotropy.

14.8.1 Relevant Energies

The *possible* effects of a magnetic field on phase stability, microstructure and texture development in materials processing, can be categorized by the nature of the magnetic material and the form of the magnetic contribution to the energy. In ferromagnets, it is possible to favor the formation of a magnetic phase by virtue of the Zeeman term, $-\mu_0 \boldsymbol{M} \cdot \boldsymbol{H}$, in the free energy density. The Zeeman energy alone does not select a particular shape or crystallographic direction for the microstructure because it is *isotropic* with respect to the phase boundaries and crystal axes of the phase formed. *Anisotropic* microstructure development can arise from the magnetostatic energy or from the magnetocrystalline anisotropy energy. The magnetostatic energy associated with certain microstructural shapes takes the form $\mu_0 \Delta M_s^2 \cos^2\theta$, where $\Delta M \cos\theta$, is the change in the perpendicular component of magnetization across a phase boundary and θ is the angle between the magnetization and the surface normal. For phases having different magnetizations, this energy favors boundaries that are parallel to H. The magnetocrystalline anisotropy energy density, $K_u \sin^2\phi$ (where ϕ is the angle between the magnetization and the uniaxial easy axis), favors development of certain crystallographic orientations relative to the field direction.

Analogous but weaker effects are possible in paramagnets and diamagnets (including superconductors). *Isotropic* effects can occur as a result of Zeeman terms $-\mu_0 \chi H^2$ in the free energy. *Anisotropic* effects can arise from anisotropy in the susceptibility, such as $\mu_0(\chi_{\parallel} - \chi_{\perp})H^2$, or simply as a result of magnetostatic effects associated with the shape of microstructural features.

In order for these magnetic energies to bring about a change in structure, there must be atomic mobility (which is a matter of sufficient thermal energy for the requisite motions in reasonable times). Further, the magnitude of the magnetic energy must be significant compared to the other energies involved. Of particular concern are the free energies of the phases present as well as the surface and interface energies which play an important role in microstructural changes.

It is helpful to refer to Figure 14.16 in understanding the thermodynamics and kinetics of magnetic effects in materials processing. This figure compares the magnitudes of some of the important surface and volume energies important for microstructural growth.

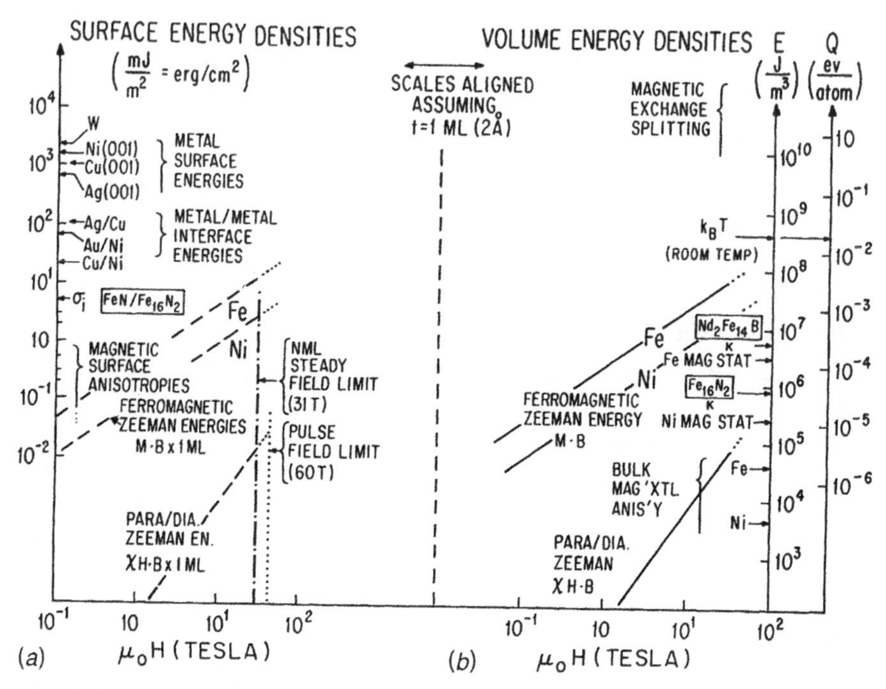

Figure 14.16 Comparison of surface energy densities (*a*) and volume energy densities (*b*) relevant to magnetic field processing of materials. The left and right scales are aligned only for samples of thickness, $t_0 = 1$ monolayer (ML). For samples of some larger thickness t normal to the dominant surface, the left-hand scale should be shifted down relative to the right in proportion to the thickness ratio t/t_0.

The right-hand scales indicate volume energy densities (in units of J/m^3 and eV/atom). Bulk magnetocrystalline anisotropy energies K_1 are shown for Fe, Ni, $Nd_2Fe_{14}B$, and $Fe_{16}N_2$. Magnetostatic energies are shown for Fe and Ni. The straight lines show the magnitude of the Zeeman energy densities: $-\mu_0 \mathbf{M} \cdot \mathbf{H}$ for ferromagnetic Fe and Ni samples in H fields given by the horizontal axis. The lines are solid up to the steady field value of 32 T = 320000 G (available at high-magnetic-field facilities), and they are dotted up to the pulse field value 60 T (available at such facilities). Also included is a line showing the Zeeman energy density for a material with susceptibility $\chi = 10^{-4}$. Note that this energy increases quadratically with the applied field.

The left-hand scale shows surface and interface energy densities (in units of mJ/m^2) for some typical metallic systems as well as the range of magnetic surface anisotropies measured for thin-film magnetic materials. The dashed lines indicating Zeeman surface energies have been aligned with the right-hand scale (volume energy densities) assuming a thickness of 0.2 nm (approximately one monolayer). The rationale for this scaling is as follows.

Consider a position-dependent *volume* energy density $\gamma(z)$, composed of a uniform contribution γ^v plus a contribution which is sharply localized near the surface $\gamma^s(z)$. These energy densities have units of energy per unit volume. The Dirac delta function may be used to approximate the surface region as a sheet of zero thickness at $z = 0$: $\gamma^s(z) \to \sigma\delta(z)$, where σ has dimensions of energy/area and the Dirac delta function has units of inverse length. The contribution of the surface energy to the volume energy averaged over a thickness t, normal to the surface of a thin sample, then becomes

$$\langle f \rangle = \gamma^v + \frac{1}{t}\int \sigma\delta(z)dz = \gamma^v + \frac{\sigma}{t} \tag{14.21}$$

Thus, volume energy densities are related to surface energy densities by the material thickness normal to the surface of interest. (In three dimensions, the scale factor between surface and volume energy densities is not t^{-1}, but rather the surface-to-volume ratio of the particle.) The left and right scales of Figure 14.16 have been aligned assuming $t = 0.2$ nm ≈ 1 monolayer (ML). The dashed Zeeman energies at left apply only for sheets of 1 ML thickness. Surface and volume energies can be compared for samples of other thicknesses by sliding the left half of the figure down or the right half up by the value of the appropriate sample thickness relative to 1 ML.

An example of how Figure 14.16 can be used in estimating magnetic field processing effects on microstructure is now given. Consider the formation in the solid state of a new phase B in A where B has lower free energy at a temperature T, $\Delta g = g_B - g_A < 0$. The interface energy of the A–B boundary is σ. It is assumed that a magnetic field can make a difference in the magnitude of σ and/or Δg. In order for B to grow, a cluster of atoms must acquire enough energy to overcome a nucleation barrier, $G^* = 64\sigma^3/(3\Delta g^2)$, and the cluster size must exceed a critical radius, $r^* = 2\sigma/\Delta g$ (below which the surface energy cost is greater than the volume energy saved). In metallic solids, interface energies are on the order of 10 mJ/m^2 and free energy differences between phases are of order 10^7 J/m^3 (14.4 cal/mol) close to the transformation temperature. Thus, $r^* \approx 1$ nm. In ferromagnets, $\mu_0 \boldsymbol{M} \cdot \boldsymbol{H}$ is on the order of 10^6 J/m^3 for $\mu_0 \boldsymbol{H} = 1$ T (fields in excess of 30 T are available at high-magnetic-field facilities). This magnetic energy is small but not insignificant relative to Δg. From Eq. (14.21), it can be seen that the magnetic field energy density ($\approx 10^6$ J/m^3) exceeds the volume-averaged surface energy ($10^{-2}/t$ J/m^3) only for samples of thickness greater than 10 nm. This thickness is greater than the critical size for a typical nucleation process ($r^* \approx 1$ nm). Hence fields of order 1 T are not expected to have significant effects on nucleation in metallic solids. Stronger magnetic fields, such as the 30 T available at special facilities, could influence nucleation processes for which $r^* \approx 1$ nm. The magnetic exchange energy ($\approx 10^{10}$ J/m^3) could influence nucleation at dimensions under 1 nm. Generally, magnetic fields of order 1 T are unlikely to have a significant effect on solid-state nucleation. There must be other processes that respond to magnetic fields of order 1 T.

14.8.2 Coarsening

The microstructural effects shown in Figure 14.15 are manifestations of coarsening (or Ostwald ripening). Coarsening is the process by which a system of two or more phases change after removal of the supersaturation conditions that existed during growth. During coarsening, larger particles of one phase grow at the expense of smaller ones such that there is no net change in the relative fraction of the two phases. Coarsening is driven by differences in surface energy or capillarity (the energy associated with the radius of curvature of a surface) and the concentration gradient that it creates. Surfaces having high convex curvature are composed of atoms with fewer bonds to their native phase and hence a greater probability of being thermally activated into the adjacent phase (with concave surface curvature and hence more bonds per surface atom). The energy difference ΔG between two idealized spherical particles due to their surface tension γ can be expressed

$$\Delta G = \gamma\Omega\left(\frac{3}{r^2}\right) dr \tag{14.22}$$

where Ω is the atomic volume and r is the radius of curvature of the spherical particle. These energy differences are small for two particles of similar size (dr approaches zero). For $r = 20$ nm and $dr = 1.0$ nm the difference in energy is of order 10^{-4} eV/atom (3×10^6 J/m^3). This is comparable to the magnetostatic or anisotropy energy differences between ferromagnetic particles (Fig. 14.17). Thus, while nucleation in solids appears to be beyond the range of conventional magnetic field energies, the coarsening process is not.

The field-induced reorientation of $Fe_{16}N_2$ platelets mentioned above (Fig. 14.15 and related text) is an example of such coarsening. The relevant interfacial energy, $\sigma_i(Fe(N)/Fe_{16}N_2) \approx 5$ mJ/m^2, becomes comparable to the volume magnetic anisotropy energy, $K_1(Fe_{16}N_2) = 0.8$ MJ/m^3, for platelets 6 nm thick or more. The platelets in Figure 14.15 approach 1 μm in thickness.

The energy considerations for magnetic-field-induced coarsening in high-aspect-ratio particles are outlined in Figure 14.17. Here, a competition exists between shape anisotropy and a uniaxial crystal anisotropy of the form $K \sin^2\theta (K > 0$ implies easy axis normal to the platelet).

The magnetic energy densities for platelets 1 and 2 are

$$\begin{aligned} E_1 &= \left(\frac{\mu_0 M_s^2}{2} - K_u\right)\cos^2\theta - \mu_0 M_s H \cos\theta \\ E_2 &= \left(\frac{\mu_0 M_s^2}{2} - K_u\right)\sin^2\phi - \mu_0 M_s H \cos\phi \end{aligned} \tag{14.23}$$

θ and ϕ are defined in Figure 14.17. The first two terms in Eq. (14.23) reflect the competition between shape anisotropy (magnetostatic energy) and the perpendicular magnetocrystalline anisotropy. [If the medium outside the plate

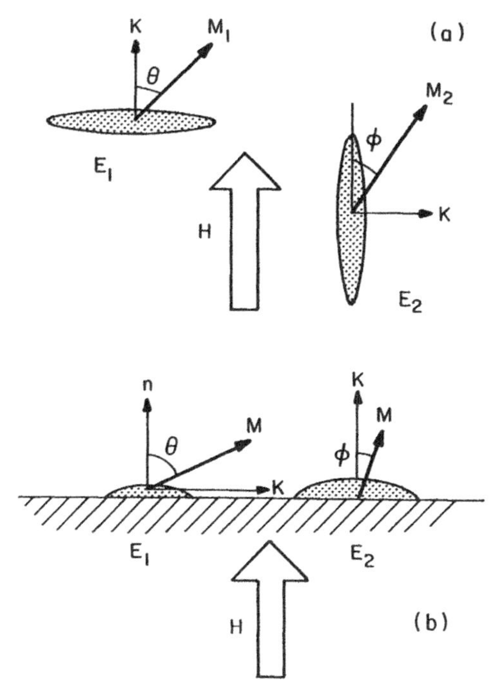

Figure 14.17 Schematic representation of magnetic particles having uniaxial shape and intrinsic anisotropy in an applied magnetic field $\boldsymbol{H}$.

is also ferromagnetic, then M_s in the first term of Eq. (14.23) should be the *difference* in magnetization between the two media, and the standard boundary conditions—Eqs. (2.6) and (2.9)—would have to apply across the interface. Also, if the platelet is not infinitesimally thin, a demagnetization factor $N < 1$ must be applied to the first term.] The last term is the Zeeman energy. In the high-anisotropy limit, $\theta = 0$ and $\phi = \pi/2$, so Eqs. (14.23) give

$$\Delta E = -\mu_0 M_s H \tag{14.24}$$

and configuration (a) in Figure 14.17 is preferred. This is the case for $Fe_{16}N_2$ illustrated in Figure 14.15. In the high-field limit where $\theta = \phi = 0$, the difference in energy between the two plate orientations is

$$\Delta E = E_1 - E_2 = \frac{\mu_0 M_s^2}{2} - K_u \tag{14.25}$$

For $K > 0$ either platelet orientation in Figure 14.17 can be stabilized depending on which term, $\mu_0 M_s^2/2$ or K_u, is stronger. If the magnitude of the shape anisotropy exceeds that of the crystal anisotropy, $\Delta E > 0$ and E_2 is the lower energy state. If, on the other hand, the crystal anisotropy exceeds the shape anisotropy, $\Delta E < 0$ and E_1 is more stable. If the particles have shape and crystal anisotropy that reinforce each other [$K_u < 0$ in Eq. (14.25)], one orientation of $\boldsymbol{M}$ is always favored.

Paramagnets and diamagnets can also respond to fields applied during processing as described below.

Isotropic Effects. In these cases there is no spontaneous moment but only one induced by the external field $\boldsymbol{M} = \chi \boldsymbol{H}$; χ is typically of order 10^{-6} but can be much larger in certain paramagnets such as Pd or in diamagnets with spatially delocalized electron states—a notable example is superconductors. Because of the small magnitude of $\mu_0 \chi H^2$, these effects require very strong fields.

Anisotropic Effects. Many molecules or microstructures with strong shape anisotropy also exhibit strong anisotropy in their magnetic susceptibility χ. Application of a field will bias the particle toward an orientation which minimizes the magnetic energy density $-\mu_0 H^2(\chi_{\parallel} - \chi_{\perp})$. Thus, for paramagnetic systems ($\chi > 0$) the high-susceptibility direction aligned parallel to the field is favored, and for diamagnetic systems ($\chi < 0$) the high susceptibility direction antiparallel to the field is favored. The high-susceptibility direction is not necessarily the long shape axis; in fact, for diamagnets, it is often the short axis if the electron states are sufficiently delocalized because the greatest contribution to the susceptibility comes from the largest orbital area [see Eq. (3.43)].

14.9 SUMMARY

In this chapter, examples have been given to show how local atomic order can influence magnetic properties, in particular magnetic anisotropy. The directional ordering of even a small fraction of interacting magnetic atomic pairs can cause a significant induced magnetic anisotropy. Further, if some of the atoms in the solid are mobile at room temperature, they can respond to changes in applied magnetic field or stress by seeking local configurations that minimize their energy with respect to the direction of magnetization and the sense of the strain. Snoek's mechanism based on magnetostriction, $\varepsilon \Sigma e_{ii} C_i$, $\varepsilon < 0$, puts the interstitials on sites parallel to the strain; Néel's magnetic interaction, $D \Sigma \alpha_i^2 C_i$, $D > 0$, puts interstitials on a plane perpendicular to the magnetization.

Figure 14.18 summarizes schematically what has been described about field annealing. Assume that an initially random distribution of atom pairs exists at room temperature, even in the presence of a saturating field. At elevated temperatures, provided that the material is still magnetic (i.e., $T_a < T_C$), the pairs redistribute so that more pair directions are perpendicular to the

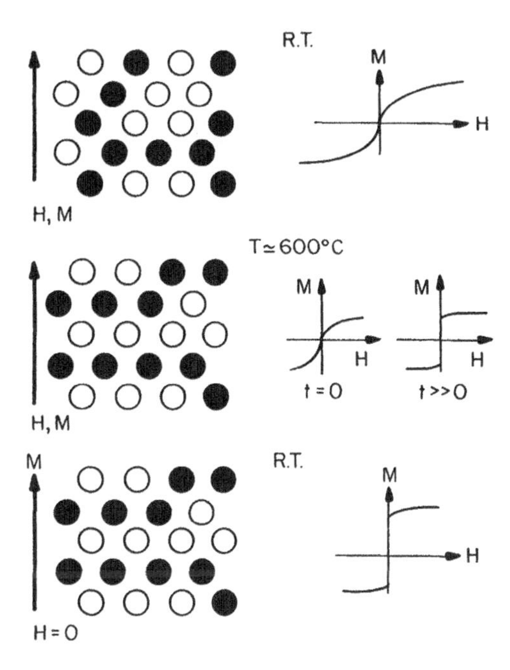

Figure 14.18 Illustration of concept of directional pair ordering (left) and consequent effect on M–H loop (right) for starting material (top), during field heat treatment (center), and after heat treatment (below).

magnetization in each domain [if $D > 0$ in Eq. (14.15)]. This requires short-range diffusion and therefore takes time. After the material is cooled to room temperature in the field, removal of the field leaves a frozen-in, anisotropic distribution of interacting atom pairs. The main effect of this distribution on magnetization is a larger remanent magnetization in the field direction than before the anneal. The energy of the directed pairs is now working against the tendency of the material to demagnetize itself. The induced anisotropy energy density is $(\frac{1}{2})M_s[H_a' - H_a]$, where H_a' is the anisotropy field after annealing.

APPENDIX: ORDER–DISORDER TRANSFORMATIONS

The chemical interactions between species in an alloy can lead to ordering or segregation depending on whether the interaction favors unlike or like pairs, respectively. In certain compounds this interaction can lead to a first-order transformation between an ordered state and a disordered one, stable below and above an ordering temperature. The ordered state is generally of lower symmetry than the disordered state.

In magnetic systems, order–disorder transformations can lead to important differences in magnetic as well as other properties. Two typical examples are

Ni_3Fe and FeCo. Ordering is also important for the development of magnetic hardness in CoPt (Chapter 13, Section 13.7). It was seen in Chapter s 6 and 7 that the ordered Ni_3Fe phase is characterized by *a more positive* value of K_1 and that the magnetostriction constants are similarly, but less sharply, affected. In this material, heat treatment can be used to bring the zero anisotropy condition closer to the zero magnetostriction condition. In the FeCo system, the disordered alloy (obtained by quenching from above the ordering temperature) is characterized by lower magnetocrystalline anisotropy, and it is more readily rolled compared to the ordered phase. Here, the formalism used to describe order-disorder transformations is described.

Consider the Ni_3Fe intermetallic phase, which is known to exist in either a disordered or an ordered structure as illustrated in Figure 14A.1.

In the *disordered* structure, the average occupancy of any given site is 75% Ni and 25% Fe. Since every site is equivalently occupied, the *Bravais lattice is* FCC. But in the *ordered structure*, the basis unit is Fe at $(0, 0, 0)$ and Ni at $(0, \frac{1}{2}, \frac{1}{2})$, $(\frac{1}{2}, 0, \frac{1}{2})$ and $(\frac{1}{2}, \frac{1}{2}, 0)$ and this unit is arranged on the lattice of a *simple cubic structure.*

The structure factors in the two cases are written as follows:

Disordered (FCC)		Ordered (SC)	
$F_{hkl} = f_{av}[1 + e^{\pi i(h+k)} + e^{\pi i(h+l)} + e^{\pi i(k+l)}]$		$F_{hkl} = f_{Fe}e^0 + f_{Ni}[e^{\pi i(h+k)} + e^{\pi i(h+l)} + e^{\pi i(k+l)}]$	
$F_{hkl} = 4f_{av} = f_{Fe} + 3f_{Ni}$	h, k, l unmixed	$F_{hkl} = f_{Fe} + 3f_{Ni}$	h, k, l unmixed
$F_{hkl} = 0$	h, k, l mixed	$F_{hkl} = f_{Fe} - f_{Ni}$	h, k, l mixed

The scattering intensity goes as $|F_{hkl}|^2$ plus other factors, including X-ray polarization and multiplicity of the family of planes. Thus, of all the possible

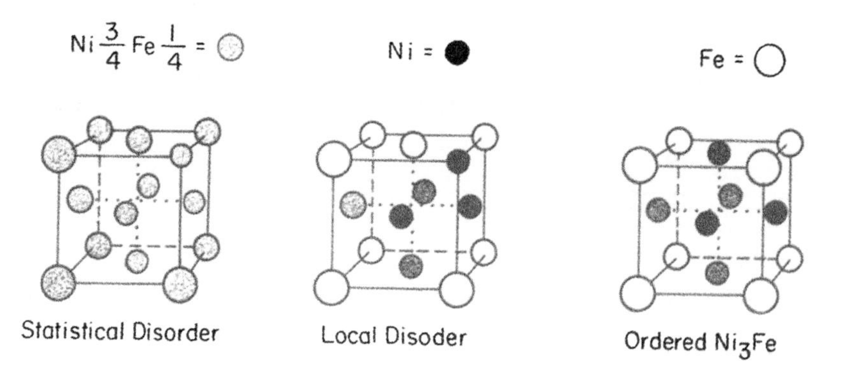

Figure 14A.1 Makeup of unit cell in disordered (left and center panels) and ordered Ni_3Fe (right panel) illustrating the change in Bravais lattice type from FCC for the disordered phase to simple cubic for the ordered phase. The Ni atoms on cube faces that are out of the line of sight are shown as lightly shaded.

diffraction peak positions on the basis of Bragg's law, only those satisfying the selection rules above appear in the disordered phase. In the disordered phase, the intensities of the superlattice lines (*h*, *k*, *l* mixed) are lost because they depend on the difference in the atomic scattering factors of the two species (Fig. 14A.2).

Note that the disordered structure exhibits a diffraction pattern identical to that of an FCC structure with an atomic scattering factor that is the weighted average of those for the Ni and Fe present. The diffraction pattern of the ordered structure is more complex. The lines corresponding to the FCC structure still appear; they are *fundamental* to both phases. In addition, the ordered structure shows lines at positions allowed in the simple cubic structure, but their intensity is weak because their atomic scattering factor is $f_{Fe} - f_{Ni}$. These are referred to as *superlattice* lines.

Structures that exhibit order–disorder phenomena may exist in intermediate states where the degree of ordering is partial. The structural order parameter S is defined as

$$S = \frac{P_A - x_A}{1 - x_A} \tag{14A.1}$$

where x_A is the fraction of species A in the alloy $A_{xA}B_{xB}$ and P_A is the probability of finding species A on the correct site. $S = 1$ for perfect ordering and $S = 0$ for complete disorder. For partial order, $0 < S < 1$, the intensity of the superlattice lines is less than it is for $S = 1$. Hence, the structure factors of Ni_3Fe can be written in general as

$$F_f = f_{Fe} + 3f_{Ni} \qquad \text{unmixed} \tag{14A.2a}$$

$$F_s = S(f_{Fe} - f_{Ni}) \qquad \text{mixed} \tag{14A.2b}$$

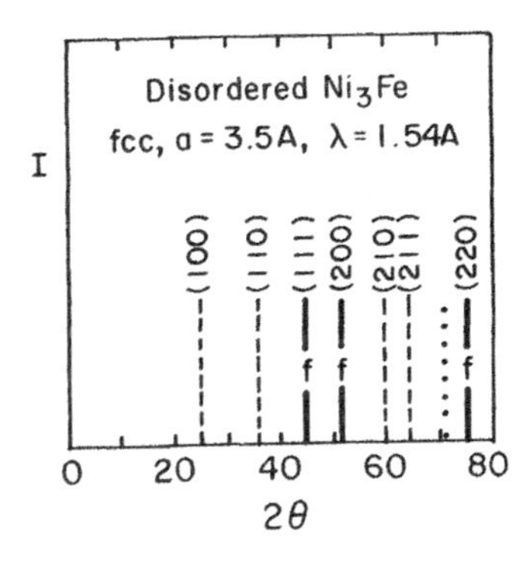

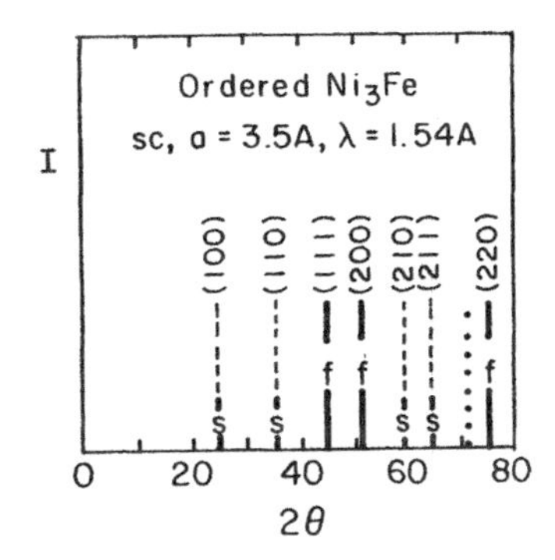

Figure 14A.2 Positions of diffracted lines for disordered (left) and ordered (right) Ni_3Fe showing the extinctions in the disordered fcc structure and the appearance of weak superlattice lines (s) at the positions of SC diffraction lines in the ordered state. The lines common to both structures are designated as fundamental (f).

These expressions determine the relative intensities of fundamental and superlattice lines.

Because measurements at elevated temperatures are difficult, the high-temperature structure may be studied by quenching a sample after annealing at an elevated temperature. In this way, the degree of order appropriate to the elevated temperature may be measured at room temperature. On the basis of Eqs. (14A.2), the order parameter S (see Fig. 14A.3) may be determined experimentally from the ratio of a superlattice peak intensity I_s to that of a nearby fundamental line I_f:

$$S = \sqrt{\frac{m_f LP_f}{m_s LP_s}} \cdot \frac{F_f}{F_s} \cdot \sqrt{\frac{I_s}{I_f}} \tag{14A.3}$$

Here m is the multiplicity and LP the Lorentz polarization factor of the reflection of interest.

Structural and Magnetic Order-Disorder Transitions Important parallels exist between structural and magnetic order–disorder transitions. Figure 14A.4 illustrates long-range order (above) and short-range order (below) in a structural sense (left) and in a magnetic sense (right).

In *disorder*, at elevated temperatures, entropy $-TS$ dominates the free energy and causes the disordered state to be favored over the ordered one (Fig. 14A.5). The local internal energy diagram for these two systems can be represented as shown in Figure 14A.6. An alternative energy representation of the systems is displayed in Figure 14A.7.

The order parameters for the two systems can be defined as the occupation of the low-energy state (ordered) minus that of the high-energy state (disordered) normalized to the total number of particles in the system (where # = number):

$$\text{Order} = S = \frac{\#\ \text{right} - \#\ \text{wrong}}{\text{total}\ \#}, \qquad M = \frac{\#\uparrow - \#\downarrow}{\text{total}\ \#} \tag{14A.4}$$

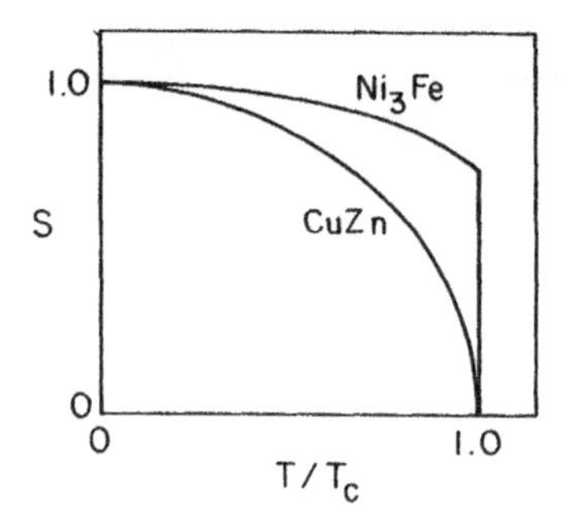

Figure 14A.3 Temperature dependence of order parameter S for Ni_3Fe and CuZn systems.

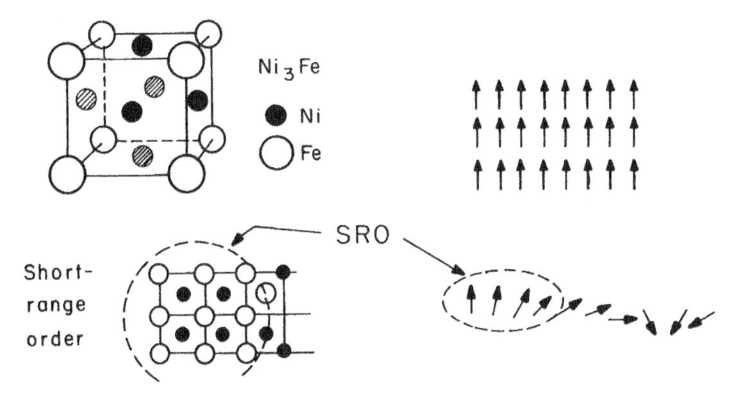

Figure 14A.4 Illustration of long-range order in crystal, above left, and magnetic system, above right. Below, the corresponding representation of order on a local scale in these two systems that is lost beyond a range defined by the two ellipses.

The relative occupations of right and wrong states can be described by means of their Boltzmann factors as was done for the quantum paramagnet in Chapter 3:

$$S = \frac{e^{-(-E_0/kT)} - e^{-(E_0/kT)}}{e^{-(-E_0/kT)} + e^{-(E_0/kT)}}, \qquad M = \frac{e^{-(-E_0/kT)} - e^{-(E_0/kT)}}{e^{-(-E_0/kT)} + e^{-(E_0/kT)}} \tag{14A.5}$$

These identical functions are simply hyperbolic tangent functions:

$$M \text{ or } S = \tanh \frac{E_0}{kT} \tag{14A.6}$$

The dependences of the order parameters on T and on E_0 are sketched in Figure 14A.8.

These graphical forms make sense; as temperature increases, entropy makes it increasingly more difficult to maintain the order that is preferred at lower temperatures (Fig. 14A.8, left). However, as the tendency to structural or

Figure 14A.5 Illustration of disorder in crystal, left, and magnetic system, right.

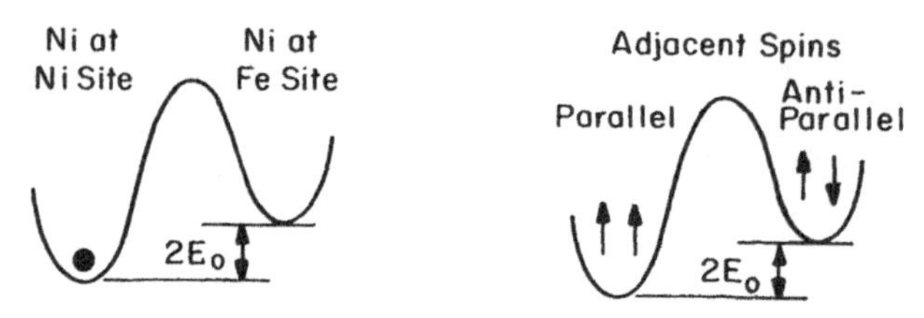

Figure 14A.6 Representation of the energy difference between ordered and disordered states as a double-well potential for Ni_3Fe structure, left, and spin system, right.

magnetic ordering (which is expressed by the magnitude of the internal energy E_0) increases, the order at a given temperature is greater (Fig. 14A.8, right).

But E_0 is not really a constant. As either of these systems proceeds to disorder, it becomes easier for further disorder to occur. Thus the model has to be modified to account for the dependence of E_0 on the degree of order.

STRUCTURE

Consider equiatomic NiFe:

MAGNETISM

Energy of spin **M** in exchange field, $\mathbf{H}_{exch}$ is $-\mathbf{M}\cdot\mathbf{H}_{exch}$

Ni in Fe Site $+E_0$ ↓

$E = 0$ ↑ H_{exch}

Ni in Ni Site $-E_0$ ↑

Figure 14A.7 Energy-level diagrams for structural and exchange systems shown in Figure 14A.6.

Structure	Magnetism
When Ni moves to an Fe site, strain makes it easier for other Ni atoms to move; thus $E_0 = E(S)$	Putting a second spin in the excited state (↓) costs less energy than the first spin because now there are two ↓ parallel to each other

$$S = \tanh \frac{E(S)}{kT} \qquad M = \tanh \frac{E(M)}{kT} \tag{14A.7}$$

where the assumption of linear dependence is made:

$$E(S) = E_0 \cdot S \qquad E(M) = -\mu_0 H_{\text{exch}} \cdot M$$

The form chosen above for the energy, $E = E_0 \cdot S$ or $E = E_0 \cdot M$), is such that the energy barrier to disorder decreases as disordering proceeds. Thus, the disordering process cascades precipitously on approaching the critical temperature T_C.

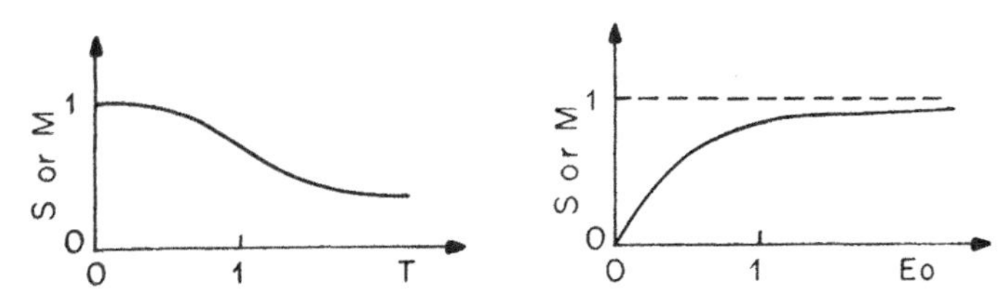

Figure 14A.8 Temperature and ordering energy dependence of the order parameters S or M.

Equations (14A.7) are transcendental equations: S (or M) appears on both sides. They can be solved graphically as was done for the quantum paramagnet in Chapter 3. We define $x = E \cdot (S \text{ or } M)/k_B T$ and plot each side of Eq. (14A.7) versus x. In Figure 14A.9, M or $S = \tanh x$ versus x is plotted at left and $M = xk_B T/(\mu_0 H_{\text{exch}})$, or $S = xk_B T/V_0$ versus x at right.

Combining the two plots above to get their common solutions, M or $S = \tanh x$, (Fig. 14A.10 left) gives a modified plot of the order parameter versus temperature (Fig. 14A.10 right) that now includes the cooperative effect of the dependence of the ordering energy on the order parameter itself.

The temperature dependence of the chemical order parameter resembles that of the Brillouin function calculated in Chapter 3. There, the energy dependence was built in by representing the energy of the magnetic system by the Zeeman energy of the *average* magnetization. The appearance of the order parameter on both sides of the equation, in either the chemical or magnetic case, indicates that the entities, atoms, or spins, act *cooperatively*. It is the cooperative nature of these phase transitions that makes the order parameter vanish so abruptly when approaching T_C from below.

The critical temperature can be expressed in terms of the energy by setting the slope of the straight line, $\partial M/\partial x = k_B T/(\mu_0 H_{\text{exch}})$ (Fig. 14A.10, left), equal to the slope of the tanh function at small $x, \partial(\tanh x)/\partial x = \text{sech}^2 x \approx (1 - x^2/2)^2 \approx 1 - x^2$. Hence T_c is proportional to V_0/k_B or H_{exch}/k_B. These forms for the critical temperature can be substituted into the small-x solution of $\partial M/\partial x = \partial(\tanh x)/\partial x$, giving

$$M^2 \text{ or } S^2 \propto T_c - T \qquad (14A.8)$$

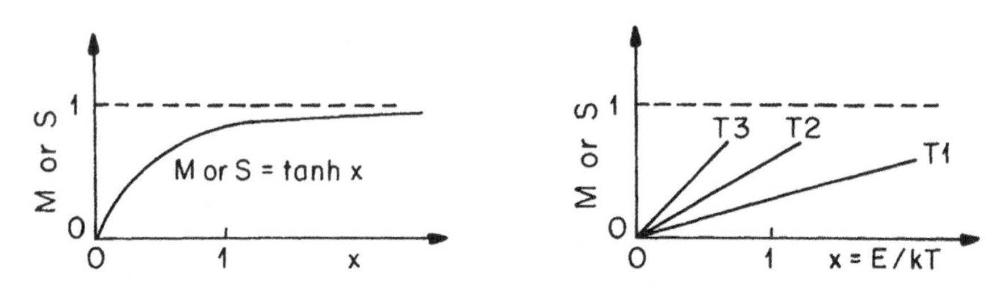

Figure 14A.9 Dependence of the order parameter on the normalized variable x according to the transcendental equation for the cooperative system, Eq. (14A.9).

Thus, the squares of the order parameters vanish linearly with temperature on approaching the critical temperature. More detailed relations were derived in Chapter 3 [Eq. (3.54)] for the behavior of the magnetization near the critical temperature. They are consistent with Eq. (14A.8). The behavior of the chemical order parameter is expressed more explicitly by the Bragg–Williams theory (Barrett and Massalski 1980). The critical temperature in that case is given by

$$T_c = \frac{V_0}{4k_B} \tag{14A.9}$$

Near the critical temperature, Bragg–Williams theory indicates

$$S^2 = \frac{3(T_c - T)}{T_c} \tag{14A.10}$$

The important lessons here are the parallels between these two superficially different ordering phenomena and the manner in which the cooperative nature of the orderings are introduced by making the ordering energy depend linearly on the order parameter itself.

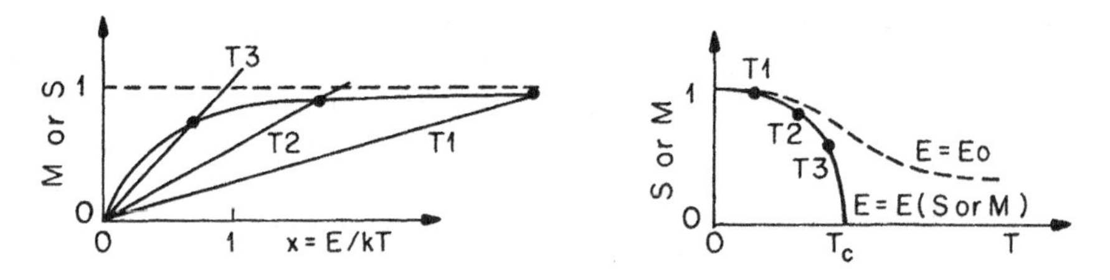

Figure 14A.10 Left, graphical solution of Eq. (14A.9); right, temperature dependence of the order parameters for cooperative systems.

PROBLEMS

14.1 Carry out the steps to derive Eqs. (14.16) and (14.17).

14.2 Verify Eqs. (14A.9) and (14A.10).

BIBLIOGRAPHY

Chikazumi, S. *Physics of Ferromagnetism* (Oxford Univ. Press, Oxford, 1997) p. 299 et seq.

REFERENCES

Barrett, C., and T. B. Massalski, *Structure of Metals*, Pergamon Press, Oxford, 1980, p. 287.

Berry, B., and W. C. Pritchett, *Phys. Rev. Lett.* **34**, 1022 (1975).

Bozorth, R. M., *Ferromagnetism*, IEEE Press, Piscataway, NJ, 1994.

Cahn, J., *Acta. Met.* **10**, 179 (1962); *J. Appl. Phys.* **34**, 3581 (1963); *Trans. Met. Soc. AIME* **242**, 166 (1968).

Chikazumi, S., *J. Phys. Soc. Jpn.* **5**, 327, 333 (1950).

Chikazumi, S. *J. Phys. Soc. Jpn.* **11**, 551 (1956).

Chikazumi, S., and C. D. Graham, Jr., in *Magnetism and Metallurgy*, A. E. Berkowitz and E. Kneller, eds., Academic Press, New York, 1969), p. 577.

Chikazumi, S., and T. Oomura, *J. Phys. Soc. Jpn.* **10**, 842 (1955).

Chin, G. Y., and J. H. Wernick, in *Ferromagnetic Materials*, Vol. 2, E. P. Wohlfarth (ed.), North Holland, New York, 1980, p. 55.

deVries, G., D. W. Van Geest, R. Gersdorf, and G. W. Rathenau, *Physica* **25**, 1131 (1959).

Ferguson, E. T., *Compt. Rend. Acad. Sci.* **244**, 2363 (1957); *J. Appl. Phys.* **29**, 252 (1958).

Fujimori, H., in *Amorphous Metallic Alloys*, F. E. Luborsky, ed., Butterworths, London, 1983.

Heindenreich, R. D., E. A. Nesbitt, and R. D. Burbank, *J. Appl. Phys.* **30**, 995 (1959); E. A. Nesbitt and R. D. Heidenreich, ibid. **30**, 1000 (1959); E. A. Nesbitt, B. W. Batterman, L. D. Fullerton, and A. J. Williams, ibid. **36**, 1235 (1965).

Iwata, T., *Sci. Rept. Res. Inst. Tohoku Univ.* **13**, 356 (1961).

Kim, T. K., and M. Takahashi, *Appl. Phys. Lett.* **20**, 492 (1972).

Kim, C. K. and R. C. O'Handley, *Metall. and Mater. Trans.* **28A**, 423 (1996).

Kronmüller, H., Springer Tracts in Natural Philos. Vol. 12 (Springer Verlag, Berlin, 1968).

Kronmüller, H., *Philos. Mag. B* **48**, 127 (1983).

Levy, P. M., *J. Phys. Chem. Solids* **26**, 415 (1965).

Luborsky, F. E., and J. L. Walter, *IEEE Trans.* **MAG-13**, 953 (1977); **MAG-13**, 1695 (1977).

Masters, R., and M. Wuttig, *Scripta Met.* **8**, 825 (1974).

Néel, L., *J. Phys. Radium* **12**, 339 (1952); **13**, 290 (1952).

Ratheneau, G. W., and G. deVries, in *Magnetism and Metallurgy*, A. E. Berkowitz and E. Kneller, eds., Academic Press, New York, 1969, p. 749.

Sauthoff, G., and W. Pitsch, *Phil. Mag. B* **56**, 471 (1987).

Slonczewski, J., "Magnetic annealing," in *Magnetism*, Vol. I, G. T. Rado and H. Suhl, eds., Academic Press, New York, 1963, p. 205.

Snoek, J. L., *Physica* **8**, 711 (1941).

Yamamoto, M., S. Taniguchi, and K. Aoyagi, *Sci. Rept. Res. Inst. Tohoku Univ.* *A***13**, 117 (1961).

CHAPTER 15

ELECTRONIC TRANSPORT IN MAGNETIC MATERIALS

15.1 INTRODUCTION

Electrical transport properties reflect the character of the valence electronic states in a material. The electrons in metals that support conduction typically consist of states with *s* or *p* character, while *f* states participate hardly at all in conduction because they are highly localized atomic states. However, *d* states are somewhere in between; they can participate in conduction to some extent. While oxides are generally not metallic conductors (the early transition metal oxides, such as TiO, are the notable exceptions), their transport properties are often governed by thermally activated electrons from *s*–*d* or *p*–*d* bonds.

In transition metals, it is the *d* states that connect magnetism with electrical transport properties. Electrical transport in metals can be affected by magnetism in many ways. Hybridization of *s* and *d* states brings a degree of orbital angular momentum to the conduction process. Empty *d* states can be occupied temporarily by conduction electrons, providing a spin-dependent and orbital-angular-momentum-dependent scattering process.

In rare-earth metals and alloys, the conduction process is carried by the 5*d* and 6*s* electrons while magnetism resides mainly in the atom-like 4*f* states. However, the conduction electrons are significantly polarized by exchange with *f* states, and in this way, magnetism affects transport in these metals.

In oxides, the conduction process is generally thermally activated and the spin of the valence states, as well as the energy gap for thermal activation, are intimately connected with magnetism by exchange and crystal field interactions.

Magnetic transport phenomena are not only valuable probes of magnetism in materials; they have a long history and a bright future in many important applications. This chapter treats some of the basic observations and important concepts connecting magnetism with electrical transport processes.

Three important observations serve well to introduce some of the effects of magnetism on electronic transport properties.

15.1.1 Observations

Temperature Dependence of Resistivity in Metals Nonmagnetic metals show a linear increase in electrical resistivity above the Debye temperature: $\rho(T) = \rho_0 + \alpha T$. The temperature dependence of electrical resistivity of a ferromagnet can show an anomaly near a magnetic transition (Fig. 15.1). Approaching the Curie temperature from below, the resistivity shows an anomalous increase. Above T_C, the resistivity increases more gradually and is almost linear in temperature but extrapolates to a low-temperature value indicative of an anomalously high residual resistivity. The electronic structure of Pd is similar to that of Ni but it is paramagnetic at all temperatures. The difference in the temperature dependence of the resistivities of these two metals suggests that when spins are disordered (Ni above T_C and Pd at all temperatures), an electron is more likely to scatter than if it moved in a medium of uniform magnetization. Thus, the high electrical resistivity of the paramagnetic state is attributed to electron scattering from the disorder in the spin system in addition to that from lattice vibrations. Spin-disorder scattering increases as the magnetic long-range order vanishes at and above the Curie temperature.

Temperature Dependence of Resistivity in Oxides For insulators, electrical transport is a thermally activated process: conductivity generally increases exponentially with increasing temperature: $\sigma \propto \sigma_0 \exp[-2E_g/k_B T]$,

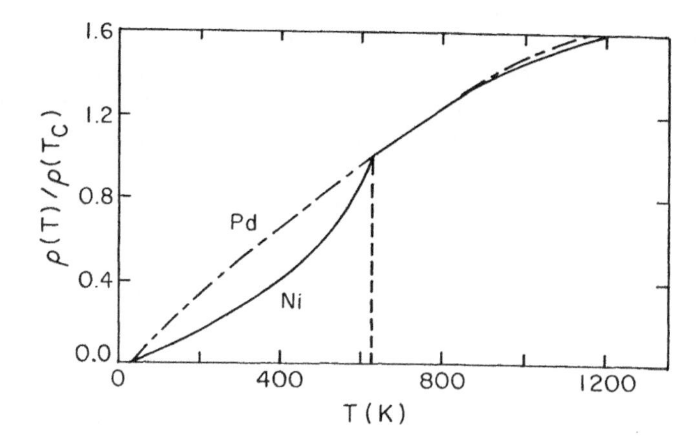

Figure 15.1 Resistivities of Ni and Pd normalized to their values at T_C of Ni, 631 K, versus temperature. [After Gerritsen (1956).]

where E_g is the energy gap between the occupied valence states and the empty (at 0 K) conduction states.

Epitaxial thin films of the doped perovskite compound $La_{1-x}Ca_xMnO_2$ have shown very large changes in their electrical resistivity with an applied field [Fig. 15.2, after Jin et al. (1994)]. Note that the system goes through a metal–semiconductor transition at a temperature where a weak magnetic moment still exists. (The low-temperature metallic phase shows a resistivity that increases with increasing temperature. The opposite is true when the Fermi energy lies in a gap as in a semiconductor or insulator). The resistivity peaks at this metal–insulator transition but the field-induced resistance change, $\Delta R/R(H)$, peaks about 25 degrees below this transition. The resistance change here is orders of magnitude greater than that observed in magnetic metals which are presently used in a variety of field sensors.

Interesting questions arise concerning the possible connection between magnetism and the metal–insulator transition, and whether this is the mechanism for the strong field dependence of the resistivity in certain oxides.

Resistivity Due to Dilute Magnetic Impurities The addition of transition metal impurities to noble metal hosts typically causes the electrical resistivity to increase linearly with the impurity concentration x:

$$\rho(x) = \rho(0) + \frac{d\rho}{dx} x$$

where $\rho(0)$ is the resistivity of the pure noble metal host.

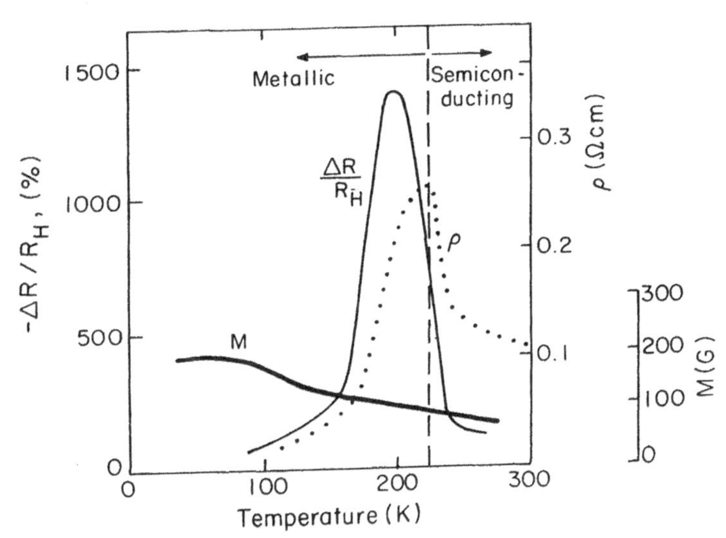

Figure 15.2 Variation of MRR, resistivity, and magnetization in laser-deposited $(La_{2/3}Ca_{1/3})MnO_3$ films. [After Jin *et al.* (1994).]

Figure 15.3 is a plot of the slope $d\rho/dx$ for transition metal impurities in Cu as a function of impurity type (or valence). The slope $d\rho/dx$ shows peaks near two separate values of the valence electron concentration of the magnetic impurity. This suggests that there is a splitting in energy of the impurity $3d^{\uparrow}$ and $3d^{\downarrow}$ states. When the $3d^{\uparrow}$ or $3d^{\downarrow}$ impurity states coincide with the conduction band Fermi level, there is enhanced scattering of charge carriers into these states. Increased scattering of conduction electrons increases the resistivity. The inserted state densities in Figure 15.3 show schematically the dominant features of the band structure for various impurities. The impurity state densities are exaggerated relative to those of Cu; 4*s* states are not shown. Conduction electron scattering with localized *d* states depends on (1) the relative spin of the two electronic states involved and (2) the relative number of initial and final states for scattering, specifically, the density of 3*d* spin up and down states.

The data of Figures. 15.1 and 15.3 indicate that spin disorder and the density of magnetic states at E_F, respectively, are both important factors in the resistivity of ferromagnetic metals. These magnetic effects are superimposed on the ordinary phenomena associated with electrical resistivity, namely, electron scattering from impurities and lattice vibrations (phonons). The data of Figure 15.2, as will be shown below, result from a more complex interplay of magnetism and electronic structure.

In order to understand these strong and provocative phenomena, the role of magnetism in electronic transport will be described. Particular attention will be given to conceptualizing the scattering processes in magnetic metals.

15.2 ELECTRICAL RESISTIVITY

Figure 15.4 compares the values of low-temperature electrical resistivity and the shape of the density of states in various metals. Alkali metals (*a*) with a valence electron configuration s^1, and noble metals (*b*), $d^{10}s^1$, have quite low values of resistivity, and their valence electronic structures near E_F are

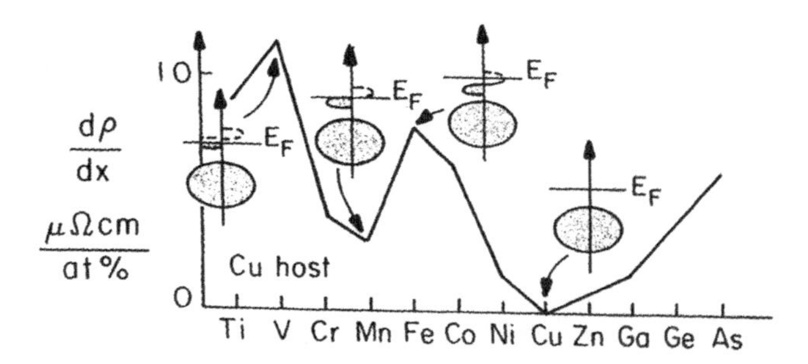

Figure 15.3 Variation with impurity type for the slope $d\rho/dx$ of Cu with transitio metal impurities in Cu: $Cu_{1-x}X_x$. [Adapted from Kittel (1963).]

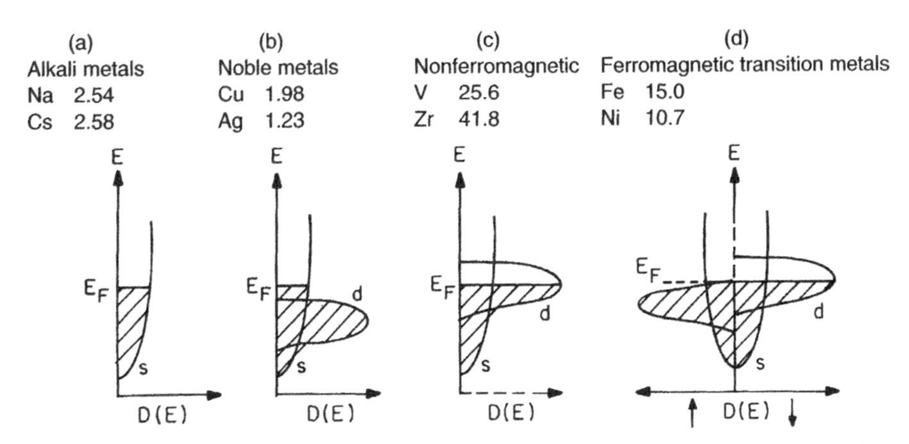

Figure 15.4 Examples of electrical resistivity at the Debye temperature (ρ in $\mu\Omega \cdot$cm) for four classes of metals (below, schematic state densities for each class: (*a*) alkali metals; (*b*) noble metals; (*c*) nonferromagnetic transition metals; (*d*) ferromagnetic transition metals.

characterized by free-electron-like bands. In the latter case, a filled *d* band exists not far below the Fermi energy. Nonmagnetic transition metals (*c*), with valence configuration close to $d^n s^1$, have much higher resistivities and both have *d* states as well as *s* states at the Fermi level. Magnetic transition metals (*d*), $d^{n\uparrow} d^{n\downarrow} s^1$, have both *d* and *s* states at the Fermi level and the densities of spin-up and spin-down *d* states at E_F are not necessarily equal.

The general trends in Figure 15.4 can be understood by starting with the simple Drude model for *free* electrons. In the Drude model, the electrical resistivity is expressed as

$$\frac{1}{\rho} = \sigma = \frac{ne^2\tau}{m^*} \tag{15.1}$$

where *n* is the volume concentration of free carriers, *e* is the electronic charge, τ is the relaxation time, and m^* is the effective mass of the charge carriers (inversely proportional to band curvature, $m^* \propto (\partial^2 E(k)/\partial k^2)^{-1}$. Equation (15.1) is applicable to *s*-electron metals and it does not hold when *d* bands intersect the Fermi surface. However, it can be used to infer the effect of *d* states at E_F on conductivity suggested in Figure 15.4. How does the presence of *d*-states near E_F affect charge transport properties?

1. The overlapping of *s* and *d* states at E_F leads to hybridization of these states so that the free electrons become partially localized (*n* decreases) and the *d* electrons become slightly delocalized (*n* increases). The net effect on free electron concentration is small.
2. When *s* and *d* states hybridize, the parabolic *s* states $[E(k) \propto k^2]$ acquire characteristics of the flatter *d* states and the effective mass of the

conduction electrons increases. This reduces the mobility, $\mu = e\tau/m^*$, of the carriers in a transition metal compared to a more free-electron-like metal.

3. Most importantly, the overlap of the *s* and *d* states allows otherwise free electrons to get scattered into more localized *d* states of the same energy. A large density of these *d* states at the Fermi energy enhances the probability of such scattering. This enhanced scattering into more localized states decreases the relaxation time, and the mobility is further suppressed. This *s–d* scattering was first described by Mott (1936, 1964). These *s–d* and other scattering effects may be expressed in terms of the strength of the scattering potential which controls the relaxation time, τ.

$$\tau^{-1} = |V_{\text{scat}}|^2 N(E_F) \tag{15.2}$$

Here, $N(E_F)$ is the density of scattering states at E_F. The strong decrease in the ratio τ/m^* when *d* bands intersect the Fermi level (Figs. 15.4*c* and *d*) results in the large increase in resistivity evidenced in the data shown with Figure 15.4.

It is interesting to note that in single crystals, the electrical resistivity can be anisotropic. This is most pronounced in cobalt where $\rho_c = 10.3\ \mu\Omega\cdot\text{cm}$ and $\rho_{ab} = 5.5\ \mu\Omega\cdot\text{cm}$ at room temperature. Thus, the degree of texture in polycrystals can affect the measured receptivity.

15.2.1 Two-Current Model for Transition Metals

In order to understand the differences between ferromagnetic and nonferromagnetic transition metals (Figs. 15.4*c* and 15.4*d*), Mott recognized that at temperatures well below T_C, the spin direction of the charge carriers is conserved during most scattering events. This is because spin waves, which mix spin-up and spin-down states, are not strongly excited at low temperature. Thus, the charge carriers having spin up and spin down can be represented as two parallel paths along which conduction can take place. (This assumption breaks down near and above T_C.) The two-current model may be represented simply by a parallel circuit with the resistivity of the two types of carrier represented by $\rho^{\uparrow}$ and $\rho^{\downarrow}$ (Fig. 15.5, left). In a single-element conductor, the resistivity in one channel is the sum of the phonon, impurity, *s–d*, and other scattering contributions. Further, $\rho^{\uparrow}$ is not necessarily equal to $\rho^{\downarrow}$ because of the difference in the density of spin-up and spin-down states at E_F. If *s–d* scattering is negligible in one of these subbands, that subband carries more of the current and the total resistance decreases toward its nonmagnetic value. This is evidently the case in Ni, where the majority *d* band is full and hence does not trap conduction electrons. Conduction in the majority-spin band of Ni tends to be favored and to short-circuit the higher-resistivity, minority-band process. The resistivity of a metal in the two-current model (low

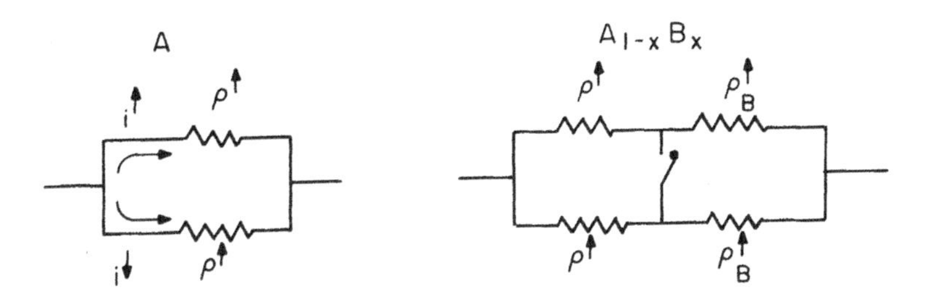

Figure 15.5 Equivalent circuits for the two-current model of resistivity in pure transition metals, left, and dilute transition metals alloys, right. The two populations of spin scatter independent of each other, that is, in parallel. When impurities or alloying elements are added, the new scattering events still contribute independently to each subband.

temperature)

$$\rho_{\text{low }T} = \frac{\rho^{\uparrow}\rho^{\downarrow}}{\rho^{\uparrow} + \rho^{\downarrow}}, \qquad (\text{low } T) \tag{15.3}$$

is always less than or equal to the resistivity of either path alone. Each resistivity in Eq. (15.3) is described by Eqs. (15.1) and (15.2) with different values of n, m^*, τ, and $N(E_F)$ for each subband. It is common to define the parameter α as the ratio of spin-down to spin-up resistivities:

$$\alpha = \frac{\rho^{\downarrow}}{\rho^{\uparrow}} \tag{15.4}$$

For Ni and Co as well as many *strongly* magnetic alloys, $\alpha \gg 1$ because the localized d states appear at E_F only in the minority-spin band.

15.2.2 Impurities

If impurity or alloying elements B are added to the metal, such as $A_{1-x}B_x$, the independence of the two currents is maintained provided there is little spin-flip scattering. The new scattering process due to the presence of the B species, ρ_B^i, may be spin-dependent and occur independently in each current path as represented in Figure 15.5, right. In this case the resistivity is given by

$$\rho = \frac{\{(1-x)\rho^{\uparrow} + x\rho_B^{\uparrow}\}\{(1-x)\rho^{\downarrow} + x\rho_B^{\downarrow}\}}{(1-x)\rho^{\uparrow} + x\rho_B^{\uparrow} + (1-x)\rho^{\downarrow} + x\rho_B^{\downarrow}} \tag{15.5}$$

If spin mixing occurs (e.g., as above T_C or when spin-flip scattering can occur), the behavior is represented by closing the switch in this circuit (Fig. 15.5, right), effectively reducing the total resistivity to the following value:

$$\rho = (1 - x)\frac{\rho^{\uparrow}\rho^{\downarrow}}{\rho^{\uparrow} + \rho^{\downarrow}} + x\frac{\rho_B^{\uparrow}\rho_B^{\downarrow}}{\rho_B^{\uparrow} + \rho_B^{\downarrow}} \tag{15.6}$$

In this case, the carriers always have a choice between the two parallel paths and can take the path of least resistance. The resistivity in this case is always lower than the two-current case without spin mixing, Eq. (15.5).

This two-current model describes a change in resistance that is opposite in sign and independent of the spin disorder scattering responsible for the behavior of $\rho(T)$ in Figure 15.1.

Equations (15.5) and (15.6) account well for the large positive deviations from Matthiessen's rule, $\rho = (1 - x)\rho_A + x\rho_B$, observed in ferromagnetic alloys. The resistivity of a two-current, ferromagnetic alloy, Eq. (15.5), is greater than that of the same alloy with ferromagnetism turned off, Eq. (15.6). Dorleijn (1976) gives a good review of the two-current model as applied to impurities in transition metals and shows that it is possible to determine consistent values of $\rho^{\uparrow}$ and $\rho^{\downarrow}$ for various impurities in specific transition metal hosts. Essentially, when either of the $\uparrow$ or $\downarrow$ d states of the impurity coincide with the Fermi energy, the $\uparrow$ or $\downarrow$ resistivity is enhanced. This is exactly the effect illustrated in Figure 15.3 for various impurities in a Cu matrix.

Data similar to those in Figure 15.3 are shown in Figure 15.6 for $3d$ impurities in Ni. These data can be interpreted within the two-current model to reveal the resistivities of the two spin-subbands of the impurities. The data show $\rho^{\downarrow}$ to be relatively insensitive to the energy of the $3d^{\downarrow}$ impurity states because $\rho^{\downarrow}$ is dominated by the large density of Ni $3d^{\downarrow}$ states at E_F. The density of states curve in Figure 15.6 shows the energy of the majority-spin $3d$ impurity

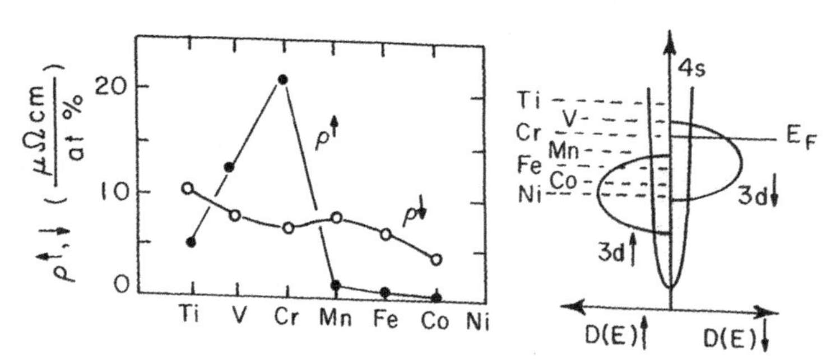

Figure 15.6 Left, resistivities determined for the spin-up and spin-down impurity bands determined from measurements on Ni host (Campbell and Fert 1982). Illustration at right shows a schematic of the band structure of the Ni host with the centroid of the majority-spin d-states of the impurities indicated.

states relative to the Ni Fermi level. When the impurity $3d^{\uparrow}$ states reside at E_F, $\rho^{\uparrow}$ increases as can be seen for Cr in Ni.

Further, Cr impurities in Ni strongly suppress the Ni magnetic moment because μ_{Cr} is antiferromagnetically coupled to μ_{Ni}. This coupling is a result of exchange of the Cr majority-spin electrons (Cr has no minority-spin electrons) with the empty $3d$ minority spin states of the Ni host. Thus the host spin-up electrons see a strong repulsive potential at the Cr sites. It is this repulsive potential [see Eq. (15.2)] that shortens the host majority-spin relaxation time.

15.2.3 Temperature Dependence

Figure 15.1 shows that the resistivity of a ferromagnet increases on approaching T_C and shows a greater value in the paramagnetic state than in the ferromagnetic state. There can be two effects contributing here: (1) conduction electron spin scattering from disorder in M that is no longer saturated at elevated temperatures and (2) magnon creation or annihilation at elevated temperatures, which flip the spin direction. These are described briefly.

1. Spin-disorder scattering implies that there is a term in the resistivity of a paramagnet or disordered ferromagnet governed by the exchange interaction between the spin of the charge carrier s and the local, paramagnetic moment, proportional to $\sqrt{J(J+1)}$. Well above T_C, it can be shown that this interaction contributes a temperature-independent, paramagnetic resistivity (Campbell and Fert 1982)

$$\rho_{\text{para}} = \frac{k_F(m\mathscr{J})^2}{4\pi e^2 Z h^2} J(J+1)$$

where k_F is the Fermi wavevector, Z is the atomic number, e/m is the charge:mass ratio of the carriers, and $\mathscr{J}$ is the appropriate exchange interaction. At and below the Curie temperature this contribution is frozen out as the moments align, giving

$$\rho_{\text{ferro}} = \rho_{\text{para}} \frac{(J - |\langle J \rangle|)(J + 1 + |\langle J \rangle|)}{J(J+1)} \approx \rho_{\text{para}} \left[1 - \left(\frac{M_s(T)}{M_s(0)} \right)^2 \right] \qquad (15.7)$$

The spin disorder resistivity is illustrated in Figure 15.7, and must be added to the impurity, phonon, and other scattering contributions.

2. The increased concentration of spin waves as T approaches T_C from below causes mixing of the spin-up and spin-down channels. A spin-up conduction electron can be scattered to a spin-down state by the annihilation of a magnon and vice versa. Mixing of the two spin channels tends to equalize the resistivities; thus, α approaches unity. This necessarily increases the net resistivity because scattering in the lower-resistivity channel increases. (This is different from the two component system depleted in Fig. 15.5, right.) When

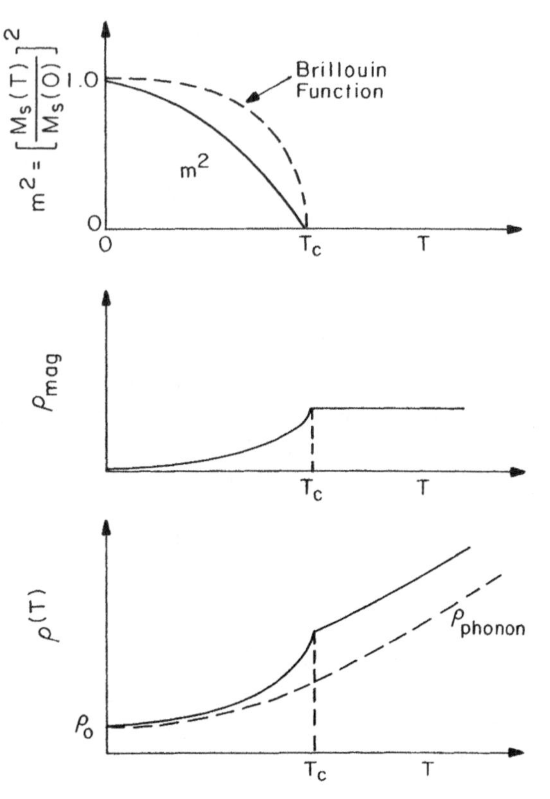

Figure 15.7 Above, temperature dependence of reduced magnetization squared. Center, temperature dependence of spin disorder scattering that goes as 1-$m^2(T)$. Below, addition of spin disorder resistivity to the residual and phonon contributions to electrical resistivity.

spin mixing occurs, the low-temperature form of ρ, Eq. (15.3), becomes (Fert and Campbell 1972):

$$\rho_{\text{high }T} = \frac{\rho^{\uparrow}\rho^{\downarrow} + \rho^{\uparrow\downarrow}(\rho^{\uparrow} + \rho^{\downarrow})}{\rho^{\uparrow} + \rho^{\downarrow} + 4\rho^{\uparrow\downarrow}} \qquad \text{high } T \tag{15.8}$$

were $\rho^{\uparrow\downarrow}$ is the spin mixing resistivity. Resistivity $\rho_{\text{high }T}$ is always greater than $\rho_{\text{low }T}$ [(Eq. 15.3)] when $\rho^{\uparrow} \neq \rho^{\downarrow}$ ($\alpha \neq 1$). At low temperatures where there is no spin mixing, $\rho^{\uparrow\downarrow} = 0$ and Eq. (15.8) reverts to Eq. (15.3). At very high temperatures where mixing dominates, $\rho^{\uparrow\downarrow} \gg \rho^{\uparrow}$ or $\rho^{\downarrow}$, the resistivity becomes simply

$$\rho_{\text{high }T} \xrightarrow{\rho^{\uparrow\downarrow} \gg \rho^{\uparrow},\rho^{\downarrow}} \frac{\rho^{\uparrow} + \rho^{\downarrow}}{4}$$

Campbell and Fert (1982) give a more thorough review of the various scattering processes contributing to the resistivity at different temperatures.

15.3 GALVANOMAGNETIC EFFECTS

We will now consider the effects of a magnetic field on transport properties in normal and ferromagnetic materials.

15.3.1 Nonmagnetic Materials: Ordinary Hall Effect and Magnetoresistance

Ordinary Hall Effect Electronic transport properties involving magnetic fields are called *galvanomagnetic effects*. The ordinary Hall effect (Chien and Westgate 1980) is a familiar phenomenon in which a transverse electric field E_H appears across a sample when an applied magnetic field $\boldsymbol{H}$ has a component perpendicular to the current density $\boldsymbol{J}$:

$$\boldsymbol{E}_H = R_H(\boldsymbol{J} \times \mu_0 \boldsymbol{H}) \tag{15.9}$$

Figure 15.8 depicts the most common geometry for measurement of the Hall effect—the applied field is normal to a sample in which E_H is measured transverse to the current direction. The ordinary Hall effect comes from the Lorentz force, $\boldsymbol{F} = \mu_0 q(v \times H)$, acting on the charge carrier. This mechanism is depicted at the right in Figure 15.8 for different charge carriers. Note how the sign of the Hall voltage changes with the nature of the charge carriers (electrons or holes).

The Hall coefficient describes the strength of the effect. From the expression for the Lorentz force and $J_y = ne\langle v_y \rangle$ (free charge carriers), it can be shown by comparison with Eq. (15.9) that

$$R_H = (ne)^{-1} \tag{15.10}$$

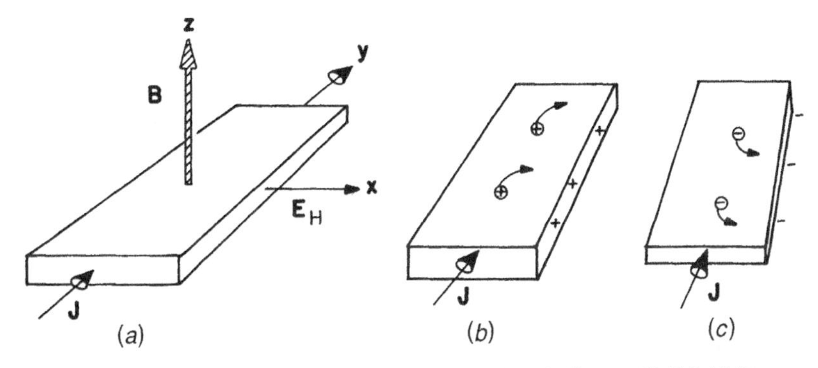

Figure 15.8 (*a*) Sample geometry for observation of Hall effect—Hall field $\boldsymbol{E}_H$ appears at right angles to current density $\boldsymbol{J}$ and magnetic field $\boldsymbol{H}$; (*b*, *c*) mechanism for appearance of positive Hall voltage when current carriers are positively charged (holes) (*b*) or negative Hall voltage for electrons (*c*).

where n is the carrier density. The material in which the Hall effect is observed need not be magnetic. Note that the sign of the Hall field reverses if the sign of the factor (JH/e) reverses; it is of odd symmetry in the applied field, the current direction and the sign of the charge of the carriers. The Hall effect is often described by the Hall resistivity, $\rho_H = E_H/J_x = R_H\mu_0 H$.

The "ordinary" Hall effect is most noticeable in semiconductors where the carrier mobility is high but the conductivity is low. In metals the high conductivity essentially short-circuits the Hall voltage; a metal does not easily support a potential difference.

Magnetoresistance The presence of an external field also causes a change in resistance of the material because the Hall effect deflects charge carriers from the current direction. Once a charge carrier begins to orbit around the magnetic field, it does not contribute to the current density ($\langle v_x \rangle = 0$ over a complete cyclotron orbit) until it is scattered. After scattering, it begins its next cyclotron orbit with an initial velocity biased toward the applied field. Thus, the longer the relaxation time (lower resistivity), the larger can be the effect of the field on the resistance. Kohler discovered this fact analytically and expressed it as $\Delta\rho/\rho = f(H/\rho)$. This description of magnetoresistance as functionally scaling with H/ρ is known as *Kohler's rule*. Because a deflection of a charge carrier in *either* direction away from J_x increases ρ, the change in resistance must be an even power of H, thus the magnetoresistance ratio to lowest order obeys.

$$\frac{\Delta\rho}{\rho} \propto \left(\frac{H}{\rho}\right)^2. \tag{15.11}$$

(The Hall resistivity $\rho_H = E_H/J$ is linearly proportional to H). Kohler's rule, Eq. (15.11), is general and applies also to ferromagnetic materials with the substitution $H \to B$.

A simple derivation of Kohler's rule can be made by considering that in the absence of a magnetic field, $\rho(0) = mv/ne^2\lambda(0)$, where $\lambda(0)$ is the zero-field mean-free path. On application of a field, the electron trajectory can be approximated as circular, tracing out an arc of length $\lambda(0)$ before scattering (Fig. 15.9). The radius of curvature of the orbit is simply $r = mv/eB$, and $\lambda(0) = r\theta$ where θ is the angle subtended by the path of the electron. But this path has a projection along the field direction of $\lambda(H) = r\sin[\lambda(0)/r]$, or, in the weak-field approximation, $\lambda(H) \approx \lambda(0)[1 - \lambda(0)^2/6r^2 \cdots] = \lambda(0)[1 - a[H/\rho(0)]$, where $a = \mu_0^2/(6n^2e^2)$. Taking $[\rho(H) - \rho(0)]/\rho(0)$ gives $\Delta\rho/\rho = a[H/\rho(0)]$, to second order in H, which is Kohler's rule.

Table 15.1 lists some representative room temperature values of R_H and $\Delta\rho/\rho$ for various ferromagnetic materials. While hexagonal cobalt exhibits an appreciable anisotropy in its electrical resistivity, there is no significant corresponding effect for Fe and Ni. The electrical resistivity of amorphous metallic alloys is very large because of the lack of long-range crystalline order (Chapter 11). This is also the reason for the negligible anisotropic magneto-

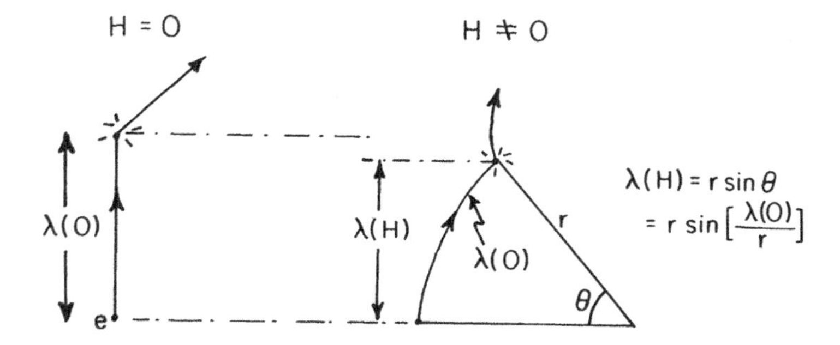

Figure 15.9 Classical depiction of magnetoresistance mechanism for justification of Kohler's rule. Mean free path is represented in zero field, left, and nonzero field right. While the relaxation time is the same in both cases, the component of the mean free path along the direction of the applied voltage is smaller for $H \neq 0$.

resistance in these materials. The sign of the ordinary Hall coefficients of Fe, Ni, and Co are difficult to interpret in terms of Eq. (15.10). Note the very large values of the spontaneous Hall coefficient for amorphous materials. This is explained below to be due to their large resistivities.

15.3.2 Galvanomagnetic Effects in Ferromagnetic Materials

It has been shown that in nonmagnetic materials, the application of a magnetic field perpendicular to the current can alter the longitudinal resistance $\Delta\rho/\rho$ and can also induce a Hall voltage in a direction orthogonal to both the current and magnetic field. These effects can be understood as consequences of the classical Lorentz force on, and cyclotron orbits of, the current carriers. In nonmagnetic materials the galvanomagnetic effects are called "ordinary" to distinguish them from the stronger effects observed in ferromagnets.

TABLE 15.1 Electrical and Galvanomagnetic Properties of Some Magnetic Materials

	ρ (RT) ($\mu\Omega \cdot$cm)	$\Delta\rho/\rho$ (%)	R_0 ($10^{-12}\,\Omega\cdot$cm/G)	R_s ($10^{-12}\,\Omega\cdot$cm/G)
Ni	10.7	+2.5	−0.6	−0.6
Fe	15	+0.8	+0.23	2.8, 7.2
Co	10.3 (*c* axis)	+3.0	−0.84	+0.6
	5.5 (basal plane)	—	—	—
Gd	—	—	—	500
Amorphous				
$Fe_{80}B_{20}$	120	≈ 0	$+2 \pm 1$	500

In ferromagnetic materials the "ordinary" anisotropic transport effects are present but are accompanied by stronger phenomena having similar geometrical dependences and symmetries. The galvanomagnetic effects unique to ferromagnets are called "extraordinary," "spontaneous" or "anomalous" because of their greater strength relative to the "ordinary" effects. The extraordinary galvanomagnetic effects derive their strength from the fact that the role of the external field is replaced by an internal field proportional to the magnetization, which is generally much stronger than an applied field. The mechanism by which the microscopic internal field associated with $\boldsymbol{M}$ couples to the current density in ferromagnets is the spin–orbit interaction between the electron trajectory (orbit) and the magnetization (spin). Thus, while the ordinary effects are classical, the spontaneous effects are quantum mechanical in their origin.

In brief, the ordinary galvanomagnetic effect arise from the macroscopic part of the flux density $\mu_0 \boldsymbol{H}$ while the extraordinary effects come from the microscopic magnetic part of the flux density $\mu_0 \boldsymbol{M}$. The appropriate expressions for the spontaneous Hall effect and magnetoresistance observed in ferromagnetic materials are obtained by replacing $\boldsymbol{B} = \mu_0 \boldsymbol{H}$ with $\mu_0 \boldsymbol{M}$ in the relations derived so far for the corresponding ordinary effects.

The challenge of a diversity of units in the magnetics literature is compounded in the case of galvanomagnetic effects by different usages for the field responsible for galvanomagnetic effects. Many references use $\boldsymbol{B}$ in nonferromagnetic materials, meaning $\mu_0 \boldsymbol{H}$. It seems more appropriate to use the latter notation in nonmagnetic materials and reserve the flux density, $\boldsymbol{B}$, for systems in which there appear both ordinary effects, which are functions of $\boldsymbol{H}$, and ferromagnetic effects, which are functions of $\boldsymbol{M}$.

Anomalous Hall Effect In ferromagnetic materials the Hall resistivity must be written

$$\begin{aligned} \rho_H &= \frac{E_H}{J} = \rho_{oH} + \rho_{sH} \\ &= \mu_o(R_o H + R_s M) \quad \text{(mks)} \qquad (15.12) \\ \rho_H &= R_o H + 4\pi R_s M \quad \text{(cgs)} \end{aligned}$$

where the first term is the ordinary Hall resistivity proportional to the external field and the second term is the spontaneous effect, proportional to the magnetization. These two contributions to ρ_H can be written as a sum because the symmetry of the two Hall effects is the same. The vector symmetry of the spin–orbit interaction, $\boldsymbol{L} \cdot \boldsymbol{s}$, responsible for the spontaneous Hall effect, is simply related to the radial component of the Lorentz force $\boldsymbol{r} \cdot (\boldsymbol{v} \times \boldsymbol{B})$, which governs the ordinary effects [Eq. (15.9)], as follows: $\boldsymbol{L} \cdot \boldsymbol{s} = (\boldsymbol{r} \times \boldsymbol{p}) \cdot \boldsymbol{s} \propto \boldsymbol{r} \cdot (\boldsymbol{p} \times \boldsymbol{M}) \propto \boldsymbol{r} \cdot (\boldsymbol{J} \times \boldsymbol{M})$. Thus the microscopic picture for the spontaneous Hall effect is similar to that shown in Figure 15.8 but with $\boldsymbol{H}$ replaced by $\boldsymbol{M}$ or $\boldsymbol{S}$.

Figure 15.10 shows the dependence of ρ_H in Ni on Al impurity content (and hence on resistivity). The spontaneous Hall resistivity is the value of the high-field Hall resistivity extrapolated to zero field. The ordinary Hall effect is responsible for the high-field slope in the data. Both the ordinary and the extraordinary effects are negative here.

Note first that the ordinary Hall coefficient (high-field slope of ρ_H) is unaffected by the Al impurities ($R_0 < 0$ implies conduction by electrons, not holes). However, the spontaneous Hall contribution is very small in Ni (low resistivity) and increases dramatically as dilute Al additions increase the electrical resistivity. The explanation for this effect is found in a microscopic model for the spontaneous Hall effect.

The spontaneous Hall resistivity is found to vary with overall resistivity as $\rho_{sH} = a\rho_\perp + b\rho_\perp^2$. This is more often expressed in terms of the spontaneous Hall angle:

$$\phi_{sH} = \frac{\rho_{sH}}{\rho_\perp} = \phi_{sk} + b\rho_\perp \tag{15.13}$$

A physical interpretation of these two terms (the first two terms in an expansion of the Hall angle in the resistivity of the material) can be given based on the ideas of Smit (1958) and Berger (1970, 1972). The first term can be considered to be a skew scattering angle that describes the average deflection of the trajectory of a charge carrier in a scattering event (Fig. 15.11*a*). The second term in Eq. (15.13), ascribed to a side jump mechanism, displaces the trajectory from its original path through the scattering center (see Fig. 15.11*b*). The Hall angle due to the side jump will be inversely proportional to the mean free path $\Delta x/\lambda$, and thus proportional to the resistivity. The side jump

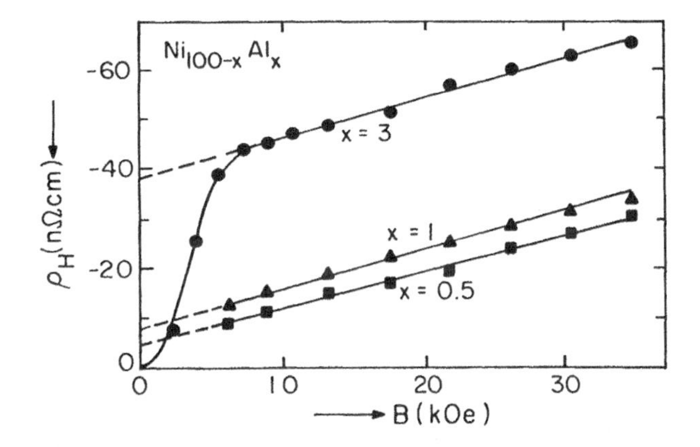

Figure 15.10 Hall resistivity of Ni with dilute Al additions. The high field slope is the ordinary Hall coefficient (Dorleijn 1976).

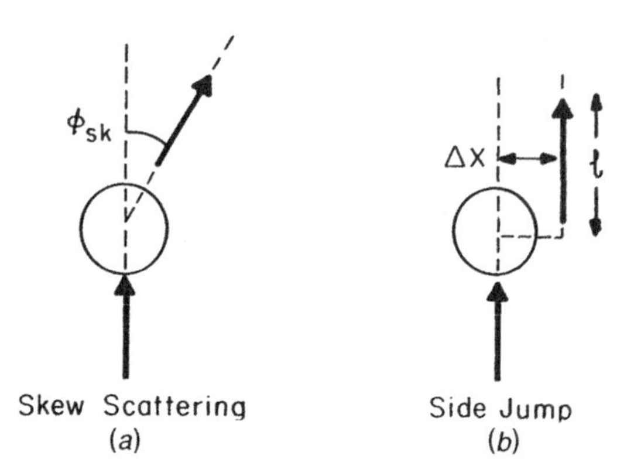

Figure 15.11 Description of skew scattering and side jump processes based on concepts of Berger (1970, 1972).

mechanism suggests that the angular momentum of the carrier (or its classical impact parameter) has been changed during the scattering event.

Another interpretation is that the skew scattering is a result of carrier scattering from the host *d* states and thus independent of the concentration of a dilute impurity. The second term in Eq. (15.13) is the result of scattering from impurity sites and thus is proportional to impurity concentration, which goes as $\rho_{\perp}$.

Thus, the reason for the dramatic increase in ρ_{sH} shown in Figure 15.10 on addition of small amounts of Al to Ni, is the side jump mechanism. The nonmagnetic impurities shorten the mean free path and allow the side jumps to accumulate more rapidly than if they occurred separated by longer mean free paths.

Another consequence of the relation in Eq. (15.13) is that the spontaneous Hall resistivity becomes very small at absolute zero. Also, at elevated temperatures where ρ is large, the side jump mechanism often dominates the skew scattering mechanism. Further, since the spontaneous Hall effect must vanish above T_C, the Hall resistivity is expected to peak at some temperature below T_C. In fact, the temperature dependence of R_s is observed to be nearly quadratic in temperature (because ρ is nearly linear in T) before plunging to zero at the Curie temperature (Fig. 15.12).

The quadratic dependence of the ρ_{sH} on $\rho_{\perp}$ leads to very large spontaneous Hall effects in amorphous ferromagnetic alloys (O'Handley 1978). However, in amorphous alloys, the resistivity is only weakly dependent on temperature (it generally remains in the range $100-140\,\mu\Omega\cdot\text{cm}$), so ρ_{sH} is nearly constant down to cryogenic temperatures.

It is interesting to note the case of ferrimagnetic rare-earth–transition metal intermetallics such as GdCo. This system shows a compensation temperature below (above) which the rare-earth (transition metal) moments dominate the magnetization (cf. Figs. 4.9, 4.11). At the compensation temperature the

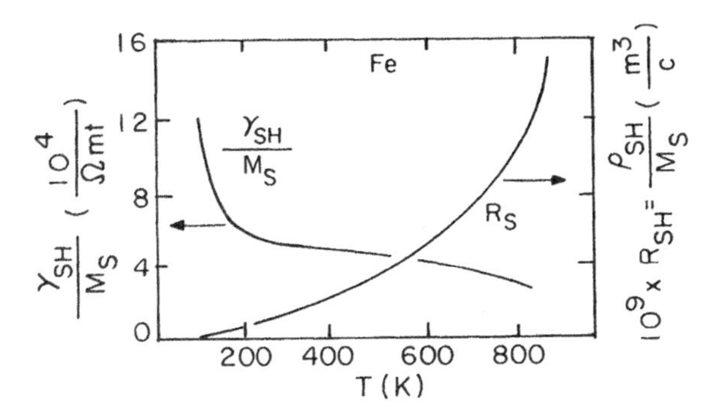

Figure 15.12 Temperature dependence of the spontaneous Hall resistivity per unit magnetization and the spontaneous Hall conductivity per unit magnetization for Fe (Jan 1958).

antiferromagnetically coupled moments of the two sublattices switch orientations if an applied field is present. Figure 15.13 shows how the Hall resistivity $R_s M_s$ changes sign through the compensation temperature in amorphous $Gd_{17}Co_{83}$ (Shirakawa 1976). For the same reason, a sign change is seen at fixed temperature as the composition is varied through T_{comp} in ferrimagnetic alloys (Stobieki 1978).

Anisotropic Magnetoresistance Just as the Hall resistivity in ferromagnets has the form of Eq. (15.12) with both ordinary and spontaneous terms (linearly dependent on H and M, respectively), so too the magnetoresistance in a

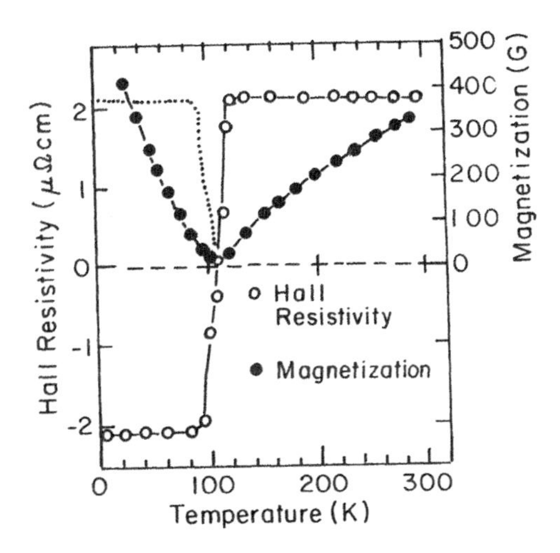

Figure 15.13 Temperature dependence of the magnetization and spontaneous Hall resistivity in $Gd_{17}Co_{83}$ (Shirakawa et al. 1976).

ferromagnet is expected to have a spontaneous contribution that depends on the orientation of magnetization. It is referred to as the *anisotropic magnetoresistance*. Thus, Kohler's rule can be generalized to read for a ferromagnet:

$$\frac{\Delta\rho}{\rho} \propto a\left(\frac{H}{\rho}\right)^2 + b\left(\frac{M}{\rho}\right)^2 \tag{15.14}$$

The first term describes the ordinary magnetoresistance and the second, the spontaneous or anisotropic magnetoresistance.

Figure 15.14*a* shows the field dependence of resistance in a Ni-Co alloy. The rapid decrease in resistance for a field applied perpendicular to the current direction is the ferromagnetic part of $\Delta\rho/\rho$. The anisotropy field for this alloy is of order 4 kOe. Above that field the magnetization is saturated and the ferromagnetic contribution to the MR is saturated. The value of the anisotropic magnetoresistance (AMR) may be determined by extrapolation of high field MR data to $H = 0$. Above H_a, all that remains is the ordinary MR which goes as H^2. The negative slope of the high-field resistance (superimposed on the low-field ferromagnetic effect) is due to the increase in magnetization order or spin order through the high-field susceptibility and the applied field (cf. Fig. 15.7). This high-field susceptibility effect overcomes the ordinary magnetoresistance due to the Lorentz force, which is positive. At cryogenic temperatures, the high-field susceptibility is much smaller and the positive Lorentz-force magnetoresistance dominates at high fields. At cryogenic temperatures the average resistance of the NiCo alloy also decreases, $\Delta\rho$ decreases, and the anisotropic MR ratio remains about the same (McGuire 1975).

Figure 15.14*b* shows the anisotropic magnetoresistance of cobalt thin films in relatively weak fields. Here the even symmetry in H is evident. Also, the presence of hysteresis (which correlates with the hysteresis the M–H loop)

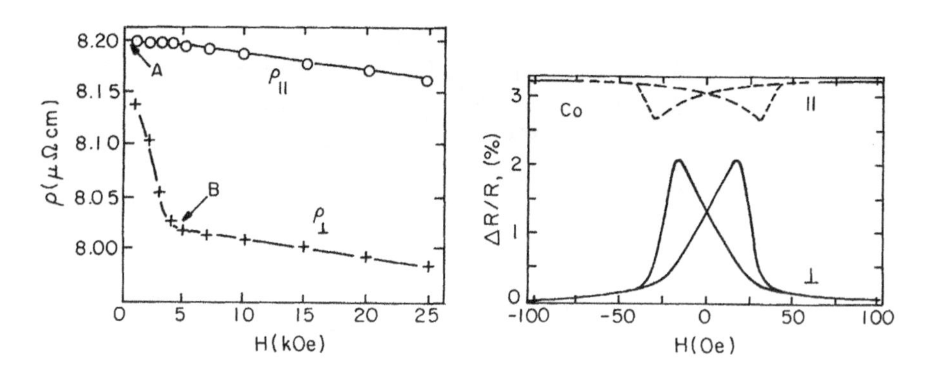

Figure 15.14 (*a*) Resistivity of $Ni_{0.9942}Co_{0.0058}$ at room temperature versus applied field (McGuire, 1975); (*b*) low-field magnetoresistance for cobalt thin film showing even field symmetry and hysteresis. [After Parkin (1994).]

confirms the origin of AMR in the state of magnetization rather than the applied field.

The anisotropic MR effect is observed in many systems to vary as

$$\frac{\Delta\rho(H)}{\rho_{av}} = \frac{\Delta\rho}{\rho_{av}}\left(\cos^2\theta - \frac{1}{3}\right) \tag{15.15}$$

where θ is the angle between $\boldsymbol{J}$ and $\boldsymbol{M}$. Note that Eq. 15.15 is even in the magnetization orientation, $\cos\theta$, as observed and predicted by Kohler's rule. The detailed consequences of this form were considered in Chapter 7 (see Fig. 7.5).

Because the state of magnetization of a demagnetized multidomain sample is not unique, the magnetoresistance in zero field, $\rho(0)$ is not well defined. Thus ρ_{av} is not necessarily equal to $\rho(0)$, and $\rho(0)$ can have different values depending on the domain structure in $H = 0$. This is shown in Figure 15.15. If the demagnetized state has a preferred direction of magnetization, then the measured anisotropic magnetoresistance in that direction is unaffected by application of a field. On the other hand, the resistivity measured in a field perpendicular to an "easy axis" is quadratic in the field. Thus the anisotropic MR effect is a measure of sample anisotropy or domain magnetization distribution. Compare this figure with the corresponding one for magnetostriction (Fig. 7.5).

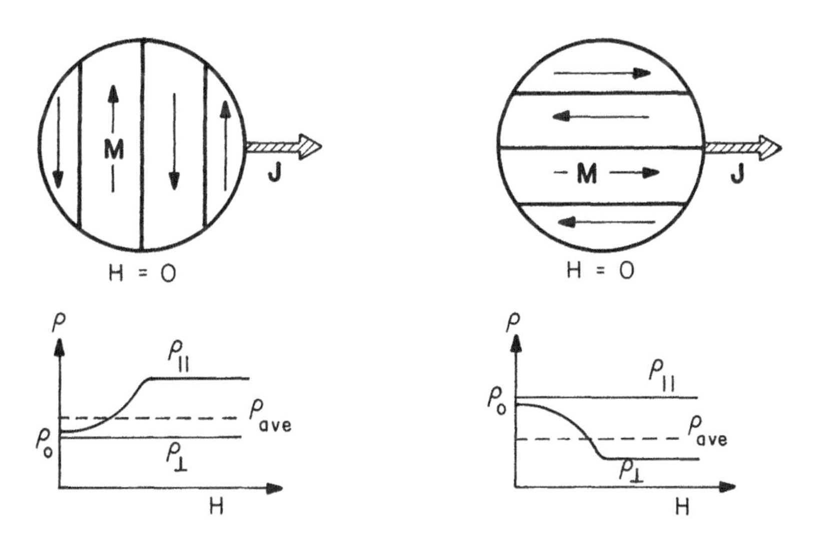

Figure 15.15 Field dependence of resistivity in fields parallel and perpendicular to $\boldsymbol{J}$ reveals the extraordinary or anisotropic magnetoresistance effect $\Delta\rho = \rho_{\parallel} - \rho_{\perp}$ at low fields superimposed on the ordinary effects (quadratic in H) at higher fields. Note that in zero field, the resistance may be larger or smaller than ρ_{av} depending on the equilibrium domain structure. (Compare with Fig. 7.5 for magnetostriction.)

Comparison with Figure 15.15 suggests that for the data of Figure 15.14, the current was applied in a direction parallel to the zero-field domain orientation (as in Fig. 15.15, right). Hence the resistivity is large at point A. At point B most of the magnetization has been rotated away from the current direction by a perpendicular field.

The low-field dependence of the anisotropic magnetoresistance [Eq. (15.15)] resembles that of the magnetostriction [Eq. (7.1)]. Both effects are quadratic in a hard-axis applied field below the anisotropy field, H_a, and both show the same zero field dependence on domain distribution. Thus, the same domain configuration that gives a large MR effect, specifically, magnetization by a 90° rotation process as opposed to a wall motion process, also results in the largest possible magnetostrictive strains. When determining magnetoresistance, care must be taken to measure the resistivity in fields of both orientations relative to $\boldsymbol{J}$ unless the quiescent domain configuration is known.

Because of the technical importance of an anisotropic magnetoresistance (see Chapter 18), some additional data on this phenomenon are included.

Figure 15.16 shows the composition dependence of the magnetoresistance ratio (MRR) in Fe-Ni and Ni-Co alloys. Note that the MRR in FeNi alloys peaks at 90% Ni, not far from the permalloy composition where λ_s and K_1 are close to zero. This is advantageous because at the permalloy composition the magnetization is easily rotated by an external field, i.e., H_a is small.

There may be more than chance to the compositional proximity of peak MR ratios and vanishing K_1 and λ_s. These features—anisotropy, magnetoresistance, and magnetostriction (as well as spontaneous Hall effect)—have their

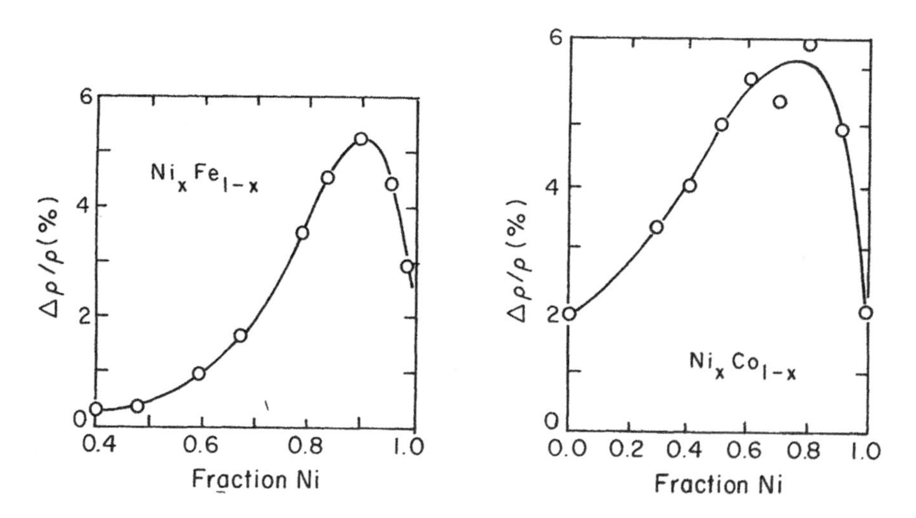

Figure 15.16 Composition dependence of anisotropic magnetoresistance at room temperature. Left, Ni_xFe_{1-x} [after Bozorth (1946)]. Right, Ni_xCo_{1-x}, [after Smit (1951) and van Elst (1959)].

origin in the spin–orbit interaction. Their composition dependence can be explained to some extent in terms of the split-band model outlined in Chapter 7 ($\langle L_z \rangle = 0$ at the minimum in the split-band density of states). When E_F for the composition coincides with a gap or minimum in the density of states, it is clear that $\boldsymbol{L} \cdot \boldsymbol{s}$ should vanish. Thus, while anisotropy and magnetostriction might vanish at nearly the same composition, it is not immediately clear why the anisotropic MR ratio should reach a maximum near the same composition, and appear to track the $K = 0$ or $\lambda = 0$ lines in ternary compounds.

Another important pattern in the compositional dependence of the MR ratio results from combining the data in Figure 15.16 with those for other alloy systems and plotting $\Delta\rho/\rho$ against magneton number as shown in Figure 15.17. While the MR ratio appears to peak for alloys with a magnetic moment per atom slightly less than one, the single-element data for Ni, Co, and Fe as well as for the ordered compound Ni_3Fe, show noticeable departures from the trend. Their smaller MR ratios reflect the diminished scattering in these pure materials. This effect is initially counterintuitive because it might be expected that the reduced resistivity in these pure materials would result in an *increase* in $\Delta\rho/\rho$.

Figure 15.18 shows the composition dependence of the electrical resistivity in the crystalline FeNi, CoNi, and CuNi alloy systems. Note that the resistivity is relatively flat in the vicinity of the MR peaks for FeNi and CoNi systems. This indicates that near 90% Ni in Figure 15.16, it is the anisotropic change in resistance $\Delta\rho$ that peaks rather than the resistivity presenting a minimum.

These ordinary and spontaneous galvanomagnetic effects can be summarized with a phenomenological model that expresses their common symmetry. Ohm's law can be written in a general form that expresses the symmetry of the Hall and magnetoresistance effects. The electric field inside an isotropic

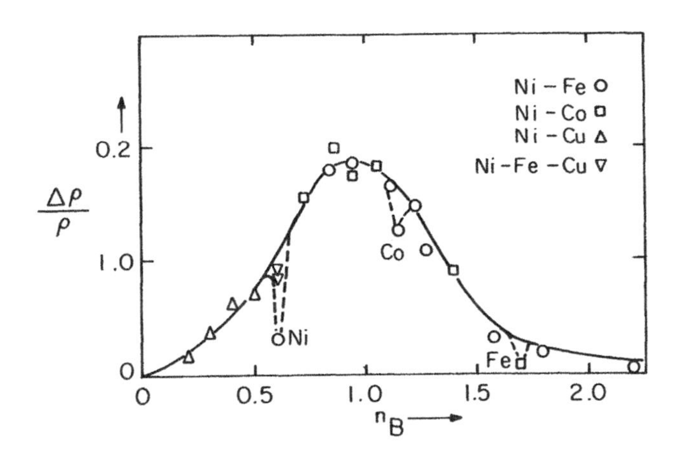

Figure 15.17 Anisotropic magnetoresistance ratio versus average Bohr magneton number for various metals and alloys. [After Smit (1951).]

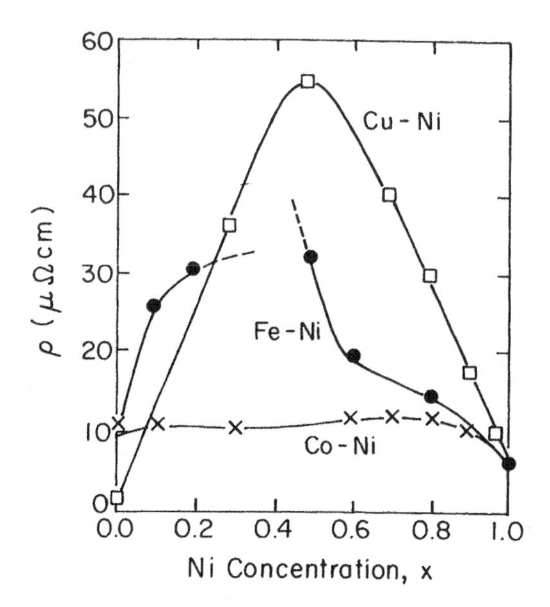

Figure 15.18 Resistivity of the binary alloy systems Ni_xFe_{1-x}, Ni_xCo_{1-x}, and Ni_xCu_{1-x}. [After Bozorth (1946).]

material is related to the current density by Ohm's law (McGuire and Potter 1975):

$$\boldsymbol{E} = \rho \boldsymbol{J} \tag{15.16}$$

In general, ρ is a tensor given by

$$\rho(H) = \begin{pmatrix} \rho_\perp(H) & -\rho_H(H) & 0 \\ \rho_H(H) & \rho_\perp(H) & 0 \\ 0 & 0 & \rho_{\parallel}(H) \end{pmatrix} \tag{15.17}$$

for $\boldsymbol{H}$ in the z direction. Here $\rho_\perp(\rho_\parallel)$ is the resistivity in a direction perpendicular (parallel) to the $\boldsymbol{H}$ field (which, in the case of a ferromagnet, is replaced by the magnetization) and ρ_H is the Hall resistivity in the geometry of Figure 15.8. All of the components of the resistivity tensor can depend on the magnitude of H. Ohm's law [Eq. (15.16)] can then be written as follows:

$$\boldsymbol{E} = \boldsymbol{h}(\boldsymbol{J} \cdot \boldsymbol{h})[\rho_\parallel - \rho_\perp] + \rho_\perp \boldsymbol{J} + \rho_H \boldsymbol{h} \times \boldsymbol{J} \tag{15.18}$$

where $\boldsymbol{h}$ is a unit vector in the direction of the applied field or, for a ferromagnet, the magnetization. Because magnetoresistance is quadratic in H, it is generally a smaller effect than the Hall effect.

In the geometry of Figure 15.8, the current density can be written as $\boldsymbol{J} = (0, J_y, 0)$, so Eqs. (15.16) and (15.17) give

$$E_x = -\rho_H J_y, \quad E_y = \rho_\perp J_y, \quad E_z = 0$$

Thus the voltage measured in the x direction is the Hall voltage and that measured in the y direction samples the resistivity perpendicular to $\boldsymbol{h}$. To measure the magnetoresistance anisotropy, the direction of the magnetization would have to be rotated into the y direction so that $E_y = \rho_{||} J_y$ could be measured. Then the difference in E_y for the two magnetization directions would give the magnetoresistance: $\Delta\rho = \Delta E_y / J_y$.

From Eq. (15.18), the spontaneous magnetoresistivity is a function of the relative direction of $\boldsymbol{M}$ and $\boldsymbol{J}$. Because the resistivity is measured in the direction of current, ρ can be written

$$\rho = \boldsymbol{E} \cdot \boldsymbol{J}/|J|^2$$

which leads to

$$\rho = \rho_\perp + [\rho_{||} - \rho_\perp]\cos^2\theta \tag{15.19}$$

where θ is the angle between $\boldsymbol{H}$ and $\boldsymbol{J}$. Equation (15.19) may be written

$$\rho(H) = \tfrac{1}{3}[\rho_{||} + 2\rho_\perp] + [\rho_{||} - \rho_\perp](\cos^2\theta - \tfrac{1}{3})$$

Subtracting $\rho_{\text{av}} = [\rho_{||} + 2\rho_\perp]/3$ from both sides and dividing both sides by ρ_{av} gives Eq. (15.15) using the definitions $(\rho_{||} - \rho_\perp) \equiv \Delta\rho$ and $\Delta\rho(H) = \rho(H) - \rho_{\text{av}}$.

Equation (15.15) or (15.19) describes the anisotropic magnetoresistance that is observed in materials. The anisotropic MR effect has been used in magnetic recording read heads for hard disk drives since the early 1990s (see Chapter 17).

15.3.3 Mechanism of AMR

It is tempting to try to understand the anisotropic ferromagnetic magnetoresistance from the classical picture for the ordinary MR effect in which a conduction electron trapped in a cyclotron orbit does not contribute to the current until it is scattered. It might be suggested by the generalization of Kohler's rule [Eq. (15.14)] that the internal field associated with $\boldsymbol{M}$ can cause a charge carrier to become localized. However, this mechanism suggests that when the current density is orthogonal to the magnetization in a ferromagnet, the electrons could become localized in cyclotron orbits about $\boldsymbol{M}$ and the resistance should increase. Just the opposite is generally the case for the anisotropic magnetoresistance; the resistivity is smaller when $\boldsymbol{J}$ and $\boldsymbol{M}$ are orthogonal (Fig. 15.14). So while the generalization of Kohler's rule may be valid, the sign of b in Eq. (15.14) is negative and the mechanism is not simple.

A satisfactory microscopic model of the anisotropic magnetoresistance effects has not been given. Here, an attempt is made to give a physical picture of the ingredients that are central to the more successful models.

The interaction of the conduction electrons with the lattice potential and impurities can be described by various contributions: the Coulomb attraction to the ion core, the spin–orbit and exchange interactions:

$$V_{\text{scat}} = V_{\text{Coul}} + V_{so} + V_{\text{exch}} + \cdots \tag{15.20}$$

The Coulomb interaction, $-Ze^2/r$, is the strongest of the three terms. The spin–orbit interaction (SOI) is the scattering mechanism that governs anisotropic MR, just as it controls the spontaneous Hall effect (Smit 1975, McGuire and Potter 1975). The exchange contribution is ignored here. The Coulomb potential and the spin–orbit scattering potential, $\xi \boldsymbol{L} \cdot \boldsymbol{S}$, are depicted in Figure 15.19.

The role of SOI in anisotropic magnetoresistance is suggested by the data for dilute rare-earth impurities (Fert et al. 1977). The asymmetric (i.e., $\langle L \rangle \neq 0$) $4f$ charge distribution of an impurity presents different scattering cross sections to an incident conduction electron of momentum, $\boldsymbol{k}$, depending on the orientation of $\boldsymbol{k}$ relative to the $4f$ moment, $g\mu_B m_J$, which is largely collinear with $\boldsymbol{L}$ and $\boldsymbol{S}$. Thus, the resistance acquires an anisotropy that is a function of the direction of the current relative to that of the magnetization. In this model, $\rho_{\parallel} > \rho_{\perp}$.

For $3d$ transition metals, Mott's two-current model must be used. While the operative scattering mechanism is recognized to be the SOI, this interaction is also known to be much weaker in $3d$ metals because the d states are strongly

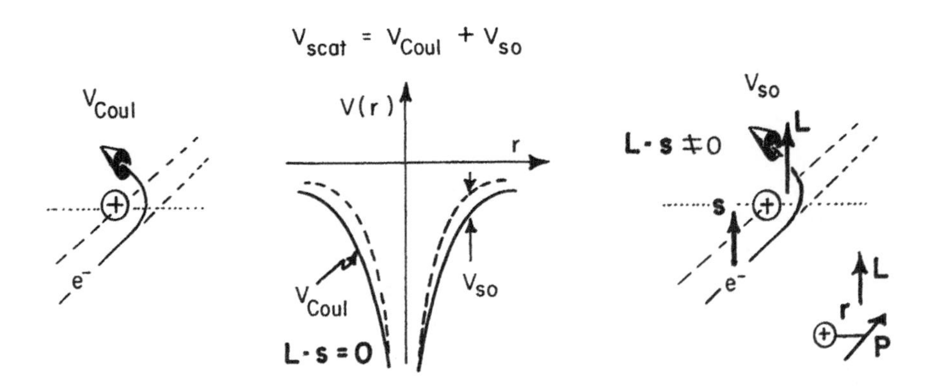

Figure 15.19 Simplified view of Coulomb scattering, left, and spin–orbit plus Coulomb scattering, right. The spin–orbit interaction alters the Coulomb scattering potential, center, for relative orientations of the magnetization and current that satisfy $\boldsymbol{L} \cdot \boldsymbol{s} \neq 0$.

perturbed by the crystal field which quenches their orbital angular momentum.

It is best to consider first the strong ferromagnetic alloys based on Ni, for which there are no $3d^{\uparrow}$ holes. This being the case, there is negligible s–d scattering in the majority spin channel and $\alpha \equiv \rho^{\downarrow}/\rho^{\uparrow}$ is very large. Alloys having large values of α generally show large, positive MR ratios (Fig. 15.21). Smit (1951) points out that in such cases with negligible majority-spin $s \to d$ scattering, a small increase in $\rho_{sd}^{\uparrow}$ could have a significant impact on the net resistivity. The SOI provides a way to mix spin-up and spin-down states so that $s^{\uparrow}$ electrons can be scattered into empty d states. The quantum mechanical reason for this mixing is outlined briefly below.

The operator in the SOI [Eq. (3.31)] can be represented as

$$\boldsymbol{L} \cdot \boldsymbol{S} = L_x S_x + L_y S_y + L_z S_z = L_z S_z + (L^+ S^- + L^- S^+)/2 \qquad (15.21)$$

by defining $L^{\pm} = L_x \pm iL_y$. Orbital, spin, or total angular momentum components perpendicular to the quantization direction can be represented as such $A^{\pm}$ operators. These complex operators are important because of what they do to the wavefunctions on which they operate. $L^{\pm}$ has the effect on the wavefunctions for angular momentum of raising or lowering (L^+ or L^-, respectively) the m_L value of the state described by the initial wavefunction

$$L^{\pm}\psi(m_l) \to \psi(m_l \pm 1) \qquad (15.22)$$

(See Problem 15.3).

The effect of operators of the form $L^+S^- + L^-S^+$ is to first lower or raise (S^- or S^+) the spin quantum number of a state [e.g., $S^-\chi(+\frac{1}{2}) \Rightarrow \chi(-\frac{1}{2})$, but $S^-\chi(-\frac{1}{2}) \Rightarrow 0$], then raise or lower its angular momentum component along z. The spin-flip operator, (15.21), therefore takes $3d^{\uparrow}(m_l)$ states into $3d^{\downarrow}(m_l + 1)$ states or $3d^{\downarrow}(m_l)$ states into $3d^{\uparrow}(m_l - 1)$ states. Thus $\boldsymbol{L} \cdot \boldsymbol{S}$ mixes spinup and spindown channels.

A simple picture is given in Figure 15.20 to illustrate how the SOI opens up new paths for s–d scattering that then contribute a resistivity anisotropy. When the spin–orbit interaction is inoperative (left), there is no s–d scattering in the majority-spin channel. In this case, the resistivity can be written

$$\rho = \frac{\rho_s(\rho_s + \rho_{sd}^{\downarrow})}{2\rho_s + \rho_{sd}^{\downarrow}} \equiv \rho_0 \qquad (15.23)$$

When the SOI is turned on, $s^{\uparrow}$ electrons can scatter to the $3d^{\downarrow}$ hole states, adding to the total resistivity. The SOI also allows $d^{\uparrow} \to s^{\downarrow}$ transitions that open $3d^{\uparrow}$ hole states, providing further channels for s–d scattering by $s^{\uparrow}$ (no spin flip) or $s^{\downarrow}$ (spin flip) electrons. However, s electrons can only scatter into the $3d$ hole states if the conduction electron momentum $\boldsymbol{k}$ is in the plane of the classical orbit of the *empty* d state. This is depicted in Figure 15.20 where, for clarity, only some of the five $3d$ wavefunctions are shown. The $3d$ hole states

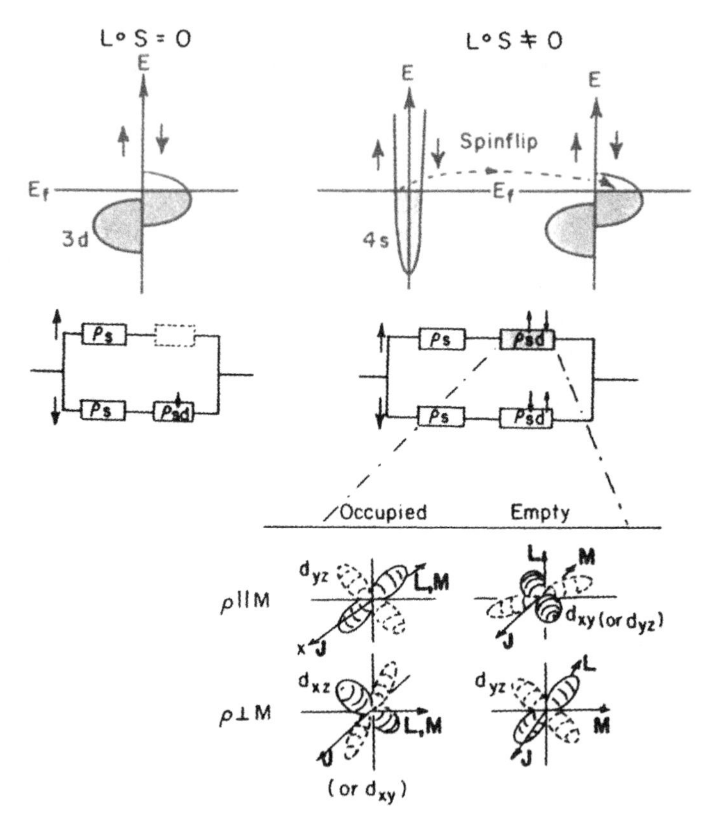

Figure 15.20 Upper left, $3d$ density of states for strong ferromagnet and equivalent circuit showing absence of $s^{\uparrow}$–d scattering. Upper right, when SOI is active, spin-flip s–d processes are allowed, increasing the resistivity in the majority spin channel. (Holes are also created in $3d^{\uparrow}$ by $d^{\uparrow}$–$d^{\downarrow}$ processes; these processes are not shown.) Below, selected unperturbed $3d$ wavefunctions are sketched to illustrate that when $J \parallel M$, the vacant d states have a component of $\boldsymbol{L}$ orthogonal to $\boldsymbol{M}$ and therefore have classical orbits $(k_x^2 + k_y^2)$ compatible with the conduction electron momentum k_x. When $\boldsymbol{J} \perp \boldsymbol{M}$ the likelihood of hole states having $\boldsymbol{J} \| \boldsymbol{L}$, and therefore incompatible with k_x, is increased. Thus $\boldsymbol{J} \| \boldsymbol{M}$ favors s–d transitions.

necessarily have different values of $\langle L_z \rangle$ than the $3d$ states that are occupied. Thus, $\boldsymbol{L}$ for a $3d^{\uparrow}$ hole state is, in general, not parallel to $\boldsymbol{L}$ of the occupied $3d^{\uparrow}$ states, which accounts for the magnetic moment. The figure illustrates that the new s–d scattering channels are more likely when $\boldsymbol{L} \| \boldsymbol{M}$, which places a greater fraction of the empty $3d$ states in the plane of the current.

The important consequence here of the $L^+S^- + L^-S^+$ mixing term is to allow for spin mixing (when $\boldsymbol{J} \| \boldsymbol{M}$), which always increases ρ in the two-current model. Hence, this model gives $\Delta\rho/\rho > 0$ as is generally observed for Ni-based alloys.

The orientation dependence of the $s^{\uparrow}$–$d^{\downarrow}$ scattering depicted in Figure 15.20 is such that $\rho_{sd}^{\uparrow}$ is nonzero when $\boldsymbol{J} \| \boldsymbol{M}$. If the angle between $\boldsymbol{J}$ and $\boldsymbol{M}$ is defined as θ, and the relative $s^{\uparrow}$–$d^{\downarrow}$ scattering strength is defined as $\delta = |V_{\mathrm{SOI}}/V_{\mathrm{Coul}}|^2$, the new SOI scattering processes can be represented as

$$\rho_{sd}^{\uparrow} = \delta \cos^2\theta \tag{15.24}$$

(The angular dependence must be at least quadratic because the effect works whether $\boldsymbol{J}$ and $\boldsymbol{M}$ are parallel or antiparallel). Thus, the resistivity from the circuit in Figure 15.20, right, can be expressed as

$$\rho = \frac{(\rho_s + \rho_{sd}^{\uparrow})(\rho_s + \rho_{sd}^{\downarrow})}{2\rho_s + \rho_{sd}^{\downarrow} + \rho_{sd}^{\uparrow}} = \frac{(\rho_s + \delta\cos^2\theta)(\rho_s + \rho_{sd}^{\downarrow})}{2\rho_s + \rho_{sd}^{\downarrow} + \delta\cos^2\theta} \tag{15.25}$$

Using the definition for ρ_0 in Eq. (15.23), Eq. (15.25) gives to lowest order:

$$\rho(M) \approx \rho_0\left[1 + \frac{\rho_{sd}^{\uparrow}}{2\rho_s}\cos^2\theta\right] \tag{15.26}$$

which is consistent with Eq. (15.15) or (15.19) to within an additive constant.

The fact that the anisotropic MR effect appears in second-order perturbation theory, specifically, $\Delta\rho/\rho \propto |V_{\mathrm{SOI}}|^2$, can be appreciated conceptually by considering that the SOI must operate twice for this mechanism to work. First, spin mixing must be turned by the SOI (creating some $3d^{\uparrow}$ holes) and second, the new s–d scattering is governed by the SOI as indicated by the symmetry of the effect.

Essential to a proper understanding of anisotropic MR are the following concepts:

1. The two-current model
2. $\rho^{\uparrow} \neq \rho^{\downarrow}$
3. The SOI results in spin mixing, creating $3d^{\uparrow}$ holes
4. $s^{\uparrow}$ electrons make use of new paths for s–d scattering via the SOI.

The reader can find more thorough and more quantitative treatments in Smit (1951), Campbell and Fert (1982), Potter (1974), and McGuire and Potter (1975).

One of the consequences of Smit's treatment is that $\Delta\rho/\rho$ is predicted to vary as $\alpha - 1$ where $\alpha = \rho^{\downarrow}/\rho^{\uparrow}$ (Fert and Campbell, 1972). That is, the larger the disparity between the s–d scattering in the two bands, the more a slight increase in $\rho^{\uparrow}$ can have a significant effect on the total resistivity. The collection of AMR ratios versus α in Figure 15.21 supports this result for Ni-based alloys. Alloys having large values of $\alpha = \rho^{\downarrow}/\rho^{\uparrow}$ generally show larger, positive MR ratios. Small values of α lead to small or even negative MR ratios.

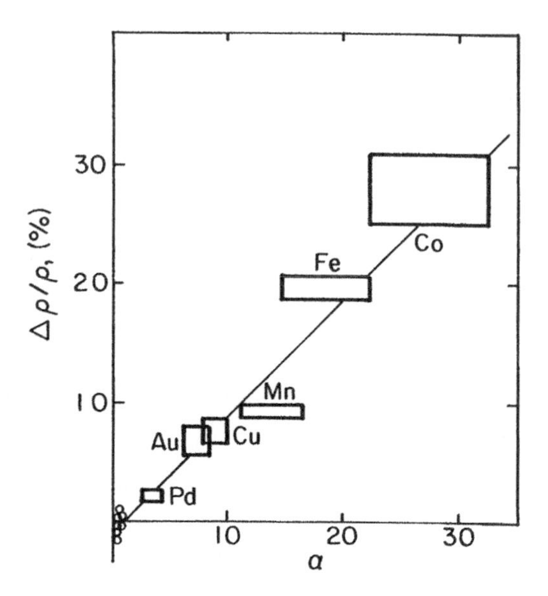

Figure 15.21 Anisotropic magnetoresistance in Ni-based alloys at 4.2 K as a function of $\alpha = 0.01(\alpha - 1)$ (Jaoul et al. 1977).

The more complicated case of Fe-based alloys (Dorleijn and Miedema 1976) is not described here.

15.4 GIANT MAGNETORESISTANCE

In 1988, Baibich et al. reported an MR ratio of order 50% at 4.2 K in FeCr multilayers that exhibit the oscillatory exchange described in Chapter 12. This magnetoresistance was approximately an order of magnitude greater than the highest values known to that time.

The Fe layers in these experiments were typically 30–60 Å thick and separated by Cr layers from 9 to 60 Å. Figure 15.22, left shows the relative orientation of the current density, the applied fields, and the crystal axes in the iron layers. The iron layers are antiferromagnetically coupled through the Cr layers and hence are difficult to saturate (Fig. 15.22, right).

The field needed to saturate the magnetoresistance is of order 20 kOe in this system (Fig. 15.23*a*). It is also clear from Figure 15.23*a* that the strength of the antiferromagnetic coupling varies with the Cr layer thickness. When the layers are coupled antiferromagnetically, the resistivity is observed to be larger than when an applied field brings the moments into alignment (Fig. 15.23*b*). The strength of this dependence of ρ on alignment of $\boldsymbol{M}_1$ and $\boldsymbol{M}_2$ is greatest when the antiferromagnetic coupling is strongest (0.9 nm Cr, Fig. 15.23), that is, when the moments of adjacent layers in zero field are almost completely antiparallel.

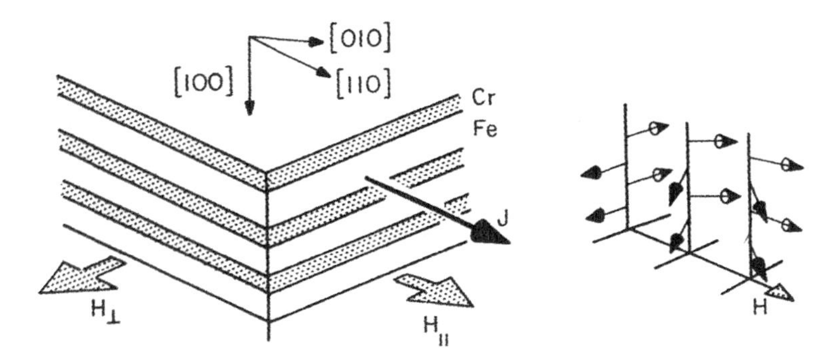

Figure 15.22 Left, relative orientation of iron easy axes, current and field for FeCr multilayers [after Baibich et al. (1988)]; right, schematic of magnetization process in the multilayers with increasing field strength. Magnetocrystalline anisotropy has been ignored.

There are two conventions for definition of the giant magnetoresistance (GMR) ratio. One refers the change in resistance ΔR to its high-field value; the other, to its low-field value. These two conventions lead to increasingly different results as ΔR becomes greater. For example, for the 9 Å Cu multilayer in Figure 15.20, right $\Delta R/R_{H=0} = -46\%$ and $\Delta R/R_{\mathrm{Hsat}} = +85\%$. Referring ΔR

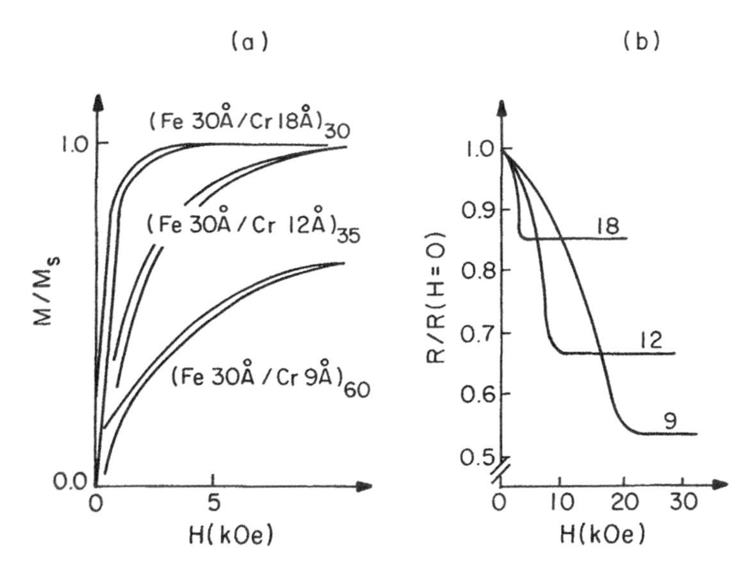

Figure 15.23 (*a*) Hysteresis loops for samples for three different FeCr multilayers at 4.2 K; (*b*) relative change in resistance with field parallel to current at 4.2 K. The numbers labeling the curves are the thicknesses (in angstroms) of the Cr layer in each case. [After Baibich et al. (1988).]

to the resistance of the antiferromagnetic or high-resistivity state limits the GMR ratios to values less than 100% but has the disadvantage that the $H = 0$ state is not always one of complete antiferromagnetic coupling or $\boldsymbol{M} = 0$. This convention is more common with experimental data. The other convention is to refer ΔR to the resistance of the ferromagnetic state R_F, which is well defined. This convention, $\Delta R/R = (R_{AF} - R_F)/R_F$ is more often used in calculations.

Parkin et al. (1990) observed that the MR ratio in Fe/Cr multilayers oscillate with the exchange coupling as shown in Figure 15.24. As the thickness of the Cr layer is increased, the oscillations in magnitude of the field needed to saturate the magnetization (Fig. 12.12) and in the magnitude of the MR ratio are observed to be in phase. *The magnetoresistance maxima occur at Cr layer thicknesses for which the magnetic layers are coupled antiferromagnetically.* The first two maxima in AF coupling and hence also in GMR ratio occur at spacer thicknesses of 9 and 24 Å for the Fe/Cr system. For Co/Cu multilayers (Parkin et al. 1991), the first two maxima of magnetoresistance occur at 8 and 19 Å. At least four oscillations in magnetoresistance are observable throughout the range of temperatures from 1 to 400 K. GMR ratios of 80% at 4.2 K and 65% at 300 K were observed in Si/Fe(40 Å)/[Co(7.5 Å)/Cu(9.3 Å)]$_{16}$.

The MR ratio in these multilayer systems is not a function of the angle between $\boldsymbol{J}$ and $\boldsymbol{M}$ as it is for AMR, but rather depends on the relative orientation of $\boldsymbol{M}$ in adjacent layers. Thus the mechanism for MR in these multilayers is different from that responsible for the anisotropic MR effect described in Sections 15.3.2 and 15.3.3. Because of the large magnitude of this new magnetoresistance, it has become known as giant magnetoresistance (GMR). GMR has been observed (Dieny et al. 1991a, 1991b) to obey the

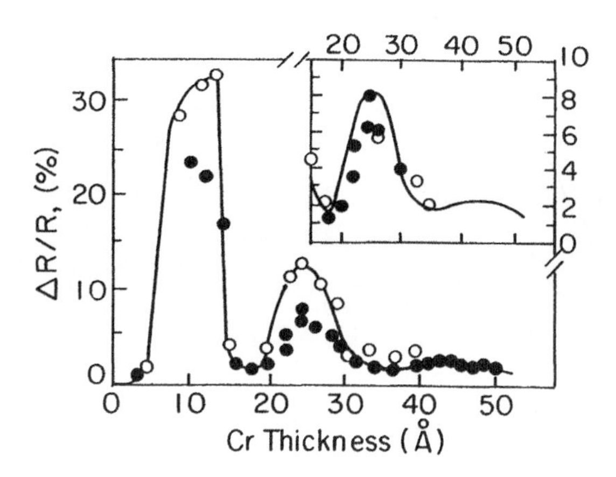

Figure 15.24 Transverse saturation magnetoresistance at 4.2 K versus layer thickness. Films deposited at 40°C, $N = 30$ (closed squares) and at 125°C, $N = 20$ (open squares). These oscillations in GMR strength are to be compared to those in the strength of the antiferromagnetic exchange coupling of this system, shown in Figure 12.12. [After Parkin et al. (1990).]

equation

$$\frac{\Delta\rho(\psi)}{\rho} = \left(\frac{\Delta\rho}{\rho}\right)_{\mathrm{GMR}} \frac{[1 - \cos\psi]}{2} \tag{15.27}$$

where ψ is the angle between the magnetizations in the two sets of layers. For the scissor-action process illustrated in Figure 15.22, right, the angle between either magnetization and the applied field, $\theta = \psi/2$, is important because it indicates the component of magnetization parallel to the field. Because $1 - \cos(2\theta) = 2\sin^2(\theta)$, Eq. (15.27) can be written

$$\frac{\Delta\rho(\theta)}{\rho} = \left(\frac{\Delta\rho}{\rho}\right)_{\mathrm{GMR}} \sin^2\theta \tag{15.28}$$

For hard axis magnetization, to a first approximation, $M/M_s = \cos\theta = H/H_a$; thus Eq. (15.27) gives for the field dependence of the MR ratio

$$\frac{\Delta\rho(\theta)}{\rho} = \left(\frac{\Delta\rho}{\rho}\right)_{\mathrm{GMR}} \left[1 - \left(\frac{H}{H_a}\right)^2\right] \tag{15.29}$$

This is the *approximate* form of the field dependence sometimes observed for GMR in antiferromagnetically coupled multilayers when the field is applied along the hard axis. It is consistent with the data in Figure 15.23*b*. Any nonlinearity of M with H would complicate the dependence of GMR on H. The field dependence of GMR is much more complicated than implied by this simple illustration.

A controversy exists concerning whether the scattering is occurring within the layers or predominantly at the interfaces. In an attempt to answer this question, Parkin (1992) studied $Ni_{81}Fe_{19}$/Cu superlattices where the permalloy/Cu interface was systematically decorated with various thicknesses of cobalt. He found that increasing the Co thickness at the interface from 0 to 4.4Å caused the GMR effect to increase from about 1% to nearly 20% at room temperature (Fig. 15.25). It was found also that if there are at least 4 angstroms of cobalt at the interfaces, the magnitude of the magnetoresistance is insensitive to the nature of the interior magnetic layer, i.e., [Co/NiFe/Co/Cu] shows the same GMR ratio as [Co/Cu]. This is true at room temperature or at 4.2 K (Fig. 15.25). Further, the permalloy/cobalt/copper superlattices show a fourfold decrease in the field required to saturate the GMR effect (to 15 Oe) compared to that for Co/Cu. These experiments suggest that interfacial scattering is important in GMR.

The interfacial nature of GMR is illustrated further in Figure 15.26, which compares the magnetic layer thickness dependence of the strength of GMR in various spin valve structures (see Section 15.5) with that for AMR in permalloy. Anisotropic MR decreases monotonically with decreasing film thickness

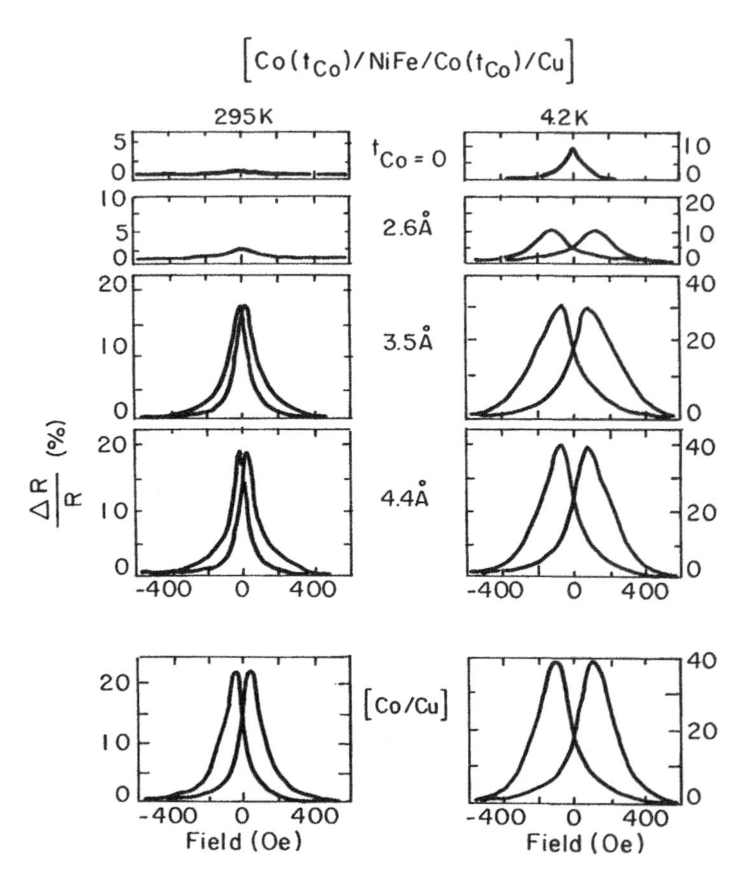

Figure 15.25 Top six panels, GMR ratio versus field for various thicknesses of an interfacial layer of cobalt in FeNi multilayers at room temperature and at 4.2 K. Lower two panels, GMR versus field for [Co/Cu] multilayers. [Adapted from Parkin (1992).]

and saturates at a bulk value for large NiFe thickness. GMR, on the other hand, shows a peak at small thicknesses and vanishes at larger magnetic layer thicknesses.

While the exchange coupling, which can play a role in GMR, is only weakly dependent on temperature, the GMR ratio at low temperatures is typically about four times that observed at room temperature.

15.4.1 Mechanism of GMR

The important observations that characterize GMR are summarized. GMR does not show the dependence on current direction relative to $\boldsymbol{M}$ that characterizes anisotropic magnetoresistance. Instead, GMR does depend on the relative orientation of the magnetization in adjacent magnetic layers. The

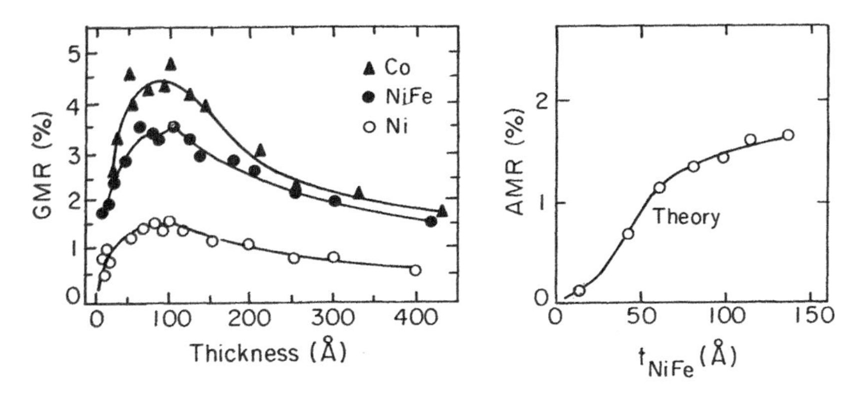

Figure 15.26 Left, variation of GMR with magnetic layer thickness in various GMR spin valve structures (Section 15.5.1) [after Dieny *et al.* (1991b)]; right, AMR versus NiFe thickness [after Daughton et al. (1992)].

GMR effect vanishes if the spacer layer thicknesses are much greater than the mean free path of the electrons (of order 10 nm); carriers of a given spin must be able to sample the spin-dependent scattering mechanisms in adjacent magnetic layers. This indicates that the important scattering events are those that depend on the relative spin of the carrier and the scattering site; spin–orbit scattering (which involves only one spin) is less important for GMR. A simple example of spin–spin scattering is exchange scattering, proportional to $\mathcal{J}\mathbf{s}\cdot\mathbf{S}$, where $\mathbf{s}$ and $\mathbf{S}$ are the spins of the charge carrier and the scattering site, respectively.

Although a complete understanding of giant magnetoresistance is not yet available, the effect may be understood by considering the preceding observations in the context of the two-current model introduced earlier. Assume that spin-dependent scattering is more likely when a carrier of one spin encounters a scattering site of opposite spin. (Recall that in spin–orbit scattering, it is the spin of the conduction electron, not that of the scattering site, that is involved). In simple terms for *antiferromagnetically coupled multilayers* (Mathon 1991, White 1992), conduction electrons of either spin having sufficiently long mean free paths will thermally sample a series of strong and weak scattering layers as they drift about the electric field direction from one magnetic layer to another. Thus carriers of either spin direction have comparable mean free paths and resistivities (Fig. 15.27*a*). However, when the layers are ferromagnetically coupled, carriers of the same spin direction as that of the magnetic layers will sample a series of weakly scattering, parallel-spin layers. Hence, their mean free path is longer and their resistivity smaller compared to those of carriers having spin opposite that of the magnetic layers (Fig. 15.27*b*).

This situation can be represented by the parallel resistance circuit shown in Figure 15.28. The nonmagnetic part of the resistivity is omitted for clarity; it is included in this discussion. The resistances R_1 and R_2 represent scattering of

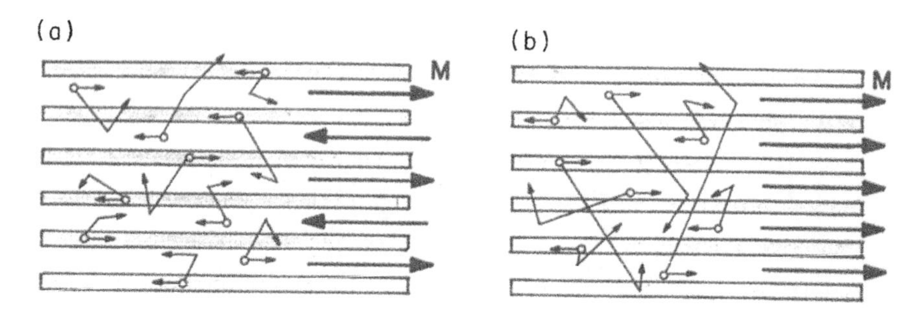

Figure 15.27 Schematic representation of magnetic multilayers with AF (*a*) and F (*b*) coupling. Charge carriers of majority and minority spin are shown as well as hypothetical trajectories with different scattering lengths. In the F case, the charge carriers with spin parallel to the direction of magnetization have a longer mean free path.

carriers of one spin in layers of the same spin (longer mean free path, smaller resistivity) and opposite spin, respectively.

Assuming that the carrier mean free path is greater than the separation between magnetic layers in the superlattice, carriers of a given spin experience a series of resistivities due to the nonmagnetic, $R_n = \rho_n l_n / A_n$, and magnetic, $R_m^i = \rho_m^\sigma l_m / A_m$, layers they traverse. The index i ($= \uparrow$ or $\downarrow$) designates the spin-up or spin-down channel and σ (open arrows in Figure 15.28) designate the spin in the magnetic layers *relative* to that of the carriers in that channel. Two nonmagnetic and two magnetic layers are assumed to be sampled for each channel. If the magnetic layers are coupled *ferromagnetically* (Fig. 15.27, left),

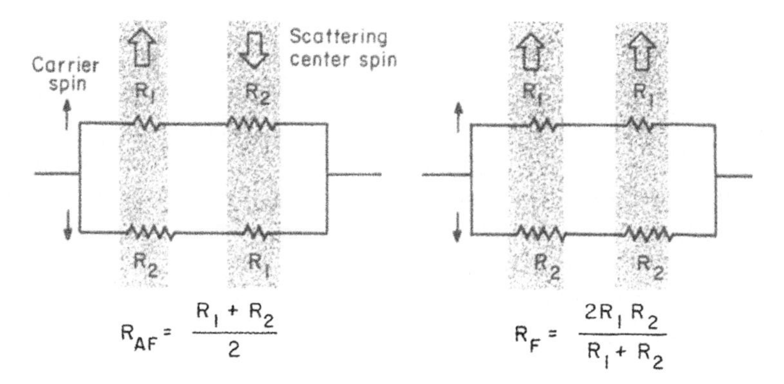

Figure 15.28 Equivalent circuits for multilayers depicted in Figure 15.27. In each circuit, the upper path represents the resistance due to spin-dependent scattering of upward-spin electrons, the lower path that of downward-spin electrons. The shaded regions indicate the magnetic layers with their direction of magnetization indicated by the open arrow. R_1 is for like-spin scattering events and R_2 for unlike-spin scattering. The spin-independent scattering in the spacer layers is omitted for simplicity.

the resistance in one channel is $R_m^i = 2\rho_n l_n/A_n + 2\rho_m^\sigma l_m/A_m$, where l_i is the path length in the n or m layer and σ is up for one value of i and down for the other. In this case, the two parallel spin channels add to give

$$R_F = \frac{(\alpha + 1)(\beta + 1)}{\alpha + \beta + 2} \frac{\rho_n l_n}{A_m}$$

where the definitions $\alpha = (\rho_m^\downarrow/\rho_n)$, and $\beta = (\rho_m^\uparrow/\rho_n)$ have been made. If the magnetic layers are coupled *antiferromagnetically*, the resistance in each of the two channels is the same, $R_n^\sigma = 2\rho_n l_n/A_n + (\rho_m^\uparrow + \rho_m^\downarrow) l_m/A_m$:

$$R_{AF} = (\alpha + \beta + 2) \frac{\rho_n l_n}{2} A_m$$

The resistance in the ferromagnetic case is always lower than that in the antiferromagnetic case as long as $\rho_m^\uparrow \neq \rho_m^\downarrow$, because the low resistance channel tends to short out the high resistance channel. The GMR ratio can then be shown to be given by

$$\frac{\Delta R}{R} \equiv \frac{R_F - R_{AF}}{R_{AF}} = -\frac{(\alpha - \beta)^2}{(\alpha + \beta + 2l_n/l_m)^2} \tag{15.30}$$

Thus, the GMR ratio referred to the antiferromagnetic, zero-field state, indicates a decrease in resistance limited in magnitude to 100%. This ratio depends quadratically on the difference between the scattering of current carriers from sites of like and unlike spin. Further, the GMR effect vanishes in the limit that the nonmagnetic layers are much thicker than the magnetic layers. This expression differs from that derived by Mathon (1991), who referred the GMR ratio to the ferromagnetic state.

It is possible to extend this phenomenological model to include microscopic interactions. Consider that the scattering of conduction electrons is due principally to the Coulomb interaction with the lattice. A small exchage interaction between the electron spin $\boldsymbol{s}$ and the local spin $\boldsymbol{S}$ of the scattering center, is added to the Coulomb scattering cross section:

$$V_{\text{scat}} = V_{\text{Coul}} - 2\mathcal{J} sS \cos\psi \tag{15.31}$$

The angle ψ is measured between the direction of $\boldsymbol{s}$ and $\boldsymbol{S}$. Depending on the sign of $\boldsymbol{S}\cdot\boldsymbol{s}$, the exchange strengthens or weakens the Coulomb scattering potential (Fig. 15.29). Substitution of the forms for $\rho^\uparrow$ and $\rho^\downarrow$ from Figure 15.29 in Eq. (15.30) indicates that the GMR ratio is quadric in $\delta = \mathcal{J} sS/V_{\text{Coul}}$.

Wang and Xiao (1994) discuss a more detailed physical model of the exchange scattering resistivity. They find that the resistivity ratio should be quadratic in the exchange interaction and also vary as $(M/M_s)^2$:

$$\frac{\Delta\rho}{\rho} = -\frac{4\gamma^2}{(1+\gamma^2)^2\left(1 + \dfrac{\rho_n}{\rho_0}\right)} \left(\frac{M}{M_s}\right)^2 \tag{15.32}$$

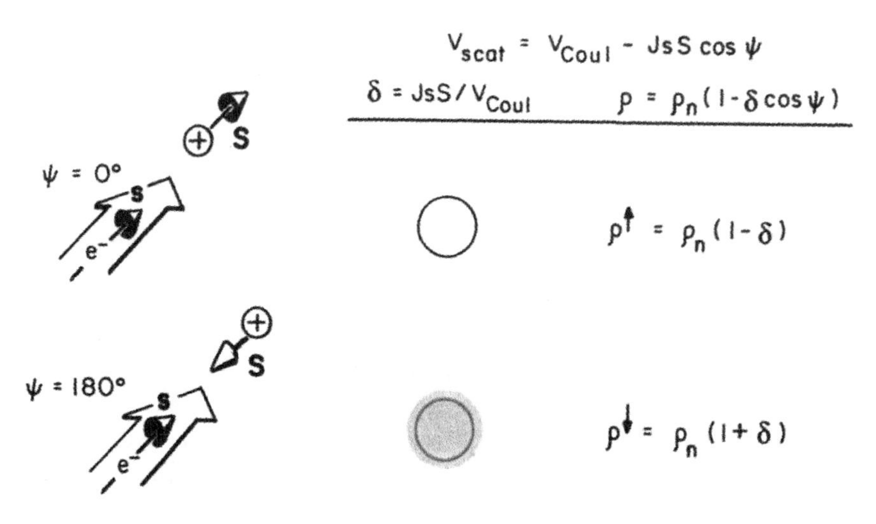

Figure 15.29 Simple model of exchange scattering exhibiting the symmetry observed for GMR. An electron of a given spin is modeled as scattering from an ion of like spin, top, and unlike spin, bottom.

where $\gamma = \mathscr{J}S/V_{\text{Coul}}$. In this equation, ρ_n is the nonmagnetic part of the thin-film resistivity and ρ_0 is the bulk classical resistivity [Eq. (15.1)]. Wang and Xiao (1994) calculate $\Delta\rho/\rho$ versus γ as shown in Figure 15.30. Their theoretical limit of GMR, $\Delta\rho/\rho = 1.0$, has not yet been reached. It would presumably be achieved in materials for which $\gamma = 1$ and ρ_r/ρ_0 is zero. So far cobalt gives the strongest GMR effect.

These observations are for the current in-plane (CIP) geometry (Fig. 15.31), in which the spin-bearing charge carriers sample different magnetic layers by their thermal drift. Experiments have also been done in the current perpendicu-

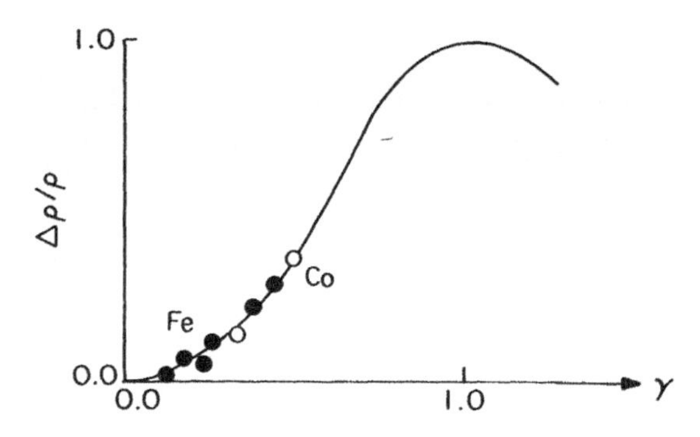

Figure 15.30 Variation of GMR ratio in granular films with parameter $\gamma = JS/V_{\text{Coul}}$. [After Wang and Xiao (1994).]

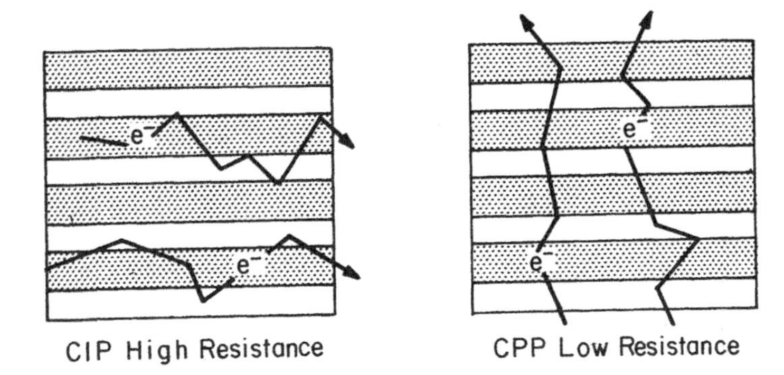

Figure 15.31 Schematic illustrations of CIP and CPP GMR geometries. In both cases the film's thickness is vertical and its lateral extent is truncated in the drawing.

lar to the plane (CPP) geometry (Pratt et al. 1993). The advantage of the CPP geometry is that electrons are forced to traverse many interfaces whereas in the CIP geometry, it is only the *thermal* motion of the electrons that causes them to move transverse to the drift direction dictated by the applied electric field. The disadvantage of the CPP geometry is that the electrical resistance of the film is extremely small normal to its plane, making measurements difficult. In CPP experiments, the magnetic field is still applied in-plane and the MRR saturates in a relatively weak field, 200 Oe. A very large GMR effect is observed in CPP.

GMR can also be observed in systems with *three* small dimensions to the magnetic component, namely, the granular magnetic films introduced in Chapter 12. [When two immiscible species are codeposited, the result is often a composite of fine ($d \leqslant 10$ nm) magnetic particles in a nonmagnetic matrix.] Chien et al. (1993) studied the magnetic properties of granular films based on the immiscible systems FCC Fe_xCu_{1-x} and FCC $Co_x\ Cu_{1-x}$. The interesting composition range, $x = 20$ to 30 at%, straddles the percolation threshold at which particle connectivity sets in.

Giant magnetoresistive effects have been studied in such granular media. At 5 K, Chien et al. (1993) measure a magnetoresistance of 9% in $Co_{16}Cu_{84}$ and 13% in $Co_{38}Cu_{62}$. At 300 K, an MR ratio of 25% has been observed in $Co_{30}Cu_{70}$. Alloys that are homogeneous (e.g., $Co_{80}Cu_{20}$) show no MR. In the $Co_{26}Ag_{74}$ system, a GMR ratio of 20% is found at room temperature (Berkowitz et al. 1992).

The characteristics of magnetoresistance observed in these granular systems are similar in some ways to those observed in the layered materials. In both cases, the maximum resistance occurs when H is equal to H_c, at which field the net magnetization is zero. In layered films, this occurs when adjacent layers are aligned antiferromagnetically. In granular systems at $M = 0$, the moments of the magnetic particles are randomly dispersed. However, the resistivity in these

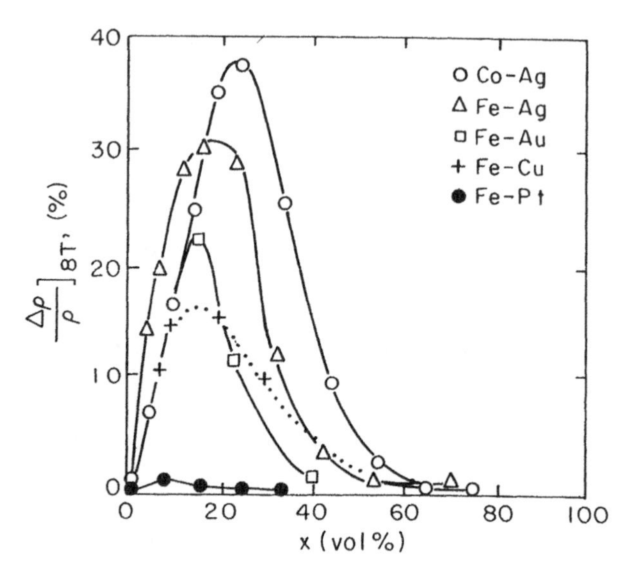

Figure 15.32 Variation of GMR at 8 T, $(\Delta\rho/\rho)_{8T}$, with volume fraction of magnetic components in several granular systems (Wang 1994).

granular systems often continues to decrease with increasing fields above saturation.

GMR in multilayers is maximized for thin layers (below which thickness ρ becomes very large and thus $\Delta\rho/\rho$ diminishes; see Fig. 15.26). In granular systems, GMR shows an optimum concentration for a variety of alloy systems (Fig. 15.32). For x too small, conduction electrons cannot communicate spin information by drifting from one grain to the next without scattering. For excessively large x (above the percolation threshold), the disorder in the particle moments decreases as they become ferromagnetically exchange-coupled.

15.5 APPLICATIONS

15.5.1 Spin Valves

A spin valve is a simple embodiment of the GMR effect in which there are only two magnetic layers separated by a nonmagnetic conductor. The magnetic layers are uncoupled or *weakly* coupled in contrast to the generally *strong* AF exchange operating in Fe-Cr-like multilayer systems. Thus the magnetoresistance can be made to change in fields of a few tens of Oersteds rather than tens of kiloersteds. One of the layers is magnetically soft and the other is magnetically hard or pinned. Thus, a modest field can cause a change in the angle between the moments of these two magnetic layers.

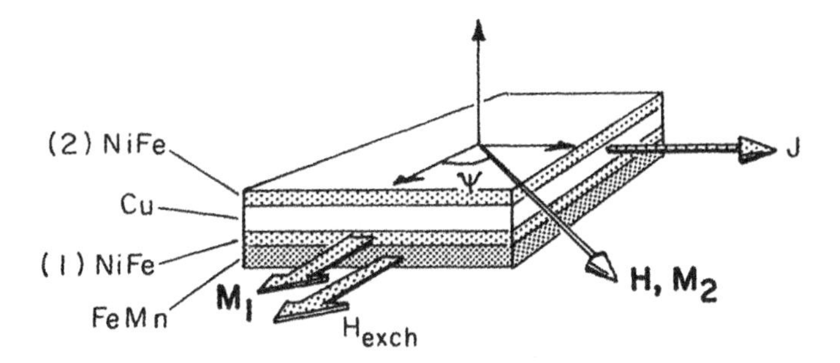

Figure 15.33 Typical composite film structure for spin valve effect. Permalloy layer (1) is exchange-coupled to the FeMn layer. Permalloy layer (2) is weakly coupled to layer (1). The magnetization of layer (2), M_2, can be arbitrarily oriented by an external field that is too weak to significantly perturb the orientation of M_1.

Such devices have been termed "spin valves" (Dieny et al. 1991a). In the case illustrated in Figure 15.33, the multilayer is Si/(NiFe 150 Å)/(Cu 26 Å)/(NiFe 150 Å)/(FeMn 100 Å)/(Ag 20 Å). The antiferromagnetic FeMn layer is exchange coupled to the adjacent soft permalloy layer. This tends to pin the direction of magnetization of this layer, NiFe(1). The second permalloy layer is weakly coupled to NiFe(1) across the 26-Å Cu spacer layer.

In an illustrative demonstration of the operation of the spin valve, the applied field is directed parallel to the exchange field and cycled in magnitude. The resulting M–H loops are shown schematically in Figure 15.34. The sharp magnetization reversal near $H = 2$ Oe is the switching of NiFe layer 2 in the presence of its weak coupling to layer 1. The more rounded magnetization reversal near 100 Oe is the switching of layer 1, overcoming its exchange coupling to the FeMn layer. The relative orientations of layers 1 and 2 are indicated by the pairs of arrows in each region of the M–H curve. Note that by cycling the field, M_1 and M_2 can be made to lie antiparallel or parallel to each other, just as adjacent layers in a GMR multilayer can be aligned by applying external field to break the strong AF coupling. In the lower panel, the change in resistance during the same magnetization cycling is shown. The resistance is larger for antiparallel alignment of the two magnetic layers, whereas the classic AMR of permalloy shows $\rho_{\|} > \rho_{\perp}$.

It should be emphasized here that a spin valve makes use of two different exchange couplings. The first is the strong exchange coupling of the pinned layer to the antiferromagnetic FeMn layer. This exchange coupling is a function of the uniaxial anisotropy K_u of the antiferromagnet and the interfacial coupling energy $\mathcal{J}_{F-AF}$ as described in Chapter 12. The second coupling is the weaker coupling between the two ferromagnetic layers, $\mathcal{J}_{F-F}$. This ferromagnetic coupling is generally balanced by the antiparallel dipole coupling between the two layers.

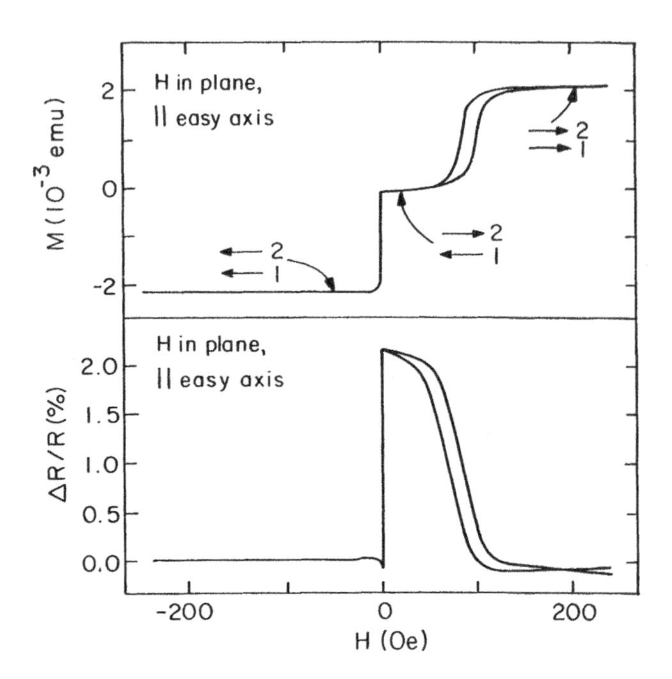

Figure 15.34 Room temperature magnetization and relative change in resistance for Si/(NiFe 150 Å)/(Cu 26 Å)/(NiFe 150 Å)/(FeMn 100 Å)/(Ag 20 Å). Current is perpendicular to the easy axis determined by the FeMn film, which is exchange-coupled to the adjacent permalloy layer. [After Dieny *et al.* (1991).]

An elegant set of experiments on this effect reveals the sharp contrast between conventional AMR and GMR as realized in a spin valve. A composite magnetic film like that shown in Figure 15.33 was fabricated [Si/($Ni_{81}Fe_{19}$ 60 Å)/(Cu 26 Å)/($Ni_{81}Fe_{19}$ 30 Å)/(FeMn 60 Å)/(Ag 20 Å)]. The exchange-coupled reference layer (1) (NiFe-30 Å) has an exchange coupled switching field of order 170 Oe. In this case, the NiFe layer (2) (60 Å) has a net weak parallel coupling to the NiFe–FeMn exchange-coupled layers in $H = 0$. The permalloy layer (2) can be rotated to any in-plane orientation, ψ by an applied field of order 10 Oe, leaving the coupled moment (1) pinned at $\psi = 0$. The GMR ratio, $\Delta R/R$, is observed to vary as $\cos\psi$ when H is rotated in plane as shown in Figure 15.35 (Speriosu et al. 1991). The exact form of the angular dependence is that given in Eq. (15.27).

Also shown is the conventional AMR ($\Delta\rho/\rho$ proportional to $\cos^2\theta$), which was subtracted from the total magnetoresistance measured. Whereas the conventional AMR peaks at $\psi = 90°$ (M_2 parallel to J), the GMR peaks for $\psi = 180°$ ($\boldsymbol{M}_2$ antiparallel to $\boldsymbol{M}_1$).

When spin valves are used as field sensors, there is a choice of directions for the pinned layer (1) relative to the field to be sensed. The quiescent state of the free layer (2) can also be designed (by sample shape or field-induced aniso-

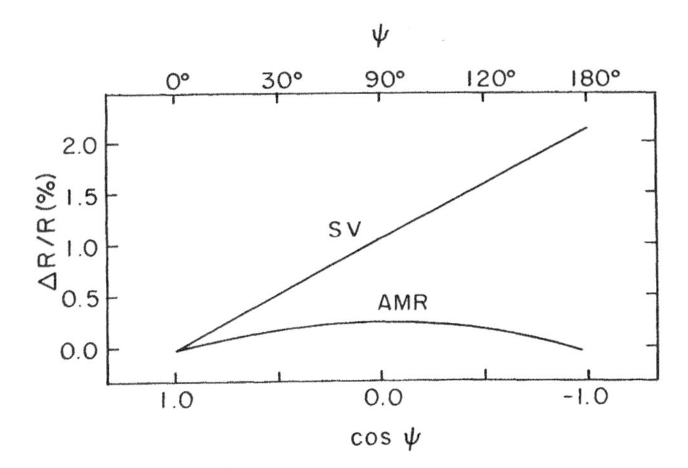

Figure 15.35 Variation of spin valve (SV) resistance and anisotropic magnetoresistance (AMR) with the cosine of the angle between $\boldsymbol{M}_1$ and $\boldsymbol{M}_2$. (Dieny et al. 1991).

tropy) to have different orientations. Figure 15.36 shows two cases: left, $M_1 \parallel H_y$ and $M_2 \perp M_1$ in $H_y = 0$ and right, $M_1 \perp H_y$ and $M_1 \parallel M_2$ in $H_y = 0$. In the first case, the M_2 magnetization process is described by $M/M_s = \cos\psi = H/H_a$. The field dependence for a spin valve with easy axis for M_1 perpendicular to the reference magnetization is then given by

$$\frac{\Delta R}{R} = \frac{\Delta\rho}{\rho}\left(\frac{1 - H/H_a}{2}\right) \tag{15.33}$$

Thus, the MRR for a spin valve in this configuration is *linear* in the applied field (Fig. 15.36*a*, lower panel).

In the other case, the M_2 magnetization process is $M/M_s = \sin\psi = H/H_a$ and the magnetoresistance is *quadratic* in the field (Fig. 15.36*b*, lower panel):

$$\frac{\Delta R}{R} = \frac{\Delta\rho}{\rho}\left(\frac{1 - \sqrt{1 - (H/H_a)^2}}{2}\right) \tag{15.34}$$

These magnetization processes account for only half of the change in magnetization in a 180° rotation, hence the factor $\frac{1}{2}$.

Even though the two magnetic layers in a spin valve are described as uncoupled, these devices actually can show a weak oscillatory coupling as Cu thickness is varied (Lottis et al. 1993). These authors also showed that if the reference layer is inadequately pinned, it can respond to the external field reducing ψ and hence reducing the GMR ratio.

Spin valves have been shown to be effective when used as a baselike control element in a spin valve transistor (Monsma et al. 1995).

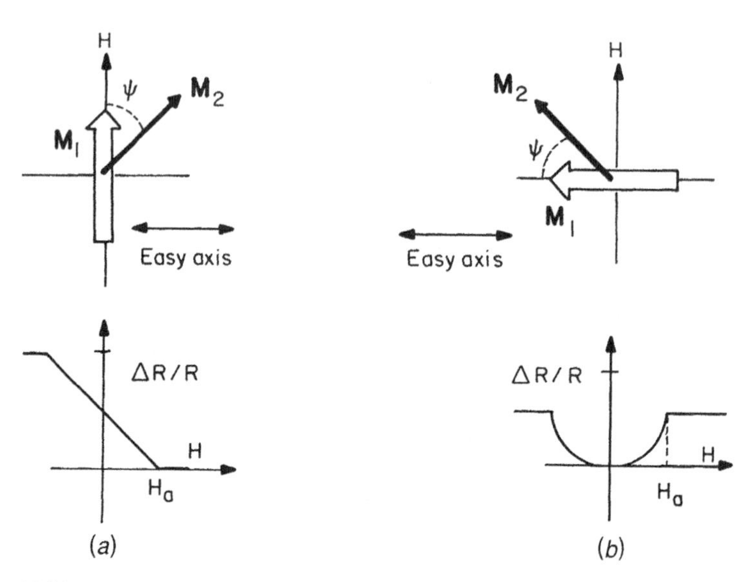

Figure 15.36 (*a*) Hard-axis magnetization process and MR for easy axis perpendicular to the reference magnetization direction (open arrow) in a spin valve structure; (*b*) hard-axis magnetization process and MR for spin valve with easy axis parallel to the reference magnetization direction.

15.5.2 Spin Switches

A spin switch is a magnetic thin film device that is configured very much like a spin valve: two thin magnetic layers sandwiching a nonmagnetic metal layer. However, the current in a spin switch passes through the thin direction of the sandwich. The terminology used to describe the functioning of the spin switch is quite different from that for a spin valve, largely for historical reasons. The development of the spin switch began in 1985, when Johnson and Silsbee (1988) began studying spin injection from ferromagnetic to paramagnetic thin films. The concept of spin injection is illustrated in Figure 15.37 for a ferromagnetic–paramagnetic bilayer. The ferromagnetic layer F_1 is chosen to be a strong ferromagnet so that electrical current is of predominantly one spin type. The effect of driving electrons of mostly one spin into the paramagnetic material is illustrated at the right. Charge neutrality in P is maintained by a net efflux of electrons. When a steady state is established, the paramagnet P gains a small nonequilibrium magnetic moment M_p at the expense of the ferromagnet; thus, the Fermi energies of the two subbands in the paramagnet now differ, whereas before the circuit was closed, they were the same. M_p relaxes with a characteristic time that appears to be the transverse spin relaxation time T_2 from magnetic resonance. The establishment of a steady-state, nonequilibrium magnetization in P implies an impedance to further charge and spin transport from F_1 to P.

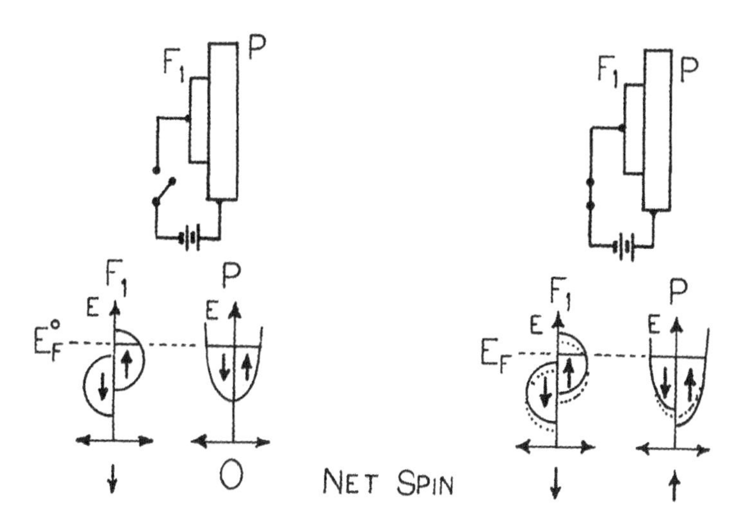

Figure 15.37 Schematic of a spin injection bilayer and state density representations that explain its operation. At left, the circuit is open; at right, closed. For open circuit, the Fermi levels of the two metals are equal, and only the ferromagnetic material exhibits a moment. For a closed circuit with the polarity shown, Fermi energy electrons from the ferromagnet (minority spin) are injected into the paramagnetic metal, which then acquires a moment at the expense of the ferromagnetic film.

A spin switch adds a second ferromagnetic layer, F_2, to the bilayer in Figure 15.37. F_2 should also be a strong ferromagnet. F_1 and F_2 are not exchange-coupled through the intermediate paramagnetic layer. The P layer thickness d must be less than the average distance δ_s that an electron could travel without loss of its original spin direction: $\delta_s = (2D_s T_2)^{1/2}$, where D_s is the spin diffusion constant. There are different ways of calculating δ_s, but in all cases the spin diffusion length, as it is called, exceeds the mean free path of charge transport by a few orders of magnitude. This is because spin-altering collisions occur much less frequently than do momentum-altering collisions.

The flow of magnetic moment into P is proportional to the charge current I scaled by (μ_B/e) to account for spin transport rather than charge transport. It also contains a current polarization factor $\eta = (J^\uparrow - J^\downarrow)/(J^\uparrow + J^\downarrow)$ to describe the extent to which the current density J is dominated by one spin type or the other. Thus the magnetic moment current is

$$I_M = \frac{\eta I \mu_B}{e} \tag{15.35}$$

The nonequilibrium magnetization buildup in P, M_p, is given by the magnetic moment current into P, I_M, times T_2 divided by the volume of P:

$$M_P = \frac{I_M T_2}{Ad} \tag{15.36}$$

The buildup of spin in P impedes further flow of current across the F_1–P interface. This is represented by the F_1–P interface spin resistance R_s.

To make a device out of this bilayer, it is helpful to consider how the spin-up electrons that make up M_P can be drained from P or trapped there. This can be accomplished by placing a second ferromagnetic layer F_2 in contact with P (see Fig. 15.38, below). For M_2 parallel M_1, spin-up states exist in F_2 at E_F, which can accept spin-up carriers from P. In this case a charge current and a magnetic current flow across P to F_2, and $E_F^{\uparrow}$ in F_2 increases. For M_2 antiparallel to M_1, no such path exists and the impedance of the device is increased; E_F in F_2 decreases. The Fermi level shifts that occur in F_2, depending on the sign of M_2/M_1, are on the order of 10^{-8} V. Essentially the impedance of the device is a function of the relative orientation of $\mathbf{M}_1$ and $\mathbf{M}_2$. Thus, these devices give weak measurable voltages and/or resistance changes that depend on the relative orientation of the magnetizations in the two magnetic layers.

Figure 15.39 shows results for a permalloy/gold/permalloy spin switch. The compositions and thicknesses of the permalloy layer are chosen to give them different coercivities. Thus, the difference in the M–H loops of F_1, the emitter (e), and F_2, the collector (c), leaves a narrow field range over which the magnetizations of the two layers are antiparallel to each other as indicated by the inset representations of the spin switch structure. Below this is shown the

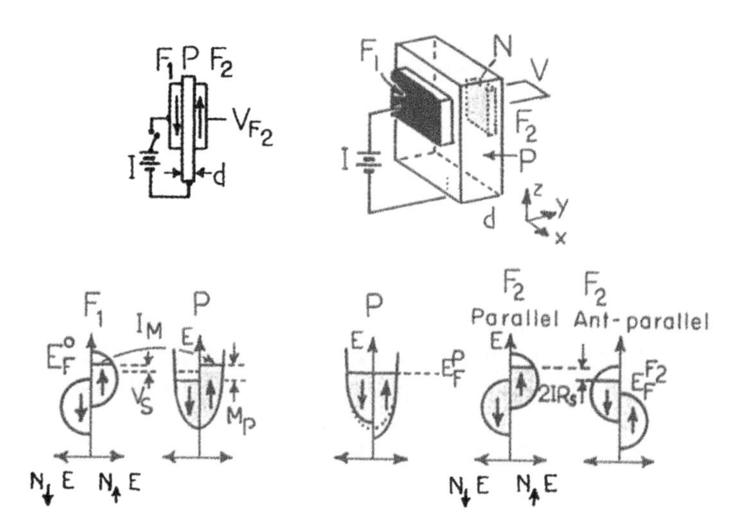

Figure 15.38 Above left, schematic of three-layer sandwich of spin switch, right, perspective view. Below left, configuration of ferromagnetic and paramagnetic layers showing dynamic process of spin transfer from the F_1 minority-spin band to the P band of same spin direction. Below right, schematic state densities for the F_2 parallel and antiparallel to F_1 and voltage difference between these two configurations referenced to paramagnetic Fermi level.

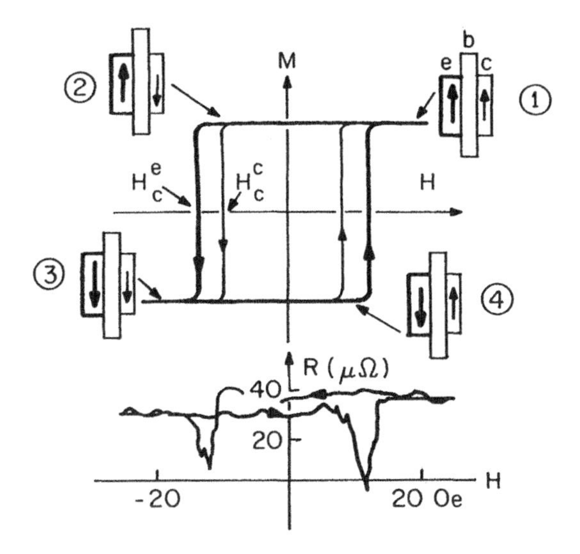

Figure 15.39 M–H loops of the emitter (e) and collector (c) and configuration of their magnetizations at different points relative to the loops. Below, the impedance versus field, H, measured by Johnson for such a device. [Adapted from Johnson (1993).]

change in resistance of the device, measured as a function of field. The impedance varies as $\cos \psi$, where ψ is the angle between the two magnetizations, just as for a spin valve. In terms of magnetoresistance, $\Delta R/R$ is 100% and the resistance change occurs over a field range of only a few oersteds. However, because of its geometry, the spin switch is a very low-impedance device and signal detection is difficult.

Johnson (1991) draws the useful analogy between the spin switch and a semiconductor device. Semiconductors derive their utility from the fact that two distinct carrier populations, electrons (N) and holes (P), exist and have Fermi energies that can differ by about an electronvolt. When a P–N junction is formed, there is a net flow of carriers across the interface until this energy difference is neutralized and the Fermi energies come to a common value. This shift in band structure across a P–N junction is the basis for many semiconductor devices including diodes, transistors, and photovoltaic cells. In ferromagnetic materials, the two carrier populations are obviously the spin-up and spin-down carriers, also with differences in their Fermi energies of order 1 eV. It has been shown that these two populations of carriers are often independent of each other. They form the basis for a number of devices, including the spin switch. Unlike semiconductor devices, the performance of spin switches improves as their thickness decreases because the spin diffusion lengths are considerably shorter than electron or hole diffusion lengths.

15.6 SPIN TUNNELING

Spin tunneling junctions have some similarities to spin valves and spin switches in their structure and field dependence. However, in a tunnel junction, the nonmagnetic spacer layer is an insulator. In order to understand the phenomena and devices associated with spin tunneling, it is helpful to begin with a review of the concepts of superconducting tunneling. An excellent review is given by Merservey and Tedrow (1996).

A tunnel junction is conveniently formed from a crossed pair of metal film stripes. The first deposited stripe, for example, aluminum, may be oxidized partially to form a barrier before deposition of the second electrode (Fig. 15.40). Alternatively, the metal and oxide layers can be deposited and patterned independently. A voltage applied across such a junction can result in a current if there are occupied states in one electrode at the same energy as unoccupied states in the other plus or minus $k_B T$. Hence varying the voltage slides the state densities of the two junctions over each other. The $I(V)$ tunneling behavior is therefore a convolution of the densities of states $D(E)$ of the two electrodes, the differences in their Fermi functions $f(E)$, and a tunneling probability function (related to the barrier characteristics) (Wolf 1985):

$$I(V) \propto \int_{-\infty}^{\infty} D_1(E - V) D_2(E)[f(E - V) - f(E)] dE \qquad (15.37)$$

The appearance of the difference Fermi function in this equation reflects the fact that electrons in states above E_F (or holes in states below E_F) cannot tunnel across the barrier unless they are already at the metal/insulator interface. If they are more than a mean free path from the interface, they have no means of transport to the interface unless they are within $k_B T$ of E_F. When one of the electrodes is a normal metal (with a relatively flat density of states

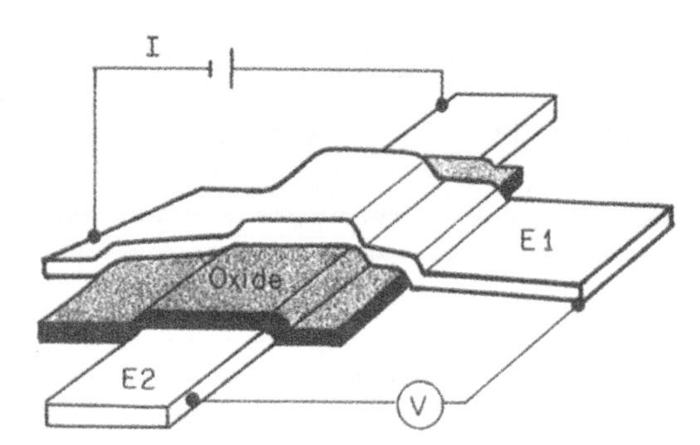

Figure 15.40 Schematic of a patterned tunnel junction. Electrode 2, E2, is either oxidized or coated with an oxide. Electrode one, E1, is deposited over the oxide on E2. A current is driven from E1 to E2 by a voltage.

near E_F), Eq. (15.37) becomes

$$I(V) \propto D_n(E_F) \int_{-\infty}^{\infty} D_s(E)[f(E+V) - f(E)]dE \qquad (15.38)$$

Here $D_i(E)$ for $i = n$ or s is the density of states for the normal and superconducting electrodes, respectively. The voltage derivative of Eq. (15.38) gives the conductance

$$\frac{dI}{dV} \propto \int_{-\infty}^{\infty} D_s(E) f'(E+V)dE \qquad (15.39)$$

where $f'(E)$ is the derivative of the Fermi distribution.

Figure 15.41*a* shows how the characteristic energy gap, $E_g = 2\Delta$, of the superconducting density of states is reproduced by a measurement of dI/dV versus voltage (measured from the Fermi energy) in a superconducting–normal (S–N) tunnel junction. Very little current flows when $|V| < \Delta$ because then at any given energy, either the density of empty or filled states is negligible.

When a magnetic field is applied parallel to the films in an S–N junction, a Zeeman splitting occurs in the energies of the electrons forming the Cooper pairs: $E(k) \pm 2\mu H$, where μ is the magnetic moment of the electron. This gives rise to a splitting of the peaks in the density of states of the superconductor. Thus the tunneling conductance splits as depicted in Figure 15.41*b*. When $\mu H < \Delta$, the Zeeman splitting of the Bardeen–Cooper–Schreiffer (BCS) density of states provides a basis for spin-polarized tunneling because predominantly the spin-up (spin-down) superconducting states are found in the energy range between $\pm(\Delta$ and $\Delta - \mu H)$. Both spin-up and spin-down states are available for $|E| > |\Delta + \mu H|$. Thus, when the junction voltage brings the metallic $f'(E)$ into the energy ranges $\pm(\Delta$ and $\Delta - \mu H)$, only the spin-up or spin-down metal states, respectively, contribute to the current.

Ferromagnetic–Superconducting Tunneling If the normal electrode of the S–N junction is ferromagnetic, tunneling measurements provide a reliable measure of electron spin polarization.

Figure 15.41*c* provides an initial basis for understanding tunneling in ferromagnetic–insulating–superconducting (F–I–S) junctions. Just as in the two-current model of ferromagnetic conduction, it is assumed here that electron spin is conserved during the tunneling process. Hence the Fermi function derivative for spin-up electrons is convoluted only with the quasiparticle density of spin-up states, and the Fermi function derivative for spin-down F states is convoluted with the density of spin-down S states. The conductances for both spin channels add in parallel. In a positive magnetic field, a positive voltage brings the $3d^{\uparrow}$ Fermi derivative function into coincidence with the spin-up quasiparticle density of states and tunneling occurs. For higher

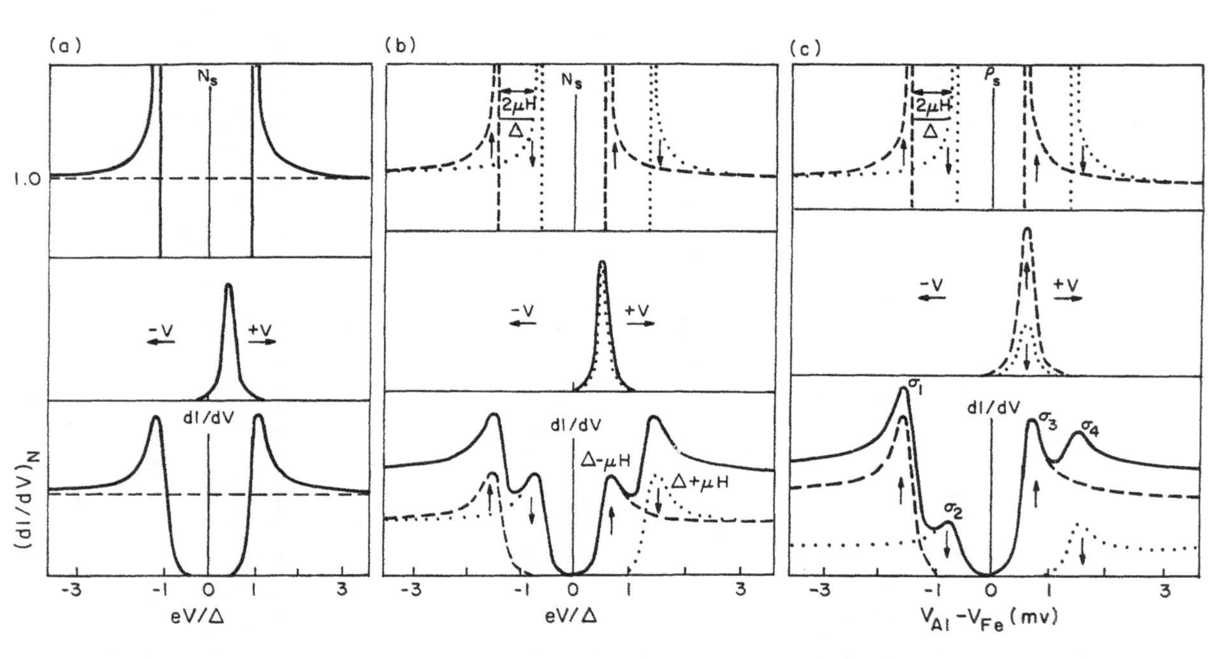

Figure 15.41 Three panels (*a*), (*b*), and (*c*), show the following: top, the superconducting density of states; middle, the metal electrode density of states, and lower, the convolution of these two densities of states for (*a*) superconductor–normal metal junctions, (*b*) superconductor–normal metal in an applied field, (*c*) superconductor-ferromagnetic junction. [After Meservey and Tedrow (1996).]

voltages, both $3d^{\uparrow}$ and $3d^{\downarrow}$ electrons can tunnel because states of both spin directions are available in the superconducting electrode. Thus the four peaks in Figure 15.41*c* correspond roughly to $\sigma_1 \approx 3d^{\uparrow}$ and $3d^{\downarrow}$, $\sigma_2 \approx 3d^{\downarrow}$, $\sigma_3 \approx 3d^{\uparrow}$, $\sigma_4 \approx 3d^{\uparrow}$, and $3d^{\downarrow}$.

Figure 15.42 shows the tunneling data for $Al-Al_2O_3-Ni$ junctions. The fact that spin-up electrons predominate in the experimental data ($\sigma_3 > \sigma_2$) whereas $D^{\uparrow}(E_F) < D^{\downarrow}(E_F)$ for Ni, indicates that tunneling is not a measure of polarization only at E_F. It is believed that mainly itinerant 3*d* states contribute to the tunnel current. There is also some evidence that the polarization measured is more representative of the surface than the bulk of the ferromagnet.

The results of tunneling experiments on several ferromagnetic materials are summarized in Table 15.2. In addition to the experimentally determined tunneling spin polarization, the table shows the polarization of the 3*d* and 4*f* electrons as determined by magnetic moment measurements, the polarization of the Gd 5*d*6*s* conduction electrons, and the band structure analysis of Stearns (1977). The 3*d* magnetic moment polarizations in Table 15.2 are based on 7.05, 8.1, and 9.4 3*d* electrons per atom for Fe, Co, and Ni, respectively.

A model by Stearns (1977) indicated that the large positive tunneling polarization was due mainly to the *s*–*d*-hybridized electrons. [Stearns had also suggested that these itinerant electrons could account for the ferromagnetism of iron in an RKKY context (Chapter 5, Section 5.6).] The tunneling probability

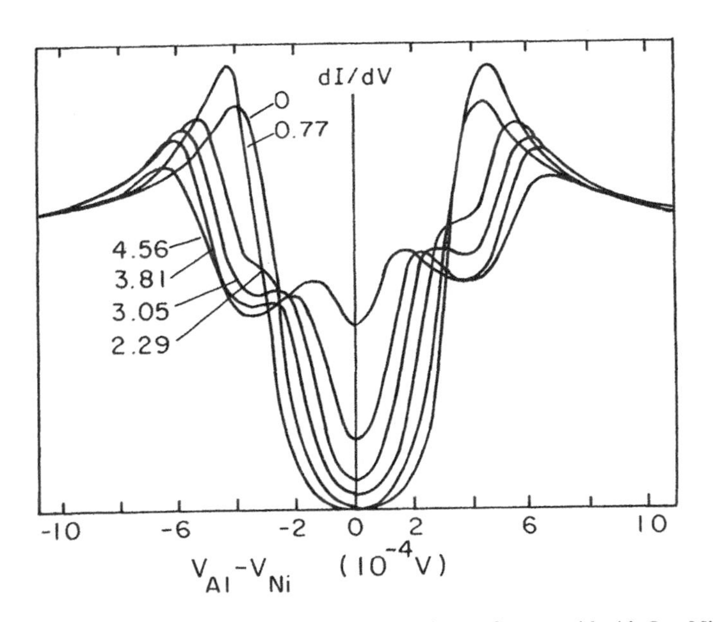

Figure 15.42 Measured conductance versus voltage for an $Al-Al_2O_3-Ni$ tunnel junction for several values of applied field (in kOe) (Messervy and Tedrow 1996).

TABLE 15.2 Spin Polarization Measured by F–I–F Tunneling [After Meservey and Tedrow (1994)] Compared with Polarizations Calculated in Different Ways

	Spin-Polarized Tunneling (%)	$P_{3d} = \frac{n^{\uparrow} - n^{\downarrow}}{n^{\uparrow} + n^{\downarrow}}$	P_{4f}	P_{5d6s}	Band Structure (%) (Stearns)
Fe	+40	31	—	—	40
Co	+35	21	—	—	≈35
Ni	+23	6.3	—	—	12
Gd	+14	—	100	19%	—

of electrons is believed to be proportional to their k vector, giving

$$P = \frac{k_f^{\uparrow} - k_f^{\downarrow}}{k_f^{\uparrow} + k_f^{\downarrow}} \tag{15.40}$$

In the case of Fe, the mobile s–d hybridized states dominate the band structure near E_F (see Fig. 5.15). Stearns found that k_f values averaged over the three crystallographic directions, [100], [110], and [111], from a calculated band structure (Callaway and Wang 1977) give $P = 40\%$ from Eq. (15.40). This model underestimates the tunneling polarization observed for Ni and gives a reasonable value for Co.

The tunneling polarization for the rare-earth metals cannot be due to the $4f$ electrons (100% polarized in Gd) because they are highly localized. It must come from the polarization of the $5d$ and $6s$ electrons. The conduction electron polarization of Gd has been indicated by other experiments and calculations (Freeman) to be of order 3–5%. A logical extension of these measurements is to investigate tunneling in F–I–F junctions.

Ferromagnetic-Ferromagnetic Tunneling Julliere (1975) studied tunneling in Fe–Ge–Co junctions and analyzed the results in terms of a two current model (i.e., no spin flip during the tunneling process). He assumed the tunneling conductance for each spin to be proportional to the product of the spin-resolved densities of states in each electrode. (There is no gap in the state densities here as there is when one of the electrodes is a superconductor. Hence, a junction bias voltage need not be applied to bring the two same-spin state densities into energy coincidence. A voltage is still needed for tunneling across the barrier.) Thus, as is the case for GMR, the current should be larger when the moments of the two magnetic layers are aligned parallel (F) rather than antiparallel (A) to each other. The result of the experiment then is a magnetoconductance ratio, $\Delta G/G$, where $\Delta G = G_F - G_A$ is analogous to the MR ratio, $(\rho_{\|} - \rho_{\perp})/\rho_{av}$. Julliere showed that

$$\frac{\Delta G}{G} = \frac{2P_1P_2}{1 + P_1P_2} \tag{15.41}$$

where P_1 and P_2 are the polarizations of the two ferromagnetic electrodes. He measured $\Delta G/G = 14\%$ at 4.2 K for an Fe–Ge–Co junction. This value seems low in light of Eq. (15.41) and the measured polarizations in Table 15.2. It is consistent with the 3*d* moment polarizations in that table.

A more recent theory by Slonczewski (1989) considered charge and spin tunneling through a rectangular barrier. In this model the spin polarization was found to be influenced by the barrier height, a feature consistent with the small values of $\Delta G/G$ found in early work. MacLaren (1977) discusses the assumptions and regimes of validity of Eq. (15.41) and Slonczewski's result.

This early work on F–I–F tunnel junctions has taken a new direction in the experiments of Moodera et al. (1996). They have shown that by using thin magnetic films of different coercivities, the F–I–F tunnel junction can be a sensitive magnetic field sensor. Figure 15.43 shows the fractional resistance change (ΔR normalized to the high-field value of resistance) in FeCo–Al_2O_3–Co junctions reported by Moodera et al. (1996). Also shown is the anisotropic magnetoresistance measured in each individual electrode (cf. Fig. 15.14). These AMR measurements show that the small value of the AMR effect contributes little to the tunneling MR effect, and they also clearly indicate the coercivities of the two uncoupled ferromagnetic layers. The tunneling MR ratio then

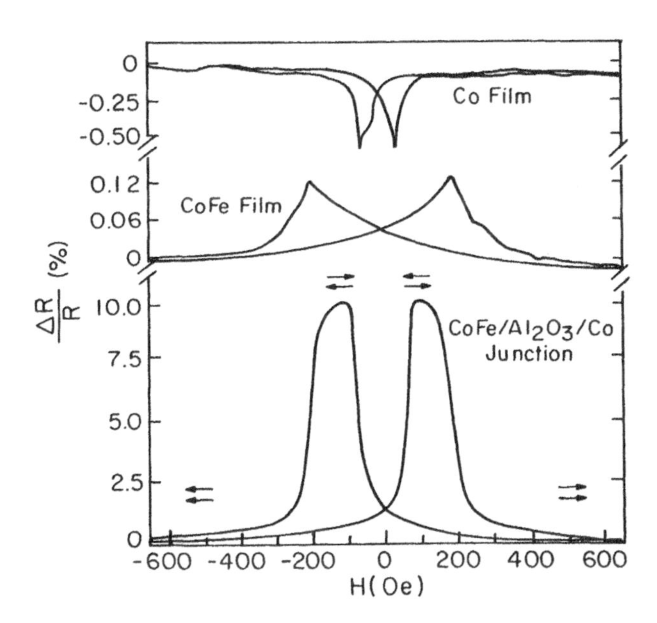

Figure 15.43 Above, anisotropic magnetoresistance in each individual CoFe and Co electrode; below, junction magnetoresistance in CoFe–Al_2O_3–Co spin tunnel junction versus applied field. Measurements done at room temperature and arrows indicate the relative directions of magnetization in the two magnetic layers. [After Moodera et al. (1996).]

appears much like that of a spin valve or a spin switch with higher resistance when the two ferromagnetic electrodes are magnetized antiparallel to each other.

The angular dependence of the junction magnetoresistance was shown by Moodera and Kinder (1996) to closely follow that of spin valve [Eq. (15.27)], which is also consistent with Slonczewski's theory.

Just as for spin valves, the highest tunneling MR ratios, about 20% and 27% at room temperature and at 77 K, respectively (Moodera et al. 1998), are obtained when one of the electrodes is Co. [The 27% spin polarization value at 77 K is close to what Eq. (15.35) predicts using the polarizations in Table 15.2). The large spin-dependent effect with a cobalt electrode can be understood to be a result of its large magnetic moment and the fact that it is a strong ferromagnet, that is, holes in the minority-spin band only.

Tunneling magnetoresistance was found to be independent of bias voltage over a range of a few millivolts and to decrease with increasing bias field, dropping to about half its zero-bias value for fields of about 0.3–0.4 V.

The junction magnetoresistance (JMR) decreases with increasing temperature. Moodera et al. (1998) and Shang et al. (1998) have shown that this temperature dependence is due primarily to the decrease in *surface* magnetization of the electrodes with increasing temperature. This temperature dependence of surface magnetization is probably due to the generation of spin waves [Eq. (3.47)]: $M^{surf}(T) = M_s^{surf}(1 - [T/T_C]^{3/2})$. The spin waves break down the independence of the two spin conduction channels, thus increasing the total conductance and reducing the JMR effect [see Eq. (15.41)]. The junction resistance is typically of order $10^4 \Omega$ and increases with decreasing temperature (see Fig. 15.44).

The measured junction resistance is less than the true junction resistance when $R_{jct} \leqslant R_{electrode}$. The reason for this is that for small junction resistance relative to electrode resistance, the current does not flow uniformly across the junction area from one electrode to the other. Instead, the current tends to take the shortest path between the two crossed electrodes, concentrating itself at the inside corner of the junction. This effect has been confirmed by finite element modeling (van de Veerdonk et al. 1997). When the current distribution is nonuniform, the apparent junction resistance is smaller than that for uniform current flow, and the JMR appears to be larger.

Junction resistance can be affected also by the quality of the junction oxide. Moodera et al. (1996) found that in poor quality junctions, where ρ_{oxide} was presumably low, the JMR was artificially enhanced. This is due to the geometrical effect mentioned above.

On the other hand, in microfabricated tunnel junctions, $R_{jct}/R_{electrode}$ increases as junction dimensions are reduced simply by the scaling of the dimensions. Kamugai et al. (1997) observed consistently smaller JMR values in microfabricated junctions due to this geometric effect. Nevertheless, spin–tunnel junctions show promise as elements in magnetic random access memories (MRAMs) (see Chapter 17).

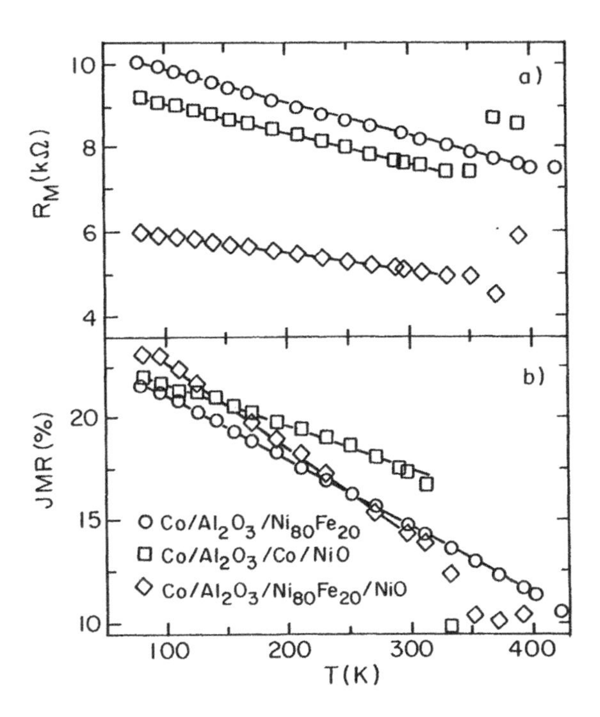

Figure 15.44 Above, temperature dependence of junction resistance for three Co/Al_2O_3/soft magnetic layer tunnel junctions; below, temperature dependence of junction magnetoresistance (Shang et al. 1998).

15.7 MAGNETIC OXIDES AND PHASE TRANSFORMATIONS: "COLOSSAL" MR

The dramatic changes in resistivity that accompany the metal–insulator transition in doped $LaMnO_3$ (Fig. 15.2) were presented as motivation in the introduction to this chapter. Intense interest in this phenomenon was stirred by the observation that application of a field could capture a large fraction of this resistivity change. This gave rise to the term "colossal magnetoresistance" for this effect. Similar very large MR ratios have been observed in a number of naturally occurring layered compounds. These materials include the perovskites NdPbMnO (Kusters et al. 1989), LaBaMnO (von Helmholt et al. 1993), $(La_{1-x}Ca_x)MnO_3$ (Chahara et al. 1993, Jin et al. 1994), and $(La_{1-x}Sr_x)MnO_3$ as well as the intermetallic compound SmMnGe (vanDover, 1993). Many of these systems have been the object of ongoing research for other reasons, and only recently have their electrical properties been examined in the context of magnetoresistance.

In the doped perovskite, lanthanum–strontium manganate, $(La_{1-x}Sr_x)$-MnO_3, it was known since at least 1950 (Jonker and van Santen 1950) that a change in electrical resistivity of six orders of magnitude (Fig. 15.45) occurs at the insulator-metal transition. Wollan and Koehler (1955) found ferromagnetic ordering in the *ab* planes (MnO layers) and antiferromagnetic ordering along the *c* axis where the MnO layers are separated by La(X)O layers, where X is a divalent alkaline-earth ion (Tokura et al. 1994).

The strong dependence of resistivity at $T = 100$ K on Sr doping level shown in Figure 15.45 can be understood in light of the phase diagram in Figure 15.46 to be associated with changes in the nature of the magnetic ordering in these materials. For Sr substitutions in lanthanum manganate, the stable phase for $x < 17\%$ is a tetragonal insulator and, below 100 K, it is either a canted spin (CS) antiferromagnet ($x < 7\%$) or a ferromagnet ($7\% < x < 17\%$). Above this Sr range, the material transforms to a cubic, ferromagnetic metal below T_C and a paramagnetic insulator or metal above T_C (Urushibara et al. 1995). Thus, with increasing Sr doping at $T = 100$ K, it can be seen that the material transforms from an insulating state to a metallic state near 17% Sr. This accounts for the six orders of magnitude drop in resistivity observed by Jonker and Van Santen (1950) (Fig. 15.45). The range of interest for magnetic-field-induced resistivity changes is between 17% and approximately 26% Sr where a metal insulator transition occurs near the Curie temperature, which is in the vicinity of room temperature.

The interest in hole-doped lanthanum manganate parallels a similar rediscovery of lanthanum cuprates, $La_{1-x}X_xCuO_3$, in 1987. The doped lanthanum cuprates, long known for their interesting electrical and magnetic properties and strong Jahn–Teller effects, were investigated by Bednorz and Muller in 1986 who discovered high-temperature superconductivity in them.

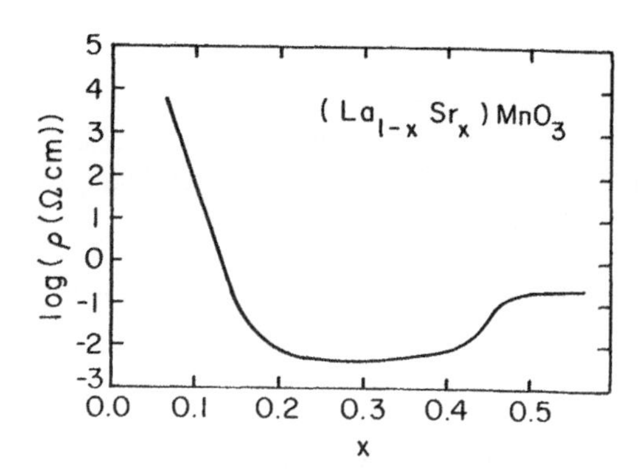

Figure 15.45 Log resistivity versus Sr concentration at 100 K in lanthanum–strontium manganate (Jonker and Van Santen 1950). The six orders of magnitude drop in resistivity near $x = 0.1$ corresponds to the insulator-to-metal transition.

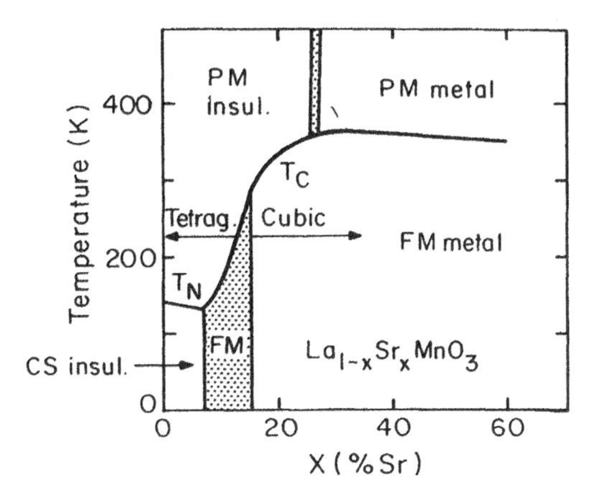

Figure 15.46 Magnetic phase diagram of $La_{1-x}Sr_xMnO_3$ [after Urushibara et al. (1995)]. PM is paramagnetic and FM is ferromagnetic. The drop in resistivity near $x = 0.1$ in Figure 15.45 corresponds here to the transition from a canted spin (CS) insulator to a ferromagnetic metal at 100 K.

The perovskite structure of these doped lanthanum manganates is illustrated in Figure 15.47.

The existence of local Mn moments in undoped lanthanum manganate below 100 K indicates that the intra-atomic (Hund) exchange energy is greater than the Coulomb crystal field splitting of the t_{2g} and e_g valence orbitals. The four valence electrons of the Mn^{3+} ions in $LaMnO_3$ are in a $t_{2g}^3 e_g^1$ configuration having $S = 2$. The three t_{2g} electrons, of lower energy than the e_g states

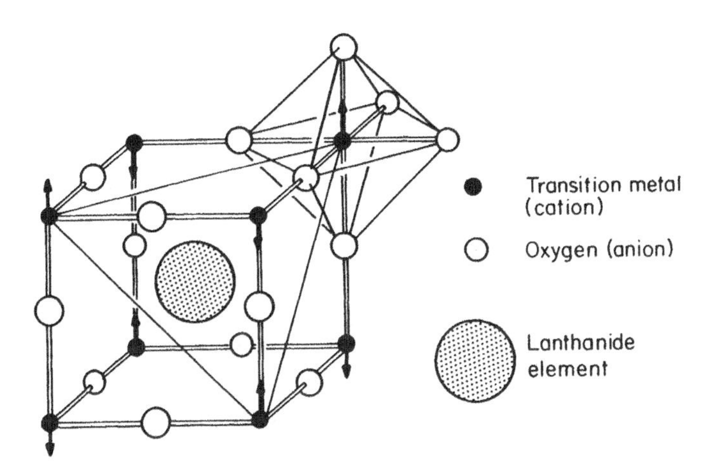

Figure 15.47 Model of the perovskite structure common to many heavy-metal/transition metal oxides such as $SrTiO_3$.

in the octahedral site, may be considered to be completely localized with a spin $S = \frac{3}{2}$. It is the lone e_g electron that plays a role in bonding and in the insulator–metal transformation. The e_g orbitals have the appropriate symmetry (Fig. 15.48) to bond with the oxygen p orbitals. The Mn spins are coupled antiferromagnetically by superexchange (Chapter 4) to their nearest neighbor Mn ions along the $\langle 100 \rangle$ directions. This leaves ferromagnetic coupling within a given $\{111\}$ plane. In lanthanum manganate, the e_g electron is highly correlated (atom-like) and the material is an insulator. Substituting divalent alkali earth ions (Sr^{2+}, Ca^{2+}, Ba^{2+}, etc.) for trivalent La, as represented by the formula $La^{3+}_{1-x}X^{2+}_{x}MnO_3$, drives the Mn^{3+} to a Mn^{4+} valence state. This leads to holes in the e_g band, which are responsible for the onset of metallic behavior.

The conducting e_g states experience competing tendencies between ferromagnetic coupling with the localized t_{2g} states and a tendency toward antiferromagnetism due to their hopping from site to site (Tokura et al. 1994). The double-exchange theory (Zener 1951, de Gennes, 1960) indicates that electron or hole transfer from site to site depends on the relative angle of the spins at the two sites, θ_{ij}—hopping goes as $\cos(\Delta\theta_{ij})$ and is most likely for parallel spins. Thus, the spin structure shown in Figure 15.47 becomes ferromagnetic at appropriate doping levels and temperatures. The colossal MR effect then comes from the applied field reducing the spin misalignment, thus decreasing the resistivity. The parallels of this field dependence of conductivity with that of F–I–F tunneling, spin valves, and spin switches is noteworthy.

The perovskite compounds studied by Jin et al. (1994) and whose properties are shown in Figure 15.2, were grown epitaxially on $LaAlO_3$ to a thickness of 100–200 nm by laser deposition. The structure is composed of metallic MnO

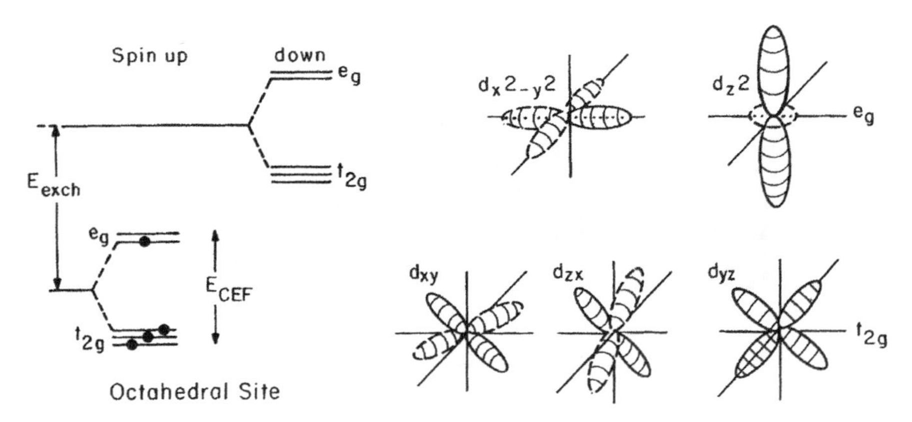

Figure 15.48 Left, crystal field splittings for transition metal ions d-levels in sites of octahedral symmetry. Occupied states are indicated with dots. The energy of the crystal-field-split d orbitals is determined by the octahedral symmetry of the negative oxygen ions coordinating the cation. Right, topology of the triplet, t_{2g} and the doublet, e_g orbitals.

layers with a lattice constant of 0.39 nm, separated by (LaCa)O insulating layers, $a = 0.38$ nm. The MnO layers show a spontaneous magnetization oriented in their planes with antiferromagnetic coupling between nearest planes in $La_{1-x}Ca_xMnO_3$.

The field dependence of the MRR is such that almost all of the resistivity change occurs over a range from 0–20 kG. Jin et al. argue that their data show this MR effect to be of an origin different from that observed in GMR. In GMR the MRR peaks at $M = 0$; here, it peaks near a transition in the electronic nature of the compound. With increasing field near the ferromagnetic–antiferromagnetic transition, ferromagnetism is favored over antiferromagnetism and consequently hopping conductivity increases ($\Delta R/R$ is negative in Fig. 15.2).

The phase diagram for $La_{1-x}Ca_xMnO_3$ (Ramirez et al. 1997) is similar to that for Sr doping in Figure 15.46. The data of Jin in Figure 15.2 occurs in a region where the metal–insulator transition coincides with the ferromagnetic–paramagnetic transition. The resistivity increases with increasing temperature in the metallic magnetic phase; the metal–insulator transformation for 33% Ca is near 250°C. Application of a field expands the ferromagnetic phase, displacing to higher temperatures the M–I transition and hence displacing the sharp increase in metallic resistivity. Thus the MR ratio is proportional to the temperature derivative of the $R(T)$ curve times the field derivative of the metal–insulator transition by a derivative chain rule:

$$\frac{\Delta R(H)}{R} \approx \frac{1}{R}\frac{dR}{dT}\frac{dT_{\text{crit}}}{dH}\,dH \tag{15.42}$$

In other words, the sharper the resistivity transition and the stronger the field dependence of that transition, the greater will be the CMR ratio.

CMR is a different physical effect than GMR but has a similar formal dependence on magnetization orientation; aligning the moments on adjacent cation sites (AF to F) causes the resistance to decrease. This is due to the increase in hopping conductivity of the cation e_g electrons for parallel spins. Fields of tens of kGauss are needed to saturate the effect because they are working against thermal energy, which is breaking down the ferromagnetic exchange at the Curie temperature. The field dependence of the CMR effect is not a function of the M–H curve; many of these compounds show appreciable magnetization in fields of order 100 Oe, but the CMR effect still requires tens of kilogauss.

Because the electrical and magnetic properties of these materials depend strongly on the degree of overlap between oxygen-p orbitals and transition metal d orbitals, these properties are sensitive functions of the size of the alkali earth metal dopant. Hwang et al. (1995) examined a series of samples for which the dopant size r_A was varied while the dopant concentration was fixed at 30% (Mn^{3+}:Mn^{2+} ratio at 7:3). They found that for decreasing $\langle r_A \rangle$, T_C decreases and the MR ratio increases dramatically. This was attributed to a departure from the ideal 180° Mn—O—Mn bond angle as $\langle r_A \rangle$ decreases (oxygen edge sites in Figure 15.47 are pushed outward from the body center for ions larger than La).

Large magnetoresistance due to phase change is not limited to oxides. The compound FeRh shows a transformation from a low-temperature, high-resistivity, AF phase to a lower-resistivity F phase near room temperature (Lommel, 1996; Vinokurova et al. 1988; Ohtani and Hatakeyama 1993). $\Delta R/R$ or order 100% occurs near T_N. This transformation can also be driven by an applied field: $dT_N/dH \approx -10^{-3}$ K/G, not particularly attractive for devices.

15.8 SUMMARY

This chapter may be well reviewed by returning to Figure 15.4, which gives an overview of the major types of electrical conduction in metals from simple models of their density of states. The resistivity associated with the scattering of itinerant electrons into localized *d* states is seen in panels (*c*) and (*d*). The spontaneous Hall effect arises from the spin–orbit interaction, which has the same symmetry as the classical Lorentz force. The explanation for anisotropic resistance in ferromagnets is more subtle, involving the breakdown of the two-current model due to spin–orbit interaction causing band mixing.

Figure 15.49 is a summary comparison of the characteristics of the four spin–spin device-related phenomena described in this chapter. Giant magnetoresistance, as manifest in spin valves and the spin switch, can be described in terms of spin–spin exchange scattering. Spin-polarized tunneling between ferromagnetic films depends on a convolution of same-spin state densities. The spin–spin devices based on these effects show low resistivity when the moments of the two magnetic layers are parallel. For colossal magnetoresistance in doped lanthanum manganates, the hopping conductivity increases for ferromagnetic coupling between adjacent MnO layers. In all these cases, lower electrical resistivity occurs between parallel-spin elements.

This chapter has only scratched the surface of the rich variety of phenomenon associated with transport effects in magnetic materials. The promise of new technology appears to be driving the increased research activity in this field, and new science should follow.

PROBLEMS

15.1 **(a)** Write out the three equations for the electric field components in Eq. (15.12), in terms of the tensor elements in Eq. (15.13) using the coordinates in Figure 15.8.

(b) Verify Eq. (15.14) by expressing its field components.

15.2 Show that for empty *d* states of only one spin direction (e.g., Ni), the

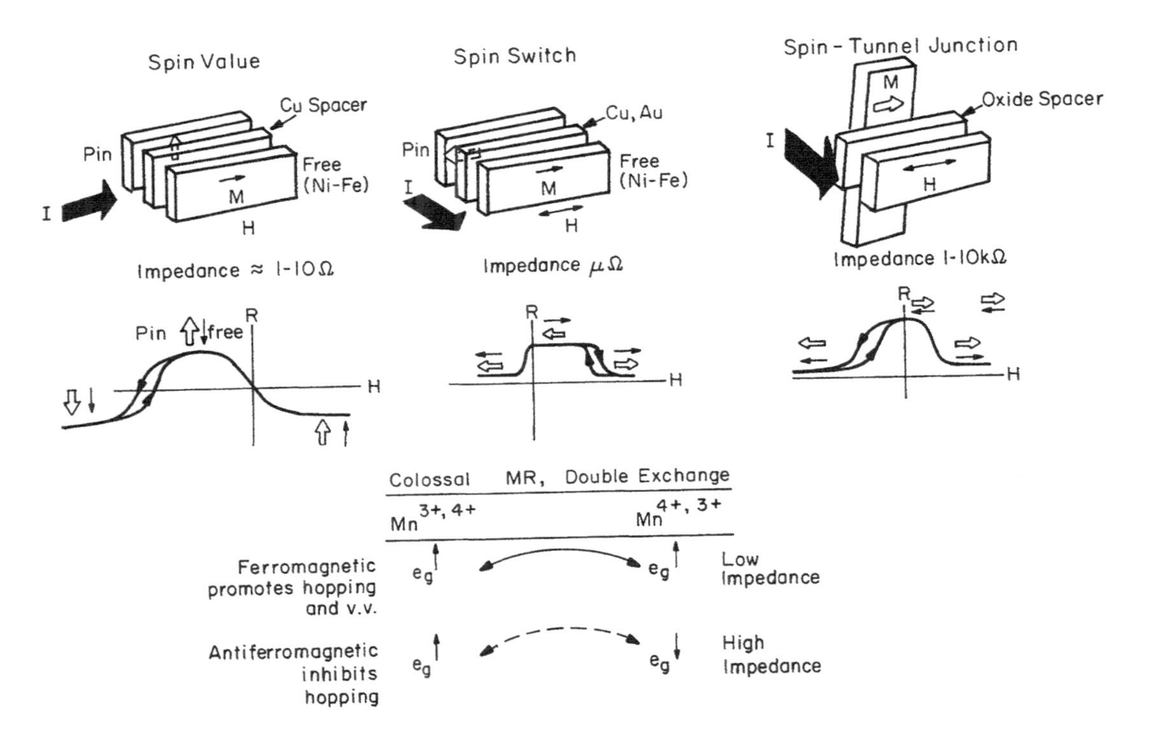

Figure 15.49 Summary diagram comparing the structure, current directions, and magnetizations (pinned layers indicated by open arrows), above, as well as galvanomagnetic characteristics of spin valves, spin switches, and spin tunnel junctions, middle. Below, the essence of the mechanism of colossal magnetoresistance is represented with the spin directions of hopping sites indicated.

form of the magnetoresistance derived is similar to that in Eq. (15.19), that is, second-order in d and decreasing for θ approaching $\pi/2$. (In this case $\rho^{\uparrow}$ is unaffected and only $\rho^{\downarrow}$ experiences spin–orbit scattering.)

15.3 Prove Eq. (15.22) using the commutation relations for the angular momentum components.

REFERENCES

Galvanomagnetic Effects

Berger, L., *Phys. Rev. B* **2**, 4559 (1970); *B* **5**, 1862 (1972).

Bozorth, R. M., *Phys. Rev.* **70**, 923 (1946).

Campbell, I. A., and A. Fert, in *Ferromagnetic Materials*, Vol. 3., E. P. Wohlfarth, ed., North Holland, Amsterdam, 1982, p. 747.

Chien, C.-L., and C. R. Westgate, eds., *The Hall Effect*, Plenum Press, New York, 1980.

Dorleijn, J. W. F., *Philips Res. Rept.* **31**, 287 (1976).

Dorleijn, J. W. F., and A. R. Miedema, *AIP Conf. Proc.* **34**, 50 (1976).

Fert, A., and I. A. Campbell, *J. Phys. F. Met. Phys.* **6**, 849 (1972).

Fert, A., R. Asomoza, D. Sanchez, D. Spanjaard, and A. Friederich, *Phys. Rev. B* **16**, 5040 (1977).

Gerritsen, A. N., *Handbuch der Physik*, vol. xix, 137 (1956).

Jan, J. P., in *Solid State Physics*, Vol. 5, F. Seitz and D. Turnbull, eds., Academic Press, New York, 1958, p. 1.

Jaoul, O., I. A. Campbell, and A. Fert, *J. Magn. Magn. Mater.* **5**, 23 (1977).

Kittel, C., *Quantum Theory of Solids*, Wiley, New York, 1963, p. 358.

McGuire, T. R., and R. I. Potter, *IEEE Trans.* **MAG-11**, 1018 (1975).

McGuire, T. R., *AIP Conf. Proc.* **24**, 435 (1975).

Mott, N. F., *Proc. Roy Soc.* **153**, 699 (1936); ibid. **156**, 368 (1936); *Adv. Phys.* **13**, 325 (1964).

O'Handley, R. C., *Phys. Rev. B* **18**, 2577 (1978).

Shirakawa, T., Y. Nakajima, K. Okamoto, S. Matsushita, and T. Sakurao, AIP Cont. Proc. No. 34, Amer. Inst. of Phys., New York, 1976, p.349.

Potter, R. I., *Phys. Rev. B* **10**, 4626(1974).

Smit, J., *Physica* **17**, 612 (1951); see also *Physica* **21**, 877 (1955), **24**, 39 (1958).

Stobieki, T., *Thin Solid Films* **15**, 197 (1978).

van Elst, H. C., *Physica* **25**, 708 (1959).

Giant Magnetoresistance

Baibich, M. N., J. M. Broto, A. Fert, F. Nguyen Van Dau, F. Petroff, P. Etienne, G. Creuzet, A. Friedrich, and J. Chazelas, *Phys. Rev. Lett.* **61**, 2472 (1988).

Berkowitz, A. I., J. R. Mitchell, M. J. Carey, A. P. Young, S. Zhang, F. E. Spada, F. T. Parker, A. Hutten, and G. Thomas, *Phys. Rev. Lett.* **68** 3745 (1992).

Chien, C. L., J. Q. Xiao, and J. S. Jiang, *J. Appl. Phys.* **73**, 5309 (1993).

Daughton, J. et al., *IEEE Trans.* **MAG-28**, 2488 (1992).

Dieny, B., V. Speriosu, S. S. P. Parkin, B. A. Gurney, D. R. Wilhoit, and D. Mauri, *Phys. Rev. B* **43**, 1297 (1991a).

Dieny, B., V. Speriosu, S. S. P. Parkin, B. A. Gurney, P. Baumgart, and D. R. Wilhoit, *J. Appl. Phys.* **69**, 4774 (1991b).

Lottis, D. et al., *J. Appl. Phys.* **73**, 5515 (1993).

Mathon, J., *Contemp. Phys.* **32**, 143 (1991).

Monsma, D. J. J. C. Lodder, Th. J. A. Popma, and B. Dieny, *Phys. Rev. Lett.* **74**, 5260 (1995).

Parkin, S. S. P., N. More, and K. P. Roche, *Phys. Rev. Lett.* **64**, 2304 (1990).

Parkin, S. S. P., R. Bhadra, and K. P. Roche, *Phys. Rev. Lett.* **66**, 2152 (1991).

Parkin, S. S. P., *Appl. Phys. Lett.* **61**, 1358 (1992).

Parkin, S. S. P., in *Ultrathin Magnetic Structures*, Vol. II, B. Heinrich and J. A. C. Bland, eds., Springer-Verlag, Berlin, 1994, p. 148.

Parkin, S. S. S. P., in *Ultrathin Magnetic Structures* vol. II, B. Heinrick and J. A. C. Bland, ed., Springer Verlag, Berlin, 1994, p. 148.

Pratt Jr., W. P., S. F. Lee, R. Laloee, P. A. Schroeder, and J. Bass, *J. Appl. Phys.* **73**, 5326 (1993).

Wang, J. Q., and G. Xiao, *Phys. Rev. B* **49**, 3982 (1994).

White, R. L. *IEEE Trans.* **MAG-28**, 2482 (1992).

Spin Switches

Johnson, M., and R. H. Silsbee, *Phys. Rev. B* **37**, 5326 (1988).

Johnson, M., *Phys. Rev. Lett.* **67**, 3594 (1991).

Johnson, M., *Phys. Rev. Lett.* **70**, 2142 (1993).

Tunneling

Callaway, J. and C. S. Wang, *Phys. Rev. B***16**, 2095 (1977).

Julliere, M., *Phys. Lett.* **54A**, 225 (1975).

Kamugai, S., .T. Yaoi, and T. Miyazaki, *J. Magn. Magn. Mater.* **166**, 71 (1997).

MacLaren, J. M., X. G. Zhang, and W. H. Butler, *Phys. Rev. B***56**, 11827 (1997).

Meservey, R., and P. Tedrow, *Phys. Rept.* **238**, 174 (1996).

Moodera, J. S., and L. Kinder, *J. Appl. Phys.* **79**, 4724 (1996).

Moodera, J. S., L. R. Kinder, J. Nowak, P. LeClaire, and R. Meservey, *Appl. Phys. Lett.* **69**, 708 (1996).

Moodera, J. S., J. Nowak, and Rene J. M. van de Veerdonk, *Phys. Rev. Lett.* **80**, 2941 (1998).

Shang, C. H., J. Nowak, R. Jansen, and J. S. Moodera, *Phys. Rev., Rapid Commun.* **58**, R2919 (1998).

Slonczewski, J., *Phys. Rev. B* **39**, 6995 (1989).

Stearns, M. B., *J. Magn. Magn. Mater.* **5**, 167 (1977).

van de Veerdonk, R. J. M., J. Nowak, R. Meservey, J. S. Moodera, and W. J. M. de Jonge, *Appl. Phys. Lett.* **71**, 2839 (1997).

Wolf, E. L., *Principles of Tunneling Spectroscopy*, Clarendon Press, Oxford, 1985.

Colossal Magnetoresistance

Chahara, K., T. Ohno, M. Kasai, and Y. Konzono, *Appl. Phys. Lett.* **63**, 1990 (1993).

de Gennes, P.-G., *Phys. Rev.* **118**, 141 (1960).

Hwang, H. Y., S.-W. Cheong, P. G. Radaelli, M. Marezio, and B. Batlogg, *Phys. Rev. Lett.* **75**, 914 (1995).

Jin, S. et al. *Science* **264**, 413 (1994).

Jonker, G. H., and J. H. Van Santen, *Physica* **16**, 337, 599 (1950).

Kusters, R. M., J. Singleton, D. A. Keen, R. McGreevy, and W. Hayes, *Phys. Rev. B* **155**, 362 (1989).

Lommel, J., *J. Appl. Phys.* **37**, 1483 (1966).

Otani, Y., and I. Hatakeyama, *J. Appl. Phys.* **75**, 3328 (1993).

Ramirez, A. P., S.-W. Cheong, and P. Schiffer, *J. Appl. Phys.* **81**, 5337 (1997).

Tokura, Y., Urushibara, A., Y. Morotomo, T. Arima, A. Asamitsu, G. Kido, and N. Furukawa, *J. Phys. Soc. Jpn.* **63**, (11), 3931 (1994).

Urushibara, A., Y. Moritomo, T. Arima, A. Asamitsu, G. Kido, and Y. Tokura, *Phys. Rev. B* **51** 14103 (1995).

vanDover, B., *Phys. Rev. B* **47**, 6134 (1993).

Vinokurova, A. et al., *J. Physique C* **49**, 99 (1988).

von Helmolt, R., J. Wecker, B. Holzapfel, L. Schultz, and K. Samwer, *Phys. Rev. Lett.* **71**, 2331 (1993).

Wollan, E. O., and W. C. Koehler, *Phys. Rev.* **100**, 545 (1955).

Zener, C., *Phys. Rev.* **82**, 403 (1951).

CHAPTER 16

SURFACE AND THIN-FILM MAGNETISM

The atoms near a surface of a crystal show a different lattice constant normal to the surface, usually smaller. There is some evidence that magnetic atoms near a surface or interface show moments that are different from their bulk values. There is abundant evidence that the magnetization near a surface or in a very thin film can have an easy axis that is different from that in the bulk. This chapter is about these and other surface phenomena in magnetic materials.

Interest in surface and thin film magnetism is driven both by scientific curiosity and technical applications. Fundamental magnetic properties depend on the local environment: the *symmetry*, *number*, *type*, and *distance* of an atom's neighbors. The symmetry at the surface is radically altered relative to the bulk because of the missing neighbors. Also, the number of nearest-neighbor atoms changes relative to the bulk. If the material is an alloy or a compound, the ratio of the different types of neighbor about an atom near the surface may be different from that about an atom in the interior. Finally, the lattice parameter contraction perpendicular to the surface, further lowers the symmetry. So it is not surprising that the magnetic moment, the Curie temperature, the magnetic anisotropy, and the magnetoelastic coupling may all be different at a surface or in a thin film compared to their bulk values.

In technical terms, the vast magnetic information technology industry adds a practical focus to research in surface magnetism (see Chapter 17 on magnetic recording). As the areal density at which information is stored increases, data reside in an increasingly thin layer of the recording medium (because all dimensions of the recording process must be scaled down as bit size decreases). Bit lengths in magnetic recording are presently a few hundreds of nanometers, and the information depth is about one-tenth of the bit length. The magnetic layers in the sensors that read magnetically recorded information can be as thin as 10 nm (MR heads), and newer sensors (spin valves) having magnetic layers 3–6 nm thick are under development. The transport phenomenon responsible for the operation of spin valves, namely, giant magnetoresistance, was described in Chapter 15. Surface and thin-film phenomena have also been covered in Chapter 12 (exchange coupling, superparamagnetism, and single-domain behavior). Although *intrinsic* surface effects (based on electronic interactions) vanish a few monolayers beneath the surface of a metal because of the short screening length, it will be seen that magnetic exchange coupling may cause surface anisotropy to affect processes several nanometers into the material.

16.1 ELECTRONIC STRUCTURE AT SURFACES

At a surface, reduced coordination and reduced bonding lead to significant changes in electronic structure that explain some of the unusual magnetic properties observed there. One early surface electronic structure calculation (Tersoff and Falicov 1982) provides excellent insight into the physical principles that govern surface magnetism. Electronic structure was calculated for eight monolayers (ML) of Ni (001) on a copper (001) substrate. Magnetic properties and electronic structure were distinguished at each layer.

Figure 16.1 illustrates the structure of the computer-generated sample and shows the variation of calculated magnetic moment from layer to layer.

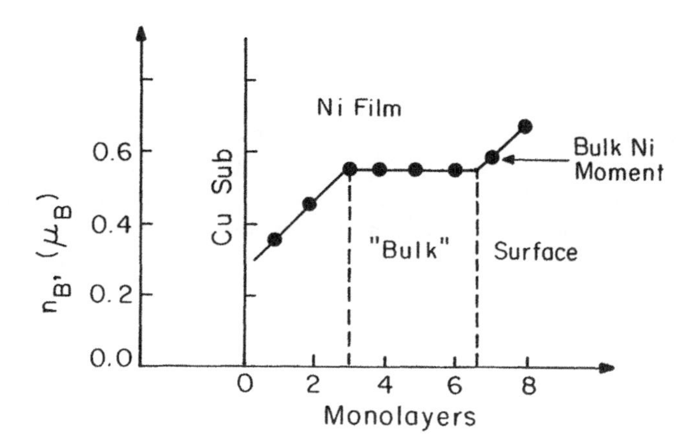

Figure 16.1 Variation of magnetic moment calculated by layer in an 8 ML Ni/Cu (001) film. The calculated bulk and surface moments are 0.56 μ_B/Ni and 0.74 μ_B/Ni, respectively. These are to be compared with a measured bulk moment of 0.6 μ_B/Ni. [After Tersoff and Falicov, (1982).]

Interaction of the Ni and Cu atoms near the Ni/Cu interface decreases the Ni moment by mixing Cu itinerant electron character with the Ni d bands. The central Ni layers are bulklike and show a moment of 0.56 μ_B/Ni atom, close to the bulk value of 0.6 μ_B/Ni atom. The surface layers show an enhanced moment of 0.73 μ_B/Ni.

The variation in Ni moment shown in Figure 16.1 can be explained in terms of the differences in the density of valence electron states for the interior, bulklike layers (numbered 3–6) and the surface layers (numbered 7–8). Figure 16.2 shows that the calculated density of states is higher (d band narrower in energy) for the surface layers (bold line, right) than for the bulk layers (fine line, left and dashed, right). This reflects the lower, 8-fold Ni bond coordination for surface atoms, compared to the 12-fold coordination of Ni in the bulklike layers. The lower coordination leaves the surface d states more localized, more atomlike. Thus intraatomic exchange is more effective at the surface. As a result, the exchange splitting, and hence the magnetic moment, can be larger near a surface.

Ni atoms at (111) and (100) surfaces have 9 and 8 nearest neighbors, respectively, in contrast to nickel atoms in the bulk which have 12 neighbors.

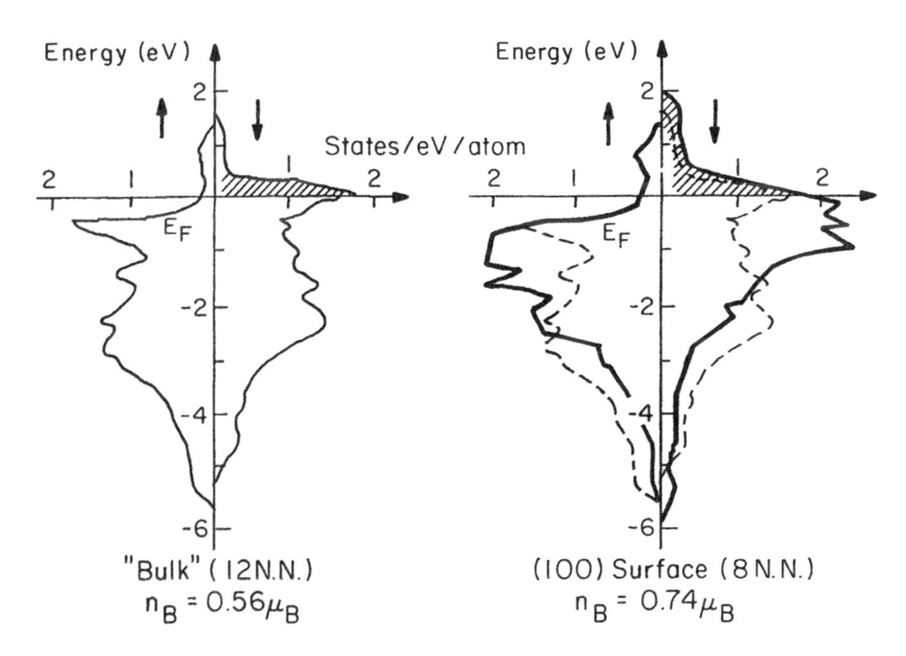

Figure 16.2 Spin-resolved density of states for 8 ML Ni (001) film on Cu. At left is the state density for Ni atoms in the interior, bulklike layers (layers 3–6 from Cu). At right is the state density for the surfacelike layers (7 and 8 from Cu). For comparison, the bulklike density of states is shown by the dashed line. [After Tersoff and Falicov (1982).]

Consequently the (111) and (100) surfaces show different state densities and magnetic moments relative to each other as well as to the bulk; the *d* band at a (111) surface is not as narrow as is that at a lower-coordination (100) surface.

Similar moment enhancement at Fe (001) surfaces has been calculated by Onishi et al. (1985) using a full-potential, linearized augmented plane-wave method (FLAPW) on a seven-layer iron structure. The magnetic moment per iron atom decreased from 2.98 μ_B at the surface to 2.25 μ_B in the central layer.

These differences in electronic structure and wavefunctions imply changes in charge distribution at a surface. The calculated electronic charge density at the surface of BCC Fe and FCC Ni are shown in Figure 16.3 (Freeman et al. 1991). Note the extension of the charge density into the vacuum; it is particularly strong in iron. The surface charge that extends into the vacuum is mostly of *s* and *p* character; the *d* electrons of surface atoms become slightly more localized about their atomic sites (consistent with the narrower surface *d* states calculated in Fig. 16.2). This spatial separation of *d* and *s*–*p* electrons has several consequences. First, there is reduced *s*–*d* hybridization in the surface layer; this further enhances *d*-band localization and increases surface moments. Second, the increased atomic character of surface *d* states enhances the orbital angular momentum there, making stronger anisotropy possible. [To the extent that the more-atomic-like surface *d* states feel the low-symmetry crystal field there, angular momentum normal to the surface (***k*** in-plane) is favored.] Third, the decreased *s* electronic charge density at surface sites plays a role in the surface lattice constant relaxation described below.

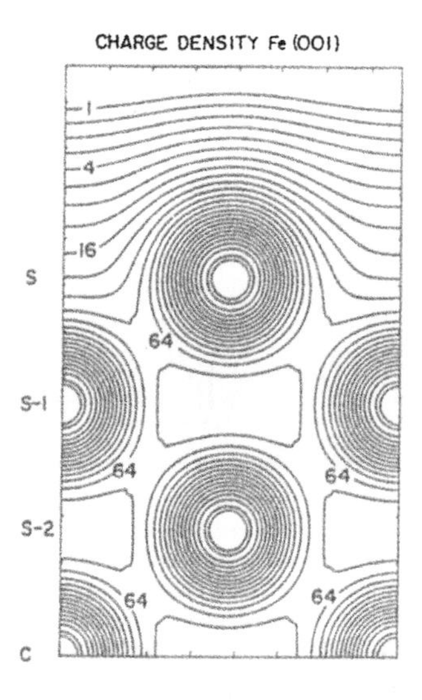

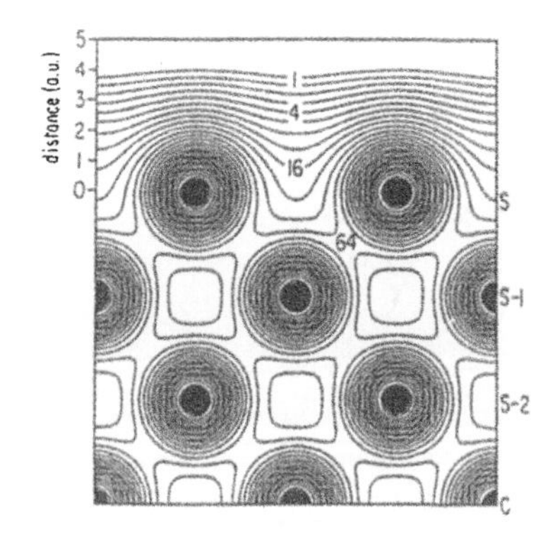

Figure 16.3 Electronic charge density at surface of seven-layer Fe (001) and Ni (001) films [Fe from Onishi et al. (1983) and Ni from Wimmer et al. (1984)]. Units are 10^{-3} e/(atomic unit)^3; each contour line differs by a factor $\sqrt{2}$.

The spin density calculated for the same model structures is depicted in Figure 16.4. Note that in the case of iron, the *majority*-spin density is predicted to extend well beyond the surface whereas in the case of Ni, the calculated electronic structure suggests that the surface acquires a thin layer of *minority*-spin density. These *total* electron polarizations are consistent with the fact that for Fe, the density of states at E_F is mostly of majority spin whereas for Ni it is mostly of minority spin. There is some evidence for this spin dipole layer from electron capture spectroscopy experiments (Rau et al. 1986). Nevertheless, experimental confirmation of such highly localized effects presents a challenge both in terms of film preparation and characterization.

The enhanced localization of the surface d states strengthens the conditions for moment formation there. This can be understood in terms of the Stoner criterion, Eq. (5.1). A number of electronic structure calculations have predicted enhanced surface moments in Fe, Co, and Ni as well as possible surface moment formation in nonmagnetic metals such as V, Ru, and Rh. See, for example, Freeman and Wu (1991), Blügel (1994), or Gay and Richter (1986, 1994). The experimental difficulty of measuring surface magnetic moment differences relative to the bulk have left this issue largely unresolved. There is

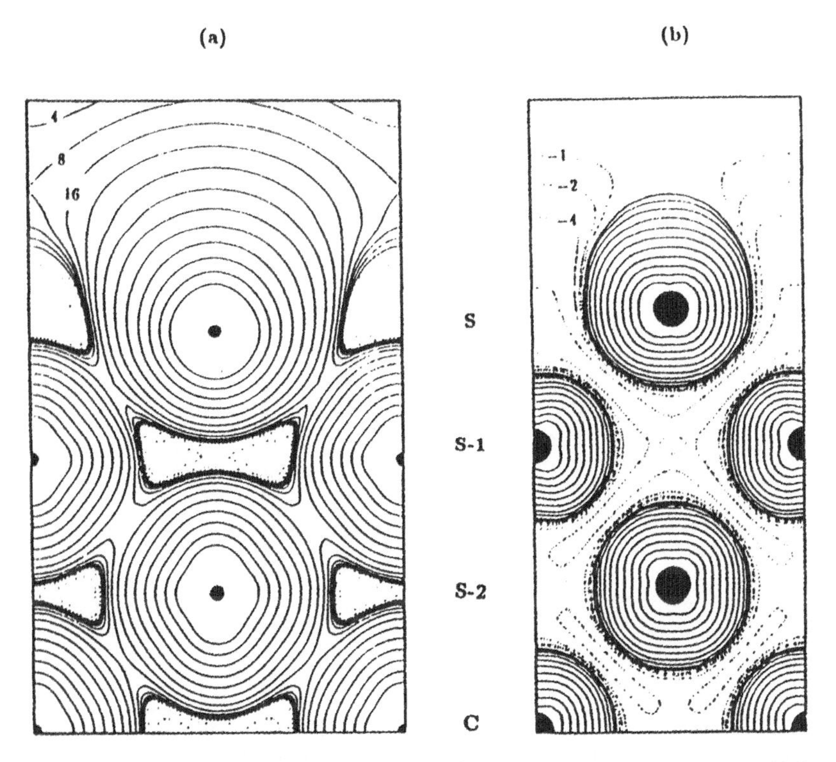

Figure 16.4 Spin density in units 10^{-4} e/(au)^3 at surface of seven layer Fe (001) and Ni (001) films; dotted lines represent negative spin densities (Freeman et al. 1991).

some evidence that the Curie temperature of a surface layer can be lower than that of the underlying layers. The enhanced surface density of states would tend to increase T_C while the reduced coordination [Eq. 4.21)] would decrease T_C.

16.2 SURFACE LATTICE CONSTANT RELAXATION

Atomic planes near a surface generally show a spacing normal to the surface that is significantly smaller than that in the interior. The surface relaxation strain can amount to several percent. Figure 16.5 displays some surface strain data expressed as the change in lattice constant between the first two layers (1–2), normalized to the number of nearest-neighbor atoms on that surface, $\Delta d(1-2)/NN$. These values are plotted against the equilibrium lattice spacing d_e, also normalized to the number of nearest neighbors on the surface in study. Most surfaces show negative (inward) relaxation. Note that the relaxation is greater for lower-atom-density faces (which are missing more bonds as a result of creation of the surface). The vertical lines draw attention to the fact that the equilibrium interplanar spacings are strongly correlated with the number of nearest neighbors in a given plane, especially for dense atomic planes. The key to understanding this surface lattice relaxation is in the opposing roles played

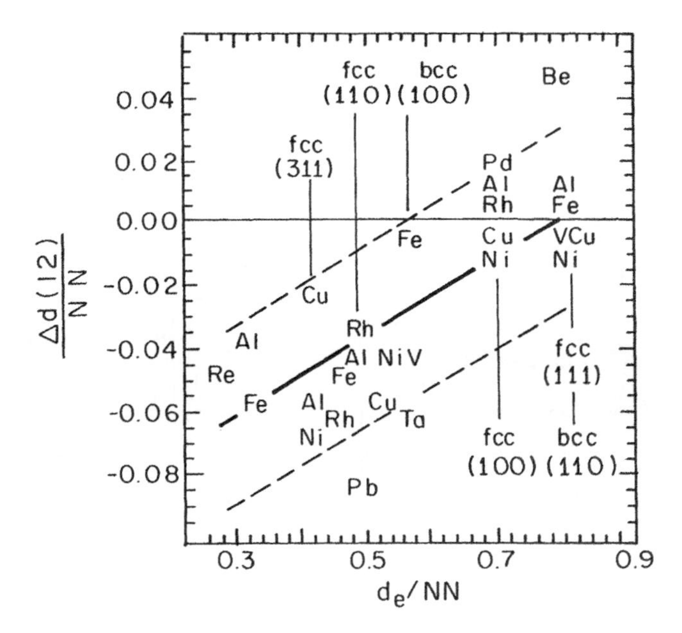

Figure 16.5 Variation of interplanar spacing between the first two surface layers, $d(1-2)$, as a function of bulk layer spacing: $\Delta d(1-2)/NN$ is $d(1-2) - d_e$, the equilibrium spacing, normalized to the number of nearest neighbors for each surface; sloped lines indicate trend of data (Davis et al. 1992).

by $3d$ and $4s$ charge densities in bonding of transition metals, described by Figure 5.21.

The major electronic structure change near a surface is the redistribution of d and s electronic charge described above. At the surface, the compression of the s electrons is relieved because of the missing neighbors. Conduction electrons tends to leak into the vacuum (because of their high kinetic energy and associated high internal pressure), decreasing s charge density ρ_s at the atomic sites in the outer atomic plane(s). Because of the reduced ρ_s in the outer atomic planes, the d electrons of surface atoms are screened less effectively from the nuclear potential, and, if they change at all, they become slightly more localized. Thus, the outermost atomic layer has an increased ratio of $3d:4s$ charge and just outside that layer, conduction electron density is increased. As a result, atoms in the surface layer experience increased attraction and diminished repulsion with their remaining neighbors (Fig. 5.21). The surface atoms thus assume a smaller equilibrium atomic volume than bulk atoms. They cannot accommodate this by relaxation *parallel* to the surface plane without breaking bonds with other surface atoms. Thus they relax inward, drawing closer to the subsurface atomic layers. Figure 16.6 summarizes these changes in electronic structure and consequent relaxation. The upper part of the structure shown in two dimensions represents the bulk material. Above the structure, the periodic bulk s and d charge densities are depicted. The lower part of the structure represents a surface and shows the inward relaxation of the outer atomic layer(s) without change in the lattice constants in the plane of the surface. Below this structure, the s and d charge densities are depicted as changing near the surface (solid lines) relative to their bulk values (dotted) in the manner just described.

It is important to note that the electronic effects associated with a surface in a metal are negligible at three or four atomic layers into the material. It will be seen later than exchange coupling can carry *magnetic* surface effects much farther into the interior.

16.3 STRAIN IN MAGNETIC THIN FILMS

In the previous section, it was pointed out that the perpendicular lattice constant near the surface of a metal generally relaxes toward smaller values. In addition to this surface relaxation, thin films can show large biaxial, in-plane strains (Koch 1994). Differences in thermal expansion between a film and its substrate, $\alpha_f - \alpha_s$, may give rise to film strain, $e = (\alpha_f - \alpha_s)\Delta T$, associated with a temperature change ΔT after deposition or during use of the film. Also, if a film grows epitaxially on a substrate from grain to grain, or especially if it is a single-crystal epitaxial film, very large lattice mismatch strains may be experienced by the film. The misfit strain in the surface plane also gives rise to a perpendicular Poisson strain.

It is possible for highly stressed films to cause the substrate on which they are grown to bend in response to the stress at the film–substrate interface. This

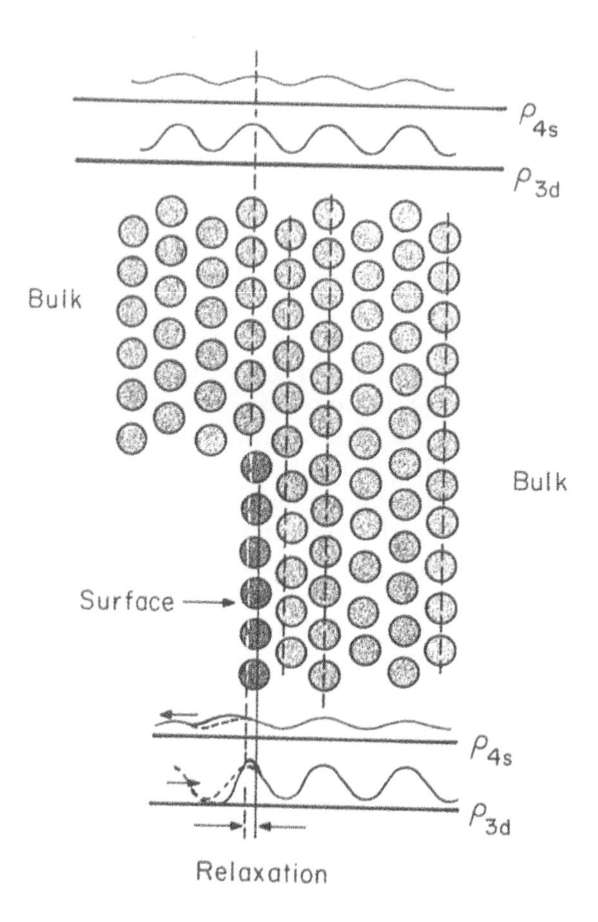

Figure 16.6 Schematic representation of bulk material (top half of structure) above which are depicted the 4*s* and 3*d* charge densities varying periodically along a line through the solid. The bottom half of the structure represents a surface with the outermost layer of atoms (shaded) relaxed. Below, the charge density near the surface (solid line), altered relative to that of the bulk (dotted), is responsible for the relaxation.

bending can be appreciable and can be used to measure the film stress. However, the bending does not result in a significant reduction of film stress or strain unless the substrate is very compliant relative to the film. This problem is analyzed in Appendix 16.A.

If a thin-film material and its substrate have a lattice mismatch $\eta = (a_s - a_f)/a_s$ (a_f and a_s are the film and substrate in-plane lattice constants, respectively), the layer with smaller product of (thickness × stiffness), usually the film, will absorb most of the strain to retain atomic coherency at the interface. (It is important to realize that the lattice constants of a free-standing thin film may differ appreciably from the bulk lattice constant of the same material because of the surface relaxation effects described in Section 16.2).

When a film grows coherently on a substrate with which there is a lattice misfit η, the elastic energy per unit area, proportional to $\eta^2 d/2$, increases as the film thickness d increases. At a critical thickness d_c, the strain energy exceeds the energy cost for formation of dislocations which relieve stress. Beyond this critical thickness, the formation of misfit dislocations (MDs) is thermodynamically favored. Thermodynamic theory (Hirth and Löthe 1982, Fitzgerald 1991) indicates that the critical thickness varies inversely with the misfit strain. As film thickness increases beyond d_c, the equilibrium dislocation density increases (Fig. 16.7, left) and the average film strain is predicted to decrease approximately as $1/d$ (Fig. 16.7, right). Above the critical thickness for misfit dislocation formation, the strain is calculated (Tsao 1993) to vary with film thickness as

$$\langle e \rangle = \frac{1}{8\pi d b_1^2}\left\{\left[b_1^2 + b_2^2 + (1-\upsilon)b_3^2\right]\left[\ln\left(\alpha\frac{d}{b}\right) - \frac{1}{2}(b_1^2 + b_2^2)\right]\right\} \quad (16.1)$$

The b_i terms are the Burgers vectors of the dislocations, υ is Poisson's ratio, and the α in the log term is a scale factor that depends on the energy of the core of the dislocation ($\alpha \approx 4$ for semiconductors and is near unity for metals). The strain is often *assumed* to decrease as d^{-1} (i.e., the log dependence is neglected), but this is rarely, if ever, observed (Matthews and Blakeslee 1974, 1975). Approximation of the log term leads to a strain dependence close to $d^{-2/3}$, and this has been observed in Cu/Ni/Cu sandwich films (Ha et al. 1999).

Figure 16.8 shows the variation of the lattice constants versus Ni thickness in a series of epitaxial Ni (001) films sandwiched by Cu. The in-plane and film–normal lattice constants were determined by grazing incidence and Bragg diffraction, respectively, using synchrotron X rays of wavelength 0.115 nm (Ha et al. 1999). The data show first that the critical thickness for the Cu/Ni/Cu system is 2.7 nm. The calculated value for a single Cu/Ni interface is 1.8 nm;

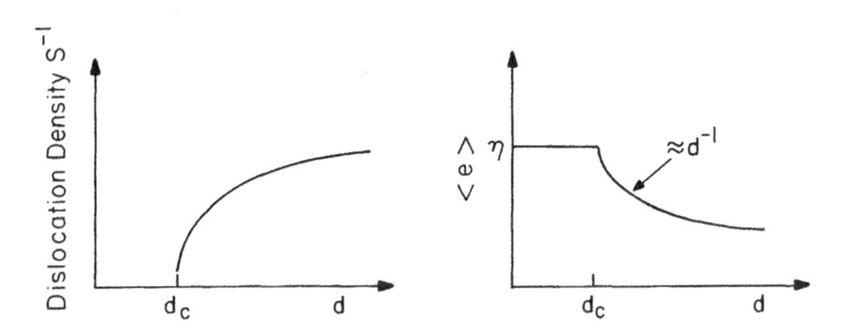

Figure 16.7 Schematic variation of dislocation density per unit length (inverse of dislocation line spacing S) and average film strain, both versus film thickness d for thermodynamic equilibrium.

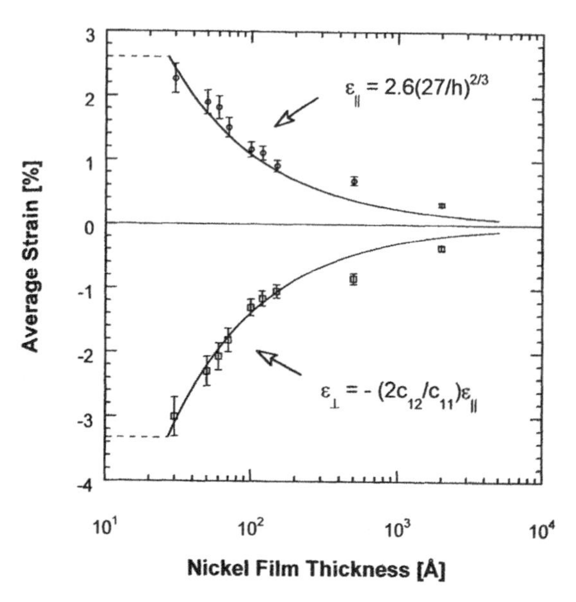

Figure 16.8 Strain versus Ni thickness in Cu/Ni/Cu sandwiches determined by synchrotron X-ray scattering. Above, strain in the in-plane lattice constant determined by grazing incidence diffraction. Below, strain in perpendicular lattice constant determined by Bragg diffraction. The two sets of data are related by the Poisson ratio of bulk Ni, and the in-plane data are well fit by $\langle e(d)\rangle = \eta(d_c/d)^{2/3}$, where $\eta = 0.026$ is the misfit strain and $d_c = 2.7$ nm is the critical thickness above which misfit dislocations are thermodynamically stable (Ha et al. 1999).

adding a Cu capping layer to a film that is already dislocated would increase the *thermodynamic* critical thickness to 2.7 nm, but *kinetics* could inhibit the film from returning completely to the equilibrium state, a process that would require many misfit dislocations to be removed. The data also show that the in-plane and perpendicular lattice constants are related, as expected, by Poisson's ratio. Finally, the strains are reasonably well fit over the Ni thickness range 3–15 nm by the function $\langle e(d)\rangle = 0.026(d_c/d)^{2/3}$ as indicated by the solid lines over the data.

Thus, in an epitaxial thin film at thicknesses above d_c, the average magnetoelastic anisotropy can be a strong function of film thickness because of its dependence on $\langle e\rangle$, which varies approximately as $\eta(d_c/d)^{2/3}$.

16.4 METASTABLE PHASES

Thin films are generally in a state of constrained equilibrium by virtue of their bonding to a substrate that often has a different lattice constant than the film. Further, even the surface of a pure metal (epitaxially joined to its bulk phase) has a different electronic structure because of its reduced symmetry and charge redistribution. As indicated above, the surface of a transition metal generally relaxes by reducing its perpendicular lattice constant. Thus a cubic phase suffers a symmetry reduction to tetragonal at its surface and a thin epitaxial film, lattice mismatched relative to its substrate, also undergoes a tetragonal distortion to maintain coherence with its substrate.

Our interest here is in possible stabilization of metastable phases in thin films [see Prinz (1985, 1991a, b)] as a result of epitaxial strains.

Figure 16.9 shows the pressure–atomic number–temperature (P–Z–T) phase diagrams for Fe, Co, and Ni. The stable phase of iron at standard temperature and pressure is BCC. This normally high-temperature phase is stabilized at room temperature by the magnetism of Fe. Close-packed phases of iron appear at high pressure and temperature. Ni is very stable in the FCC phase, and no other phases are shown until it is alloyed strongly with cobalt. Cobalt shows a transformation to a BCC phase at room temperature and negative pressure.

The FCC phase of iron can be accessed at room temperature by deposition of iron on single-crystal copper substrates. The moment and Curie temperature of FCC iron are reduced relative to their BCC iron values. On the other hand, FCC cobalt is relatively easy to produce on a variety of substrates and in widely different preparation conditions. Even bulk FCC cobalt is readily made because the marginal stability of the HCP phase (0.036 eV) is easily upset by small impurity content, stress, or other constraints).

Perhaps more unexpected is the metastable magnetic BCC phase of cobalt; it occurs on 6% lattice *expansion* of the HCP phase (Fig. 16.10 insert). BCC

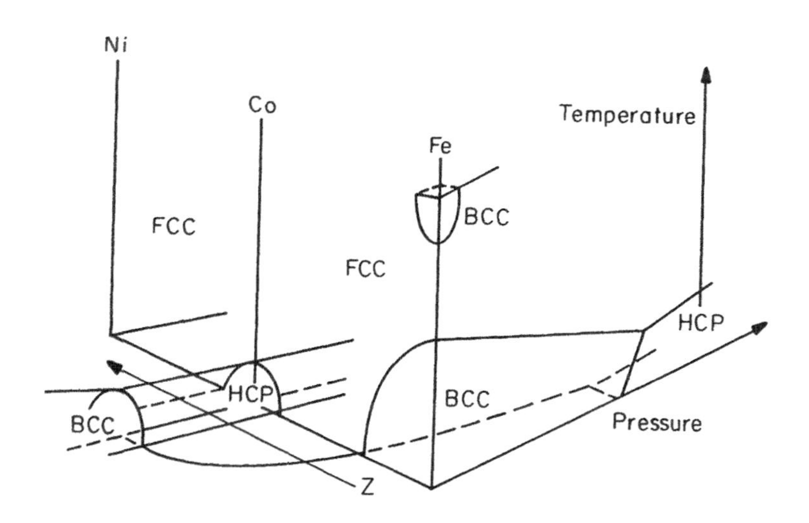

Figure 16.9 P–Z–T phase diagram (Z = atomic number) for iron, cobalt, and nickel showing fields of stability for various structures. BCC cobalt is stable at negative pressure. [After Prinz (1991b).]

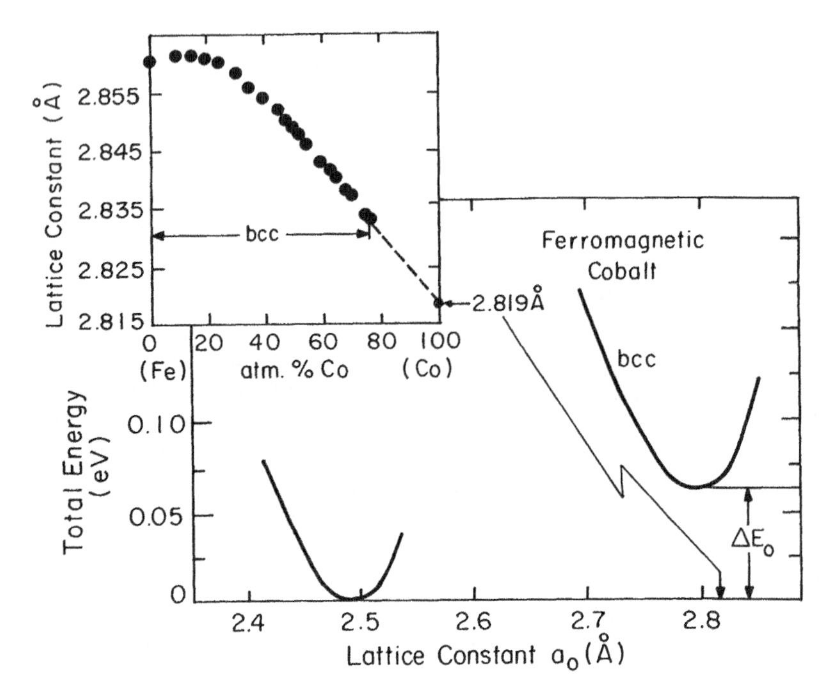

Figure 16.10 Total energy calculated for cobalt at different atomic volumes. Insert shows the lattice constants of BCC FeCo alloys (Prinz 1991a).

cobalt is calculated to lie 0.03 eV above the HCP phase and at a lattice constant of 2.819 Å (Fig. 16.10). Despite the lattice expansion, BCC cobalt has a reduced moment (1.53 μ_B/Co) relative to close-packed cobalt (1.7 μ_B/Co). Note that even though the lattice is expanded and the atomic volume (Wigner–Seitz radius) is increased, the nearest-neighbor distance along [111] is smaller than that in HCP cobalt. Also BCC cobalt shows a negative anisotropy $K_1 \approx -10^5$ J/m^3 ($\langle 111 \rangle$ easy axes), with magnitude twice that of BCC iron. Clearly, many exciting opportunities exist in this new field; the number of potential new materials that can be created by different deposition conditions and different combinations of film and substrate (including insulators as well as metals and semiconductors) is quite large.

16.5 SECONDARY ELECTRON SPIN POLARIZATION

Electrons can be ejected from the surface of a material by photoemission, field emission, or the Auger process. Analysis of the energy, temperature, and angle dependence of the spin polarization of electrons ejected from magnetic materials represents a powerful technique for studying surface magnetism. Knowledge of the spin polarization is useful both in an average way (electrons ejected from an area representative of one or more domains) and, with focused primary beams, as a local probe (electrons ejected from an area much smaller than a domain). In the latter case the local polarizations can be combined to generate a magnetic domain image of a surface. One example of this is scanning electron microscopy with spin polarization analysis (Section 16.7.1). The use of polarized *primary* electron beams is less common and will not be described here. Here we present selected results of spin-polarized electron studies on magnetic materials.

When an energetic primary electron or photon enters a ferromagnetic material, electrons can be excited from their ground state by an Auger process or by a photoemission process. The electrons excited may come from the core levels or the valence states of the material depending mainly on the energy of the probe. Once an electron is freed from an atom, it must reach the surface to escape from the material. As the excited or hot electron passes through the material, a cascade of collisions may result, producing more and more secondary electrons at progressively lower energies. These abundant low-energy, cascade electrons come primarily from the valence band.

Inside the material the valence electrons are characterized by a polarization P':

$$P' = \frac{n\uparrow - n\downarrow}{n\uparrow + n\downarrow} = \frac{n_B}{n_v} \tag{16.2}$$

where $n^{\uparrow}$ and $n^{\downarrow}$ are the numbers of spin-up and spin-down valence electrons per atom, n_B is the magneton number, and n_v is the number of valence electrons per atom. The polarization takes on values from 0 to 1. A certain fraction of the "hot" electrons created by the probe will be emitted from the solid depending on their energy, momentum, and starting depth and the scattering center density. It is generally a safe approximation to assume that electron spin is conserved during scattering and emission. This is because pure Coulomb scattering does not flip spins, exchange and spin–orbit scattering events can flip spins, and the energies of the latter are generally small relative to Coulomb energies (see Chapter 15).

Once the hot electron is transmitted through the surface barrier it is called a secondary electron. The electrons emitted from a ferromagnet may be characterized by a polarization, $P = (N^{\uparrow} - N^{\downarrow})/(N^{\uparrow} + N^{\downarrow})$, where $N^{\uparrow}$ and $N^{\downarrow}$ represent the number of spin-up and spin-down electrons *in the emitted beam.* The polarization of emitted cascade electrons can be taken as an indication of valence electron polarization: $P \propto P'$ (Unguris et al. 1982). However, the polarization P of secondary electrons is a strong function of their kinetic energy, so it is generally not true that the cascade electrons emerge from the material with a polarization P equal to their polarization inside the material P'. The reason for this is that as the hot $n^{\uparrow}$ and $n^{\downarrow}$ electrons are moving toward the surface, they have different scattering probabilities along the way [Eq. (15.2)]. Those hot electrons whose spin and energy corresponds to the band with the greater density of unoccupied states (that is, minority spin electrons) are more likely to be scattered and thus have a reduced probability of reaching the surface. Hence, the cascade polarization is enhanced in proportion to the distance traveled from excitation to the surface: $P > P'$. This filtering process can be accounted for quantitatively (Siegmann, 1994) to make cascade electron magnetometry more reliable.

The spin polarization of the secondary electrons can be measured by doing a scattering experiment with them. This experiment makes use of spin-dependent scattering based on exchange or spin–orbit interactions; the latter is more commonly used (Kirschner 1985). These spin-dependent scattering processes are the same ones that, in a ferromagnetic solid, account for anisotropic magnetoresistance (spin–orbit) and giant magnetoresistance (exchange scattering) (see Chapter 15). In spin-polarization analysis, the conditions for those atomic scattering processes are created in the vacuum chamber. Spin-dependent scattering events are generally a small fraction of the total number of scattering events. One way to enhance asymmetric spin–orbit scattering is to direct the secondary electrons toward a heavy metal, such as gold or tungsten, that exhibits a strong spin–orbit interaction. One could also scatter the secondary electrons from a ferromagnet to make use of the strong exchange interaction in measuring secondary electron polarization.

The principles of spin polarization analysis of secondary electrons are briefly described. The secondary electrons from the surface of interest may be accelerated in ultrahigh vacuum to either 100 keV (Mott scattering from Au foil), 35 keV (Mott scattering in a "mini-Mott" polarimeter), 150 eV (diffuse scattering from the core potential in a polycrystalline gold film), or to a few tens of volts (LEED scattering from W single crystal). As a result of the asymmetry in the scattering process, more electrons of one spin will be scattered to the left or right of the plane containing the k vector and the spin quantization axis of the secondary electrons. For the spin detector represented in Figure 16.11, more or fewer electrons will be detected in quadrant 4 than quadrant 2 of a detector array. The scattering potential can be represented by Eq. (15.20).

The scattering asymmetry A measured at the four-quadrant detector, $A = (I_4 - I_2)/(I_4 + I_2)$, is weak, departing from unity by only a few percent because $\xi\langle \boldsymbol{L} \cdot \boldsymbol{S} \rangle \ll V_{\text{Coul}}$. This makes these detectors inherently inefficient. Their spin analyzing efficiency, represented by the Sherman function S, can be calibrated according to

$$S = \frac{A}{P} \tag{16.3}$$

Thus, S is the asymmetry produced for a given incident beam polarization; typically $0.1 < S < 0.2$. The quantities A and P are functions of scattering angle and so, generally, is S. Once a polarimeter is calibrated, that is, its Sherman function is known, then the secondary electron spin polarization P can be calculated from the measured scattering asymmetry A, using Eq. (16.3).

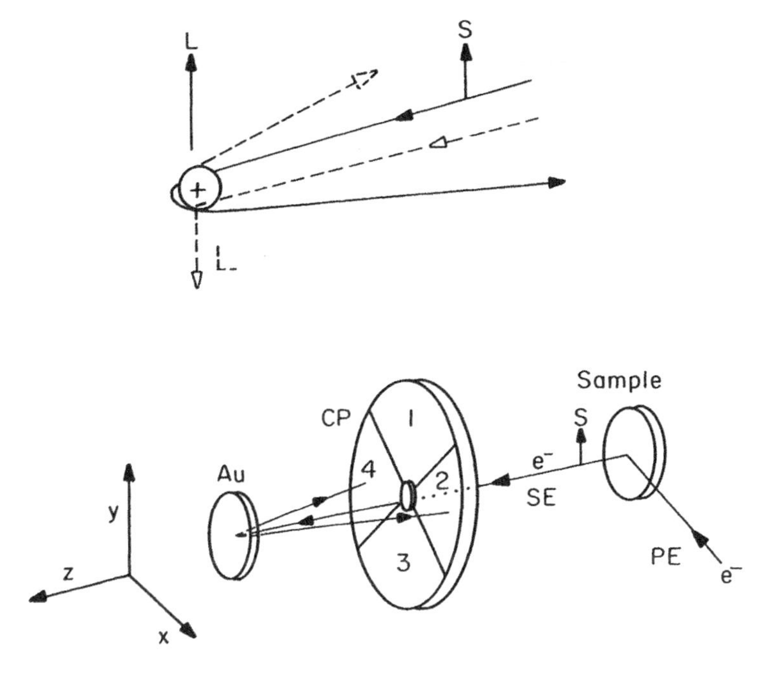

Figure 16.11 Above, incoming spin-up electron is scattered from positive core potential in a process that includes spin–orbit interaction ($\langle \boldsymbol{L}\cdot\boldsymbol{S}\rangle > 0$, solid line). The dotted line shows spin-up scattering with $\langle \boldsymbol{L}\cdot\boldsymbol{S}\rangle < 0$. The total scattering potentials [Eq. (15.20)] for left and right scattering therefore differ between the two cases by $2\xi(\langle \boldsymbol{L}\cdot\boldsymbol{S}\rangle)$. Below, from the same perspective, the major components of a secondary-electron polarization experiment. Unpolarized primary electrons (PE) impinge on the surface of a sample, causing it to emit secondary electrons (SE) having a net spin indicative of the sample polarization. These SEs are scattered from a diffuse scattering, thin gold film (Au) spin polarimeter. The scattering asymmetry from the gold film is detected as a difference signal in quadrants 2 and 4 of the four-quadrant, channel plate (CP) detector. Spins polarized in the x direction produce a 1–3 asymmetry.

The spin polarization of the cascade electrons has been used as a surface-sensitive probe of magnetism because these electrons come from the outermost 5 Å or so of the material. When the incident electron beam has a diameter of order one millimeter, the surface magnetization represented by the secondary electrons is averaged over several domains. Secondary electron spin polarization has been used to probe the temperature dependence and critical exponents of magnetization in thin films (Pierce et al. 1982), to study exchange coupling phenomena (Donath et al. 1991), and to explore the variation of magnetization with distance from the surface (Mauri et al. 1985).

A number of experiments have measured the spin polarization of field-emitted (Landolt 1977, 1978), photoemitted (Kisker 1985), and Auger electrons (Landolt and Mauri 1982). Some of these results are described briefly to illustrate the power and utility of these techniques for magnetic surface analysis as well as to draw a comparison between these methods and spin-polarization measurements made by tunneling (Chapter 15). Photoemitted electrons come from localized (high density of states) bands in contrast to tunneling electrons, which come from itinerant states near E_F. The different results of these various probes, although not thoroughly understood, point to important aspects of magnetism in materials.

Landolt et al. (1977, 1978) measured the polarization of field-emitted electrons from Ni and Fe crystals as summarized in Table 16.1. Their results are consistent with expectations based on calculated band structures (cf. Fig. 16.4).

Some representative results for polarization of electrons from photoemission experiments include the following.

Single crystals: Fe, +54%; Co, +21%; and Ni, 15% (Busch et al. 1971).

Polycrystalline Ni, +15.5% (Bänninger et al. 1970);

Gd, +5.5% (Busch et al. 1969).

The positive spin polarization measured for Ni was particularly provoking because Ni is a strong ferromagnet with only minority states at E_F. In photoemission, electrons are emitted from states within an energy $h\nu - W$ of the Fermi energy. Here $h\nu$ is the energy of the incident radiation, and W is the work function of the material being probed. Eib and Alvarado (1976) reasoned that by lowering the energy of the incident radiation, they could sample the polarization closer to E_F and should see $P < 0$ for Ni. In fact, they observed $P < 0$ for electron energies only within 0.3 eV of the Fermi energy. This value is smaller than the expected "Stoner gap" (about 0.8 eV) between the top of the Ni majority-spin band and E_F. It is now understood that the more positive polarization measured for Ni (and also for Co) is a result of the spin filtering effect referred to above. Minority spin electrons see a greater scattering probability during transport to the surface because of the large number of empty minority-spin 3*d* states in the material. Hence, majority spin hot electrons have a greater probability of escape and P is more positive than P' (Seigmann 1994).

TABLE 16.1 Spin Polarization of Field-Emitted Electrons from Different Surfaces of Fe and Ni Crystals

	Surface		
	(100)	(110)	(111)
Ni	$-3 \pm 1\%$	$+5 \pm 2\%$	—
Fe	$+25 \pm 5\%$	$-5 \pm 10\%$	$+20 \pm 5\%$

The spin polarization of the low energy, cascade electrons from clean transition metal surfaces has been found always to be positive (Unguris et al. 1982). Further, its magnitude is enhanced at very low energies and at larger energies approaches a value close to the average valence band polarization for Fe, Co, or Ni. The cascade (threshold) polarizations are Fe, Co, and Ni are $44\% \pm 2\%$, $+34 \pm 2\%$, and $24 \pm 3\%$, respectively. These authors pointed out that with a *scanning* primary beam, it should be possible to use the cascade electron polarization to create high-resolution, surface magnetic domain images. This technique is briefly described in Section 16.7.1.

16.6 SURFACE MAGNETIC ANISOTROPY

Néel (1954) pointed out that the reduced symmetry at the surface of a cubic solid changes the usual form of the anisotropy that has lowest-order terms of fourth order in the direction cosines: $K_1(\alpha_1^2\alpha_2^2 + \alpha_2^2\alpha_3^2 + \alpha_3^2\alpha_1^2)$. At a surface, uniaxial terms of second and higher order apply. Hence the surface anisotropy energy per unit area may be written (see Chapter 6 Appendix 1):

$$\sigma = K_{s1}\alpha_3^2 + K_{s2}\alpha_3^4 + K_{s3}\alpha_1^2\alpha_2^2 + \cdots \tag{16.4}$$

In other words, in the plane of the surface there remains a fourfold anisotropy, $K_{s3}\alpha_1^2\alpha_2^2 = K_{s3}\sin^4\theta\sin^2\phi\cos^2\phi$, which can also be written as $K_{sp}\sin^2 2\phi$ when $\theta = \pi/2$ (Gradmann 1986). It is now supplemented by uniaxial surface terms, $K_{s1}\cos^2\theta + K_{s2}\cos^4\theta\cdots$ which to first order favor magnetization perpendicular to the surface if $K_{s1} < 0$ and favor magnetization in the plane of the surface if $K_{s1} > 0$. The first term in Eq. (16.4) is often written as $K^s\sin^2\theta$, in which case $K^s > 0$ implies perpendicular magnetization. The absence of terms of the form $K_a(\alpha_1^2 + \alpha_2^2)$ or $K_b(\alpha_1^2 + \alpha_2^2)\alpha_3^2$ is the subject of Problem 16.2. The Néel model does not fully specify a microscopic mechanism; it assumes the surface energy to be a function of magnetization orientation relative to bond directions.

The frequent occurrence of perpendicular magnetization in a variety of thin-film systems [e.g., Fe/Ag (001), Ni/Cu (001), and Co/Pd] can be understood qualitatively in terms of the difference between electronic structure in the bulk and at the surface. The *d* electrons in the *bulk* may have components of momentum in any direction, although certain directions may be more likely on the basis of orbital topology; this gives rise to the bulk magnetic anisotropy based on the spin–orbit interaction $(\boldsymbol{r} \times \boldsymbol{p}) \cdot \boldsymbol{S}$. However, at the *surface*, electron momentum components perpendicular to the surface must be significantly reduced because the *d* electrons have a reduced probability of being found outside the surface. Velocity in the plane of the surface is associated with angular momentum perpendicular to the surface plane. Thus near a surface, the ratio $L_z^2/(L_x^2 + L_y^2)$ must increase. If the spin–orbit interaction is significant, then the z component of spin perpendicular to the surface will also be increased and perpendicular magnetization may be favored.

16.6.1 Iron Films

Gay and Richter calculated the surface anisotropy for unsupported monolayers of iron and nickel (1986) and found a strong tendency for perpendicular magnetization at an iron surface: $K^s = 0.77$ erg/cm^2 (0.2 meV/atom). This value for the Fe monolayer is two orders of magnitude stronger than the bulk cubic anisotropy of iron, $K_1 \approx 1.3 \times 10^{-2}$ erg/cm^2 per ML (monolayer) (4 μV/atom). More recently, Freeman et al. (1991) have calculated the anisotropy of an unsupported Fe monolayer and found it to favor in-plane magnetization. Their calculations for one ML of Fe on Au, Pt, and Ag show perpendicular anisotropy. These calculations were motivated in part by numerous observations of perpendicular magnetization in iron thin films. Here, one particularly instructive experiment will be described.

Stampanoni et al. (1987) grew BCC Fe films on Ag (001) at room temperature in ultrahigh vacuum (UHV). (There are often improvements in film quality with advances in preparation conditions and deposition technique. These films may not have grown layer by layer below 3 or 4 ML. Nevertheless, the results are instructive.) In order to determine the anisotropy of the films at different thicknesses, the spin polarization of photoemitted electrons was measured at $T = 30$ K. A strong field could be applied perpendicular to the films for the measurements. Figure 16.12, above represents the experiment, and Figure 16.12, below summarizes the results. For iron thicknesses up to 1.5 ML, the polarization indicates that the magnetization is hard to saturate perpendicular to the film and the remanence is zero. This indicates no spontaneous magnetization perpendicular to the film. In fact, the shape of the M–H curves at these thicknesses resembles a Langevin function suggestive of paramagnetism with some short-range rather than long-range ferromagnetic order. (Measurements would have to be done in a parallel field to determine whether there is any long-range magnetic order at all.) However, for 3.5 and 5 ML a relatively square M–H loop is observed, and, in the former case, an appreciable perpendicular polarization remains after removal of the field. As film thickness increases further (10 ML), the magnetostatic energy of the films increases and the magnetization reverts to its in-plane orientation. This result suggests that for these Fe/Ag (001) films, a perpendicular surface anisotropy exists that is able to produce a spontaneous perpendicular moment, provided the magnetostatic energy is not too large (i.e., if the film is not too thick).

If there is an anisotropy K^s localized at the surface that differs from that of the bulk, the effective anisotropy energy density measured for a film of thickness d, may be described as

$$K^{\text{eff}} = \frac{1}{d}\int_0^d (K^v + K^s\delta(z))dz \qquad (16.5)$$

where K^v is the bulk anisotropy energy density operating uniformly throughout the film and K^s is an energy per unit area localized at the surface ($z = 0$)

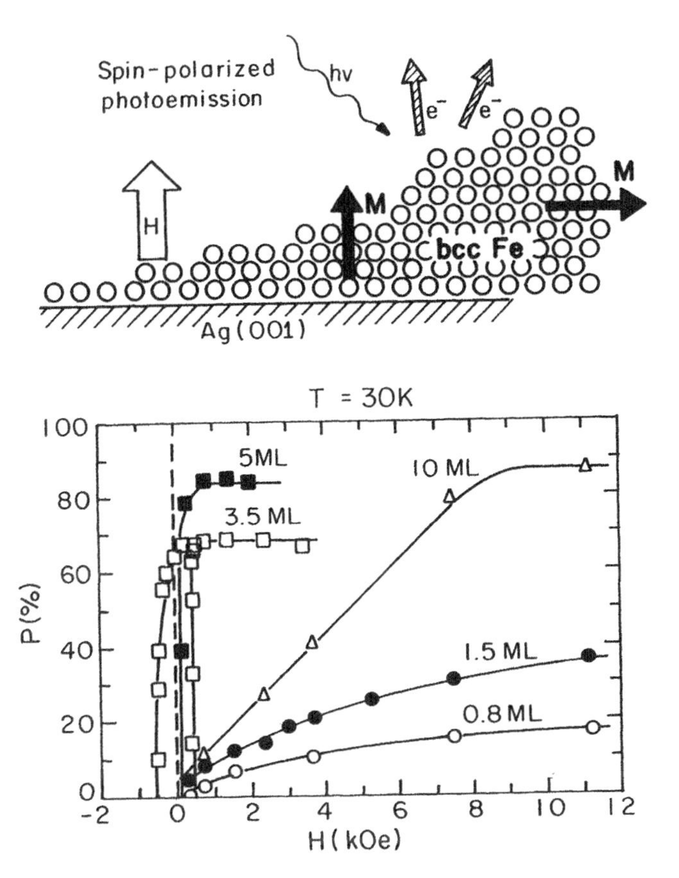

Figure 16.12 Above, schematic of experiment measuring the spin polarization of photoemitted electrons as a function of film growth. Below, results of valence electron spin polarization in BCC Fe films on Ag (001) substrates. Fe shows perpendicular magnetization over a narrow thickness range. [After Stampanoni et al. (1987).]

by the Dirac delta function $\delta(z)$. Thus the effective anisotropy energy density is expressed by carrying out the integration, giving

$$K^{\text{eff}} = K^{v} + \frac{K^{s}}{d} \tag{16.6}$$

The *volume* anisotropy term contains the magnetocrystalline anisotropy, a magnetostatic term, and magnetoelastic terms arising from strains that are uniform throughout the film. The *surface* anisotropy term in Eq. (16.6) may arise from Néel spin–orbit contributions or from strains that are localized at the surface. Restoring the angular dependence, it is possible to write the anisotropic free energy density for a film under biaxial strain, as

$$f_{\text{anis}} = K^{\text{eff}} \sin^2\theta$$

where

$$K^{\text{eff}} = K_u^{mc} + 2B_1 e - \frac{\mu_0 M_s^2}{2} + \frac{K^s}{d} \tag{16.7}$$

Here, uniaxial magnetocrystalline anisotropy, magnetoelastic anisotropy [cf. Eq. (7.8)], and magnetostatic anisotropy ($2\pi M_s^2$ in cgs units) have been included as bulk contributions, along with the surface anisotropy term, K^s/d. Bulk *cubic* anisotropy has been neglected because it neither favors nor hinders perpendicular magnetization. (In the case of Fe or Ni, K_1 is approximately two orders of magnitude smaller than the magnetostatic energy.) If two surfaces are present (e.g., in a thin film), a factor of 2 should appear before K^s in Eqs. (16.6) and (16.7). When $K^{\text{eff}} > 0$, perpendicular magnetization is favored. A uniform magnetization, uniaxial model such as this does not allow for the existence of canted spin states having $0 < \theta < \pi/2$.

The inverse dependence of the surface anisotropy on film thickness is a natural consequence of the Néel model [see Chuang et al. (1994)]. As was seen in Chapters 6 and 7, magnetic anisotropy can be justified on the basis of the anisotropic atom–pair interaction [Eq. (6.8)]. The anisotropic interaction energy w_{ij} can be summed over neighboring atoms to model the macroscopic anisotropy of a material. For a bulk material the result is w_{ij} summed over nearest neighbors and multiplied by the ratio of the number of such identical nearest-neighbor sites to the volume of the material, $N^b/V = 1/\Omega$, where Ω is the atomic volume (Fig. 16.13):

$$f_a^{\text{bulk}} = \frac{N^{\text{bulk}}}{V} \sum_{NN}^{\text{bulk}} w_{ij}(r, \psi) = \frac{1}{\Omega} \sum_{NN}^{\text{bulk}} w_{ij}(r, \psi) \tag{16.8}$$

When the nearest-neighbor environment has missing atoms, as at a surface, the sum must be taken over the atoms actually present about a surface atom or, equivalently, the anisotropy energy of the atoms missing from the local environment of a surface site must be subtracted from the energy in Eq. (16.8). The result must then be multiplied by the ratio of the number of such surface sites to the volume of the material, $N^s/V = 2/\alpha d$, where α is the area per surface atom and d is the sample thickness (Fig. 16.13). The factor of 2 reflects the fact that there are two surface sites in this volume αd:

$$f_a^{\text{surf}} = \frac{N^{\text{surf}}}{V} \sum_{NN}^{\text{surf}} w_{ij}(r, \psi) = \frac{2}{\alpha d}\left(\sum_{NN}^{\text{bulk}} w_{ij}(r, \psi) - \sum_{NN}^{\text{missing}} w_{ij}(r, \psi) \right) \tag{16.9}$$

The quantity in parentheses divided by α is the surface anisotropy energy per

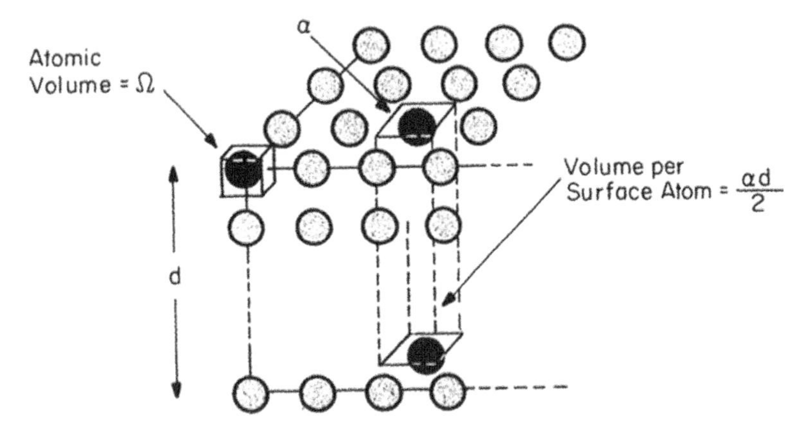

Figure 16.13 Cross section of a thin film showing that the volume per surface atom is a much larger quantity than the atomic volume when film thickness d is large. Normalization of the surface energy to the volume per surface atom thus gives surface anisotropy an inverse dependence on film thickness.

unit area, K^s. Thus Eqs. (16.8) and (16.9) combine to give Eq. (16.6). Surface terms become less significant relative to volume contributions as film thickness increases.

The value of the surface anisotropy coefficient can be determined to first order from the film thickness at which the perpendicular ($K^{\text{eff}} > 0$) to in-plane ($K^{\text{eff}} < 0$) magnetization transition occurs, $K^{\text{eff}} = 0$. Thus, from Eq. (16.7), we obtain

$$K^s = d\left(\frac{\mu_0 M_s^2}{2} - 2B_1 e\right) \tag{16.10}$$

In the case of the Fe/Ag (001) data cited above, the surface anisotropy can be estimated by noting that the perpendicular to in-plane transition occurs near 1 nm of iron. Assuming the saturation magnetization to be suppressed from its bulk value $\mu_0 M_s = 2.2\,\text{T}$ ($M_s = 1750\,\text{G}$) to approximately 1.8 T gives $\mu_0 M_s^2/2 \approx 1.2 \times 10^6\,\text{J/m}^3$ ($2\pi M_s^2 \approx 1.2 \times 10^7\,\text{erg/cm}^3$). Because the strain was not measured on these films, it is assumed to be given by the lattice mismatch between BCC Fe ($a = 2.866\,\text{Å}$) and FCC Ag ($a = 4.09\,\text{Å}$), which is 0.9% (Fe [010] grows parallel to Ag [100]). From Table 7.1, $B_1 = -2.9 \times 10^6\,\text{N/m}^2$ for Fe. Thus, the magnetoelastic energy density, of order $10^4\,\text{J/m}^3$, can be neglected relative to the magnetostatic energy. Equation (16.10) then indicates a surface anisotropy energy density of $\mu_0 M_s^2 d/2 \approx +0.9\,\text{mJ/m}^2$, which is consistent with other measurements. A more careful analysis is required when the magnetoelastic energy is significant (see Section 16.6.3, below). Gay and Richter calculated $0.77\,\text{mJ/m}^2$ for a freestanding iron film. (Note that such surface anisotropies are comparable in magnitude to the surface energy of a typical domain wall, described in Chapter 8.)

16.6.2 Cobalt Films

Another striking manifestation of perpendicular magnetic anisotropy arising from an interface effect is the data of den Broeder et al. (1988) on CoAu multilayers having a period of approximately 10 Å. Multilayered films prepared by ion-beam sputtering showed rather diffuse interfaces at which there appeared significant intermixing of Co and Au species. VSM measurements in fields applied either parallel or perpendicular to the film plane gave the loops shown in Figure 16.14, upper right. These loops indicate a clear preference for in-plane magnetization. When these multilayers are annealed at 275°C, the immiscible Co and Au atoms tend to segregate, markedly sharpening the composition profile. The preferred direction of magnetization is now observed from VSM measurements to be directed out of the plane of the multilayers

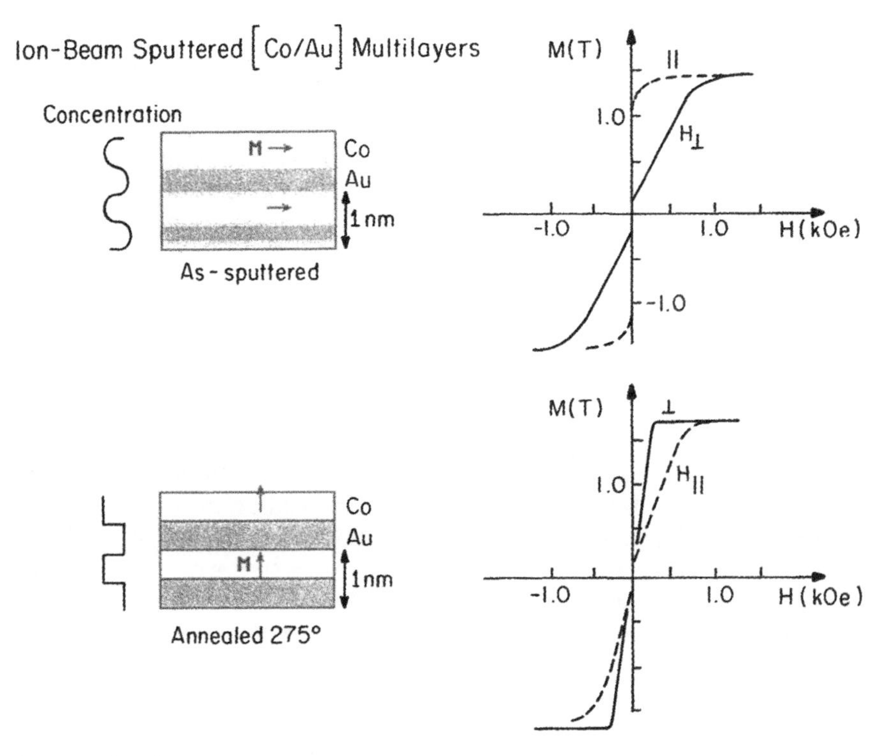

Figure 16.14 Structure and magnetization of ion-beam sputtered Co/Au multilayers concentration profiles and M–H loops of the as-deposited films are shown above, and the effects of annealing on the definition of the Co/Au interfaces and M–H loops are depicted below (den Broeder et al. 1988).

(Fig. 16.14, lower right). This result may be taken to be indicative of the anisotropy of a Co/Au film; the multilayer merely increases the number of surfaces measured. It is also possible that the state of strain is altered by the annealing process.

In the previous section, the surface anisotropy of Fe films was *estimated* from the thickness of the reorientation transition ($K^{\text{eff}} = 0$) using Eq. (16.7) and making the assumptions (valid there) that magnetocrystalline $|K_1|$ and magnetoelastic $|2B_1 e|$ energy densities are much less than the magnetostatic energy density, $2\pi M_s^2$. These assumptions are not valid in cobalt films.

Effective anisotropy data for the Co/Pd system (taken from VSM, SQUID, and torque magnetometry) are plotted for three different crystallographic orientations as well as for polycrystalline multilayers of FCC $[\text{Co/Pd}]_N$ in Figure 16.15 (Engle et al. 1991). These superlattices were grown by seeded-layer epitaxy on GaAs substrates (Lee et al. 1989, 1990) in an MBE chamber.

The [111] superlattices were grown on Pd (50 nm)/Co (0.6 nm)/GaAs (110).

The [100] films were grown on Ag (50 nm)/Co (0.6–1.0 nm)/GaAs (001).

The (110) superlattices were grown on Pd (20 nm)/Ag (40 nm)/GaAs (001).

All depositions were done at room temperature at rates of 0.1 to 0.25 Å/s.

The notation in Engle et al. used to analyze this data is different from that used in Eq. (16.7): Engle—$K_u^{\text{Co}} \times t = (K^{\text{eff}} - 2\pi M_s^2)t + 2K^s$ with $K^{\text{eff}} = K^{mc} + K^{me}$; here— $K^{\text{eff}} \times d = (K^{mc} + 2B_1 e - 2\pi M_s^2)d + 2K^s$. These equations predict a linear dependence of $K^{\text{eff}} \times t$ on t, and that is observed for all four series of

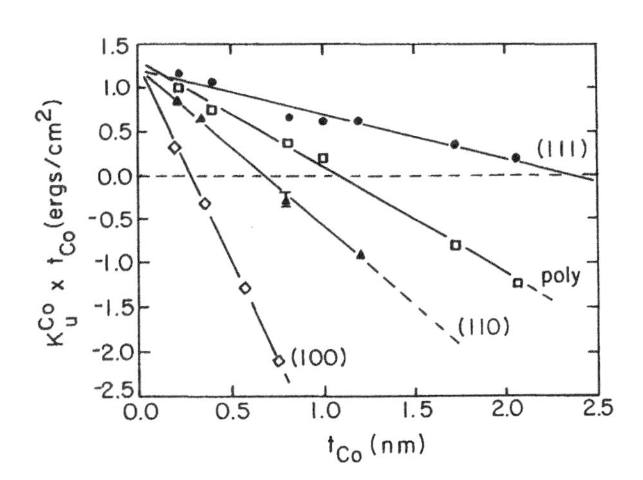

Figure 16.15 Effective anisotropy times Co thickness versus cobalt thickness for [Co/Pd] superlattices. Epitaxial structures having different orientations were grown on GaAs and the polycrystalline multilayers were grown on Si. [After Engle et al. 1991).]

superlattices in Figure 16.15. The data surprisingly indicate the same Co/Pd interface anisotropy regardless of crystal structure: $K^s = +0.65 \pm 0.05\,\text{mJ/m}^2$, consistent with other measurements (Lee et al. 1989). The differences in slope for the different orientations must be due to differences in the volume anisotropy terms associated with crystal structure K^{mc} and strain $K^{me} = 2B_1e$. The strongest perpendicular anisotropy is observed for the (111) superlattices for which a slope of $K^{mc} + 2B_1e - 2\pi M_s^2 = -5.4 \times 10^5\,\text{J/m}^3$ is observed. Given the magnetostatic energy of cobalt as $1.27 \times 10^6\,\text{J/m}^3$, Eq. (16.7) suggests $K^{mc} + 2B_1e \approx 0.75 \times 10^5\,\text{J/m}^3$. It is not possible to carry the analysis further without more information on the cobalt layer structure and strain. It is possible that stacking faults are present putting the films in a structure that may be a mixture of FCC and HCP characteristics. These structures have very different values for K_1 and B_1 (Table 7.2). It is instructive to seek a simpler material system in which it might be possible to carry out a more complete quantitative analysis of the magnetic behavior. The Cu/Ni/Cu system, treated next, moves in this direction.

16.6.3 Nickel Films

Ballentine et al. (1989) found results in thin epitaxial Ni/Cu (001) films that were similar to those of Stampanoni et al. for epitaxial Fe/Ag (001) in some respects. In the Ni/Cu case, the film magnetization was measured at 100 K by the magnetooptic Kerr effect. Because the Kerr effect is more sensitive to perpendicular magnetization (Bader and Erskine 1994), the MOKE signal from a film magnetized at 45° to the normal is enhanced in a perpendicular field relative to that in a parallel field. Figure 16.16 summarizes the results on Ni/Cu (001). For Ni thicknesses below about 4 ML, the remanent state shows a preference for in plane, not perpendicular magnetization. At 4.7 ML, the remanent state shows that a significant component of the magnetization remains perpendicular to the plane of the film. This work did not determine the orientation of magnetization in films thicker than 10 ML. A similar result had been obtained earlier by Gradmann (1966) and has been confirmed more recently by Huang et al. (1994) and Schulz and Baberschke (1994). These data alone seem to suggest a negative surface anisotropy for Ni because $K^{\text{eff}} < 0$ at small d [Eq. (16.7)].

Room temperature measurements on similar epitaxial Ni/Cu (001) films were continued to much larger Ni thicknesses by Bochi et al. (1995a). They also used MOKE to measure the films in situ, applying fields either perpendicular or parallel to the film. In those measurements perpendicular magnetization was observed for nickel thicknesses up to approximately 6 nm. Above that thickness, magnetostatic energy drives the magnetization back in plane. Using the method of analysis applied above to the iron magnetostatic reorientation transition, Eq. (16.7) suggests a net surface anisotropy of order 0.8 mJ/m^2, neglecting the magnetocrystalline and the magnetoelastic contributions (it will be seen later that neglecting $2B_1e$ is a bad assumption for Ni).

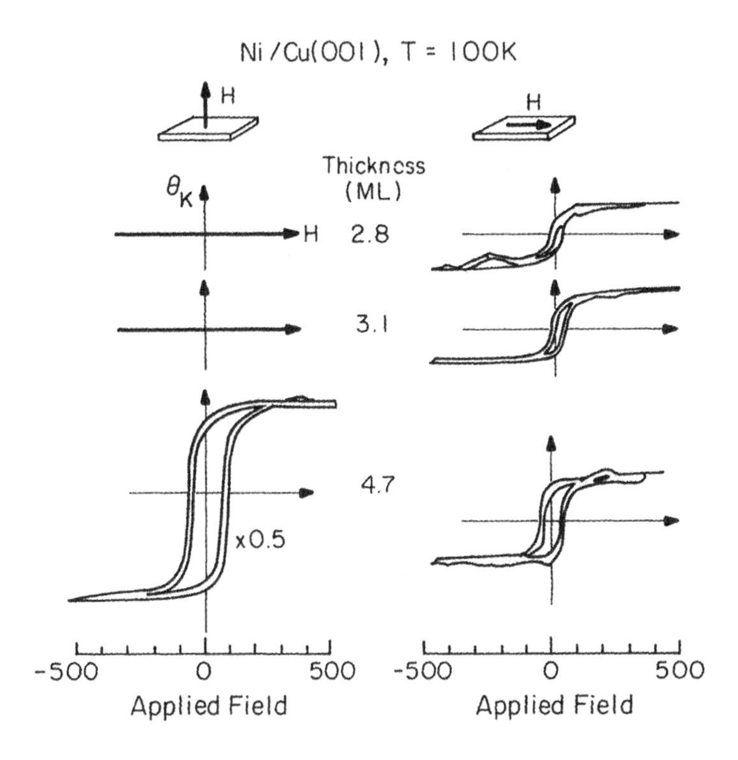

Figure 16.16 Magnetooptic Kerr effect measurements of magnetization in ultrathin Ni/Cu (001). The results for fields applied either perpendicular or parallel to the film plane indicate that above about 4 ML, Ni films prefer to be magnetized normal to their plane. [Adapted from Ballentine et al. (1990).]

Cu-capped Ni/Cu (001) epitaxial films (two Cu–Ni interfaces) have been studied by Jungblut and O'Handley (1994) using in situ MOKE, Bochi et al. (1995a) using ex situ VSM, and Ha and O'Handley (1999) using ex situ torque magnetometry. One advantage of capping with Cu is that the two Ni interfaces are now identical, in principle. Also, the Cu-capped films can be removed from vacuum for more extensive testing, with no significant change in magnetic properties at least over a year. From the MOKE and magnetometer measurements, the energy needed to saturate the films in perpendicular and parallel fields were subtracted to give a quantitative determination of the effective anisotropy K^{eff}. (The energy needed to saturate is calculated from an anhysteretic loop.) The torque measurements give a direct measure of the uniaxial as well as higher-order components of K^{eff}. The results of these three sets of measurements are shown in Figure 16.17 in terms of Eq. (16.7) plotted as $K^{eff}d$ versus d. The first important conclusion from this data is that the range of perpendicular magnetization is now observed to extend to about 13.0 nm, roughly double the range for Ni/Cu. This suggests that a major source of the

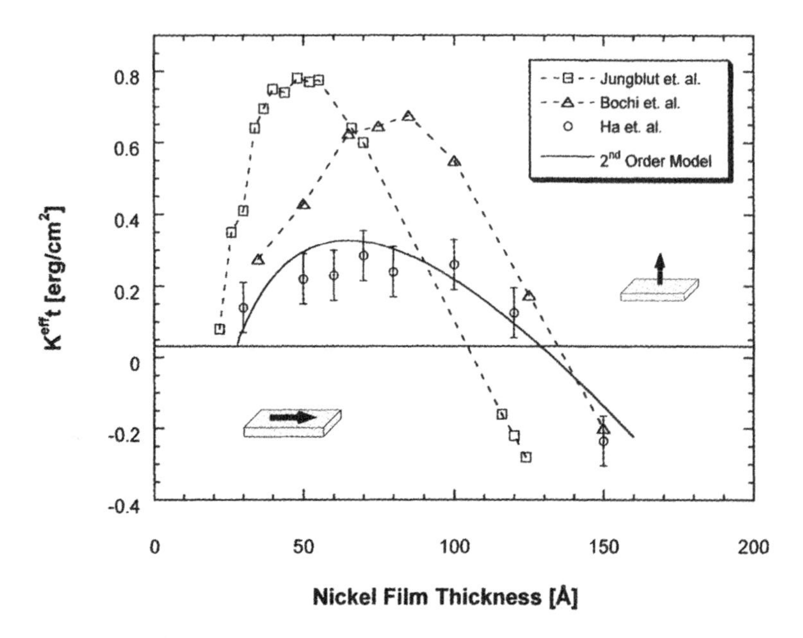

Figure 16.17 Variation of effective anisotropy times Ni thickness ($K^{\text{eff}} \times d$) versus Ni thickness in Cu/Ni/Cu (001) sandwiches after Jungblut et al. (1994), Bochi et al. (1995a), and Ha and O'Handley (1999), using in situ MOKE, ex situ VSM, and ex situ torque magnetometry, respectively, to determine the effective anisotropy. $K^{\text{eff}} \times d > 0$ implies perpendicular magnetization in the uniaxial approximation.

perpendicular anisotropy in this system is the Cu/Ni interface [i.e., K^s(Cu/Ni) > 0)] because in the capped films the number of Cu/Ni interfaces has doubled relative to that in the uncapped films. The Ni strain in the last set of films has been found to vary as $0.026(2.7/d)^{2/3}$ (Fig. 16.8). Using this form for the strain, neglecting the small magnetocrystalline anisotropy of Ni, and using $B_1 = 6.2 \times 10^6\ \text{N/m}^2$ for Ni, Eq. (16.7) predicts a nearly linear dependence of $K^{\text{eff}}d$ versus d and no reversion to in-plane magnetization. [If the strain were found to vary as d^{-1} for $d > d_c \approx 2.8$ nm, it is clear from Eq. (16.7) that the magnitude of the slope would change from $2B_1e_0 - 2\pi M_s^2$ below d_c to $-2\pi M_s^2$ above d_c (Chappert and Bruno 1988)]. The changes in magnetization with film thickness are small and are included in the model fit. Clearly, the model is missing some important energy contributions for these films. The next section indicates what that contribution is.

16.6.4 Magnetoelastic Coupling in Thin Films

Caution should be exercised in applying the bulk values of the magnetoelastic coupling coefficients in Table 7.1 to thin films. There is growing evidence that magnetoelastic coupling constants can deviate from bulk values near a surface (Sun and O'Handley 1991). This should not be surprising because it is well established that lattice constants, electronic structure, symmetry, spin–orbit interactions, and the fundamental magnetic properties derived therefrom, can take on unique values at a surface. The same is true for thin films.

In a thin film bonded to a substrate, it is not appropriate to speak of magnetostriction in terms of the strains produced as $\boldsymbol{M}$ is rotated because there is no free anisotropic strain in a constrained sample. A film cannot even show a true magnetostrictive strain normal to its plane because magnetostrictive strains conserve volume to first order and the film cannot strain in the other two directions to conserve volume. The magnetostrictive stress B_i associated with a change in the direction of $\boldsymbol{M}$, can be measured through its tendency to cause the film–substrate couple to bend, expressing a small fraction of the magnetostrictive strain.

It has recently been reported that the magnetoelastic coupling coefficient of a thin film can also depart sharply from the value observed in thicker films and in bulk. Figure 16.18 shows the measured values for the ME coupling coefficient of a polycrystalline permalloy film (slightly iron rich, $Ni_{79}Fe_{21}$) plotted to show that the data follow the Néel form, $B^{\text{eff}}(t) = B^{\text{bulk}} + B^{\text{surf}}/(t - t_0)$, with $B^{\text{bulk}} = -0.78 \times 10^5\ \text{N/m}^2$, $B^{\text{surf}} = 1.4 \times 10^{-4}\ \text{N/m}$, and $t_0 = 0.7\ \text{nm}$. The accepted bulk value for this composition is $|B^{\text{bulk}}| < 1.0 \times 10^5\ \text{N/m}^2$ which corresponds to a magnetostriction coefficient of $+0.3 \times 10^{-6}$. Similar results have

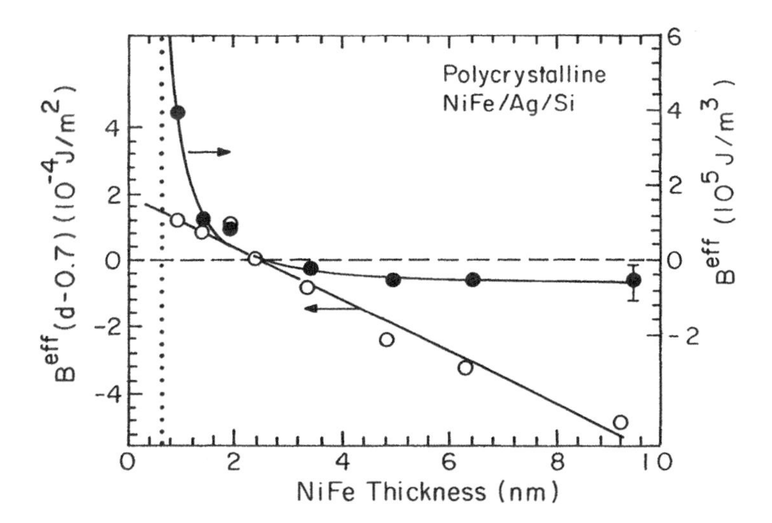

Figure 16.18 Solid data points: effective ME coupling coefficient measured in situ for polycrystalline NiFe/Ag/Si versus NiFe thickness. Open data points: $B^{\text{eff}}(t - 0.7)$ versus t show quality of fit to Néel model. Note that the vertical scale unit, $B^{\text{eff}} = 1 \times 10^5\ \text{J/m}^3$, corresponds to a magnetostrictive strain of approximately 0.5×10^{-6}. The value of t_0 is shown by vertical dotted line (Song and O'Handley 1994).

been observed in various other permalloy compositions (Gurney 1997) as well as in rare-earth intermetallics (del Moral et al. 1998).

The trend in the effective ME coupling shown in Figure 16.18 is probably not all due to an intrinsic Néel, surface effect. It is likely that changes in the structure, chemistry, and state of stress of the film with thickness are also affecting the data. Koch et al. (1996) have shown that the effective ME coupling coefficient of thick, epitaxial Fe/W also changes sign as a function of the state of stress in the film. This supports the notion that such changes can be second-order magnetoelastic interactions.

Extension of the Néel model to include second-order strain, as well as other interactions, indicates that in addition to the terms in Eq. (16.7), new terms should be considered. These new contributions to the effective anisotropy include terms of the form (Ha and O'Handley 1999):

$$De^2\alpha_i^2, \quad \frac{B^s}{d}e\alpha_i^2, \quad \text{and} \quad \frac{D^s}{d}e^2\alpha_i^2 \tag{16.11}$$

The first term is a second-order, *bulk* magnetoelastic interaction. The second and third are *surface* magnetoelastic terms (analogous to K^s) of first order and second order in strain. The Néel model relates the magnitude of these new coefficients to K_1, B_1, B_2, and so on, which in turn can be described in terms of the magnetic dipole interaction strength l and its spatial derivatives l', l'', and so forth. This pair interaction model indicates that the bulk second-order term, $De^2\alpha_i^2$, is more important than either of the surface magnetoelastic terms for strains greater than a few tenths of a percent. This second-order term shows up as a strain dependence in the effective magnetoelastic coefficients (O'Handley and Sun 1992):

$$B^{\text{eff}} = \frac{\partial K^{\text{eff}}}{\partial e} = (B_1 + 2De) \tag{16.12}$$

Measurements of such interactions in ultrathin films and multilayers are just beginning to appear. Koch et al. (1996) find for epitaxial BCC Fe (100)/W, $D = -1.1 \times 10^{10}\,\text{mJ/m}^3$, Sander et al. (1999) find for BCC Fe (100)/MgO, $D = -1.1 \times 10^{10}\,\text{mJ/m}^3$. With the inclusion of the second-order ME interaction in Eq. (16.7), the torque data of Figure 16.17 can be fit to give $K^s = 0.8\,\text{mJ/m}^2$ and $D = -1.1 \times 10^{10}\,\text{mJ/m}^3$. The effective ME coupling coefficient, Eq. (16.12), should replace B_1 in Eq. (7.8) to describe the ME contribution to the anisotropy energy density when strain is appreciable.

The implications of the thickness and/or strain dependence of B^{eff} are important for surface and thin-film magnetism. If a device is to be designed with a specified range of magnetoelastic coupling, bulk data may not serve as a reliable guide for designing the appropriate alloy. Further, the combination of large strains and ME coupling coefficients out of the specified range compounds the problem of property control. Stresses due to interfaces, dislocations, and differential thermal expansion could alter the strength of the effective coupling. On the positive side, there is potential for using stress to control magnetic anisotropy through the ME contribution to anisotropy as well as for development of highly sensitive strain sensors, accelerometers and other devices based on multiple layers of very thin magnetic films.

16.7 MAGNETIC DOMAINS IN THIN FILMS

Several images of magnetic domain patterns in crystalline (Figs. 1.13, 8.1*a*, 8.13, 8.14, 8.25, 12.14) and noncrystalline magnetic materials (Fig. 11.8) have been presented and described. In thin films, the energies that determine magnetic domain structures may be quite different than in bulk materials because of the proximity of the interfaces (cf. Fig. 8.14). In Chapter 8 it was shown that a 180° Bloch domain wall generally has a Néel cap (***M*** parallel to the surface) where it intersects the surface. Further, in thin films, magnetostatic energy can turn the entire Bloch wall into a Néel wall.

This section treats magnetic domains in thin films for which the anisotropy is either in plane or out of plane. Two measurement techniques well-suited to these cases are scanning electron microscopy with spin polarization analysis and magnetic force microscopy. Each of these powerful tools will be described and examples given.

16.7.1 Scanning Electron Microscopy with Polarization Analysis (SEMPA)

If a primary electron beam is highly focused and scans the surface as in a scanning electron microscope, it is possible to analyze the secondary electron spin polarization to construct a high-resolution image of the surface magnetization (Celotta and Pierce 1982). This technique is known as scanning electron microscopy with (spin) polarization analysis (SEMPA). The first SEMPA images (10-μm resolution) were made by adding a scanning electron gun to a 100 kV gold foil Mott detector (Koike and Hayakawa 1984). Higher-resolution images ($<1\ \mu$m) soon followed [Unguris et al. (1985), Celotta and Pierce (1986), Hembree et al. (1987), using a diffuse scattering, gold thin film detector] as well as detailed descriptions and analysis of the technique (Unguris et al. 1986, Celotta and Pierce 1986, Scheinfein et al. 1990). High resolution SEMPA has been applied to a variety of problems in surface magnetism, including surface domain profile determination [Scheinfein et al. (1989, 1991) and Oepen and Kirschner (1989), using a low-energy electron diffraction spin polarimeter] and exchange coupling of Fe layers through Cr spacers (Unguris et al. 1991).

The principles of SEMPA are an extension of the concepts of SEM, but they are practiced in ultrahigh vacuum to maintain surface cleanliness and to ensure that the mean free path of the secondary electrons is greater than the distance from the sample to the gold scatterer. SEMPA also makes use of electron spin polarization analysis, which was outlined above. The implementation of SEMPA is not trivial, largely because of the demands placed on the electron optics by the need for high-resolution and integrity of the secondary electron beam polarization.

The time τ required to accumulate a pixel for an image with resolvable contrast relative to an adjacent pixel (Hembree et al. 1987) is given by

$$\tau = C\left\{\left(\frac{N}{N_0}\right) TYi_{PE}\right\}^{-1} \tag{16.13}$$

Here, C is a constant describing the electron count needed for detectable asymmetry threshold, N/N_0 measures the gold backscattering efficiency, T is the transmission of the electron optics between the sample and the polarimeter ($T > 0.5$), Y is the secondary electron yield of the sample ($Y \approx 0.2$), and i_{PE} is the current in the primary scanning electron beam (of order 10^{-9} A). It is clear that many instrumental factors must be simultaneously optimized if the image time is to be reasonably short. Depending on resolution requirements, SEMPA systems have image times of order 20 min.

Figure 1.13 is a SEMPA image taken from the surface of a 3% SiFe crystal illustrating some of the advantages of this technique. The upper panel shows the intensity image collected by summing the signals from all four quadrants of the detector $I_1 + I_2 + I_3 + I_4$ (Fig. 16.11). The intensity image reveals surface topography due to polishing roughness. The next two panels display the asymmetry in the two orthogonal in-plane directions. The images in all three panels were assembled from the *same set of data*, four independent intensities for each pixel, taken during one scan. The left panel shows the strongest contrast between domains magnetized to the left and right (M_x), while the right panel shows the strongest contrast for $\pm M_y$. The panels differ only in the data combinations represented; for instance, P_x is proportional to $(I_1 - I_3)/(I_1 + I_3)$ in Figure 16.11. There is no ambiguity in assignment of vector components of $\mathbf{M}$. Also note that topographic features evident in the intensity image are largely normalized out of the polarization images by virtue of the definition of polarization, which involves division by the total intensity. SEMPA allows high-resolution imaging of surface domain structures with minimal sample preparation. Surface roughness generally is not a problem because of the favorable depth of field afforded by SEM imaging.

Figure 8.1 is a high-resolution image and line scan of the spin-polarization variation across the Néel surface termination of a 180° Bloch wall. This result should be compared with those in Figure 8.12 for permalloy and iron whisker surfaces. Note that the Néel cap wall is imaged by SEMPA, not the underlying Bloch wall. The Néel cap on a Bloch wall may be calculated by adding magnetoelastic terms to the usual domain wall energy and thickness calculation (Chapter 8). The thickness of a Néel cap is greater than that of the Bloch wall that it terminates and the magnetization distribution through it is asymmetric.

As a final example of the elegance and power of SEMPA, refer to the thin film structure and the corresponding SEMPA image in Figure 12.16. These figures describe the domains in a thin iron film deposited over a tapered Cr layer on an iron whisker. The two iron layers are coupled by an exchange interaction through the interposed Cr layer. In this one experiment, the dependence of iron–iron exchange coupling on chromium layer thickness is beautifully and unambiguously revealed. The exchange coupling oscillates from ferromagnetic to antiferromagnetic repeatedly as the chromium layer thickness increases from zero. Above one lengthwise domain in the iron whisker, the iron film shows a periodically varying contrast due to the different orientation of its magnetization, reflecting the F–AF–F··· exchange coupling. When the iron depositions on the chromium wedge are made at elevated temperatures and at a slower rate, the quality of films and interfaces is much improved. In this case, even shorter-period oscillations, which are superimposed on the longer-period oscillations, can be resolved by SEMPA (Unguris et al. 1991).

16.7.2 Magnetic Force Microscopy (MFM)

The 1983 Nobel Prize in Physics was awarded to Binnig and Rohrer (Binnig et al. 1982) for inventing the scanning tunneling microscope (STM). They demonstrated the amazing resolution achieved by scanning a metal tip over a surface at a height of a few angstroms and measuring surface topography via the tunneling of electrons between the tip and sample. The tunneling current varies exponentially with tip height above the surface. Thus, very small changes in surface elevation can produce strong changes in tunnel current. The pioneering demonstration of STM was followed by the development of scanning force microscopy (SFM) (Binnig et al. 1986, 1987) and later, magnetic force microscopy (MFM) (Martin and Wickramasinghe 1987, Saenz et al. 1987). The latter provides a powerful means of characterizing magnetic surfaces that in many respects complements SEMPA. Figure 8.14 is an example of an MFM image of the perpendicular magnetization domain pattern in an epitaxial, 2-nm-thick Ni (001) film capped with Cu.

An MFM, like an STM, consists of a micrometer-scale tip attached to the end of a flexible cantilever (Fig. 16.19). The tip is scanned close to the surface, but at a height considerably greater than that used for an STM. The tip is magnetic or is coated with a thin film of a hard or soft magnetic material. The magnetoelastic interaction between the tip and the fringe field above the sample results in cantilever deflection and/or a change in its resonance frequency and phase. If the MFM tip is too close to the sample, it can snap into contact in an attractive force field. Cantilever deflection can be measured by optical interferometry or other means in the "static" mode of operation. In

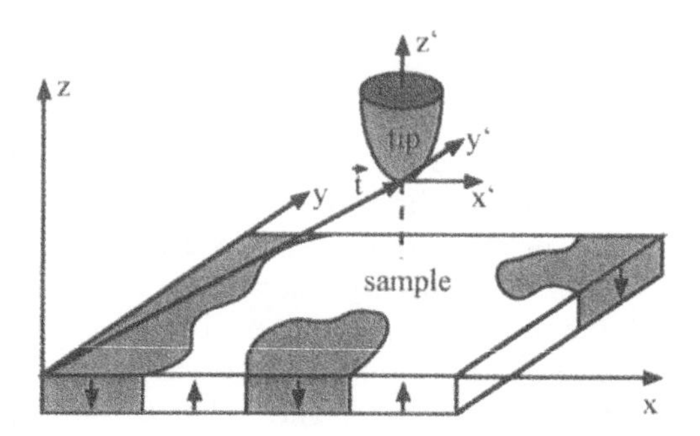

Figure 16.19 Geometry for description of MFM technique. A tip scans at a height z above a thin sample whose surfaces are charged in a pattern determined by the normal component of the magnetization distribution, $M_z(x, y)$.

the "dynamic" mode, the cantilever frequency is a measure of the attractive or repulsive force on the cantilever due to the fringe field.

The conventional tip/cantilever is most sensitive to magnetic forces normal to the surface. While this mode of MFM is most suitable for imaging domains on perpendicularly magnetized surfaces, fringe fields above domain walls in longitudinally magnetized materials can also be imaged. An example considered below is the quantitative analysis of the domain structure in perpendicular-magnetization, epitaxial films of Ni (001) sandwiched by Cu (001) layers. In MFM the tip is usually scanned a few tens of nanometers above the sample surface; in AFM the tip is a few tenths of a nanometer above the surface. The same tip can be used to measure surface topography, as well as magnetic force. Thus, MFM allows a detailed correlation of magnetic and structural features near a surface.

It is not possible, in general, to calculate the perpendicular magnetization distribution $M_z(x, y)$ in a sample from the MFM force data $F(x, y, z)$. The principal reason for this is that the same stray field pattern can be generated from many different magnetization configurations. That is, the mapping from field to magnetization is not unique. One case in which the mapping is unique is that of completely perpendicular magnetization. In this case it is important to have a good signal-to-noise ratio.

The features in a perpendicular magnetization pattern decay with increasing height above the surface like e^{-kz} where $k = 2\pi/\lambda$. Thus, long wavelength changes in $M_z(x, y)$ produce stronger forces at a given height, z, than do sharp, short wavelength features. In order to do *quantitative* MFM, it has been found useful to Fourier analyze the force field, fringe field and magnetization patterns and relate their Fourier components by transfer functions that scale the different spatial frequency components (Fourier components) appropriately (Hug et al. 1998). The Fourier components of an arbitrary surface magnetization charge distribution $\sigma(r)$ are given by

$$A_M(\boldsymbol{k}) = \int_{-\infty}^{\infty} \sigma(r) e^{i\boldsymbol{k}\cdot\boldsymbol{r}} dr \tag{16.14}$$

where $\boldsymbol{r} = (x, y)$ and $\boldsymbol{k} = (k_x, k_y)$. The Fourier coefficients of the z component of the fringe field $A_H^{z,h}(\boldsymbol{k})$ and that of the charges on the two surfaces of a thin film are simply related by a factor called the *transfer function*:

$$A_{H_z}^{z,h}(k) = \frac{e^{-kz}(1 - e^{-kh})}{2} A_M(k) \tag{16.15}$$

The first term in the transfer function, $e^{-kz}/2$, describes the exponential decay of the field with height z above the top surface (at $z = 0$), while the second term, $-e^{-k(z+h)}/2$, describes the signal from magnetic features at the lower surface of the sample (at $z = -h$) (Fig. 16.19). For an infinitely thick sample, the field from the bottom surface is zero.

The implications of this spatial frequency dependence of signal decay are graphically represented in Figure 16.20. Here the field transfer function, $A_{H_z}^{z,h}(k)/A_M(k)$ from Eq. (16.15), is shown for a thin film [solid line, both terms in Eq. (16.15) apply] and for a bulk surface (dotted line, second term vanishes), over more than four decades in k (dimensionless units of $2\pi/\lambda$). For a single surface, long wavelength surface charge features scale with equal strength to give the long wavelength field components. At short wavelengths, the field at a given height z drops off exponentially. For a perpendicularly magnetized thin film (two charged surfaces a distance h apart), the short wavelength magnetization features show up less strongly in the fringe field, as is the case for bulk samples. However, at long wavelengths the field also decays because the mirror-image charge distribution at the bottom surface tends to cancel the field from the top surface charge.

The lower panel of the figure shows how magnetic thin-film surface charge distributions, having three different spatial frequencies but equal amplitudes, are scaled to give different fringe fields. The features having $\lambda \approx \lambda_{\max} \approx 6.6$ (dimensionless units of $2\pi/k$) contribute most strongly to the MFM image while longer and shorter wavelength features are attenuated.

Figure 16.21 shows a commercial pyramid scanning probe tip with an electrodeposited needle intentionally grown on its end. This needle is coated with a 25-nm-thick Co film to form the high-resolution MFM tip. The images described below on Cu/Ni/Cu/Si (001) films were taken using such a tip and analyzed using the Fourier transfer function method.

It is instructive to consider the domain patterns in the series of epitaxial Ni/Cu/Si (001) films, capped with 2 nm of Cu (Fig. 16.22). These films show perpendicular magnetization over a wide thickness range (Fig. 16.17). The domain patterns were taken using a magnetic force microscope (MFM) at the

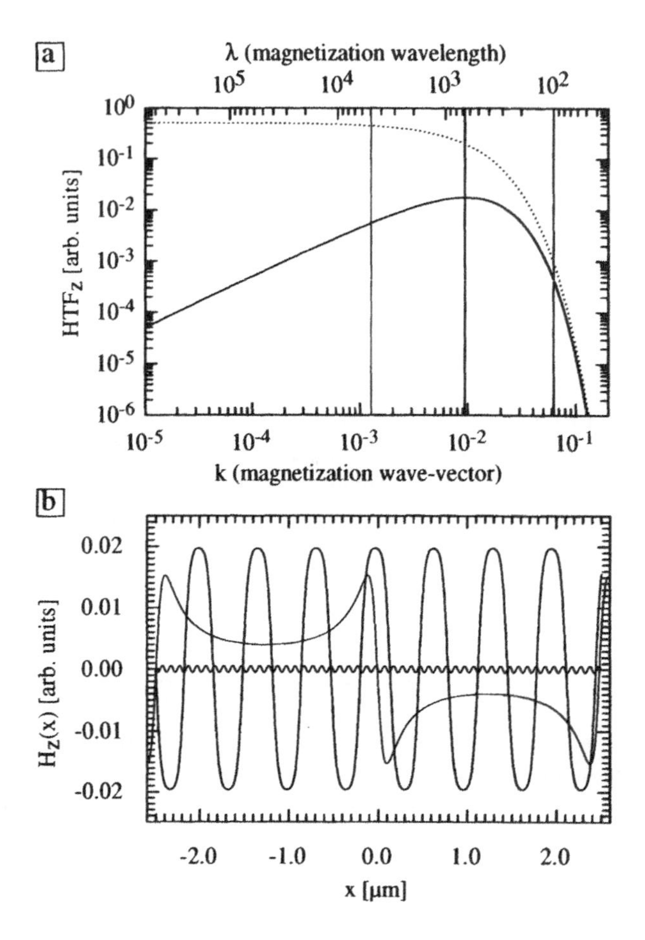

Figure 16.20 (*a*) Dependence of the field transfer function, $A^{z,h}H_z(k)/A_M(k)$, Eq. (16.15), on magnetization wavenumber and wavelength λ. The transfer function is shown for a sample with one charged surface (bulk sample, dotted line) and for two charged surfaces (thin film, solid line). (*b*) Spatial dependence of the z component of the fringe field for three different thin-film magnetization patterns of k vectors indicated by the three vertical lines in (*a*). [After Hug et al. (1998).]

University of Basel (Bochi et al. 1995a, b). The strong contrast of the MFM images shows dramatically that the magnetization is indeed predominantly perpendicular to the plane of the films as indicated by the positive effective anisotropy in Figure 16.17. Magnetic domains are shown for Ni layer thicknesses from 2 to 12.5 nm in Figures 16.22a–16.22d. The overall length scale D of the domains is refined with increasing Ni thickness. However, the length scale is not well defined, especially at smaller Ni thicknesses; the dispersion in D values is larger there. Further, a change in domain character occurs between

Figure 16.21 Scanning electron microscopy image of tip used in MFM imaging. The commercial pyramid tip has a nonmagnetic, electron-beam-induced tip deposited formed on its surface in vacuum. This tip is made magnetic by deposition of a thin magnetic film on one of its long sides.

8.5 and 10 nm Ni thickness. For the thicker Ni films, the larger domains are broken up by an increasing proliferation of finer bubble domains. This suggests that as the Ni film grows, the domain walls stable at smaller Ni thickness cannot move, as Ni thickness increases, in order to give a domain pattern that represents an energetic minimum; these are not equilibrium domain patterns.

It must first be determined whether the force images in Figure 16.22 can be interpreted as magnetization images. Figure 16.23*a* is a reproduction of Figure 16.22*c* for 10 nm of Ni. The line scan (lower left) through the MFM image gives the position dependence of the measured force and field. Using a discrimination routine, the fringe field pattern in Figure 16.23*a* was used to generate the trial magnetization pattern (Fig. 16.23*b*). From this trial $M_z(x, y)$ data, the force field in panel (*c*) was calculated. Not only is the calculated force-field image in excellent agreement with the MFM data in panel (*a*), but also the details shown in the calculated line scan below panel (*c*) match very well with those in panel (*a*). This analysis (Hug et al. 1998) is evidence that, at least for this perpendicular magnetization case, the MFM data are a good representation of the magnetization pattern in the film.

It is striking that the domain walls in Figure 16.22 show no orientational correlation with the easy in-plane $\langle 110 \rangle$ crystallographic directions in these epitaxial Ni films. The domain walls are Bloch walls, not Néel walls, because the magnetization rotates from $+M_z$ in one domain to $-M_z$ in the adjacent domain, by rotating essentially in the plane of the wall. Thus, the wall magnetization runs along the wall length with no cost in magnetostatic energy. (Néel walls are preferred only in thin films exhibiting in-plane magnetization.) On the basis of the anisotropy measurements for these films (Fig. 16.17), the wall width δ_{dw} should be of order 30 nm for a Ni thickness of 8.5 nm and increases for thinner or thicker films (because of the thickness dependence of

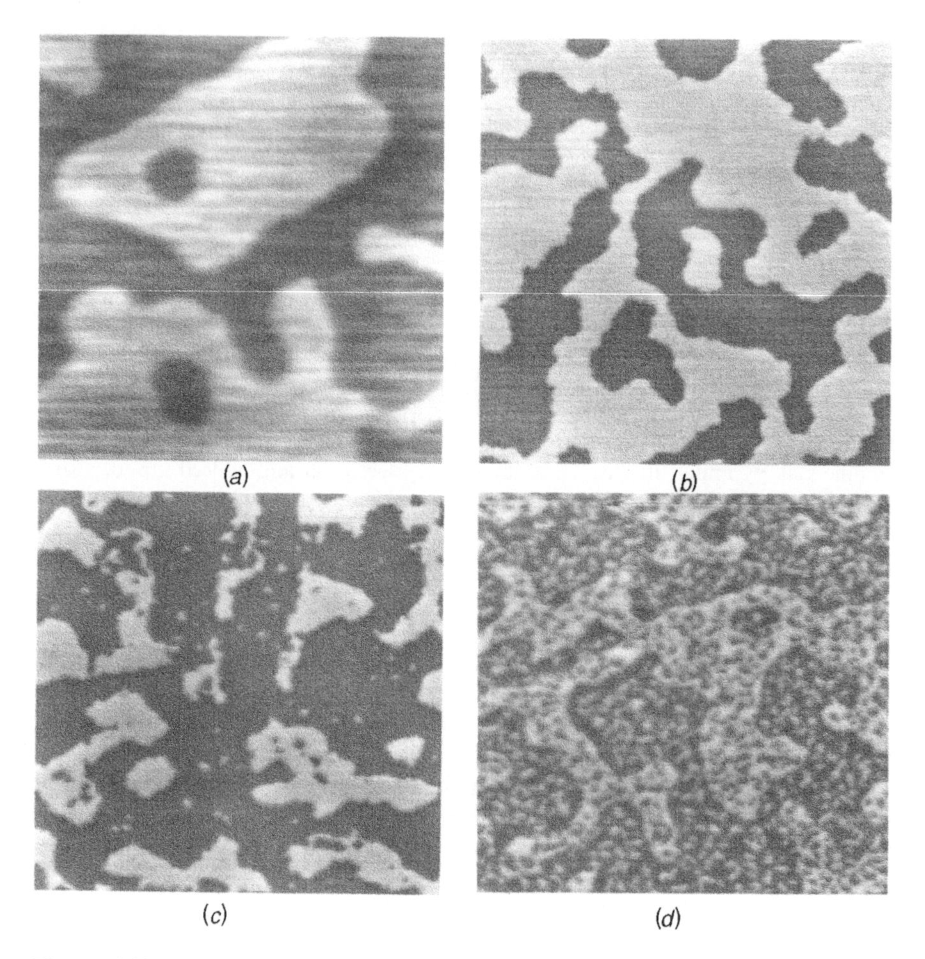

Figure 16.22 Domain structure of epitaxial Cu/t_{Ni}Ni/Cu (001) films imaged by MFM over a 12-μm square: (*a*) 2 nm Ni, (*b*) 8.5 nm Ni, (*c*) 10 nm Ni, (*d*) 12.5 nm Ni (Bochi et al. 1996).

K^{eff}). The energy saved by having the magnetization at the center of the Bloch wall follow the in-plane easy crystallographic directions, is small compared with the wall energy saved by minimizing ratio of the wall length to domain area. Thus, the walls follow more curved paths rather than rectilinear ones along $\langle 110 \rangle$.

A simple domain model that explains some aspects of the length scale D of these domain patterns is now considered. The energy density for a stripe domain structure

$$u = 16M_s^2 \frac{D}{\pi^2} \sum_{n,\,\text{odd}} \frac{1 - \exp(-n\pi d/D)}{n^3} + \frac{\sigma_{dw} d}{D} \tag{16.16}$$

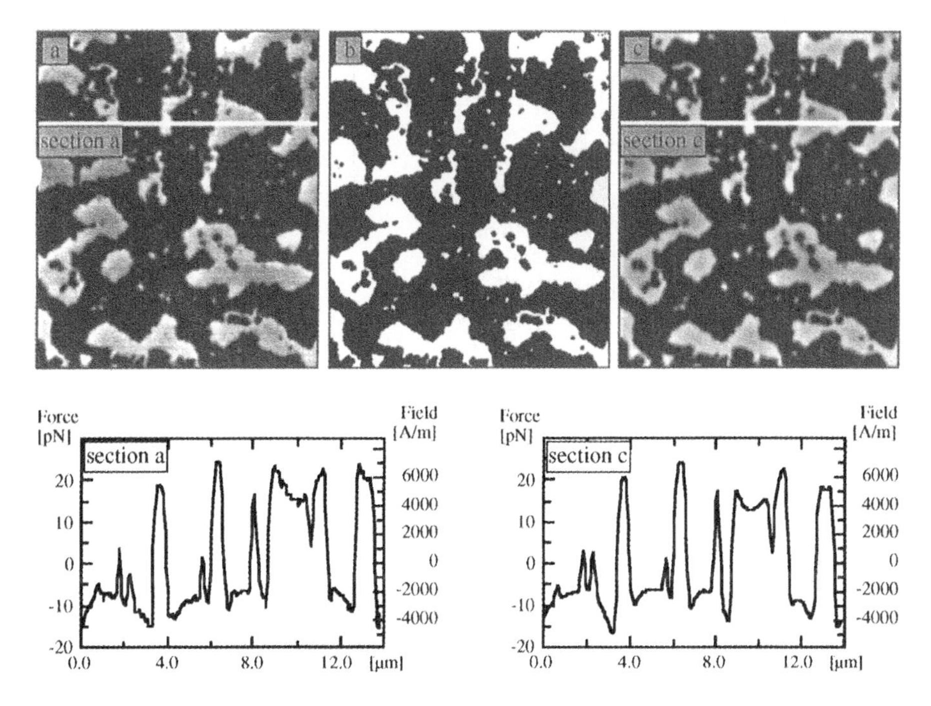

Figure 16.23 Above (*a*) MFM force-field image taken from epitaxial Cu/10 nm Ni/Cu (001) film of Figure 16.22*c*; (*b*) trial perpendicular magnetization distribution generated from (*a*) by discrimination routine; (*c*) field pattern calculated from magnetization distribution (*b*) using Eq. (16.16)); below, measured (left) and calculated (right) force and field line scans across panels (*a*) and (*c*) above (Hug et al. 1996).

has been numerically minimized with respect to D, by Paul (Bochi et al. 1995b), using the methods of Malek and Kambersky (1958). The energy difference between stripe and checker domain patterns is small.

The equilibrium spacing D of a perpendicular domain pattern periodic in one direction is found, to first order, to be given by

$$D \approx \frac{27M_s^2d^2}{9.17M_s^2d - \sigma_{dw}(d)} \tag{16.17}$$

where M_s is the magnetization density, d is the thickness of the magnetic layer, and σ_{dw} is the domain wall energy density. The domain wall energy density is thickness-dependent when the effective anisotropy energy is thickness-dependent [see Eq. (16.7) or Fig. 16.17)]. Figure 16.24 shows schematically the form of Eq. (16.17). When the magnetostatic energy per unit film area, $9.17M_s^2d$, is much greater than the wall energy density, that is, at very *large film thicknesses*, Eq. (16.17) indicates that the length scale of the domain structure should

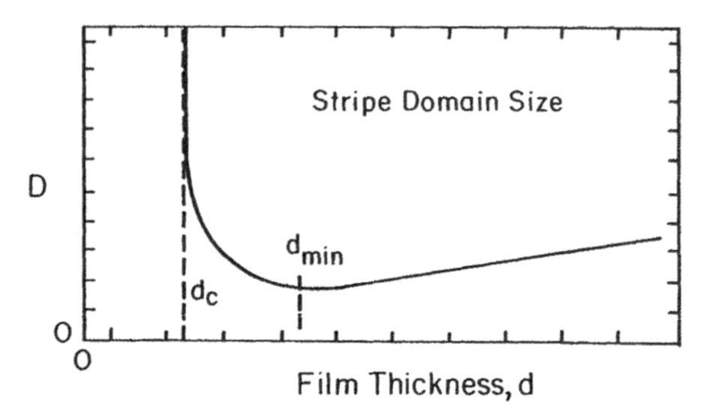

Figure 16.24 Schematic representation of the variation of domain size with magnetic film thickness according to Paul's first-order solution, Eq. (16.17) (Bochi et al. 1995b).

increase linearly with increasing film thickness. At very *small film thicknesses*, D diverges at a critical film thickness d_c, given by $d_c = \sigma_{dw}/9.17M_s^2$. For $\sigma_{dw} \approx 1.0\,\mathrm{erg/cm^2}$ and $M_s \approx 800\,\mathrm{emu/cm^3}$, $d_c \approx 20\,\text{Å}$. Below this critical thickness, a magnetic film should consist of a single domain; that is, the domain diameter D should be much greater than the film thickness. Closure domains may still exist near the edges of such films where the magnetoelastic fields are greater than in the interior of the film. In between these film thickness limits, D shows a minimum value, $d_{min} \approx 20\,\mathrm{nm}$.

The exact form of the numerical solution to Eq. (16.17) is graphed in Figure 16.25 using the experimentally measured effective anisotropy K^{eff} in σ_{dw}. The inserts show schematically the shape of the energy minima at different film

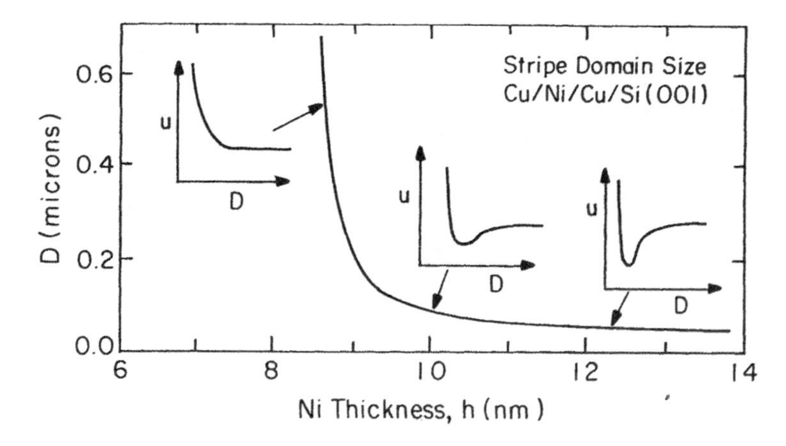

Figure 16.25 Ni-thickness dependence of domain size by numerical minimization of the energy in Eq. (16.16) using independently measured effective anisotropy; the energy versus D inserts are schematic (Bochi et al. 1995b).

thicknesses. These u/D curves explain why the domain structure is characterized by a well-defined length scale at large film thickness but not at small thicknesses. For large Ni thickness, the energy minima are deep and define a narrow range of D values, consistent with the observations of well-defined domain sizes in Figures. 16.22*c* and 16.22*d*. At smaller Ni thicknesses approaching the critical value, the minima are neither deep nor well defined: the energy difference between a wall spacing of 5 and 20 μm is less than $k_B T$. At small film thickness, a wide range of domain sizes results in almost the same total energy for the structure.

16.8 INHOMOGENEOUS MAGNETIZATION IN FILMS

It has been shown that when the effective uniaxial anisotropy can be described by a first-order, uniaxial, positive term, $K^{\text{eff}} > 0$, uniform, perpendicular magnetization is stable. For $K^{\text{eff}} < 0$, if the magnetization is assumed to be uniform and no higher-order anisotropy terms are present, the stable state is one of in-plane magnetization. Higher-order anisotropy terms (e.g., $\sin^4\theta$, Baberschke 1996) can stabilize uniform *canted* states of magnetization ($0 < \theta < \pi/2$). In addition, for uniaxial K^{eff} slightly less than zero, inhomogeneous states of magnetization having significant out-of-plane components, are possible. Finally, in films of thickness greater than the exchange length, the magnetization may be inhomogeneous through the film thickness. These cases are reviewed briefly.

16.8.1 Magnetization Variations Normal to the Film Surface

It is common for a magnetic film having a thickness much less than the domain size, to be dominated by magnetostatic energy and thus to be magnetized predominantly in plane. However, if the surface is also characterized by a strong positive surface anisotropy, the magnetization there may cant out of the film plane while remaining more-nearly parallel to the film plane in the interior, provided the film is thicker than an exchange length. The mathematical form of $\theta(z)$ for perpendicular surface anisotropy and in-plane bulk anisotropy can be described analytically for a semiinfinite medium (Mills 1989, O'Handley and Woods 1990). The energy density for a semiinfinite sample with a surface at $z = 0$ ($M \neq 0$ for $z > 0$), and with $\theta(z) = \pi/2$ at $z = +\infty$ measured from the surface normal (see Fig. 16.26*a*) is

$$f = \int_0^d \left[A\left(\frac{d\theta}{dz}\right)^2 + (K^v - 2\pi M_s^2)\sin^2\theta + \delta(z)K^s \sin^2\theta \right] dz \qquad (16.18)$$

Here, A is the exchange stiffness, K^v is a uniaxial, volume anisotropy (crystal-

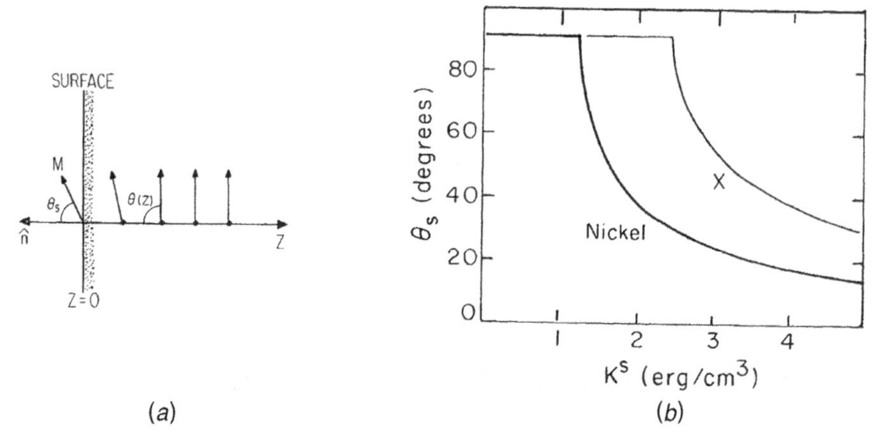

Figure 16.26 (*a*) Coordinate system for the perpendicular anisotropy problem with in-plane bulk anisotropy in a semiinfinite medium. The magnetization direction, $\theta(z)$, is defined relative to the film surface normal. (*b*) Plot of Eq. (16.22) showing that the magnetization remains in plane unless the surface anisotropy exceeds a critical anisotropy threshold, $K_c = 1.22$ and $2.5\,\text{erg/cm}^3$ for Ni and X, respectively.

line or magnetoelastic), and the Dirac delta function $\delta(z)$ confines the anisotropy K^s to the surface layer $z = 0$.

Applying the variational principle to Eq. (16.18), as was done in Chapter 8 for the 180° Bloch wall problem, gives two equations. The first is an Euler equation identical to Eq. (8.9) but with f_a now given by $K^v - 2\pi M_s^2$, and the second is a boundary condition coming from the integration of the Dirac delta function:

$$2A\frac{d\theta_s}{dz} - K^s \sin(2\theta_s) = 0 \tag{16.19}$$

Here θ_s is the value of $\theta(z)$ at the surface. The solution, $\theta(z)$, is similar to Eq. (8.14), but shifted on the θ scale to meet the boundary condition $\theta = \pi/2$ at $z = \infty$:

$$\theta(z) = 2\arctan\left[\sinh\left(\frac{a_0 + z}{\xi}\right)\right] \tag{16.20}$$

Here, $\xi = [A/(2\pi M_s^2 - K^v)]^{1/2}$ is the exchange length given in Eq. (12.11b). The energy density $2\pi M_s^2 - K^v$ provides the torque that drives the surface moments in plane as z increases from zero. Equations (16.19) and (16.20) combine to give for a_0:

$$\tanh(a_0) = \frac{\sqrt{(2\pi M_s^2 - K^v)A}}{K^s} \equiv \frac{K_c}{K^s} \tag{16.21}$$

The parameter a_0 shifts the domain-wall-like solution along the z axis so that the magnetization at the surface satisfies [from Eqs. (16.20) and (16.21)]

$$\csc(\theta_s) = \frac{K^s}{K_c} \equiv \kappa \qquad (16.22)$$

This equation is plotted in Figure 16.26*b*, where it is clear that the surface anisotropy must exceed a threshold K_c before the surface magnetization pops out of the film plane. K_c is the energy per unit surface area associated with the variation in spin orientation near the surface and is analogous to a domain wall energy density σ_{dw}, but here, K_c is the energy of a partial wall that is pinned at the surface. For surface anisotropy greater than this threshold, that is, for $\kappa > 1$, the form of $\theta(z)$ [Eq. (16.20)] is shown in Figure 16.27 for parameters characteristic of Ni ($M_s = 480\,\text{emu/cm}^3$, $\xi = 8\,\text{nm}$, $K_c = 1.22\,\text{erg/cm}^3$) and an arbitrary material, X ($M_s = 1000\,\text{emu/cm}^3$, $\xi = 4\,\text{nm}$, $K_c = 2.5\,\text{erg/cm}^3$). In both cases illustrated in Figure 16.27, $K^v = 0$ has been assumed. The magnetization at the surface takes on an orientation that satisfies Eq. (16.22) and $\theta(z)$ decays (with the form of a Bloch wall) toward $\pi/2$ with increasing z. The weaker magnetization of Ni allows the surface moments to lie closer to the surface normal, $\theta_s = 0$. The longer exchange length of Ni allows the surface perturbation to stretch deeper into the material. It should be emphasized that just beneath the surface, there is no surface anisotropy, only the volume anisotropy that favors in-plane magnetization. It is the exchange coupling that extends the influence of the surface anisotropy in the material over a range ξ.

The values of surface anisotropy in Figures 16.26*b* and 16.27 are large compared to the results of most measurements, $K^s \approx 0.1$–$0.6\,\text{erg/cm}^2$. The

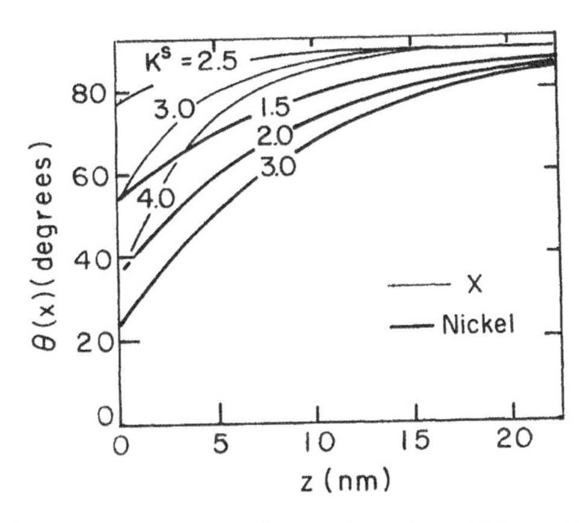

Figure 16.27 Variation of magnetization orientation $\theta(z)$ with distance from the surface in a semiinfinite medium having perpendicular anisotropy according to Eq. (16.20). The surface angle is defined by Eq. (16.22). Curves are shown for Ni parameters ($M_s = 480\,\text{emu/cm}^3$, $\xi = 8\,\text{nm}$, $K_c = 1.22\,\text{erg/cm}^3$) and for material X ($M_s = 1000\,\text{emu/cm}^3$, $\xi = 4\,\text{nm}$, $K_c = 2.5\,\text{erg/cm}^3$). $K^v = 0$ has been assumed in both cases.

calculations leading to these curves were made for $K^v = 0$. If there is a volume anisotropy favoring perpendicular magnetization, $K^v > 0$ (as is the case for Cu/Ni/Cu sandwiches because of the positive strain in the Ni layers), the exchange length is increased and K_c is decreased. In such cases, less surface anisotropy is required to give a perpendicular component of magnetization at the surface and the decay to in-plane magnetization is more gradual.

The field dependence of these solutions is illustrated numerically (O'Handley and Woods 1990), and analytic solutions are also available (Aharoni 1993).

16.8.2 Magnetization Configurations with Two Perpendicular-Anisotropy Surfaces

The preceding analysis is for a semiinfinite material for which $\theta(z = \infty) = \pi/2$. It is important to know the form of the solutions, $\theta(z)$, for a film where the magnetization in the interior may be pulled away from its preferred in-plane orientation by exchange coupling to both surfaces. There appears to be no simple analytic solution for $\theta(z)$ for this case. However, it is possible to derive a phase diagram for films with perpendicular surface anisotropy that indicates the nature of $\theta(z)$ in reduced surface anisotropy, $\kappa = K^s/K_c$ versus film-thickness space (Thiaville and Fert 1992, Hu and Kawazoe 1995, Thomas 1995, Bertram and Paul 1997). The phase diagram, shown in Figure 16.28 indicates three regimes of behavior: (1) at large film thickness and small reduced surface anisotropy, the magnetization is predicted to be uniformly in plane; (2) at small film thickness and κ greater than a thickness-dependent critical value, the

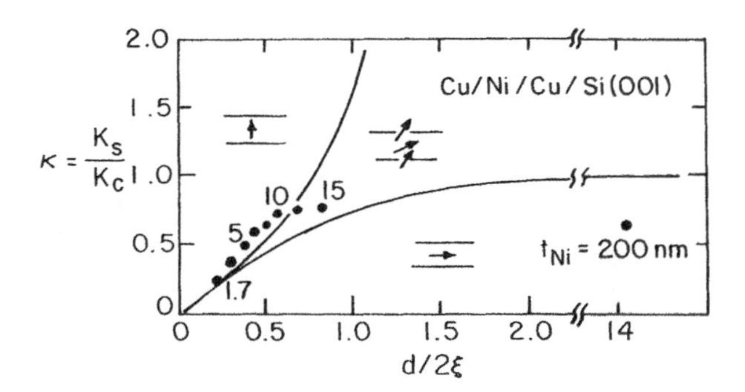

Figure 16.28 Phase diagram for magnetization structure in films of thickness d normalized to twice the exchange length ξ. The vertical axis, $\kappa = K^s/K_c$, is the surface anisotropy energy density K^s normalized by the energy per unit area due to the surface magnetization structure $(AK^v)^{1/2}$. Phase diagram shows three stable, zero-field magnetization configurations in a thin film with bulk anisotropy in plane (dominated by magnetostatic energy) and perpendicular anisotropy at each surface. (After H. Thomas, unpublished). The data points are from the effective anisotropy measurements of Ha et al. (1999) on epitaxial Cu/Ni/Cu (001) films.

magnetization is perpendicular throughout the film; (3) at intermediate thicknesses and surface anisotropies, the magnetization orientation is a function of depth in the film, with the magnetization more nearly in plane at the film center plane and more nearly perpendicular at its surfaces. The data shown in Figure 16.28 will be explained below. (The magnetization in the film with 200 nm of Ni turns out to be not uniformly in plane; see Section 16.8.3.)

A simple and insightful stability argument due to Thomas (1995) leads to the analytic forms of these phase boundaries. Assume a z-dependent energy density in the film similar to that of Eq. (16.18) (independent of coordinates in the film plane, a large domain approximation). This energy density is integrated over half the film thickness (from the origin at the center of the film to $+d/2$). Also, the delta function in Eq. (16.18) must have the argument $z - d/2$ to put the surface anisotropy at the film surface. The energy density function then takes the form

$$f = 2A \int_0^{d/2} \left[\left(\frac{d\theta}{dz} \right)^2 - \frac{1}{\xi^2} \sin^2\theta + \frac{\delta(z - d/2)}{D} \sin^2\theta \right] dz \qquad (16.23)$$

Here, a generalized exchange length ξ [Eq. (12.11b)] and surface anisotropy length D have been defined:

$$\xi^2 = \frac{A}{|2\pi M_s^2 - 2B_1 e|} \qquad \text{and} \qquad D = \frac{A}{K^s}$$

The volume anisotropy is perpendicular and of magnetoelastic origin. While analytic solutions to Eq. (16.23) do not appear to exist, it is still possible to test the stability of the out of plane state, $\theta_0 = 0$ everywhere, by considering perturbations of the form $\theta_p(z) = 0 + \Delta\theta_0 \cos(z/\xi)$, and the stability of the in-plane state, $\theta_0 = \pi/2$ everywhere, by considering perturbations of the form, $\theta_p(z) = \pi/2 - \Delta\theta_0 \cosh(z/\xi)$. Substitution of these forms into the energy function, Eq. (16.23), leads to conditions for the instability of the perturbation, $f(\theta_p) > 0$, or equivalently, the stability of the unperturbed state. The method is outlined in Figure 16.29.

These stability conditions are plotted as the phase boundaries in Figure 16.28. They indicate that the fully perpendicular magnetization state is stable only in the thin-film, strong-surface-anisotropy regime indicated, and fully in-plane magnetization is stable only in the thick-film, weak-surface-anisotropy regime indicated. In between these regions, states of inhomogeneous magnetization having different canting angles at different values of z across the film thickness are allowed.

Torque measurements on CuNiCu (001) epitaxial films (Ha and O'Handley 1999) give the experimental points in Figure 16.28. As the Ni layer thickness increases from 1.7 nm, the measured effective anisotropy indicates that the films should evolve from being magnetized fully perpendicular (1.7–10 nm), to

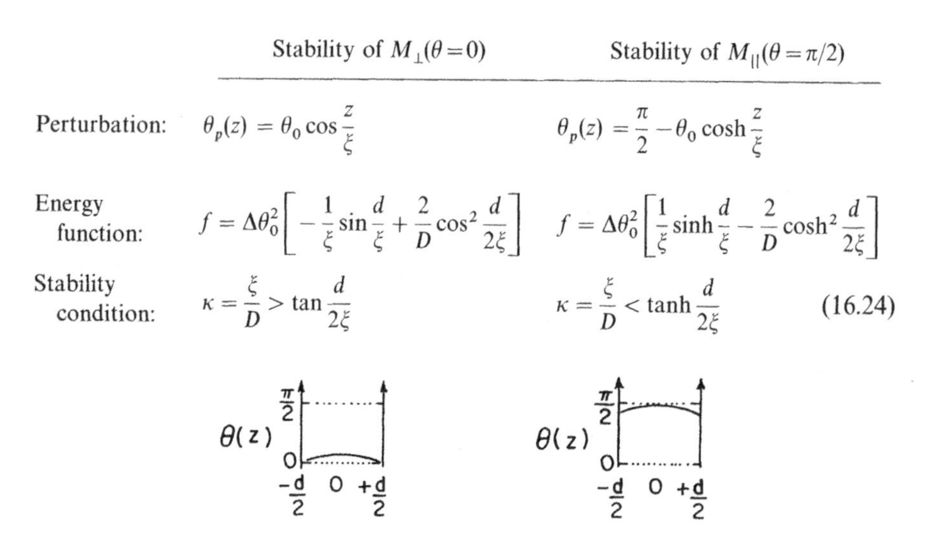

Figure 16.29 Outline of the method and results for determining thickness-dependent magnetization configuration in thin films in the presence of surface anisotropy and in-plane bulk anisotropy, but neglecting in-plane variations in magnetization. The form of the perturbation to the energy functional is sketched for the two limiting cases, and the solutions for the phase boundaries of these regions of stability are given.

inhomogeneously magnetized (11–15 nm), and finally to magnetized fully in plane. The data plotted on this phase diagram also suggest that for Ni thicknesses smaller than 1.7 nm, the magnetization should revert to an in-plane orientation, as has been observed by Gradmann (1986) and Ballentine (1989). Finally, this phase diagram predicts that the magnetization should be uniformly oriented in the film plane for Ni thicknesses above about 15 nm, as was suggested by bulk magnetometry measurements. However, it will be shown below that this model, which assumes uniform magnetization in the film plane, still does not explain all the observations on this system; a 200 nm Ni film shows a ripple domain pattern (see below).

16.8.3 Magnetization Variations in the Film Plane: Ripple Domains

It has been shown that epitaxial Cu/Ni/Cu (001) films exhibit a strong perpendicular component of magnetization for Ni layer thickness in the range 1.7–14 nm. Above this thickness range, magnetometry indicates that most of the remanent magnetization lies in the film plane ($K^{\text{eff}} < 0$). However, even up to 200 nm of Ni, a maze domain pattern with an appreciable perpendicular fringe field is observed. Figure 16.30*a* shows the domain pattern in epitaxial Cu/200 nm Ni/Cu (001) measured by magnetic force microscopy (Section 16.7). The width of the domains in this maze pattern is approximately 110 nm. Below the MFM image are in-plane and out-of-plane magnetization loops that clearly show this film to prefer in-plane magnetization (Hug et al. 1999). This domain pattern has been shown to be due to magnetization ripple (Muller 1961, Spain 1963, Saito et al. 1964). In such domains, $\boldsymbol{M}$ is largely in the film plane (because $|2\pi M_s^2| > K_\perp$), but a perpendicular component of the magnetization M_z exists which oscillates *periodically* in x or y. When the magnetization vector ripples in and out of the film plane, the secondary tendency for perpendicular magnetization can be accommodated (energy decreases like $-K_\perp \sin^2\theta_r$, where θ_r is the ripple angle measured from the film plane) without paying a price in magnetostatic energy as large as $2\pi M_s^2$, because of the small perpendicular component of magnetization and the alternating sign of the magnetostatic field on the charged surfaces (see Fig. 2.20). Ripple domains are distinct from the magnetization patterns observed in films that are magnetized either fully in-plane or perpendicular to the film plane inside the domains (see Section 16.7, Fig. 16.22).

Ripple domain states occur in certain ranges of film thickness and normalized perpendicular anisotropy as depicted in Figure 16.31 (Hubert and Schäfer

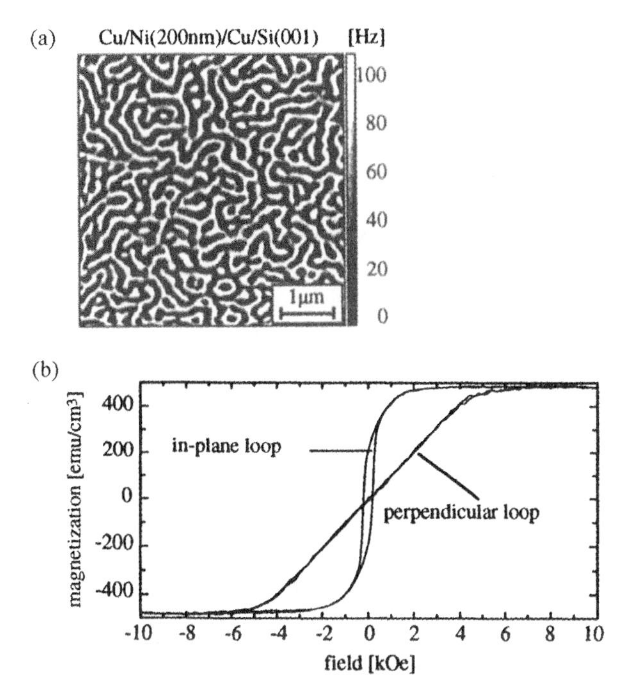

Figure 16.30 (*a*) Domain pattern as measured by MFM from the fringe field above the surface of an epitaxial Cu/200 nm Ni/Cu (001) film. (*b*) vibrating sample magnetometry M–H loop of the film in (*a*) showing that the quiescent state of magnetization is predominantly in the film plane despite the strong fringe field observed in (*a*). [After Hug et al. (1999).]

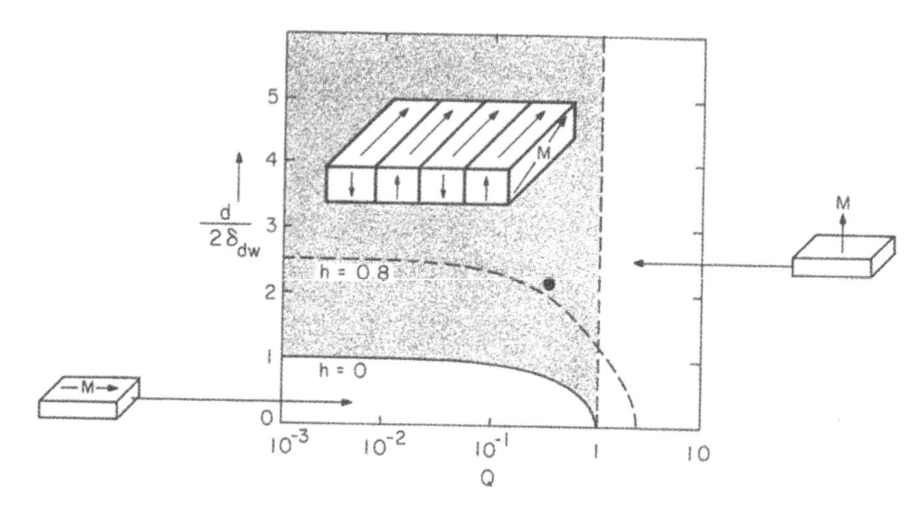

Figure 16.31 Phase diagram showing stable regimes of zero-field magnetization distributions in a thin film with perpendicular bulk anisotropy (exclusive of magnetostatic energy). The vertical axis is film thickness normalized to twice the Bloch wall width. The horizontal axis is $Q = K^{\mathrm{eff}}/(2\pi M_s^2) + 1$. Only the $h = H/H_a = 0$ and $h = 0.8$ critical lines are shown. [Adapted from Hubert and Schäfer (1998).]

1998). Here, the vertical axis is film thickness normalized to twice the domain wall thickness and the horizontal axis is $Q = K^{\mathrm{eff}}/(2\pi M_s^2) + 1$. When $Q > 1$ ($K^{\mathrm{eff}} > 0$), the uniform, perpendicular magnetization state is stable regardless of the value of $d/2\delta_{dw}$. When $Q < 1$ ($K^{\mathrm{eff}} < 0$), uniform in-plane magnetization is stable for normalized film thickness less than a critical value, $d/2\delta < d_{\mathrm{crit}}$. Above this critical thickness, the ripple domain state is stable. As Q approaches unity, there is an increasing thickness range over which a ripple state is stable. The phase boundary between the in-plane state and the ripple state is a function of applied, in-plane field. The field dependence of this boundary at $Q \approx 0$ is given by $d_{\mathrm{crit}} = 2\delta_{dw}(1 - h)^{-1/2}$, where $h = H/H_a = M_s H/2K_u$. The width w of the ripple domains has the field dependence at small Q given by $w = 2\delta_{dw}(1 + h)^{-1/2}$. In the ripple domain state, as in-plane field increases, the thickness range over which ripple domains are stable, shrinks, and the width of the domains increases.

The data point in Figure 16.31 represents the 200 nm thick Ni film shown in Figure 16.29 (Hug et al. 1999). For this film, the perpendicular anisotropy is due mostly to its tensile strain, measured to be +0.355% (Ha et al. 1999): $K_\sigma = 4.6 \times 10^5\ \mathrm{erg/cm^3}$. The magnetostatic energy is $1.5 \times 10^6\ \mathrm{erg/cm^3}$, so $Q = 0.31$. The domain wall width, calculated using the total anisotropy, is 47 nm, so the ratio $d/2\delta_{dw} = 2.13$. These coordinates place this film well within the range predicted for ripple domain formation in zero field. For these values, ripple theory predicts a zero-field domain width of 94 nm, which compares well with the observed value of 110 nm. Further, the phase diagram predicts that the ripple state of this 200-nm-thick Ni film should be destabilized by an in plane field of magnitude $H \approx 0.85H_a \approx 1.7$ kOe. The $M-H$ loop in Figure 16.30 shows the 200 nm Ni film to be saturated by an in-plane field of 2 kOe.

The ripple state is distinct from the inhomogeneous magnetization state shown in the phase diagram in Figure 16.28. Magnetization ripple is stabilized by a perpendicular *volume* anisotropy whereas the inhomogeneous state in Figure 16.28 is stabilized by perpendicular *surface* anisotropy.

16.9 SUMMARY

Electronic structure calculations have tended to focus on magnetic moment enhancement at surfaces because band structure calculations are reasonably good at determining the difference in spin-up and spin-down populations. They are not yet accurate enough to determine magnetic anisotropy with as much reliability. It is extremely difficult, experimentally, to measure a moment enhancement localized in a few atomic layers at a surface. The difficulty comes from the measurement accuracy itself, which depends on a knowledge of the weight or volume of the material giving the moment. It also comes from the difficulty of creating a clean and well-defined surface. Magnetic anisotropy is an easier parameter for experiments to determine. Magnetooptic Kerr loops, vibrating sample magnetometry, magnetic resonance, magnetic force microscopy (MFM), and a variety of polarized electron techniques are capable of revealing the preferred direction of magnetization in a thin film or at a surface. This is revealed in the shape of the $M-H$ (or θ_K-H or polarization vs. H) loops for different field orientations.

Despite these difficulties, it is becoming clear that magnetism at a surface or in an ultrathin film can be very different from bulk magnetism of the same material. This is a result of the altered coordination and symmetry at a surface, leading to Néel surface anisotropy, and in some cases to strong magnetoelastic interactions. These effects are most clearly manifest in the appearance of perpendicular magnetic anisotropy in many thin film systems. In Fe-based thin-film structures, the magnetostatic energy dominates except for thicknesses below about 1 nm, where a positive Néel surface anisotropy becomes important. In cobalt-based thin films, the bulk magnetocrystalline anisotropy and magnetoelastic anisotropy cannot usually be neglected and surface anisotropy may be important below about 2.5 nm. In Ni-based thin-film structures, bulk magnetocrystalline anisotropy is negligible but not bulk magnetoelastic anisotropy. A Néel surface anisotropy (apparently positive for Cu/Ni interfaces) and magnetoelastic anisotropy up to at least second order in strain, can dominate the magnetostatic energy for Ni thicknesses up to about 13 nm.

Magnetic domain structures in thin films show many features that distinguish them from domains in bulk materials. Scanning electron microscopy with spin polarization analysis and magnetic force microscopy are widely used techniques for imaging surface magnetic domains. Table 16.2

TABLE 16.2 Comparison of the Capabilities and Limitations of SEMPA and MFM

	SEMPA	MFM
Quantity measured	Spin polarization	Stray field
Atmosphere	Vacuum is essential	Vacuum not necessary but does enhance stability and sensitivity
Surface topography	Not an issue	Can be a problem without adequate tip feedback controls
Sample conductivity	Metals required[a]	Insulators as well as metals can be imaged
Vector components imaged	x, y, z	Mainly z; governed by tip orientation
Present resolution	40 nm	40 nm
Limitations on resolution	Stray fields, electron optics, spin detector efficiency, beam diameter and current	Tip size, tip height above sample, and instrumental sensitivity
Other limitations	Images cannot be collected in the presence of magnetic fields	Magnetic tip may interact with very soft magnetic materials, changing the domain pattern during imaging
Advantages	Topography and magnetism can be imaged independently Scanning at different magnifications is controlled by SEM electronics	Topography and magnetism can be separated; images can be taken in magnetic fields; vacuum not necessary

[a] Domains in insulators can be imaged if samples are coated with a thin magnetic, metallic film.

compares some of the capabilities, advantages, and disadvantages of SEMPA and MFM.

The magnetization configurations that are stable in thin films having volume and/or surface anisotropies that favor out-of plane magnetization can be predicted with reasonable accuracy in various approximations. For uniaxial anisotropy and homogeneous magnetization, perpendicular magnetization is favored for $K^{\text{eff}} > 0$ or $Q > 1$. When surface anisotropy is present, the surface moments are confined to the surface plane of a semiinfinite medium unless K^s exceeds a critical value, $K_c = A/\xi$ where ξ is the exchange length. For a thin film, as opposed to a semiinfinite medium, perpendicular magnetization states can be stable for K^s values less than K_c at small film thickness. In addition, states of magnetization inhomogeneous in z can exist. When the constraint of having magnetization uniform in the film plane is relaxed, ripple states can exist even for $Q < 1$. Epitaxial Cu/Ni/Cu sandwiches provide a useful model system in which these various magnetization configurations have been observed.

APPENDIX: STRESS IN THIN FILMS ON SUBSTRATES

Thin films are generally in a state of *biaxial* stress: $\sigma_x = \sigma_y$ and $\sigma_z = 0$. The combined effects of σ_x and σ_y can be understood by applying σ_x first: $e_x = \sigma_x/E$ and $e_y = -\nu\sigma_x/E$. Next, apply σ_y to the deformed y direction, getting $e_y = \sigma_y/E - \nu\sigma_x/E$. But because $\sigma_x = \sigma_y$, the strain for biaxial stress is given by

$$e_y = \sigma_x \frac{1-\nu}{E} = e_x \tag{16A.1}$$

When a film on a substrate is in a state of stress, the film–substrate couple will respond by bending so that the film stress on the interface is balanced by the substrate stress on the interface. This may be expressed as a force balance: the force per unit film width (in the y direction, Fig. 16.A.1) between the film and substrate is

$$\sigma_f h_f = -\sigma_s h_s \tag{16A.2}$$

or, from Eq. (16A.1)

$$e_f h_f \frac{E_f}{1-\nu_f} = -e_s h_s \frac{E_s}{1-\nu_s} \tag{16A.3}$$

Because $h_s \gg h_f$, most of the strain appears in the film. The substrate experiences both tension and compression with an average strain close to zero. (These assertions assume comparable stiffnesses for the film and substrate).

In order to determine how much the film–substrate couple bends under this force, it is necessary to consider the bending moments per unit film width, $M = Fz$, about the center of the substrate. For the film, the bending moment is given by

$$M_f = F_f z = \sigma_f h_f \frac{h_s}{2} \tag{16A.4}$$

Now consider the bending moment in the substrate where the stress is not a constant but varies with z (Fig. 16A.1, right). If the substrate curvature is assumed to be circular (and it is not, see below), similar triangles indicate that the substrate strain, $e_x = \delta x/x$, is related to the distance, z, from the center of

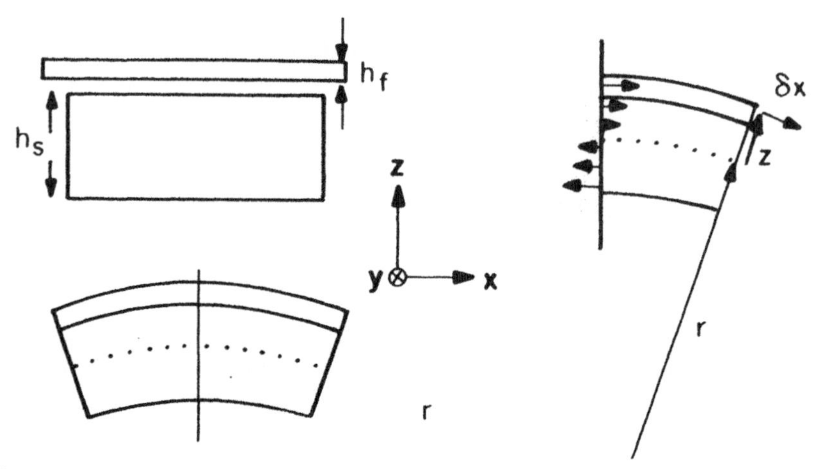

Figure 16.A.1 Upper left, film that would have larger dimensions were it not bonded to the substrate, is in compression when bonded. This exerts a bending moment on the substrate, lower left. The bending moments, which are zero on the symmetry plane, are shown at right for half of the x-symmetric, film–substrate couple. The strain, δx, increases with z, measured from the median plane of the substrate.

the substrate, by

$$e_s(z) = \frac{\delta x}{x} = \frac{z}{r}$$

or, from Eq. (16A.1),

$$e_s(z) = \frac{z}{r} = \sigma_s(z)\,\frac{1 - \nu_s}{E_s}$$

Thus, the stress and hence the bending moment is a function of z. The bending moment of the substrate is the z integral over the stress in the substrate times its moment arm, z:

$$M_s = \int_{-h_s/2}^{h_s/2} \sigma_s(z) z\,dz = \frac{E_s}{1 - \nu_s}\frac{h_s^3}{12r} \tag{16A.5}$$

The moment on the substrate, Eq. (16A.5), must be equated to the moment on the film, Eq. (16A.4), giving

$$\sigma_f = \frac{E_s}{1 - \nu_s}\frac{h_s^2}{6rh_f} \tag{16A.6}$$

This is known as the *Stoney equation.* It is applicable only near the center of a curved substrate where the curvature is circular. Equation (16A.6) shows that the stress in a film can be calculated from the radius of curvature, $K = 1/r$, knowing the elastic properties of the substrate and the film and the substrate thicknesses. The Stoney equation is used extensively in a variety of thin film measurements. The strain in the film can be determined from the radius of curvature and the substrate thickness or from σ_f and Eq. (16A.3).

A more exact relation between substrate bending and film stress (derived by Ha, unpublished), expresses film stress not in terms of a radius of curvature but rather in terms of the deflection y at a distance L from the center of symmetry:

$$\sigma_f = \frac{E_s}{1 - v_s} \frac{h_s^2}{3h_f L^2} y \tag{16A.7}$$

Comparison of this equation with the Stoney equation [Eq. (16A.6)] shows that the film stress is actually about 50% greater for a given substrate deflection than suggested by the Stoney equation.

It is important to note that even though stress causes the film to strain and bend the substrate, the bending strain in the film is far from sufficient to relieve fully the stress in the film. The reason for this is that it remains clamped to the much thicker substrate. If the film were not constrained by the substrate, it would strain much more. The effect of thin film strain on magnetic properties is covered in Chapter 7 text and Appendix 7B.

PROBLEMS

16.1 Show that the quantum mechanical polarization is given by $P = (N^+ - N^-)/(N^+ + N^-)$ by evaluating the expectation value of the Pauli spin matrix $\sigma_z = (1/0\ 0/-1)$ for multi-electron spin state that is a weighted sum of spin up $\alpha = (1/0)$ and spin down $\beta = (0/1)$ functions:

$$\chi = a_1|\alpha\rangle + a_2\langle\beta\rangle$$

16.2 Write the second- and fourth-order terms in α_i for the anisotropy energy expansion in tetragonal symmetry and show that they all reduce to those in Eq. (16.4).

16.3 **(a)** Verify that the function in Eq. (16.20) solves the Euler equation (8.9) and boundary condition in Eq. (16.19).

(b) Derive Eq. (16.22) from the two preceding equations.

BIBLIOGRAPHY

Freeman, A. J., C. L. Fu, S. Onishi, and M. Weinert, in *Polarized Electrons in Surface Physics*, R. Feder, ed., World Scientific, Singapore, 1985.

Gradman, U., in *Handbook of Magnetic Materials*, Vol. 7, Elsevier Science, Amsterdam, 1993, p. 1.

Hopster, H., in *Ultrathin Magnetic Structures*, Vol. 1, J. A. C. Bland and B. Heinrich, eds., Springer-Verlag, Berlin, 1994, p. 123.

Johnson, M. T., P. J. H. Bloemen, F. J. A. den Broeder, and J. J. de Vries, *Rep. Prog. Phys.* **59**, 1409 (1996).

Landolt, M., in *Polarized Electrons in Surface Physics*, R. Feder, ed., World Scientific, Singapore, p. 385.

REFERENCES

Aharoni, A., *Phys. Rev. B* **47**, 8296 (1993).

Baberschke, K., *Appl. Phys. A* **62**, 417 (1996).

Bader, S. M., and J. L. Erskine, in *Ultrathin Magnetic Structures*, Vol. 2, Springer-Verlag, Berlin, 1994, p. 297.

Ballentine, C. A. Ph.D. thesis, Univ. Texas at Austin, 1989.

Ballentine, C. A., R. L. Fink, J. Araya-Pochet, and J. L. Erskine, *Appl. Phys. A.* **49**, 459 (1989).

Banninger, U., G. Busch, M. Campagna, and H. C. Siegmann, *Phys. Rev. Lett.* **25**, 585 (1970).

Bertram, H. N., and D. I. Paul, *J. Appl. Phys.* **82**, 2439 (1997).

Binnig, G., H. Rohrer, Ch. Gerber, and E. Weibel, *Phys. Rev. Lett.* **49**, 57 (1982).

Binnig, G., C. F. Quate, and Ch. Gerber, *Phys. Rev. Lett.* **56**, 930 (1986).

Binnig, G., Ch. Gerber, E. Stoll, R. T. Albrecht, and C. F. Quate, *Eur. Phys. Lett.* **3**, 1281 (1987).

Bland, J. A. C., and B. Heinrich, eds., *Ultrathin Magnetic Structures*, 2 vols., Springer-Verlag, Berlin, 1994.

Blügel, S., *Phys. Rev. Lett.* **68**, 851 (1994).

Bochi, G., C. A. Ballentine, H. E. Inglefield, C. V. Thompson, R. C. O'Handley, H.-J. Hug, B. Steifel, A. Moser, and H.-J. Güntherodt, *Phys. Rev. B* **52**, 7311 (1995a).

Bochi, G., H. Hug, B. Stiefel, A. Moser, A. Paroshikov, H.-J. Güntherodt, D. I. Paul, and R. C. O'Handley, *Phys. Rev. Lett.* **25**, 1839 (1995b).

Bochi, G., C. A. Ballentine, H. E. Inglefield, C. V. Thompson, and R. C. O'Handley, *Phys. Rev. B* **53**, R1729 (1996).

Busch, G., M. Campagna, P. Cotti, and H. C. Siegmann, *Phys. Rev. Lett.* **22**, 597 (1969).

Busch, G., M. Campagna, and H. C. Siegmann, *Phys. Rev. B* **4**, 746 (1971).

Celotta, R. J., and D. T. Pierce, in *Microbeam Analysis*, K. F. J. Heinrich, ed., San Francisco Press, San Francisco, CA, 1982, p. 469.

Celotta, R. J., and D. T. Pierce, *Science* **234**, 249 (1986).

Chappert, C., and P. Bruno, *J. Appl. Phys.* **64**, 5736 (1988).

Chuang, D. S., C. A. Ballentine, and R. C. O'Handley, *Phys. Rev. B* **49**, 15084 (1994).

Davis, H. L. et al., *Phys. Rev. Lett.* **68**, 2632 (1992).

den Broeder, F. J. A. et al., *Phys. Rev. Lett.* **60**, 2769 (1988).

Donath, M., D. Scholl, H. C. Seigmann, and E. Kay, *Phys. Rev. B* **43**, 3164 (1991).

Eib, W., and S. F. Alvarado, *Phys. Rev. Lett.* **37**, 444 (1976).

Enders, E., D. Sander, and J. Kirschner, *J. Appl. Phys.* **85**, 5279 (1999).

Engle, B. N., C. D. England, R. A. van Leeuwen, M. H. Wiedmann, and C. M. Falco, *Phys. Rev. Lett.* **67**, 1910 (1991).

Fitzgerald, E. A., *Mater. Sci. Rept.* **7**, 87 (1991).

Freeman, A. J. et al., *J. Magn. Magn. Mater.* **100**, 497 (1991).

Freeman, A. J., and R. Wu, *J. Magn. Magn. Mater.* **100**, 497 (1991).

Gay, J. G., and R. Richter, *Phys. Rev. Lett.* **56**, 2728 (1986).

Gay, J. G., and R. Richter, in *Ultrathin Magnetic Structures*, Vol. 1, J. A. C. Bland and B. Heinrich, eds., Springer-Verlag, Berlin, 1994, p. 21.

Gradmann, U., *Ann. Physik* **7**, 91 (1966).

Gradmann, U., *J. Magn. Magn. Mater.* **54–57**, 733 (1986).

Gurney, B. et al., INTERMAG abstract, unpublished (1997).

Ha, K., M. A. Ciria, R. C. O'Handley, P. W. Stephens, and S. Pagola, *Phys. Rev. B*, in press (1999).

Ha, K. and R. C. O'Handley, *J. Appl. Phys.* **85**, 5282 (1999).

Ha, K. and R. C. O'Handley, *J. Appl. Phys.* (2000).

Hembree, G. G., J. Unguris, R. J. Celotta, and D. T. Pierce, *Scan. Microsc. Suppl.* **1**, 229 (1987).

Hirth, J. P., and J. Lothe, *Theory of Dislocations*, 2nd ed., Krieger, 1992.

Hu, X., and Y. Kawazoe, *Phys. Rev. B* **51**, 311 (1995).

Huang, F., M. T. Kief, G. J. Mankey, and R. F. Willis, *Phys. Rev. B* **49**, 3962 (1994).

Hubert, A., and R. Schäfer, *Magnetic Domains*, Springer, Berlin, 1998, p. 297.

Hug, H. J., G. Bochi, D. I. Paul, B. Steifel, A. Moser, I. Parashikov, A. Klicznik, D. Lipp, H.-J. Güntherodt, and R. C. O'Handley, *J. Appl. Phys.* **79**, 5609 (1996).

Hug, H. J., B. Stiefel, P. J. A. van Schendel, A. Moser, R. Hofer, S. Martin, H.-J. Güntherodt, S. Porthun, L. Abelman, J. C. Lodder, G. Bochi, and R. C. O'Handley, *J. Appl. Phys.* **83**, 5609 (1998).

Hug, H. J., P. J. A. van Schendel, B. Steifel, S. Martin, H.-J. Güntherodt, K. Ha, M. A. Ciria, and R. C. O'Handley, unpublished (1999).

Jungblut, R., M. T. Johnson, J. van de Stegge, A. Reinders, and F. J. A. den Broeder, *J. Appl. Phys.* **75**, 6424 (1994).

Kirschner, J., in *Polarized Electrons in Surface Physics*, R. Feder, ed., World Scientific, Singapore, 1985, p. 245.

Kisker, E., in *Polarized Electrons in Surface Physics*, R. Feder, ed., World Scientific, Singapore, 1985.

Koch, R., *J. Phys: Condens. Matter* **6**, 9519 (1994).

Koch, R., M. Weber, K. Thurmer, and K. H. Reider, *J. Magn. Magn. Mater.* **159**, L11 (1996).

Koike, K., and K. Hayakawa, *Jpn. J. Appl. Phys.* **23**, L187 (1984); *Appl. Phys. Lett.* **45**, 585 (1984).

Landolt, M., and M. Campagna, *Phys. Rev. Lett.* **38**, 663 (1977).

Landolt, M., and Y. Yafet, *Phys. Rev. Lett.* **40**, 1401 (1978).

Landolt, M., and D. Mauri, *Phys. Rev. Lett.* **49**, 1783 (1982).

Lee, C. H., H. He, F. Lamelas, W. Vavra, C. Uher, and R. Clark, *Phys. Rev. Lett.* **62**, 653 (1989).

Lee, C. H., R. Farrow, C. J. Lin, E. E. Marinero, and C. J. Chien, *Phys. Rev. B* **42**, 11384 (1990).

Malek, Z., and V. Kambersky, *Czech. J. Phys.* **8**, 416 (1958).

Martin, Y., and H. K. Wickramasinghe, *Appl. Phys. Lett.* **50**, 1455 (1987).

Matthews, J. W., and A. E. Blakeslee, *J. Cryst. Growth* **27**, 118 (1974); **29**, 273 (1975).

Mauri, D., R. Allenspach, and M. Landolt, *J. Appl. Phys.* **58**, 906 (1985).

Mills, D. L. *Phys. Rev. B* **39**, 12306 (1989).

Moral, A. del, M. A. Ciria, J. I. Arnaudas, and C. de la Fuente, *Phys. Rev. B* **57**, R9471 (1998).

Muller, M. W., *Phys. Rev.* **122**, 1485 (1961).

Néel, L., *J. Phys. Rad.* **15**, 225 (1954).

Oepen, H. P., and J. Kirschner, *Phys. Rev. Lett.* **62**, 819 (1989).

O'Handley, R. C., and J. P. Woods, *Phys. Rev. B* **42**, 6568 (1990).

O'Handley, R. C., and S. W. Sun, *J. Magn. Magn. Mater.* **104–107**, 1717 (1992).

Onishi, S., A. J. Freeman, and M. Weinert, *Phys. Rev. B* **28**, 6741 (1983).

Onishi, S., C. L. Fu, and A. J. Freeman, *J. Magn. Magn. Mater.* **50**, 161 (1985).

Pierce, D. T., R. J. Celotta, J. Unguris, and H. C. Seigmann, *Phys. Rev. B* **26**, 2566 (1982).

Prinz, G., *Phys. Rev. Lett.* **54**, 1051 (1985).

Prinz, G., *Science* **250**, 1092 (1990).

Prinz, G., *J. Magn. Magn. Mater.* **100**, 469 (1991a).

Prinz, G. A., in *Magnetism in the '90's*, A. J. Freeman, ed., Elsevier, Amsterdam, 1991b.

Prinz, G. A., in *Ultrathin Magnetic Structures*, Vol. 2, J. A. C. Bland and B. Heinrich, eds., Springer-Verlag, Berlin, 1994, p. 1.

Rau, C., C. Schneider, G. Zing, and K. Jamison, *Phys. Rev. Lett.* **57**, 3221 (1986).

Saenz, J. J., N. Garcia, P. Grutter, E. Nayer, H. Hinzelmann, R. Wiesendanger, L. Rosenthaler, H. R. Hidber, and H.-J. Guentherodt, *J. Appl. Phys.* **62**, 4293 (1987).

Saito, N., H. Fujiwara, and Y. Sugita, *J. Phys. Soc. Jpn.* **19**, 421 (1964).

Scheinfien, M. R., J. Unguris, M. H. Kelley, D. T. Pierce, and R. J. Celotta, *Rev. Sci. Instrum.* **61**, 2501 (1990).

Scheinfien, M. R., J. Unguris, R. J. Celotta, and D. T. Pierce, *Phys. Rev. Lett.* **63**, 668 (1989); *Phys. Rev. B* **43**, 3395 (1991).

Schulz, B., and C. Baberschke, *Phys. Rev. B* **50**, 13467 (1994).

Siegmann, H. C., in *Core Level Spectroscopies for Magnetic Phenomena*, P. S. Bagus, G. Pacchione, and F. Parmigiani, eds., Plenum Press, New York, 1994.

Song, O., and R. C. O'Handley, *Appl. Phys. Lett.* **64**, 2593 (1994).

Spain, R. J., *Appl. Phys. Lett.* **3**, 208 (1963).

Stampanoni, M., A. Vaterlaus, M. Aeschlimann, and F. Meier, *Phys. Rev. Lett.* **59**, 2483 (1987).

Sun, S. W., and R. C. O'Handley, *Phys. Rev. Lett.* **66**, 2798 (1991).

Tersoff, J., and Falicov, L., *Phys. Rev. B* **26**, 6186 (1982).

Thiaville, A., and A. Fert, *J. Magn. and Magn. Mater.* **113**, 161 (1972).

Thomas, H., unpublished (1995).

Tsao, J. Y., *Materials Fundamentals of Molecular Beam Epitaxy*, Academic Press, Boston, 1993.

Unguris, J., D. T. Pierce, A. Galejs, and R. J. Celotta, *Phys. Rev. Lett.* **49**, 72 (1982).

Unguris, J., G. G. Hembree, R. J. Celotta, and D. T. Pierce, *J. Microsc.* **139**, RP1 (1985).

Unguris, J., D. T. Pierce, and R. J. Celotta, *Rev. Sci. Instr.* **57**, 1314 (1986).

Unguris, J., R. J. Celotta, and D. T. Pierce, *Phys. Rev. Lett.* **67**, 140 (1991).

Wimmer, E. et al., *Phys. Rev. B* **30**, 3113 (1984).

CHAPTER 17

MAGNETIC RECORDING

17.1 INTRODUCTION

Magnetic recording products include magnetic storage media for tape and hard disks, read and write heads, consumer audio and video equipment, floppy disks, and credit cards. Although information storage can be accomplished by any of several competing technologies (e.g., thermoplastic or phase change memories, magnetic bubbles, semiconductor memories, Josephson memories, and magnetooptic recording), none has been able to match the combination of information areal density and access time of hard disk magnetic storage.

The density at which information can be stored in high-end magnetic disk files has doubled every 2–3 years since the early 1960s (a 30% annual growth rate). In 1991, IBM demonstrated a recording system with a storage density of 1 Gb/in^2 (gigabit per square inch): 158 kfci (thousands of flux changes per inch) and 7470 tpi (tracks per inch) corresponding to a 3.4-μm track pitch. This system makes use of advanced thin-film media and new read heads based on anisotropic magnetoresistance (MR heads). With the introduction of these new MR heads and of higher-density thin-film media, the rate of growth in information density has doubled to 60% per year (see Fig. 17.1). In 1997, Tsang et al. demonstrated a 5 Gb/in^2 hard disk system employing spin-valve read head (see below) based on giant magnetoresistance (Chapter 15) and Co-Cr-Pt-Ta thin-film media. The bit size was 1400 Å long (7 bits/μm) on a 0.7-μm track width. The demonstration achieved a data rate of 10^7 bits per second. As of this writing, densities of 22 Gb/in^2 are available.

However, past performance is no guarantee of future results. It will be shown that increases in recording density are achieved only through a reduction of *all* the critical dimensions involved in the recording process as well

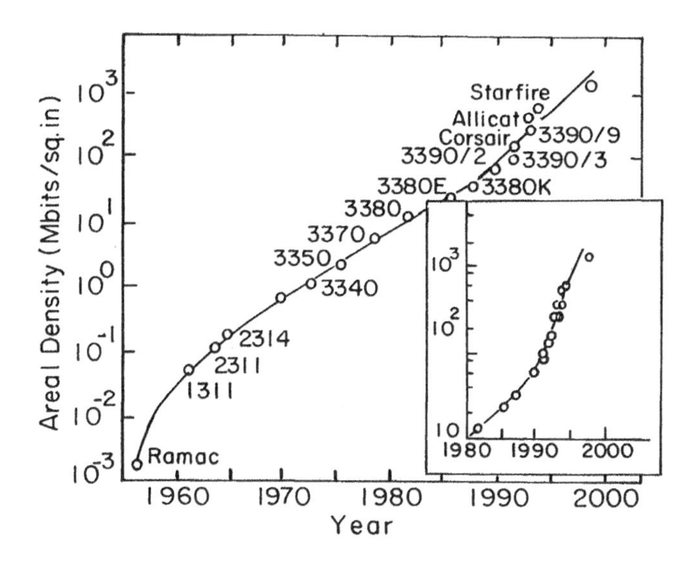

Figure 17.1 Areal density history of the hard disk drive. Insert: more detail of the upturn in areal density growth in 1991. The present compound growth rate is about 60%. [After Grachowski and Thompson (1994).]

as improvements in signal processing, head-media tribology, and track servo-control. The ability to control the fidelity and stability of the magnetic recording process at still smaller dimensions is limited by our ability to control the processing and properties of the materials used in magnetic recording heads and media. Ultimately, information storage density will be limited by physical factors such as superparamagnetism (zero coercivity implies loss of memory) in small storage elements and by magnetic resonance limitations at high read and write rates.

This chapter aims to connect the basic principles of magnetic materials with the needs of magnetic recording. It has six parts:

1. An overview of the principles of magnetic recording (Section 17.2)
2. A description of particulate recording media (Section 17.3)
3. A description of thin-film recording media (Section 17.4)
4. A treatment of magnetic recording heads and materials (Section 17.5)
5. An overview of magnetic random access memories (MRAMs) (Section 17.6)
6. A look forward at developing trends and fundamental limitations (Section 17.7)

The interested reader is referred to several in-depth texts, reviews, and articles found in the list of references at the end of the chapter. Magnetooptic recording and materials are not covered here.

17.2 MAGNETIC RECORDING OVERVIEW

The original Ampex wire recording system made use of an iron wire, drawn to have a strong fiber texture. This texture produced a uniaxial magnetic anisotropy and a coercivity of a few hundred oersteds. The information was stored in a magnetization pattern along the wire length whose amplitude and frequency replicated the recorded sound. This is *analog recording.* The magnetization pattern was written with a small ring head having a gap that provided a fringe field to the moving wire. When the written wire passed over the passive head gap, a voltage was induced in the windings that could be read acoustically by a speaker coil to produce the recorded sound.

The same principles used then apply to analog audio magnetic recording today. However, far more information is stored by *digital recording* on magnetic tapes, floppy disks, and hard disks. The digital recording process is outlined in Figure 17.2. The magnetic recording medium (tape or disk) moves relative to an electromagnetic transducer, which is essentially a magnetic circuit with a gap. When a current passes through windings about the head, the head is magnetized and a fringe field appears in the gap. In the write process, the fringe field in the gap magnetizes the medium alternately in one direction or the other as the drive current changes polarity. Because the head and the medium move relative to each other, information can be described in the head reference frame in terms of the variable ωt (e.g., $e^{-i\omega t}$) or in the medium reference frame by the variable kx where $k = 2\pi/\lambda$. Thus, the full variable is $kx - \omega t$ and the head and medium have a relative velocity given by $v = \omega/k$. In analog recording, the spatial *waveform* written on the recording medium

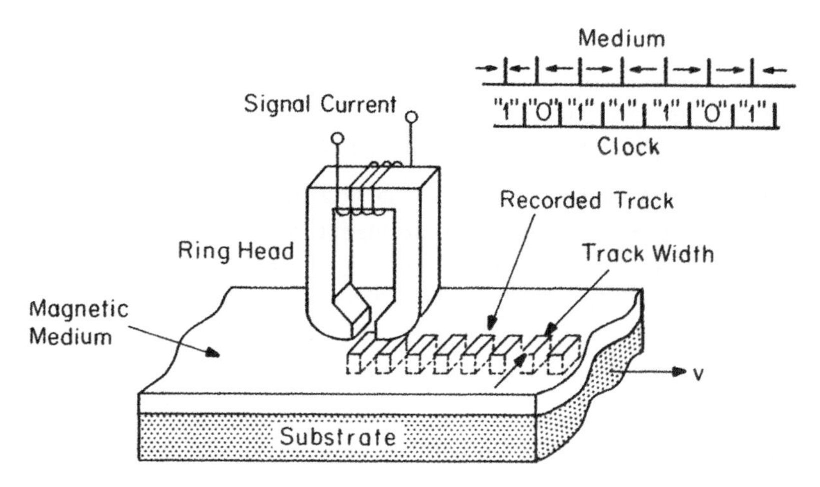

Figure 17.2 Schematic representation of longitudinal, digital magnetic recording write processes. Insert, upper right, sequence of transitions constitute the bits that are read as binary information.

replicates the temporal waveform put into the medium by the write current. In digital recording, the spatial *sequence* of the magnetized bits replicates the temporal sequence of current pulses. The sequence of binary states has information significance (Fig. 17.2, inset). A clock sets the system frequency, indicating when or where a transition might occur. The presence or absence of a transition at expected intervals (called *bits*) is read as a "one" or a "zero" to represent binary coded information.

When the recorded medium is moved across the gap of a passive magnetic circuit, the fringe field above the written bits induces a flux change in the head. That flux change corresponds to the spatial magnetization pattern on the medium. The flux change in the head induces a proportional voltage in the pickup windings of the read head. This voltage is amplified and read with an electronic signal processor to make use of the information (sounds or data) that has been recorded.

If one direction of magnetization is written in a longitudinal medium immediately adjacent to an oppositely directed domain, the two bits see each others' magnetostatic fields. If the coercivity of the medium is small, the bits may demagnetize each other, and the information is lost. For stable information storage, even in the absence of external fields, high-coercivity recording media are required. It is not enough that the domain walls separating bits be stable against demagnetization. For high-frequency analog recording or for high-density digital recording, the domain walls should appear as sharp transitions to a read head that spans the track width. The domain walls should not assume a sawtooth pattern between head-to-head domains (see Problem 17.5).

The write head should produce a large flux density in the gap when it is activated and none when the write current is zero. That is, $\mu \gg 1$, M_s large, and $B_r = 0$. The read head should have a very large permeability at low fields and no coercivity: $\mu_i \gg 1$ and $H_c = 0$. Figure 17.3 contrasts the ideal, square loop of a digital recording medium with that of a recording head where the presence of a gap shears the loop over.

17.2.1 The Write Head

The current through the N turns enclosing the core of the write head provides the magnetic potential or magnetomotive force, $V_m = NI$ (Appendix, Chapter 2), which generates the field in the gap. The head efficiency η is the fraction of V_m that appears as field in the gap:

$$\eta = \frac{2gH_g}{2gH_g + \ell_c H_c}$$

where the gap length is $2g$ and the flux path length in the core is l_c. From the conservation of magnetic current, or flux, about the circuit, the efficiency can be expressed in terms of the reluctance and the head parameters (core

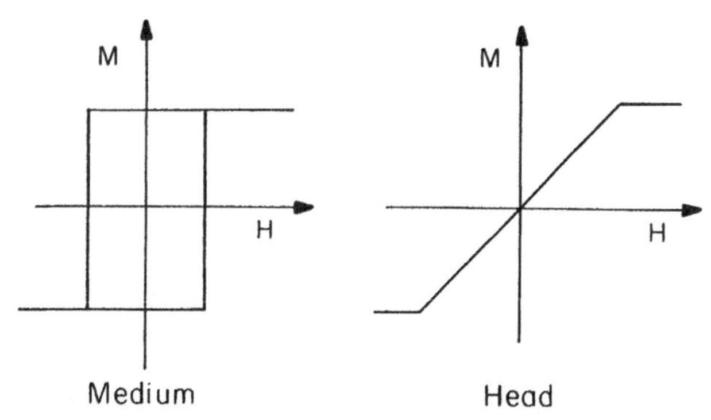

Figure 17.3 Schematic M–H loops for ideal magnetic recording medium and head material.

permeability μ and cross-sectional areas of the core and gap A_c and A_g):

$$\eta = \frac{R_g}{R_g + R_c} = \frac{1}{1 + (R_c/R_g)} \approx 1 - \frac{\ell_c A_g}{2g\mu A_c} \tag{17.1}$$

The efficiency drops below 100% by the amount of the ratio of the core reluctance to the gap reluctance. If the gap length is too small and the cross-sectional area there too large, the head field remains in the gap rather than fringing out toward the medium.

The field about the gap of a recording head has the approximate form

$$H_x(x, y) = \frac{H_g}{\pi}\left[\arctan\left(\frac{x+g}{y}\right) - \arctan\left(\frac{x-g}{y}\right)\right]$$

$$H_y(x, y) = -\frac{H_g}{2\pi}\log\frac{(x+g)^2 + y^2}{(x-g)^2 + y^2} \tag{17.2}$$

where H_g is the field in the gap at $x = y = 0$. These equations for the head field were derived by Karlqvist and are known by his name. He assumed that the field in the gap at $y = 0$ is a constant equal to $(4\pi)M$, the magnetization of the pole pieces. Actually, this is true only well inside the gap, $y < -2g$, as in Figure 17.4. The true field is weaker than these Karlqvist solutions. Note that these fields drop off sharply with distance y from the head and that the field strengths at a given height y, also decrease with decreasing gap length $2g$ (see Problem 17.1).

The Mathematica program to plot the Karlkvist solutions at a distance $y = 1$ from the gap, shown in Figure 17.4, is given next.

```
hx=(ArcTan[(x+1)/y]-ArcTan[(x-1)/y])/Pi
hy=-(Log[((x+1)^2+y^2)/((x-1)^2+y^2)])/(2 Pi)
y=1
```

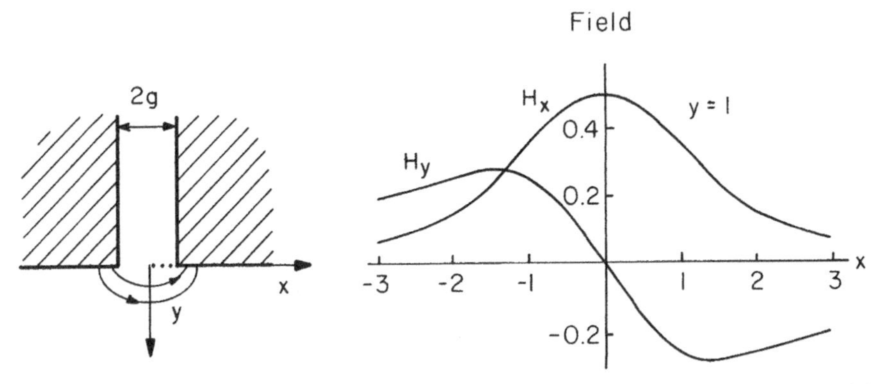

Figure 17.4 Schematic cross section of the gap in a magnetic recording head at left showing coordinate system for calculation of gap fields. The recording medium moves in the x direction at a distance y from the face of the head; x and y are normalized to g. At right, the forms of the Karlkvist field solutions normalized to H_g are shown for $y = 1$.

```
Plot[{hx,hy}, {x, −3,3}, AxesLabel→{"x", "Field"},
  PlotRange→{−0.3, 0.6}]
```

To compare the decay in H_x with increasing y with an exponential decay and with $0.6/y$ we can follow the first two Mathematica expressions above with the following statements:

```
x=0
Plot[{hx, Exp[−y/2], 0.6/y}, {y, 0, 5},AxesLabel→{"y", "Field"},
PlotRange→{0, 1}]
```

The full x and y dependence of H_x and H_y can be generated by the following simple Mathematica plot commands after defining the fields as above:

```
Plot3D[hx, {x, −3, 3}, {y, 0, 3}, PlotRange→(0, 1.2},
  AxesLabel→{"x", "y", "Hx"}]
Plot3D[hy, {x, −3, 3}, {y, 0, 3}, PlotRange→{−.8, .8},
  AxesLabel→{"x", "y", "Hy"}]
```

The x component of the gap field drops off exponentially at first and then goes as $1/y$. The loss of signal at the medium from the write head field [Eq. (17.2), Fig. 17.5] is slightly less than exponential; it represents a 45 dB loss at $y = 1$. (See also Fig. 17.6.)

17.2.2 The Recording Medium

Magnetic recording media are ideally composed of a regular array of isolated single-domain magnetic elements. These elements should be bistable; that is, they should be capable of being magnetized using a reasonable field strength.

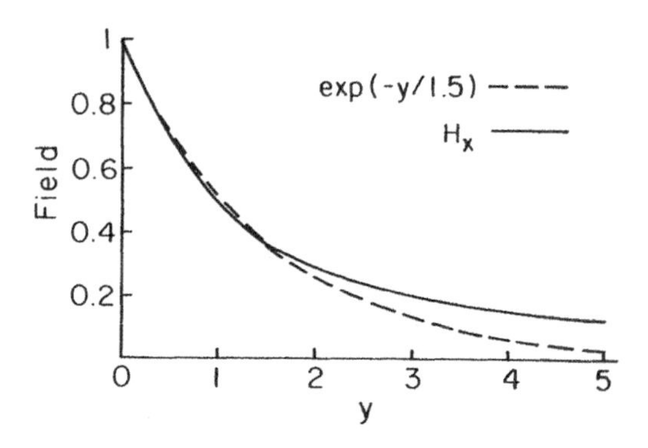

Figure 17.5 Karlkvist field H_x at $x = 0$ normalized to H_g as a function of distance y from the gap, compared with exponential function. For $y > 1.8$, the field drops off more like $0.6/y$, which plots almost directly over H_x. Coordinates x and y are normalized to g.

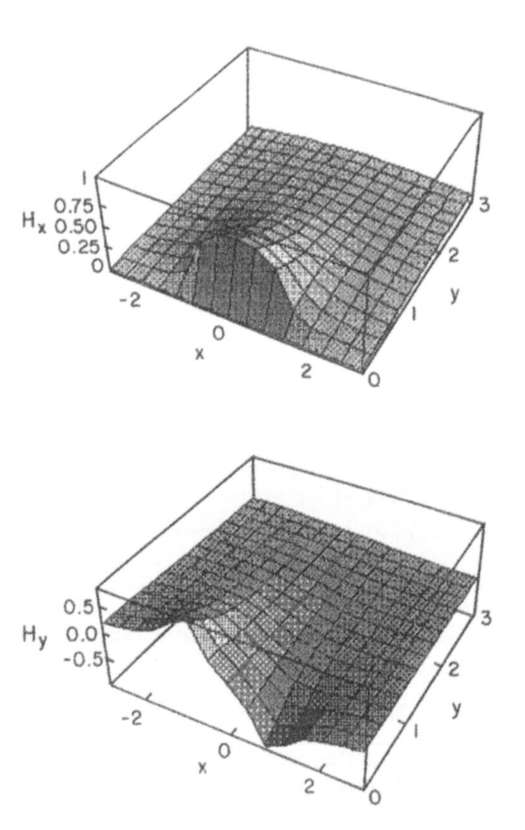

Figure 17.6 Full x and y dependence of the Karlkvist solutions H_x and H_y normalized to the gap field H_g. View is from the head gap centered at $y = 0$, $x = 0$ looking toward the medium in $y > 0$.

When the field is removed, the elements should have a large remanent magnetization. The simplest embodiment of these criteria is the ferrite ring-core memory (Section 17.6). But uniformly sized pieces of magnetic material, large enough to assemble, were not able to provide sufficient bit density. For higher information storage densities, smaller-sized pieces of magnetic material were less regular and harder to arrange periodically. A continuous, two-dimensional magnetic medium that met these criteria had to be developed. The path of development of continuous magnetic media began with slurries or paints filled with magnetic particles. Such a tape-cast recording medium is essentially an array of microscopic, independent magnetic dipoles. A single bit consisted of hundreds or thousands of particles with a net magnetization in one direction. Magnetic independence of the particles is required so that orienting a given dipole or cluster of dipoles will not affect the orientation of adjacent regions. The small size of the particles and their magnetic isolation from each other allows that, on the scale of the head dimensions, the transition between domains appears sharp (see Fig. 17.2).

At higher recording densities, the number of particles per bit decreases unless the particle size decreases correspondingly. When there are fewer particles per bit, the transition between domains becomes less sharp and pickup signal decreases. Why not make each recorded region a single-domain particle or grain? This is not possible by simple film deposition alone. The difficulty comes from the strict periodicity requirement: the bits must pass the head at a regular clock speed so that, at a constant relative velocity between the head and recording medium, they must be arranged with strict spatial periodicity. It is now possible to create nanoscale patterned media by high-resolution lithography (Section 17.7).

The magnetostatic field due to a *sharp*, longitudinal, head-to-head transition can be derived from magnetic potential theory (White 1985, Bertram 1994). Consider a recording medium of thickness δ in the y direction and a track width of w in the z direction. The field along the track, h_x for $w \gg \delta$, is identical to what is derived from Eq. (2.3):

$$H_x(x, y) = \frac{(4\pi)M_r}{\pi}\left[\arctan\left(\frac{y + \delta/2}{x}\right) - \arctan\left(\frac{y - \delta/2}{x}\right)\right] \quad (17.3)$$

Along the midline of the recording medium, specifically , at a height $y = 0$, Eq. (17.3) becomes

$$H_x(x, 0) = \frac{(4\pi)2M_r}{\pi}\arctan\left(\frac{\delta}{2x}\right) \approx \frac{(4\pi)M_r\delta}{\pi x}$$

The last approximation applies close to the transition at $x = 0$.

Thus, the strength of the field produced by the recorded bits is proportional to the product of the bit remanence and the thickness of the recorded information, $M_r\delta$. The fringe field of a single magnetic transition is plotted in Figure 17.7 for $y = 0$ and 0.5 (fine lines).

In practice, the magnetostatic energy at the transition is minimized by a smearing of the transition over a length a along the track direction, x. The form of the magnetization through the transition is generally assumed to be that of an arctangent function (Fig. 17.7, inset):

$$M_x(x) = \frac{2M_r}{\pi} \arctan\left(\frac{x}{a}\right)$$

This change in transition shape decreases the fringe field, Eq. (17.3), which now takes the form (White 1985, Bertram 1994):

$$H_x(x, y) = \frac{(4\pi)M_r}{\pi}\left[\arctan\left(\frac{y + a + \delta/2}{x}\right) - \arctan\left(\frac{y - a - \delta/2}{x}\right) - 2\arctan\left(\frac{a}{x}\right)\right] \qquad (17.4)$$

The form of the fringe field for $a = 0.5$ and $\delta = 1$ is shown in Figure 17.7 at $y = 0$ (center of the medium thickness) and $y = 0.5$ (the top surface of the recording medium).

This fringe field of a sinusoidal magnetization pattern varies sinusoidally with position along the track and drops off exponentially with distance above

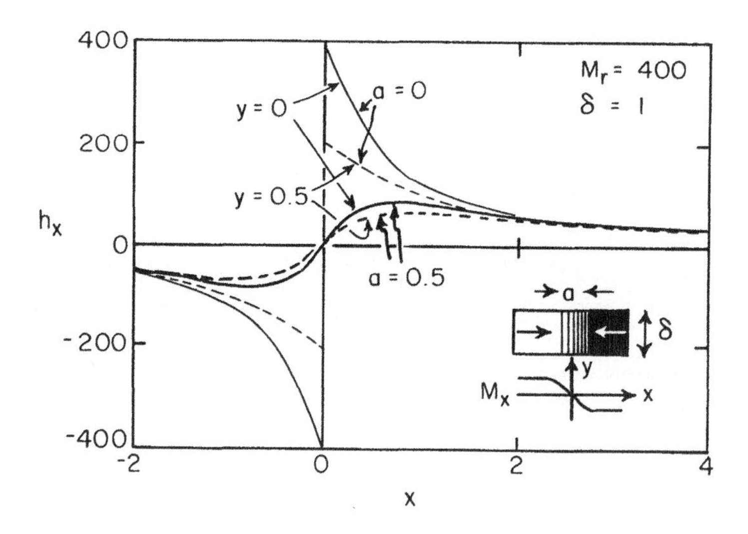

Figure 17.7 Horizontal fringe field h_x for a longitudinal transition of zero width ($a = 0$), Eq. (17.3) (fine lines), and for $a = 0.5$, Eq. (17.4) (bold lines), at $y = 0$ (solid lines) and $y = 0.5$ (dashed lines). A magnetization of 400 emu/cm^3 is assumed and medium thickness, $\delta = 1$.

the medium. This exponential signal loss, $H = H_0e^{-ky}$, where $k = 2\pi/\lambda$, corresponds to nearly a 55 dB loss for $y = \lambda$, specifically, $H(\lambda)/H(0) = 1.9 \times 10^{-3}$. Here λ is the wavelength of the recorded bits. Thus, in the coordinates of Figure 17.8 with $x = 0$ at a transition:

$$H_x \propto e^{-ky}\sin(kx) \quad \text{and} \quad H_y \propto e^{-ky}\cos(kx) \tag{17.5}$$

The H_x field is shown schematically in Figure 17.8. The origin of the sinusoidal H_x field is displaced to larger values of y as H_x is evaluated at larger values of y.

More complex magnetization patterns in analog recording, or bit sequences in digital recording, must be Fourier-analyzed. Each Fourier component, corresponding to a different wavenumber, k, decays like e^{-ky}. Thus, for weak signals or larger head-to-medium distance, the higher-frequency information is lost first.

The field loss in the write process [Eq. (17.2)] and in the read process [Eq. (17.5)] account for 100 dB or five orders of magnitude loss. Clearly, at higher recording density (smaller λ) the head–medium spacing must decrease or these losses will increase to the detriment of the recording process.

Because of the experimental loss of signal with distance of the head above the medium, or more precisely, above different strata within the medium, it turns out that most of the signal to be read comes from a depth of a little more than a third of the recorded wavelength. This can be shown by integrating the signal from the medium [Eq. (17.5)] to get the normalized voltage V^d from a depth, d (Fig. 17.9), and setting it equal to, for example, 90%:

$$\frac{V^d}{V^\infty} = \frac{\int_0^d \exp(-ky)dy}{\int_0^\infty \exp(-ky)dy} = 1 - e^{-kd} = 0.9$$

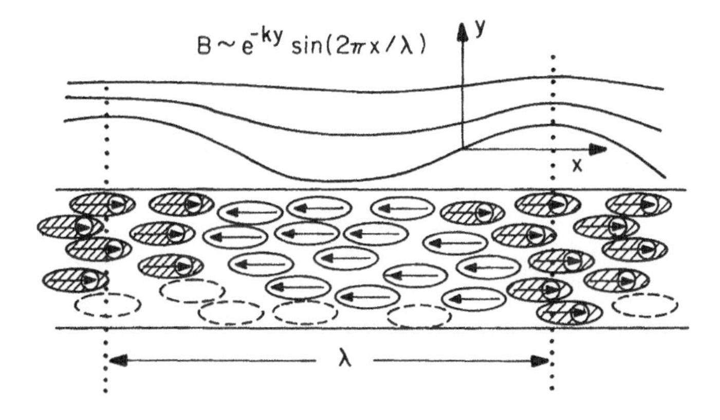

Figure 17.8 Schematic representation of field above a longitudinal recording medium. The sinusoidal field variation is represented at three increasing values of y.

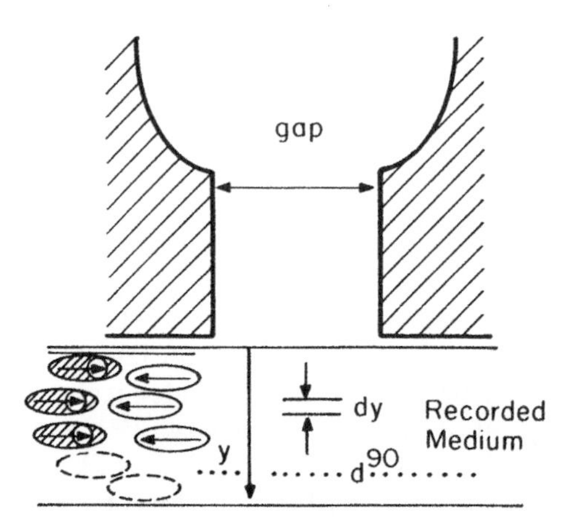

Figure 17.9 Coordinates describing integration over recorded medium to obtain field strength at read head.

Thus, the depth d^{90} from which 90% of the read signal originates, is given by

$$d^{90} = \frac{2.3\lambda}{2\pi} \approx 0.37\lambda \tag{17.6}$$

17.2.3 The Read Head

The read function can be accomplished with a write head operated in its demagnetized state ($i_{\text{coil}} = 0$). The fringe field of the transition, Eq. (17.4), is seen by the moving head as time dependent, and therefore a voltage is induced in the windings. We will see below that advantages come from the use of a different, more sensitive read transducer based on the MR effect or the GMR effect. This has the added benefit of allowing the write head to be optimized for only writing and the read head for reading.

Tape heads operate in contact with a protective layer over the magnetic recording tape medium. In hard disk drives, the head flies on an air cushion at a height of about 40 nm above the disk surface. The hard disk recording medium is protected from head crashes by a thin diamond-like carbon (DLC) coating. The air bearing surface of the head may be uncoated but some heads also have a thin DLC protective layer. To understand the importance of head–medium separation, it may be useful to consider an analogy. A scanning tunnel microscope is extremely sensitive to atomic-scale variations in its separation from a sample because the tunneling current varies exponentially with the tip height above the surface. Similarly, a read head (and to a lesser extent a write head) is extremely sensitive to variations in its height above the recording medium. The variation here also is exponential with separation. This places severe restrictions on substrate flatness, medium thickness uniformity, head aerodynamics (or tribology for contact reading), and mechanical stability of the head–medium physical supports.

17.2.4 Material Requirements

Let us summarize some of the requirements placed on magnetic materials used in recording heads and media.

The *write head* must have adequate magnetic permeability at high frequency so that it can be driven to saturation with minimal current.

The write head must have a sufficiently high saturation magnetization so that its fringe field exceeds the coercivity of the medium (500–3000 Oe). $Ni_{81}Fe_{19}$ ($4\pi M_s \approx 10$ kG) is generally used in thin-film write heads but higher-induction permalloys, such as $Ni_{50}Fe_{50}$, and iron nitrides based on $Fe_{16}N_2$($\mu_0 M_s \approx 3$ T), are being considered.

The *recording medium* should respond to the field of the write head (H_c not too high) and retain the sense of magnetization in spite of the magnetostatic fields of adjacent bits, stray fields and ambient temperature fluctuations (H_c not too low). The coercivity is typically in the range 500–3000 Oe. These values place recording media in the low end of the coercivity range of hard magnetic materials.

(*Note*: The medium should be composed of small, independent magnetic entities (grains or single-domain particles) that can retain their direction of magnetization across a sharp transition. A bit ideally should be composed of a single-domain, isolated magnetic particle. Because this is generally not practical, approximately $N = 10^3$ particles should constitute a bit in order to insure a sharp transition.)

The medium must have adequate remanent magnetization so that the fringe field due to the written domain sequence extends above the surface with sufficient strength (several Oe) to be detected by a read head. A magnetic recording medium requires a saturation magnetization of at least 500 G ($\mu_0 M_s \approx 0.63$ T), and typical values range up to 1000 G.

The *read head* must have low coercivity, low noise, and extremely high permeability in order to respond with a substantial change in flux to the weak fringe field above the medium. Near-zero-magnetostriction permalloy is generally used in thin-film read heads.

The read and write functions can be filled by the same inductive head but there are advantages to separating these functions.

The flying height should be as small as possible but not to the extent that head crashes are frequent or friction and wear become problems.

17.2.5 Longitudinal versus Perpendicular Recording

Having reviewed the basics of magnetic recording, the reader is in a position to distinguish between longitudinal, perpendicular, and isotropic recording media.

A *longitudinal medium* is one where the easy axis of magnetization lies in the plane of the recording layer. This geometry takes advantage of the strength of the in-plane component of the write head fringe field [Eq. (17.2) or Fig. 17.4]. Longitudinal media may be textured so that the easy axis of the grains lies predominantly along one direction in the plane. This is advantageous for tape media where information is recorded linearly, but it is ineffectual for floppy disk media where the tracks are circumferential. As recording density increases in a longitudinal medium, the demagnetization factor of the recorded bits, proportional to $M_r t/\lambda$, becomes more unfavorable unless the depth of information storage is proportionally reduced. But reduced thickness reduces the read signal strength, which is also proportional to $M_r t$. Longitudinal media constitute the bulk of the tape, floppy, and hard-disk media. They have demonstrated linear bit densities in the range of 10^5 bits per inch (bpi) or $\lambda = 0.5\,\mu$m.

Perpendicular media are those for which the preferred direction of magnetization is perpendicular to the recording layer. For a perpendicular medium, higher information density stabilizes the bits against demagnetization (the demagnetization factor in this case goes as $M_r \lambda/t$). Thus, recorded information can be packed with greater density in a perpendicular medium than in a longitudinal medium. This is depicted schematically in Figure 17.10. Densities in the range of 100,000–500,000 bpi are achievable in perpendicular media. However, at increased densities the fringe field of a perpendicular medium is confined closer to the medium [compare Eqs. (16.14) and (17.5)]. This makes inductive reading of perpendicular media more difficult. See Suzuki (1980) for the case in favor of perpendicular media and Mallinson (1981) for the case against. Perpendicular media for inductive recording systems have long been of interest but are not yet in production. Magneto-optic media are preferably perpendicular because that geometry optimizes the Kerr rotation in normal-incidence reading.

Perpendicular magnetic recording requires a different kind of head than that used on longitudinal recording. Figure 17.11 shows one configuration for a single-pole tip for perpendicular recording. The medium is backed by a high-permeability layer, which creates an image of the pole and focuses its field. The single pole is joined to a return pole whose larger cross-sectional area

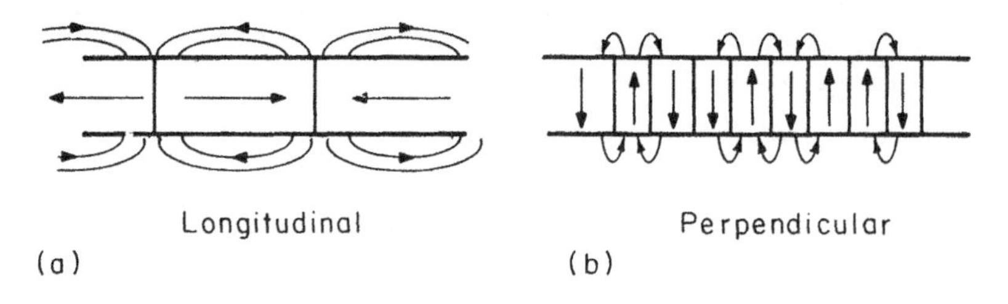

Figure 17.10 Comparison of recorded bits in longitudinal (*a*) and perpendicular (*b*) media.

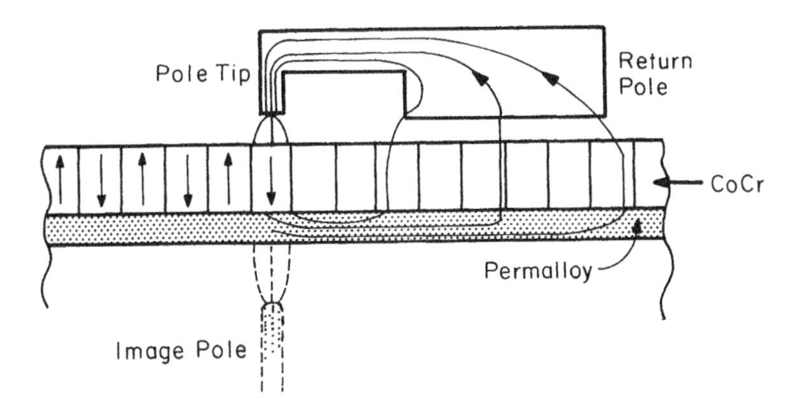

Figure 17.11 Perpendicular recording using flux closure layer beneath the medium.

allows the returning flux to pass back through the medium at lower flux density, thus not interfering with stored information.

There has been interest also in media that are neither longitudinal nor perpendicular, but rather, isotropic. An isotropic recording medium should respond to the fact that the field of an inductive ring head has both longitudinal and perpendicular field components in its gap. An isotropic recording medium presumably would respond optimally to the entire head field rather than just to its longitudinal or perpendicular field component. This is referred to as vector magnetic recording.

17.3 PARTICULATE RECORDING MEDIA

Particulate media generally consist of single-domain particles because of their high coercivity. The magnetization process in single-domain particles is described in Chapter 9. H_c reaches a maximum between the single-domain and superparamagnetic limits (Fig. 12.2). It was shown in Chapter 8 that the volume of a single-domain particle can be increased if the particle is elongated; this reduces the magnetostatic energy that is always the driving force for domain formation. Acicular particles are most often used because of their strong shape anisotropy.

The particles are suspended in a polymer matrix (including binders, plasticizers, solvents, and wetting agents) that allows the medium to be painted on a substrate then cured to a flexible magnetic layer. Thus, particulate media are more suitable than metal films when soft substrates such as mylar tape or polyester floppy disks are used.

The process of coating a particulate recording medium on a substrate tends to align acicular particles along the application direction. As the solvents are baked off, a field can be applied to enhance alignment of the particles. This is desirable if the medium is to be magnetized in the alignment direction because it squares up the hysteresis loop. Unidirectional texturing is used in magnetic tape.

A square $M-H$ loop implies that the material is bistable, that is, that either direction of magnetization is stable up to fields of H_C. Further, the material should be bistable on a submicron scale. A quantitative measure of loop squareness that is widely used outside the recording media community is the remanence ratio

$$S = \frac{M_r}{M_s} = m_r$$

This parameter is an indicator mainly of the *strength* of the read signal, proportional to M_r, in either of the two bistable states. It says nothing about stability, which is related to H_c. Another measure of squareness that includes the field needed to switch the magnetization is the ratio of M_r/H_c. This parameter measures the average susceptibility in the second quadrant. Because a magnetic medium is generally near a state of net demagnetization, a more appropriate susceptibility is $\chi_0 = [\partial M/\partial H]_{H_c}$ at $M = 0$. It is usually the case that $\chi_0 > M_r/H_c$ so local squareness is proportional to the magnitude of the ratio $\chi_0/(M_r/H_c)$. To express this ratio as a squareness parameter that varies from 0 (not square) to 1 (most square), the coercivity squareness parameter S^* is defined as:

$$S^* = 1 - \frac{M_r}{\chi_0 H_c} \tag{17.7}$$

These squareness parameters are illustrated in Figure 17.12. A value of S or S^* approaching unity indicates an $M-H$ loop with a sharp second quadrant change in magnetization with changing field. More importantly, with applied field varying in time over a moving medium, a large value of S^* implies that a spatially sharp magnetization transition can be written and sustained in the

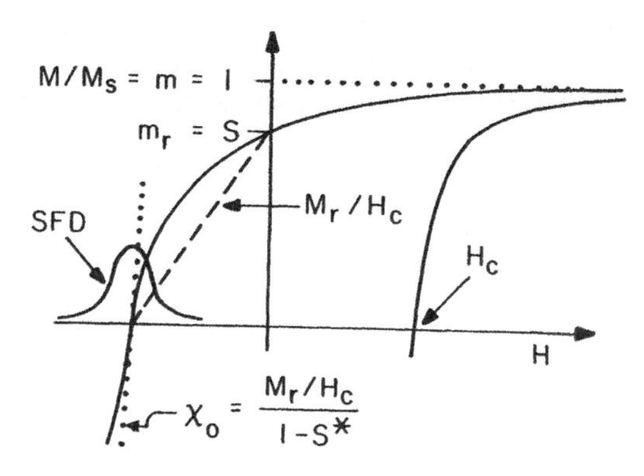

Figure 17.12 Part of $M-H$ loop showing various measures of loop squareness.

medium. The spatial sharpness or *acuity* of the transition is enhanced if the medium consists of a large number of small magnetizable units, grains, or particles. These units should be as weakly interacting magnetically as possible in order to allow a spatially abrupt transition. At different points across the track width the transition fluctuates in position by about half the particle dimension along the track length. A read head averaging the transition across the track width will see more noise and less signal as the spatial fluctuations in the transition position approaches the bit length.

The switching field distribution (SFD) is defined (with reference to Fig. 17.12) as

$$\mathrm{SFD} = \frac{\Delta H}{H_c}$$

where ΔH is the full width at half maximum of the differential susceptibility, $\chi = dM/dH$, near H_c. The SFD is a measure of the transition fluctuations. High SFD requires a narrow particle size distribution.

A particulate magnetic recording medium should be composed of rod-shaped (acicular) particles. While a favorable shape factor may contribute to a bistable loop, a strong uniaxial magnetic anisotropy in the particles also enhances loop squareness, provided the easy axes are well aligned. If the magnetic easy axes do not show a strong texture but instead are spatially dispersed, then the magnetostatic energy at the transition is reduced and the transition width may be narrow. However, in this case, the signal is also reduced because M_r is reduced. On the other hand, if the grains show a strong texture (high squareness) the magnetostatic energy at the transition is large and acuity may be degraded by a zig-zag transition (especially in thin film media). In this case the signal from the transition may be noisier. The best media lie between these extremes.

The magnetization density of particulate media is reduced relative to that of thin-film media because of the presence of binders and plasticizers in the former. The magnetic particle loading in particulate media is typically 20–50% by volume. Tape media (low density) typically have a magnetic coating thickness of order 10 μm, while floppy disks (higher density) are coated only to a few microns. Particles are used in media for recording systems in which the bit length is typically greater than one micron.

17.3.1 Gamma Iron Oxide

The composition Fe_2O_3 is chemically stable because the iron ions are in the fully oxidized state, Fe^{3+}. At room temperature the stable phase of this composition is the hexagonal corundum phase, α-Fe_2O_3 or hematite, which is antiferromagnetic and therefore not suitable as a recording medium. A metastable phase, gamma ferric oxide or maghemite, γ-Fe_2O_3, has a spinel structure similar to that shown in Figure 4.6, but with the divalent iron ions missing and the two trivalent irons unequally distributed over the A and B sites making it

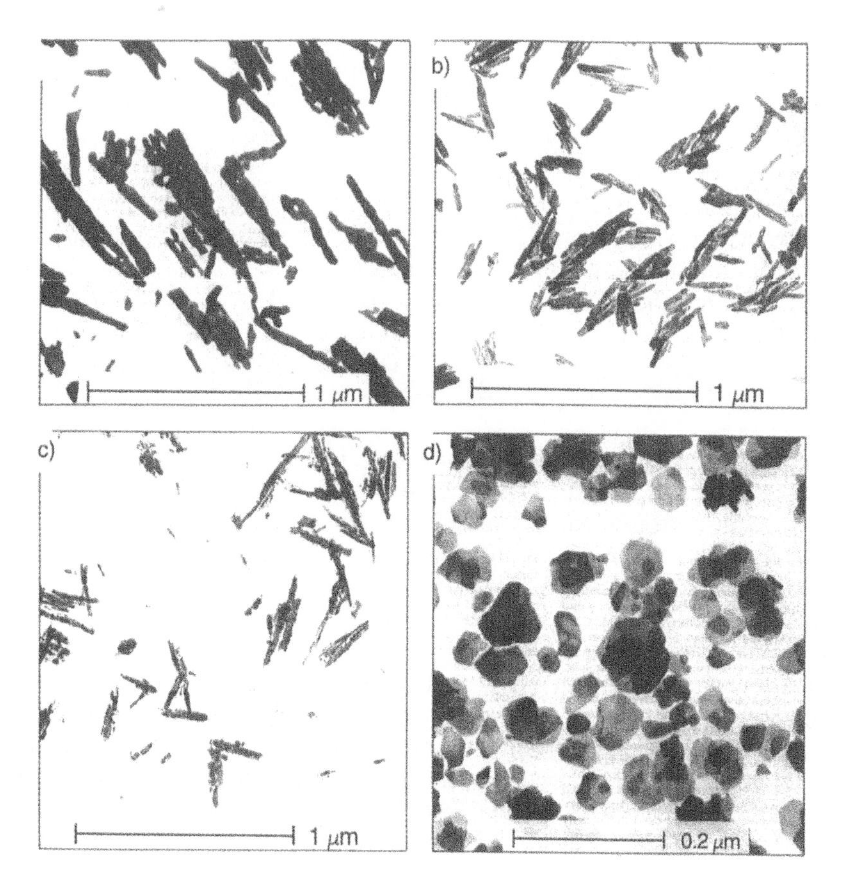

Figure 17.13 Magnetic particles used in recording: (*a*) iron oxide particles with surface-deposited cobalt of coercivity 700–750 Oe; (*b*) iron oxide with Co-treated surface having coercivity of order 900 Oe; (*c*) iron metal particles with coercivity of 1500 Oe; (*d*) barium ferrite particles (note different scale) (Sharrock 1990).

a ferrimagnet. The room temperature saturation magnetization of γ-Fe_2O_3 is approximately 400 G, but only about 350 G in fine particles. Acicular particles of γ-Fe_2O_3 with roughly a 10:1 aspect ratio (typically 100 nm long and 10 nm in diameter), have been the most widely used recording medium since the late 1940s. Thus the bit length is of order ten particle lengths. Because K_1 of γ-Fe_2O_3 is of order only $10^4\,\mathrm{erg/cm^3}$, the particle shape dominates the anisotropy and controls the coercivity. From Eqs. (2.17) and (2.18), the demagnetization factor of a prolate ellipsoidal particle is $\Delta N \approx \frac{1}{2} - 1/m^2$, where $m = a/b \approx 10$, so H_d is given by $2\pi M_s \approx 2200\,\mathrm{Oe}$. Thus the shape anisotropy is of order $K_u \approx M_s H_d/2 = \pi M_s^2 \approx 4 \times 10^5\,\mathrm{erg/cm}$ for γ-Fe_2O_3. Acicular γ-Fe_2O_3 particles exhibit a coercivity of order 350 Oe, much reduced from the limiting value, H_d. The first hard disks (those in IBMs Winchester drives) were coated with an iron–oxide particulate medium.

The fabrication of acicular γ-Fe_2O_3 is described in detail by Bate in Wohlfarth (1980). Two paths can be followed from a starting iron salt solution: nucleation and growth of FeOOH (which is dehydrated to give α-Fe_2O_3), or direct precipitation of hematite. Hematite must be reduced to form the spinel, magnetite. Oxidation of magnetite gives maghemite, a metastable spinel ferrimagnet, γ-Fe_2O_3.

17.3.2 $Co^{2+}\gamma$-Fe_2O_3

Cobalt-modified ferric oxide media (Fig. 17.13, *a* and *b*) presently make up the most widely used class of particulate recording materials. They were developed during the late 1960s to improve on the low coercivity of γ-Fe_2O_3. The cobalt creates a magnetically harder phase that results in a higher coercivity. Co^{2+} ions preferentially occupy the B sites of the spinel structure. The energy-level splitting in that octahedral site (see Chapter 6) gives a partially occupied $t_{2g}(d_\varepsilon)$ valence configuration. Thus $\langle L_z \rangle \neq 0$ and the spin–orbit interaction can give rise to magnetic anisotropy. The higher anisotropy enhances H_c. Early development of cobalt-treated iron oxide particles, which focused on uniform doping, has given way to surface doping which gives more stable magnetic properties.

1. *Uniform Co^{2+} Doping.* The increased contribution of magnetocrystalline anisotropy to the total anisotropy in Co-doped ferric oxide means that H_c is more sensitive to temperature and stress compared to γ-Fe_2O_3. Also, Co^{2+} ions in the presence of Fe^{2+} ions experience enhanced mobility. Thus, time dependence of magnetic properties can be a problem in slightly reduced, uniformly doped cobalt-iron oxide. A further disadvantage of uniform doping comes about if the Co^{2+} ions are randomly distributed over different sites in a particle. This results in a random crystalline anisotropy which detracts from, rather than enhances, the shape anisotropy.

2. *Co^{2+} Surface Treatment.* When the cobalt is confined to a thin surface layer on the γ-Fe_2O_3 particles, the composition there can approach $CoOFe_2O_3$, which has no Fe^{2+} ions. This reduces the instability problem encountered with uniformly doped γ-Fe_2O_3. Also, the anisotropy arising from the Co^{2+} rich surface layer is less random and reinforces the shape anisotropy.

17.3.3 CrO_2

Chromic oxide is metallic and ferromagnetic, the only known ferromagnetic oxide at room temperature. It was developed during the late 1960s specifically to improve on the relatively low coercivity of γ-Fe_2O_3. Its coercivity is in fact significantly improved (500–600 Oe) and its magnetization (350 G) is comparable to that of γ-Fe_2O_3.

CrO_2 is made by a relatively expensive, high-pressure, hydrothermal process. A clear advantage of this process is its simplicity; only one step is involved. Sb, Fe, or Te is sometimes added to control the nucleation and growth of the needle-shaped particles. Because of their shape uniformity, CrO_2 particles give highly oriented, high-SFD, media. Chromic oxide media compete with Co^{2+}-γ-Fe_2O_3 for market share in the 500–900 Oe media range.

CrO_2 particles having coercivities approaching 3000 Oe have been made in the laboratory by using Ir as a growth inhibitor. The resulting particles have smaller diameters and the magnetization reversal mechanism probably is dominated by coherent rotation rather than curling (which appears to be a factor in larger particles).

The low Curie temperature of CrO_2 (125°C) makes it the material of choice in thermomagnetic copying, a process for contact duplication of large amounts of high-density information by cooling a CrO_2 medium from above its Curie temperature while in contact with a higher T_C master.

17.3.4 Metal Particles

Acicular metal particles were developed for magnetic recording in the late 1970s. The use of pure metal particles instead of oxides results in significant increase in magnetization density. The large magnetization and relatively low crystal anisotropy of iron ($2\pi M_s^2/K_1 \approx 10^2$) demand that the particles be acutely acicular (Fig. 17.13*c*) to avoid demagnetization. Larger magnetization enhances the shape-induced anisotropy (proportional to M_s^2) and hence increases the coercivity (as M_s). The most widely used metal particle recording medium is based on iron.

Transmission electron microscopy (TEM) studies show oxide layers of 30–40 Å thickness on iron particles. This accounts for most of the loss in magnetization of particulate iron relative to pure iron. The surface oxide is generally inhomogeneous and reflects an expected oxygen gradient: outer layers of Fe_2O_3, and underlying layers of Fe_3O_4, then sometimes FeO before reaching the Fe core. While pure iron has a magnetization density in excess of 1700 G, particles for recording media have a high surface to volume ratio and generally exhibit only 50–60% of this value. This is still a significant improvement over M_s for oxide particles. The tendency of metal particles to oxidize can be diminished by alloying additives to the pure metal, by the polymer binder used in the medium, or by surface passivation with selected oxides.

Commercial production of iron particles begins much like the production of γ-Fe_2O_3. The α or γ phase of (FeOOH) is nucleated and grown from an iron salt solution. Al, Si, or P additions to the FeOOH are sometimes used to control morphology and minimize sintering in subsequent steps. Dehydration and reduction then result in the final metal needles. Their shape is similar to that of γ-Fe_2O_3 but they are slightly smaller.

Iron particles are widely used in 8-mm video cassette and other high-density media. The mechanism of magnetization reversal is consistent with the chain-of-spheres model (see Chapter 9). In particles that show increasing coercivity with decreasing diameter, a curling model seems to apply.

17.3.5 Barium Ferrite

$BaO \cdot 6Fe_2O_3$, developed in the 1980s, is unique among magnetic particles inasmuch as its anisotropy is not dominated by shape. Rather, these hexagonal platelets (approximately 10 nm thick and 100 nm in diameter; Fig. 17.13*d*) are magnetized normal to their thin dimension because of a strong crystalline anisotropy (3.2×10^5 J/m^3) (see Chapter 13).

Barium ferrite has the hexagonal magnetoplumbite structure assumed by the class of materials of formula $MO \cdot 6Fe_2O_3$, where M = Ba, Pb, or Sr. Barium ferrite is ferrimagnetic with $M_s \leqslant 400$ G. Barium and strontium ferrites are used as high coercivity permanent magnets because their very large magnetocrystalline anisotropy favors *c*-axis magnetization. These ferrites grow most rapidly in their *a*—*b* plane to form hexagonal platelets and the crystal anisotropy overwhelms the shape anisotropy ($2\pi M_s^2 \approx 10^5$ J/m^3), so the particles remain magnetized perpendicular to their thin dimension. Barium ferrite particles are well suited for use in perpendicular particulate media; the flat particles generally lie with their plate normals perpendicular to the plane of the coating. Despite demonstrations of high recording density, barium ferrite media have not captured a sizable share of the particulate media market. This is due in part to their low magnetization.

The large magnetocrystalline anisotropy gives the particles a coercivity too high for magnetic recording. Consequently, Ti^{2+}, Co^{2+}, or Fe^{2+} ions are sometimes substituted for some of the Ba^{2+} to reduce K_u and bring H_c into the range 500–1200 Oe. Barium ferrite media show a very sharp switching field distribution despite some evidence of strong interparticle magnetic interactions.

Small platelets typically 100 nm in diameter and 10 nm thick can be made by a hydrothermal process or by devitrification of glass.

In summary, the prototype particulate recording medium, acicular γ-Fe_2O_3, is the product of a mature and cost-effective processing technology. However, it is characterized by low magnetization density and low coercivity. Cobalt-treated ferric oxide and CrO_2 particles were developed to improve on the H_c of iron oxide. Metal particles were developed for their enhanced magnetization. Barium ferrite particles were developed as a perpendicular recording medium. Smaller particle sizes, desirable for high-density recording, bring problems of lower H_c and reduced M_s due to the higher surface to volume ratio. Development of new compositions for particulate media have focused on rare earth–transition metal intermetallic compounds, among others. See Table 17.1 for a summary of properties of several particulate media.

17.4 THIN-FILM RECORDING MATERIALS

The challenge for thin-film media is to achieve high coercivity to insure a sharp transition with low noise, while at the same time maintaining adequate signal strength, proportional to $M_r t$. In the case of particulate media, the anisotropy needed for high coercivity is generally provided by particle shape. Thin film

TABLE 17.1 Summary of Characteristics of Various Particulate Media

	Dimensions (Length, mm)	Source of Anisotropy	M_s (G)	H_c (Oe)	Application
γ-Fe_2O_3	10:1 acicular	Shape	350	350	Audio and low-density data
CrO_2	Acicular	Shape and crystal	350 ± 50–90	550 ± 50	Audio/video and data tape
Co^{2+}-γFe_2O_3	10:1 acicular (0.1–0.25)	Shape	350	900 ± 100	Audio/video
α-Fe	10:1 acicular (0.1–0.25)	Shape	750–900	1500	8-mm video and digital audio
$BaO \cdot 6Fe_2O_3$	Hexagonal platelets (0.01 × 0.1)	Crystal	300	Broad	range, typically 500–1200

media rely more on intrinsic crystal anisotropy. In both cases, single-domain particles are required to eliminate domain walls, which lower coercivity. Thin-film magnetic recording media, just as particulate media, can be either longitudinal and perpendicular.

Typical values of S^* in thin film media are 0.5–0.9. At low recording densities, it is desirable to maximize the coercive squareness. High squareness implies a large demagnetizing field at the transition. At high recording density, square media often reduce their magnetostatic energy at the sharp transition by forming a zig-zag domain wall. Such media show increased noise.

17.4.1 Noise in Thin-Film Media

Thin-film media are more prone to noise than are particulate media. The reason for this is that in the latter case, the particles are fully isolated from each other by the polymer matrix so the transition fluctuations are limited by the particle size. In thin film media, the grain boundaries are narrow enough and often sufficiently magnetic to allow the particles to couple by exchange or dipole fields. Thus several particles can act in unison (an interaction domain) effectively increasing particle size and transition thickness. The more irregular transitions in thin-film media are referred to as zigzag or sawtooth transitions. Noise in thin-film media is due primarily to the formation of zigzag transitions between bits.

As is demonstrated in Problem 17.5, the amplitude of the sawtooth pattern scales roughly as $M_s^2/K^{1/2}$. This has been qualitatively confirmed by experiment. One solution is to reduce the magnetostatic energy by decreasing the $M_r t$ product of the thin film. However, this reduces signal strength. Another solution would be to increase the magnetic anisotropy. This is not always easy and must be limited by the ability of the write head to perform its function. Zigzag domain wall amplitude can be reduced also by use of a soft magnetic underlayer. This alloys flux closure and hence reduces the magnetostatic energy from the head-to-head transition. Media noise increases with increasing bit density in longitudinal films having no underlayer. It decreases with increasing bit density in perpendicular media because they have no zigzag domain walls. Zigzag or sawtooth transitions were identified in longitudinal Co-Cr films in the early 1970s. Bertram (1994) shows that a change in transition width from a delta function to an arctan-like transition of length a along the track $[M_x(x) = M_r \tan^{-1}(x/a)]$, results in a loss of signal as if the head-to-medium distance were increased by the amount a.

Signal-to-noise ratio (SNR) is also a function of statistical counting noise in a measurement. Assume that there are N independent grains per bit. The transition position will be ill-defined on the scale of the grain size, thus noise goes as $w/N^{1/2}$. Signal is proportional to w giving SNR $\propto N^{1/2}$. Equivalently, random walk considerations show that fluctuations (noise) from N random events increase like $N^{-1/2}$ as more events are averaged. The signal is proportional to the number of measured events or particles per bit, N. Hence

$$\mathrm{SNR} \propto \frac{N}{N^{1/2}} = N^{1/2} \tag{17.8}$$

An SNR of 20 dB [dB = 20 log (ratio of amplitudes)] is equivalent to $N^{1/2} = 10$, $N = 100$, or 10 particles by 10 particles on the surface of a bit array. It is desirable to reduce particle size as long as H_c does not drop too severely on approaching the superparamagnetic regime. Superparamagnetism becomes a problem for particles smaller than about 10 nm in diameter, corresponding to areal densities of about 20 Gb/in^2.

Lambeth (1998) argues that for an ideal thin-film medium (completely noninteracting particles), the signal-power to noise-power, proportional to N, is important.

For either of these measures, SNR increases with increasing number of particles in a bit. If the particle easy axes are distributed in orientation, those orthogonal to the read and write axis are essentially inactive and contribute little to the signal. This cause of reduced signal can be minimized by texturing the thin film medium so that the c-axes of the cobalt-rich grains are more favorably aligned with the track direction. Directional roughening of the aluminum substrate is also useful.

Interaction domains consist of clusters of coupled grains up to several microns wide that switch as a unit. They have been observed in high-noise thin-film media but not in low-noise media. Interaction domains result from either magnetostatic or exchange interactions between single-domain particles. Thus it is desirable to try to isolate the magnetic grains by a nonmagnetic intergranular layer. Particulate media use nonmagnetic, organic binders to isolate the separate particles. This reduces the likelihood of interaction domains and promotes a spatially sharp magnetization transition region. Grain isolation in thin film media is difficult to achieve but can be approached by alloy selection, processing conditions, and buffer layers.

17.4.2 Longitudinal Thin-Film Media

The first electrochemically deposited cobalt thin films for magnetic recording (1952) had a coercivity of less than 300 Oe. The addition of phosphorus to the electrochemical solution increased H_c nearly sixfold. Apparently, phosphorus segregates to the grain boundaries, isolating the grains and rendering them single-domain particles.

The addition of nickel to Co-P led to films having smaller grain sizes and better corrosion resistance. It was studies of Co-Ni-P-plated media that first identified zigzag domain walls at bit transitions as a source of noise. Low noise was found to be associated with single-domain particles.

Evaporation of Co, Co-Ni, or Co-Ni-Cr on polyester at low angle of incidence (70° from normal) and high deposition rates (1–10 μm/s), has been used to make longitudinal media called *metal evaporated tape* (MET). These processing conditions lead to a tilted columnar microstructure due to shadowing effects during deposition. The shape anisotropy of this microstructure and the presence of cobalt oxide at the particle boundaries are responsible for the high coercivity (1500 Oe) of MET media.

Studies of electroless deposition of Co-P showed that under some conditions an HCP structure resulted with its *c* axis normal to the substrate plane. In these films, H_c was of order 300 Oe for fields applied in plane (see Fig. 9.3*a*) and exceeded 1000 Oe for fields applied along the perpendicular easy axis (see Fig. 9.3*b*). These media having a significant perpendicular component of anisotropy showed lower noise for longitudinal recording. The reason for this may be that the magnetization near the transition is allowed to rotate up and out of the medium toward the head. This would reduce magnetostatic energy at the interface, decreasing the likelihood of zigzag transition formation. It may also increase signal strength. However, increased perpendicular anisotropy can also reduce H_c.

The most widely used substrate for hard disk media is presently an Al-Mg alloy with an electroless Ni-P coating. Harder glass substrates are also finding use in disk drives for personal computers because of their shock resistance.

As a method of thin-film media deposition, sputtering allows for high deposition rates, the ability to deposit a wide range of complex compositions (metals and insulators) and good adhesion. The use of bias sputtering helps reduce the amount of oxygen trapped in films. γ-Fe_2O_3 and its cobalt-doped variant have been successfully sputtered on Ni-P/Al disks. These media show low noise and a coercivity of 1000 Oe. RF-sputtered Co-Ni-Pt films show a coercivity of nearly 900 Oe along with good corrosion and wear resistance and a high remanent magnetization of 800 G.

Presently used high-density, longitudinal thin film media are based on Co-Cr with Pt and Ta additions. Pt is used to increase the magnetic anisotropy of the cobalt-rich film. It also improves the epitaxial relation between the cobalt film and the Cr underlayer. Cr seems to play a role in isolating the magnetic grains. The use of Ta as an alloying addition in Co-Cr longitudinal media is found to enhance segregation of Cr to the grain boundaries as well as improving epitaxy to the Cr underlayer. Co-Cr-Pt-Ta longitudinal media show improved grain isolation, increased H_c, and significantly lower media noise, particularly when there is a significant perpendicular component of anisotropy compared to Co-Cr alone.

The natural growth mode for BCC Cr is (110). Co-Cr or Co-Ni media grow with their HCP *c* axis 28° out of plane on Cr (110) surfaces. From Figure 9.7 it is clear that coercivity should be increased by enhancing the alignment of the *c* axes of the grains with the track length. Hexagonal Co-based films grow with their *c* axis in plane on Cr buffer layers having (002) or (112) surfaces. These modes of growth can be induced in Cr by deposition at elevated temperatures and carrying the growth to greater thicknesses, respectively (Lambeth et al. 1998). The epitaxial relations between HCP Co-based films and BCC Cr buffer layers are illustrated in Figure 17.14.

In the case of Co $(11\bar{2}0)$ on Cr (002), there are two possible epitaxial Co *c*-axis orientations. As a result, such media show a bicrystal-type structure with a reduced remanence (sharper transitions). There is only one axis for alignment of the HCP *c* axis on a Cr (112) surface. The difficulty of achieving Cr (112) growth on glass substrates led to the development of Ni-Al (Lee et al. 1995) as an improved buffer layer favoring Cr (112) growth. Ni-Al offers the same advantages of adhesion, flatness, and promotion of desired texture in thin-film media. In addition, it promotes a finer, more uniform grain size in the magnetic thin-film medium.

Table 17.2 summarizes the relevant properties of some thin film systems used in magnetic recording media.

Hard disks are generally coated with a 10–20-nm-thick, sputtered, diamondlike carbon (DLC) layer. This provides protection against head crashes but increases the distance between the head and the magnetic medium. In the future, overcoats may have to be reduced to thicknesses of 5 nm.

Although the head flies on an air bearing when the disk is spinning, accidental contact between the head and the hard disk is unavoidable. In addition, the head may come to rest on the disk on shutdown. In order to decrease the likelihood of the head sticking to the medium, the substrate is

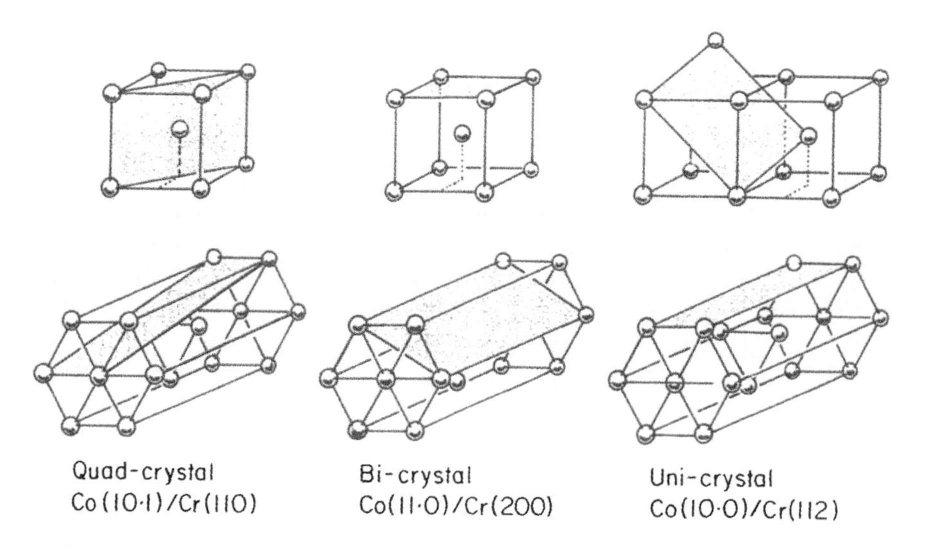

Figure 17.14 Epitaxial relations between Co-rich media and Cr underlayers for in-plane and 28°-out-of-plane orientation of the Co *c* axis. [From Lambeth et al. (1998).]

TABLE 17.2 Comparison of Properties of Various Thin-Film Compositions for Media

	H_c	Substrate	M_s	Thickness (mm)	Method or Application
CoP	1000	Plastic	—	0.3	Plate
MET[a] CoNiCr	1500	Polyester	—	—	Evaporated Sony 8-mm video
γ-Fe_2O_3	1000	NiP/Al	250	0.12	Sputter
CoNiPt	900	NiP/Al	800	0.03	Sputter
Co	1000	Cr/NiP/Al	—	—	Sputter
CoCrTa	1400	Cr/NiP/Al	—	—	Sputter
CoCrM (M = Pt, Ta, Zr)	—	Cr/NiP/Al	—	0.05	Sputter
CoNiCr	2000	CrGd/NiP/Al	—	—	RF-biased sputter

[a]Metal evaporated tape.

often roughened on a scale finer than the head dimensions. This can be done by mechanical polishing, by laser texturing of dedicated, circumferential landing zones, or by deposition of a small amount of low-melting-temperature metal such as In or Ga (which tends to bead up to reduce its surface energy). The latter, so-called "transient metal underlayer" process was developed for glass substrates where mechanical roughening is more difficult. Al-N coatings can also be used to roughen the disk surface to reduce striation.

Materials challenges in thin-film media include:

1. Decreased grain size for increased SNR without loss of H_c
2. Improved grain isolation for low noise and higher H_c
3. Higher H_c in smaller, single domain (particle) grains

Future media may involve higher anisotropy materials such as SmCo, barium ferrite, or CoPt.

17.4.3 Perpendicular Media

Perpendicular media are generally deposited on a thin layer of high-permeability, longitudinal material, such as NiFe, which provides a flux closure path for a single pole tip record head. This high-permeability layer effectively creates an image pole tip opposite the active one, focusing its flux (see Fig. 17.10).

When Co-Cr films having 18 at% Cr are RF sputtered on polyester substrates, a strong perpendicular anisotropy results. Iwasaki et al. (1979, 1980) developed this idea, which became the foundation of a class of magnetic media designed specifically for perpendicular recording. This is to be contrasted with the longitudinal media that are processed to exhibit a degree of perpendicular anisotropy for reduced noise.

There is a complex interplay between thin-film processing conditions, microstructure, and magnetic properties for RF-sputtered Co-Cr. Increasing the magnetization by increasing Co/Cr ratio leads to more domain walls and hence lower H_c. Increasing film thickness results in larger, multidomain particles, also having lower H_c. Optimal H_c is obtained by choosing the correct Co/Cr ratio and film thickness so that single-domain particles result with coherent rotation as the dominant magnetization process.

In a laboratory demonstration of high-density perpendicular recording, a floppy disk coated with 100 nm of Co-Cr on 0.5 μm of Ni-Fe achieved a density of 680 kbpi (thousands of bits per inch) when written with a 0.4-μm-thick single-pole head.

Magnetooptic recording uses perpendicular media based on rare-earth transition metal alloys such as Fe-Tb-Dy (Gambino and Suzuki, 1999). Magnetooptic recording is thoroughly discussed by Mansuripur (1993) and Gambino and Suzuki (1999) and is not described here.

17.5 RECORDING HEADS

Magnetic recording heads are basically transducers that convert electrical signals into a magnetic field (write head) or that sense a magnetic field and convert it to an electrical signal (read head). All write heads make use of Ampère's law using electrical windings around a high-permeability pole piece. Read heads can be inductive (Faraday's law), in which case the strength of the electric signal depends on the speed at which the fringe field is read. They can also be magnetoresistive, in which case the signal is independent of reading or scanning speed.

From the calculated field dependence on head height [Eq. (17.1) or (17.2)], it is clear that the head must be as close to the medium as possible. This puts constraints on head wear resistance. Thin-film heads will become more prevalent as demands for higher recording density increase.

We review the general material requirements, survey those materials presently in use, and outline what is needed in the future.

17.5.1 Inductive Heads

Inductive heads can be ring heads (Fig. 17.2) or single-pole heads (Fig. 17.10).

Table 17.3 summarizes the present classes of inductive recording head materials and their properties.

Bulk recording heads may soon become obsolete because they cannot achieve the dimensional refinements needed for high-density recording. Nickel

TABLE 17.3 Properties of Various Materials for Inductive Recording Heads

Material	$4\pi M_s$ (kG)	μ	H_c (Oe)	λ_s	ρ ($\mu\Omega \cdot$cm)	Comments
Bulk						
Sendust[a]	12	2000-		0	10^6	Poor WR[b]
MnZn ferrite	5.5	5000		≈ 0	10^6	Poor CR[c]
NiZn ferrite	4.5	100–200		$\neq 0$	10^{10}	1% the wear of permalloy
Thin Film						
81–19 Permalloy	9.0			0	40	
(PI)	15					
50-50 Permalloy						
(P2)						
Sendust,				0		
amorphous	10–16		0.01	0	130	

[a]Sendust composition: 85% Fe + 9.6% Si + 5.4 Al.
[b]Wear resistance.
[c]Corrosion resistance.

zinc ferrite heads lose their sensitivity with use because the contact friction with the medium causes wear and strain on the head surface. The appreciable magnetostriction of the compositions used allows the strain to create a very strong magnetoelastic anisotropy field at the surface, which eventually pins the magnetization there and reduces the permeability. This problem is referred to as the "dead layer" problem, which means that the remaining active part of the head is further removed from the recording medium and so the signal written or read is weaker. To correct this problem, MnZn ferrites can be used instead. They generally show higher permeability than NiZn ferrites, but their resistivity (of order $1\,\Omega \cdot$cm) makes them less suitable for high-frequency use (NiZn ferrite has $\rho = 10^4\,\Omega \cdot$cm). Ferrites also suffer from relatively low saturation flux density. For this reason some heads are made of Sendust (see Chapter 11), which has a saturation magnetization $4\pi M_s$ of 12 kG.

Another means of enhancing the field strength of a head is to cap or coat the pole tips with a high saturation magnetization metal. These are called *metal-in-gap* (MIG) heads.

Presently, 81-19 permalloy is the most widely used thin-film head material. Permalloy thin-film heads show improved recording density and resolution relative to bulk ferrite and Sendust heads. A thin-film head is shown in Figure 17.15. The film thickness is typically 2–3 μm, and the gap of order 200 nm for high-density recording. Reduction of permeability due to corrosion of permalloy could be reduced by the addition of small concentrations of Cr, Ti, Ir, or Rh. However, Cr causes unacceptable degradation of magnetic properties.

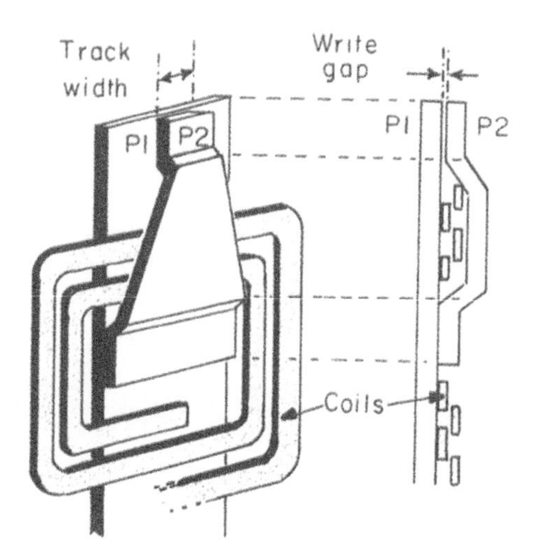

Figure 17.15 Thin-film recording head. Left, layout of pole pieces and windings; right, enlarged, cross-sectional view of magnetic pole pieces.

Mumetal (77% Ni, 14% Fe, 5% Cu, 4% Mo (wt%)] and Sendust thin films have also been used in thin-film heads.

Single-pole heads for perpendicular recording are limited in their resolution by the pole tip thickness. For higher resolution, thickness must decrease; to maintain write field strength, $4\pi M_s$ of the tip material must increase correspondingly.

A write head is driven to near saturation so its domain structure is not important. A read head, on the other hand, operates from its quiescent or demagnetized state. This state is very sensitive to the domain structure. A read head should respond to the fringe field of the medium by magnetization rotation rather than wall motion (wall motion generates noise). Thus the head material should be able to be field annealed to develop a weak, cross-track uniaxial anisotropy in order to define the demagnetized domain state.

A number of materials are under consideration for future inductive head applications either for their higher saturation magnetization or for their good high-frequency response. Most notable among these are the amorphous alloys based on Co-Zr: $4\pi M_s = 14\,\text{kG}$, $H_c < 0.5\,\text{Oe}$, and $\mu = 3500$. These films also show exceptional hardness. Lamination with SiO_2 allows good permeability to be maintained to higher frequencies: $\mu = 1000$ at $f = 100\,\text{MHz}$ in amorphous $Co_{87}Nb_8Zr_5$.

Metastable iron nitride films composed of the $Fe_{16}N_2$ phase show high saturation flux density (nearly 3 T) but also show strong negative magnetostriction. The large iron moment in this system is believed to be due to a tetragonal expansion of the iron lattice by nitrogen, resulting in an iron moment increase from $2.2\mu_B$ to nearly $3.0\mu_B$. The concept of lattice expansion could possibly be exploited in multilayers to achieve high $4\pi M_s$ and low λ_s in more stable compositions.

Iron–carbon multilayers can be tailored to achieve near zero magnetostriction and show coercitivies under 1 Oe. Ni/Fe multilayers having thickness ratios of 1:5 and a period of 22 nm have shown low magnetostriction and $4\pi M_s$ approaching 20 kG. The low magnetostriction is not of the same origin as that found in 81-19 permalloy but may be related to strain-induced changes in electronic structure. Insulation layers of SiO_2 or Al_2O_3 are sometimes used between the metal layers to reduce eddy currents.

Nanocrystalline materials made by devitrification of amorphous alloys or simply by underquenching (e.g., Fe-B-Cu-Si) are promising and show lower H_c the finer the grain size (see Chapter 12). Nanocrystalline Fe-Co-B-Si films 1 μm thick have exhibited $\mu = 1000$ up to 10^9 Hz (Fig. 17.16).

Because higher recording densities require higher-frequency head operation, new head materials should be designed with an objective of pushing operating frequencies toward 10^9 Hz. Again this implies that the magnetization process must be dominated by rotation. Hence a weak uniaxial anisotropy should be present or be able to be induced. Thin films of high electrical resistivity will

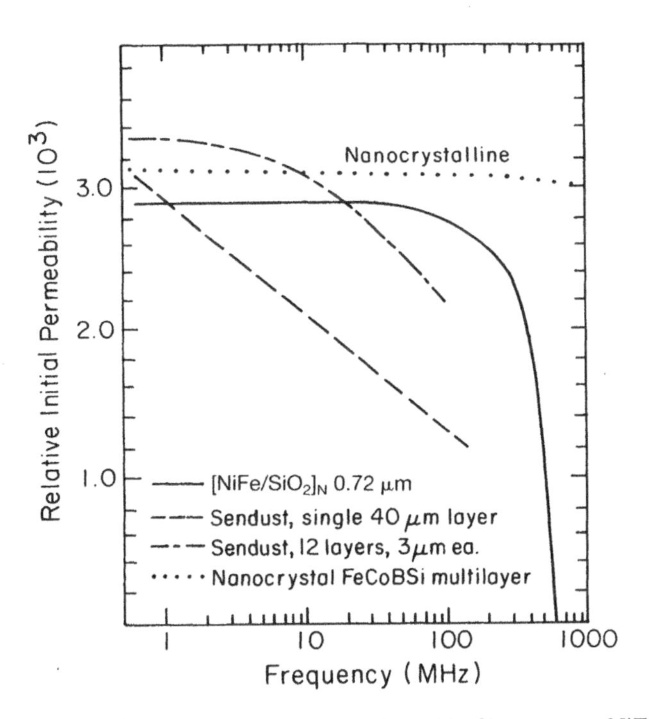

Figure 17.16 Permeability versus frequency for four thin-film systems; $NiFe/SiO_2$ and Sendust results from Jagielinski (1990).

probably play a dominant role. Multilayers will allow eddy-current suppression while maintaining high flux-carrying capability.

Ni-Fe/SiO_2 multilayers have shown permeabilities in excess of 2500 that are sustained up to $f = 300$ MHz (Jagielinski 1990) (Fig. 17.16). Amorphous alloys allow for field-induced anisotropy but lack the temperature stability to permit high-temperature glass bonding.

17.5.2 Magnetoresistive Heads

In 1975, Thompson *et al.* described the use of the magnetoresistance effect in magnetic recording heads. They described the need for bias field to allow the head to operate on the steep, nearly linear portion of the curve. The resistance versus field for the anisotropic magnetoresistance effect follows the general form shown in Figure 17.17: $\Delta\rho(H)/\rho = (\Delta\rho/\rho)(\cos\theta - \frac{1}{2})$. In the case of transverse anisotropy and y-directed field, $M_y/M_s = H_y/H_a = \sin\theta$. Thus, this form of $\Delta\rho/\rho$ leads to a quadratic field dependence below saturation:

$$\frac{\Delta\rho}{\rho} = \left(\frac{\Delta\rho}{\rho}\right)_{\max}\left[\frac{2}{3} - \left(\frac{H_y}{H_a}\right)^2\right]$$

Shield layers on either side of the MR element were found to increase its sensitivity and reduce signal pickup from adjacent transitions. Further, Thompson *et al.* found that the MR ratio increases with increasing film thickness, saturating at about 2–3% for 100 nm of permalloy.

Consideration of Figure 17.17 showing $\Delta\rho/\rho$ vs. H makes it clear that MR sensitivity is greatest near the inflection point of the curve. Thus it is desirable to apply a bias field in the direction of the sensed field (hard axis). Optimal sensitivity occurs for a bias that holds $\boldsymbol{M}$ at approximately 45° from the current direction. This is usually accomplished by the incorporation with the MR element a means of applying a bias field in the $\pm y$ direction.

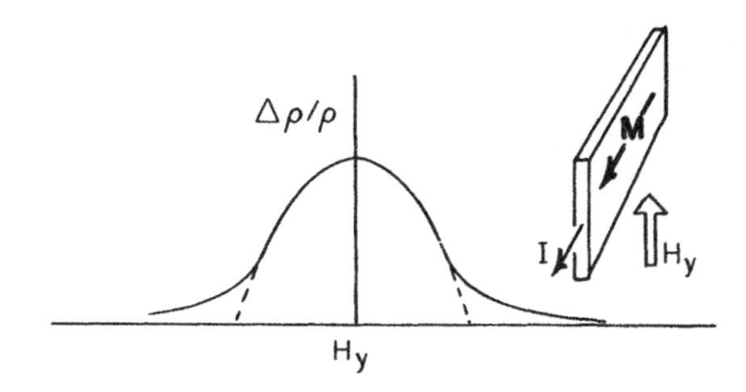

Figure 17.17 Field dependence of magnetoresistance (solid line) for uniform response to a uniform field H_y. Dotted line shows idealistic, quadratic MR response.

The sensitivity of an MR head depends not only on the magnitude of the magnetoresistance ratio $\Delta\rho/\rho$ but also inversely on the field change over which $\Delta\rho$ occurs:

$$\text{Sensitivity} \approx \frac{\Delta\rho}{\rho H_a} \tag{17.9}$$

where $H_a = H_u + H_d$ is the sum of the uniaxial anisotropy field and the demagnetizing field.

Figure 17.18 is a schematic diagram of an MR head (bias and shields not shown). Typical MR head parameters are $h = 1\text{–}2\,\mu\text{m}$, $w = 2\text{–}4\,\mu\text{m}$, $t = 10\text{–}20\,\text{nm}$, and $\Delta\rho/\rho = 2.0\%$ ($Ni_{81}Fe_{19}$), in which case $H_d \gg H_u$, where the demagnetizing field is $H_d \approx 4\pi M_s\,(t/2h) \approx 40\,\text{Oe}$. Permalloy is currently the most widely used material for MR heads.

We need to know the mathematical form of the resistance change with applied field H_y. The earliest description of MR heads, found in the classical paper by Hunt (1971), shows explicitly the role of *bias field* and nonuniform h_y from the transition. The magnetization responds to the field in the y direction, $H_{\text{bias}} + h_{\text{ext}}$, by rotation through an angle θ measured from the easy horizontal axis. Thus, in terms of the θ in Figure 17.18, we have

$$\frac{M_y}{M_s} = \sin\theta = \frac{H_{\text{bias}} + h_y}{H_a}$$

If the field from the transition is small compared to the bias field, the MR signal ($\Delta\rho/\rho \propto \frac{2}{3} - \sin^2\theta$) can be linearized when operating at the bias point:

$$\frac{\Delta R}{R} \approx \left(\frac{\Delta\rho}{\rho}\right)_{\text{max}} \left[\text{constant} + 2\frac{H_{\text{bias}} h_y}{H_a^2}\right]. \tag{17.10}$$

The fractional change in voltage generated by this effect is given by

$$\frac{\Delta V}{V} = 2\left(\frac{\Delta\rho}{\rho}\right)_{\text{max}} \frac{H_b}{H_a^2}\langle h_{\text{ext}}\rangle \tag{17.11}$$

where $\langle h_{\text{ext}}\rangle$ is the field from the medium, $h(x'y')$, averaged over the MR head.

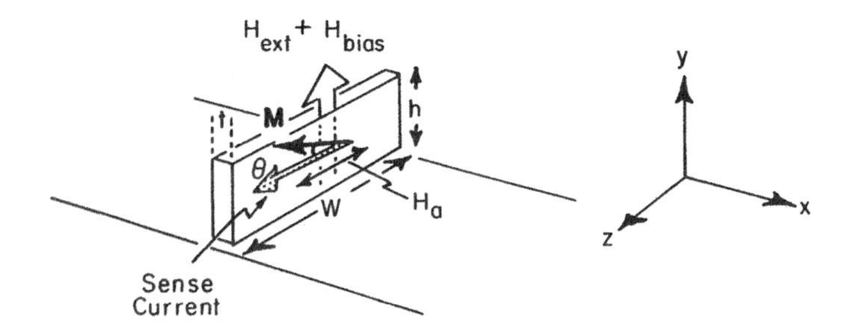

Figure 17.18 Geometry of magnetoresistive sensor showing sense current, anisotropy field, and external or fringe field of medium, and their effect on magnetization.

With the field of the medium given by Eq. (17.4) with a factor from Eq. (17.5), the voltage across the MR head is

$$\Delta V = J\rho w\left(\frac{\Delta\rho}{\rho}\right)\frac{4\pi M_r H_b}{H_a^2}e^{-kd}(1 - e^{-k\delta})\frac{1 - e^{-kh}}{kh}\cos kh \qquad (17.12)$$

where d is the head–medium spacing, δ is the medium thickness, M_r is the remanence of the recording medium that sets the strength of k_y, and $k = 2\pi/\lambda$ as before. The term e^{-kd} accounts for the falloff of the field above the recorded transition, and the next term describes the effects of medium thickness. The last two factors come from integration of the fringe field over the MR element height, h. Note that the medium does not induce a voltage in the head by Faraday's law of induction $V = -Nd\phi/dt$. The MR signal is independent of the relative speed of the head and medium; the voltage or resistance change is a result of the amount of rotation of $\boldsymbol{M}$, which depends on the strength of the fringe field above the recorded medium and on the head characteristics.

The strength of the signal from an MR head is usually expressed per unit track width, $\Delta V/w$. This is seen from Eq. (17.12) to vary as $J\Delta\rho H_b M_r/H_a^2$ times factors related to the length scales of the recording process. The signal strength is limited practically by the ability of the head to dissipate heat generated by the sense current.

A completely passive MR element bias is provided by a soft magnetic layer adjacent to the MR strip. The soft adjacent layer (SAL) is magnetized by the primary current in the MR strip. In turn, the magnetization of the bias layer causes a dipole field that provides the necessary bias. To a first approximation, the flux per unit track width Mt is closed through the magnetic circuit of the SAL layer and the MR element: $M_s t_{SAL} \approx M_r t_{MR}$.

Barkhausen noise results from irregular domain wall motion in the head, shields or SAL layers, and generally becomes more severe as the element width (which scales with track width) is decreased and as the aspect ratio is more favorable to domain formation. Figure 17.19 illustrates the reduction of Barkhausen noise by use of an applied easy-axis bias field in an otherwise unbiased MR head. This transverse bias field is often referred to as a *stabilization field.* Note that as the stabilization field H_x increases, the noise associated with Barkhausen jumps of domain walls vanishes.

Adjacent strips of hard magnetic material can apply a stabilization field. Exchange coupling between the MR strip and a suitable layer in intimate contact can also provide the bias needed for low noise. Permanent magnet materials such as CoPt, as well as the more familiar γ-FeMn or the magnetic oxide CoO, are known to apply stabilization fields to neighboring magnetic layers.

It is useful to summarize the response of an MR head to the various effective fields governing the direction of its magnetization:

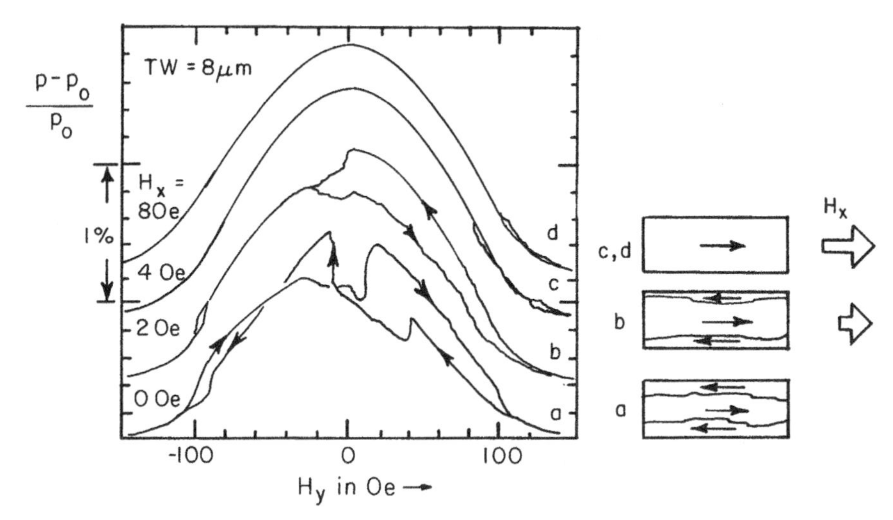

Figure 17.19 Measured MR ratio in a 24-μm-wide element versus excitation field for different bias fields directed along the track width (easy) direction (Jeffers et al. 1987). Insert at right shows schematic of domain elimination by application of stabilization field, H_x.

1. Uniaxial anisotropy of the MR element due to shape and processing, $K_u \sin^2\theta$
2. Vertical bias field for linear operation, $-M_s H_b \sin\theta$
3. Exchange or other cross-track bias field to reduce noise, $-M_s H_{ex} \cos\theta$
4. Fringe field from the medium, $-M_s H_y \sin\theta$

Figure 17.20 illustrates the effects of these terms on the *m-h* and $\Delta R/R$-*h* curves. The shape of the MR transfer curves can be generated to first order by recalling that $\Delta\rho/\rho$ goes as $\cos^2\theta - \frac{1}{3}$ and for hard-axis magnetization, $\cos\theta = m = h$, the reduced magnetization and reduced field, respectively.

The familiar shearing effect of the anisotropy field, H_a, is shown first in Figure 17.20 as a reference. The bias field H_b cants the MR magnetic moment from its horizontal easy axis and consequently shifts the loop along the field axis (first two panels, Figure 17.20). The domain stabilization field H_{ex} shears the loop, as does H_a (first panel), but it also adds curvature in the approach to saturation (third panel) because of its different field dependence. The final panel shows the additive effects of exchange and anisotropy. In real MR elements, the demagnetizing field is not uniform and so is not described by a uniaxial anisotropy. Thus, even without H_{ex}, the M–H loop is not linear but curves toward saturation; the $\Delta\rho/\rho$ curve does not show a break at H_a but also curves toward saturation.

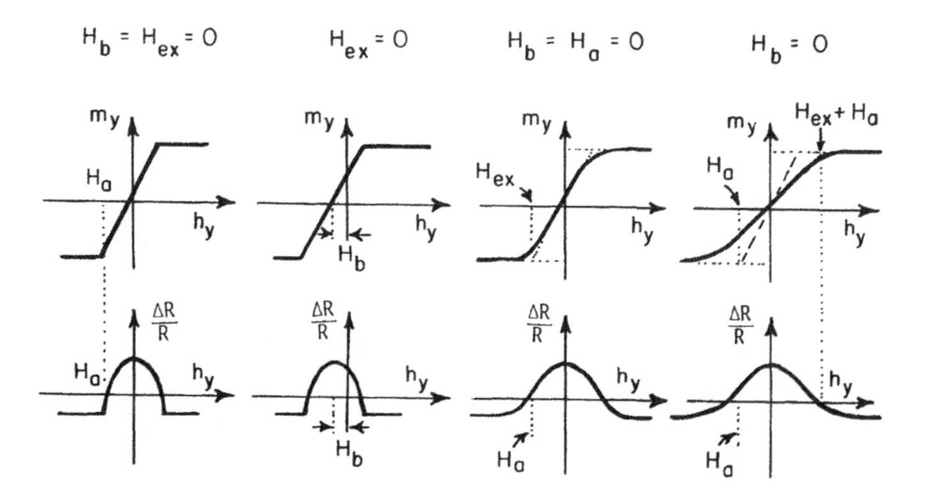

Figure 17.20 Schematic reduced magnetization versus external field h_y, showing the different effects of anisotropy field, bias field, exchange field, and exchange plus anisotropy. Lower part shows the transfer functions $\Delta R/R$ corresponding to each magnetic effect above.

A quantitative expression of these effects provides a useful review of the magnetization process and of the MR transfer function. In the geometry of Figure 17.21, the free energy is written

$$f = K_u \sin^2\theta + (N_2 - N_1)\mu_0 M_s^2 \sin^2\theta - M_s H_{\mathrm{exch}} \cos\theta - M_s H_{\mathrm{bias}} \sin\theta - M_s H_y \sin\theta \tag{17.13}$$

The uniaxial anisotropy and shape anisotropy terms can be combined to a single effective uniaxial anisotropy, $K_u^{\mathrm{eff}} \sin^2\theta$. Equation (17.13) leads to a zero-net-torque condition that can be simplified by the substitutions $m = \sin\theta$, $(1 - m^2)^{1/2} = \cos\theta$:

$$H_a^{\mathrm{eff}} m(1 - m^2)^{1/2} + H_{\mathrm{exch}} m - (H_{\mathrm{bias}} + H_y)(1 - m^2)^{1/2} = 0 \tag{17.14}$$

The M–H loops in Figure 17.21 were calculated for $H_a^{\mathrm{eff}} = 10$ Oe, $H_{\mathrm{bias}} = 4$ Oe, and $H_{\mathrm{exch}} = 1$ and 6 Oe. Without an exchange field, the M–H response below saturation would be linear with a slope M_s/H_a^{eff}. The exchange field strengthens the effective uniaxial anisotropy with a unidirectional anisotropy that also retards the approach to saturation of the M–H curve.

To the lower right in Figure 17.21 is displayed the MR transfer functions using the same parameters that generated the M–H loop at left. Because $\Delta R/R$ goes as $\cos^2\theta - \frac{1}{3}$, it varies also as $\frac{2}{3} - \sin^2\theta = \frac{2}{3} - m^2$, which is displayed here.

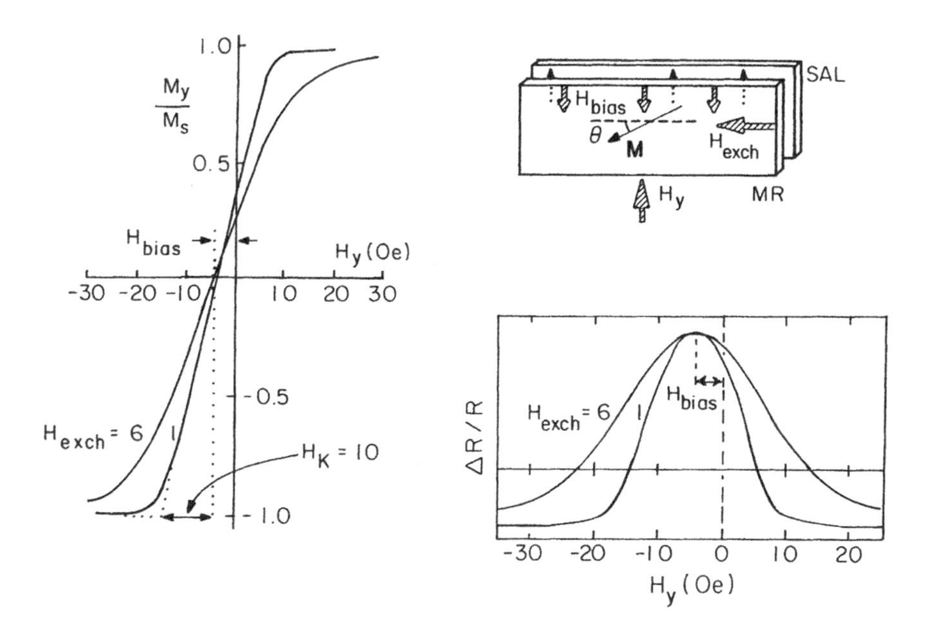

Figure 17.21 Upper right, geometry of MR head and various fields acting on its magnetization. Left, M–H loop calculated from simple theory in text using exchange fields of 6 and 1 Oe, respectively, $H_a = 10$ and a 4 Oe bias. Lower right, MR transfer functions calculated from same model for the parameters used in M–H.

Note that this head is underbiased in the sense that a larger bias field would be required to shift the transfer function so that the zero-external-field point is at the steepest, most linear part of the curve.

The interested reader is encouraged to generate the MR transfer curves for these effects by incorporating the energies of Eq. (17.13) in Eq. (17.10) (Problem 17.6).

Figure 17.22*a* shows the arrangement of the exchange tabs or hard magnet domain stabilization layers viewed from the air-bearing surface (ABS). The three layers making up the portion of the MR head over the track width are the SAL layer, a spacer (often Ta) and the MR element itself. Permalloy is most often used for the MR layer. Certain Co-Fe-Ni alloys show larger $\Delta\rho/\rho$ but have larger anisotropy and electrical resistivity.

MR heads are generally integrated with the inductive write head. The configuration of shields in a dual head, that is, one including both thin film write and MR read functions, is shown in Figure 17.22*b*. The two elements of the inductive write head, P1 and P2, are shown with the write gap between them. P1 also serves as one of the shields for the MR read sensor, shown schematically between the two shields. This design, with the current leads defining a narrower read track than the write track defined by P1, is called "write wide, read narrow."

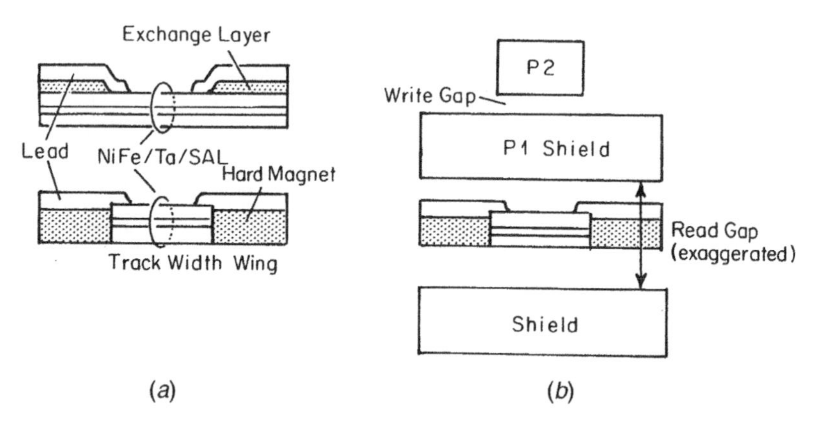

Figure 17.22 (*a*) Cross-sectional views of exchange-stabilized (above) and permanent-magnet-stabilized MR elements (Ishiwata et al. (1995); (*b*) Arrangement of components in a dual-function head. The P1 element of the inductive write head also serves as one shield for the MR element.

The sensitivity of an MR head is often expressed in terms of voltage output per unit track width w [Eq. (17.12)]. Clearly, as track width decreases to accommodate higher recording density, the signal from MR heads decreases. This has driven the development of spin valve read heads, based on the GMR effect described in Chapter 15.

17.5.3 Spin-Valve Read Heads

It will be recalled from Chapter 15 that a spin valve is composed of two magnetic layers separated by a conducting spacer. The resistance of the trilayer depends on the relative orientation of the magnetizations in the two layers. Exchange coupling is often used to pin the direction of magnetization in one of the layers. A favored configuration of the device is for the free layer to have its quiescent orientation orthogonal to the direction of the field to be sensed; the pinned layer should be magnetized in the sense field direction. Spin valve (SV) structures, because of the weak magnetic coupling between their two magnetic layers, are easily changed from the quiescent state ($\uparrow\rightarrow$) by a magnetic field to either the $\uparrow\uparrow$ or $\uparrow\downarrow$ state. The structure of a simple spin valve sensor is illustrated in Figure 17.23. The magnetization of the reference layer is pinned in the vertical direction by deposition in a field with an adjacent FeMn exchange-coupled layer. The sensitivity of the device is improved by interposing a thin layer of Co at the interfaces between the magnetic layers and the Cu spacer.

Note the structural similarity between this spin valve and an MR head. The SAL layer in an MR head is essentially saturated in the vertical direction, similar to the pinned layer in a spin valve. Both devices make use of a thin, conducting spacer layer to decouple the moments of the two magnetic films.

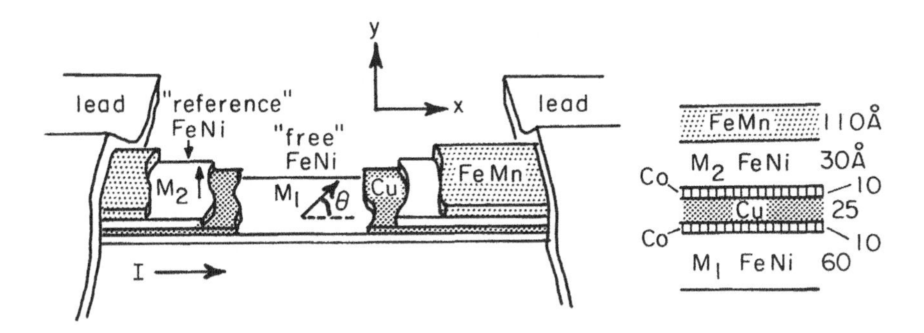

Figure 17.23 Structure of a simple spin valve. Note that free layer is deposited first, reference layer is magnetized in positive y direction, and sense current is in positive x direction; the device dimensions are approximately $h = 2$ to $6\ \mu m$ and $w = 10\ \mu m$ (Heim et al. 1994).

In an MR head, the active layer is biased at 45° to the current direction and its angular range for near-linear response is bounded well within 90°. In a spin valve, the free layer is magnetized parallel to the sense current and its quasilinear range is bounded within 180°. Besides being based on different physical interactions (Chapter 15) there is a further functional difference between these two devices. In the MR head, all of the MR effect occurs within the MR element; the SAL and spacer layers ideally should carry no current. The operation of a spin valve, on the other hand, depends critically on the ability of the charge carriers to drift between the free and pinned layers. In terms of applications, the defining difference is that the MR ratio of an MR head decreases monotonically with decreasing thickness of the MR element. The spin valve, on the other hand, shows improved performance with decreasing thickness of the three layers down to a limit that seems to depend on our ability to mass-fabricate high-quality ultrathin films.

Transfer curves for finished, unshielded sensors measured in uniform fields are shown in Figure 17.24 for a 2-μm-high device (Heim et al. 1994). Note that a positive field leads to saturation of the signal. A negative field leads first to a negative saturation near -100 Oe then the signal returns through zero toward positive saturation. Of the three resistance states implied by Figure 17.24, the middle one corresponds to antiparallel $\boldsymbol{M}_1$ and $\boldsymbol{M}_2$. Thus, the signal here is inverted relative to $\Delta R/R$ of the device. Note that changes in the direction of the sense current give different quiescent orientations for $\boldsymbol{M}_1$.

The magnetization profile in the free layer is a result of four fields acting on $\boldsymbol{M}_1$ (Fig. 17.25):

1. Magnetostatic field due to the poles on the pinned layers. This field is strongest at the top and bottom of the free layer.
2. Exchange field favoring $\boldsymbol{M}_1 \| \boldsymbol{M}_2$. This field is roughly uniform in y.

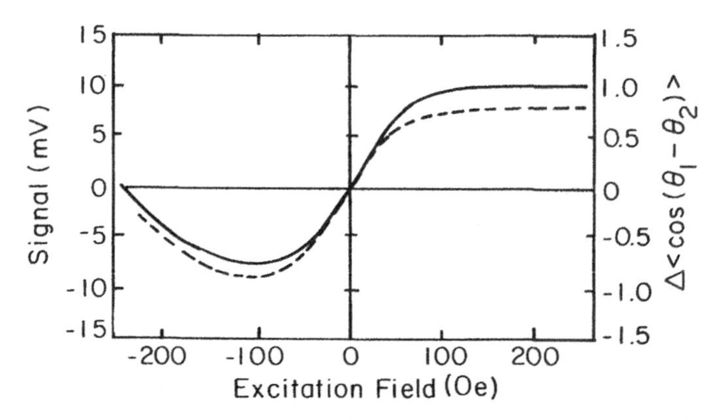

Figure 17.24 Experimental transfer curve for a 2-μm high-spin valve sensor for $+5$ mA (solid) and -5 mA (dashed) sense current. Computed transfer curves follow the data within experimental error. [Adapted from Heim et al. 1994).]

3. The magnetostatic field of the free layer itself. This field is greatest near the top and bottom surfaces.
4. The field due to sense current distribution. This field is greatest near the center of the free layer.

The spin valve should be designed and operated so that these fields nearly cancel, leaving the orientation of $\boldsymbol{M}_1$ governed by the horizontal uniaxial anisotropy and the vertical field to be sensed.

It is critical for the operation of a spin valve that the pinned layer not respond to the transition field. A less than fully pinned $\boldsymbol{M}_2$ would reduce the field-induced change in relative orientation between $\boldsymbol{M}_1$ and $\boldsymbol{M}_2$ that determines the resistance change of the device. Consequently, considerable attention is given to the exchange coupling layer used to pin $\boldsymbol{M}_2$. Two cases are worth mentioning: FeMn "top" spin valve (pinned FeMn/NiFe layers deposited after free NiFe layer) and NiO "bottom" spin valve (pinned FeNi/NiO before free

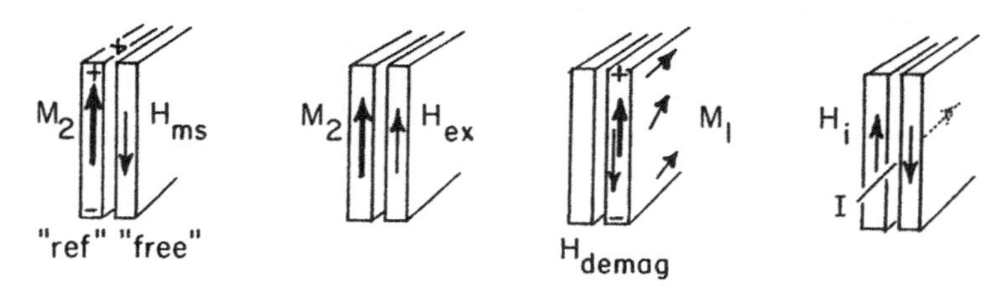

Figure 17.25 Representation of four contributors to the field (fine arrows) acting on $\boldsymbol{M}_1$ of the spin valve in Figure 17.23 to establish its quiescent orientation. Bold arrows give orientation of $\boldsymbol{M}_2$ or $\boldsymbol{M}_1$.

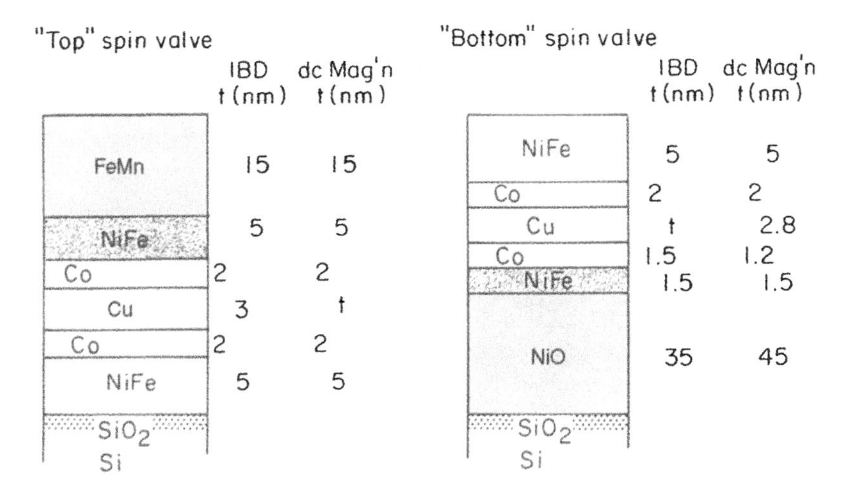

Figure 17.26 Structure of spin valves made by Wang *et al.* (1997) to compare effects of ion-beam versus dc magnetron sputter deposition and the relative merits of NiO versus FeMn exchange biasing layers.

NiFe layer) (see Fig. 17.26). The reason for the difference in sequence of these depositions is largely related to the effects of processing and substrate on the quality of the deposited film. FeMn is a good antiferromagnetic exchange coupling layer only in its metastable FCC phase. This phase is best achieved by deposition on an FCC substrate such as permalloy. NiO, on the other hand, has no such structural constraints and further provides a suitable substrate for the growth of the pinned permalloy layer.

17.6 MAGNETIC RANDOM ACCESS MEMORIES

17.6.1 Ferrite Ring Core Memories

The ideal magnetic recording medium was described above as consisting of a regular array of noninteracting, single-domain, bistable magnetic elements. Such memories were manufactured in the 1950s and 1960s using an array of small (<1-mm-diameter) ferrite toroids laced together by a grid of x and y conducting wires. Each toroid was a bit that could be written by sending a current pulse through the two wires that defined the coordinates of the toroid, its address. The current in each wire is chosen so that its field alone is insufficient to switch any of the toroids along its path:

$$H_{\text{toroid}} = \left(\frac{\sqrt{2}}{2}\right) H_i < H_c$$

But two intersecting currents do produce a field that exceeds the switching threshold:

$$H_{\text{toroid}} = \sqrt{2}\, H_i > H_c$$

Reading the bit at a given address is a matter of testing the address to see if it switches under a given pulse. If the toroid is not switched by a pair of inquiry current pulses of a given polarity, the impedance of the x and y wires is purely resistive ($\mu_r = 1$) and the state of the bit is known. If the toroid is switched, its permeability gives it an appreciable inductance causing a back EMF. Again, the state is known as a result of the inquiry but it must be reset to its original state. Thus either one or two current pulses are required to query an address and leave the memory unchanged.

Ferrite core memories obviously have areal density limitations based on the ability to make and assemble the microscopic components. However, they have advantages based on the fact that there are no moving parts and there is no separate head required.

These memories have been used in a few special applications for which nonvolatile, robust information storage is more important than areal density. The bulk of the storage market is owned by hard disk drives, tape drives, and floppy disks.

17.6.2 Thin-Film, High-Density MRAM

Some of the advantageous concepts of ferrite core array memories are used today in a class of storage devices called *magnetic random access memories* (MRAMs). Those advantages are no moving parts, no heads, and the ability to access information at an arbitrary sequence of addresses (random access) as opposed to sequential access as in tape and disk storage. The individual ferrite cores and their tedious assembly have been replaced by magnetoresistive storage elements made by thin-film technology and high-resolution lithography. Magnetoresistive random access memories were pioneered by Schwee (1972), Pohm et al. (1991), and Daughton (1992).

Present MRAM information storage densities measure in the tens of kilobits per chip using an array of MR elements. Several orders of magnitude increase in storage density are expected in the near future in devices using spin-valve-like (SVL) storage elements and improved design and lithography. How does an MR random access memory element work?

We describe one recently proposed, spin-valve-like MRAM structure that embodies the essential concepts of the technology (Irie et al. 1995). The MRAM design retains the grid of conductors that defines the bit addresses in the ferrite core memory. However, the ferrite cores are replaced by SVL devices, and, most importantly, the dimensions of the thin-film structure can be reduced dramatically, limited by the lithography, film quality, and the peak response of the GMR effect (Fig. 15.26).

A conventional spin valve would show destructive readout (DRO) because of the strong pinning of the reference layer. A pseudo-spin-valve (PSV) structure was developed that shows nondestructive readout (NDRO). Figure 17.27 provides a basis for description of the device.

A PSV structure consists of a free layer, a spacer, and a semihard layer (unlike a spin valve that uses an exchange-coupled layer as a reference). Irie et al. use a multilayered PSV device totaling 130 nm in thickness: $[Ni_{80}Fe_{10}Co_{10}/Cu/Co_{75}Pt_{25}/Cu]_N$. The magnetization and relative resistance change of such a device is shown schematically below the structure in Figure 17.27. The coercivities H_{c1} and H_{c2} correspond to the switching of the free and semihard layers, respectively. The form of the output signal of these storage devices is closer to that of a spin switch described in Section 15.2.2 (Fig. 15.35) than to that of a spin valve. However, here the current is in the plane of the layers. The array of storage elements is connected by conducting shunts to form a series of lines called sense lines. An insulating layer of polyimide or SiN is deposited

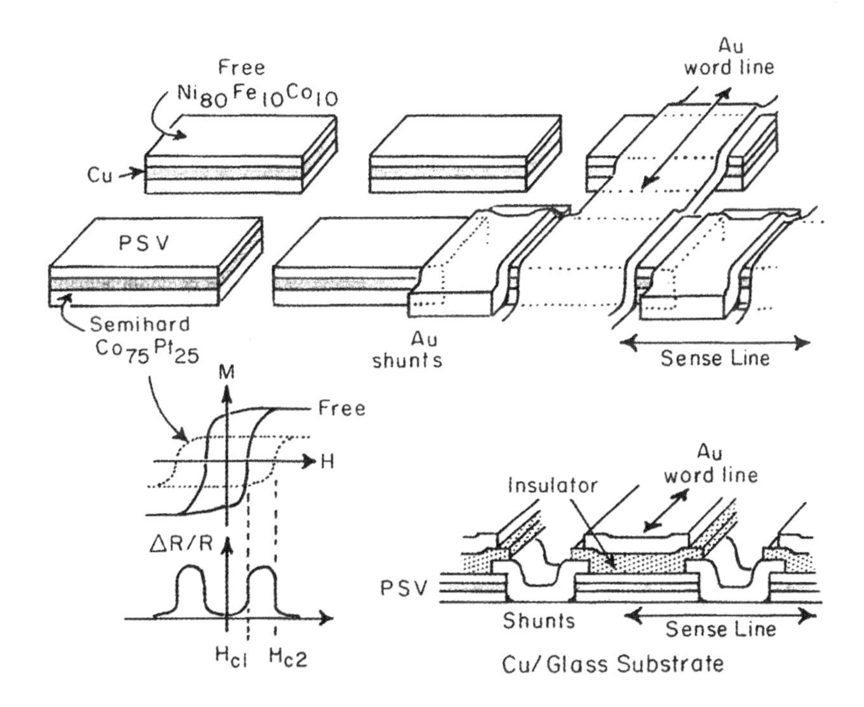

Figure 17.27 Above, simplified depiction of assembly of an MRAM. At left, the pseudo-spin-valve (PSV) structure elements are deposited on a substrate. Gold shunts connect the PSVs to form the sense line. After application of an insulation layer, the gold word lines are deposited over the PSV elements. Below right, cross section through the middle of the sense line showing structure. Below left, M–H and $\Delta R/R$ characteristics of the PSV.

TABLE 17.4 Composition and Dimensions of the Principal Layers in a Current Representative MRAM Device

Device Element	Composition (at%)	Thickness (nm)	Line Width (μm)
Free layer	$Ni_{80}Fe_{10}Co_{10}$	3.0	40
PSV			
Spacer	Cu	2.3	40
Semihard	$Co_{75}Pt_{25}$	7.4	40
Sense line shunts	Au	300	>40
Word line	Au	200	80

to electrically isolate this array of storage elements from succeeding layers. After the deposition of the insulator, an array of "word lines" (typically gold) orthogonal to the sense lines is deposited. The layer dimensions for the device made by Irie et al. are summarized in Table 17.4. Clearly, a current in a word line generates a field parallel to the underlying sense lines. Similarly, a current through a sense line generates a field that cants the free layer magnetization away from its easy axis; this makes it easier for the word line field to switch only the storage element at the intersection of the two current pulses.

As shown in Figure 17.28, the write process consists of magnetizing *both* the free and semihard layers in one direction or another by an appropriately directed word current pulse (and simultaneous sense line current). After the write process, the two layers are in their remanent states and the resistance of either state has the same minimum value ($H = 0$ in Fig. 17.28, lower left).

The read process consists of applying a bipolar current pulse to the word line. This pulse produces a field sufficient to switch the soft layer but not the hard layer: $H_{c1} < H_{\text{word}} < H_{c2}$. Thus, depending on the state of the element, "0" or "1," the resistance in the sense line changes in phase or out-of-phase, respectively, with the word current pulse. After application of the read pulse, the MR element reverts to its original remanent state ($\uparrow\uparrow$ or $\downarrow\downarrow$). This is possible only if the two magnetic layers are ferromagnetically exchange-coupled through the Cu spacer, or if the read pulse is followed by a small reset pulse of opposite polarity. Thus, this device exhibits NDRO.

Some MRAM devices have been proposed making use of symmetric PSV storage elements (Everitt and Pohm 1998).

If MRAM storage density were to approach the Gb/in^2 range, it could possibly displace tape and disk drive storage in some applications.

17.7 OUTLOOK AND FUNDAMENTAL LIMITS TO RECORDING

An areal density of 1 Gb/in^2 is realized at 158 kfci (thousands of flux changes

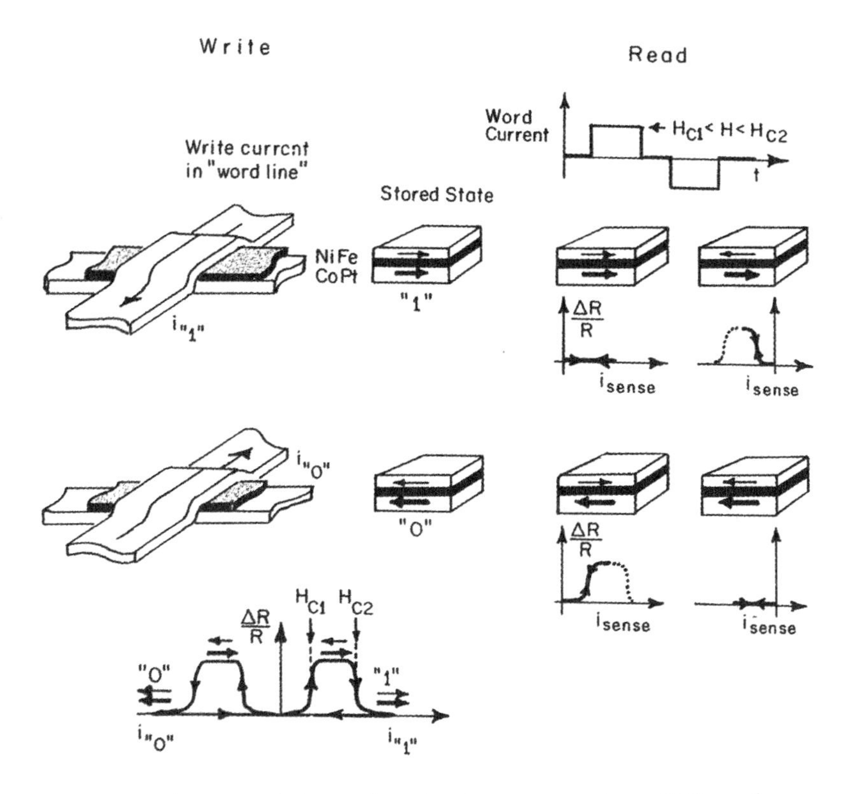

Figure 17.28 Schematic of the read and write processes in a PSV random access memory. The write process involves current pulses through both the word line and the sense line such that the field at the PSV exceeds H_{c2}. The read process involves a field pulse that takes the device to the high-resistance state without switching the semihard layer ($H_{c1} < H < H_{c2}$).

per inch) and 7470 tpi (tracks per inch). This areal density corresponds to a bit size that is approximately 2 times greater than the diffraction limit of visible optical recording. The bit density limit of thin-film media is estimated to be approximately of order 100 Gb/in^2.

It is desirable that the density with which the information is recorded be as great as possible without sacrificing signal-to-noise ratio (SNR). Media alone do not limit recording density. It should be clear from Eqs. (17.2) and (17.5) that if bit size, $\lambda/2$, in Eq. (17.5) is to decrease, the write gap, g must decrease, Eq. (17.2), and thus the write head must be closer to the medium. The first 1-Gb/in^2 system had a write gap of 200 nm and the head flew at 50 nm (2μ in). Also, smaller λ demands that the depth to which the signal is recorded in the medium be reduced [Eq. (17.6)]. This means that there is less volume magnetized and smaller flux changes at the transitions. The fringe field above the medium decreases and signal strength drops even more. The read head, then, must be either more sensitive or closer to the medium. Thus, all of the relevant dimensions of the recording process need to be scaled down together to achieve high recording density. In addition to reduced dimensions of the recording process, all the issues associated with friction, wear, head-track alignment (tracking), and reliability must advance apace. It is amazing that despite these demanding conditions, the density of information storage has been able to increase so dramatically in recent years.

17.7.1 Fundamental Limits

Optical recording is diffraction limited in its density. (Near-field optical techniques are being developed that circumvent the diffraction limit.) Magnetic recording is not so limited and presently achieves higher information density than magnetooptical recording. The information density in magnetic recording systems based on currently used concepts is fundamentally limited by the minimum particle or grain size which is magnetically stable against ambient thermal demagnetization. For stable recorded information, the particle volume must exceed the superparamagnetic limit [see Eq. (8.34)]. For $K_u = 10^5\,\mathrm{J/m^3}$, the critical volume of an isolated particle is $V = 1.5 \times 10^{-24}\,\mathrm{m^3}$. However, at head-to-head transitions in thin-film media, magnetostatic energy reduces the effective anisotropy of the grains. The grain size in longitudinal, thin-film media, below which recorded information is unstable, is given approximately by (Lu and Charap 1995)

$$r_0^{1\,\mathrm{yr}} \approx \left(\frac{15\,k_B T}{K_u}\right)^{1/3} \approx 28\,\mathrm{nm} \tag{17.15}$$

Below the superparamagnetic limit a particle has no memory. But Eq. (17.15) looks at only one aspect of the medium limitations on recording density. Note that a larger value of anisotropy reduces the size below which particles are superparamagnetic. Thus, for higher-density recording, the medium anisotropy should be increased. Increased areal density therefore puts a *lower* limit on medium anisotropy of $60\,k_B T/V$. But increased anisotropy increases medium coercivity. The medium coercivity should not be so high that information cannot be written (write head fields are limited by the saturation flux density of the head material). This places an *upper* limit on media anisotropy of $M_s H_a/2$. Lu and Charap (1995) estimate that these limits will converge for recording densities of order $40\,\mathrm{Gb/in^2}$. The fuller results of their calculations are graphically illustrated in Figure 17.29. Here the region of stable (lower limit to K_u) and recordable (upper limit to K_u) media are bounded on an anisotropy–magnetization plane. In region 1 the signal strength is weak because M_s is small. In region 2, the signal is weak because magnetostatic energy smears the transition. Region 5 is optimal because for higher magnetization media, higher anisotropy is required.

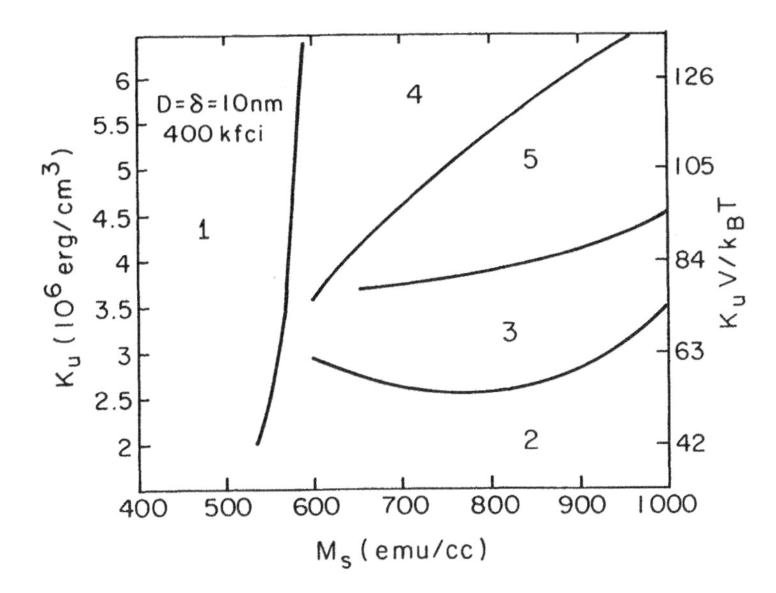

Figure 17.29 Recording performance field in $K_u - M_s$ space for grain size D and thickness δ, each measuring 10 nm. See text. [After Lu and Charap (1995).]

It should be noted that for higher writing rates (shorter write times) the factor of 15 in Eq. (17.14) must increase (see Chapter 13). Thus, the effective coercivity increases at higher data rates and the write process requires larger fields. Also, as data rates increase the head is effectively operating at higher frequencies. Presently, frequencies are in the range of 100 MHz, and higher frequencies will become typical for high-density recording. The speed with which magnetic materials can respond to a field is limited by resonance phenomena, and in metals, by eddy-current damping. (These effects were touched on in Chapter 9. Eddy currents are associated with the decrease in AC field intensity with depth inside a magnetic material. If the skin depth is appreciably smaller than the sample dimensions, then a significant volume fraction of the magnetic material is unresponsive to the field and, effectively, the permeability is reduced.) The thickest film dimensions in high-end heads is of order 2 μm. The head permeability should hold its value up to these high frequencies. The skin depth for permalloy ($\rho = 20\,\mu\Omega \cdot \text{cm}$) at 10^8 Hz is of order 2 μm if the permeability is 100. Thus, the entire thickness of a 2-μm-thick write head is being magnetized and eddy currents should not cause a loss of permeability.

The natural resonance frequency for magnetic materials, the Larmor precession frequency [Eq. (3.2)], is of order 14 GHz per tesla of flux density. At higher frequency, the moments cannot keep up with the drive field and the permeability decreases sharply. It is not simply the data rate (e.g., 100 MHz) that needs to be considered in relation to the resonance frequency (see Section

10.8). A sequence of transitions at a fundamental frequency of 10^8 Hz has higher harmonics associated with the sharpness of the transition. If these high-frequency components are suppressed by the resonance limit, even a spatially sharp transition in the recording medium will be read as less sharp and signal strength will decrease. There are some material effects that can alter the resonance frequency and the relaxation time, but these generally involve decreased magnetic moment. When metals are used in the recording process, the ferromagnetic resonance line width is broadened to several kilogauss. Attention to the issue of resonance imitations is increasing.

17.7.2 Patterned Media

One way to achieve low-noise, high-density media is to make each bit consist of a single piece or grain of magnetic material. Such bits should be arranged periodically to be synchronized with the signal channel. This can be achieved using high-resolution lithography.

The term *patterned media* is used to refer to media for which each bit consists of a single, lithographically defined grain. Such a recording medium eliminates the random $\sqrt{N}$ noise associated with multigrain bits. It also eliminates the noise associated with irregular or sawtooth transitions that cause noise in thin-film media. Patterned media will allow for high bit densities because the superparamagnetic limit applies to a single bit, not to each of the many grains in a multigrain bit. Finally, the patterning process defines a sharper transition between bits, and dispersion of easy axes can be minimized relative to that in thin-film media. Thus patterned media have relaxed conditions on coercivity and $M_r t$ product.

Figure 17.30 shows an array of Ni pillars grown by electrodeposition on a lithographically defined pattern of holes in a removable template. The substrate is coated with a conductive plating base, an antireflection coating, a silica etch mask, and, finally, photoresist. A pattern of holes is formed in the mask using laser interferometric lithography. After electrodeposition of the Ni, the template is removed (Ross and Smith 1998). The magnetic properties if nanoscaled arrays of particles such as these represents a possible path toward higher-density, lower-noise media.

Patterned media with bit sizes on the scale of 100 nm square will require new read and write head technology. Inductive MR and GMR heads produce signals proportional to track width; track width will be reduced tenfold in patterned media. Further, the ability of the servo system that allows the head to follow a given information track will have to be improved.

17.8 SUMMARY

Magnetic recording technology is driven by the desire to increase the areal density and access speed of stored information. The quantitative description of the write field and the signal from the written bits makes it clear that all

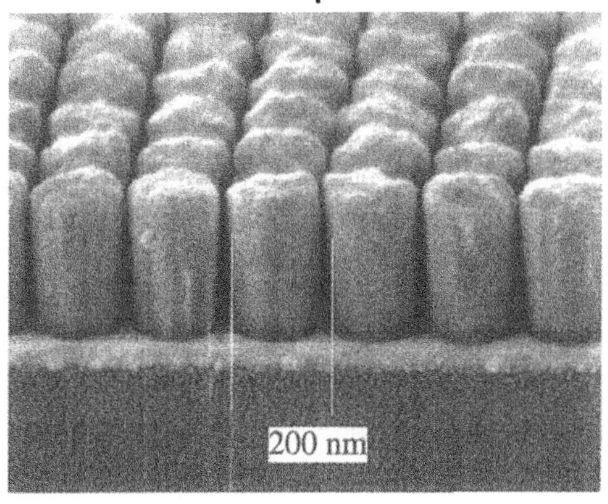

Figure 17.30 Scanning electron microscopy image of a square array of electrodeposited Ni pillars of height 300 nm and period, 200 nm. Courtesy of Ross et al. 1999.

dimensions of the recording process must be reduced proportionally to achieve increased density. Other aspects of the recording system must also be improved to accommodate the reduced length scales and higher data rates. These include issues of tribology, track servo mechanics, and signal processing.

From the point of view of materials, magnetic recording offers a rich range of challenges in terms of thin-film processing, microstructure, and interface control, all within strict limitations of reliability and high operating temperatures.

Present modes of magnetic recording face hard fundamental limits to materials performance at higher areal densities. These arise from loss of magnetization in small, superparamagnetic particles and the magnetic resonance relaxation limit to read head response time. Nevertheless, fascinating devices (such as spin valves) and improved thin film media (perhaps patterned media) incorporating advanced materials science, thin-film processing, and high-resolution lithography, hold promise for information densities approaching 100 Gb/in^2.

PROBLEMS

17.1 Evaluate the reduced Karlkvist field $h_x = H_x(xy)/H_g$ at midgap ($x = 0$) for $y = 0$, g and $2g$.

17.2 Derive the vector field components beneath the gap of a write head by replacing the head with a current flowing in the cross-track direction in the gap at $x = 0$ and $y = 2g$. Choose the strength of the current so that $H_x(0, 0) = H_g$.

17.3 Derive the form of the vertical and horizontal components of the fringe field above a sharp head-to-head transition using Eqs. (2.2) and (2.3).

17.4 Calculate the vertical component of the field above a sharp head-to-head transition in a thin-film recording medium characterized by $M_r t = 2 \times 10^{-3}$ erg/cm^2. Assume infinite track width.

17.5 Consider a rectangular area (see accompanying diagram) on the surface of a recording medium measuring a by $2a$ magnetized to a depth t into the medium. Compare the sum of magnetostatic plus domain wall energy for the case of a linear wall separating head-to-head domains with that of a sawtooth of peak-to-peak amplitude $a' = a/10$ and period $\lambda = a/10$.

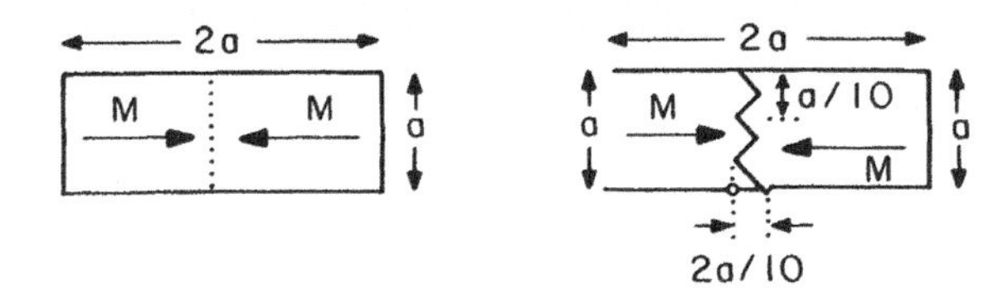

17.6 Generate the MR transfer curves corresponding to the four M–H loops shown in Figure 17.20, using the energies in Eq. (17.14) and $\Delta R/R = (\Delta\rho/\rho)\,(1 - \cos^2\theta - \frac{1}{3})$.

BIBLIOGRAPHY

Bertram, N. H., *Proc. IEEE* **74**, 1949 (1986).

Bloomberg, D. S., and G. A. N. Connell, in *Magnetic Recording Handbook*, C. D. Mee and E. D. Daniel, eds., McGraw-Hill, New York, 1981, p. 530.

Daughton, J., *IEEE Trans.* **MAG-29**, 2705 (1993).

Derbyshire, K., and E. Korczynski, *Solid State Technol.*, p. 57 (Sept. 1995).

Dorner, M. F., and Richard L. White, *Mater. Bull.* **21**, 28 (1996).

Fontana, R. E., *IEEE Trans.* **MAG-31** 2579 (1995).

Gambino, R. J., and T. Suzuki, *Magneto-Optical Materials*, IEEE Press, Piscataway, NJ, 1999.

Lodder, J. C., in *Handbook of Ferromagnetic Materials*, ed. K. H. J. Buschow, North Holland, Amsterdam, 1988, p. 291.

REFERENCES

Bate, G., in *Ferromagnetic Materials*, Vol. 2, E. P. Wohlfarth, ed., North Holland, Amsterdam, 1980, p. 381.

Bertram, H. N., *Theory of Magnetic Recording*, Cambridge Univ. Press, Cambridge, UK, 1994.

Daughton, J., *Thin Solid Films* **216**, 162 (1992).

Everitt, B. A., and A. Y. Pohm, preprint (1998).

Grachowski, E., and D. A. Thompson, *IEEE Trans.* **MAG-30**, 3797 (1994).

Heim, D. E., R. E. Fontana, C. Tsang, V. S. Speriosu, B. A. Gurney, and M. L. Williams, *IEEE Trans.* **MAG-30**, 316 (1994).

Hunt, R. P., *IEEE Trans.* **MAG-7**, 150 (1971).

Ishiwata, N., H. Matsutera and K. Yamada, Paper AS, TMRE Pittsburgh PA, July, 1995.

Iwasaki, S., Y. Nakamura, and K. Ouchi, *IEEE Trans.* **MAG-15**, 1456 (1979).

Iwasaki, S., Y. Nakamura, and N. Honda, *IEEE Trans.* **MAG-16**, 1111 (1980).

Irie, Y., H. Sakakima, M. Satomi, and Y. Kawawake, *Jpn. J. Appl. Phys.* **34**, 415 (1995).

Jagielinski, T., *MRS Bull.* **15**, 36 (1990).

Jeffers, F., D. Wachenschwanz, D. Phelps, and J. Freeman, *IEEE Trans.* **MAG-23**, 2088 (1987).

Jorgensen, F. *The Complete Handbook of Magnetic Recording*, TAB Books, Blue Ridge Summit, PA, 1988.

Kryder, M. H., *IEEE Trans.* **MAG-25**, p. 4358 (1989).

Lambeth, D. L. et al., *MRS Conf. Proc.* (1998).

Lee, L. L., D. E. Loughlin, L. Fang, and D. L. Lambeth, *IEEE Trans.* **MAG-31**, 2728 (1995).

Lemke, J., Magnetic Recording Materials, M.R.S. Bulletin, March 1990.

Lu, P. L., and S. H. Charap, *IEEE Trans.* **MAG-31**, 2767 (1995).

Mallinson, J., *IEEE Trans.* **MAG-21**, 1217 (1981).

Mallinson, J., *Foundations of Magnetic Recording*, Academic Press, San Diego, 1987.

Mansuripur, M., *Physical Principles of Magneto-optical Recordings*, Cambridge Univ. Press, Cambridge, UK, 1993.

Mee, C. D., and E. D. Daniel, eds., *Magnetic Recording Handbook*, McGraw Hill, NY, 1989.

Mee, C. C., and E. D. Daniel, Second edition, IEEE Press, 1995.

Onodera, S., H. Kondo, and T. Kawana, *Mater. Bull.* **21**, 35 (1996).

Pohm, A. V., C. S. Comstock, G. B. Granley, and J. M. Daughton, *IEEE Trans.* **MAG-27**, 5520 (1991).

Ross, C. A., H. I. Smith, T. Savas, M. Schattenburg, M. Farhoud, M. Hwang, M. Walsh, M. Abraham, and R. Ram, *J. Vac. Sci. Technol.* in press (1999).

Schwee, L. J., *IEEE Trans.* **MAG-8**, 405 (1972).

Sharrock, M. P., *MRS Bull.* **15**, 53 (1990).

Speriosu, V. S., D. A. Herman Jr., I. L. Sanders and T. Yogi, *IBM Jour. of R&D* **34**, 884 (1990).

Suzuki, T., *IEEE Trans.* **MAG-20**, 675 (1980); **MAG-24**, 675 (1984).

Suzuki, T., *Mater. Bull.* **21**, 42 (1996).

Tang, L. et al., *J. Appl. Phys.* **81**, 4906 (1997).

Thompson, D. et al., *IEEE Trans.* **MAG-11**, 1036 (1975).

Tsang, C., M. Chen. T. Yogi, and K. Ju, *IEEE Trans.* **MAG-26**, 1689 (1990).

Tsang, C., R. Fontana, T. Lin, D. Heim, V. Speriosu, B. Gurney, and M. Williams, *IEEE Trans.* **MAG-30**, 3801 (1994).

Tsang, C., T. Lin, S. MacDonald, M. Pinarbasi, N. Robertson, H. Santini, M. Doerner, T. Reith, Lang Vo, T. Diola, and P. Arnett, *IEEE Trans.* **MAG-33**, 2866 (1997).

Wang, S. X., W. E. Bailey, and C. Sürgers, *IEEE Trans.* **MAG-33**, 2369 (1997).

White, R. L., *IEEE Trans.* **MAG-28**, 2482 (1992).

White, R. M., *Physics of Magnetic Recording*, IEEE Press, New York, 1985.

Wood, R., IEEE Spectrum, May 1990, p. 32.

Yang, W., and D. L. Lambeth, *IEEE Trans. Magn.* **33**, 2965 (1997).

APPENDIX A

TABLE OF CONSTANTS

		MKS/SI	CGS-EMU
A	Exchange stiffness	10^{-11} J/m	10^{-6} erg/cm
a_0	First Bohr radius	5.29×10^{-11} m	5.29×10^{-9} cm
c	Speed of light	2.998×10^{8} m/s	2.998×10^{10} cm/s
e	Electronic charge	1.602×10^{-19} coulombs	1.602×10^{-20} emu 4.8×10^{-10} esμ
ε_0	Dielectric constant	8.85×10^{-12} A^2 s^2 m^{-2}	
γ	Gyromagnetic ratio of electron e/m	1.76×10^{11} C/kg (=1/T·s)	1.76×10^{7} emu/g (=1/Oe·s)
h	Planck's constant	6.625×10^{-34} J·s	6.625×10^{-27} erg·s
$\hbar$	Planck's constant/2π	1.054×10^{-34} J·s	1.054×10^{-27} erg·s
k_B	Boltzmann constant	1.38×10^{-23} J/deg	1.38×10^{-16} erg/deg
m_e	Mass of the electron	9.108×10^{-31} kg	9.108×10^{-28} g
m_{proton}	Mass of the proton	$1832 \times m_e$	$1832 \times m_e$
μ_o	Free space permeability	$4\pi \times 10^{-7}$ $= 1.26 \times 10^{-6}$ NA^{-2}	—
μ_B	Bohr magneton	0.927×10^{-23} J/T (=Am^2)	0.927×10^{-20} erg/Oe (=emu)
N_A	Avogadro's number	—	6.025×10^{23} (g·mol)$^{-1}$
$R = N_A k_B$	Gas constant	8.317 J/mol·deg)	8.317×10^{7} erg/(mol·deg)

APPENDIX B

CONVERSION FACTORS AND USEFUL RELATIONS

Symbol	MKS		Gaussian
Q	1 coulomb	=	3×10^9 esu
I	1 ampere	=	0.1 abamperes
E	1 joule	=	10^7 erg
L	1 henry (Vs/A)	=	10^9 esu henrys
F	1 newton	=	10^5 dynes
Ω	1 ohm (Ω)	=	$10^{11}/9$ esu ohms
V	1 volt	=	1/300 esu volts
ϕ	1 weber (V · s)	=	10^8 maxwells
B	1 tesla (W/m^2)	=	10^4 gauss
H	1 A/m	=	$4\pi/10^3 \approx 80$ oersteds

1 abcoulomb = 1 statcoulomb $\times\, c$ (cm/s)

$1/(\mu_o \varepsilon_o) = c^2$ (m^2/s^2)

1 calorie = 4.184 J

1 eV = 1.6×10^{-12} erg

1 Calorie = 4184 J

$k_B T = 0.025$ eV

1 bar = 10^6 dyne/cm^2

1 atm = 1.013×10^6 dyn/cm^2

1 mm Hg = 1 torr = 1333 dyne/cm^2

1 lb/in^2 = 6.895×10^4 dyn/cm^2

Electron spin precession frequency = 2.8 GHz/kOe

INDEX

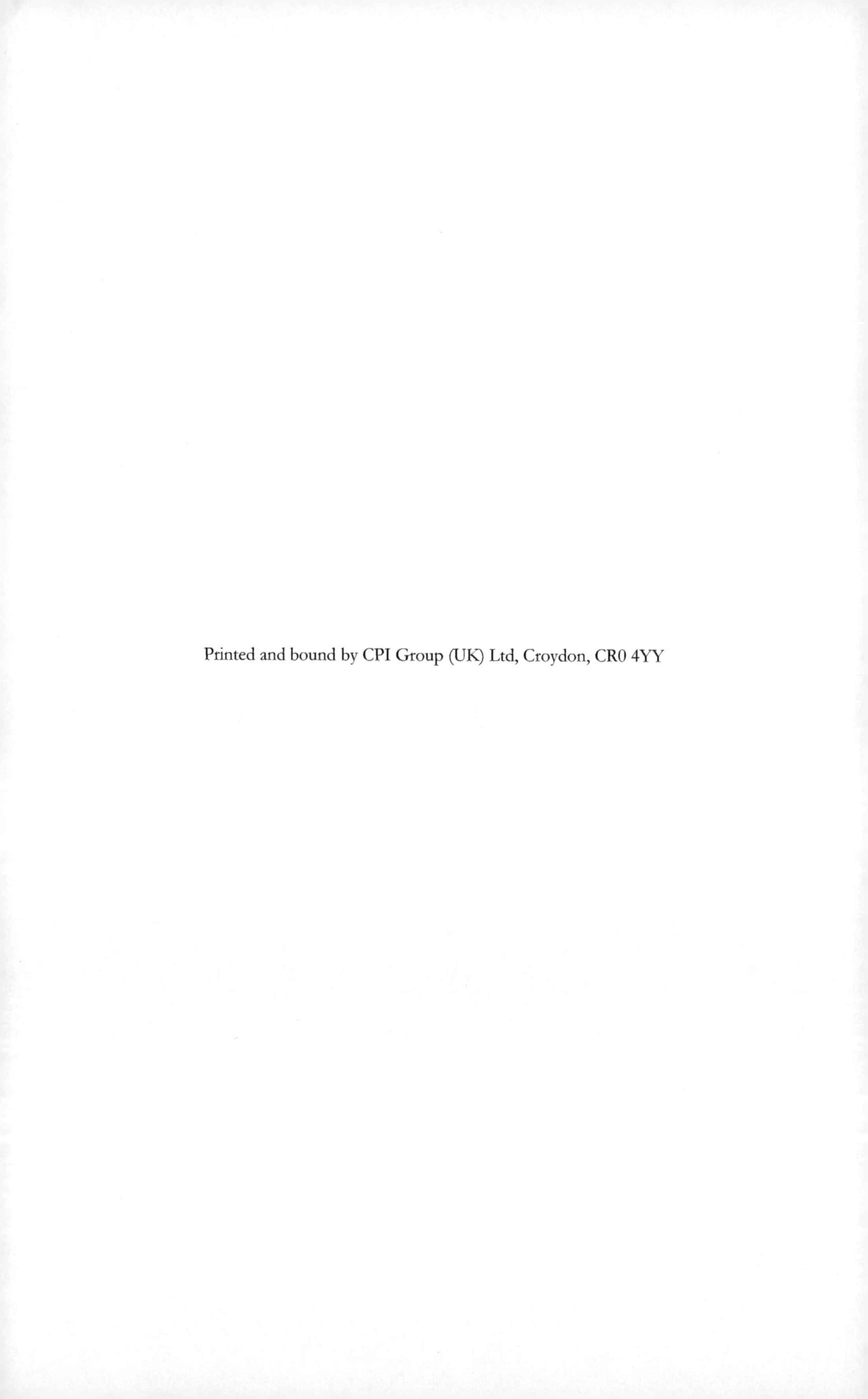

Printed and bound by CPI Group (UK) Ltd, Croydon, CR0 4YY

Printed and bound by CPI Group (UK) Ltd, Croydon, CR0 4YY
10/03/2022
03116443-0001